EXCLUSIVE BOTANY FOR

NEET

National Eligibility-cum-Entrance Test

ABOUT THE BOOK

The book "EXCLUSIVE BOTANY FOR NEET" was prepared as per the latest syllabus and the patterns of NEET. It contains nineteen chapters of Botany out of Biology incorporating all relevant topics. Special focus has been given to difficult topics. The comprehensive information in all chapters is followed by a limited number of most probable multiple-choice questions. Simple and easy multiple-choice questions were not included. The book contains all the relevant materials from the competition's point of view by avoiding unnecessary verbosity and repetitions. The book was written systematically in lucid language. Short and long answer-type questions are given at the end of each chapter for intermediate students. The book also contains five model papers, each containing 50 probable multiple-choice questions at the end of chapters. I am convinced that the students will be able to perform well with greater confidence in the NEET and other related examinations. This book will also cater to the needs of all students associated with Botany.

ABOUT THE AUTHOR

Dr. Khwaja Salahuddin is an Assistant Professor at the University Department of Botany, LN. Mithila University, Darbhanga, Bihar. He has been JRF and SRF of CSIR, New Delhi. He has been highly efficient in teaching the students of UG and PG levels. His knowledge of the subject is par excellent. He started teaching from Bahauddin Science College, Junagadh under the Government of Gujarat to brand coaching institutes of India to impart his quality education to the students who want to compete in the prestigious examination of NEET. To his credit, he has published over 20 research articles in noted, International and UGC-CARE-listed journals, and several book chapters. He has supervised 1 PhD student successfully and is currently guiding 2 research scholars. He has been on the editorial board of several international journals. He has almost fifteen years of experience in teaching.

EXCLUSIVE BOTANY FOR NEET

National Eligibility-cum-Entrance Test

(As per Latest Syllabus of NEET)

by

Khwaja Salahuddin

ACADEMIC ENCLAVE

First Edition, 2025

ISBN: 978-93-48642-29-5 (Paperback)

Printed at Replika Press Pvt. Ltd.

Published by
ACADEMIC ENCLAVE
House No. 10/15, Ground Floor,
Block-A, Street No. 14/3, Kamal Vihar,
Kamal Pur, Burari, Delhi-110084
Email: info@academicenclave.com
Mobile: +91-7042423166
Website: www.academicenclave.com

In memory of my father

the late Prof. Khwaja Aminuddin

PREFACE

It gives me immense pleasure to write this book in response to requests from a large number of students and friends keeping the requirements of medical aspirants in mind. I have been teaching Botany for the last several years and realized the problems of students in Botany. It has taken more than three years to bring out this book and I am sure that the teachers and students will like the book. It helps students to understand the subject more precisely. The subject of Biology is packed with a plethora of information and there are several books on Botany in the market, each book has its shortcomings, so students are unable to purchase all. Even if students have access to all the books, they usually may not have the time to consult them all. I have incorporated all the relevant materials from the competition's point of view by avoiding unnecessary verbosity and repetitions.

I have singled out the entire content of Biology into Botany and Zoology. I have divided the entire content of this book into 19 chapters. I have designed the book according to the latest syllabus of NEET and further I have incorporated two chapters viz. 'Biotechnology: Principles and processes', and 'Biotechnology and its applications' in this book. Special focus has been given to those topics which are usually tougher. The thorough and comprehensive information on the said topics followed by the limited number of most probable questions is given so students are advised to go through theories first and then after solving the multiple-choice questions. Each chapter contains short-type questions and long-type questions at the end for plus two students. The book has five model papers each containing 50 multiple-choice questions for the sake of NEET aspirants.

I have included all the materials containing relevant information systematically in lucid language in a readable and interesting manner without increasing the volume of the book. I am convinced that the students will be able to perform well with greater confidence in competitive medical examinations such as NEET. This book will also cater to the needs of all students associated with Botany.

I have taken utmost care to make this book flawless and indispensable for the students but the chance of some minor errors cannot be ruled out. I will welcome your suggestions for improvement.

I am sure this book is my ambitious venture and shall prove more useful to the students who aspire to be a doctor. I wish the medical aspirants all the best in their endeavours.

Khwaja Salahuddin
University Dept. of Botany
L. N. Mithila University, Darbhanga, Bihar

ACKNOWLEDGEMENTS

I want to extend my heartfelt thanks to the Hon'ble Vice Chancellor of L.N.Mithila University, Darbhanga, Prof. Sanjay Kumar Chowdhry, and a galaxy of erudite scholars of University Dept. of Botany, viz. Prof. Indra Nath Mishra, Dr. K.K. Sahu, Dr. R.N. Jha, Dr. Shahnaz Jamil for showering extraordinary kindness and evincing keen interest in this book.

I am thankful to my colleagues at LNMU, Dr. Gajendra Prasad, Dr. Sabita Verma, Dr. Ankit Kumar Singh, Dr. Roop Kala Sinha, Dr. Ghazala Urfi, Dr. Naushad Alam, Prof. Manju Roy, Dr. Nisha Saxena, Dr. Chitra Jha, Dr. Shagufta Khanam, Dr. Neelam Sen, Dr. Kiran, Dr. Amit Gupta, Dr. Manoj Kumar, Dr. Anupama, Dr. Amit Kumar Jha, Dr. Sangita Kumari, Dr. Sarita Kumari, Dr. Shailja Mishra, Dr. Garima Pathak, Dr. Dheeraj Gautam for making several useful suggestions that helped me in completing this book.

I am thankful to my well-wishers Prof. S.M. Zofair, Col. Ishrat Nabi, Ms. Mohsena Ahmad, Ms. Sabiha Yasmin, Mr. Hashmat Nabi, Ms. Farah Naz, Mr. Shakeb Nabi, Dr. Asadullah Khan, Mr. Shafqat Nabi, Mr. Azmat Nabi, Dr. Shambhu Nath Sharma, Prof. Basant Mishra, Mr. Jawed Humayun, Dr. Ahmad Ali, Dr. Ram Prasad, Md. Umar, Dr. Fauzia Farheen, Dr. Amin Arif, Dr. Deokant, Dr. Salamat Ali, Mr. Amar Thakur, Dr. Abhishekh Priyaranjan, Mr. Dipen Ghimiray, Mr. Firoz Ahmad, Dr. Nizamuddin Ansari, Dr. Chandan Kumar Singh, Dr. P.K Patel, Dr. Virendra Kumar Soni, Mr. Pawan Kumar, Mr. Trilochan Mishra, Mr. Manoj Singh, Mr. Pankaj Mathur & Mrs. Sonali Bhattacharya for encouraging me from time to time.

I submit myself to my mother and offer if I have anything at my disposal at her feet with most supplication and reverence. I have the privilege of her blessings day in and day out. This book is nothing but the fruition of her blessings and blessings alone. I pray, may she live long.

I appreciate my wife Shabana Nabi and daughters, Sara and Simra who always supported my venture even when they were greatly inconvenienced due to my preoccupations with the manuscript. I also want to thank my nephews Mr. Syed Hussain and Hassan and niece, Ms. Adiva Faisal for giving useful suggestions.

I dedicate it to my father, the late Prof. Khwaja Aminuddin who did all that he could in his lifetime for the well-being, prosperous living and all-round betterment of my kin including myself. I pray to God, may his soul rest in eternity.

Last but not least I would like to express my appreciation to my publishers, M/s Academic Enclave, New Delhi for bringing out the book in nice form.

Dr. Khwaja Salahuddin

CONTENTS

SYLLABUS OF BIOLOGY (BOTANY AND ZOOLOGY)

UNIT L: DIVERSITY IN LIVING WORLD

- **What is living?; Biodiversity; Need for classification;; Taxonomy & Systematics; Concept of species and taxonomical hierarchy; Binomial nomenclature;**
- **Five kingdom classifications: salient features and classification of Monera; Protista and Fungi into major groups: Lichens; Viruses and Viroids.**
- **Salient features and classification of plants into major groups: Algae, Bryophytes, Pteridophytes, Gymnosperms (three to five salient and distinguishing features and at least two examples of each category);**
- Salient features and classification of animals-nonchordate up to phyla level and chordate up to classes level (three to five salient features and at least two examples)'

UNIT 2: STRUCTURAL ORGANISATION IN ANIMALS AND PLANTS

- **Morphology and modifications; Tissues; Anatomy and functions of different parts of flowering plants: Root, stem, leaf, inflorescence- cymose and racemose, flower' fruit and seed (To be dealt along with the relevant practical of the Practical Syllabus) Family (Malvaceae, Cruciferae, Leguminosae, Compositae, Graminae).**
- Animal tissues; Morphology, anatomy and functions of different systems (digestive, circulatory, respiratory, nervous and reproductive) of an insect (Frog) (Brief account only)

UNIT 3: CELL STRUCTURE AND FUNCTION

- **Cell theory and cell as the basic unit of life; Structure of prokaryotic and eukaryotic cell; Plant cell and animal cell; Cell envelope' cell membrane, cell wall; Cell organelles structure and function; Endomembrane system-endoplasmic reticulum' Golgi bodies' lysosomes, vacuoles; mitochondria, ribosomes, plastids, microbodies; Cytoskeleton' cilia flagella centrioles (ultrastructure and function); Nucleus-nuclear membrane, chromatin, nucleolus.**
- Chemical constituents of living cells: Biomolecules-structure and function of proteins, carbohydrates. lipids, nucleic acids; Enzymes-types, properties' enzyme action' classification and nomenclature of enzymes
- **Cell division: Cell cycle, mitosis, meiosis and their significance**

UNIT 4: PLANT PHYSIOLOGY

- **Photosynthesis: Photosynthesis as a means of Autotrophic nutrition; Site of photosynthesis takes place; pigments involved in Photosynthesis (Elementary idea); Photochemical and biosynthetic phases of photosynthesis; Cyclic and non-cyclic and ll photophosphorylation; chemiosmotic hypothesis; photorespiration c3 and c4 pathways; Factors affecting photosynthesis.**
- **Respiration: Exchange gases; cellular respiration-glycolysis. fermentation (anaerobic), TCA cycle and electron transport system (aerobic); Energy relations- Number of ATP molecules generated; Amphibolic pathways; Respiratory quotient.**
- **Plant growth and development: Seed germination; phases of plant growth and plant growth rate; Conditions of growth; Differentiation, dedifferentiation and redifferentiation; Sequence of developmental process in a plant cell; Growth regulators: auxin, gibberellin, cytokinin, ethylene, ABA;**

UNIT 5: HUMAN PHYSIOLOGY

- Breathing and Respiration: Respiratory organs in animals; Respiratory system in humans; Mechanism of breathing and its regulation in humans- Exchange of gases, transport of gases and regulation of respiration

Respiratory volumes; Disorders related to respiration-Asthma, Emphysema, Occupational respiratory disorders.

- Body fluids and circulation: composition of blood, blood groups, coagulation of blood; composition of lymph and its function; Human circulatory system- structure of human heart and blood vessels; cardiac cycle, cardiac output. ECG. Double circulation; Regulation of cardiac activity; Disorders of circulatory system-Hypertension, coronary artery disease, Angina pectoris, Heart failure.
- Excretory products and their elimination: Modes of excretion- Ammonotelism, ureotelism, uricotelism; Human excretory system- structure and function; Urine formation, osmoregulation; Regulation of kidney function-Renin-angiotensin, Atrial Natriuretic Factor' ADH and Diabetes insipidus; Role of other organs in excretion; Disorders; Uraemia, Renal failure, Renal calculi, Nephritis; Dialysis and artificial kidney.
- Locomotion and Movement: Types of movement- ciliary, flagellar, muscular; Skeletal muscle- contractile proteins and muscle contraction; Skeletal system and its functions (To be dealt with the relevant practical of practical syllabus); Joints; Disorders of muscular and skeletal system-Myasthenia gravis, Tetany, Muscular dystrophy, Arthritis, Osteoporosis, Gout.
- Neural control and coordination: Neuron and nerves; Nervous system in humans: central nervous system, peripheral nervous system and visceral nervous system; Generation and conduction of nerve impulses;
- Chemical coordination and regulation: Endocrine glands and hormones; Human endocrine system-Hypothalamus, pituitary, pineal, Thyroid, parathyroid, Adrenal, Pancreas, Gonads; Mechanism of hormone action (Elementary idea); Role of hormones as messengers and regulators, Hypo-and hyperactivity and related disorders (common disorders e.g. Dwarfism, Acromegaly, Cretinism, goitre, exophthalmic goitre, diabetes, Addison's disease). (Important diseases and disorders mentioned above to be dealt in brief.)

UNIT 6: REPRODUCTION

- **Sexual reproduction in flowering plants: Flower structure; Development of male and female gametophytes; Pollination-types, agencies and examples; Outbreeding devices; Pollen-Pistil interaction; Double fertilization; Post fertilization events- Development of endosperm and embryo, Development of seed and formation of fruit; Special modes: apomixis, parthenocarpy, polyembryony; Significance of seed and fruit formation.**
- Human Reproduction: Male and female reproductive systems; Microscopic anatomy of testis and ovary; Gametogenesis-spermatogenesis & oogenesis; Menstrual cycle; Fertilisation, embryo development up to blastocyst formation, implantation: Pregnancy and placenta formation (Elementary idea); Parturition (Elementary idea); Lactation (Elementary idea).
- Reproductive health: Need for reproductive health and prevention of sexually transmitted diseases (STD); Birth control- Need and Methods, Contraception and Medical Termination of Pregnancy (MTP); Amniocentesis; Infertility and assisted reproductive technologies - IVF, ZIFT, GIFT (Elementary idea for general awareness).

UNIT 7: GENETICS AND EVOLUTION

- **Heredity and variation: Mendelian inheritance; Deviations from Mendelism- Incomplete dominance, Co-dominance, Multiple alleles and Inheritance of blood groups, Pleiotropy; Elementary idea of polygenic inheritance; Chromosome theory of inheritance; Chromosomes and genes; Sex determination in humans, birds, honey bee; Linkage and crossing over; Sex-linked inheritance-Haemophilia, colour blindness; Mendelian disorders in humans-Thalassemia; chromosomal disorders in humans; Down's syndrome, Tumer's and Klinefelter's syndromes.**
- **Molecular basis of Inheritance: Search for genetic material and DNA as genetic material; Structure of DNA and RNA; DNA packaging; DNA replication; Central dogma; Transcription, Genetic code, translation; Gene expression and regulation- Lac Operon; Genome and human genome project; DNA finger printing, Protein biosynthesis.**
- Evolution: Origin of life; Biological evolution and evidence for biological evolution from Palaeontology, comparative anatomy, embryology and molecular evidence); Darwin's contribution, Modem Synthetic theory of Evolution; Mechanism of evolution- Variation (Mutation and Recombination) and Natural Selection with examples, types of natural selection; gene flow and genetic drift; Hardy-Weinberg's principle; Adaptive Radiation; Human evolution.

UNIT 8: BIOLOGY AND HUMAN WELFARE

- Health and Disease; Pathogens; parasites causing human diseases (Malaria, Filariasis, Ascariasis. Typhoid, Pneumonia, common cold, amoebiasis, ring worm, dengue, chikungunya); Basic concepts of immunology-vaccines; Cancer, HIV and AIDS; Adolescence, drug and alcohol abuse, Tobacco abuse
- **Microbes in human welfare: In household, food processing, industrial production, sewage treatment, energy generation and as biocontrol agents and biofertilizers.**

UNIT 9: BIOTECHNOLOGY AND ITS APPLICATIONS

- **Principles and process of Biotechnology: Genetic engineering (Recombinant DNA technology).**
- **Application of Biotechnology in health and agriculture: Human insulin and vaccine production, gene therapy Genetically modified organisms-Bt crops: Transgenic Animals, Biosafety issues-Biopiracy and patents.**

UNIT L0: ECOLOGY AND ENVIRONMENT

- **Organisms and environment: Population interactions-mutualism, competition, predation, parasitism; Population attributes-growth, birth rate and death rate, age distribution.**
- **Ecosystem: Patterns, components; productivity and decomposition: Energy flow: Pyramids of number, biomass. energy**
- **Biodiversity and its conservation: Concept of Biodiversity; Patterns of Biodiversity: Importance of Biodiversity; Loss of Biodiversity Biodiversity Conservation; Hotspots, endangered organisms. extinction; Red Data Book, Biosphere reserves, National parks and sanctuaries, Sacred Groves.**

Note: The Bold portion of the syllabus is concerned with Botany

CHAPTER-1

THE LIVING WORLD

Biology (Gr. *Bios*=life, logos=study) is a branch of science dealing with the study of living organisms. The term 'Biology' was coined by Jean Baptiste de Lamarck (1801). Aristotle (384-322 B.C.) is regarded as the father of Biology.

Botany (Gr. *Botane* = herb) is the study of plants. Theophrastus (370-287 B.C.) is regarded as the Father of Botany.

DIVERSITY

An unlimited variety of organisms of different shapes, sizes and structures are estimated to exist right from minute bacteria to big banyan trees and elephants. During nearly 3.5 billion years of existence of life on earth, so many types of organisms have become extinct and their dead remains or impressions are now represented as fossils. Now Biologists have been able to describe near about 1.7 million types of organisms on the planet Earth in which 1.2 million animals and 0.5 million plants are found at different geographical areas. Of these, insects show maximum diversity with 7,50,000 species. Due to diversity, it is very difficult to study every plant and animal so these are required to be classified for further study.

TAXONOMY

It is the arrangement of different types of organisms on the basis of shared characteristics. It is the study of characterisation (description), identification (correct placement), nomenclature (naming) and classification of living organisms. The term 'Taxonomy' was coined by A.P. de Candolle (1813). He composed '*Theorie elementare de la botanique*'. The aim of taxonomy is to

(a) Classify the organisms on the basis of morphological and phylogenetic relationship

(b) Name and identify the organisms

(c) Assign them ranks and positions

The Father of Taxonomy is Carolus Linnaeus (Full name: Carl Von Linne) and the Father of Indian Taxonomy is Shantapau.

Linnaeus composed several books:

(i) Philosophia Botanica (1751): The principles of binomial nomenclature were mentioned.

(ii) Species Plantarum (1753): The nomenclature of 5900 plants was given and

(iii) Systema Naturae (1758): The description of 4326 animals was made.

Systematics: It was coined by Linnaeus. It is the process of organising taxonomic information about organisms into a logical classification. It deals with the kinds and diversity of all organisms and their existing relationship. It is taxonomy plus phylogeny (evolutionary history of organisms). It includes those areas of biology concerned primarily with similarities and differences between organisms.

Biosystematics: It was coined by Julian Huxley. It is the study of the diversification of living forms, both past and present and the relationships among living things through time. It includes discovery, characterization (description), identification, nomenclature, classification, and cataloguing, with their diversity, life histories, habits, role in an ecosystem and geographical distributions. Biosystematics aims to

(a) Provide information regarding habit, habitat, characters, affinities, and diversity of organisms

(b) Provide information of the evolution of different types of organisms

(c) Identify the newly discovered organisms and fossils

TYPES OF TAXONOMY

Turril divided taxonomy into 3 types:

1. **Alpha taxonomy** (Classical taxonomy): It is based on morphological characters, compilation of flora and monograph. Classical taxonomists are Theophrastus and Aristotle.
2. **Beta Taxonomy** (Modern or Explorative taxonomy): It is based on morphological, anatomical, embryological, palynological, physiological, cytological and biochemical characteristics. It deals with the identification of plants.

   It is also known as Biosystematics.
3. **Omega taxonomy** (Encyclopaedic taxonomy): It is based on all available known information about plants. It deals with evolutionary patterns.

   It is also known as Neosystematics (a term coined by Julian Huxley, 1940).

NEW BRANCHES OF TAXONOMY

1. **Chemotaxonomy**: It is based on studies of chemical constituents such as polysaccharides, amino acids, proteins, fatty acids, nucleic acids (DNA sequences), alkaloids, pigments (betacyanin in beetroot), aromatic compounds, tannins, crystals in the form of cystolith (Calcium carbonates) and raphides (Calcium oxalates).
2. **Cytotaxonomy**: It is based on cytological studies such as number, structure and meiotic behaviour of chromosomes.
3. **Karyotaxonomy**: It is based on studies of chromosomal bands (light and dark bands).
4. **Numerical Taxonomy** (Phenetics / Taximetrics or computer-aided taxonomy): It was proposed by Adanson. It is based on numerical or statistical methods in which a large number of characters are observed and recorded. It determines the similarities and dissimilarities or primitiveness and advancement of several characters at the same time. It has the following features:

   (a) If the content of information in taxa is increased then the classification will be better.

   (b) Every character is of equal weight in creating natural taxa.

   (c) Classification is based on the percentage similarity coefficient for each operational taxonomic unit (OTU).

   Dendrogram: It is a family tree of organisms based on numerical taxonomy.

   Cladogram: It is a family tree of organisms based on phylogeny (showing ancestry).

TAXONOMIC CATEGORY

It is the position or status or level in a hierarchal classification of organisms.

Taxon: It was coined by Adolf Mayer (1926). It is the scientific term of category. It is a taxonomic group or unit of organisms of any rank in any category. It is a group of genetically related individuals having distinct characteristics from those of other groups. It was introduced in 1956 by ICBN. All fishes or birds form taxon. Cats, mammals and animals represent taxa at different levels or categories.

Classification of Mango and wheat

Category	Taxon	Taxon
Kingdom	Plantae	Plantae
Division	Angiospermae	Angiospermae
Class	Dicotyledonae	Monocotyledonae
Order	Sapindales	Poales
Family	Anacardiaceae	Poaceae
Genus	*Mangifera*	*Triticum*
Species	*indica*	*aestivum*

Classification of Human and Housefly

Category	Taxon	Taxon
Kingdom	Animalia	Animalia
Phylum	Chordata	Arthropoda
Class	Mammalia	Insecta
Order	Primata	Diptera
Family	Hominidae	Muscidae
Genus	*Homo*	*Musca*
Species	*sapiens*	*domestica*

TAXONOMIC HIERARCHY

It is a classification of organisms in a definite but descending sequence of categories. Hence all categories constitute a hierarchy. There are two types of Hierarchies:

1. **Linnean Hierarchy:** Linnaeus first proposed a hierarchy (Ranking system) in which 5 categories were mentioned. But he did not incorporate Division and Family.

   Class----- Order----- Genus----Species----Variety

2. **Obligate Hierarchy:** It mentions 7 obligate categories.

   Kingdom---Division/Phylum---Class----Order---Family---Genus---Species

1. **Kingdom**: It is the highest taxonomic category. It is a group of related divisions. All plants are placed in kingdom-Plantae.
2. **Division**: It is a group of related classes in plants. It was coined by Eichler. It mentions suffix-phyta. If subdivision exists then it will be -phytina.

   Phylum: It is a group of related classes in animals. The term 'Phylum' was coined by Cuvier.
3. **Class**: It is a group of related orders. It was coined by Linnaeus. It mentions suffix-phyceae (algae) or -opsida (plants such as Bryophytes, Pteridophytes and Gymnosperms). If a subclass exists then it will mention the suffix -idea for plants. The orders such as Ranales, Polemoniales and Personales have been placed in a class, Dicotyledonae.
4. **Order**: It is a group of related families having a few similar features. It was coined by Linnaeus. It mentions suffix-ales. If a suborder exists then it mentions suffix-ineae. The families such as Solanaceae and Convolvulaceae have been placed in the order, Polemoniales.
5. **Family**: It is a group of related genera having some similar or common features. It was coined by John Ray. It mentions the suffix -aceae (plants) or –idae (animals). If there is a sub-family then the suffix will be –oideae. The three different genera such as *Solanum, Atropa* and *Nicotiana* have been placed in the family, Solanaceae.

   Tribe: It is between family and genus. It mentions the suffix -eae (plants) and -ini (animals)
6. **Genus**: It is a group of closely related species having many similar or common features. It was coined by Tournefort. It occupies a special position in classification hence species cannot be named without assigning it to a genus. Plants such as Potato (*Solanum tuberosum*) and Brinjal (*Solanum melongena*) are two species but both belong to the same genus-*Solanum*. The genus is of two types:

   (a) Monotypic genus: Some genera have only one species. e.g., *Ginkgo biloba*, *Homo sapiens.*

   (b) Polytypic genus: Some genera have more than one species. e.g., *Solanum* has *tuberosum* and *melongena*; *Panthera* has *leo* and *tigris.*
7. **Species**: It is the basic (smallest) unit of biological classification. It was coined by John Ray. It is a group of naturally interbreeding populations with the ability to produce fertile offspring but reproductively isolated. It is a group of closely related individuals with similar morphological, anatomical, biochemical, and cytological characteristics. It is of two types:

(a) Monotypic species (Microspecies): The species without differentiation of sub-species or variety. It is the species in which variations are less but it reproduces asexually. e.g., Bacteria (*Escherichia coli*), Cyanobacteria (*Anabaena azollae*).

(b) Polytypic species: (Macrospecies): The species in which more than one variety or subspecies are found. It is of three types:

(i) Biotype: The members of the same species inhabit a similar environment and they show some permanent genetic variations. The members cannot breed among themselves. e.g., *Brassica oleracea* var. *botrytis* (Cauliflower), *Brassica oleracea* var. *capitata* (Cabbage) and *Brassica oleracea* var. *pristis* (Knol kohl).

(ii) Ecotype: The members of the same species inhabit different environments and they show some permanent genetic variations. The members can breed themselves but due to geographical barriers they cannot interbreed or it is a genetically different population. e.g., *Corvus splendens splendens* (Indian crow), *Corvus splendens insolence* (Myanmarese crow) and *Corvus splendens perotegatus* (Srilankan crow).

(iii) Ecades (Ecophenes): The members of the same species show some temporary non-genetic variations due to the environment or It is a form of a plant modified by its habitat but not heritable. e.g., Money plant is dark green (in shade) and yellow (in sunlight).

TYPES OF SPECIES

1. Morphospecies: The species which are based on morphological characters only.
2. Taxonomic species: The species which have definite binomial names.
3. Sympatric species: The species that live in same geographical area.
4. Allopatric species: The species that live in different geographical areas.
5. Parapatric species: The species that live in the narrow overlap of two different areas.
6. Synchronic species: The species which belong to the same period. e.g., *Archaeopteryx and Dinosaur*
7. Allochronic species: The species which belong to different periods. e.g., Dinosaur and Man.
8. Sibling species (Cryptic species): True species that do not breed (reproductively isolated).
9. Palaeospecies: These species which are known from fossils only. e.g., Dinosaurs.
10. Neontological species: The species which are living.

Features of Categories:

Kingdom---Division---Class---Order--Family--Genus---Species

(Less)←**Specific characters**→(more)

(Less)←**Common or Similar characters**→(more)

(More)←**Diversity**→(less)

Institutions for Nomenclature:

1. International Code of Botanical Nomenclature (**ICBN**)
2. International Code of Zoological Nomenclature (ICZN)
3. International Code of Viral Nomenclature (ICVN)
4. International Code for Nomenclature of Bacteria (ICNB)
5. International Code of Nomenclature for Cultivated Plants (ICNCP)

ICBN

It was established at 12th International Botanical Congress held at Leningrad in 1975 to provide uniformity and to avoid confusion regarding nomenclature of plants. It came into existence in 1978.

ICBN is set of rules and recommendations dealing with the botanical names given to plants. The problem of internationally acceptable nomenclature was discussed in the "First International Botanical congress" held in Paris in August 1887. The name of ICBN was changed to "International Code of Nomenclature (ICN) for Algae, Fungi and Plants at the International Botanical Congress held at Melbourne, Australia in July, 2011. The current version of the code is "Shenzhen code' adopted by the IBC held in Shenzhen, China in July 2017. The documentation of the code was published in its final form on 26 June, 2018.

ICBN of St. Louis code (2000) is a book of law containing the rules and recommendations prescribed for naming the plants in an international context and comprise three divisions:

1. **Principles:** These are basic points on which code is based. They do not give detailed rules about nomenclature but show the main ideas that have guided the compilers of the Code and should be kept in view by any botanist attempting to publish a new taxon.
   (a) Botanical nomenclature is an independent of zoological and bacteriological nomenclature. Plant names became effective from 1st May, 1753 and animal names from 1st August, 1758.
   (b) The application of names of taxonomic group is determined by means of nomenclature type.
   (c) The nomenclature of taxonomic group is based on the priority of publication.
   (d) Each taxonomic group with a particular circumscription, position and rank can bear only one current name, the earliest that is in accordance with the rules, except in specified cases.
   (e) Scientific names of taxonomic groups are treated as Latin, regardless of their derivation.
   (f) The rules of nomenclature are retroactive unless expressly limited.
2. **Rules**: These give detailed account on all the points concerned with the naming of plants. These were distributed in 62 articles. The names which contradict any one or more of the above rules are illegitimate.
3. **Recommendations:** These are practical applications of the rules and deal with subsidiary points. The objective is to bring about greater uniformity in nomenclature. The names contrary to Recommendations need not be rejected.

BINOMIAL NOMENCLATURE

It was first introduced by Kasper Bauhin (1623) in his book 'Pinax theatri botanici' (Illustrated Exposition of Plants) but it was finally proposed by Linnaeus (1753). There are several **rules** which are as follows:

1. The name of an organism is composed of two words: generic name (identifying genus) and specific name (identifying species).
2. The name of the genus should be written first followed by the name of the species.
3. The first letter of the name of the genus should be in capital and that of the species in small.
4. The name of the genus and species should contain between 3 and 12 letters.
5. The botanical name should be taken from the Latin language because it was the language of scholars at the time of Linnaeus.
6. The botanical name (genus and species) should be printed in italics. If it is handwritten then it should be underlined separately.
7. The name of the discoverer of a plant should be written after the species in abbreviation. e.g., *Zea mays* L.
8. The name of the family and other categories should be based on the type genus.
9. The name of the family and other categories should not be printed in italics.
10. If a species name is transferred or revised, then the name of the original worker should be written but in parentheses. e.g., *Syzygium cumini* (L) Skeels
11. If the name of the species is compound (having two words) then a hyphen should be mentioned between these words.
12. In case of two or more names, the name given first is recognised as a valid name and all other names will be called synonyms.

Trinomial Nomenclature: It was proposed by Lamarck. In the case of plants, it includes the name of genus, species and variety. e.g.,*Brassica oleracea botrytis* (cauliflower), *Brassica oleracea capitata* (cabbage) but in the case of animals, it includes the name of genus, species and sub-species. e.g., *Homo sapiens sapiens* (Human being), *Corvus splendens splendens* (Indian crow)

Nomenclature Types: It acts as the source for naming a certain taxonomic group. The ICBN recognised the following several types:

1. Holotype: It is the original description of the type specimen by the author.
2. Isotype: It is a duplicate of the holotype so this is another specimen collected at the same place, same time by the same author.
3. Neotype: It is a new type of specimen when the original description (holotype) is missing.
4. Paratype: It is an actual description of the holotype with some additional information.
5. Syntype: It is two or more type specimens when the original description (holotype) is missing.
6. Lectotype: It is used as a substitute when holotype is lost.

Nomenclature Terms:

1. Autonym: It is a name with the same species and subspecies. e.g., *Corvus splendens splendens*
2. Synonym: It is more than one name of a taxon. e.g., *Catharanthus roseus* (=*Vinca rosea*) for periwinkle and *Malus pumila* (=*Pyrus malus*) for apple.
3. Homonym: It is the same name for two different plants. e.g., *Prunus dulcis* for almond and plum.
4. Basionym: It is the name on which the new combination is based.
5. Tautonym: It is a name with the same genus and species. It is only for animals. e.g., *Bison bison, Rattus rattus, Naja naja, Gorilla gorilla, Catla catla.*

SHORT ANSWER TYPE QUESTIONS

1. Distinguish between taxonomy and systematics.
2. How does alpha taxonomy differ from beta taxonomy?
3. What do you mean by numerical taxonomy?
4. What is the difference between a dendrogram and a cladogram?
5. Explain species with examples.
6. Distinguish between sympatric and allopatric species.
7. Distinguish between taxon and category.
8. Explain the Linnean hierarchy.
9. Explain the monotypic genus and polytypic genus.
10. Distinguish between binomial and trinomial nomenclature.

LONG ANSWER TYPE QUESTIONS

1. Discuss the different rules of binomial nomenclature.
2. Describe all the categories of taxonomic hierarchy.
3. Explain obligate categories.
4. Write notes on ICBN.
5. Write the classification of mango.

MULTIPLE CHOICE QUESTIONS

1. **The term 'Taxonomy 'was coined by**
   (a) John Ray (b) Julian Huxley
   (c) A.P. de Candolle (d) Linnaeus
2. **Systematics involves**
   (a) Taxonomy and ontogeny
   (b) Taxonomy and phylogeny
   (c) Numerical taxonomy and karyotaxonomy
   (d) Cytotaxonomy and karyotaxonomy
3. **The term 'Neosystematics' was coined by**
   (a) Linnaeus (b) A.P. de Candolle
   (c) Julian Huxley (d) Adanson
4. **Which of the following features is not included in Taxonomy?**
   (a) Nomenclature (b) Phylogeny
   (c) Identification (d) Classification
5. **The family tree of organisms on the basis of phylogeny is called**
   (a) Dendrogram (b) Histogram
   (c) Karyogram (d) Cladogram
6. **Linnean hierarchy does not include**
   (a) Class (b) Order
   (c) Family (d) Variety
7. **In animals, the category 'Division' of plants is equivalent to**
   (a) Domain (b) Phylum
   (c) Class (d) Variety
8. **Which of the following categories in animals is equivalent to 'Subspecies' of plants?**
   (a) Form (b) Section
   (c) Series (d) Variety
9. **The suffix of 'Family' is**
   (a) –ae (b) –eae
   (c) -ceae (d) -aceae
10. **The suffix of 'Tribe' is**
    (a) –ae (b) –eae
    (c) –ceae (d) –aceae
11. **The 'Tribe' is placed between**
    (a) Order and family (b) Family and Genus
    (c) Genus and Species (d) Species and Variety
12. **Which is not correct with respect to taxonomy?**
    (a) Numerical taxonomy is also known as phenetics
    (b) Beta taxonomy is also known as biosystematics
    (c) All birds form a taxon
    (d) Taxonomy includes phylogeny
13. **The exceptions to the 'Biological species concept' in natural condition are**
    (a) Mule and Tigon (b) Tigon and Liger
    (c) Mule and Hinny (d) Hinny and Tigon
14. **The specific characters are more in**
    (a) Class (b) Order
    (c) Family (d) Genus
15. **The diversity is less in**
    (a) Class (b) Order
    (c) Family (d) Genus
16. **The polytypic species is shown by the genus**
    (a) Solanum *(b) Brassica*
    (c) Lilium (d) All of the above
17. **Dinosaurs and *Archeopteryx* belong to**
    (a) Allochronic species (b) Synchronic species
    (c) Allopatric species (d) Sympatric species
18. **The name of plants was effective from**
    (a) 1-5-1751 (b) 1-5-1753
    (c) 1-5-1758 (d) 1-5-1759
19. **The botanical name, *Prunus dulcis*, is homonym for**
    (a) Plum and Almond (b) Plum and Walnut
    (c) Almond and Blackberry
    (d) Cashew and Walnut
20. **If a particular specimen was originally described by the author, it is called**
    (a) Isotype (b) Holotype
    (c) Paratype (d) Lectotype
21. **The botanical name of cauliflower is**
    (a) Brassica oleracea var. *botrytis*
    (b) Brassica oleracea var. *capitata*
    (c) Brassica oleracea var. *pristis*
    (d) Brassica oleracea var. *glomerata*
22. **Which of the following terms of nomenclature is used especially for animals?**
    (a) Autonym (b) Basionym
    (c) Homonym (d) Tautonym

23. Dinosaurs and human beings belong to
(a) Sympatric species (b) Allopatric species
(c) Synchronic species (d) Allochronic species

24. Biological epithet includes
(a) Generic name and specific name
(b) Specific name and variety name
(c) Generic name, specific name and author citation
(d) Specific name, variety name and author citation

25. The ICBN came into existence in the year
(a) 1975 (b) 1978
(c) 1981 (d) 1986

26. ICBN does not include
(a) Principles (b) Rules
(c) Recommendation (d) Hypothesis

27. Potato and brinjal belong to
(a) Same genus but different species
(b) Same species but different varieties
(c) Same families but a different genus
(d) Different families but same genus

28. The species which live in the same geographical areas are known as
(a) Allopatric species (b) Sympatric species
(c) Parapatric species (d) Synchronic species

29. The phylogenetic relationship can be established by
(a) Isoenzymes
(b) Alkaloids
(c) Alcoholic dehydrogenase
(d) Morphology

30. The phylogenetic relationship can be precisely established by
(a) Amino acid sequences
(b) DNA
(c) rRNA
(d) mRNA

31. The molecular clock of evolution could be traced on the basis of
(a) Comparison of the short arm of 16 S rRNA
(b) DNA fingerprinting
(c) Fossil study
(d) Substitution in amino acids of polypeptides due to mutation

32. When two species are morphologically almost identical but reproductively isolated, are termed as
(a) Taxonomic species (b) Ecotypes
(c) Sibling species (d) Morphospecies

33. A group of species that are phylogenetically closer but they are lacking common ancestor. Such group is regarded as
(a) Monophyletic (b) Polyphyletic
(c) Paraphyletic (d) Sympatric

34. Which of the following techniques is used among plants for genetic relationships?
(a) Amino acid sequence
(b) Serological investigation
(c) Isoenzyme profiling
(d) Chromosome morphology

35. ICBN is not applicable for
(a) Algae (b) Bryophytes
(c) Angiosperms (d) Cultivated plants

36. Besides the nomenclature of plants in the wild, ICBN also gives binomial names for
(a) Fungi (b) Cultivated plants
(c) Bacteria and fungi (d) Fungi & cultivated plants

37. ICBN was established at
(a) 10th International Botanical Congress
(b) 11th International Botanical Congress
(c) 12th International Botanical Congress
(d) 13th International Botanical Congress

38. Which of the following is capitalised in a printed scientific name?
(a) Species (b) Genus
(c) Family (d) Order

39. Each category of taxonomic hierarchy refers to as a unit of
(a) Systematics (b) Nomenclature
(c) Identification (d) Classification

40. The common characteristic between tomato and potato will be maximum at the level of
(a) Genus (b) Family
(c) Order (d) Division

41. The true species are
(a) Interbreeding
(b) Sharing the same niche
(c) Feeding on the same food
(d) Reproductively isolated

42. Which of the following taxons is 'Domain'?

(a) Bacteria (b) Eubacteria
(c) Archaebacteria (d) Protista

43. The diversity is maximum in

(a) Class (b) Order
(c) Family (d) Genus

44. The term 'family' was coined by

(a) Linnaeus (b) John Ray
(c) Eichler (d) Julian Huxley

45. House fly belongs to the order

(a) Isoptera (b) Orthoptera
(c) Hymenoptera (d) Diptera

ANSWERS

1.c	11.b	21.a	31.d	41.d
2.b	12.d	22.d	32.c	42.a
3.c	13.c	23.d	33.b	43.a
4.b	14.d	24.c	34.a	44.b
5.d	15.d	25.b	35.a	45.d
6.c	16.b	26.d	36.c	
7.b	17.b	27.a	37.c	
8.d	18.b	28.b	38.b	
9.d	19.a	29.a	39.d	
10.b	20.b	30.b	40.b	

CHAPTER-2

BIOLOGICAL CLASSIFICATION

Biological classification is the study of arranging organisms into groups on the basis of their similarities and dissimilarities. It is very important because

(a) It helps in the easy identification of organisms.

(b) It gives an idea of the diversity of living organisms.

(c) It gives an idea of the origin of the organism and their interrelationship

(d) It gives an order of evolution of different types of organisms.

Several classifications have been proposed which are as follows:

1. **Scientific Classification:** It was proposed by Aristotle. He classified plants into Herbs, Shrubs and Trees based on morphological characters. He also classified animals on the basis of availability of RBCs into **Anaima** (animals without red blood. e.g., Invertebrates) and **Enaima** (animals having red blood. e.g., Vertebrates).

2. **Two-kingdom system**: It was proposed by Linnaeus (1758). He divided living organisms into two kingdoms on the basis of (a) Presence or absence of cell wall, (b) Mode of nutrition, (c) Locomotion and (d) Response to external stimuli

Plantae	Animalia

Demerits:

(a) The organisms are very much inconsistent.

(b) It failed to place a few dual organisms such as *Chlamydomonas, Euglena*, Slime moulds, *Rhizopus*, and Mushroom which showed their disputed status.

(c) The position of fungi is not clear.

(d) It failed to distinguish between unicellular and multicellular organisms.

(e) It failed to distinguish between prokaryotic and eukaryotic organisms.

(f) It failed to distinguish between algae (photosynthetic) and fungi (non-photosynthetic).

3. **Three Kingdom system**: It was proposed by Haeckel (1866). He included Protista for the first time. He divided living organisms into three kingdoms:

Protista	Plantae	Animalia

Demerits:

(a) Bacteria and Cyanobacteria were placed in Protista.

(b) Slime moulds, Protozoans, Algae and Fungi were placed in Protista in spite of dissimilarities between these organisms.

4. **Four Kingdom System**: It was proposed by Copeland (1956). He included Monera for the first time. He divided living organisms into four kingdoms:

Monera	Protista	Plantae	Animalia

Demerits:

(a) Slime moulds, and protozoans were placed in Protista in spite of dissimilarities

(b) Algae and fungi were included in Protista.

5. **Five Kingdom System**: It was proposed by Whittaker (1969). He included Fungi for the first time. He divided living organisms into five kingdoms on the basis of the following features:

Features	Monera	Protista	Fungi	Plantae	Animalia
Body organisation	Unicellular	Unicellular	Multicellular/ Loose tissue	Multicellular	Multicellular/ organ system
Cell structure	Prokaryotic	eukaryotic	Eukaryotic	Eukaryotic	Eukaryotic
The presence or absence of cell wall	Present	Mostly present	Present	Present	Absent
Composition of cell wall	Peptidoglycan	Cellulose	Chitin	Cellulose	Absent
Mode of nutrition	Mostly heterotrophic	Autotrophic and Heterotrophic	Heterotrophic (Absorptive)	Autotrophic	Heterotrophic (Ingestive)
Ecological Lifestyle	Mostly decomposer	Producers, Consumers, Decomposers	Decomposers	Producers	Consumers

Merits:

(a) It is natural and more scientific

(b) It follows a phylogenetic system.

(c) It gives more emphasis on cellular organisation and mode of nutrition.

(d) Fungi have been placed separately due to their heterotrophic mode of nutrition.

(e) The disputed organisms such as *Euglena* and slime moulds have been placed in Protista.

(f) Plantae have been more homogeneous with the exclusion of fungi and some unicellular algae.

(g) Animalia has become more homogeneous with the exclusion of Protozoans.

Demerits:

(a) Viruses have not been placed properly.

(b) Archaebacteria and bacteria have not been separated.

(c) Archaebacteria and *Mycoplasma* have been placed in the kingdom Monera.

(d) Monera and Protista do not show monophyletic origin (developing from a single ancestor).

(e) Slime moulds showing fungal features have been placed in Protista.

(f) Protista is an artificial group and diverse.

(g) The unicellular algae such as *Volvox* and *Chlamydomonas* have been placed in the kingdom Plantae.

(h) Algae have been placed in three kingdoms such as Monera, Protista and Plantae.

6. **Six kingdom system**: It was proposed by Carl Woese (1990). He divided all organisms into three Domains on the basis of 16S ribosomal RNA.

Origin	Domains	Kingdoms
Progenitor	Archea	Archaebacteria
	Bacteria	Eubacteria
	Eukarya	Protista
		Fungi
		Plantae
		Animalia

He proposed that Archaea, Bacteria and Eukarya each developed separately from an ancestor with less developed genetic machinery called progenote. The ribosomal RNA unlike the genome, has changed more slowly hence there was a slow evolution of rRNA. So, it is used for comparing the ancestry of different strains of bacteria. The sequencing of 23S rRNA provided more information for classification because the evolutionary changes in rRNA occurred in different segments at different times. The rRNA has been referred to as a molecular clock because of the independent rate of changes in sequences.

Thus, he separated Archaebacteria from Eubacteria on the basis of the presence of peptidoglycan in the cell wall and the occurrence of branched-chain lipids in the membrane of Archaebacteria.

ARCHAEBACTERIA:

These are ancient bacteria present in extreme or hostile conditions of the environment that prevailed on the primitive earth. These appeared to be the oldest living fossils. These are free-living, non-motile, and Gram-negative. These are a heterogeneous group of prokaryotes that are different from bacteria. These are of mainly three types:

1. **Methanogens:** e.g., *Methanococcus, Methanobacterium, Methanosarcina.*
   (a) These are present in marshy areas, water-logged fields, wetlands, gobar gas plants, lake mud, marine sediment, anaerobic sludge digester, the intestinal tract of animals and humans, and rumen of cattle.
   (b) These are rounded in shape (cocci form).
   (c) These are chemoautotrophs and obligate anaerobes.
   (d) These convert CO_2, formic acid and acetic acid (fermentation products of other anaerobes) to methane.
   (e) These help in the digestion of cellulose in ruminants.

2. **Halophiles:** e.g., *Halococcus, Halobacterium.*

These are present in sea, salted lake, salted fish, salted hides etc.

(a) These are rounded in shape (cocci form).
(b) These are chemoheterotrophs and facultative anaerobes.
(c) These tolerate high salt concentrations due to the presence of special lipids in membrane and mucilage covering.
(d) They form a purplish pigmented membrane (made up of bacteriorhodopsin) which protects from strong sunlight.
(e) They utilise light energy to run their metabolism and produce ATP but they cannot utilise this ATP for the synthesis of food.

3. **Thermoacidophiles:** e.g., *Pyrococcus, Thermoplasma, Sulpholobus*

(a) These are present in hot sulphur springs, and thermal vents.
(b) These grow at high temperature (80^{0}C) and acidity (pH=2).
(c) These are chemoautotrophs facultative anaerobes.
(d) These tolerate high temperatures due to the presence of homopolar bonds in their protein.
(e) These convert sulphur to sulphur dioxide under aerobic conditions and further oxidise to form sulphuric acid which makes the environment acidic.
(f) These convert sulphur to hydrogen sulphide under anaerobic conditions.

Archaebacteria	Bacteria
These are extremophiles.	These are mesophiles
Their cell wall is not made up of peptidoglycan. Instead, protein and non-cellulosic polysaccharides are present.	Their cell wall is made up of peptidoglycan.
Their membrane lipids have branched chain hydrocarbons linked to glycerol by ether linkage.	Their membrane lipid has straight-chain hydrocarbons linked to glycerol by ester linkage.

They have introns.	They do not have introns.
They have several RNA polymerase enzymes	They have only one RNA polymerase enzyme.
They have an initial amino acid, methionine, for protein synthesis.	They have formylated methionine for protein synthesis.
They are not sensitive to streptomycin and chloramphenicol.	They are sensitive to streptomycin and chloramphenicol.

BACTERIA

Bacteria are the simplest, most primitive and microscopic organisms. Bacterium was discovered by Antonie Van Leeuwenhoek (1675) from rainwater using a compound microscope of his design and called it 'Wild animalcule'. The term 'Bacterium' was coined by Ehrenberg (1838).

Father of Microbiology (study of microbes): Leuwenhoek

Father of Modern Microbiology: Louis Pasteur

Father of Bacteriology (study of bacteria): Robert Koch

Smallest bacterium	*Mycoplasma gallisepticum* (0.2μm). It is equal to the largest virus.
Largest bacterium	*Thiomargarita namibiensis* (100 μm). It is visible to unaided eye

Contributions of Pasteur:

(a) He proposed the germ theory of disease.

(b) He discovered that microbes can cause fermentation

(b) He developed a vaccine for Rabies (Immunisation technique).

(b) He disproved the theory of spontaneous generation.

(b) He discovered the existence of life in the absence of oxygen.

(b) He developed the Pasteurisation technique.

Contributions of Robert Koch:

(a) He proved the germ theory of disease.

(b) He isolated *Bacillus anthracis, Mycobacterium tuberculosis* and *Vibrio cholerae.*

(b) He cultured bacteria for the first time.

(b) He developed a staining method for bacteria.

(b) He proposed **Koch's postulates** to establish a causal relationship between a microbe and a disease. His postulates are as follows:

(i) The suspected microorganism must always be found in diseased individuals but never in healthy ones.

(ii) The microorganism must be isolated in pure culture on a nutrient medium.

(iii) The same disease must develop when the isolated microorganism is inoculated into a healthy individual.

(iv) The same organism must be reisolated from the experimentally infected host.

Features of Bacteria:

(a) These are found in every habitat (ubiquitous) but mostly in soil (one gram of soil contains 40 million bacterial cells and 1 ml. of water contains one million bacterial cells).

(b) These show simple structure but their behaviour is complex.

(c) These are prokaryotic and unicellular but some show colonial or filamentous structure. It is one-tenth the size of a eukaryotic cell.

(d) The cell wall is made up of peptidoglycan or murein or mucopeptide.

(e) The cell has a single envelope type of organisation.

(f) The flagella are single-stranded and are made up of flagellin protein.

(g) The cell lacks membrane-bound cell organelles such as mitochondria, chloroplast, endoplasmic reticulum, Golgi bodies etc.

(h) The respiratory enzymes are present along the infolding of the plasma membrane called mesosome.

(i) The ribosome is 70S and found freely in cytoplasm.

(j) These do not show cyclosis or cytoplasmic streaming.

(k) Their DNA is single (haploid), circular, double-stranded, and naked (without histone proteins).

(l) These show the most extensive metabolic diversity.

(m) Their mode of nutrition is autotrophic as well as heterotrophic.

(n) Their mode of respiration is aerobic as well as anaerobic.

(o) These undergo asexual reproduction by transverse binary fission, or endospore or conidia.

(p) These do not undergo mitosis and meiosis.

(q) Their cell division is amitotic type.

(r) These lack sex organs hence there is no formation of gametes.

(s) These do not undergo sexual reproduction.

Shapes of Bacteria:

1. **Bacillus (Pl. Bacilli):** It is rod-shaped. It is most common.
   (a) Monobacillus: It is a single rod-like cell. e.g., *Rhodospirillum*
   (b) Diplobacillus: It divides in one plane and remains paired. e.g., *Bacillus subtilis*
   (c) Streptobacillus: It divides several times in one plane forming several bacteria attached in a chain. e.g., *Bacillus anthracis*
   (d) Trichome: Cells of chains have large areas of contact with each other. e.g., *Beggiatoa*
   (e) Palisade-like: Cells are lined side by side and at angles to one another. e.g., *Corynebacterium*
   (f) Hyphae: Cells are arranged in long-branched filaments. e.g., *Streptomyces*

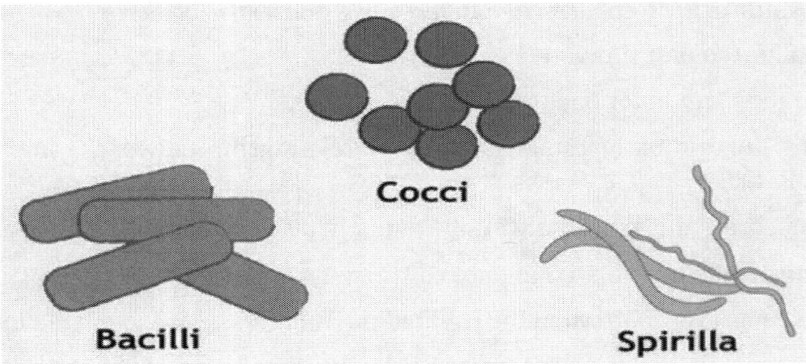

2. **Coccus (Pl. Cocci):** It is spherical.
   (a) Micrococcus (Monococcus): It is single. e.g., *Micrococcus flavus*
   (b) Diplococcus: It divides in one plane and remains paired. e.g., *Diplococcus pneumoniae*
   (c) Tetracoccus: It divides into two planes at a right angle to one other forming a group of four cells. e.g., *Tetracoccus cechii*
   (d) Streptococcus: It divides several times in one plane forming several bacteria attached in the chain. e.g., *Streptococcus lactis*

(e) Staphylococcus: It divides several times irregularly and all are attached together just like a bunch of grapes. e.g., *Staphylococcus aureus*

(f) Sarcinae: It divides into three planes at a right angle to one another forming a cubical packet of 8, 16, 32. e, g., *Sarcina*

3. **Spirillum (Pl. Spirilla)**: It is spiral-shaped. It has several twists with several turns. e.g., *Heliobacter pylori, Spirillum.*
4. **Vibrion (Pl. Vibrio):** It is comma-shaped or slightly curved. It shows less than one complete twist or turn. e.g., *Vibrio cholerae*
5. **Spirochaetes**: It is highly flexible and tightly curved e.g., *Treponema pallidum.*
6. **Pleomorphic**: It has a different shape and form. e.g., *Rhizobium, Arthrobacter, Azotobacter*

Mode of nutrition:

1. **Autotrophs**: They utilise CO_2 as a carbon source for the synthesis of food.

(a) Photoautotrophs (Photolithotrophs) or Photosynthetic: These absorb and utilise light energy to prepare organic food from CO_2.

Carbon source	CO_2
Energy source	Light
Electron donor	$H_2S/Na_2S_2O_3$

$$6CO_2 + 12H_2S \xrightarrow{\text{Light, bacterioviridin}} C_6H_{12}O_6 + 6H_2O + 12S$$

e.g., *Chlorobium* (Green sulphur bacteria)

$$6CO_2 + 3Na_2S_2O_3 + 15H_2O \xrightarrow{\text{Light, bacteriochlorophyll}} C_6H_{12}O_6 + 6NaHSO_4 + 6H_2O$$

e.g., *Chromatium* (Purple sulphur bacteria)

(b) Chemoautotrophs (Chemolithotrophs) or Chemosynthetic: These capture energy released by oxidation of inorganic compounds to prepare food from CO_2. These bacteria help in recycling of nutrients.

Carbon source	CO_2
Energy source	Oxidation of inorganic compounds (NH_3, NO_2^-, NO_3^-, Fe^{+3}, H, S)
Electron donor	Inorganic compound

(i) $2NH_3 + 3O_2 \longrightarrow 2NO_2^- + 2H_2O + 2H^+ + \text{Energy (78kcal)}$

e.g., *Nitrosococcus, Nitrosomonas*

(ii) $2NO_2^- + O_2 \longrightarrow 2NO_3^- + \text{Energy (17 kcal)}$

e.g., *Nitrococcus, Nitrobacter*

(iii) $4FeCO_3 + 6H_2O + O_2 \longrightarrow 4Fe(OH)_3 + 4CO_2 + \text{Energy}$

e.g., *Ferrobacillus* (Iron bacteria)

(iv) $2H_2 + O_2 \longrightarrow 2H_2O + \text{Energy.}$

e.g., *Hydrogenomonas* (Hydrogen bacteria)

(v) $2H_2S + O_2 \longrightarrow 2H_2O + 2S + \text{Energy}$

e.g., *Beggiatoa* (Sulphur bacteria)

2. **Heterotrophs**: These require organic compounds as carbon sources.

(a) Photoheterotrophs (Photoorganotrophs):

Carbon source	Organic compound
Energy source	Light
Electron donor	Organic compound (alcohol/acid)

$$6CO_2 + CH_3CH_2CH_2OH \xrightarrow{\text{Light, bacteriorhodopsin}} C_6H_{12}O_6 + 12 \text{ acetone} + 6H_2O$$

e.g., *Chloronema, Chloroflexus* (Green non-sulphur bacteria)

Rhodopseudomonas, Rhodospirillum (Purple non-sulphur bacteria)

(b) Chemoheterotrophs (Chemoorganotrophs):

Carbon source	Organic compound
Energy source	Oxidation-reduction of organic compound
Electron donor	Organic compound

These are of three types:

(i) Parasites: These obtain nutrition for growth from living plants and animals. These may be **Obligate parasites** (These organisms survive and grow only in living cells) and **Facultative parasites** (These are basically saprotrophs but become parasitic under certain conditions. e.g., *Bacillus, E. coli, Salmonella, Pseudomonas, Agrobacterium, Xanthomonas*

(ii) Saprotrophs: These obtain nutrition from dead organic matter by breaking down complex organic compounds into simpler forms with the secretion of extracellular enzymes. e.g., *Bacillus ramosus, Bacillus mycoides*. The majority of bacteria are aerobic saprotrophs.

(iii) Symbionts: These obtain nutrition from host plants and provide nitrates to plants by establishing symbiotic relationships. These produce humus from animals and plant remains and cause the decay of materials and useful to man. e.g., *Rhizobium*

Terms:

- Fermentation: It is an anaerobic breakdown of carbohydrates.
- Putrefaction: It is an anaerobic breakdown of protein.
- Decay: It is an aerobic breakdown of organic compounds.

Bacterial photosynthesis	Green plant photosynthesis
There is one photochemical reaction which involves only one pigment system.	Two photochemical reactions involve two pigment systems.
It occurs at a wavelength longer than 700 nm.	It occurs at a wavelength in a range of 400-700 nm.
The reaction centre is B-890.	The reaction centres are P-680 and P-700.
It involves only cyclic photophosphorylation.	It involves both non-cyclic and cyclic photophosphorylation.
The electron donor is H_2S.	The electron donor is H_2O.
The evolution of O_2 does not occur.	Evolution of O_2 occurs.
The reductant is $NADH + H^+$	The reductant is $NADPH + H^+$
Emerson's enhancement effect is not seen.	Emerson's enhancement effect is seen

Mode of respiration: Bacteria show the following types of respiration

1. **Obligate aerobes**: These grow only in the presence of oxygen. e.g., *Bacillus subtilis, Azotobacter, Vibrio*
2. **Obligate anaerobes**: These grow only in the absence of oxygen. e.g., *Clostridium*
3. **Facultative aerobes**: These are basically anaerobes but change to aerobic mode if oxygen is available. e.g., *Chlorobium, Chloroflexa*

4. **Facultative anaerobes**: These are basically aerobes but change to anaerobic mode if oxygen is unavailable. e.g., *Escherichia coli*

Mode of Reproduction:

1. **Vegetative reproduction** takes place by the following methods:

   (a) **Budding:** In this case, a small protuberance called bud develops at one end of the cell. Then DNA (genome) replication occurs and one copy of DNA gets into the bud. Then the bud enlarges and are filled with cytoplasm and other cellular components. The bud gets detached from the parent cell by a constriction and each grows into a new bacterial cell. It occurs in *Rhodopseudomonas*.

   (b) **Fragmentation:** It occurs during unfavourable condition. In this bacterial protoplasm undergoes compartmentalisation and forms minute bodies called gonidia. Then DNA (genome) replication occurs and one copy of DNA gets into the gonidia. Each gonidium grows to a new bacterium under favourable condition.

   (c) **Transverse binary fission**: It is the simplest and most common method which occurs during favourable condition of temperature, moisture and nutrients. It is very rapid and takes 20 to 30 minutes. It involves the following steps:

      (i) **Replication of DNA:** The genetic material of bacteria (nucleoid) replicates (divides) amitotically forming two circular nucleoids after undergoing theta mode in bidirectional manner.

      (ii) **Partitioning or separation of DNA**: The nucleoid is attached to mesosome which starts dividing because of the synthesis of membrane between nucleoid-mesosome attachment sites. Then each mesosome is pushed to the middle of daughter cell. Finally, two nucleoids get properly partitioned into two daughter cells.

      (iii) **Cross-wall formation**: The transverse septum is formed due to invagination of plasma membrane and simultaneously cross wall is formed and grows centripetally forming two equal sized organisms.

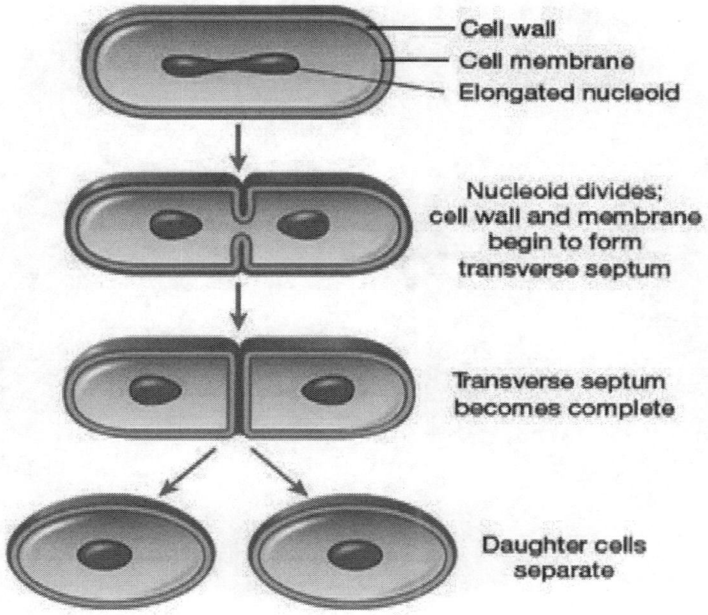

2. **Asexual reproduction** takes place by the following methods:

   (a) **Conidia:** It occurs in several filamentous bacteria such as *Streptomyces.* It is small, thin and rounded spore like structure formed exogenously at the tip of filament in chain under favourable condition. It separates from chain and forms a new bacterium.

   (b) **Endospore:** It occurs in *Bacillus, Clostridium, Thermoactinomyces* (mostly in Gram positive bacteria). It occurs in unfavourable condition (exhaustion of nutrients, high temperature and desiccation). It is thick-walled, highly resistant body (tolerates very high temperature) formed inside the cell (endogenous). It is single formed per cell. Its formation involves the following steps:

(i) Nucleoid is converted into central core containing DNA and cytoplasm.

(ii) Transverse septum (but no transverse wall) is formed towards one pole, which separates cell into small and large part. Small part containing DNA and cytoplasm is called forespore.

(iii) Prespore is formed and develops membrane outside forespore.

(iv) Cortex is formed and deposited between two membranes surrounding forespore.

(v) Spore coat (exosporium) is formed over cortex. Spore now becomes refractile.

(vi) Ca^{+2} and dipicolinic acid are synthesised to provide heat resistance to endospore.

(vii) Autolysis of parent cell occurs.

(viii) Mature endospore is released from cell on the onset of favourable condition when spore coat bursts by imbibition of water. Only one endospore is formed per cell.

(ix) It germinates and gives rise to a single bacterial cell. then endospore is a means to ensure survival (perennation).

(x) It survives high temperature, drying, freezing, toxic chemicals, radiation hence endospore is a means to ensure survival (perennation).

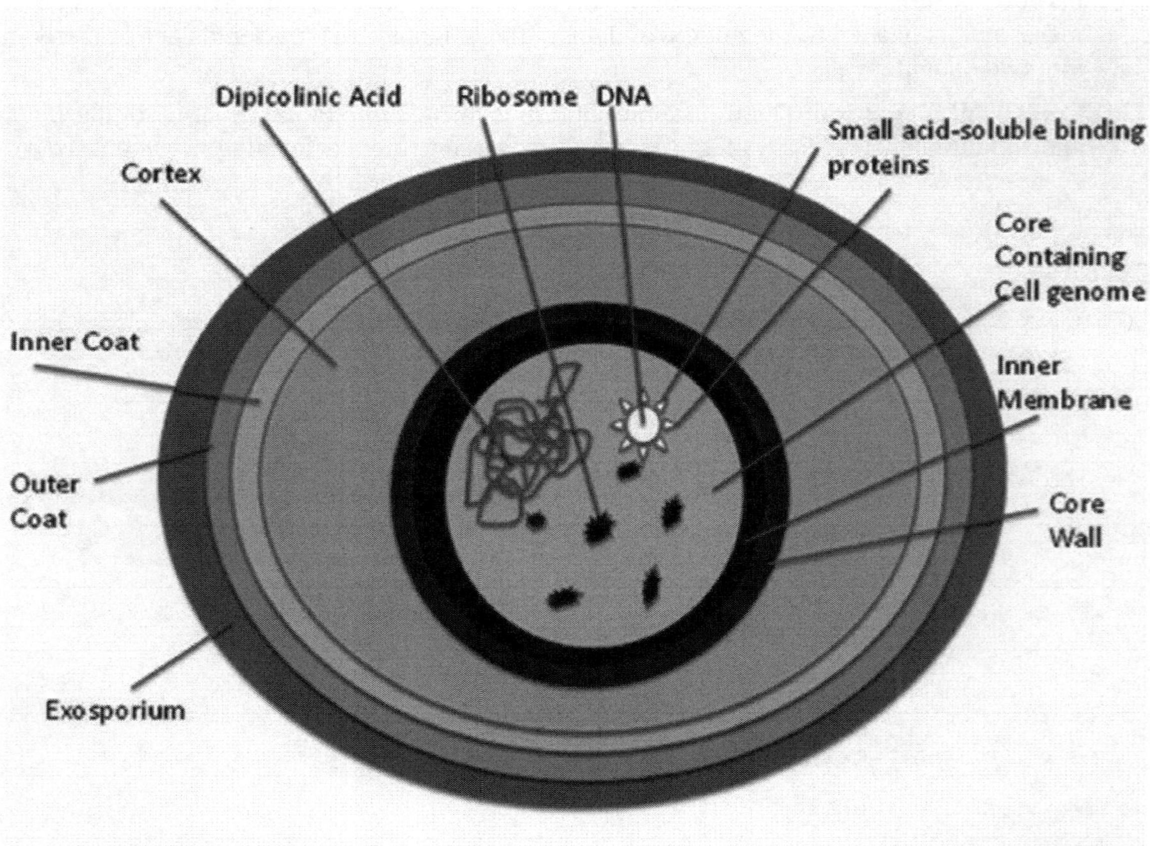

Fig. Endospore formation

3. **Sexual reproduction (Gene recombination):** It does not occur in bacteria because there is no gamete formation and fertilisation but the exchange of genetic material occurs hence there is gene recombination

which occurs in the following three methods:

(a) **Transformation:** It is the transfer of relatively small segments of naked or exogenous DNA from capsulated (donor) cell to non-capsulated (recipient) cell through the medium in which they grow but both cells are not in contact. It was discovered by Griffith (1928) in *Diplococcus pneumoniae.*

(b) **Conjugation:** It is the transfer of DNA from the donor cell to the recipient cell during a period of contact. It was discovered by Joshua Lederberg and Edward Tatum (1946) in K_{12} strain of *E.coli.*

(c) **Transduction**: It is the transfer of DNA from donor cell to recipient cell by means of a temperate λ-bacteriophage. It was discovered by Zinder and Lederberg (1952) in *Salmonella typhimurium.*

Mycoplasma: It comes under class Mollicutes. It was discovered by Nocard and Roux (1898) from pleural fluid of cattle. It is the smallest and simplest microorganism. It was earlier known as pleura Pneumonia like Organism (PPLO).

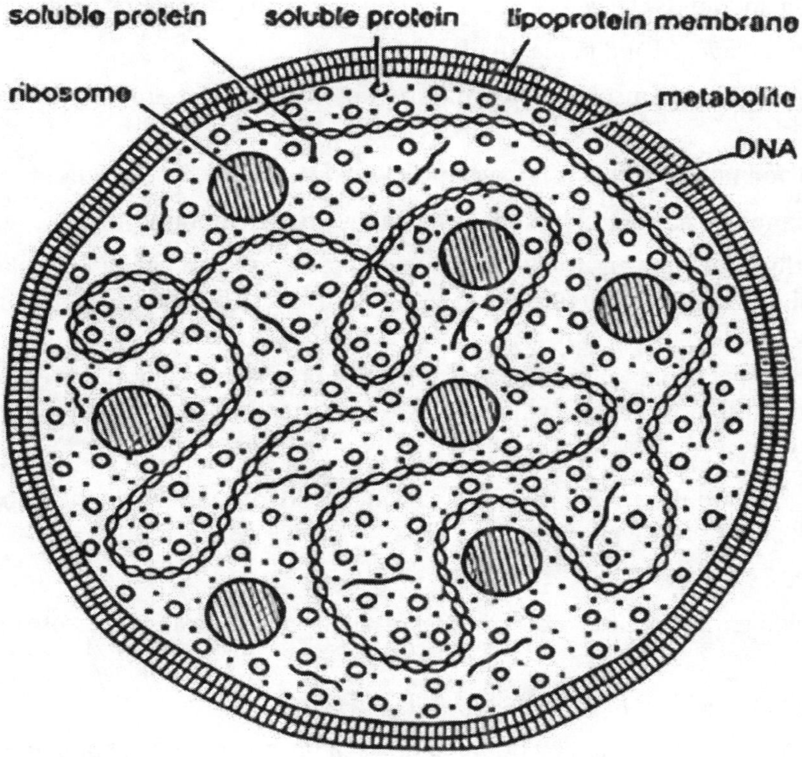

Fig: ***Mycoplasma***

Features:

1. It is spherical, non-motile, Gram negative, facultative anaerobe, chemoheterotroph (parasitic) and highly pleomorphic (of different shape hence called Joker of Plant kingdom).
2. It does not have a cell wall.
3. It is filterable through a bacterial filter (of size 450 nm)
4. It grows artificially in culture and forms fried egg-shaped colonies
5. It is resistant to penicillin because it attacks on cell wall which is absent in *Mycoplasma.*
6. It is sensitive to Erythromycin, Streptomycin and Tetracycline.
7. It has linear, double-stranded, and coiled DNA where G+C content is low.
8. It causes serious diseases in plants such as little leaf of Brinjal, Sandal spike and Witch's broom of potato, groundnut, strawberry etc.
9. It causes Pleuropneumonia in animals (by *Mycoplasma pneumoniae*), Urethritis in human beings (by *Mycoplasma genitalium*) and sterility in human beings (by *Mycoplasma hominis, Spiroplasma*).

CYANOBACTERIA:

The term 'Cyanobacteria' was coined by ICBN (1978). It is also known as Blue Green algae or Cyanophyceae. It originated in the Precambrian period.

Features:

(a) These are ubiquitous (mostly fresh water and some are terrestrial and marine).

(b) These may be unicellular (e.g., *Chrococcus, Chamaesiphon*), multicellular colonial (e.g., *Microcystis*), filamentous (e.g., *Anabaena, Nostoc, Oscillatoria, Spirulina*) and heterotrichous (e.g., *Scytonema, Tolypothrix*).

(c) It is the largest photosynthetic bacterium.

(d) Cell is extremely simple, larger than bacteria and Gram-negative.

(e) It is enclosed in a layer of thick, dense and undulating gelatinous sheath which is made up of pectin. It protects cells from desiccation and holds the cells in a colony together. Thus, it is also known as myxophyceae.

(f) Its cell wall is rigid, four-layered and made up of peptidoglycan.

(g) Its protoplast is simple without any plastid. It is differentiated into peripheral dense chromoplasm and central clear centroplasm.

(h) It is photoautotroph and photosynthesis is oxygenic.

(i) Its reserve food is cyanophycean starch (showing a negative test to iodine).

(j) Its photosynthetic thylakoids occur freely in the cytoplasm and are called chromatophores. It contains pigments such as chlorophyll-a (imparting green colour to thallus), β-carotene, xanthophyll and phycocyanin (giving blue colour to thallus).

(k) It contains cytoplasmic inclusions such as gas vacuoles (for buoyancy), α-granules (reserve food), β-granules (lipid droplet), volutin granules (storage of phosphates), polyhedral bodies (containing DNA and RNA).

(l) It shows gliding movement.

(m) Cyanobacteria show the Gaidukov phenomenon or chromatic adaptation where the colour changes depending upon the wavelength of light received.

(n) These show Nitrogen-fixing ability.

(o) These form water blooms in polluted water bodies causing the release of neurotoxins which make water undrinkable.

(p) These play a role in the recycling of nutrients like N, P, S, Fe etc.

(q) These undergo vegetative reproduction by fragmentation of hormogonia (part of filament).

(r) These undergo asexual reproduction by

(i) Akinete: It is thick-walled, rounded, large-sized, pale yellow. It lacks pigments. It tolerates desiccation and high temperature hence it is perennating structure. It germinates to form a new individual after a period of rest under unfavourable conditions.

(ii) Heterocyst: It is a clear, thick-walled, two-layered, barrel-shaped and pale yellow structure. It is terminal as in *Gloeotricha* or intercalary as in *Nostoc, Scytonema* and lateral as in *Nostochopsis*. It is absent in *Oscillatoria*. It obtains nutrition from adjacent vegetative cells. It retains PS-I activity but lacks Ps-II and RuBP carboxylase hence it neither fixes CO_2 nor releases O_2 in light. It takes part in N_2-fixation and contributes to fixed N_2 in the form of an amino acid (Glutamine) because the heterocyst provides anaerobic conditions to cyanobacteria.

(s) These do not show sexual reproduction due to a lack of sex organs.

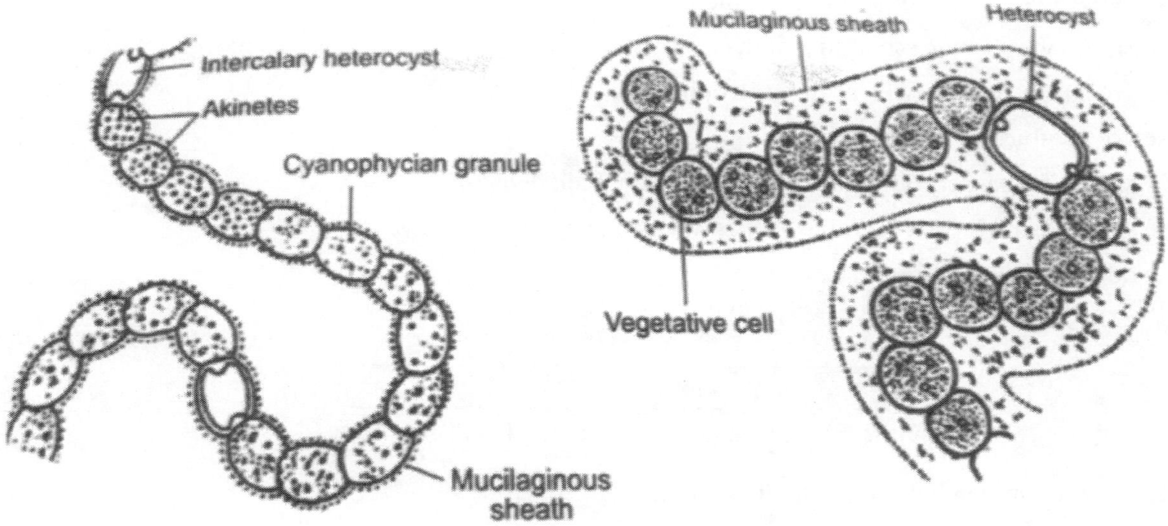

Fig: Akinete and Heterocyst

Economic importance:

(a) *Spirulina* is rich in protein (70%) and is eaten.

(b) *Nostoc commune* is eaten (edible).

(c) *Lyngbia* is a source of antibiotics.

(d) It increases soil fertility by fixing nitrogen (biofertiliser) but in the rice field, *Aulosira* is used.

(e) It helps in the reclamation of alkaline (usar) soil into fertile land. (It is grown in alkaline soil, causing a lowering of pH and enhancing N, P and organic matter).

Bacteria	Cyanobacteria
Cell is smaller in size.	Cell is larger in size.
Cell wall is one to two-layered.	Cell wall is four-layered.
Flagella are present.	Flagella are absent.
These are mostly anaerobes.	These are aerobes.
These are mostly heterotrophs (non-photosynthetic).	These are autotrophs (photosynthetic).
These have glycogen as reserve food.	These have cyanophycean starch as reserve food.
These take up H_2S as an electron donor and release Sulphur.	These take up H_2O as an electron donor and release O_2.
These do not contain heterocysts and akinetes.	These contain heterocysts and akinetes.

PROTISTA

The term 'Protista' was coined by Ernst Haeckel (1866). It is an artificial and diverse group. The protists are microscopic, unicellular and eukaryotic organisms. These are phylogenetically or evolutionary links between fungi, Plantae and Animalia. These are difficult to characterise because of their great diversity.

General features:

(a) These are mostly aquatic (plankton).

(b) These are solitary or colonial unicellular organisms.

(c) These do not show any differentiation into tissues and organs.

(d) These have no well-defined boundaries.

(e) These have membrane-bound cell organelles.

(f) These show a 9+2 arrangement of cilia and flagella; if present.

(g) These show cytoplasmic streaming or cyclosis.

(h) These are autotrophic as well as heterotrophic.

(i) These are aerobic.

(j) These show asexual reproduction by binary fission (cell division).

(k) These show sexual reproduction by isogamy, anisogamy and oogamy.

(l) These do not show the presence of an embryo.

1. Diatoms (Desmids):

Division	Chrysophyta
Class	Bacillariophyceae (Golden Brown Algae)

(a) These are known as jewels of the plant kingdom.

(b) These are freshwater as well as marine.

(c) These are passively floating hence non-flagellate.

(d) These are of Pennate type (bilaterally symmetrical or rectangular) or Centric type (radially symmetrical or circular).

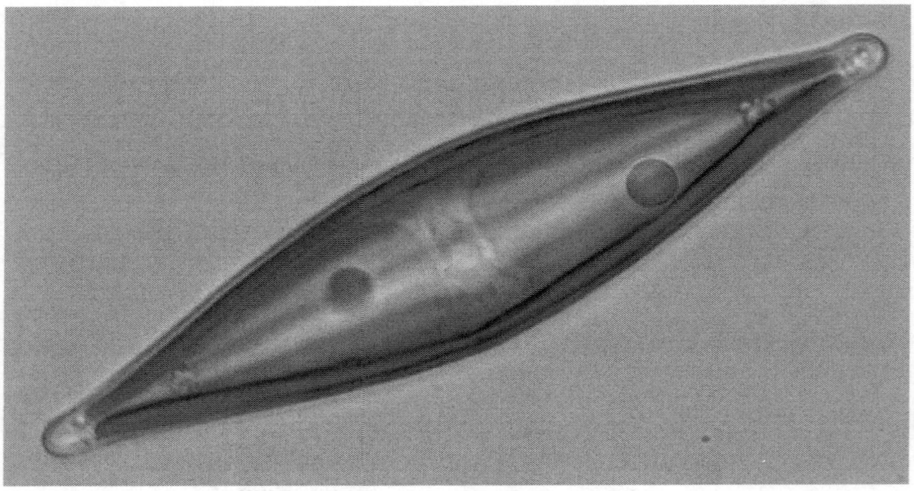

Fig: Diatoms

(e) The body is made up of two thin overlapping shells (halves)-outer and larger epitheca and younger and smaller hypotheca. These shells look like soapboxes.

(f) Cell wall is made up of cellulose and silica.

(g) Cell has several discoid or two plate-like chromatophores.

(h) These are photosynthetic due to the presence of chlorophyll a and -c.

(i) The reserve food is leucosin or chrysolaminarin (carbohydrate) and oil.

(j) These are chief producers of the ocean.

(k) Asexual reproduction takes place by binary fission (cell division).

(l) Sexual reproduction takes place by isogamy, anisogamy and oogamy. The zygote forms a diploid auxospore that develops a new set of vegetative cells.

(m) These form diatomite or diatomaceous earth or kiesulghur when they settle at the bottom of the sea after death.

(n) These are indicators of sewage pollution.

e.g., *Naviculla, Pinnularia, Milosira*

2. Dinoflagellates:

Division	Pyrrophyta
Class	Dinophyceae (Dark yellow algae)

(a) These are mostly marine.

(b) The body has a longitudinal groove (sulcus) and transverse groove (annulus/cingulus). These may be of different colours.

(c) These are biflagellate-heterokont and oriented at a right angle to each other thus they are also called whirling whips.

(d) The cell wall is made up of cellulose and pectin.

(e) These show mesokaryotic cellular organisation.

(f) These are photosynthetic due to the presence of chlorophyll-a and -c.

(g) Reserve food is starch and oil.

(h) These contain a non-contractile vacuole called pusule which helps in excretion and osmoregulation.

(i) Asexual reproduction takes place by binary fission (cell division).

(j) Sexual reproduction takes place by isogamy, anisogamy and oogamy.

(k) Life cycle is haplontic (involving zygotic meiosis) as well as diplontic (involving gametic meiosis).

(l) These cause red tide (the sea looks red) due to the rapid multiplication of *Gonyaulax.*

(m) Some dinoflagellates such as *Noctiluca, and Pyrodinium* (Fire algae) show bioluminescence (emitting light at night which makes the sea surface illuminated).

(n) These produce powerful toxins such as saxitoxin and brevitoxin that cause respiratory failure in vertebrates esp. fishes.

e.g., *Gymnodinium, Gonyaulax, Noctiluca, Pyrodinium*

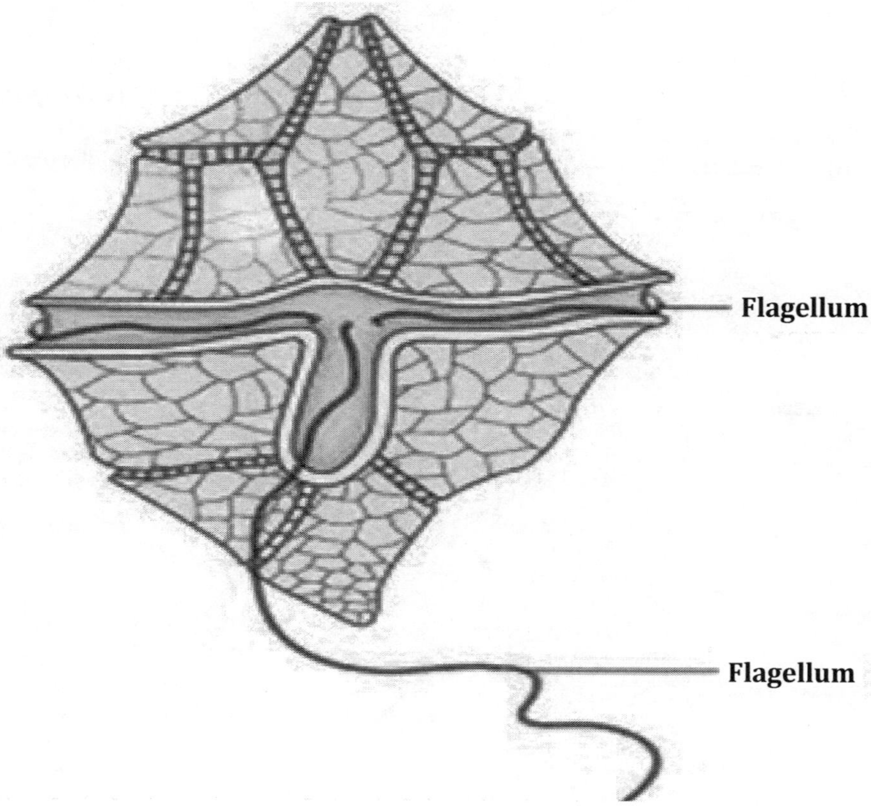

Fig: Dinoflagellate

3. Euglenoids:

Division	Euglenophyta
Class	Euglenophyceae

(a) These are freshwater esp. in stagnant water.

(b) The body is spindle-shaped having a blunt anterior end and pointed posterior end. The anterior end has invagination containing cytostome (mouth), cytopharynx (canal) and reservoir. The eye spot (stigma) is situated at the anterior end and is involved in the perception of light stimulus.

(c) The body is covered with a protein-rich layer called pellicle which makes the body elastic or flexible. The pellicle has oblique but parallel stripes called myonemes.

(d) These are biflagellate and heterokont (of different lengths).

(e) Cell wall is absent (animal feature).

(f) Chloroplast is present which contains pyrenoid.

(g) These are photosynthetic (autotrophic) in the presence of light and chlorophyll a and - b (just like higher plants). These also become heterotrophic (holozoic) in the absence of light. Hence the mode of nutrition is mixotrophic.

(h) Reserve food is paramylon starch (β-1,3 glucan).

(i) These contain contractile vacuole (at the anterior end) which helps in osmoregulation.

(j) These contain food vacuoles.

(k) Asexual reproduction takes place by longitudinal binary fission.

(l) Sexual reproduction is absent.

(m) These develop encystment to tide over unfavourable condition.

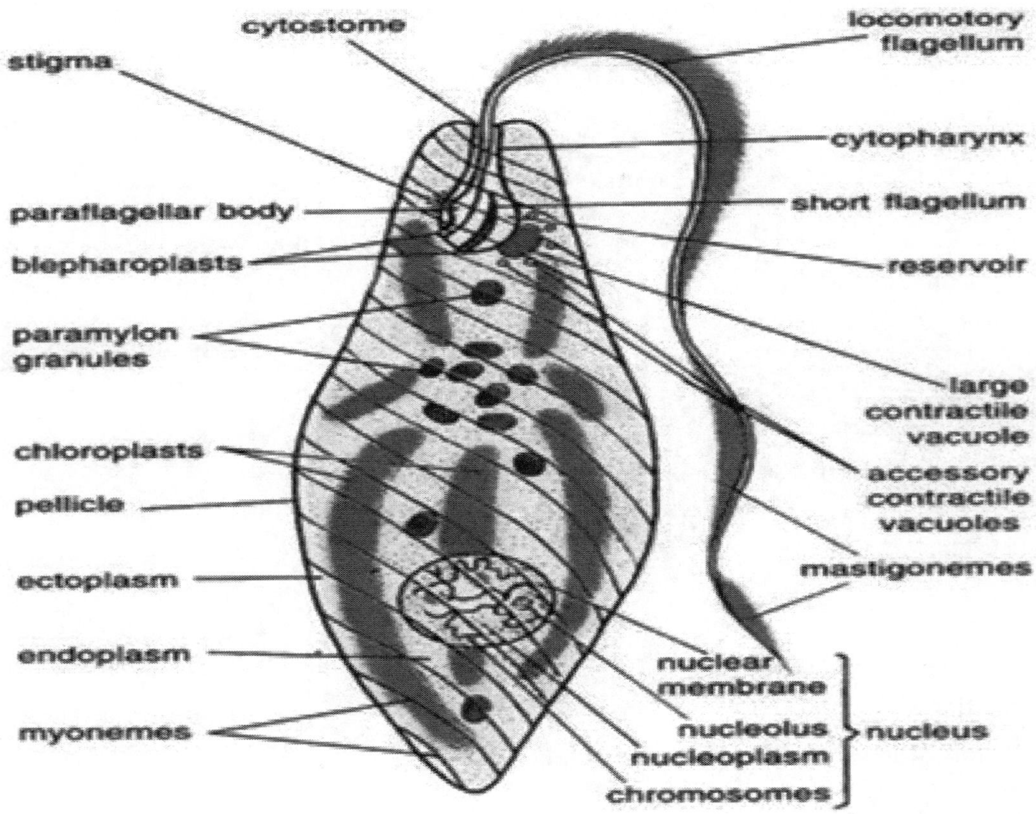

Fig: Euglena

4. Slime moulds:

Division	Myxomycota
Class	Myxomycetes

These are also known as consumer-decomposer protists or fungus-animals or saprophytic protists. These are on the transition line between fungi and Protista. These are of two types:

1. **Acellular or True slime moulds**: e.g., *Dictyostellium, Fuligo, Physarum*
   - (a) These are aquatic.
   - (b) The somatic body is made up of an active, free-living, naked, amoeboid, diploid and multinucleate mass of protoplasm called plasmodium. It is covered with slime and formed during favourable conditions. It moves over the surface of the substratum by pseudopodia (amoeboid movement).
   - (c) These are heterotrophic (holozoic or ingestive) due to the absence of chlorophyll (animal-like feature).
   - (d) These form a fruiting body called sporocarp by differentiation under unfavourable conditions. Sporocarp contains sporangia which form highly resistant spores by meiosis under unfavourable conditions. Spore has cellulosic cell walls (plant feature). Spore is dispersed by air currents and germinates to form one, biflagellate and spindle-shaped swarm cell.
   - (e) Asexual reproduction takes place by spores.
   - (f) Sexual reproduction takes place by isogamy.

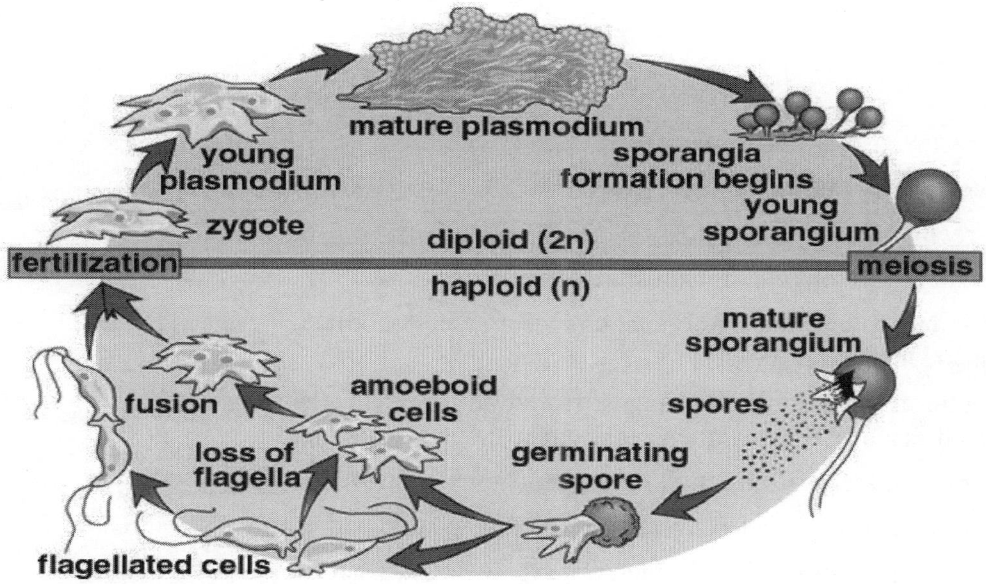

2. **Cellular slime moulds**: e.g., *Protostelium*
   - (a) These are terrestrial.
   - (b) Somatic body is made up of aggregated amoeba like organisms called pseudoplasmodium.
   - (c) These form a fruiting body called sporocarp which contains sporangia. Sporangium develops spores. Spore has cellulosic cell wall. Spore germinates to form non-flagellate, wall-less, uninucleate and haploid cell called myxamoebae which act as gamete.
   - (d) Sexual reproduction takes place by anisogamy.
   - (e) Zygote directly forms pseudoplasmodium which becomes multinucleate by repeated division.
   - (f) Life cycle is diplontic.

5. Protozoans

These are considered as 'first animals' or predator protists.

Class 1:	Rhizopoda (Sarcodina): e.g., *Amoeba, Entamaoeba*
Class 2:	Ciliata: e.g., *Paramoecium, Balantidium*
Class 3:	Flagellata (Mastigophora): e.g., *Leishmania, Trypanosoma*
Class 4:	Sporozoa: e.g., *Monocystis, Plasmodium*

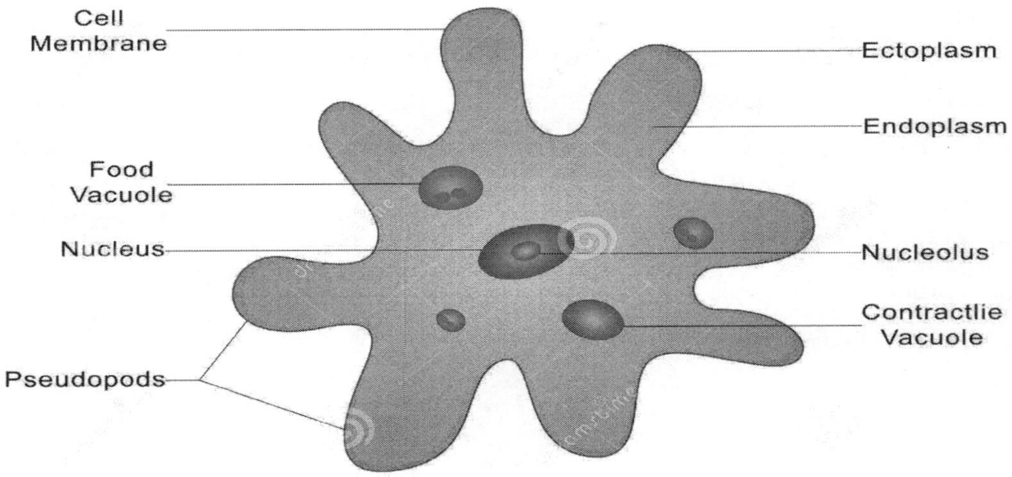

Fig. Amoeba

(a) These are freshwater as well as marine.

(b) These are solitary as well as colonial.

(c) These are free-living as well as parasites.

(d) These are animal-like microorganisms that lack cell walls.

(e) These move through locomotory organelles such as pseudopodia (false feet), cilia and flagella which help in locomotion and food capture.

(f) These are heterotrophic (holozoic or ingestive and saprozoic involving diffusion or active transport of soluble nutrients such as glucose, and amino acids across the plasma membrane.

(g) These undergo intracellular digestion through the general body surface.

(h) These undergo respiration and excretion through the general body surface by diffusion.

(i) Asexual reproduction takes place through transverse binary fission and multiple fission.

(j) Sexual reproduction takes place through conjugation, hemixis, and automixis.

(k) These form cysts (encystment) during unfavourable conditions.

FUNGI

Fungi are multicellular, haploid, nonvascular, achlorophyllous, spore-bearing and heterotrophic organisms. Fungi are major decomposers of wastes (along with bacteria) and help in the recycling of inorganic and organic materials present on earth.

The study of fungi is called Mycology (*Mykes*=Mushroom).

Father of Mycology: Pier Antonio Micheli (He wrote Nova Genera Plantarum).

Father of Modern Mycology/Plant pathology: de Bary

Father of Systemic Mycology: E.M. Fries (Linnaeus of Mycology)

Father of Indian Mycology: E.J. Butler

Smallest fungus	*Saccharomyces cerevisiae* (Yeast)
Largest fungus	*Armillaria* (Honey mushroom)

Features:

(a) These are ubiquitous and cosmopolitan. These show great diversity in morphology and habitat.

(b) These grow well at 20-30^{0}C and pH=6.

(c) The body is thalloid. Thallus is unicellular (Yeast) or made up of fine, tubular and thread-like filaments called hyphae which modify into the following structures:

(i) Hyphae become thick and root-like called rhizomorph. e.g., *Agaricus.*

(ii) Hyphae become terminally swollen to penetrate the host and cause infection. e.g., *Erysiphe.*

(iii) Hyphae become closely packed and compact as in *Claviceps.*

(iv) Hyphae develop appendages to called haustoria which penetrate the host cell for nutrition. These occur commonly.

(v) Hyphae are usually interwoven into a network called mycelium.

(vi) Mycelia also become multinucleate and non-septate hence called coenocytes. e.g., Oomycetes and Zygomycetes.

(vii) Mycelia form a chain of cells by budding forming pseudomycelium. e.g., yeast.

(viii) Mycelia often aggregate into tissue-like structure called plectenchyma which is of two types:

- Prosenchyma: Several hyphae come close together and run parallel to each other.
- Pseudoparenchyma: Several hyphae are closely packed.

(d) The cell wall is made up of chitin (a polymer of N-acetyl glucosamine). It is absent in oomycetes where cellulose is present. Chitin is also absent in yeast where mannan (polymer of mannose) and glucan (polymer of glucose) are found.

(e) Lomasome (invagination of plasma membrane) is present between cell wall and plasma membrane.

(f) Chloroplast is absent.

(g) Golgi body is unicisternal.

(h) These are heterotrophs (saprophytes, parasites and symbionts) due to the absence of chloroplast and chlorophyll molecules. These may be

- **Obligate saprophytes**: These totally depend on dead organic matter. e.g., *Mucor*
- **Facultative saprophytes**: These are normally parasites but can also be saprophytes. e.g., *Ustilago*
- **Obligate parasites**: These totally depend on the host for nutrition. e.g., *Puccinia, Albugo*
- **Facultative parasites**: These are normally saprophytes but can also be parasites. e.g., *Pythium, Fusarium*

(i) Reserve food is glycogen.

(j) These show an absorptive (assimilative) type of nutrition.

(k) Sex organs are unicellular and non-jacketed.

(l) Vegetative reproduction takes place by fragmentation of hyphae (common), budding, and fission.

(m) Asexual reproduction takes place by different spores such as

- **Zoospores** (oval, unstalked, thin, endogenous and flagellate)
- **Sporangiospore** (rounded, unstalked, thin, endogenous)
- **Conidia** (rounded, stalked, thin, exogenous and non-flagellate)
- **Chlamydospore** (rounded, unstalked, thick).

(n) Fungi are monoecious (homothallic) and dioecious (heterothallic).

(o) Fungi are mostly eucarpic (a part of the thallus is converted into a reproductive structure). Some are holocarpic (the entire thallus is converted into a reproductive structure).

(p) Sexual reproduction involves plasmogamy (fusion of two protoplasm of either two motile or non-motile gametes takes place bringing two compatible nuclei close together), karyogamy (fusion of two nuclei to form diploid nucleus) and meiosis (formation of haploid spores from zygote).

(q) Sexual reproduction takes place by following methods:

(i) Planogametic copulation: It is the fusion of naked gametes. It involves

- Isogamy: It is the fusion of morphologically similar gametes.
- Anisogamy: It is the fusion of morphologically and physiologically dissimilar gametes. In this, the male gamete is smaller and more active than the female gamete.
- Oogamy: It is the fusion of smaller, active and motile sperm (formed due to repeated division of male gametangium) with larger, passive and non-motile eggs (formed directly from female gametangium without division).

(ii) Gametangial contact: Both male and female gametangia lie in close contact with a fertilisation tube. The gametes are never released. Then one or several male nuclei shift to female gametangia where fusion takes place to form an oospore. The gametangia never fuse.

(iii) Gametangial copulation: The two isogametangia come in close contact and their contents are mixed after dissolving their walls at the point of contact. Both fuse to form a thick-walled zygospore.

(iv) Somatogamy: The two vegetative cells or two vegetative hyphae bearing nuclei of different parental strains fuse after the dissolution of the wall to form karyon formation. In this sex organs and gametes are not formed. It occurs in *Agaricus, Morchella, Peziza.*

(v) Spermatisation: It occurs when non-motile and spore-like spermatium (male gamete) releases its contents into trichogyne (female receptive hyphae or female gamete). In this sex organs are absent.

(r) Sexual reproduction shows gradual and progressive simplification from lower to higher forms.

Oogamy→Anisogamy→Isogamy→Gametangial copulation→Spermatisation→Somatogamy

(s) Embryo is not formed.

Classification: Fungi have been divided into five classes based on the following criteria.

(a) Nature of somatic hyphae: either unicellular or myceloid; if latter septate or non-septate.

(b) Types of asexual spore-zoospore, sporangiospore, conidia.

(c) Presence or absence of sexual stage

(d) Nature of life cycle: haplontic, diplontic, haplo-diplontic, dilobiontic, haploid dikaryotic

1. Oomycetes (Egg fungi) or Phycomycetes (Algal fungi):

(a) These are aquatic.

(b) Mycelia are branched and coenocytic (multinucleate and non-septate).

(c) The cell wall is made up of cellulose.

(d) Flagella (whiplash or tinsel type) are found

(e) Asexual reproduction takes place by biflagellate and heterokont zoospore.

(f) Sexual reproduction takes place by gametangial contact to form oospores.

Examples:

- *Phytophthora infestans*: It causes late blight of potatoes (Irish famine in 1845).
- *Synchytrium endobioticum*: It causes wart disease in potatoes.
- *Albugo candia*: It causes white rust on crucifers.

- *Pythium debarynum*: It causes damping off of seedlings in tobacco, chilly
- *Sclerospora*: It causes the green ear of pearl millet.
- *Saprolegnia* (Water mould): It causes salmon disease on fish gills.
- *Pilobolus*: It is a coprophilous fungus.

2. Zygomycetes (Conjugating fungi):

(a) Mycelia are branched and coenocytic.

(b) Zoospore is absent.

(c) Asexual reproduction takes place by unflagellate sporangiospore.

(d) Sexual reproduction takes place by gametangial copulation forming zygospore.

Examples:

- *Mucor* (Pin mould/Dung mould): It produces trisporic acid.
- *Rhizopus* (Bread mould): It shows heterothallism as discovered by Blakeslee where conjugation occurs between + and – strains of hyphae.

3. Ascomycetes (Sac fungi):

It is the largest class.

(a) These are mostly coprophilous (growing on cow dung) and some are parasites.

(b) Mycelia are branched and septate.

(c) Motile structure is absent.

(d) Asexual reproduction takes place by conidia.

(e) Ascocarp (fruiting body) containing asci is formed.

(f) Crozier formation occurs in the formation of ascus.

(g) Dikaryophase is small between plasmogamy and karyogamy.

(h) Sexual reproduction is isogamous in lower forms but anisogamous in higher forms.

(i) Sexual reproduction takes place by usually 8 (sometimes 4), linear and endogenous ascospores.

Examples:

- *Penicillium* (Blue green mould): Fruiting body is cleistothecium (spherical).
- *Peziza* (Cup fungus): Fruiting body is apothecium (cup-shaped).
- *Claviceps* (Ergot fungus): Fruiting body is perithecium (flask-shaped).
- *Neurospora* (Red or pink mould): *Drosophila* of Plant kingdom
- *Aspergillus* (Black mould): Guinea pig of plant kingdom or weed of laboratory.
- *Saccharomyces* (True yeast or sugar fungus): Asci are formed.
- *Candida* (False yeast): Asci are not formed.
- *Morchella* (Morels): It is edible and proteinaceous.
- *Tuberaestivum* (Truffle): It is edible.
- *Erysiphe*: It causes powdery mildew of pea.

4. Basidiomycetes (Club fungi):

(a) These are saprophytes as well as parasites.

(b) Mycelia are branched and septate.

(c) These are mostly heterothallic.

(d) Dolipore septum (centrally located and barrel-shaped pore) is present.

(e) Motile structure is absent.

(f) Mycelia may be primary (short-lived containing monokaryotic cells) and secondary (long-lived containing dikaryotic cells).

(g) Clamp connection is seen in dikaryotic hyphae. It forms a dikaryon.

(h) Vegetative reproduction takes place by fragmentation.

(i) Asexual spore is absent.

(j) Dikaryophase is large and present between plasmogamy and karyogamy.

(k) Basidiocarp (fruiting body) containing basidia is formed.

(l) Sexual reproduction takes place by 4, exogenous basidiospores.

Examples:

- *Agaricus* (Mushroom/Gill fungus): It is edible.
- *Puccinia graminis tritici* (rust fungus): It causes black stem rust in wheat
- *Ustilago tritici* (Smut fungus): It causes loose smut of wheat.
- *Tilletia*: It causes a bunt of wheat.
- *Polyporus* (Bracket fungus): It causes heart rot disease.
- *Amanita* (Toad stool): It contains hallucinogens.
- *Lycoperdon* (Puff ball): It is fleshy and edible.

5. Deuteromycetes (Fungi Imperfecti):

(a) It is also known as fungi imperfecti because only asexual stage is known.

(b) These are mostly decomposers. Some are parasitic or saprophytic.

(c) Mycelia are branched and septate.

(d) Asexual reproduction takes place by conidia.

(e) Asexual fruiting bodies such as synnema, sporodochium (hemispherical), pycnidium (flask-shaped), acervulus (saucer-shaped) are formed.

(f) Sexual reproduction is absent hence perfect stage is not known. If the sexual stage was discovered, then it was moved to either ascomycetes or basidiomycetes.

(g) These help in mineral cycling.

Examples:

- *Alternaria solanii*: It causes early blight of potato
- *Colletotrichum falcatum*: It causes red rot of sugarcane
- *Fusarium oxysporum*: It causes wilt of red gram
- *Cercospora personata*: It causes tikka disease of groundnut.
- *Helminthosporium oryzae*: It causes brown leaf spot disease of rice (Bengal famine,1942).
- *Pyrecularia oryzae*: It causes the blast disease of rice.
- *Trichoderma*: It is used in biocontrol.

MYCORRHIZA

It is a symbiotic association of fungi with roots of higher plants. It was discovered by Frank (1885).

In this association, the fungus gets nutrition (carbohydrates) from the roots of higher plants and in return, higher plants uptake water and minerals due to the spread of fungal hyphae in the soil. It occurs in most of the plants which live in nutrient (organic matter) poor soil. The association is not very specific. Mycorrhiza does not form root hair and root cap.

Types: There are two types of mycorrhizae.

Ectomycorrhiza	Endomycorrhiza
Fungal hyphae form a sheath or mantle around roots (Hartig net) because these do not penetrate the root.	Fungal hyphae do not form a sheath or mantle because these penetrate roots intercellularly and intracellularly.
The cells of the root secrete sugar into intercellular space for fungal hyphae.	The cells of the root do not secrete sugar.
It does not form rounded structures (vesicles) and branch-like structures (arbuscles).	It forms arbuscles and vesicles hence called Vesicular Arbuscular Mycorrhiza (VAM).
It is less common.	It is more common.
It is shown by *Boletus, Lacterius* (Ascomycetes) and *Amanita* (Basidiomycetes).	It is shown by *Glomus, Rhizoctonia* (Zygomycetes).
It occurs in the roots of Birch, *Eucalyptus, Pinus,* and *Quercus.*	It occurs in the roots of Tomatoes, Peppers, Coffee, Cardamom (Crop plants), Grasses, and Orchids.

Economic importance:

(a) Mycorrhiza helps in the absorption of water and minerals (esp. N, P, S, Ca) because fungus increases the surface absorbing area of roots.

(b) It protects the roots and provides resistance in plants because fungus secretes antimicrobial substances.

(c) It increases the solubilisation of humus because fungus secretes several growth-promoting substances.

(d) It increases the fertility of the soil and is used as a biofertiliser.

(e) It minimises the requirement of chemical fertilisers since it does not cause pollution.

(f) It increases the tolerance of plants to salinity and drought conditions in soil.

LICHEN

It is a composite or dual organism formed as a result of the symbiotic association of algae and fungi. The term 'Lichen' was coined by Theophrastus. The Father of Lichenology is Acharius.

Features:

(a) Lichens are cosmopolitan in distribution but in India, these are abundantly found in Eastern Himalayas.

(b) These are perennial, variously coloured and show very slow growth.

(c) These show permanent symbiotic association (consortium) of algae and fungi. The alga provides food to the fungus and in return, the fungus provides protection, water and nitrogenous substances to alga. Fungus is a controlled parasite over alga because fungus sends haustoria into algal cells, induces alga to secrete organic substance and prevents alga from developing a covering of pectin. Hence this phenomenon is helotism (master-slave relationship). Fungi thus get more benefits than algae.

(d) These can thrive on barren rocks (saxicolous), earth crust (terricolous), tree bark (corticolous), and wood (lignicolous).

(e) These have an algal component (**phycobiont**) which is shown mostly by Chlorophyceae (esp. *Trebouxia, Trentepohlia, Pleurococcus*) and some Cyanophyceae (esp. *Anabaena, Calothrix, Gloeocapsa*) as well as fungal component (**mycobiont**) which is shown by ascomycetes. The algal component is minor (5%) and the fungal component is major (95%). Both components do not change within a species even under varying habitats.

(f) Thallus shows the following structures:

- **Cyphellae:** These are small, circular, scattered depressions on the underside. These are aerating structures that help in the exchange of gases (analogous to the stomata of higher plants). These are found in *Parmelia.*
- **Cephalodia**: These are gall-like outgrowth on the upper side. These retain moisture. These are found in *Peltigera.*

(g) Vegetative reproduction takes place by fragmentation of thallus.

(h) Asexual reproduction takes place by sorecia (a small and rounded outgrowth of thallus), isidia (an elongated outgrowth of thallus) and conidia.

(i) Sexual reproduction takes place by fungal partner (Ascomycetes) through ascospore.

Types:

1. **Crustose Lichen:** This forms a thin crust on the substratum. e.g., *Dermatocarpon* (Stone mushroom), *Rhizocarpon, Graphis, Lecanora, Strigula.*
2. **Foliose lichen:** This forms a flat, branched and leaf-like thallus attached to the substratum by a rhizoid-like structure (rhizine). e.g., *Cetraria, Loboria, Parmelia, Peltigera* (Dog lichen), *Physcia, Chowdhuria*
3. **Fruticose lichen:** This forms much branched, bushy and pendant to upright. e.g., *Cladonia* (Reindeer moss), *Usnea, Ramalina*

Economic Importance:

(a) Lichens are indicators of SO_2 pollution.

(b) Crustose lichens are pioneer colonisers (communities) in xeric succession (xerosere).

(c) Lichens such as *Cetraria icelandica* (Iceland moss), *Cladonia* and *Dermatocarpon* are edible.

(d) Rocella tinctoria is a source of litmus solution and orcein dye (Orchill).

(e) Peltigera is used to cure hydrophobia.

VIRUS:

It is a simple, ultramicroscopic, non-cellular infectious agent. Its unit is called a virion which is a fully assembled mature particle capable of causing infection. The term 'Virus' was coined by Louis Pasteur. The Father of Virology is Beijerinck. It is neither prokaryotic nor eukaryotic because it has no cell structure. It is considered as a connecting link between non-living organisms and living organisms.

Non-living properties:

(a) The virus does not have protoplasm and cell organelles.

(b) It does not show any type of metabolism.

(c) It can be crystallised (as done by Stanley (1935) on TMV and Poliovirus)

(d) It does not secrete enzymes.

(e) It does not grow on culture medium.

(f) It can be precipitated, stored and dissolved in water.

(g) It is an inert chemical or molecule outside the host cell and exists as a virion.

(h) It is autocatalytic.

(i) It remains unaffected by antibiotics.

Living properties:

(a) It grows and multiplies (reproduces) inside the host cell by redirecting host metabolism. Its life cycle includes
- Attachment (adsorption) of virion to susceptible host
- Penetration (injection)
- Synthesis of nucleic acid and proteins
- Assembly of proteins and packaging of nucleic acid
- Virion (progeny) release from the host cell

(b) It has either DNA or RNA as genetic material.

(c) It shows mutation due to nucleic acid (as proved by Luria, Delbruck and Lwoff). HIV is most mutable.

(d) It is sensitive to temperature, chemicals and UV radiations.

(e) It has genetic continuity and has a definite race.
(f) It shows antigenic properties.
(g) It can direct protein synthesis by using ribosomes of a host cell.
(h) It can be killed by autoclaving
(i) It causes diseases in plants and animals hence it is an obligate intracellular parasite.

Symmetry of virus: Viruses show the following three types of symmetry:

1. Helical (Cylindrical): TMV, Influenza virus, Rabies virus
2. Polyhedral (Icosahedral): Herpes simplex virus, Polio virus, Turnip mosaic virus
3. Complex (Mixed symmetrical): Bacteriophage, Poliovirus.

Disease Symptoms of viruses:

1. Stunting (reduced growth)
2. Mosaic (dark green, light green and yellow angular areas)
3. Mottling (development of different colours)
4. Necrosis (death of tissues)
5. Leaf rolling
6. Curling
7. Yellowing
8. Vein clearing

Classification of Viruses: Lowff, Home and Tournier proposed LHT system which was adopted.

Phylum Vira: It was divided into two subphyla:

Deoxyvira: Deoxyhelica, Deoxycubica and Deoxybinala

Ribovira: Ribohelica and Ribocubica

Virus family	Nature of genome	Examples
Parvoviruses	Single-stranded DNA	Coliphage ψ x174, Canine parvovirus
Adenoviruses, Pox viruses	Double-stranded DNA	Smallpox virus, Cytomegalo virus.
Picornaviruses	Single-stranded RNA	Polio virus, Influenza virus.
Retroviruses	Double-stranded RNA	Reovirus, Rice dwarf virus

Types of viruses:

1. Plant virus: Genetic material in plant viruses is usually single-stranded RNA.

Exceptions: The ds RNA is found in Rice dwarf virus, Wound tumour virus (Reo virus)

The ds DNA is found in the Cauliflower mosaic virus.

Smallest plant virus	Tobacco necrosis virus & Alfalfa mosaic virus (17nm)
Largest plant virus	Citrus tristeza virus (2000x12 nm)

Tobacco mosaic Virus: It was the first virus to be discovered by Dmitri Ivanovsky (1893) who found it smaller than bacteria because it passed through a bacterial-proof filter. Beijerinck (1898) proposed that the filtrate of tobacco leaves caused infection or tobacco mosaic disease (dark green, light green, yellow angular areas on leaves) in healthy plants and called it '*Contagium Vivum Fluidum*' (Infectious living fluid). Adolph Mayer discovered infective nature of tobacco mosaic disease which was developed by applying sap of infected leaves onto the leaves of healthy plants. Pirie and Bawden (1937) discovered the nucleo-proteinaceous nature of the virus.

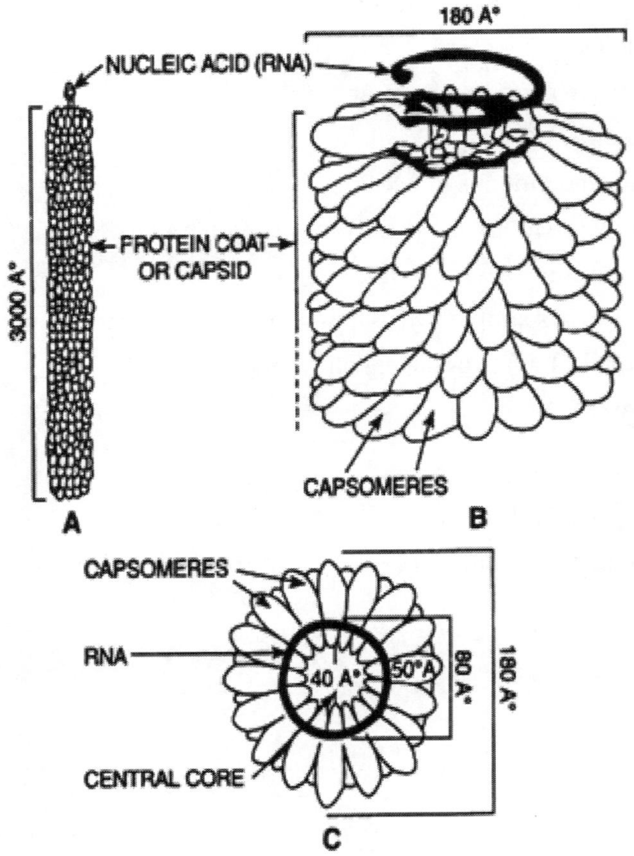

Fig: Tobacco mosaic virus

Features:

(a) It is the most thoroughly studied virus.

(b) It was the first virus to be seen under an electron microscope.

(c) It was the first virus to be purified.

(d) It was the first virus from which infectious nucleic acid was obtained.

(e) It is rod-shaped and the total length of each rod has 130 helicals.

(f) Its size is 3000 A^0 x180 A^0.

(g) It is made up of RNA (5%) and protein (95%).

(h) RNA is single-stranded, linear, helical, grooved and intertwined with protein. It has 6400 nucleotides. RNA performs self-replication and synthesis of specific proteins.

(i) Protein coat is a capsid that protects RNA and has 2130 capsomeres (subunits of protein). Each capsomere contains 158 amino acids.

(j) The ratio of nucleotides and capsomeres is 3:1.

2. Animal Virus: Genetic material is mostly double-stranded DNA.

Exceptions: The ss-DNA is found in Parvovirus

The ss-RNA is found in Retrovirus (Rous sarcoma virus, HIV), Polio virus and Influenza virus.

The ds-RNA is found in the Reovirus (Wound tumour virus).

Smallest animal virus	Foot and mouth disease of virus (20 nm)
Largest animal virus	Small pox virus (350 x 250 x 100 nm)

3. Bacterial virus: Genetic material is mostly double-stranded DNA.

Exceptions: The ss-DNA is found in ϕ x174, fd, M-13,

The ss-RNA is found in F_2, QB, R_{17}, MS_2

The ds-RNA is found in ϕ_6

Smallest bacterial virus	Coliphage P_2 (2 nm)
Largest bacterial virus	Pandora virus (11 nm)

Types of viruses:

(a) Cyanophage: It is a virus that kills cyanobacteria esp. *Lyngbya, Phormidium* and *Plectonema* (LPP-1). It was discovered by Schfferman and Morris (1962).

(b) Mycophage: It is a virus that kills *Agaricus bisporus.* It was discovered by Sinden (1957).

(c) Bacteriophage: It is a virus that kills bacteria. It was discovered by Twort and Felix D' Herelle (1957).

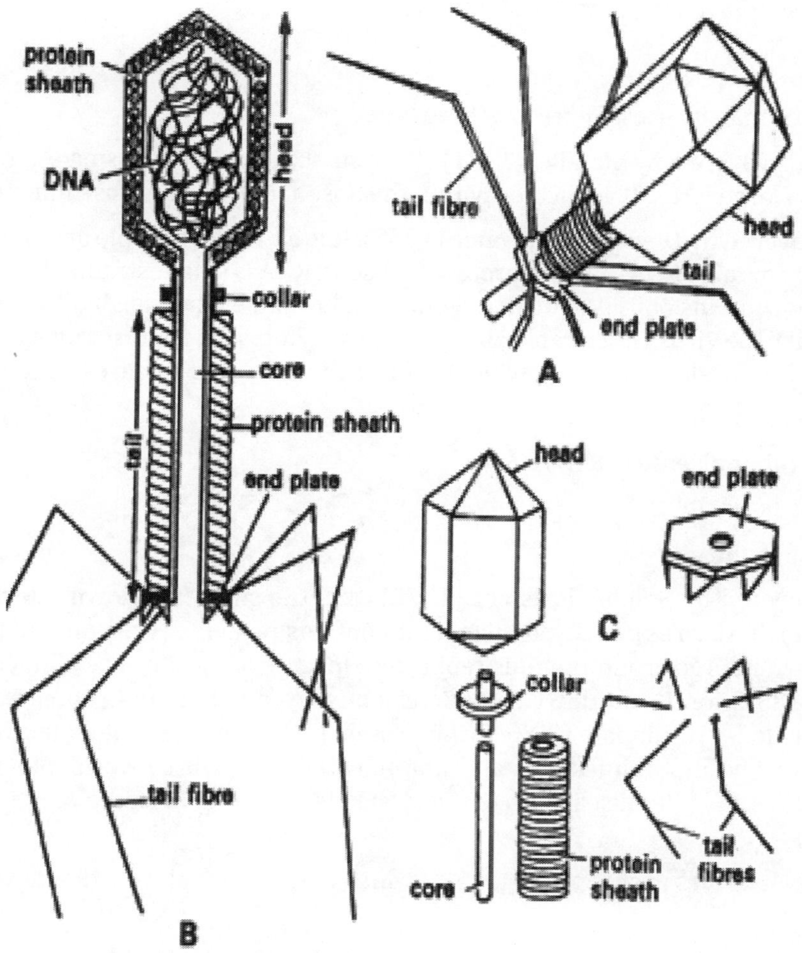

Fig: Bacteriophage

Features of bacteriophage:

(a) It is tadpole-shaped.

(b) It has a polyhedral (hexagonal) head (size=950x650A°) and a binal (helical) tail.

(c) The head has double-stranded DNA surrounded by a protein coat called capsid. The amount of DNA is 40% and that of protein is 60%. Capsid has 2130 protein subunits called capsomeres. Each capsomere has 158 amino acids. DNA of T even phage (T_2, T_4) has hydroxymethyl cytosine instead of cytosine of normal DNA.

(d) Tail is attached with Head by collar. It is proteinaceous and surrounded by a sheath having 24 annular rings (striations). At the lower end, the hexagonal base plate contains 6 small tail pins and 6 large tail fibres. Tail fibres help in the attachment of phage to host cells.

(e) It is obligate parasite.

Types of Bacteriophages:

1. Temperate phage: It is a non-infective phage. In this lysis does not occur. The viral genome gets integrated with the host genome forming prophage (lysogeny). e.g., λ-phage
2. Virulent phage: It is infective phage. In this, lysis occurs. e.g., T_2, T_4- bacteriophage.

Economic Importance:

1. T_2 bacteriophage was used by Hershey and Chase to prove that DNA was genetic material.
2. Bacteriophage was used by Zinder and Lederberg to bring about transduction in bacteria.
3. Bacteriophage was used to destroy bacteria to purify water body
4. Cyanophages were employed to control algal bloom.

Sub-viral particles

These are smaller particles than viruses. There are three types:

1. **Virusoid:** It was discovered by Randle (1981). It is small, circular and single-stranded RNA. It lives in the protein coat of a large virus. It replicates within the host and does not cause infection.
2. **Viroid:** It was discovered by Diener and Raymer (1967). It is extremely simple and one of the smallest known infectious agents. It contains RNA only but protein is absent. RNA is single-stranded, circular, covalently closed and tightly folded. It contains 260-380 nucleotides having low molecular weight. RNA replicates in the nucleus of the host cell, unlike the virus which replicates in the host cytoplasm. It causes diseases only in higher plants. It is transmitted mechanically or through wind. It causes the following diseases:
   (a) Cadang cadang of coconut
   (b) Potato spindle tuber disease (PSTD)
   (c) Citrus exocortis (stunt)
   (d) Chrysanthemum stunt
5. **Prion:** It was discovered by Stanley Prusiner (1972). It is the smallest known infectious agent (100 times smaller than the smallest virus). It is rod-shaped. It contains proteins (PrP) only but does not have nucleic acids (DNA/RNA). It undergoes autonomous replication in spite of the absence of nucleic acid. It is latent and resistant to heat (135°C for 18 min), UV radiations, ionising radiation and protease. It is sensitive to urea, phenol, and sodium dodecyl sulphate (SDS). Prion protein (PrP) is not sedimentable and cannot be separated by centrifugation. PrP occurs in infectious and non-infectious cells. It converts folded protein into PrP. It is called a slow virus because the disease occurs very slowly. It causes diseases only in animals. It causes the following diseases:
   (a) Scrapie disease: It occurs in sheep and goats in which there is loss of wool due to scrapping of the body on a rough surface.
   (b) Kuru disease: It is a degenerative disease of CNS in the human beings of New Guinea.
   (c) Mad Cow disease: It is also known as Bovine Spongiform Encephalopathy in which soft brain and cavities are developed in the brain of cattle.
   (d) Creutzfeldt-Jakob disease: It occurs in human beings in which there is a loss of memory and coordination.

SHORT ANSWER TYPE QUESTIONS

1. Discuss the importance of biological classification.
2. Mention the demerits of the 2-kingdom system of classification.

3. Distinguish between methanogens and halophiles.
4. Distinguish between archaebacteria and bacteria.
5. What are the different shapes of bacteria?
6. Write short notes on *Mycoplasma.*
7. Distinguish between bacteria and cyanobacteria.
8. Draw a well-labelled diagram of *Euglena*.
9. Discuss the five characteristic features of fungi.
10. What are the main features which put Fungi in different kingdoms?
11. Discuss the importance of mycorrhiza.
12. Discuss the economic importance of lichen
13. Discuss the salient features of tobacco mosaic virus.
14. Draw a well-labelled diagram of bacteriophage.
15. Distinguish between viroid and prion.

LONG ANSWER TYPE QUESTIONS

1. Describe the different criteria of the 5-kingdom system of classification.
2. Discuss the different modes of nutrition in bacteria.
3. Describe the important features of each group of Protista.
4. Describe the characteristic features of different classes of fungi.
5. What are the living and non-living properties of viruses?

MULTIPLE CHOICE QUESTIONS

1. Human immune deficiency virus (HIV) has a protein coat and a genetic material that is

(a) Single-stranded RNA
(b) Double-stranded RNA
(c) Single-stranded DNA
(d) Double-stranded DNA

2. Which of the following plant viruses has DNA?

(a) Tobacco mosaic virus
(b) Tomato mosaic virus
(c) Potato mosaic virus
(d) Cauliflower mosaic virus

3. Phages that show a lysogenic cycle are called

(a) Virulent phages (b) Temperate phages
(c) Avirulent phages (d) Lytic phages

4. The most complex animal virus is

(a) Poliovirus (b) Pox virus
(c) Parrot fever virus (d) Influenza virus

5. Which of the following viruses was first to be cultured *in vitro* in human cells?

(a) Poliovirus (b) Pox virus
(c) Influenza virus (d) HIV

6. Which of the following statements is incorrect about *Rhizopus*?

(a) Coenogametes are found.
(b) Trisporic acid is a hormone involved in sexual reproduction.
(c) The common bread mould is *Rhizopus stolonifer.*
(d) Coenocytic condition is not found.

7. The disease 'Aspergillosis' is caused by *Aspergillus* which infect

(a) Liver (b) Kidney
(c) Lung (d) Ear

8. Koch's postulates are not applicable to

(a) Mycobacterium leprae
(b) Mycobacterium tuberculosis
(c) Diplococcus pneumonia
(d) Vibrio cholera

9. Bacteria responsible for nitrification come under

(a) Bacillus *(b) Coccus*
(c) Spirilla *(d) Vibrio*

10. The kingdom 'Fungi' was included for the first time in

(a) Two kingdom system
(b) Three kingdom system
(c) Four kingdom system
(d) Five kingdom system

11. In five Kingdom system classification, greater emphasis was given on

(a) Cellular organisation and mode of nutrition
(b) Cellular organisation and Lifestyle
(c) Composition of cell wall and Cellular organisation
(d) Composition of cell wall and Lifestyle

12. Carl Woese divided all organisms into three domains on the basis of

(a) 23 S rRNA (b) 5S rRNA
(c) 28S rRNA (d) 16 S rRNA

13. Which of the following archaebacteria are found in hot spring?

(a) Methanogens
(b) Halophiles
(c) Thermoacidophiles
(d) All of the above

14. Which of the following is incorrect regarding archaebacteria?

(a) Methanogens are found in the rumen of cattle.
(b) Halophiles have purplish pigmented membrane
(c) Thermo-acidophiles grow at high temperatures and acidity.
(d) All archaebacteria are motile and Gram-negative.

15. Halophiles tolerate high salt concentrations due to the presence of

(a) Special lipids in membrane
(b) Mucilage covering
(c) Special protein and mucilage covering
(d) Special lipid and mucilage covering

16. Which of the following is incorrect regarding bacteria?

(a) They have a simple structure and complex behaviour
(b) They do not show cyclosis
(c) They show the most extensive metabolic diversity
(d) They form gametes

17. Most of the bacteria are

(a) Photoautotrophs
(b) Chemoautotrophs
(c) Photoheterotrophs
(d) Chemoheterotrophs

18. ***Chromatium* is also known as**

(a) Green sulphur bacteria
(b) Purple sulphur bacteria
(c) Green non-sulphur bacteria
(d) Purple non-sulphur bacteria

19. Which of the following is incorrect regarding the endospore of bacteria?

(a) It is thick-walled and highly resistant
(b) It is endogenous
(c) It occurs during unfavourable period
(d) It is several formed per cell

20. Which of the following is incorrect regarding cyanobacteria?

(a) They are Gram-negative
(b) Their photosynthesis is anoxygenic
(c) They form algal blooms
(d) They contain gas vacuoles.

21. Which of the following is incorrect regarding the heterocyst of cyanobacteria?

(a) It is a thick-walled structure
(b) It takes part in asexual reproduction
(c) It plays a role in nitrogen fixation
(d) It undergoes vegetative reproduction.

22. Which of the following is incorrect regarding Protista?

(a) These are unicellular and eukaryotic
(b) These have no well-defined boundaries
(c) These are anaerobes
(d) These show a 9+2 arrangement of cilia and flagella

23. Diatoms are photosynthetic due to the presence of

(a) Chlorophyll-a and b
(b) Chlorophyll-a and c
(c) Chlorophyll-a and d
(d) Chlorophyll-a only

24. Auxospores are found in

(a) Diatoms (b) Dinoflagellates
(c) Euglenoids (d) Slime moulds

25. **Which of the following dinoflagellates causes red tide in the sea?**
(a) *Gymnodinium* (b) *Gonyaulax*
(c) *Pyrodinium* (d) *Noctiluca*

26. **Euglenoids resemble higher plants in having**
(a) Reserve food
(b) Composition of cell wall
(c) Food vacuole
(d) Chlorophyll pigments

27. **Which of the following is not an animal feature in Euglenoids?**
(a) Presence of eyespot
(b) Absence of cell wall
(c) Heterotrophic mode of nutrition in dark
(d) Longitudinal binary fission

28. **The spore of slime moulds is made up of**
(a) Starch (b) Protein
(c) Cellulose (d) Glycogen

29. **The somatic body of slime moulds is made up of**
(a) Thallus
(b) Sporocarp
(c) Pseudoplasmodium
(d) Plasmodium

30. **The locomotary organelles are absent in**
(a) Rhizopoda (b) Ciliata
(c) Flagellata (d) Sporozoa

31. **The mycelium in lower fungi is coenocytic which is**
(a) Binucleate and septate
(b) Uninucleate and septate
(c) Multinucleate and aseptate
(d) Multinucleate and aseptate

32. **The cell wall in fungi is made up of chitin which is a polymer of**
(a) NAG (b) NAM
(c) NAG and NAM (d) Amino acids

33. **Sex organs in fungi are**
(a) Unicellular and jacketed
(b) Multicellular and non-jacketed
(c) Multicellular and jacketed
(d) Unicellular and non-jacketed

34. **Which of the following spores is not involved in asexual reproduction?**
(a) Oospore (b) Conidia
(c) Sporangiospore (d) Zoospore

35. **Irish famine is caused by**
(a) *Alternaria solanii*
(b) *Synchytrium endobioticum*
(c) *Helminthosporium oryzae*
(d) *Phytophthora infestans*

36. **Which is incorrectly matched?**
(a) *Mucor*-Pin mould
(b) *Rhizopus*-Bread mould
(c) *Aspergillus*-Black mould
(d) *Penicillium*-Red mould

37. **Which of the following fungi is involved in biocontrol?**
(a) *Alternaria* (b) *Mucor*
(c) *Colletotrichum* (d) *Trichoderma*

38. **Sexual reproduction is absent in**
(a) Oomycetes (b) Zygomycetes
(c) Ascomycetes (d) Deuteromycetes

39. **Dikaryophase is longer and present between plasmogamy and karyogamy in**
(a) Phycomycetes (b) Ascomycetes
(c) Basidiomycetes (d) Deuteromycetes

40. **Crozier formation occurs in**
(a) Phycomycetes (b) Ascomycetes
(c) Basidiomycetes (d) Deuteromycetes

41. **Clamp connection is seen in**
(a) Phycomycetes (b) Ascomycetes
(c) Basidiomycetes (d) Deuteromycetes

42. **Which of the following classes of fungi is involved in mineral cycling?**
(a) Phycomycetes (b) Ascomycetes
(c) Basidiomycetes (d) Deuteromycetes

43. **Gametangial copulation in fungi results in the formation of**
(a) Oospore (b) Zygospore
(c) Sporangiospore (d) Zoospore

44. **Endomycorrhiza is shown by**
(a) *Lacterias* (b) *Boletus*
(c) *Amanita* (d) *Glomus*

45. **The main algal and fungal components in lichens are, respectively,**
(a) Cyanophyceae and Oomycetes
(b) Chlorophyceae and Ascomycetes
(c) Chlorophyceae and Basidiomycetes
(d) Phaeophyceae and Ascomycetes

46. Virus is an inert chemical or molecule outside the host cell and exists as

(a) Viroid (b) Prion
(c) Virusoid (d) Virion

47. In bacteriophage, the nature of genetic material is

(a) ss RNA (b) ss DNA
(c) ds RNA (d) ds DNA

48. The protein subunit in TMV is known as

(a) Capsule (b) Capsid
(c) Capsomere (d) Cap

49. Which of the following is incorrect with respect to sub-viral particles?

(a) Viroid does not contain protein
(b) Prion does not contain nucleic acid
(c) Viroid causes PSTD
(d) Prion is the largest known infectious agent

50. Mad cow disease is caused by

(a) Viroid (b) Virion
(c) Prion (d) A bacterium

51. The first organisms evolved on Earth were

(a) Aerobic heterotrophs
(b) Anaerobic heterotrophs
(c) Aerobic autotrophs
(d) Anaerobic autotrophs

52. Mycorrhiza works as

(a) Modified root
(b) A root hair in an adverse condition
(c) Vegetative propagule
(d) Mechanical support

53. The electron donor in photoautotrophic bacteria is

(a) H_2O (b) H_2S
(c) Both a and b (d) Organic compound

54. Fungi imperfecti is also known as

(a) Phycomycetes (b) Ascomycetes
(c) Basidiomycetes (d) Deuteromycetes

55. The unit of a virus is known as

(a) Viroid (b) Virion
(c) Prion (d) Virusoid

56. Endomycorrhiza is seen in

(a) *Pinus* (b) *Betula*
(c) *Eucalyptus* (d) Orchids

57. The dominant pigment in cyanobacteria is

(a) Phycocyanin (b) Fucoxanthin
(c) Phycoerythrin (d) Beta carotene

58. Which of the archebacteria belongs to hot sulphur springs?

(a) Methanogens (b) Halophiles
(c) Thermoacidophiles (d) All of the above

59. The locomotary organelles in *Leischmania* and *Trypanosoma* are

(a) Cilia (b) Flagella
(c) Pseudopodia (d) None

60. Virus is also known as

(a) Obligate intracellular parasite
(b) Facultative intracellular parasite
(c) Obligate intracellular saprophyte
(d) Facultative intracellular saprophyte

61. Who classified animals into two groups on the basis of the presence or absence of RBCs?

(a) Aristotle (b) Linnaeus
(c) Copeland (d) Whittaker

62. How many bacteria are produced in 3 hours if a bacterium divides once in half an hour?

(a) 16 (b) 32
(c) 64 (d) 128

63. Which of the following processes are involved in the reproduction of protists?

(a) Binary fission and budding
(b) Cell fusion and zygote formation
(c) Spore formation and cyst formation
(d) All of the above

64. Which of the following fungi is used in biochemical and genetic research?

(a) Mucor *(b) Neurospora*
(c) Aspergillus *(d) Trichoderma*

65. Fungi show sexual reproduction by all the following spores except

(a) Conidia (b) Zygospore
(c) Oospore (d) Ascospore

66. In basidiomycetes, the vegetative reproduction takes place by

(a) Endospore (b) Conidia
(c) Akinetes (d) Fragmentation

67. The term '*Contagium vivum fluidum*' was proposed by

(a) Louis Pasteur (b) Beijerinck
(c) Stanley Prusiner (d) Leeuwenhoek

68. The algal component of lichen mostly belongs to the class

(a) Chlorophyceae (b) Phaeophyceae

(c) Rhodophyceae (d) Cyanophyceae

69. Which of the following is an incorrect statement?

(a) Diatoms do not contain locomotary organelles

(b) Euglena shows mixotrophic nutrition

(c) Slime moulds belong to class myxomycetes

(d) Sporozoa class of Protozoa contains locomotary organelles.

70. Potato spindle tuber disease is caused by

(a) Prion (b) Viroid

(c) Bacteria (d) Protozoa

ANSWERS

1.a	11.a	21.d	31.c	41.c	51.c	61.a
2.d	12.b	22.c	32.a	42.d	52.a	62.c
3.b	13.c	23.b	33.c	43.b	53.b	63.d
4.c	14.d	24.a	34.a	44.d	54.d	64.b
5.a	15.d	25.b	35.d	45.b	55.b	65.a
6.d	16.d	26.d	36.d	46.d	56.d	66.d
7.c	17.d	27.d	37.d	47.d	57.a	67.b
8.a	18.b	28.c	38.d	48.c	58.c	68.a
9.a	19.d	29.d	39.c	49.d	59.b	69.d
10..d	20.b	30.d	40.b	50.c	60.b	70.b

CHAPTER-3

PLANT KINGDOM

The plant kingdom comprises five groups which are organised as such:

Kingdom: Plants

↓ ———————————————————— ↓

Subkingdoms: **Cryptogams** (Non-seed bearing plants) **Phanerogams** (Seed bearing plants)

↓ ↓

Divisions: **Thallophyta Bryophyta Pteridophyta** **Gymnosperms Angiosperms**

General characters of Plants

1. These are multicellular and eukaryotic.
2. These have a cell wall which is made up of cellulose.
3. These are autotrophic (photosynthetic) due to the presence of pigments.
4. These store food in the form of starch.
5. These possess rhizoids or roots for fixation and absorption of water.
6. These complete the life cycle in two alternating phases- a diploid or sporophyte and a haploid or gametophyte.

ALGAE

Algae come under thallophyta (young shoot plant) which was coined by Endlichler (1836). Algae are simple, non-flowering, chlorophyllous, non-vascular and aquatic groups of plants. These originated from the Cambrian period. The study of algae is known as phycology (Gr. *Phykos*=sea weed). Algae are known as children of the sea.

The Father of phycology is Fritsch.

The Father of Indian phycology is M.O.P. Iyenger

Smallest alga	*Chlamydomonas* (a green alga)
Largest alga	*Macrocystis* (a brown alga)

Characteristic features:

1. Algae are ubiquitous (found everywhere). These are mostly aquatic (freshwater forms are called pond scum and marine forms are called seaweed).
2. Plant body is gametophyte. It is made up of thallus which does not contain organised tissues hence roots, stems and leaves are not differentiated. Thallus shows great variation.
3. Mechanical and vascular tissues are absent
4. The cell wall is made up of cellulose, galactose and mannose.
5. These are photoautotrophs due to the presence of different chlorophyll molecules.
6. Reserve food is mostly starch.
7. Vegetative reproduction takes place by fragmentation.
8. Asexual reproduction takes place by binary fission, budding, palmella stage and different spores such as zoospore-most common (thin-walled and motile), aplanospore (thin-walled and non-motile) and hypnospore (thick-walled and non-motile).

9. Sex organs are unicellular and non-jacketed; if multicellular, then all cells are fertile.
10. Sexual reproduction involves plasmogamy (fusion of cytoplasm of gametes), karyogamy (fusion of nuclei of gametes) and meiosis.
11. Sexual reproduction occurs in three ways:
    (a) **Isogamy:** It is the fusion of morphologically and physiologically similar gametes. It is most primitive method. It marks the transition from asexual reproduction to sexual reproduction. e.g., *Chlamydomonas debaryanum, Ulothrix.*
    (b) **Physiological anisogamy:** It is the fusion of morphologically similar but physiologically (behavioural) dissimilar (+ and -) gametes. e.g., *Spirogyra, Zygnema*
    (c) **Anisogamy:** It is the fusion of morphologically and physiologically dissimilar gametes. In this, the male gamete is smaller and more active while the female gamete is larger and less active. e.g., *Chlamydomonas braunii*
    (d) **Oogamous:** It is the most advanced type in which the protoplast of male gametangium undergoes repeated division to form smaller, motile and active male gamete and the protoplast of female gametangium does not divide but is directly converted into larger, non-motile and passive egg. e.g., *Chlamydomonas coccifera, Volvox, Oedogonium, Fucus.*
12. Embryo is not formed after fertilisation.
13. The process of reproduction increases in complexity from lower to higher forms.
14. Sporophyte is not dependent on gametophyte.
15. The life cycle is haplontic in most algae. The vegetative phase is haploid and zygotic meiosis occurs. Some other life cycles are as such
    (a) Diplontic: The vegetative phase is diploid and gametic meiosis occurs. e.g., *Fucus, Sargassum* etc.
    (b) Haplo-Diplontic: There is an alternation of gametophytic and sporophytic phases and it shows sporic meiosis. It may be
    (c) Isomorphic (morphologically similar): *Cladophora, Ectocarpus, Ulva*
    (d) Heteromorphic (morphologically different): *Laminaria*
    (e) Haplobiontic (diphasic): There are two haploid phases-(i) sexual individual and (ii) carpospores bearing stage alternating with diploid phase (zygote).e.g., *Nemalion*
    (f) Haplobiontic (triphasic): There is one gametophytic phase and two sporophytic phases such as carposporophyte and tetrasporophyte.
    (g) Diplobiontic: (triphasic): There are two diploid phases (i) diploid carposporophyte, and (ii) tetrasporic individual alternating with single haploid phase (sexual individual). e.g., *Polysiphonia.*

Classification: Fritsch (1935,1945) proposed classification of algae in his book 'The structure and reproduction of the Algae'. He considered algae as the rank of division and divided algae into 11 classes. He classified algae on the basis of

(a) Flagellation
(a) Types of pigment
(a) Nature of reserve food
(a) Composition of cell wall
(a) Types of plastids
(a) Nature of pyrenoid
(a) Types of sexual reproduction

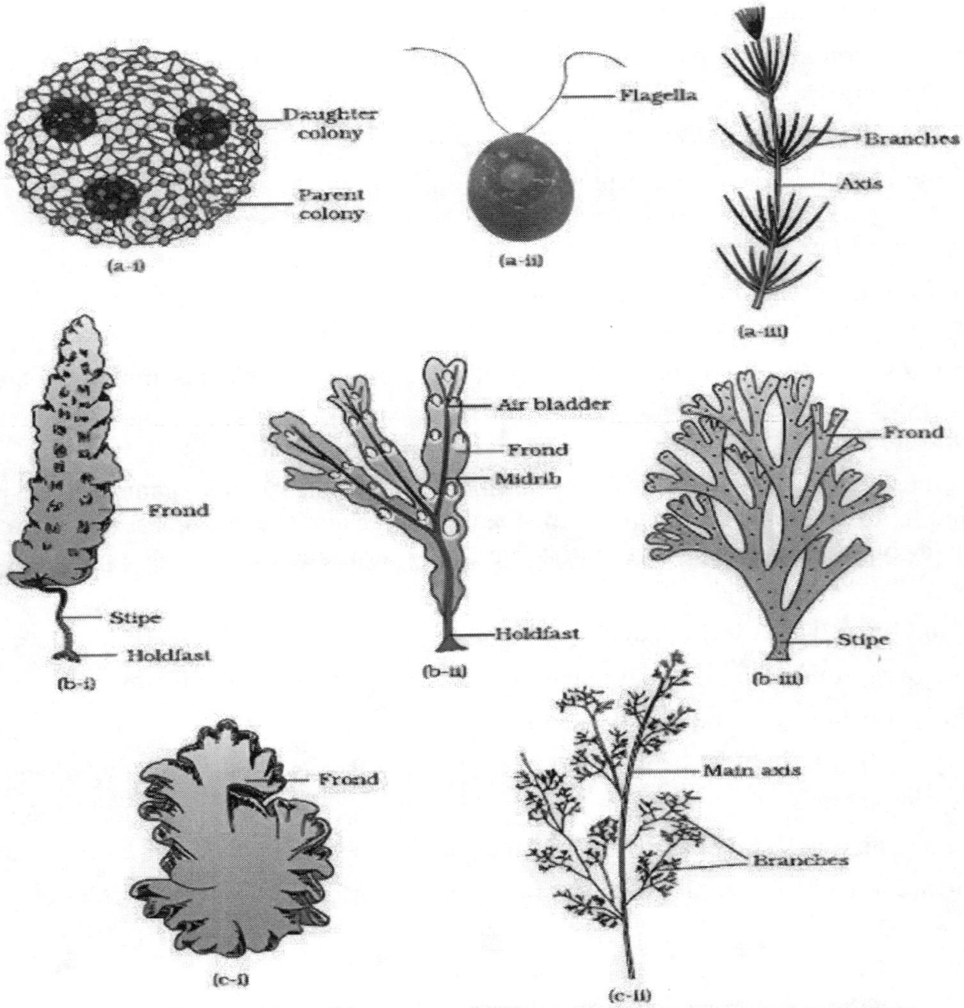

Figure Algae : (a) Green algae (i) *Volvox* (ii) *Chlamydomonas* (iii) *Chara*
(b) Brown algae (i) *Laminaria* (ii) *Fucus* (iii) *Dictyota*
(c) Red algae (i) *Porphyra* (ii) *Polysiphonia*

Class 1: **Chlorophyceae:**

(a) These are mostly fresh water (90%).

(b) Thallus may be

- Unicellular and motile- *Chlamydomonas* (the smallest gametophytic plant)
- Unicellular and non-motile- *Chlorella*
- Filamentous, unbranched and attached- *Ulothrix, Oedogonium*
- Filamentous, unbranched and free-living- *Spirogyra.*
- Filamentous and branched- *Cladophora.*
- Colonial and motile- *Volvox, Gonium*
- Colonial and non-motile- *Scenedesmus, Hydrodictyon*
- Coenocytic- *Caulerpa*
- Heterotrichous- *Coleochaete, Stigeoclonium*
- Palmella stage- *Tetraspora.*

(c) Cell wall is made up of cellulose (inner) and pectin (outer).

(d) Chloroplast is of variable shape

- Spiral or ribbon shaped: *Spirogyra*

- Cup shaped: *Chlamydomonas*
- Girdle or collar shaped: *Ulothrix*
- Discoid: *Chlorella*
- Reticulate: *Oedogonium*
- Stellate (star shaped): *Zygnema*

(e) Chlorophyll-a and b are present and β-carotene is dominant.

(f) Pyrenoid is a proteinaceous body having a sheath of starch. It is found in an envelope of chloroplast. It synthesises starch.

(g) Reserve food is starch which is in the form of grains.

(h) Flagella are isokont and acronematic (anteriorly inserted).

(i) Vegetative reproduction takes place by fragmentation.

(j) Asexual reproduction takes place by zoospores (e.g., *Chlamydomonas*, *Ulothrix*) and aplanospore (e.g., *Chlamydomonas*; and hypnospore).

(k) Sexual reproduction takes place by isogamy, anisogamy, physiological anisogamy and oogamy (advanced).

2. **Phaeophyceae** (Brown algae):

(a) These are mostly marine (found in cooler seas) and lithophytes (growing attached to rocks).

(b) Thallus is differentiated into hold fast (for attachment), stipe (stalk) and frond (leaf-like). Air vesicles are present on the frond for providing buoyancy.

(c) Thallus is heterotrichous (different types of branches such as prostrate and erect system)

(d) The cell wall is made up of cellulose and algin.

(e) Chlorophyll-a and-c are present in Chromatophore.

(f) Dominant pigment is fucoxanthin.

(g) Reserve food is laminarin (sugar) and mannitol (alcohol) in pyrenoid.

(h) Flagella are heterokont (of different lengths) and stigonematic (laterally inserted).

(i) Pyrenoid is naked (without starch sheath) and stalked.

(j) Conducting or trumpet hyphae are present in large forms for the transport of food.

(k) Vegetative reproduction takes place by fragmentation.

(l) Asexual reproduction takes place by pear-shaped biflagellate zoospores.

(m) Sexual reproduction takes place by isogamy (e.g., *Ectocarpus*). anisogamy (e.g., *Cutleria*) and oogamy (e.g., *Fucus, Dictyota, Laminaria*).

3. **Rhodophyceae** (Red algae):

(a) These are predominantly marine (found in warmer seas) and lithophytes. The only freshwater red alga is *Batrachospermum* (Frog spawn alga)

(b) (b) Thallus may be

Unicellular (e.g., *Porphyridium*)

Pseudoparenchymatous (e.g., *Gelidium*)

Filamentous -uniaxial filament (e.g., *Batrachospermum*) and Multiaxial filament (e.g., *Polysiphonia*)

(c) The cell wall is made up of cellulose and pectin.

(d) Flagella are absent.

(e) Chlorophyll a and d are present in chromatophore (photosynthetic lamellae).

(f) Dominant pigment is phycoerythrin.

(g) Pyrenoid is absent.

(h) The reserve food is Floridian starch (it stains red with iodine test and resembles amylopectin (branched).

(i) Vegetative reproduction is not very common but occurs only in unicellular forms.

(j) Asexual reproduction takes place by unflagellate spores such as tetraspores, carpospores, monospore etc.

(k) Sexual reproduction takes place by oogamy. The gametes are non-motile.

(l) Male sex organ is spermatangium which produces non-motile male gamete (spermatium) and female sex organ is carpogonium which is one celled and its prolonged distal end has an outgrowth called trichogyne.

(m) During fertilisation, the spermatium comes in contact with the trichogyne with the help of water current.

(n) Post fertilisation stage is very complex and occurs in carpogonium (flask-shaped).

(o) Post fertilisation changes develop carposporophyte. Carposporangia are developed from each carposporophyte and each carposporangium produces single carpospores.

(p) They mostly show biphasic or triphasic life cycle.

4. **Xanthophyceae** (Yellow green algae):

5. **Chrysophyceae** (Yellow green algae):

6. **Cryptophyceae**:

7. **Chloromonadinea** (Bright green algae):

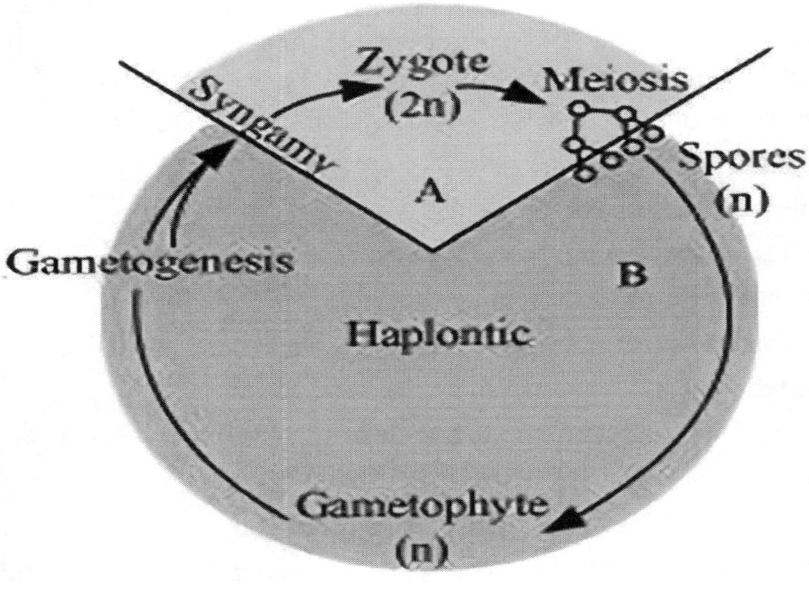

Fig: Haplontic life cycle

EVOLUTION OF PLANTS

Tetrasporine line: It is the most accepted.

Chlamydomonas (motile and unicellular) →*Tetraspora* (non-motile colony) →*Ulothrix* (filamentous) →*Draparnaldiopsis* (heterotrichous) → *Ulva* (parenchymatous) →*Chara* (branched multilayered)

Economic Importance

(a) Algae produce organic matter as a result of photosynthesis and provide oxygen to aquatic organisms.

(b) Some algae such as *Chlorella, Scenedesmus, Ulva, Chondrus, Porphyra* etc. are sources of proteins.

(c) Some brown algae such as *Fucus, Sargassum, Laminaria* etc. are used as fodder for animals.

(d) A few red algae such as *Gelidium, Gracilaria, Gigartina, Pterocladia* produce agar-agar which is dried hydrophilic and phycocolloidal polysaccharide (galactose monomer) obtained from the cell wall of algae. It is used in food, pharma and cosmetic industries. It is also used in media preparation as a solidifying agent.

(e) A few brown algae such as *Laminaria* and *Fucus* which produce algin which phycocolloidal salt of alginic acid obtained from the cell wall of algae. It is used as a thickener for preparing polish, cream, soup, sauce etc.

(f) A red alga, *Chondrus* produces carrageenan which is phycocolloidal sulphate polysaccharide present in its cell wall. It is used as a stabiliser in cosmetic and paint industries.

(g) A green alga, *Chlorella* produces an antibiotic, chlorellin.

(h) Brown algae (kelps) are the source of iodine and soda.

(i) A few green algae such as *Chlamydomonas, Chlorella, Scenedesmus, Pandorina* etc. remove organic matter and coliform bacteria by aerobic oxidation (pumping of O_2 as a result of photosynthesis).

(j) *Chlorella* has been used in space by astronauts because it has a rapid rate of photosynthesis hence it produces oxygen and food for space travellers.

(k) *Chlorella* has also been used as experimental material in photosynthesis for studying the path of carbon.

(l) *Acetabularia* has been used in cytology to prove the nucleus as a storehouse of genetic information.

BRYOPHYTA

The term 'Bryophyta' (Gr. *Bryon*=moss, phyton=plant) was coined by Robert Brown (1864). It is a group of simplest land plants. Bryophytes originated in the Devonian period. These are relatively poorly adapted to life on land.

Father of Bryology: Hedwig

Father of Indian bryology: SR. Kashyap

Smallest bryophyte	*Zoopsis argentea* (a liver wort)
Largest bryophyte (gametophyte)	*Dawsonia* (a moss) =70 cm
Fossil bryophytes	*Ricciopsis, Hepaticites, Naiadetia*
Aquatic bryophytes	*Riccia fluitans, Ricciocarpus natans, Riella*
Saprophytic bryophytes	*Bauxbaumia, Cryptothallus*

Features:

1. These grow in moist and shady places for the leading vegetative phase as well as live in water for the leading reproductive phase hence called amphibious plants. Western Himalaya is known as a gold mine of bryophytes.
2. The plant body is differentiated into root-like rhizoids for attachment with substratum, stem-like and leaf-like structures but there is no true root, stem or leaf.
3. Vascular tissues are absent.
4. The main plant body is gametophyte which is dominant, free-living, green (photosynthetic), independent and develops from spore. It is attached to the soil so it can absorb water and minerals.
5. Sex organs are multicellular and jacketed.
6. The male sex organ is antheridium which encloses the mass of sperms mother cell (androcyte) producing biflagellate and spirally coiled sperms that are released on water.
7. The female sex organ is archegonium which is flask-shaped containing a tubular neck and swollen venter. The neck encloses 6-10 rows of neck canal cells while the venter consists of a smaller and sterile ventral canal cell and a larger and fertile egg.
8. Sperm is attracted chemotactically (sugar) to an egg.
9. Water is required for fertilisation.
10. Fertilisation occurs inside archegonium.
11. The zygote does not undergo a resting period. It is the first cell of the sporophyte. It divides mitotically to form an embryo.
12. Embryo is exoscopic (apex of the embryo develops from outer wall) and later forms sporophyte.
13. Sporophyte is well-developed and differentiated into foot, seta and capsule.

1. Sporophyte is nutritionally (parasitically) dependent on gametophyte because it is aerial and has no soil contact and thus it is non-photosynthetic.
2. Sporogenous tissue (archesporium) shows progressive sterilisation from lower to higher forms.
3. Spore mother cell undergoes sporic meiosis inside the capsule to form haploid spores. Bryophytes are homosporous (only one type of spores).
4. Spores are disseminated by wind and germinate immediately.
5. Vegetative reproduction takes place by progressive death and decay of thallus, fragmentation of gemmae, tubers, buds, adventitious branches and protonema.
6. Asexual reproduction by spores is absent.
7. Life cycle is a haplo-diplontic and heteromorphic alternation of generation.
8. Bryophytes do not attain great height because (i) there is no root, (ii) there is no vascular tissue and mechanical tissue, (iii) there is no cuticle on the body, and (iv) there is requirement of water.

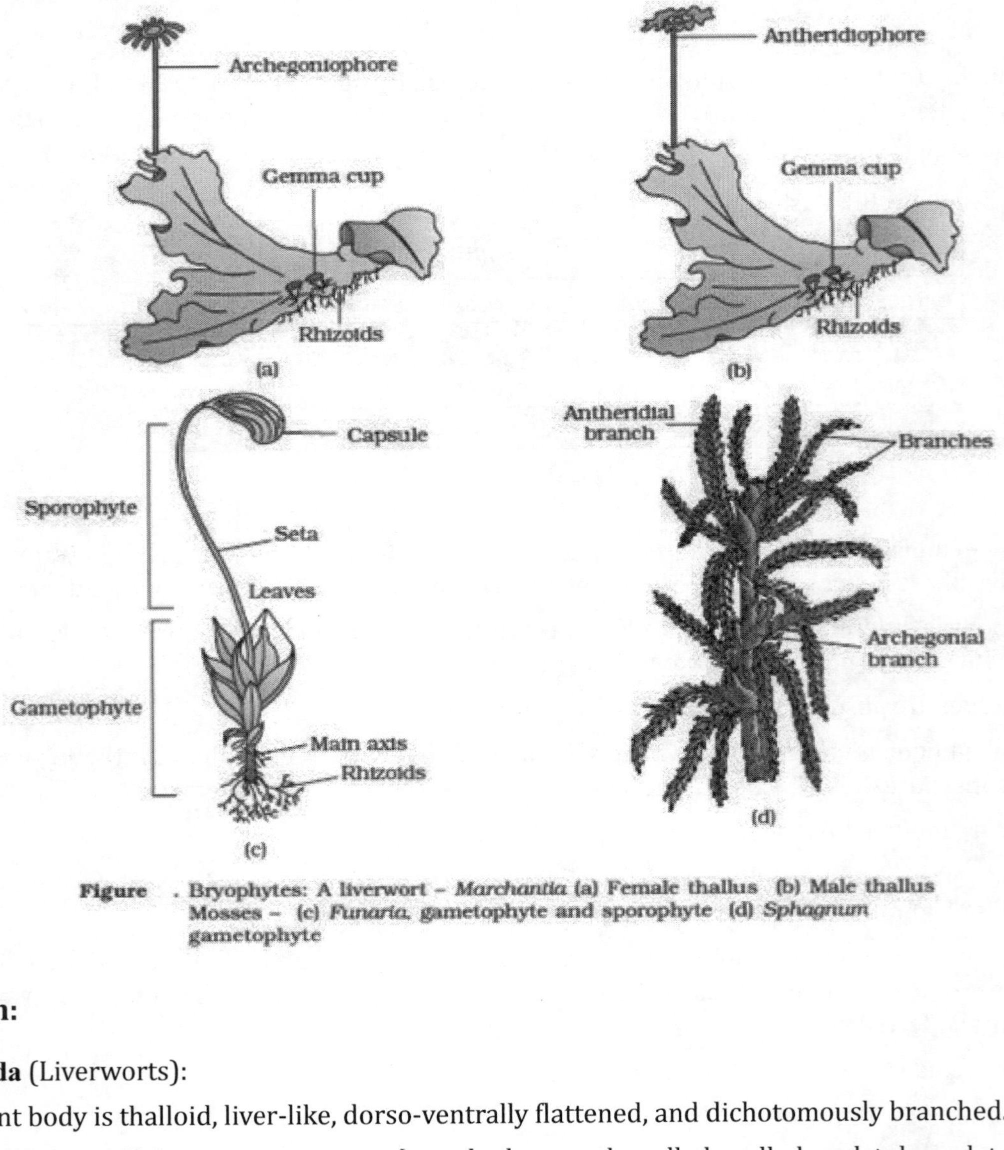

Figure . Bryophytes: A liverwort – *Marchantia* (a) Female thallus (b) Male thallus Mosses – (c) *Funaria*, gametophyte and sporophyte (d) *Sphagnum* gametophyte

Classification:

1. Hepaticopsida (Liverworts):

(a) Main plant body is thalloid, liver-like, dorso-ventrally flattened, and dichotomously branched.

(b) Rhizoids are unicellular, non-septate, unbranched, smooth-walled walled and tuberculate and found on the ventral surface of thallus.

(c) Thallus anatomically consists of an upper (dorsal) assimilatory region having air chambers and air pores and a lower (ventral) storage region.

(d) Chloroplast is without pyrenoid.

(e) Scales are multicellular and found at the ventral surface of the thallus.

(f) Gemmae are green, multicellular, stalked buds and are found on the dorsal surface of thallus.

(g) Sex organs are dioecious (e.g., *Marchantia*) and monoecious (e.g., *Riccia*).

(h) Sex organs are embedded on the dorsal surface of the thallus (homothallic-*Riccia*) and projected from the thallus in the form of stalked antheridiophore and archegoniophore (heterothallic-*Marchantia*).

(i) Sporophyte is simple, and primitive and is found at the dorsal surface of thallus.

(j) Sporophyte is completely dependent on gametophyte.

(k) Seta is delicate and short-lived.

(l) Capsule does not contain stomata and chlorophyll hence non-photosynthetic. It does not show external differentiation.

(m) Sporogenous tissue (archesporium) develops from endothecium. It forms elaters and spores. Elaters are dead, elongated and hygroscopic having spiral bands of thickening. These help in spore dispersal.

(n) Capsule splits irregularly by death and decay of spore wall.

(o) Vegetative reproduction takes place by death and decay of thallus and tubers.

(p) Asexual reproduction takes place due to the germination of gemma. Each gemma has two thalli because two notches are present on the opposite lateral side.

2. **Anthocerotopsida** (Hornworts):

(a) The plant body is thalloid. The thallus is simple in form, dorso-ventrally flattened variously lobed without midrib, air chambers and air pores.

(b) Rhizoids are unicellular and smooth-walled. Tuberculate rhizoids are absent.

(c) Scales are absent.

(d) Sex organs are embedded in the thallus on the dorsal side.

(e) Sperms are biflagellate and whiplash type.

(f) Sporophyte is elongated and is differentiated into bulbous foot, buried in thallus and long, upright, and emerged cylindrical capsule.

(g) Seta is absent.

(h) Capsule is photosynthetic and has simple functional stomata.

(i) Capsule grows continuously with the help of basal intercalary meristem.

(j) Capsule contains a central and sterile columella which develops from the entire endothecium.

(k) Sporogenous tissue (archesporium) develops from an inner layer of amphithecium.

(l) Archesporium gives rise to spores and elaters (no thickening band hence called pseudoelaters).

(m) Archesporium is dome-shaped arching over columella at the top.

3. **Bryopsida (Musci):**

(a) The main plant body is laterally branched. It shows

Primary protonema: It is green, thalloid, filamentous and branched and juvenile stage. It develops from the germination of spore. It resembles a green alga. It disappears after the establishment of a leafy gametophore.

Gametophore: It is leafy (foliose), erect, spiral and mature stage. It is differentiated into rhizoids, axes and leaves but without midrib. It is monoecious.

(b) Rhizoids are multicellular, branched and oblique septate.

(c) Air chambers and air pores are absent.

(d) Scales are absent.

(e) Gemmae are absent.

(f) Sex organs are always monoecious. These develop in the axil of the leaf. Antheridium is club-shaped and stalked and it is surrounded by perigonial leaves. Archegonium is flask-shaped and surrounded by perichaetial leaves. It arises from apical cell of the stem thereby terminating its growth (acrogynous condition). Both are intermingled with paraphysis. The largest archegonium is found in *Funaria*.

(g) Sporophyte is complex and advanced. It is partially dependent on the gametophyte. It is differentiated into foot, seta and capsule.

(h) Seta is strong and long-lived. It performs conduction.

(i) Capsule is differentiated into operculum, theca and paraphysis. It contains stomata and chlorophyll hence photosynthetic.

(j) Sporogenous tissue (archesporium) develops from an outer layer of endothecium as in *Funaria* and from an inner layer as in *Sphagnum*. The entire archesporium forms spores.

(k) Columella is a sterile structure that develops from an inner layer of endothecium.

(l) Elaters are absent.

(m) Spore sac has many small spores.

(n) Split of capsule for spore dispersal is regular. There is an elaborate mechanism of spore dispersal.

(o) Vegetative reproduction takes place by bulbil which is a rounded bud-like structure on the rhizoid. It germinates to form protonema which forms thallus. It also occurs by secondary protonema which is formed from structures other than spores such as rhizoids, stems and leaves. It breaks into some segments containing bud and each gives rise to leafy gametophores.

e.g., *Funaria* (Cord moss), *Polytrichum* (Hair cap moss), *Sphagnum* (bog/peat moss), *Pogonatum* (Maiden hair moss), *Andrea* (Granite moss), *Grimmia* (Black moss), *Tortula* (Twisted moss).

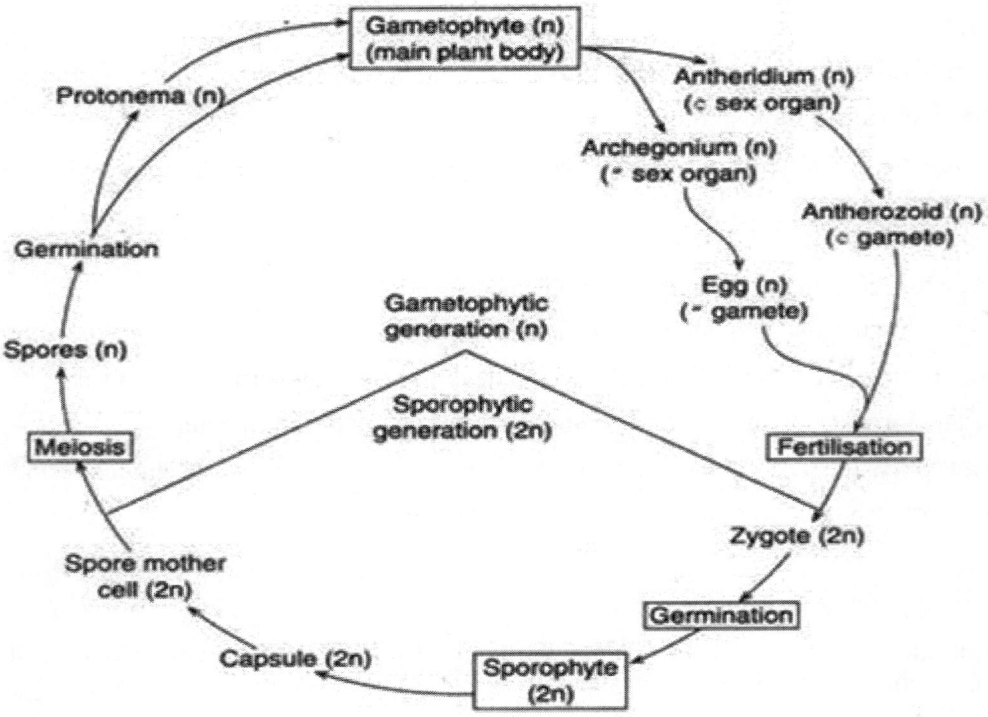

Fig: Haplo-diplontic life cycle

False moss: These plants belong to groups other than bryophyte.

1. Algae: *Chondrus crispus* (Irish moss/Sea moss), *Gracilaria* (Ceylon moss)
2. Lichen: *Cladonia rangiferina* (Reindeer moss), *Cetraria icelandica* (Iceland moss), *Evernia* (Oak moss).

3. Pteridophytes: *Lycopodium* (Club moss), *Selaginella* (Little club moss/spike moss), *Selaginella rupestris* (Bird's nest moss).
4. Angiosperms: *Silene* (Champion moss), *Tillandsia* (Spanish moss/ Florida moss).

Economic importance:

(a) Bryophytes such as *Grimmia, Tortula Polytrichum* are link or transitional communities of primary succession in xerosere.

(b) *Marchantia polymorpha* is used in the treatment of liver and lung disease.

(c) *Sphagnum* is a source of peat which is dried, compressed, spongy and accumulation of partially decomposed plants due to anaerobic and highly acidic conditions in wetland bog. It is also used as fuel and in surgical dressing due to its antiseptic properties.

(d) *Funaria* is an indicator plant for copper.

(e) *Atrichium* is used as a pollution indicator (bioindicator).

(f) Mechanism of sex determination was discovered in *Sphaerocarpus.*

PTERIDOPHYTA

The term 'pteridophyta' (Gr. *Pteris*=feather, phyton=plant) was coined by Haeckel. Tracheophyta includes pteridophyta, gymnosperms and angiosperms. Pteridophytes originated in the Silurian period of the Paleozoic era. Thus, the age of pteridophytes is the carboniferous period. These are primitive vascular cryptogams (first land plants). These are called botanical snakes due to the circinate coiling in young leaves.

Father of pteridology: Bowers

Father of Indian pteridology: S. S. Bir

Smallest pteridophyte	*Azolla*
Largest pteridophyte	*Cyathea*
Oldest fossil pteridophytes	*Rhynia, Cooksonia*

Features:

1. These are mostly herbaceous (annual and perennial).
2. These grow in moist cool and shady places (mostly terrestrial).
3. The main plant body is sporophyte which is dominant, free-living, independent, photosynthetic and develops from the zygote.
4. The plant is divided into true roots (rhizome), stem and leaves.
5. Vascular tissues are present and form a stele which is medullated (with pith) and non-medullated (without pith). The xylem lacks vessels and the phloem lacks companion cells. Cambium is absent hence there is no secondary growth.
6. Ramenta are the scales that cover and protect the stem and leaves.
7. Leaves are microphyllous (small, simple and unveined) and megaphyllous (large, pinnately compound and branched veined). Young leaves show circinnate vernation.
8. Sporophyll is a leaf-like structure that bears sporangia (spore-bearing organ) on its abaxial side. Sporophyll may be uniform as in ferns; compact (cone-like) as in *Selaginella* or form a specialised structure called sporocarp as in *Marsilea.*
9. Sporangia may be single or in a group (sori) protected by an indusium.
10. The development of sporangia is of two types:
    - **Leptosporangiate:** It develops from a single superficial cell. It divides transversely into outer and inner cells. The outer cell forms the entire sporangium. The inner cell remains inactive and forms a stalk of sporangium.

- **Eusporangiate:** It develops from a group of superficial cells that divides periclinally differentiating into an outer layer of jacket initials and an inner layer of primary sporangium cells. The jacket cells eventually produce sporangium; the primary sporogenous cell undergoes nuclear division.

11. Sporangia show the following types of arrangement:
    - Simple: All sporangia grow and mature together. e.g., *Ophioglossum*.
    - Gradate: Mature sporangia at the centre and immature at the periphery. e.g., *Marsilea*.
    - Mixed: Mature and immature sporangia are arranged irregularly. e.g., *Pteris*.
12. These are mostly homosporous (same type of spores) but heterosporous (different types of spores) condition is seen in *Selaginella, Isoetes, Azolla, Marsilea, Salvinia.*
13. Spore germinates to form small, multicellular, short-lived, independent (free-living), photosynthetic (green), thalloid and heart-shaped gametophyte called prothallus. It is monoecious and protandrous.
14. The development of gametophytes is either exosporic (outside spore wall) as in homosporous forms which are monoecious or endosporic (inside spore wall) as in heterosporous forms which are dioecious.
15. Sex organs are multicellular and jacketed and these develop from the ventral surface of the prothallus. Sex organs may be embedded or projected.
16. Antheridium (male sex organ) is round and sessile which develops multi-flagellate and spiral sperms as in ferns but it forms numerous and biflagellate sperms as in *Selaginella.*
17. Archegonium (female sex organ) contains a venter and neck. The venter divides to form smaller and sterile ventral canal cells and larger and fertile egg. There are 1-4 binucleate neck canal cells.
18. Fertilisation is internal and zooidogamous in which male gametes swim in a film of water to the female gametes (archegonium). Pteridophytes are found in narrow geographical regions due to dependence on water for fertilisation.
19. Sperm is chemotactically (malic acid/fumaric acid) attracted to archegonia. (Smallest sperm is seen in *Isoetes*).
20. The zygote produces a small multicellular embryo for its nutrition.
21. Embryo is formed but no ovule. The embryo is endoscopic (the apex is towards the base of archegonia or pushed inside the interior of the female gametophyte.
22. Embryo develops into a mature diploid sporophyte.
23. The life cycle is diplo-haplontic or heteromorphic or heterologous alternation between independent sporophyte and gametophyte generation,
24. Vegetative reproduction takes place.
25. Asexual reproduction occurs by apogamy.

Classification:

1. Psilopsida: It is primitive and the simplest.

(a) Sporophyte is rootless and leafless.

(b) It is differentiated into underground prostrate stem and aerial and erect short system.

(c) It shows dichotomous branching.

(d) Sporangia are found in the axil of small leaf-like appendages.

e.g., Whisk fern (*Psilotum*)

2. Lycopsida:

(a) Leaves are microphyllous and spirally arranged.

(b) Sporangia are found singly at the axial surface of sporophyll.

(c) Homosporous as well as heterosporous conditions are found.

e.g., Little club moss/Spike moss (*Selaginella*), Club moss (*Lycopodium*), Quill wort (*Isoetes*).

3. Sphenopsida:

(a) Sporophyte has a joint axis differentiated into roots and internodes.

(a) Leaves are microphyllous and show whorled phyllotaxy.

(a) Sporangia are found in groups at the end of branches on sporangiophore.

(a) These are homosporous.

e.g., Horse tail (*Equisetum*), Adder's tongue fern (*Ophioglossum*)

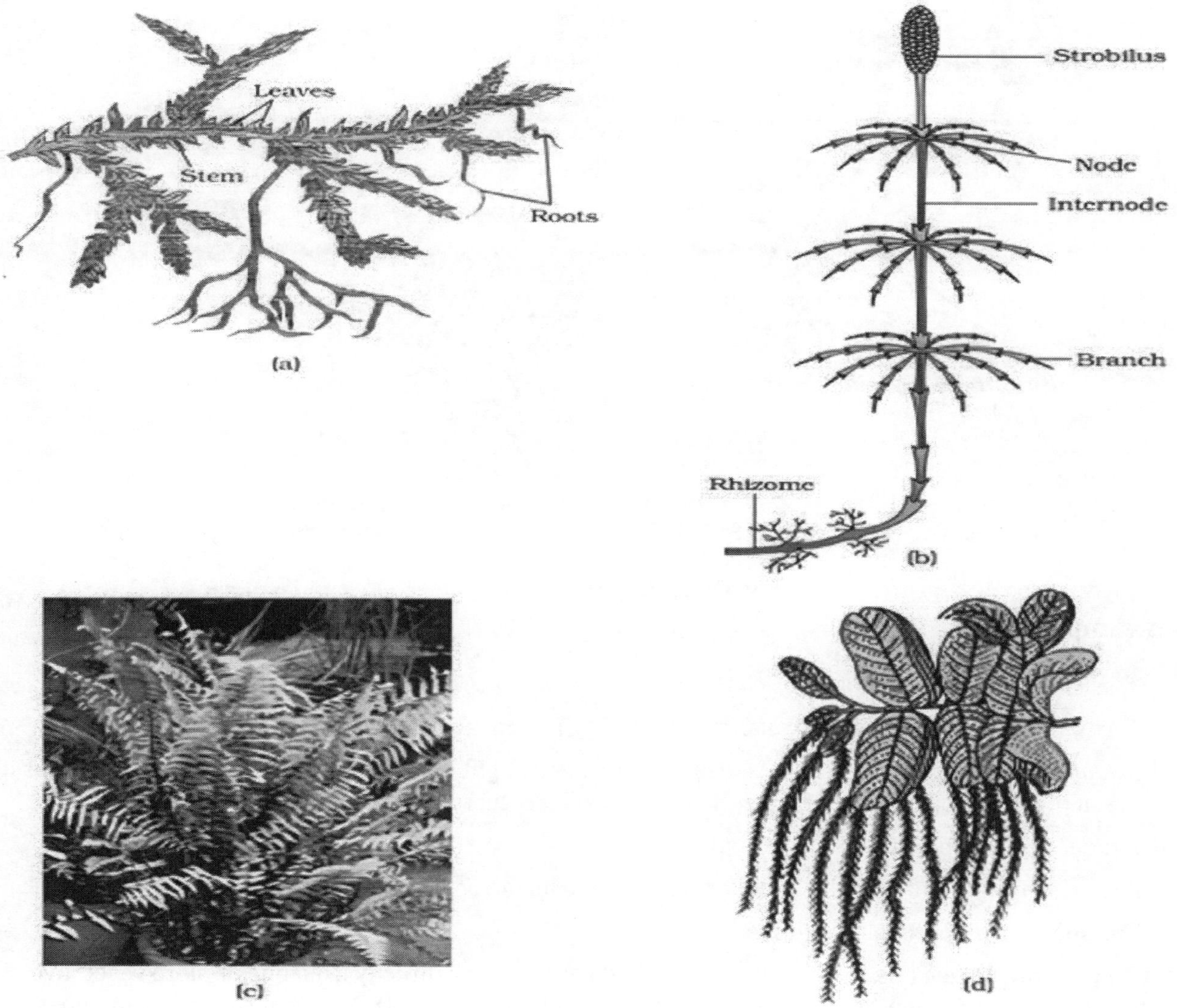

Fig: (a) *Selaginella*, (b) *Equisetum*, (c) Fern and (d) *Salvinia*

4. Pteropsida: It is the largest and most highly advanced.

(a) Sporophyte consists of an underground stem called a rhizome.

(b) Leaves are megaphyllous and pinnately compound.

(c) Sporangia are aggregated to form sori covered by indusium and at the ventral surface of sporophyll.

e.g., Male shield fern (*Dryopteris*), Sun fern (*Pteris*), Maiden hair fern or Walking fern (*Adiantum*), Tree fern (*Cyathea*) Filmy fern (*Hymenophyllum*), Water ferns (*Azolla, Marsilea, Salvinia*), Bracken fern (*Pteridium*).

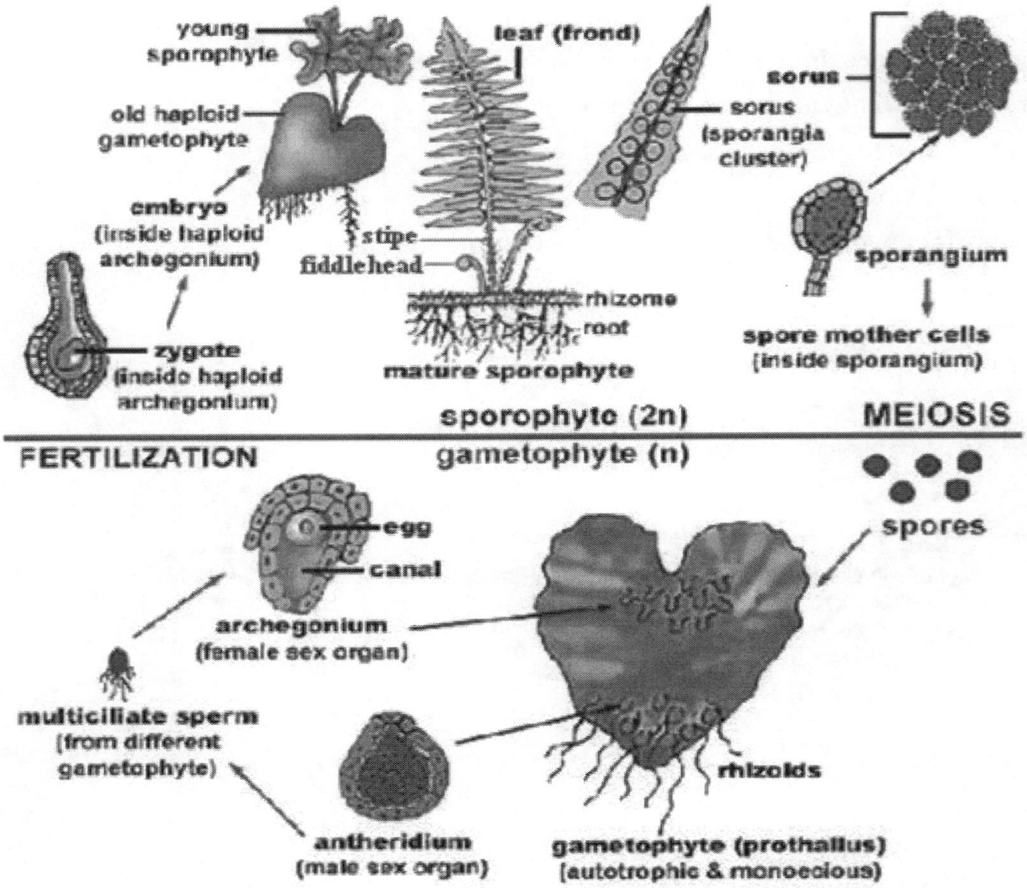

Fig: Diplo-haplontic life cycle

Economic importance:

(a) The fossil members of pteridophytes constitute coal formation.

(a) Ferns are grown as ornamental plants because of their attractive foliage.

(a) *Selaginella lepidophylla* is sold as novelty. It appears as tight ball during dry period and later opens up in green plant in the presence of water hence it is called resurrection plant or Sanjeevani.

(a) *Equisetum* is an indicator plant of gold in soil.

(a) *Azolla* is grown commercially as biofertiliser due to its nitrogen fixing ability.

(a) The sporocarp of *Marsilea* is edible.

Apospory in pteridophytes: It is the formation of diploid gametophyte directly from one or more vegetative cell of sporophyte without spore formation or meiosis. It was reported by Druery (1884) in a fern, *Arthyrium filix-femina*.

Apogamy in pteridophytes: It is the formation of haploid sporophyte directly one or more vegetative cell of gametophyte except egg without fertilisation. It was reported by Farlow (1874) in *Pteris cretica*.

Pre-requisites for seed habit: For the first time, seed habit originated in pteridophytes. The following features are required to form seeds in any plant:

(a) Production of two types of spores (heterospory).

(b) Formation of one megaspore (female gametophyte) within megasporangium.

(c) Presence of an additional protective coat called integument around megasporangium.

(d) Permanent retention of megaspore inside megasporangium.

(e) *In situ* fertilisation and development of embryo.

GYMNOSPERMS

The term 'Gymnosperm' (Gr. *Gymnos*=naked, sperma=seed) was coined by Theophrastus. These are seed-bearing plants, but seeds are not enclosed in an ovary. These are the smallest group of plants having around 900 and 70 living genera. These originated in the Carboniferous period of the Paleozoic era.

Smallest gymnosperm	*Zamia pygmaea*
Tallest gymnosperm	Giant Redwood (*Sequoia sempervirens*)
Gymnosperm of highest life span	*Macrozamia* (10,000 years)

Features:

1. The gymnosperms are mostly evergreen trees or shrubs (no herbs). These grow in temperate regions.
2. These are xerophytic in nature.
3. The main plant body is sporophyte which is divided into roots, stem and leaves.
4. Root is well developed and is of tap root. Root shows diarch to polyarch vascular bundle with exarch condition. The root of *Cycas* (coralloid root) is negatively geotropic and does not bear root hairs and root caps. It shows a symbiotic association with *Anabaena* for nitrogen fixation. However, the root of *Pinus* shows symbiotic association with fungi to form ectomycorrhiza.
5. Stem is woody and unbranched, soft and manoxylic (containing parenchyma) in *Cycas* but branched, hard and pycnoxylic (without parenchyma) in *Pinus*. The xylem contains tracheids and parenchyma with bordered pits but lacks vessels. *Ephedra* and *Gnetum* exceptionally show the presence of a vessel. Phloem contains fibres and parenchyma but lacks a sieve tube and companion cell. Instead, it contains sieve cells. *Cycas* has polyxylic vascular bundle (many cambial rings and vascular bundle) but *Pinus* has monoxylic vascular bundle (one cambial ring and vascular bundle). Secondary growth occurs and annual rings are formed.
6. The leaf is dimorphic and spirally arranged. The foliage leaf is green and pinnately compound as in *Cycas* and simple and needle-like (acicular) as in *Pinus*. The scaly leaves are minute and deciduous. There is no venation of the leaf. Leaf has a thick cuticle, thick epidermis having sunken stomata and sclerenchymatous hypodermis. Transfusion tissue is found in the endodermis around the vascular bundle. It is composed of tracheid and parenchyma cells. It helps in the lateral conduction of water.
7. Sporophyll is leaf-bearing spores. It may be microsporophyll and megasporophyll which aggregate to form male and female cones, respectively. *Cycas* has the largest male cone.

Microsporophyll	Equivalent to stamen of angiosperms
Microsporangium	Equivalent to pollen sac of angiosperms
Microspore	Equivalent to pollen grain of angiosperms
Megasporophyll	Equivalent to carpel of angiosperms
Megasporangium	Equivalent to nucellus of ovule of angiosperms
Megaspore	Equivalent to embryo sac of angiosperms

8. Eusporangiate development of sporangia occurs.
9. Heterospory condition is seen. Microspore germinates to form male gametophyte and megaspore germinates to form female gametophyte.
10. Male and female gametophytes are highly reduced and not free-living (independent). These develop from sporophytic structure.
11. Gymnosperms are mostly dioecious (e.g., *Cycas*) and few are monoecious (e.g., *Pinus*).
12. Antheridium is a male sex organ. It bears non-flagellate sperms but *Cycas* shows multiciliate and top-shaped sperm.
13. Archegonium is a female sex organ. It contains ventral canal cells and egg. *Cycas* has no neck canal cells but show the largest egg.

14. Pollination is anemophilous (through wind).
15. Pollen grains shed at the 3-celled stage as in *Cycas* and the 4-5-celled stage as in *Pinus*.
16. Pollens are deposited in a pollen chamber in the nucellus of an ovule.
17. Fertilisation does not require water. It is siphonogamous (with the formation of a pollen tube) and zooiodogamous (sperm reaches the archegonium by swimming in the fluid present in the archegonial chamber).
18. Zygote forms embryo.
19. Endosperm is haploid (develops before fertilisation).
20. Double fertilisation does not occur. It occurs in *Ephedra* only.
21. The embryo develops from a part or lower half of the zygote (meroblastic development). It shows endoscopic condition.
22. Polyembryony occurs but usually one embryo survives.
23. Embryo shows radicle, plumule, haustorium and cotyledon.
24. Ovule is orthotropous and unitegmic (one integument has outer fleshy, middle stony and inner fleshy). *Cycas thourasii* has the largest ovule (7cm in length).
25. Ovule is transformed into seed which is not covered but exposed on the surface of megasporophyll.
26. Seed has two cotyledons as in *Cycas* and three to fifteen (polycotyledonous) as in *Pinus*. Seed undergoes a resting period and during the period seeds are dispersed over vast areas. Germination of seed is hypogeal.
27. Seed represents three generations:
    (a) Parent (old) sporophytic generation is represented by nucellus and integument.
    (b) Gametophytic generation is represented by endosperm.
    (c) Future (new) sporophytic generation is represented by embryo.
28. The life cycle is diplontic.

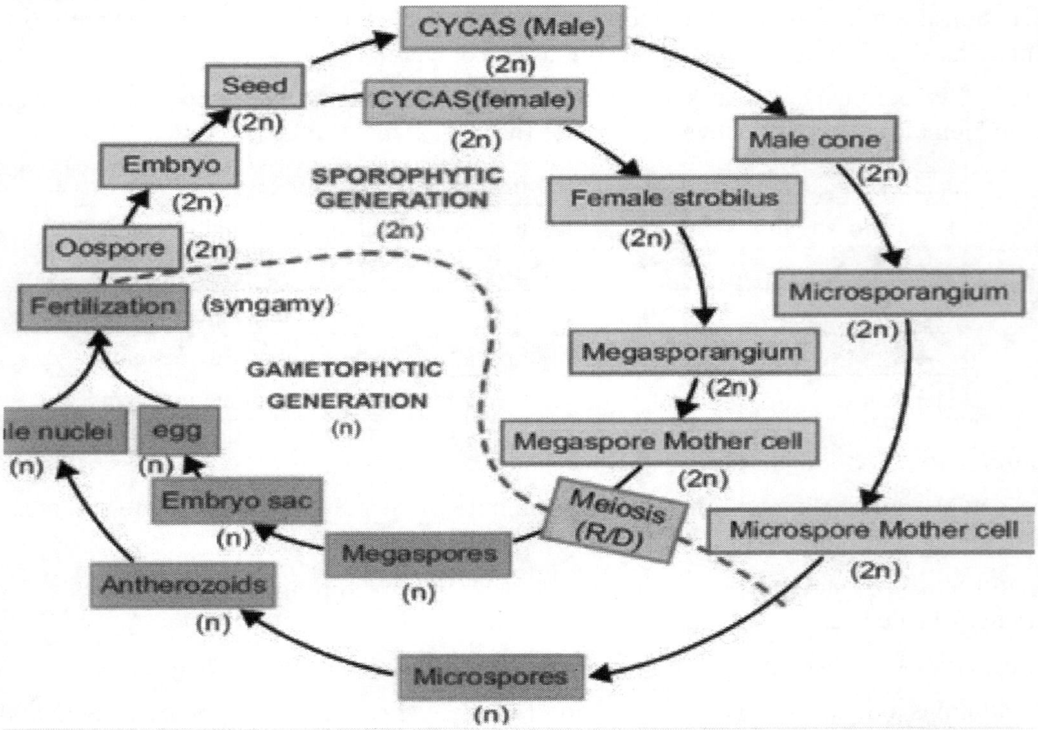

Fig: Diplontic life cycle of ***Cycas***

Classification

1. **Pteridospermopsida:**
   (a) Leaf is large, pinnately compound and frond-like having large leaf trace.
   (b) Spores are aggregated to form synangia.
   (c) Sporangia are borne on leaves which never form cones.
   e.g., *Caytonia, Glossopteris*

2. **Cycadopsida:**
   (a) Stem is unbranched and manoxylic.
   (b) Leaf is a pinnately compound having a direct and girdle-like trace.
   e.g., *Cycas* (Sago palm), *Cycadeoidea, Nilssonia, Pentoxylon.*

3. **Coniferopsida**: (largest class)
   (a) Stem is branched and pycnoxylic.
   (b) Leaf is simple.
   e.g., *Pinus, Taxus, Ginkgo* (Maidenhair tree or Pagoda tree), *Cedrus deodara* (the strongest softwood).

4. **Chlamydospermopsida:**
   (a) The xylem contains vessels.
   (b) Leaves are dimorphic.
   (c) Strobilus is a compound.
   e.g., *Ephedra* (Jointed fir), *Gnetum, Welwitchia*

Economic importance:

(a) *Cycas revoluta, Araucaria excelsa* (Christmas tree), *Araucaria imbricata* (Monkey puzzle), *Pinus, Cupressus* are grown as ornamental plants.

(b) Stem and seed of *Cycas circinnalis* yield a starch called sago.

(c) Leaves of *Cycas* are rich in nitrogen and used as green manure for rice, sweet potato and sugarcane.

(d) Wood of *Cedrus* is also durable, oily, fragrant, insect repellant and rot resistant.

(e) Wood of *Araucaria canninghana* used for plywood manufacture.

(f) Leaf extract of *Ginkgo biloba* is useful in the treatment of cerebral insufficiency and vertigo.

(g) Seed of *Pinus gerardiana* is source of chilgoza which is a dry fruit and edible.

(h) Stem of *Pinus roxburghii* (Chir), *Pinus wallichiana* (Kail) and *Pseudotsuga menzeiss* (Douglas fir) are source of timber.

(i) Stem of *Pinus roxburghii* (Chir) is source of resin which makes them invaluable in paints, varnishes, paper sizing, medicines and liquor industries and is also source of turpentine oil.

(j) Bark of *Taxus brevifolia* produces taxol which is anticancerous compound and is effective against ovarian cancer, breast cancer, and melanoma and colon cancer.

(k) The leaves of *Taxus baccata* are used in asthma, bronchitis, hiccough, epilepsy and for indigestion.

(l) Stem of *Juniperus virginniana* (Red cedar) produces red cedar wood oil used for cleaning tissues in histological works and also used with the oil immersion lenses of the microscope.

(m) Stem of *Abies balsamia* produces a mounting agent, canada balsam used in preparation of permanent slide. It has a very high refractory index hence it is extremely suitable as a mounting medium for microscopic objects and as cements for uses in optical work.

(n) Stem or whole plant of *Ephedra geraradiana* produces ephedrine used in respiratory trouble.

(o) Species of *Ephedra* is a source of an aromatic beverage, known as Mormon tea used in south western United State.

(p) Wood pulp of *Picea* (Spruce) is source of paper.

Adaptations to land plants

(a) Gametophytic generation is highly reduced and is protected inside sporophytic tissue.

(b) Pollination occurs through wind.

(c) Fertilisation does not require water.

(d) Fertilisation occurs through a pollen tube.

(e) Ovule is developed into seed.

ANGIOSPERMS

The term 'Angiosperm' (Gr. *Angeion*=covered; sperma=seed) was coined by Oswald Tippo. Angiosperms are flowering plants in which seeds are enclosed inside the fruit. These are the largest group of plants, the most advanced and showed maximum adaptability. Angiosperms originated in the Cretaceous period of the Mesozoic era and appeared about 130 million years ago.

Smallest angiosperm	*Wolffia microscopica* (0.1 mm)
Tallest angiosperm	*Eucalyptus regnans* (114 m)
Largest spreading tree	*Banyan tree (Ficus benghalensis)*-Father of forest)

Features:

1. These occur in all types of habitats (terrestrial and aquatic). Marine angiosperms are *Thalassia* and *Zostera.*
2. The main plant body is the dominant sporophyte.
3. Sporophyte is divided into root, stem and leaves
4. Mechanical and Vascular tissues are well-developed in vegetative parts.
5. Vessels are found in the xylem for long-distance and quick transport
6. A sieve tube is found in the phloem for the translocation of food material.
7. Secondary growth occurs due to the presence of cambium.
8. Bark is present which is meant for protection.
9. Sporophylls are aggregated to form flowers.
10. These are mostly monoecious and some are dioecious.
11. The male sex organ is the stamen and the female sex organ is the carpel.
12. Dispersal of fruits and seeds takes place.
13. Heterospory condition is seen.
14. Megaspore is single and functional and permanently retained in the nucellus.
15. Embryo sac is well developed.
16. Gametophyte is dependent on sporophyte.
17. Archegonia are absent.
18. Pollination occurs by several agencies.
19. Fertilisation does not require water.
20. Fertilisation is siphonogamous (presence of pollen tube).
21. Double fertilisation is seen.
22. Endosperm is triploid which occurs after fertilisation due to triple fusion.

1. Fertilised ovule ripens into seed.
1. Fertilised ovary ripens into fruit.

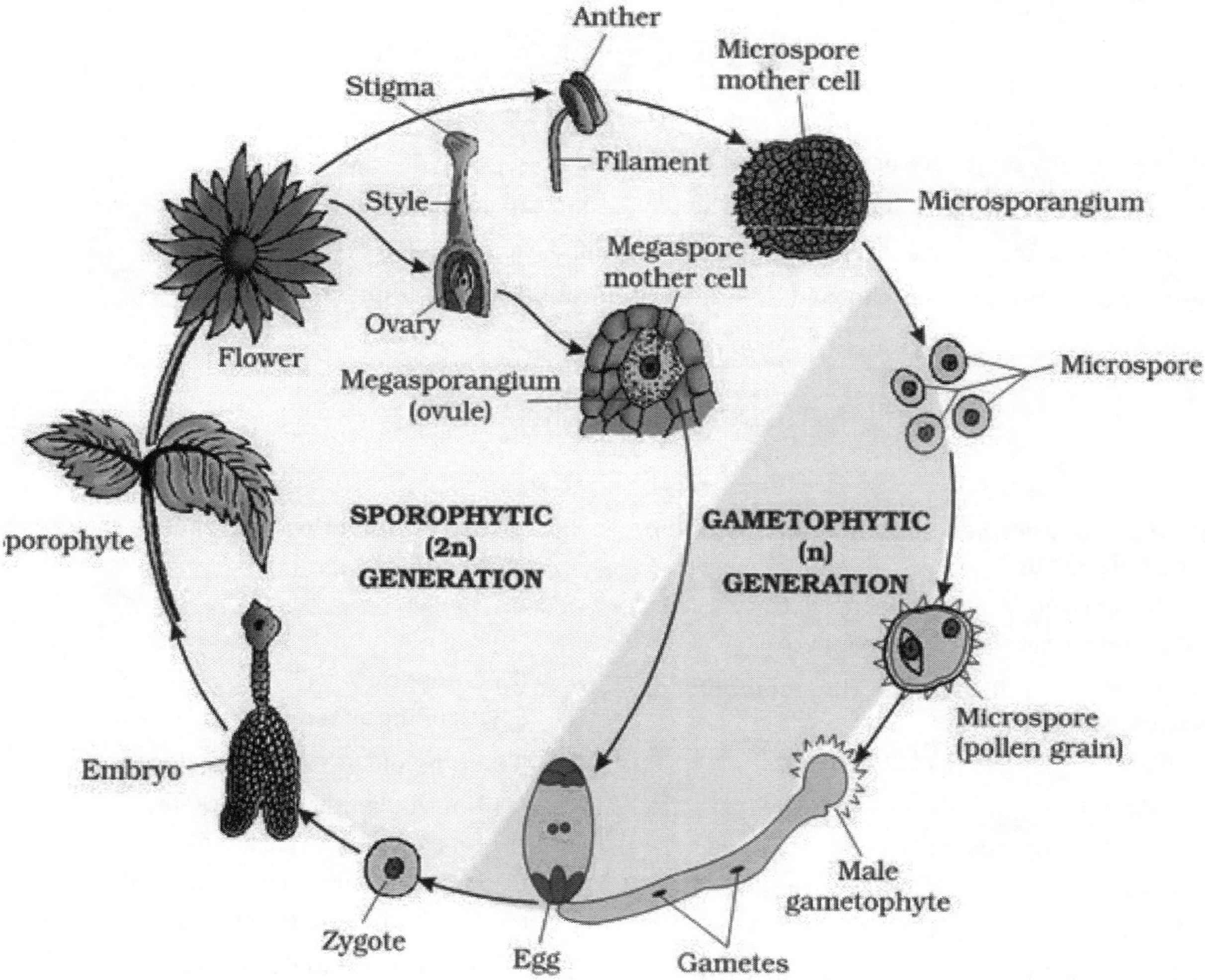

Figure Life cycle of an angiosperm

SHORT ANSWER TYPE QUESTIONS:

1. Discuss the different modes of sexual reproduction in algae.
2. What are the different criteria for classification in algae?
3. Discuss the economic importance of algae.
4. Discuss the range of thallus structure in green algae.
5. Distinguish between brown algae and red algae.
6. Explain the haplo-diplontic life cycle in bryophytes.
7. Draw a well-labelled diagram of the dioecious condition of *Marchantia*.
8. What are the prerequisites for seed habits in plants?
9. Distinguish between bryophytes and pteridophytes.
10. How does the life cycle of pteridophytes differ from bryophytes?
11. Distinguish between leptosporangiate and eusporangiate development.
12. What are the adaptations to land plants?

13. Write the characteristic features of angiosperms.
14. Distinguish between *Cycas* and *Pinus.*
15. Distinguish between gymnosperms and angiosperms.

LONG ANSWER TYPE QUESTIONS

1. Describe the characteristic features of green algae, brown algae and red alage.
2. Distinguish between liverworts and mosses.
3. Describe the characteristic features of pteridophytes and explain their life cycle.
4. Describe the characteristic features of gymnosperms and explain its life cycle.
5. Describe the outline of Bentham and Hooker's classification and mention their demerits.

MULTIPLE CHOICE QUESTIONS

1. Agar-agar is obtained mostly from *Gelidium* which belongs to

(a) Chlorophyceae (b) Phaeophyceae
(c) Rhodophyceae (d) Cyanophyceae

2. The position of flagella in the members of brown algae is

(a) Apical (b) Lateral
(c) Subapical (d) Basal

3. Motile stages are absent in

(a) Green algae (b) Brown algae
(c) Red algae (d) Mosses

4. The haplontic life cycle is common in algae but the diplontic life cycle is seen in

(a) *Chlamydomonas* (b) *Spirogyra*
(c) *Fucus* (d) *Ulothrix*

5. Heterotrichous habit is seen in

(a) *Fucus* (b) *Ectocarpus*
(c) *Chara* (d) *Chlamydomonas*

6. Which of the statements is true for red algae?

(a) Pyrenoid is present.
(b) Chlorophyll a and -c are present
(c) Reserve food material is Floridian starch
(d) The dominant pigment is fucoxanthin.

7. Which of the statements is true for brown algae?

(a) Flagella are not seen
(b) Algin is not found in the cell wall
(c) Frond is seen
(d) Pyrenoid is absent

8. Which of the following types of meiosis occurs in algae?

(a) Sporic
(b) Zygotic
(c) Gametic
(d) All of the above.

9. The type of sex organs in algae is

(a) Unicellular and non-jacketed
(b) Unicellular and jacketed
(c) Multicellular and non-jacketed
(d) Multicellular and jacketed.

10. Embryophyta does not include

(a) Algae, Bryophytes
(b) Brypohytes, pteridophytes
(c) Gymnosperms and angiosperms
(d) Algae only

11. Archegoniatae does not include

(a) Algae and Bryophytes
(b) Bryophytes and Pteridophytes
(c) Gymnosperms and angiosperms
(d) Algae and angiosperms

12. The capsule of sporophyte is photosynthetic in

(a) *Riccia* (b) *Marchantia*
(c) *Anthoceros* (d) *Sphagnum*

13. In which plant of bryophytes does archesporium arise from endothecium?

(a) *Marchantia* (b) *Sphagnum*
(c) *Funaria* (d) *Anthoceros*

14. Heterospory and ligulate leaves are the characteristic features of

(a) *Lycopodium* (b) *Selaginella*
(c) *Equisetum* (d) *Adiantum*

15. Prothallus in pteridophytes is

(a) Monoecious and protogynous
(b) Monoecious and protandrous
(c) Dioecious and protogynous
(d) Dioecious and protandrous

16. Spores of *Funaria* on germination give rise to

(a) Primary protonema
(b) Secondary protonema
(c) Gametophores
(d) All of the above

17. Sporophytes are partially dependent on gametophytes in

(a) Liverworts (b) Hornworts
(c) Mosses (d) All of the above

18. Sporophylls are compact and cone-like in

(a) Ferns (b) *Selaginella*
(c) *Azolla* (d) *Marsilea*

19. Which of the following is wrong with respect to pteridophytes?

(a) Young leaves show circinate vernation
(b) Phloem lacks companion cells
(c) Sex organs are multicellular and jacketed
(d) Life cycle is diplontic

20. Circinate vernation in ferns refers to

(a) Uncoiling of new leaves from base to apex
(b) Coiling of new leaves from base to apex
(c) Shedding of leaves from base to apex
(d) Arrangement of leaves

21. *Cycas* is vegetatively propagated by

(a) Corms (b) Bulbs
(c) Bulbils (d) Suckers

22. Which of the following is wrong with respect to gymnosperms?

(a) Phloem contains sieve cells
(b) Leaf has sunken stomata
(c) Polyembryony is seen
(d) Ovule is anatropous

23. Which of the following is wrong with respect to *Cycas*?

(a) The root of *Cycas* shows nitrogen-fixing ability
(b) Stem is unbranched
(c) Leaf is pinnately compound
(d) It is monoecious

24. The sperm of *Cycas* is

(a) Top shaped and multiciliate
(b) Pear-shaped and biflagellate
(c) Pear-shaped and unflagellate
(d) Top shaped and biflagellate

25. In Gymnosperms, the endosperm is formed by the

(a) Fusion of egg with one of the polar nuclei
(b) Fusion of egg with two polar nuclei
(c) Fusion of egg with synergid cells
(d) Germination of megaspore

26. The wing in the seed of *Pinus* develops from

(a) Nucellus (b) Ovuliferous scale
(c) Endosperm (d) Integument

27. The winged pollen grains of *Pinus* are produced in

(a) Pollen chamber (b) Anther
(c) Tapetum (d) Microsporangium

28. The edible part of the *Pinus* seed is

(a) Pericarp (b) Female gametophyte
(c) Perisperm (d) Endosperm

29. The seed of *Pinus* is

(a) Adaxial and endospermic
(b) Adaxial and non-endospermic
(c) Abaxial and endospermic
(d) Abaxial and non-endospermic

30. If the number of chromosomes in the pollen grain of *Pinus* is 12, then the number of chromosomes in endosperm cell will be

(a) 6 (b) 12
(c) 18 (d) 36

31. Which of the following sets of gymnosperms is considered as living fossils?

(a) *Cycas* and *Gnetum*
(b) *Pinus* and *Ginkgo*
(c) *Pinus* and *Ephedra*
(d) *Cycas* and *Ginkgo*

32. Which group of plants showed dominant sporophytes for the first time?

(a) Bryophyta
(b) Pteridophyta
(c) Gymnosperms
(d) Angiosperms

33. Two distinct generations present in a single life history are found in

(a) Bryophytes only
(b) Pteridophytes only
(c) Bryophytes and Angiosperms
(d) Thallophytes and Bryophytes

34. The largest archegonium is seen in

(a) *Selaginella,* (b) *Pinus,*
(c) *Funaria* (d) *Cycas*

35. Which of the following statements is wrong with respect to angiosperms?

(a) Bark is formed
(b) Archegonia are absent
(c) Double fertilisation occurs
(d) Life cycle is haplo-diplontic

36. Isomorphic alternation of generations is found in

(a) *Chara* (b) *Vaucheria*
(c) *Oedogonium* (d) *Ectocarpus*

37. Sexual reproduction in algae results in the formation of

(a) Zygospore (b) Zoospore
(c) Oospore (d) Zygote

38. Agranal chloroplast is seen in

(a) Algae (b) Bryophytes
(c) Pteridophytes (d) Angiosperms

39. The largest gametophyte is found in

(a) *Cycas* (b) *Nephrolepis*
(c) *Polytrichum* (d) *Cycas*

40. Gymnosperms are called soft wood spermatophytes because they lack

(a) Cambium (b) Tracheids
(c) Phloem fibres (d) Xylem fibres

41. Megasporophyll of *Cycas* is homologous to

(a) Sepal (b) Petal
(c) Stamen (d) Carpel

42. An ovule of *Cycas*

(a) One archegonium (b) Two archegonia
(c) 2-4 archegonia (d) 2-8 archegonia

43. The gametophyte of *Funaria* is

(a) Monoecious and autoecious
(b) Dioecious and heteroecious
(c) Monoecious and heteroecious
(d) Dioecious and autoecious

44. Meiosis does not take place in the formation of gamete from

(a) Prothallus (b) Protonema
(c) Promycelium (d) Sporangium

45. Primitive type of stomata is found in

(a) Axis of moss
(b) Leaf of moss
(c) Apophysis of moss capsule
(d) All of the above

46. Which of the following features is incorrect regarding bryophytes?

(a) Dominant stage is gametophyte
(b) Cone is not formed
(c) Antherozoids are biflagellate
(d) Prothallus is seen

47. Which of the following features is incorrect regarding pteridophytes?

(a) They are mostly heterosporous
(b) Seed habit is shown by *Selaginella*
(c) Circinnate vernation is seen
(d) Antherozoids are mostly multiflagellate

48. Elater mechanism for spore dispersal is shown by

(a) *Riccia* (b) *Marchantia*
(c) *Funaria* (d) *Sphagnum*

49. Transfusion tissue found in the leaves of gymnosperms helps in

(a) Lateral conduction of water
(b) Downward conduction of water
(c) Upward conduction of water
(d) All of the above

50. Life cycle in angiosperms is

(a) Haplontic (b) Diplontic
(c) Haplo-diplontic (d) All of the above

51. Sporophyte is very reduced in

(a) Algae (b) Bryophytes
(c) Pteridophytes (d) Angiosperms

52. Multicellular and oblique septate rhizoids are found in

(a) Liverworts (b) Hornworts
(c) Mosses (d) All of the above

53. Exoscopic embryo is found in

(a) Bryophytes only
(b) Pteridophytes only
(c) Both of the above
(d) Angiosperms only

54. Zooiodogamous fertilisation occurs in
(a) Bryophytes and pteridophytes
(b) Pteridophytes and gymnosperms
(c) Gymnosperms and angiosperms
(d) All of the above

55. *Riccia* is a bryophyte due to
(a) Thalloid habit
(b) Presence of rhizoids
(c) Jacketed sex organs
(d) Dependent sporophyte

56. The ovules in gymnosperms are
(a) Orthotropous and bitegmic
(b) Orthotropous and unitegmic
(c) Anatropous and unitegmic
(d) Anatropous and bitegmic

57. Sex organs are projected in
(a) *Riccia* (b) *Marchantia*
(c) *Funaria* (d) *Sphagnum*

58. Sperms are biflagellate in
(a) *Selaginella* (b) *Cycas*
(c) *Pinus* (d) All of the above

59. Pyrenoid in algae synthesises
(a) Cellulose (b) Protein
(c) Glycogen (d) Starch

60. Secondary protonema in mosses is not developed from
(a) Spore (b) Rhizoid
(c) Stem (d) Leaf

61. In which periods, the land plants arose?
(a) Jurassic (b) Cambrian
(c) Carboniferous (d) Ordovician

62. Cones of gymnosperms differ from flowers of angiosperms in
(a) Not having sporophyll
(b) Not having accessory reproductive structures
(c) Not having reproductive structures
(d) Not having shoot

63. Multicellular and oblique septate rhizoids are seen in
(a) *Riccia* (b) *Marchantia*
(c) Both a and b (d) *Funaria*

64. Simple sporophyte is found in
(a) *Riccia* (b) *Marchantia*
(c) *Funaria* (d) *Sphagnum*

65. Unicellular and non-jacketed sex organs are seen in
(a) Algae (b) Bryophytes
(c) Pteridophytes (d) Gymnosperms

66. Iodine and soda are produced from
(a) Green algae (b) Brown algae
(c) Red algae (d) All of the above

67. Cup-shaped chloroplast is seen in
(a) *Spirpgyra* (b) *Ulothrix*
(c) *Chlorella* (d) *Chlamydomonas*

68. The roots of *Pinus* show
(a) Symbiotic nitrogen fixation
(b) Formation of mycorrhiza
(c) Presence of xylem vessel
(d) All of the above

69. Pollination in gymnosperms takes place through
(a) Water (b) Wind
(c) Insects (d) Birds

70. Siphonogamous fertilisation occurs in
(a) Algae (b) Bryophytes
(c) Pteridophytes (d) Angiosperms

71. Aplanospores in algae are
(a) Thin walled and motile
(b) Thin walled and non-motile
(c) Thick walled and motile
(d) Thick walled and non-motile

72. Which of the following algae is filamentous, unbranched and attached?
(a) *Ulothrix* (b) *Spirogyra*
(c) *Chlorella* (d) *Volvox*

73. Globule and nocule are male sex organ and female sex organs, respectively are found in
(a) *Chara* (b) *Volvox*
(c) *Chlamydomonas* (d) *Fucus*

74. Conducting or trumpet hyphae are seen in
(a) Green algae (b) Brown algae
(c) Red algae (d) Dark yellow algae

75. Floridean starch is reserve food material in
(a) Green algae (b) Brown algae
(c) Red algae (d) Dark yellow algae

76. Which of the following bryophytes shows projected sex organs
(a) *Riccia* (b) *Marchantia*
(c) *Sphagnum* (d) *Funaria*

77. Fucoxanthin is the dominant pigment in

(a) Green algae (b) Brown algae
(c) Red algae (d) Dark yellow algae

78. Sporangia in some pteridophytes are in groups called sori which are protected by

(a) Ramenta (b) Sporophyll
(c) Rhizophore (d) Indusium

79. Ferns are the plants to appear first after a forest fire and they survive due to

(a) Spores (b) Leaves
(c) Rhizomes (d) Sporophylls

80. Fruits are not formed in gymnosperms due to absence of

(a) Ovary (b) Pollination
(c) Fertilisation (d) Seed

ANSWERS

1.c	11.d	21.c	31.d	41.d	51.a	61.d	71.b
2.b	12.d	22.d	32.b	42.d	52.c	62.b	72.a
3.c	13.c	23.d	33.c	43.a	53.a	63.d	73.a
4.c	14.b	24.a	34.c	44.a	54.b	64.a	74.b
5.b	15.b	25.d	35.d	45.c	55.d	65.a	75.c
6.c	16.a	26.b	36.d	46.d	56.b	66.b	76.b
7.c	17.c	27.d	37.a	47.a	57.b	67.d	77.b
8.b	18.b	28.d	38.a	48.b	58.a	68.b	78.d
9.a	19.d	29.a	39.c	49.a	59.d	69.b	79.c
10.d	20.b	30.b	40.d	50.b	60.a	70.d	80.a

CHAPTER-4

MORPHOLOGY OF FLOWERING PLANTS

Morphology is the study of form, size, colour, structure and relative position of various parts of plants.

Forms of Plants: Theophrastus (Father of Botany) classified plants into three categories based on their shape and size.

1. **Herbs**: These are small-sized plants. There is no single main stem and all branches emerging out of the soil are not woody in nature. The herbs may be
   (a) **Annuals**: These plants complete their life cycle (from seed germination to seed formation) in one growing season of a year lasting from a few weeks to a few months. Such plants are monocarpic (bearing flowers and fruits only once). e.g., Cereals, *Ranunculus*

   (b) **Biennials**: These plants complete their life cycle in two growing seasons in two years (the vegetative phase in one year and the reproductive phase in the next year). Such plants are monocarpic (producing flowers and fruits once). e.g., Radish, Turnip, Carrot, Henbane etc.

   (c) Perennials: These plants complete their life cycle in many years. Such plants are polycarpic (bearing flowers and fruits every year). e.g., Banana, Cynodon, Ginger, Turmeric etc.
2. **Shrubs**: These are medium-sized plants. There is no single main stem, but several woody branches are developed. These are Perennial and are known as bushes. e.g., Henna, Rose, China rose, *Capparis* etc.
3. **Trees**: These are large-sized plants. There is a single main stem, and all branches are woody arising from a certain height. These are perennial and polycarpic. The trees have the following variations:
   (a) **Caudex:** Unbranched stem or trunk having a crown of leaves at the top. e.g., Coconut palm, Date palm etc.

   (b) **Excurrent:** The main stem or trunk tapers from base to apex but lateral branches give a cone-like appearance. e.g., *Casuarina, Eucalyptus, Pinus, Polyalthia* etc.

   (c) **Deliquescent:** The main stem or trunk is distinguishable only for some distance after which it is replaced by several large branches which form a dome-shaped crown. e.g., *Dalbergia*, mango etc.

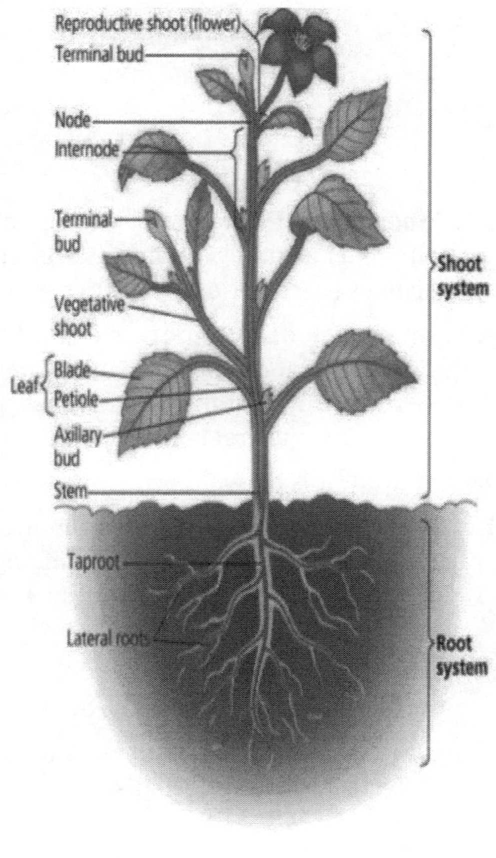

Parts of a flowering plant: A flowering plant consists of root, stem, leaf, flower, fruit and seed.

ROOT

It is an underground and cylindrical part of a plant. The longest root is found in *Adansonia digitata* but rootless angiosperms are *Aldrovanda, Wolffia, Balanophora, Utricularia, Ceratophyllum* etc.

Features:

(a) It is positively geotropic, negatively phototropic and positively hydrotropic

(b) It is usually a non-green part.

(c) It does not bear nodes and internodes.

(d) It usually does not form buds.

(e) It bears unicellular hairs.

(f) It develops endogenous branches.

(g) It is protected by a root cap at its tip.

Parts of a typical root:

1. **Zone of root cap**: It is conical and cap-like present at the apical region. Simple root cap is found in most of the plants, but multiple root cap is found in screw pine (*Pandanus*). The root cap is regenerated if damaged. It protects young apical meristematic cells against soil particles. It secretes mucilage which lubricates the passage of the root while penetrating the soil particles. Root pockets are found instead of root caps in *Eichhornia, Pistia,* and *Lemna* (hydrophytes) which function as balancers. These root pockets cannot be regenerated if damaged.
2. **Zone of Cell division**: It is found just behind the root cap hence it is subterminal or subapical. The cells are in an active state of division where cells are thin-walled having dense cytoplasm, large nucleus and reduced vacuoles or without vacuoles. This is a meristematic zone that produces root caps and other regions.

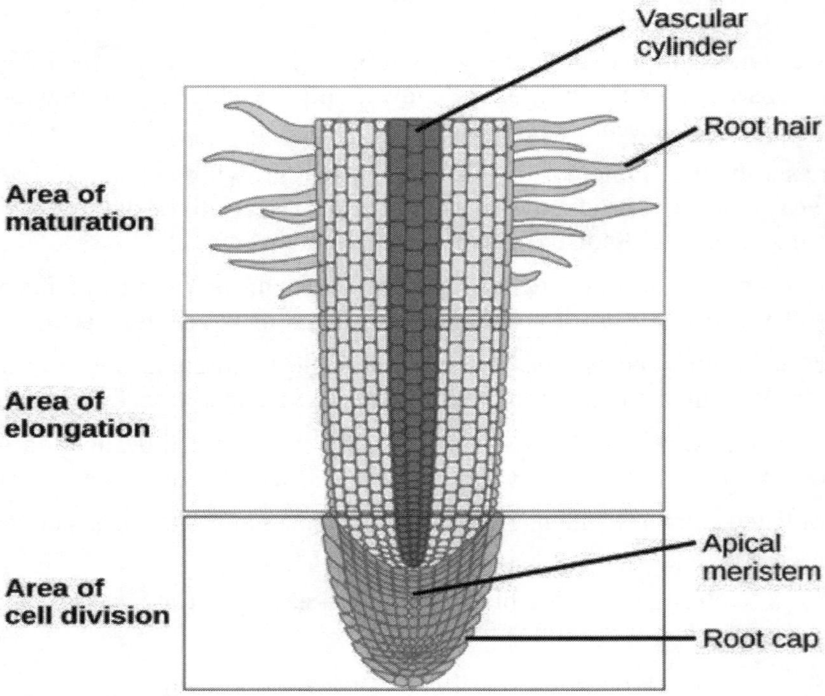

Fig: Different zones of root

3. **Zone of cell elongation**: It is a region below the meristematic zone. In this rapid growth in length occurs. The cells are elongated and show a large central vacuole filled with sap due to the fusion of a large number of small vacuoles (vacuolation).
4. **Zone of cell differentiation or maturation**: The cells are differentiated into permanent tissues of root such as cortex, endodermis and stele depending upon the functions. It bears unicellular root hairs which increase the surface area of the root for absorption. In hydrophytes, root hairs are absent because they absorb water through the general body surface.

Types of Root: It is of two types.

1. **Tap root:** It develops from continuous growth of the radicle of the embryo. Basically, it is primary root. It gives rise to secondary roots at certain angles and later forms tertiary roots. Tap roots together with secondary roots and tertiary roots constitute a tap root system. If tap root grows indefinitely it is called a racemose type of root system and if it stops growing after some time then it is known as a cymose type of root system.It is usually found in dicots.

   Modifications of Tap root: For storage of food

   (a) **Fusiform root:** It is thick in the middle and tapers towards the top and bottom. It bears secondary and tertiary roots. e.g., Radish

(b) **Napiform root**: It is broad at the top and tapers abruptly downwards. e.g., Beet, Turnip etc.

(c) **Conical root:** It is wide at the top and tapers gradually downwards. e.g., Carrot

(d) **Tuberous root:** It stores food in irregular way, and it does not have any definite form. e.g., Four o' clock plant (*Mirabilis*), *Cassava, Trichosanthes* etc.

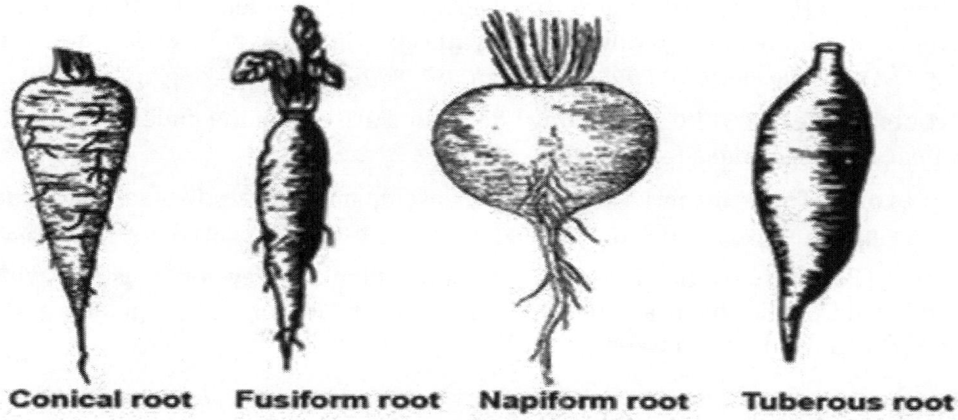

Special type:

(e) **Respiratory roots (Pneumatophores):** These come out of the soil to obtain oxygen with the help of several aerating pores called pneumathodes. These are negatively geotropic and grow vertically upwards in marshy or salty areas. e.g., *Rhizophora, Heritiera, Avicennia, Sonneratia, Penniseria, Ceriops* (Mangrove plants) etc.

2. **Adventitious root**: It grows from any part other than the radicle such as the hypocotyl, base or node of the stem because the radicle dies immediately after seed germination. It is usually found in monocots.

Modifications: For storage of food

(a) **Tuberous roots:** These store food in an irregular way and have no regular shape. e.g., Sweet potato (*Ipomoea batatas*)

(b) **Fasciculated roots:** These are fleshy and grow in clusters. e.g., *Asparagus,* Dahlia

(c) **Palmate roots:** These are fleshy and fuse to thickened structures like the palm of a human hand. e.g., *Orchis, Aconitum*

(d) **Moniliform or beaded roots:** These store food at frequent intervals looking like beads. e.g., Bitter gourd (*Momordica charantia), Portulaca, Vitis.*

(e) **Nodulose roots**: These store food at or near the apex. e.g., Arrow root (*Maranta*), Mango ginger (*Curcuma amada*).

(f) **Annulated root:** These store food in a ring-like manner. e.g., Ipecac (*Cephaelis*), *Psychotria* etc.

Modifications: For mechanical support

(a) **Prop roots:** These arise from aerial branches of stem and hang downwards in air. These become large, and woody and penetrate the soil to provide additional support. e.g., Banyan tree (*Ficus benghalensis*)

(b) **Stilt roots (Brace roots):** These arise obliquely from the lower node of stem. These move downwards and grow in clusters. e.g., Maize, Sugarcane, *Sorghum*. In Screw pine (*Pandanus*), stilt roots arise from the upper node.

(c) **Climbing roots**: These arise from the node of the stem and help in climbing by fixing to support. e.g., Money plant (*Pothos*), Betel (*Piper betle*).

(d) **Clinging roots**: These arise from the node of the stem and axil of leaves and pierce the substratum for fixation. e.g., Orchid, Ivy

(e) **Buttress root:** These arise from the base of the stem, become irregular and spread in different directions. e.g., *Ficus religiosa*, Red silk cotton (*Bombax*), *Salmalia*, *Terminalia.*

(f) **Contractile roots**: These are fleshy and thick and their apical parts contract and fix the plant in the soil in a vertical direction. e.g., *Canna*, Onion, *Crocus, Lilium*.

Modifications: For vital functions

(a) **Assimilatory roots (Photosynthetic roots):** These are long, slender, hanging and arise from branches. These develop chlorophyll and perform photosynthesis. These have finger-like attaching organs called haptera. e.g., *Trapa, Tinospora, Taeniophyllum, Podostemon.*

(b) **Floating roots:** These are white, soft, spongy, aerial and arise from the node of the stem. These store air to help in floating. e.g., *Jussiaea (=Ludwigia*).

(c) **Hygroscopic roots:** These are aerial, hanging and develop multi-layered epidermal tissue called velamen which absorbs moisture from moist air (atmosphere), dew, and rain. e.g., A few orchids such as *Vanda*

(d) **Sucking roots (Parasitic roots)**: These arise from the stem and develop haustoria which obtain water, minerals and food from the xylem and phloem of the stem and leaf of the host plant. e.g., Dodder (*Cuscuta*), Broom rape (*Orobanche*), Watch weed (*Striga*).

Special types:

(a) **Fibrous roots**: These are thin, underground, thread-like and equal in length arising from the node of the horizontal stem in groups. e.g., Grass, wheat, barley (Monocot plants).

(b) **Foliar root**: These arise from the leaf margin. e.g., *Bryophyllum*

(c) **Seminal root**: These arise from coleorhiza or base of radicle. These behave like primary root.

(d) **Lateral roots**: These arise from the pericycle region of the root. It is initially horizontal and later downwards.

Main Functions:

(a) It absorbs water and minerals and transports the different parts of the plant.

(b) It fixes the plant body in the soil (Anchorage) and provides support to the plant.

Special functions:

(a) Storage of food (b) Assimilation (c) Vegetative reproduction

STEM

Stem is the ascending and aerial part of a plant that bears leaf, flower and fruit.

Features:

(a) It develops from the plumule of an embryo.

(b) It is negatively geotropic and positively phototropic.

(c) It bears nodes, internodes, and apical buds. If a stem has a solid node and hollow internode then it is called culm e.g., Bamboo, wheat.

(d) It bears multicellular hairs.

(e) It bears an apical or terminal bud for growth in length.

(f) It bears exogenous branches.

Types of Stems:

1. **Erect stems:** These are upright and ascending. e.g., Majority of Angiosperms.
2. **Weak stems:** These are thin and delicate and cannot be upright without any support. There are several types:

   (a) **Trailers:** The plants that creep on the ground without rooting at intervals. e.g., *Eichhornia, Euphorbia prostrate*

   (b) **Creepers:** The plants that creep on the ground rooting at intervals. e.g., Grass, Strawberry, Pumpkin.

(c) **Twiners:** The plants of a weak stem in which the stem coils or twines around upright support. e.g., Bean, *Clitoria*, Railway creeper (*Ipomoea carica*), Rangoon creeper (*Quisqualis*) etc.

(d) **Climbers:** The plants that climb on upright support by special clinging structures like tendrils, roots, and hooks. e.g., Grapevine (*Vitis*), Pea

(e) **Lianas:** These are woody twiners and climbers. e.g., *Bauhinia vahlii, Aristolachia, Hyptage* etc.

Branching of stem

(a) Unbranched: Maize, sugarcane, coconut

(b) Branched: Most of the plants

Modifications of stem

1. **Underground modifications of stem**: The modified part remains inside the soil. Its function is the storage of food and also helps in perennation.

(a) **Rhizome:** Stem is modified into thick, fleshy and branched rhizome. It grows horizontally below the soil. It bears distinct nodes and internodes. It bears adventitious roots on the lower side of the node. It bears scale leaves at a node.

e.g., Ginger, Turmeric, *Canna*, Ferns

(b) **Corm:** Stem is modified into short, thick, unbranched and massively laden with food and arises from its base. It grows vertically below the soil. It bears horizontal circular nodes (lines). It bears adventitious roots on the lower side of the node. It bears scale leaves bearing axillary bud on the node.

e.g., Zaminkand (*Amorphophallus), Colocasia, Crocus, Gladiolus, Colchicum*

(c) **Tuber**: Stem is modified into an oval or spherical and unbranched tuber arising from the axil of a lower leaf (branch). It has no fixed orientation. It bears node (depression or eye) and internode. The eye is a modified axillary bud (1-3) in the axil of a small-scale leaf. It does not bear adventitious roots.

e.g., Potato, Jerusalem artichoke (*Helianthus tuberosus*)

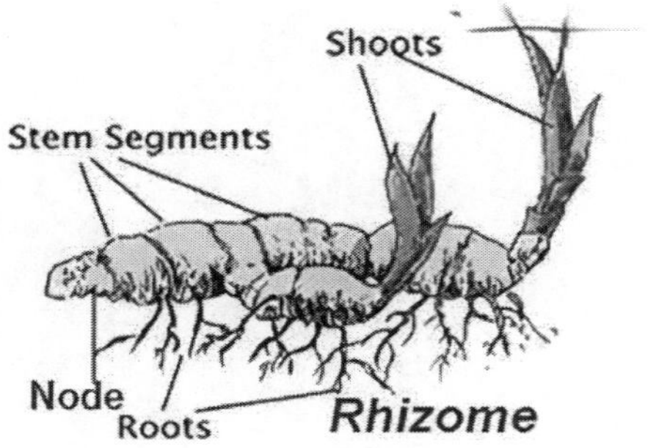

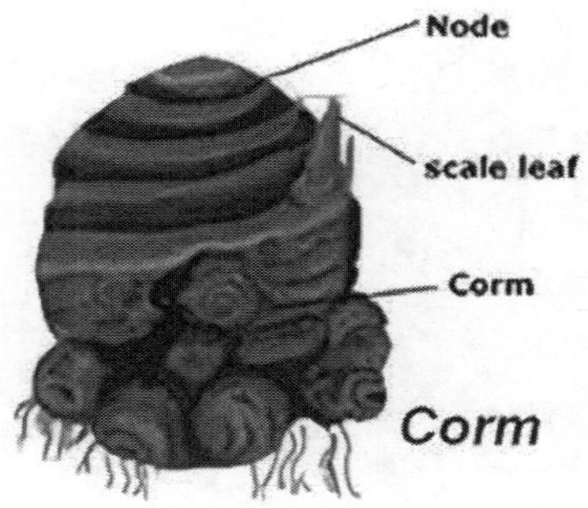

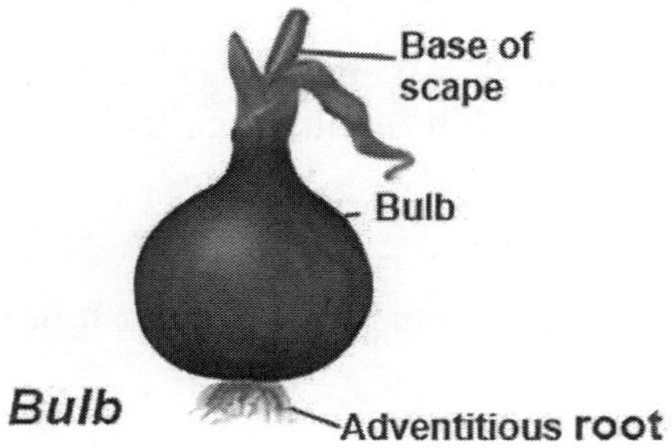

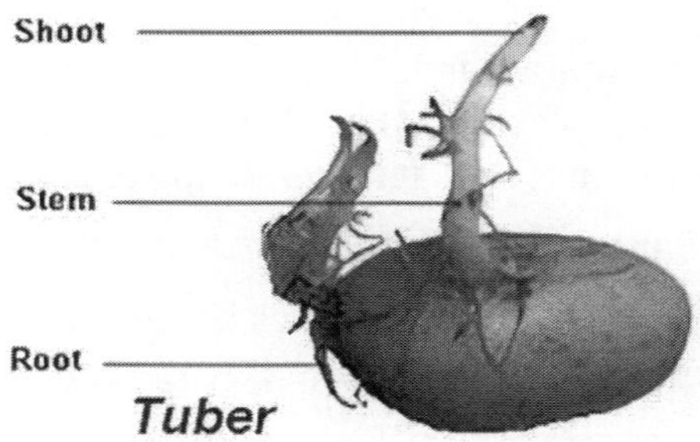

(d) **Bulb**: Stem is modified into short, condensed and discoid bulb. It bears short internodes and fleshy scale leaves in a concentric manner. It bears numerous adventitious roots from the lower side. It bears an apical bud which gives rise to aerial flowering shoots (scape). It is of three types:

(i) **Simple tunicated bulb:** It is covered with dry scaly leaves (tunic). e.g., Onion, Tulip, Daffodil etc. (The characteristic smell of Onion is due to the sulphurous compound (allyl sulphide) present in fleshy leaves).

(ii) **Compound tunicated bulb**: It contains several bulblets which are covered with a tunic (dry scaly leaves). e.g., Garlic (Garlic contains sulphur-containing amino acid, S-allyl-cystine sulphoxide).

(iii) **Scaly (Imbricate) bulb**: It is not a compact body and is not covered with a tunic. Fleshy scale leaves lie loose without forming concentric circles. e.g., Lily.

2. **Subaerial modifications**: Some part of the stem lives underground or on water and the remaining part of the stem is aerial. The branch breaks forming a new plant (vegetative reproduction). These are of the following types:

(a) **Runner**: It is a long branch (internode) running horizontally on the soil surface in different directions. It bears adventitious roots at nodes on the lower side. It bears an axillary bud. It is called sobole (underground runner/ creeping stem) if the internode is thick as in *Agropyron*.

e.g., Lawn grass (*Cynodon)*, Wood sorrel (*Oxalis), Centella,*

(b) **Stolon**: It is a slender lateral branch arising from the lower node of the stem. It grows aerially for some time and bends downwards to touch the ground. It bears adventitious roots at nodes on the lower side and new shoots from its terminal bud.

e.g., Jasmine, Strawberry, Mentha, *Vallisneria*

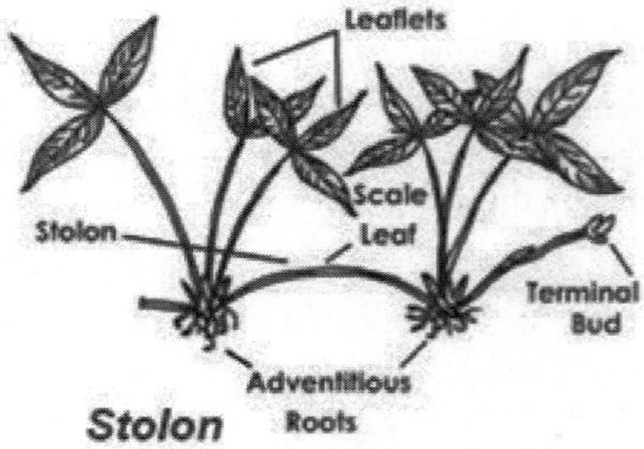

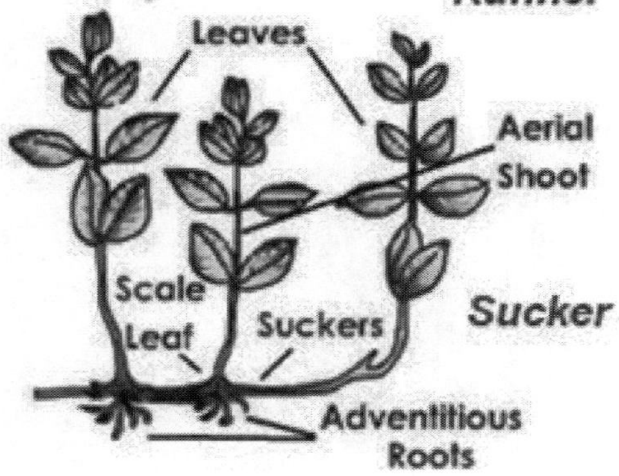

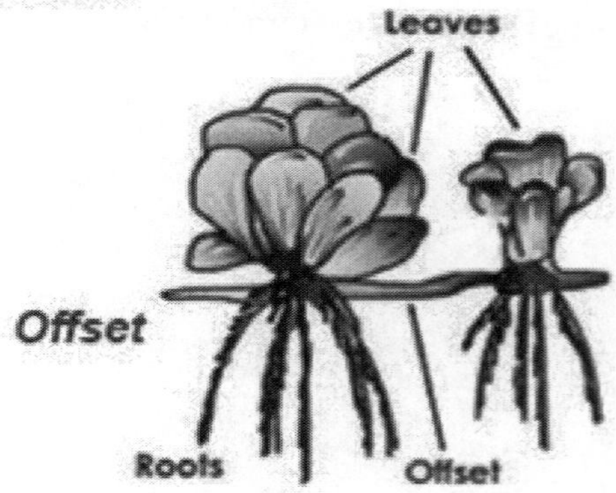

(c) **Sucker**: It is a slender lateral branch arising from the lower node of the stem. It grows horizontally below the soil surface to some distance and then comes out obliquely to form an aerial leafy shoot. It bears adventitious roots at nodes on lower side.

e.g., Banana, *Chrysanthemum*, Mint, Rose, Pineapple.

(e) **Offset**: It is a short and thick horizontal branch (internode) arising from the main stem. It bears a cluster of roots below the node and a rosette of leaves above.

e.g., Water hyacinth (*Eichhornia*), Water lettuce (*Pistia*).

3. **Aerial modification**: The entire part of the stem remains above ground. It performs specific and different functions.

(a) **Stem tendril**: It is thin and spirally coiled branch arising from node of the branch. It helps plants to climb.

Modified axillary bud: e.g., Cucumber, *Passiflora*

Modified apical bud: e.g., Grapevine (*Vitis vinifera*)

Modified floral bud or inflorescence: e.g., *Antigonon, Cardiospermum*

(b) **Stem thorn**: It is a straight, hard and pointed structure that protects from animals (defensive) and also checks transpiration. It has a vascular connection.

Modified axillary bud: e.g., Lemon, Orange, Duranta, *Bougainvillea*, Wood apple.

Modified apical (terminal) bud: e.g., *Carissa*

(c) **Phylloclade**: It is a flattened, fleshy and green structure of unlimited growth bearing several nodes and internodes. It performs photosynthesis and also stores water and food. It helps plants to grow in dry habitats (xeric or xerophytic).

Flattened: e.g., Cactus (*Opuntia),* Coccoloba (*Muhlebenckia*)

Cylindrical: e.g., *Casuarina, Euphorbia.*

(d) **Cladode**: It is a green and scaly leaf-like structure of limited growth bearing a few internodes. It performs photosynthesis.

One internode: e.g., *Asparagus*

Two internodes: e.g., *Ruscus*

Phylloclade	Cladode
It is a modified stem and branch (whole shoot).	It is a modified branch.
It has several internodes.	It has one or two internodes.
It has indefinite growth.	It has definite growth.

(e) **Bulbil**: It is a swollen and multicellular structure which is a modified fleshy bud. It helps in vegetative reproduction by being shed from the parent plant.

Modified vegetative bud (axillary bud): e.g., *Dioscorea, Lilium*

Modified floral bud (apical bud): e.g., *Agave*, Garlic, Onion

(f) **Pseudobulb**: The internode of the stem becomes fleshy and swollen for storing moisture. e.g., Orchid

(g) **Thalamus**: Stem is modified into floral axis bearing floral organs

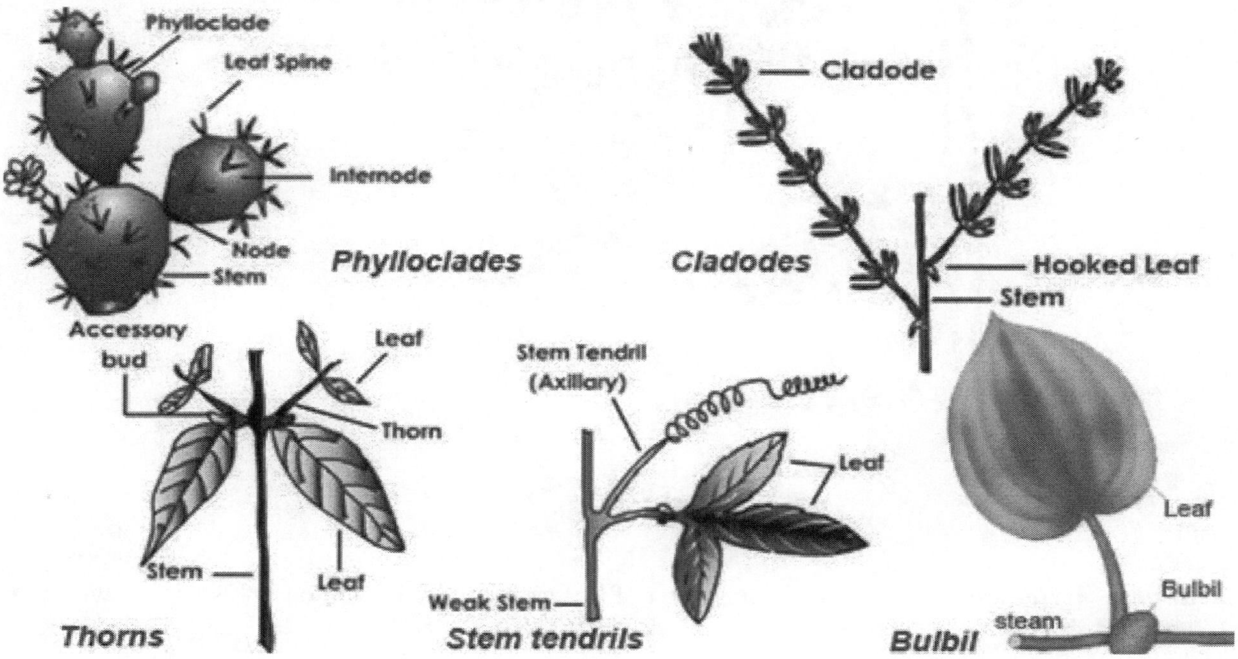

BUD

It is a young, undeveloped and condensed shoot having a growing point surrounded by closely packed immature leaves.

Types of bud: On the basis of position on the plant.

1. **Normal bud**: It is borne on the stem. It may be
   (a) Apical (Terminal) bud: It is present at the apex of the main stem and branches. It is responsible for the elongation of plants. e.g., Cabbage (the largest bud)
   (b) Lateral (Axillary) bud: It is present in the axil of leaf.
2. **Adventitious bud**: It arises from a part other than the node. It may be
   (a) Radical bud: It grows from the root. e.g., *Dalbergia, Ipomoea*
   (b) Cauline bud: It grows from a stem. e.g., *Artocarpus*, rose
   (c) Foliar bud: It grows from a leaf. e.g., *Begonia, Bryophyllum*

LEAF

It is a flat and lateral or flattened outgrowth of a stem or branch. It arises from a node and usually bears a bud in its axil. Its fresh weight is minimum at afternoon and maximum at morning. It grows to a specific pattern and shape and then stops. The green leaves are called foliage.

Smallest leaf	*Wolffia microscopica*
Largest leaf	*Victoria regia* (diameter=1.5m)
Longest leaf	*Raphia ruffia* (length=10m)

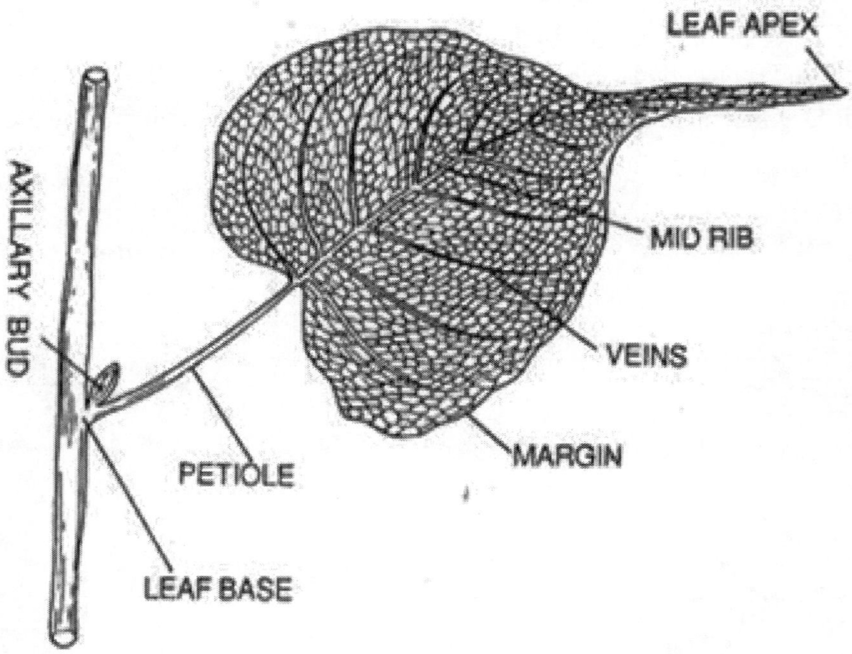

Parts of leaf

1. **Leaf base (Hypopodium):** It is lowermost part of leaf attached with stem on one side and with petiole on other side.

   It expands into sheath encircling stem in monocots. It bears two lateral appendages called stipules to protect young leaves in dicots except Solanaceae and Liliaceae.

Leaf base is swollen called pulvinus. It is responsible for shock movement (siesmonasty) or drooping of leaves. e.g., Touch-me-not (*Mimosa pudica*), mango, pea, gram, bean, *Ficus* etc.

It forms outgrowth called ligule between leaf base and leaf lamina. e.g., Grasses.

2. **Petiole (Mesopodium)**: It is long stalk which holds leaf at a certain height. It mechanically links the leaf to the stem for transport of materials.

   It is hollow (fistular) in Papaya.

   It is modified into wing in Citrus plants.

   Petiole is absent in leaf hence called sessile as in Rice, wheat, *Gloriosa, Calotropis.*

3. **Leaf lamina=Leaf blade (Epipodium)**: It is uppermost and expanded part of leaf. It produces prominent midrib at centre and a number of lateral veins. These veins form skeleton of leaf and function as channels for transport of water, minerals and food.

Surface of leaf: It may be glaucous (shinning and smooth) and glabrous (shinning and dull). There are two surfaces of leaf:

(a) Abaxial= lower or dorsal

(b) Adaxial= upper or ventral

Types of leaf: On the basis of origin and function

(a) Cotyledonary leaves: These are embryonic and nutritive leaves after seed germination.

(b) Bract leaves: These are protective leaves attached at the base of flower.

(c) Foliage leaves: These are green and photosynthetic leaves.

(d) Scale leaves: These are non-green (non-photosynthetic) and protective present on bud.

(e) Sporophylls: These are spore bearing leaves. These are reproductive in function and equivalent to stamen and carpel of angiosperms.

Types of leaf: On the basis of insertion

(a) Radical: Leaves are borne at root (ground level). e.g., Radish, Turnip.

(b) Cauline: Leaves are borne at stem. e. g., Maize

(c) Ramal: Leaves are borne at branch of stem. e.g., Plum, *Dalbergia, Ficus*

Types of leaf: On the basis of structure

1. **Simple leaf**: Leaf lamina is entire (there is no cut or incision) and if incised the incisions do not touch the midrib. It shows the presence of bud in its axil. e.g., Mango. Cucurbits, China rose, Guava.

2. **Compound leaf**: Leaf lamina shows incisions which touch the midrib or petiole. Hence leaf is broken up into several segments called leaflets which are free from one another. A leaflet appears as a leaf, but it does not bear bud in its axil hnce called stipule. The leaflet contains stipule called stipel as in bean. It is of two types:

   (a) **Pinnately compound leaf**: The leaflets are attached laterally on both sides at midrib known as rachis.

      (i) Unipinnate: The leaflets are attached directly at both sides of rachis (mid rib). It may be paripinnate (having even number of leaflets) in Cassia, Tamarind and imparipinnate (having odd number of leaflets) in Neem, Rose etc.

      (ii) Bipinnate: The leaflets are borne on secondary axis. e.g., *Acacia, Mimosa*

      (iii) Tripinnate: The leaflets borne on tertiary axis. e.g., *Ficus, Moringa*

      (iv) Decompound: Leaf lamina is divided into narrow segments because rachis divided repeatedly without any definite pattern. e.g., Carrot, Coriander, Fennel etc.

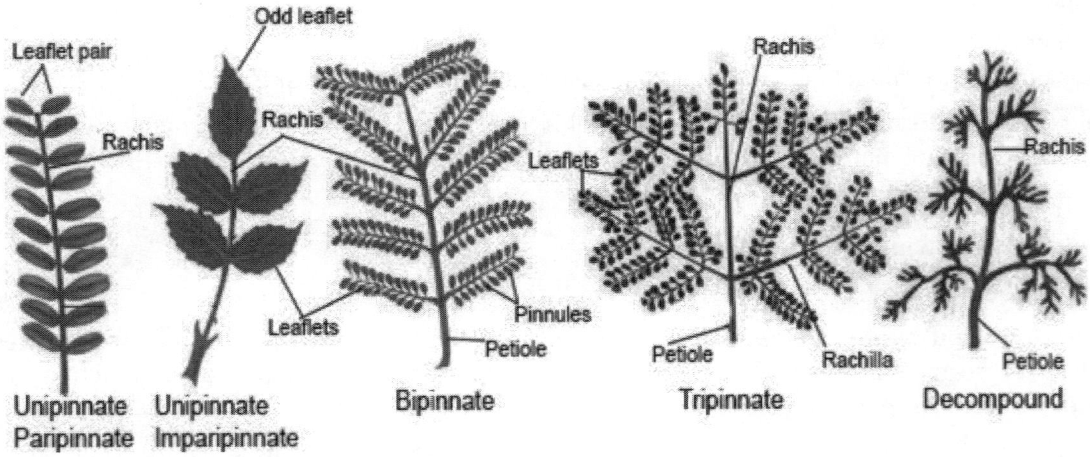

(b) **Palmately compound leaf**: The leaflets are joined at the tip of petiole. It may be

(i) Unifoliate: There is one leaflet. e.g., Citrus

(ii) Bifoliate: There are two leaflets which arise from petiole. e.g., Balanites, *Hardwickia.*

(iii) Trifoliate: There are three leaflets which arise from petiole. e.g., Oxalis, Wood apple., *Trifolium*

(iv) Quadrifoliate: There are four leaflets which arise from petiole. e.g., *Marsilea.*

(v) Multifoliate (digitate): There are five or more leaflets which arise from petiole. e.g., *Bombax*, Lupin.

Palmately compound leaf

Duration of leaf:

(a) Caducous: Leaves fall off just after their formation. e.g., *Opuntia*

(b) Deciduous: Leaves fall off at the end of the growing season. e.g., Mulberry, Poplar.

(c) Persistent (Evergreen): Leaves remain on the plant for more than one year or leaves fall off individually at different times so that plants remain evergreen. e.g., Mango

Phyllotaxy: It is the arrangement of leaves on the main stem or branches. It is three types:

1. **Alternate (Spiral):** It is most common. There is a single leaf at a node. Leaves arising at successive nodes alternate with each other at a certain angle of divergence (angle between leaves of successive nodes).

   (a) Distichous: There are two rows and the angle of divergence is 180°. e.g., *Annona*, Vanda

   (b) Tristichous: There are three rows and angle of divergence is 120°. e.g., *Cyperus*

(c) Penatstichous: There are five rows and angle of divergence is 144°. e.g., *Ficus*

(d) Octastichous: There are eight rows and angle of divergence is 135°. e.g., Papaya

2. **Opposite (Spiral)**: There is a pair of leaves at a node and all leaves are opposite to each other. It is of types:

   (a) Decussate: Leaves are present at right angle to next upper or lower pair. e.g., *Calotropis*, Guava (in horizontal branches), *Mussaenda,* Sacred basil (*Ocimum*)

   (b) **Superposed:** Leaves are present parallel to next upper or lower pair e.g., *Quisqualis, Syzygium* (=*Eugenia*), Guava (in vertical branches).

3. **Whorled/cyclic:** There are three or more leaves at a node. e.g., *Nerium,Hydrilla, Alstonia*

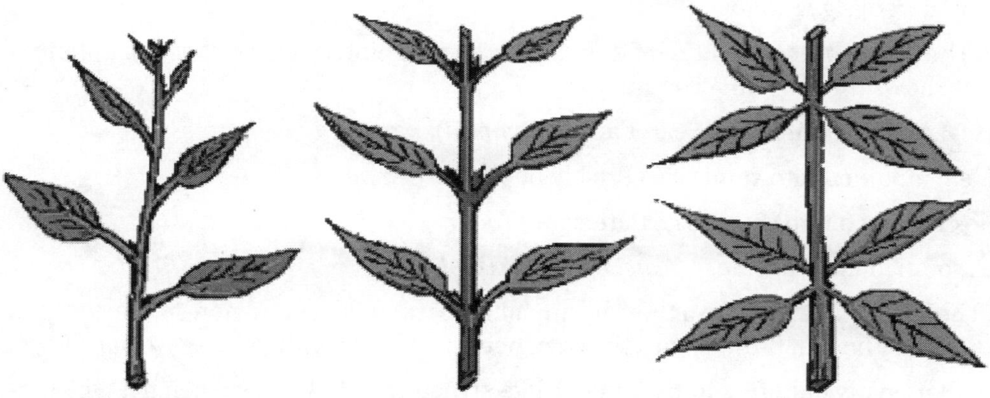

Heterophylly: It is the occurrence of different types of leaves on the same plant. It comes under plasticity of growth which is the ability of plants to follow different pathways in response to the environment. Plasticity is irreversible stretching. It is of several types.

1. Environmental heterophylly: The submerged leaves are much dissected, linear and narrow but aerial or floating leaves are expanded and broad due to changes in the environment including soil, temperature, humidity and air current. e.g., Buttercup (*Ranunculus*), *Limnophila, Sagittaria*
2. Developmental heterophylly: The young or juvenile leaves are different in shape than those in mature leaves. e.g., Cotton, Coriander, Larkspur (*Delphinium*), Mustard, Pea, Eucalyptus.
3. Habitual heterophylly. The leaves of different incisions and shapes occur at the same time. e.g., Jack tree (*Artocarpusheterophyllus*).

Venation: It is the arrangement of veins on the leaf lamina. Veins provide shape and mechanical strength to the lamina. Veins are hard structure that contains xylem and phloem. It is of two types:

1. **Reticulate**: Veins are irregularly distributed forming a network. It occurs in dicots except *Calophyllum, Corymbium, Eryngium.*

   (a) Pinnate: One single main vein or midrib gives rise to several veins on the lateral side. e.g., *Ficus*, Guava, Mango etc.

   (b) Palmate: Veins radiate outward from the base of the leaf.

   (i) Convergent: Several veins start from the base and merge at the tip. e.g., Bay leaf, plum

   (ii) Divergent: Several veins start from the base and go towards the margin. e.g., Castor, cucurbits, papaya.

2. **Parallel venation**: Veinlets are arranged parallel to each other. It occurs in Monocots except *Alocasia, Colocasia, Dioscorea, Smilax* etc.

   (a) Pinnate: One single main vein or midrib gives rise to lateral veins on both sides. e.g., Banana, *Canna*, ginger, turmeric etc.

   (b) Palmate: Veins radiate outward from the base of the leaf.

   (i) Convergent: Several main veins start from the base and go towards the tip. e.g., Bamboo, grass, rice, sugarcane, maize, wheat, *Eichhornia.*

   (ii) Divergent: Several main veins start from the base and go towards the margins. e.g., Fan palm.

Modifications of Leaf:

1. **Leaf tendril**: The leaf or part of a leaf is modified into a slender and coiled structure which helps the plant to climb. It shows the following conditions:
   (a) Whole leaf modified: e.g., Wild pea (*Lathyrus aphaca*). In this stipule, performs photosynthesis.
   (b) Terminal leaflet modified: e.g., Sweet Pea (*Lathyrus odoratus*), Garden Pea (*Pisum sativum)*
   (c) Leaf apex modified: e.g., Glory lily (*Gloriosa*)
   (d) Petiole modified: e.g., *Clematis, Nepenthes*, Garden nasturtium (*Tropaeolum)*
   (e) Stipule modified: e.g., *Smilax*
2. **Leaf spine**: The leaf or part of a leaf is modified into a sharp pointed structure called a spine which is defensive in function. It shows the following conditions:
   (a) Whole leaf modified into spine: e.g., Cactus (*Opuntia*), Barberry
   (b) Partial leaf modified into spine: e.g., Prickly poppy (*Argemone*)
   (c) Leaf apex modified into spine: e.g., Date palm, *Yucca*
   (d) Stipule modified into spine: e.g., *Acacia*, Plum (*Zizyphus*)
3. **Scale leaf**: The leaf is modified into a thin, dry, membranous, stalkless and non-photosynthetic scale leaf which protects the axillary bud. In these plants, the stem performs photosynthesis. e.g., *Asparagus, Casuarina, Ruscus.*
4. **Pitcher**: Leaf lamina is modified into a pitcher-like structure (20-30 cm in height) which captures, kills and digests insects to obtain nitrogenous compounds esp. amino acids. In this leaf apex is modified to a lid to attract insects, the rim contains nectariferous glands and the pitcher contains digestive fluid. e.g., Pitcher plant (*Nepenthes*), Venus fly trap (*Dionea), Dischidia, Darlingtonia, Sarracenia.*

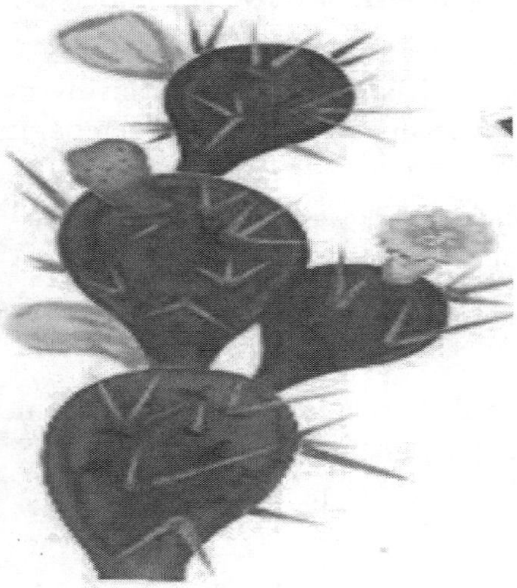

Opuntia

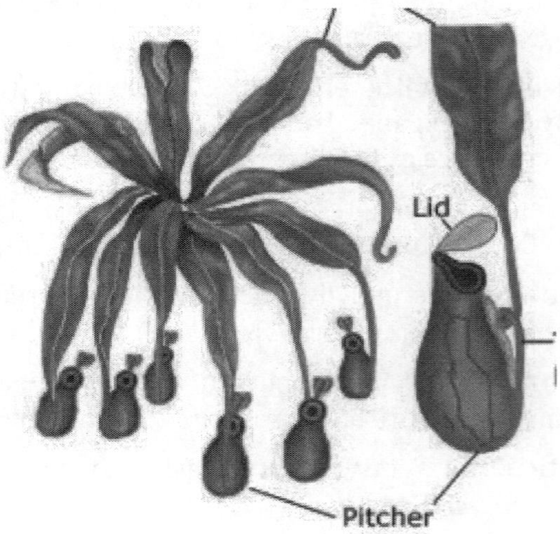

Nepenthes-pitcher shaped leaf

5. **Phyllode**: The petiole of the leaf is modified into a flattened, green and looks like a leaf to perform photosynthesis when the normal leaf of the plant falls off at the seedling stage. It is a mechanism to reduce transpiration. e.g., Australian acacia (*Acacia melanoxylon), Parkinsonia.*
6. **Bladder**: The leaf shows a segmented structure and some of the segments are modified to form small bladder showing sensitive hairs, branched bristle, trap valves and internal or external glands for trapping and killing aquatic animals such as water fleas, Cyclops etc. e.g., Bladderwort (*Utricularia*).
7. **Leaf root**: One leaf of each node is modified into a root for maintaining balance on water. e.g., *Salvinia*
8. **Leaf hook**: Three terminal leaflets of the compound leaf are modified into hooks which help in climbing. e.g., *Bignonia.*
9. **Succulent leaf:** The leaf is modified into a fleshy structure that stores water or mucilage. e.g., *Agave, Aloe vera, Bryophyllum.*

Functions of leaf:

1. Photosynthesis
2. Transpiration and
3. Exchange of gases

Terms:

1. Phyllome: Presence of all the leaves on a plant.
2. Convulate: Folding of leaves lengthwise along the midrib
3. Conduplicate: Bending of the upper half of the leaf upon the lower half
4. Ptyxis: Folding of individual leaves in bud condition.
5. Vernation: Arrangement of leaves in bud condition. It is of three types:
   (a) Circinnate vernation: Rolling (coiling) of leaves from apex to base. e.g., Fern, *Cycas*
   (b) Convolute vernation: Rolling of leaves from one margin to the other or with the lower surface outermost.
   (c) Involute vernation: Rolling of leaves from both margins on opposing sides of the leaf to form two tubes meeting at midrib.

INFLORESCENCE

It is the arrangement and distribution of a group of flowers on the mother axis called the peduncle. Morphologically it is the modified part of the shoot. The peduncle is the axis of inflorescence. The enlarged peduncle is called a cob found in maize. The flattened peduncle is called a receptacle (torus). The largest inflorescence is seen in *Agave* (12m).

Cauliflory: It is the development of flowers on an old stem as in *Artocarpus.*

Tassel: It is the staminate inflorescence of corn.

Solitary flower: Flower is single and does not form inflorescence. It may be

(a) Solitary axillary: A single flower is present in the axil of a leaf. e.g., Bean, *Petunia,* Lily

(b) Solitary terminal: A single flower is present at the tip of the main stem. e.g., Cucurbits, *Crotalaria, Magnolia,* Poppy.

(c) Solitary intercalary: A single flower is present at the intercalary position. e.g., Bottlebrush (*Callistemon*)

Types of Inflorescences:

1. **Racemose inflorescence:** Growth of the axis is continuous (indefinite/indeterminate or monopodial) as it does not end in a terminal flower. It keeps forming lateral flowers. Flowers are laterally and acropetally arranged (younger and inner flowers are at the top and older and outer flowers are at the base). Flowers open centripetally (peripheral flowers open first followed by central flowers).

The main axis is elongated:

(a) **Raceme**: Flowers are stalked (pedicellate). e.g., Radish, mustard, Tamarind, *Delphinium,* Snapdragon *(Antirrhinum).*

(b) **Spike**: Flowers are stalkless (sessile) and arise in axils of bracts. e.g., *Achyranthes, Adhatoda, Amaranthus,* barley etc.

(c) **Spikelet**: It is a small spike and arises in the axil of two fertile glumes or lemma (modified bract). At the base of the flower, there is a membranous structure called a palea. e.g., Members of Poaceae (=Graminae) such as wheat, rice etc.

(d) **Catkin (Amentum):** It is pendulous and bears only unisexual and sessile flowers. e.g., *Casuarina,* Mulberry (*Morus*), *Betula, Quercus*, Poplar, Willow.

(e) **Spadix**: The axis is fleshy, and flowers are covered with large and coloured bracts called spathe. e.g., Banana, Maize, *Colocasia,* Palm (Monocots).

The main axis is flattened or suppressed.

(f) **Head (Capitulum or Anthodium):** Sessile flowers are directly produced on the thalamus called florets which are covered by involucres of bracts. There are two types of florets on the receptacle:

- Disc florets: These have central, actinomorphic, bisexual and tubular flowers.
- Ray florets: These have peripheral, zygomorphic, unisexual, and strap-shaped (ligulate) flowers.

Head is of two types:

- Homogamous head: It does not have both ray florets and disc florets. Either ray or disc florets are found. e.g., *Sonchus, Zinnia*
- Heterogamous head: It has both ray florets and disc florets. e.g., Sunflower.

It is an advanced or perfect type of inflorescence due to aggregation of florets which ensure maximum protection to flowers. e.g., Asteraceae (=Compositae) family

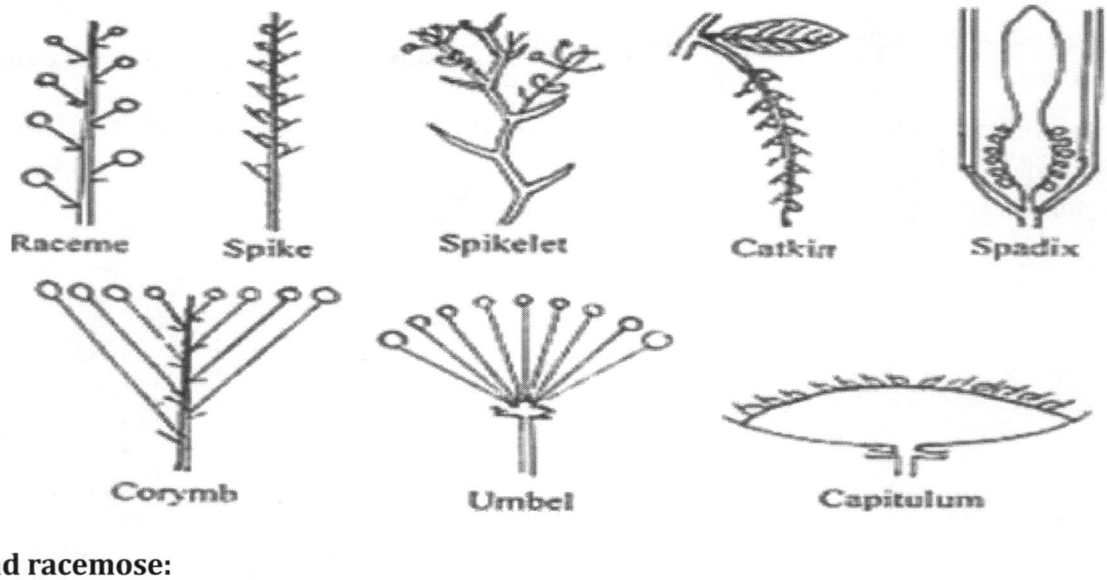

Compound racemose:

(a) Compound or branched raceme (Panicle): e.g., *Azadirachta, Cassia, Delonix*, Mango, *Yucca*

(b) Compound or branched spike or spike of spikelet: e.g., Members of Poaceae family

(c) Compound or branched spadix: e.g., Coconut palm, Date palm

(d) Compound or branched corymb: e.g., Cauliflower, Apple

(e) Compound or branched Umbel: e.g., Carrot, Coriander, Fennel

(f) Compound or branched Head: e.g., *Echinops, Sphaeranthes*

The main axis is shortened:

(g) Corymb: Flowers are produced from several points, but they reach the same levels. e.g., *Cassia*, Candytuft (*Iberis*), *Delonix, Iberis*

(h) Umbel: Flowers are produced from one point, and they reach the same levels because flowers have the same length of pedicels. e.g., Members of Apiaceae (Umbelliferae) family such as Carrot, Coriander, Cumin, Fennel etc.

2. **Cymose inflorescence:**

Growth of the main axis is definite/determinate or sympodial and ends in a terminal flower. Flowers are basipetally arranged (smaller flowers are at the base and larger flowers are at top. Thus, maturity starts from top. Flowers open centrifugally.

(a) **Uniparous (Monochasial) cyme**: Flowers are produced only on one lateral side. It shows the following two conditions:

Helicoid cyme: The successive flowers are found in a helical manner. e.g., *Begonia, Juncus, Saraca, Myostosis* etc.

Scorpoid cyme: The successive flowers are found in a zig-zag manner. e.g., Cotton, *Drosera*, Onion, *Ranunculus.*

(b) **Biparous (Dichasial cyme)**: Flowers are produced on both sides from two lateral branches which in turn end in a flower. e.g., *Ixora*, Jasmine, *Nyctanthes,* Bougainvillea, Pink, teak etc.

(c) **Multiparous (Multichasial) cyme**: Numerous flowers are produced on the lateral side. e.g., *Calotropis, Hamelia.*

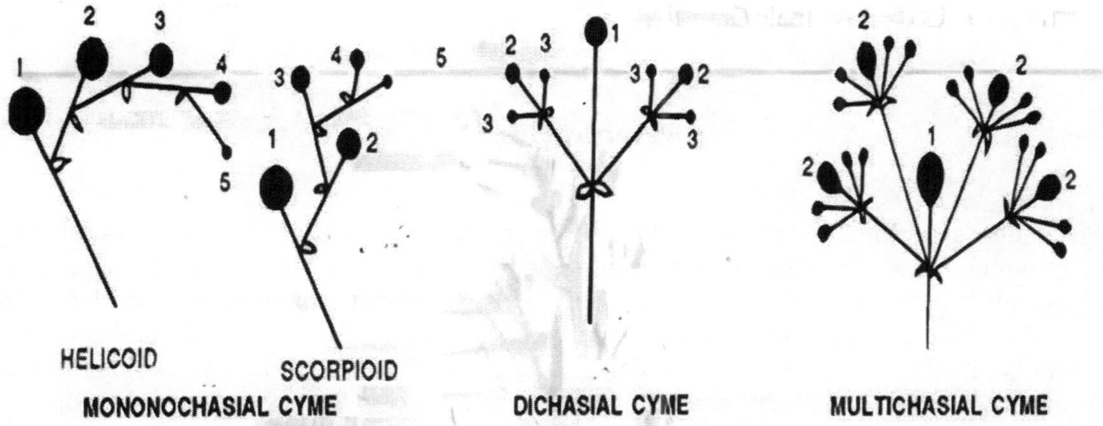

3. **Special type of Inflorescence:**

(a) **Cyathium**: It is cup-shaped and looks like a single flower. All flowers are enclosed by involucres of bract. Nectariferous glands are present. At the centre, there is a single female flower on a long stalk which is surrounded by many male flowers. It is a mechanism to reduce transpiration. e.g., Poinsettia (*Euphorbia pulcherimma), Pedilanthus.*

(b) **Hypanthodium (Syconium)**: Peduncle and receptacle are flask-shaped and fleshy. Inside there is a hollow central cavity where flowers are unisexual and not exposed. Male flowers are present towards the apical opening, female flowers are present at the base and gall flowers (female sterile & short-styled) are present below male flowers. e.g., Banyan tree (*Ficus benghalensis*), Fig (*Ficus carica*), Gular (*Ficus glomerata*) etc.

(c) **Verticillaster**: It is a condensed form of cymose (dichasial) inflorescence in which a cluster of sessile flowers borne on dichasial cyme ends in monochasial cyme (two scorpoid cyme). These sessile flowers appear in pseudowhorls (verticil) and develop from each of two opposite axillary leaves. e.g., Sacred basil (*Ocimum), Salvia, Leucas, Mentha* etc

Fig: Cyathium

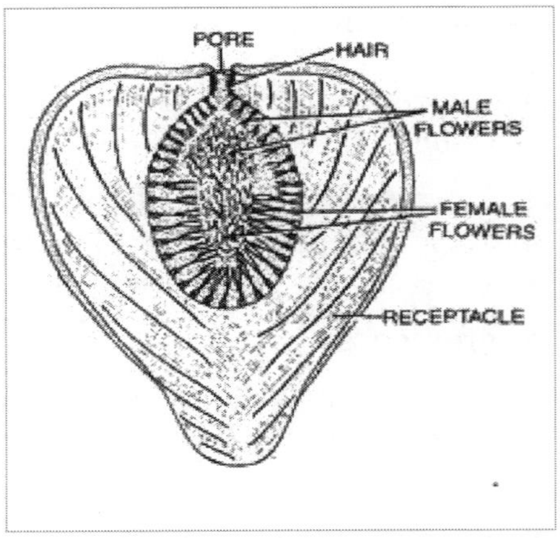

Fig: Hypanthodium

Fig: Verticillaster

FLOWER

It is a specialised condensed or modified shoot in which shoot apical meristem is converted into floral meristem. In this internode does not elongate hence the axis gets condensed. It is called a reproductive unit in angiosperms. It is meant for sexual reproduction. It is a characteristic feature of angiosperms. The study of flowers is called an anthology. Its swollen part is called the thalamus. There are usually four nodes and a highly reduced internode in the thalamus. The stalk of a flower is called a pedicel. Flowers are stalkless (without pedicel) in *Achyranthes, Adhatoda*, Mulberry (*Morus)* etc. Flowers are involved in pollination and fertilisation thus these help in sexual reproduction.

Mother axis: It is part of the shoot on which the flower is borne. It is always posterior in position. A part of a flower is called posterior when it is towards the mother axis or facing the stem and a part of a flower is anterior when it is opposite to the mother axis or facing the subtending bract.

Smallest flower	*Wolffia* (0.1mm)
Largest flower	*Rafflesia arnoldi* (1m)
Largest flower in India	*Sapria himalayensis* (a root parasite)
National flower of India	Lotus (*Nelumbo nucifera*)

Parts of flower:

1. **Accessory outer whorl**: It contains calyx and corolla. These do not take part in sexual reproduction.
   (a) Calyx: Its unit is sepal. It is green, lower and smaller. It protects other inner floral parts.
   (b) Corolla: Its unit is a petal. It is coloured other than green, upper and larger. It protects inner floral parts and attracts pollinators for pollination.

2. **Essential inner whorl**: It contains androecium and gynoecium. These take part in sexual reproduction.
   (a) Androecium: Its unit is stamen.
   (b) Gynoecium: Its unit is carpel.

   Development: The floral parts are developed in ascending order. Calyx is developed first, then corolla and androecium and in the last, gynoecium is developed.

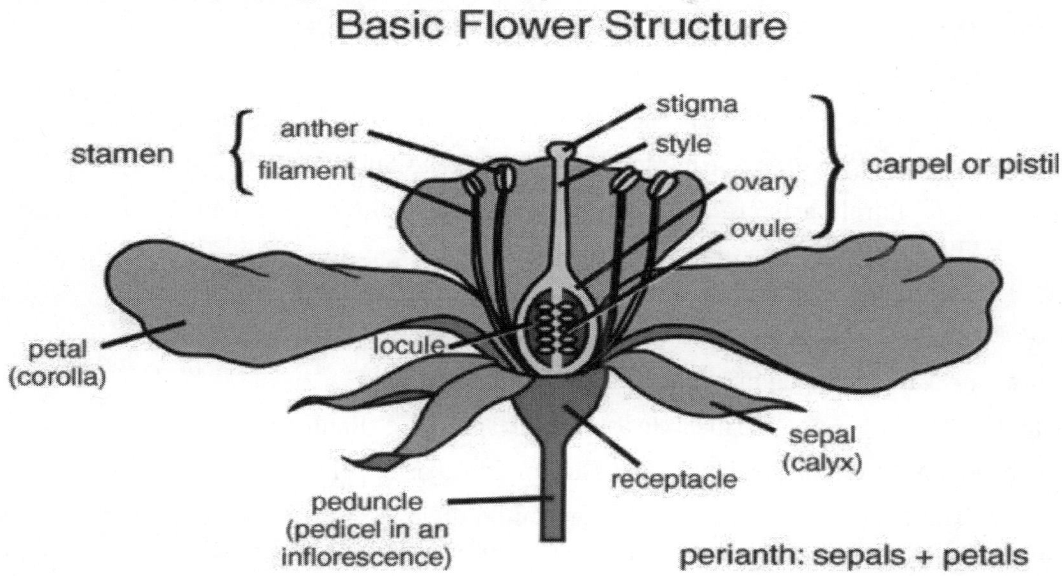

Types of a flower:

1. Achlamydeous: It does not have an accessory whorl (neither calyx nor corolla) hence flower is naked. e.g., Betel, *Euphorbia, Poinsettia.*
2. Monochlamydeous: It has only one accessory whorl (perianth). e.g., Castor
3. Dichlamydeous: It has both accessory whorls (calyx and corolla) due to differentiation of perianth. It is of two types:
   (a) Homochlamydeous: Both whorls (calyx and corolla) are similar in all respects. e.g., *Michelia*
   (b) Heterochlamydeous: Both whorls (calyx and corolla) are dissimilar in all respects. It is most common.

Terms:

1. Monoecious: Both male and female sex organs are present in the same plant.
   (a) Bisexual (Hermaphrodite): Both male and female sex organs are present in the same flower. It is common.
   (b) Unisexual: Male and female sex organs are present in different flowers. e.g., Castor, Coconut, Cucurbits, Maize, *Acalypha.*
2. Dioecious: Male and female sex organs are present in separate plants. These are always unisexual. e.g., Date palm, Mulberry, Papaya, *Cannabis*, *Coccinia*, Spinach, hollyhock.
3. Trioecious: The plant contains male, female and bisexual flowers. e.g., *Silene.*
4. Andromonoecious: The plant contains male and bisexual flowers. e.g., Lily
5. Gynomonoecious: The plant contains female and bisexual flowers. e.g., Sunflower
6. Polygamous: The plant contains male, female, bisexual and neuter flowers. e.g., Mango, Cashew, *Polygonum.*

Types of an internode: In many flowers, the internode is not seen clearly due to the condensed thalamus but in some flowers where internode is elongated and distinct. It may be

1. Anthophore: It is a long internode between calyx and corolla. e.g., *Silene*
2. Androphore: It is a long internode between the corolla and androecium. e.g., *Passiflora*

3. Gynophore: It is a long internode between androecium and gynoecium. e.g., *Capparis*
4. Gynandrophore: It shows both androphore and gynophore. e.g., *Gynandrophis*

Merosity: It is the number of floral parts in whorl. It is of two types:

1. **Isomerous**: All floral parts except the carpel are similar in shape, size, colour, and origin. Each whorl of a flower has an equal number of parts. It may be
   (a) Dimerous (having two sets or two floral parts in each whorl): e.g., Poppy
   (b) Trimerous (having three sets): e.g., Members of Liliaceae family
   (c) Tetramerous (having four sets): e.g., Members of Brassicaceae (=Cruciferae)
   (d) Pentamerous (having five sets): e.g., Members of Fabaceae, Solanaceae
2. **Heteromerous:** All floral parts are dissimilar in shape, size, colour and origin. Each whorl of a flower has an unequal number of parts.

Anthotaxy: It is the arrangement of flowers.

1. Cyclic: Floral parts are arranged in a circle. e.g., Brassicaceae, Malvaceae.
2. Spiral: Floral parts are arranged in a spiral. e.g., Magnoliaceae
3. Spirocyclic (Hemicyclic): Some are in circles and some spiral. e.g., Ranunculaceae.

Symmetry: It is the arrangement of the floral organs around the axis of a flower.

1. Actinomorphic (Radially symmetrical): A flower can be divided into two equal parts by any vertical plane passing through the centre. e.g., Solanaceae, Liliaceae, Brassicaceae (Cruciferae).
2. Zygomorphic (Bilaterally symmetrical): A flower can be divided into two equal parts by only one vertical plane passing through the centre. e.g., Fabaceae, *Ocimum, Delphinium*.
3. Asymmetrical: A flower cannot be divided into two equal parts by any plane. e.g., *Canna, Opuntia*.

Bract (Hypsophyll): It is a special leaf-like structure that occurs at the base of a flower. Generally, there is a single bract. If a bract is a small-scale structure, it is called a bracteole. Generally, there are two bracteoles.

Modifications of bract:

(a) Leafy (Foliaceous): It is green and leaf-like. e.g., *Acalypha, Adhatoda*
(b) Glume: It is small and dry. e.g., Poaceae (=Graminae)
(c) Scaly bracteole: It is present at the base of each floret. e.g., Sunflower
(d) Epicalyx: It is a whorl of bracteoles below the calyx. e.g., Malvaceae
(e) Involucre: It is green and in one or more whorl below inflorescence. e.g., Asteraceae
(f) Petaloid bract: It is brightly coloured. e.g., *Bougainvillea*, Poinsettia
(g) Spathe: It is large and encloses inflorescence. e.g., Banana, Coconut, Maize

Parts of a flower: A flower consists of four types of floral organs.

1. **Calyx:** Its unit is sepal. It may be polysepalous when sepals are free and gamosepalous when sepals are united or fused. Sepal is usually green but in *Clematis and Salvia,* it is called petaloid where it is coloured other than green to attract pollinators. Odd sepal is usually posterior when it is in line with the mother axis but anterior in the Fabaceae family. In the Malvaceae family, there is an extra series of sepal-like organs present outside the calyx called epicalyx.

Duration of sepals:

(a) Caducous: Sepals fall off at the time of opening the flower. e.g., Poppy
(b) Deciduous: Sepals fall off at the time of maturity. e.g., Mustard
(c) Persistent: Sepals remain attached to fruit to provide protection. e.g., Tomato, Brinjal.

Modifications of sepal:

(a) Pappus: Sepal is thread-like and helps in dispersal. e.g., Asteraceae

(b) Leafy: Sepal is large to attract pollinators. e.g., *Mussaenada*

(c) Bilabiate: Sepal has two lips. e.g., *Salvia*

(d) Campanulate: Sepal is bell-shaped. e.g., China rose

(e) Hooded: Sepal is hood-like. e.g., *Aconitum*

(f) Infundibuliform: Sepal is funnel like. e.g., *Atropa*

(g) Spurred: Sepal is beak-like to store nectar. e.g., *Delphinium*

(h) Tubular: Sepal appears as tube-like. e.g., *Datura*, Tobacco.

2. **Corolla:** Its unit is petal. It may be polypetalous when petals are free and gamopetalous when petals are fused or united.

Appendages of corolla:

(a) Spur: it is a tubular projection from the bottom. e.g., Larkspur (*Delphinium*)

(b) Nectary: it is a protuberance containing nectar secreting gland. e.g., *Salvia*

(c) Corona: It is an additional whorl between the corolla and stamen or on the corolla formed transverse splitting of the corolla. It may be coronary corona (lobed) in *Calotropis* and staminal corona in *Nerium*.

Modifications of petal:

(a) Cruciform: Petals are cross-like. e.g., Brassicaceae

(b) Bilabiate: Petals have two separate lips. e.g., *Ocimum, Salvia*

(c) Campanulate: Petals are bell-shaped. e.g., *Cuscuta, Physalis*

(d) Infundibular: Petals are funnel shaped. e.g., *Datura*

(e) Spurred: Petals are beak-like. e.g., Larkspur (*Delphinium*)

(f) Tubular: Petals are tube-like. e.g., Disc floret of sunflower.

(g) Ligulate: Petals are strap shaped. e.g., Ray floret of sunflower

(h) Caryophyllaceous: Petals are clawed. e.g., Pink, Carnation

(i) Rosaceous: Petals are rose-like. e.g., Rose

Perianth: Its unit is tepal. It is a structure when there is no distinction between sepal and petal. It is found in Monocots. It may be polyphyllous when tepals are free and gamophyllous when tepals are united.

(a) Tepal is sepaloid when it is green. e.g., Date palm

(b) Tepal is petaloid when it is coloured other than green. e.g., *Asphodelus.*

Aestivation: It is the arrangement of sepals and petals with respect to one another in a flower. It is of several types.

(a) **Valvate**: The margins of adjacent sepals touch each other but do not overlap. e.g., Solanaceae, Liliaceae.

(b) **Twisted (Contorted)**: The one margin of adjacent sepal overlaps each other only in one direction regularly. e.g., Members of Asteraceae (sunflower), Malvaceae (Cotton, Lady's finger).

(c) **Ascending imbricate**: The posterior petal is completely inner; one is completely outer and the rest are partly outer and partly inner. e.g., Members of Caesalpinaceae (*Cassia, Delonix*).

(d) **Quincuncial:** Two petals are completely outer; two are completely inner and the rest one petal is partly outer and partly inner. e.g., *Ranunculus*, guava, *Ipomoea.*

(e) **Vexillary (Descending imbricate):** The posterior petal (vexillum or standard) is completely outer; two anterior fused petals (keel or carina) are inner, and two lateral petals (alae or wing) are partly outer and partly inner. e.g., Members of Fabaceae

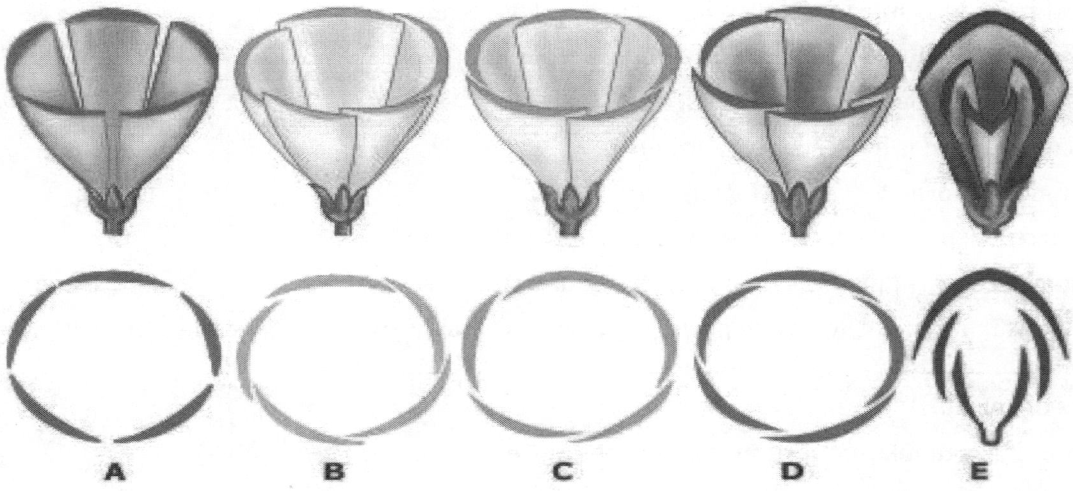

Different types of aestivation of calyx and corolla
A, Valvate, B, Twisted, C. Imbricate, D.Quincuncial, E. Vexillary

Positions of floral parts on the thalamus:

(a) Hypogynous: The floral parts such as the sepal, petal and stamen are situated below the ovary hence ovary is superior. e.g., Members of Solanaceae, Malvaceae, Brassiccaceae, Liliaceae etc.

(b) Epigynous: The floral parts such as sepal, petal and stamen are situated above the ovary hence ovary is inferior. e.g., Members of Cucurbitaceae, Ray florets of Asteraceae, Guava.

(c) Perigynous: The floral parts such as sepal, petal and stamen are in the same line with the ovary hence ovary is either half superior or half inferior. e.g., Rose, Plum, Peach.

3. **Androecium:** Its unit is stamen (microsporophyll). It is the third whorl and male sex organ of a flower. It is differentiated into fertile anther, sterile connective and long filament. Fertile stamen produces pollen grains. If stamen is sterile (does not produce pollen), then it is called staminode as seen in Cassia.

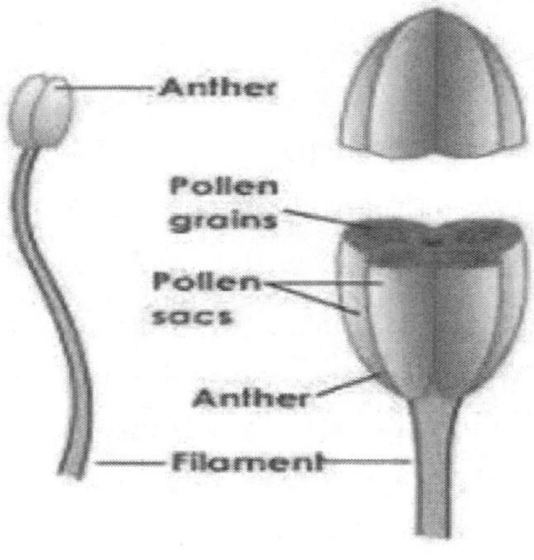

Fig: Stamen

Length of stamens:

(a) Isostemonous: Stamens are of the same length. It is common

(b) Heterostemonous: Stamens are of different length. It is of two types

(i) Didynamous: There are 4 stamens in which two are outer and shorter and two are inner and longer. e.g., Members of Lamiaceae (=Labiatae), Verbenaceae.

(ii) Tetradynamous: There are 6 stamens in which two are outer and shorter and four are inner and longer. e.g., Members of Brassicaceae (=Cruciferae).

Arrangement of stamens:

(a) Alternipetalous: Stamens are alternate with petals.

(b) Antipetalous: Stamens are opposite to petals

(c) Diplostemonous: Stamens are in two whorls-outer alternipetalous and inner antipetalous. e.g., *Cassia*

(d) Obdiplastemonous: Stamens are in two whorls- outer antipetalous and inner alternipetalous. e.g., Pink (*Dianthus*), Castor

Cohesion of stamen (Fusion among themselves):

(a) Polyandrous: Stamens (anther and filament) are free. e.g., Papaya, Poppy, *Ranunculus*

(b) Monadelphous: Filaments are fused in one bundle, but anthers are free. e.g., Members of Malvaceae (China rose, cotton, lady's finger)

(c) Diadelphous: Filaments are fused in two bundles, but anthers are free. e.g., Members of Fabaceae

(d) Polyadelphous: Filaments are fused in several bundles, but anthers are free. e.g., Citrus plants (Lemon, orange), castor, cotton, *Bombax* etc.

(e) Syngenesius: Filaments are free, but anthers are fused. e.g., Members of Asteraceae (Sunflower, Marigold).

(f) Synandrous: Filament and anther are fused throughout their length. e.g., Cucurbits

Adhesion of stamens (Fusion with other floral parts):

(a) Epipetalous: Stamens are fused with petals. e.g., Solanaceae, Asteraceae

(b) Epiphyllous: Stamens are fused with tepals. e.g., Liliaceae

(c) Gynandrous: Stamens are fused with carpels. e.g., *Calotropis.*

Condition of anther

(a) Monothecous: Anther is one lobed having two microsporangia (bisporangiate). e.g., Members of Malvaceae (Cotton, lady's finger)

(b) Dithecous: Anther is two lobed having four microsporangia (tetrasporangiate). e.g., most common.

(c) Introrse: The face of the anther is towards the centre of the flower. e.g., Tomato

(d) Extrorse: The face of the anther is towards the periphery. e.g., *Ranunculus*

Fixation of anther:

(a) Adnate: Filament is attached to the total length of another on the back. e.g., *Magnolia, Ranunculus, Michelia, Nymphaea.*

(b) Basifixed: Filament is attached to the base of another. e.g., *Datura*, Mustard

(c) Dorsifixed: Filament is attached to another by its dorsal side. e.g., *Passiflora*

(d) Versatile: Filament is attached to the back of the anther which swings freely in all directions. e.g., Members of the Poaceae family (Grasses, Wheat).

4. **Gynoecium:** Its unit is carpel. It is differentiated into ovary, style and stigma. The carpel is centrally placed in the thalamus. If it is sterile then it is called pistillode. It may be apocarpous (carpels are free) and syncarpous (carpels are fused or united). In some plants, the gynoecium shows the following types:

(a) Monocarpellary: e.g., Members of Fabaceae and Poaceae

(b) Bicarpellary and syncarpous: e.g., Members of Solanaceae

(c) Tricarpellary and syncarpous: e.g., Members of Liliaceae

(d) Pentacarpellary and syncarpous: e.g., China rose (*Hibiscus*)

(e) Multicarpellary and apocarpous: e.g., *Michelia, Ranunculus*

(f) Multicarpellary and Syncarpous: e.g., Poppy (*Papaver*)

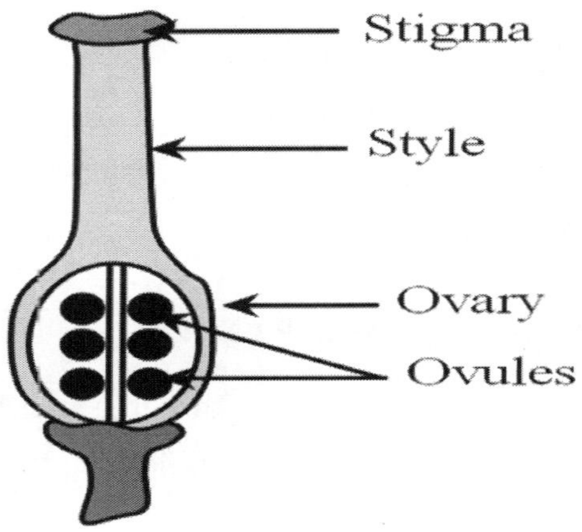

Fig: The different parts of carpel

Placentation: It is the arrangement and position of placentae and ovules in the ovary. Placenta is cushion-like structure to which ovules are attached. It is of the following types:

(a) Marginal: Ovary is unilocular and ovules arise from ventral suture (along the junction of both margins of ovary). e.g., Fabaceae.

(b) Axile: Ovary is multilocular and ovules arise from central axis. It is most common. e.g., Solanaceae, Malvaceae, Liliaceae.

(c) Parietal: Ovary is unilocular and ovules arise from inner wall of ovary. e.g., Brassicaceae, Cucurbitaceae.

(d) Free-central: Ovary is unilocular and ovules arise from central axis, but septum is absent. e.g., Pink (*Dianthus*), *Primula*, *Silene* (Members of Caryophyllaceae).

(e) Basal: Ovary is unilocular and ovules arise from base of ovary. It is considered as advanced. e.g., Asteraceae, Poaceae

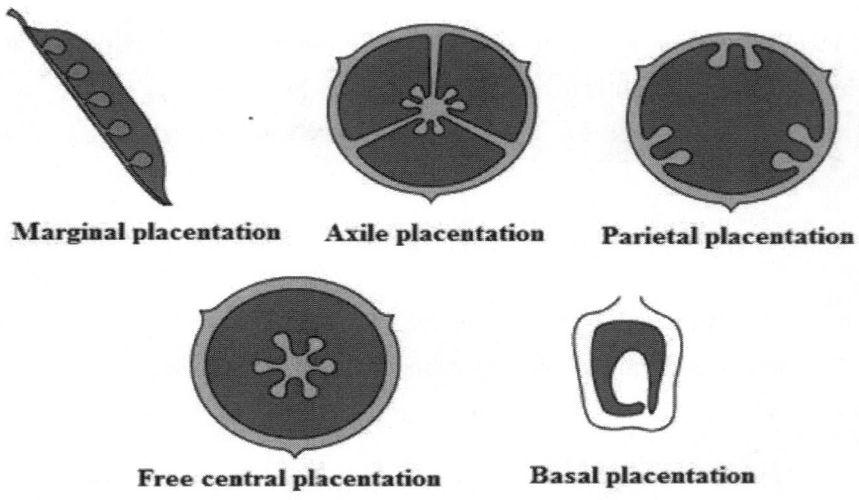

Types of style:

(a) Terminal style: Style arises from terminal part of ovary. e.g., *Datura*, *Solanum*

(b) Lateral style: Style arises from one side of ovary. e.g., Mango

(c) Gynobasic style: Style arises from the base of ovary. e.g., *Ocimum* and other members of Lamiaceae (=Labiatae).

(d) Stylopodium: Style is swollen at its base. e.g., Coriander and other members of Apiaceae (=Umbelliferae).

Types of Stigma:

(a) Capitate: It is rounded. e.g., China rose

(b) Forked: It is divided. e.g., *Tridax*

(c) Feathery: It is brush-like. e.g., Grasses

Terms:

1. Cauliflory: It is the production of flowers on old stems from dormant bud. e.g., *Artocarpus, Ficus.*
2. Resupination: It is a rotation of the ovary by 180°. e.g., Orchids

FRUIT

It is a ripened or fertilised ovary. The study of fruits is called Pomology. It consists of a pericarp (formed from the wall of an ovary) and a seed (formed from the ovule). Pericarp is differentiated into epicarp, mesocarp and endocarp. Fruit may be False when it is developed from the thalamus along with the ovary. Such fruits are called pseudocarp. e.g., Apple, pear and all composite fruits.

Largest fruit	Double coconut (*Ladoicea maldivica*) = 6kg
Smallest fruit	*Wolffia*

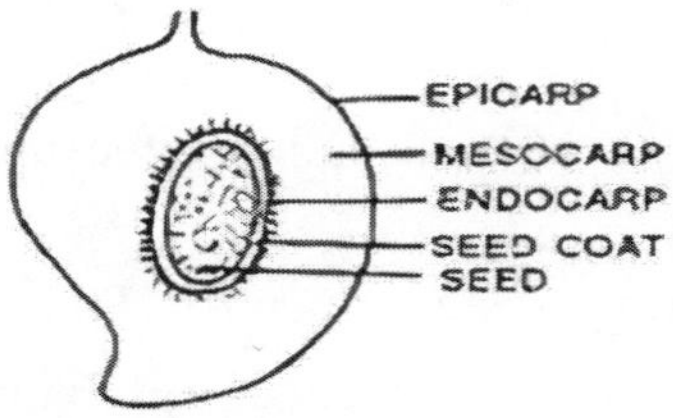

MANGO IN LS
A. TRUE FRUIT

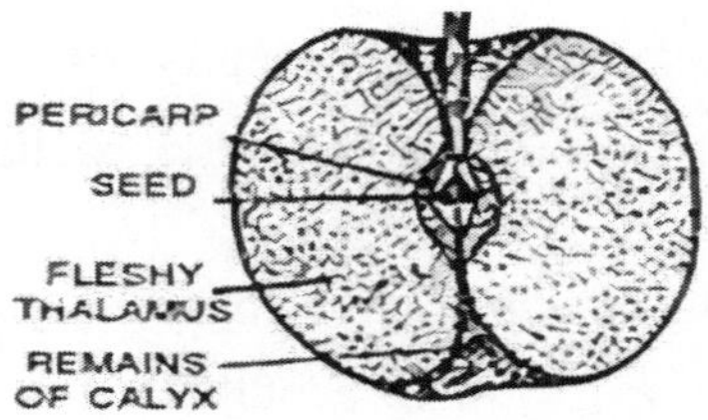

APPLE IN LS.
B. FALSE FRUIT

Fruit may be parthenocarpic (seedless) when it is developed without pollination and fertilisation. e.g., Banana, Grapes, Orange, Pineapple etc.

Types of true fruit:

1. **Simple fruits**: These develop from syncarpous (either monocarpellary or bicarpellary or polycarpellary) ovary. These are categorised into four types:

   (a) **Dry dehiscent fruits**: Fruits split open. These are of following types:

   (i) **Legume (Pod):** It develops from monocarpellary ovary and dehiscence occurs by both halves. e.g., Beans (Fabaceae family)

   (ii) **Follicle**: It develops from bicarpellary ovary and dehiscence occurs by one half. e.g., *Calotropis, Michelia*, Periwinkle (*Catharanthus*).

   (iii) **Siliqua**: It develops from bicarpellary ovary and dehiscence occurs by both halves. e.g., Brassicaceae. Flat siliqua is known as silicula. e.g., *Capsella, Iberis.*

   (iv) **Capsule**: It develops from multicarpellary ovary and dehiscence occurs by pores (Poricidal) as in *Argemone*, Poppy; by locules (Loculicidal) as in Cotton, Lady's finger; by septa (Septicidal) as in Linseed, Mustard and by fragments of septa (Septifragal) as in *Datura*.

   (b) **Dry indehiscent fruits**: These are one seeded fruit which does not split inside. These are of following types:

   (i) **Achene**: It develops from monocarpellary ovary and pericarp is not fused with seed coat. e.g., *Boerhaavia*, rose.

   (ii) **Caryopsis**: It develops from monocarpellary ovary and pericarp is fused with seed coat. e.g., Members of Poaceae (=Graminae)

(iii) **Cypsela**: It develops from bicarpellary and inferior ovary. Persistent pappus is attached with fruit. e.g., Asteraceae.

(iv) **Nut:** It develops from multicarpellary ovary and epicarp is rough or hard. e.g., Cashew nut, litchi, hazel nut, *Trapa*.

(v) **Samara**: It is winged fruit and wing develops from pericarp. e.g., *Dioscorea, Shorea*.

(c) **Dry Schizocarpic fruits**: Fruits split into segments with one or several seeded structures called mericarp. These are of following types:

(i) **Lomentum**: It develops from monocarpellary ovary and fruit splits into one seeded structure at maturity. e.g., Tamarind, Groundnut

(ii) **Cremoacarp**: It develops from bicarpellary ovary and fruit splits into two seeded structures at maturity. e.g., Carrot, Coriander

(iii) **Regma**: It develops from tricarpellary ovary and fruit splits into three seeded structures at maturity. e.g., Castor

(iv) **Carcerulus**: It develops from bicarpellary ovary and fruit splits into four seeded structures at maturity. e.g., members of Lamiaceae

(v) **Double samara**: It develops from bicarpellary ovary and fruit splits into two samaras at maturity. e.g., *Acer*

(d) **Fleshy fruits: These are of following types:**

(i) **Amphisacra**: It develops from multicarpellary ovary. It is multiseeded and epicarp is stony. e.g., Wood apple

(ii) **Balausta:** It develops from multicarpellary ovary. Testa is juicy. e.g., Pomegranate

(iii) **Berry**: It develops from bicarpellary ovary. It is multiseeded. e.g., Banana, Brinjal, Grapes, Guava, Tomato, gooseberry etc.

(iv) **Drupe**: It develops from multicarpellary ovary. It is one seeded. Endocarp is stony. e.g., Mango, almond, coconut, plum, walnut, olive, cherry, apricot etc.

(v) **Hesperidium**: It develops from multicarpellary ovary. Endocarp is segmented but epicarp and mesocarp are fused forming rind or skin. e.g., Lemon, Orange.

(vi) **Pepo**: It develops from tricarpellary inferior ovary. It is multiseeded. e.g., Cucurbits, Papaya.

(vii) **Pome**: It develops from multicarpellary ovary. e.g., Apple, Pear

2. **Aggregate Fruits**: These develop from multicarpellary and apocarpous ovary. It is a group of fruitlets (simple fruit) called etaerios. These are of following types:

(i) Etaerio of achene: Strawberry, lotus, *Ranunculus, Clematis*

(ii) Etaerio of Berry: Custard apple

(iii) Etaerio of Drupe: Raspberry, Blackberry

(iv) Etaerio of follicle: *Aconitum, Delphinium*

3. **Composite or Multiple fruits**: These develop from inflorescence. Each flower produces a fruit but these mature into a single mass. These are regarded as false fruits. These are of following types:

(i) **Sorosis**: Perianth and bract are fused and fleshy. It develops from

- Female spike: e.g., Pineapple
- Female catkin: e.g., Mulberry
- Female spadix: e.g., Jackfruit

(ii) **Syconus**: It develops from fleshy hypanthodium inflorescence. e.g., Fig, Banyan tree, Peepal.

Types of fruits with edible parts:

Plant	Botanical name	Type	Edible parts
Mango	*Mangifera indica*	Drupe	Mesocarp
Grapes	*Vitis vinifera*	Berry	Pericarp and Placenta
Banana	*Musa sapientum*	Berry	Mesocarp and Endocarp
Tomato	*Lycopersicon esculentum*	Berry	Mesocarp and Placenta
Guava	*Psidium guajava*	Berry	Thalamus and pericarp
Papaya	*Carica papaya*	Berry	Mesocarp
Date palm	*Phoenix dactylifera*	Berry	Fleshy pericarp
Cucumber	*Cucumis sativus*	Pepo	Mesocarp, endocarp and Placenta
Lemon	*Citrulus aurantifolia*	Hesperidium	Juicy endocarp (hairs)
Orange	*Citrus reticulate*	Hesperidium	Juicy endocarp
Apple	*Malus pumila (=Pyrus malus)*	Pome	Fleshy thalamus
Pear	*Pyrus communis*	Pome	Fleshy thalamus
Pineapple	*Ananas comosus*	Sorosis	Receptacle, bract & perianth
Jack fruit	*Artocarpus heterophyllus*	Sorosis	Receptacle, bract and perianth
Cashew nut	*Anacardium accidentale*	Nut	Cotyledon and stalk
Coconut	*Cocos nucifera*	Drupe	Endosperm
Walnut	*Juglans regia*	Drupe	Lobed cotyledon
Water melon	*Cucumis melo*	Pepo	Mesocarp
Plum	*Prunus domestica (=Prunus dulcis)*	Drupe	Epicarp and mesocarp
Black berry	*Syzygium cumuni*	Drupe	Mesocarp
Mulberry	*Morus alba*	Sorosis	Fleshy receptacle
Strawberry	*Fragaria indica*	Etaerio of achene	Fleshy receptacle
Raspberry	*Rubus idaeus*	Etaerio of drupe	Fleshy receptacle
Fig	*Ficus carica*	Syconus	Fleshy inflorescence
Litchi	*Litchi chinensis*	Nut	Fleshy aril
Sapodilla	*Achras sapota*	Berry	Mesocarp
Tamarind	*Tamarindus indica*	Lomentum	Mesocarp
Wood apple	*Aegle marmelos*	Amphisacra	Endocarp and placentae
Custard apple	*Annona squamosa*	Etaerio of berry	Fleshy pericarp
Pomegranate	*Punica granatum*	Berry	Fleshy testa
Pumpkin	*Cucurbita moschata*	Pepo	Mesocarp
Lady's finger	*Abelmoschus esculentus*	Capsule	Whole fruit
Ground nut	*Arachis hypogea*	Lomentum	Seed
Almond	*Prunus dulcis*	Drupe	Seed
Pista	*Pistacia vera*	Drupe	Cotyledon

SEED

It is fertilised or ripened ovule which contains embryo. Study of seed is called Spermology.

Smallest or lightest seed	Orchis (0.004mg)
Largest or heaviest seed	Double coconut (*Ladoicea maldivica*)=18 kg

Parts of seed: A typical seed consists of

1. **Seed coat**: It is outer protective covering. Its outer and hard layer is called testa and inner layer is called tegmen (usually fused with testa). It has a fine pore called micropyle on surface. It has a scar called hilum where funicle (stalk of ovule/seed) is attached. If seed coat is absent, then such seed is called kernel.
2. **Embryo**: It has young or miniature plant enclosed within seed coat. It has two main parts:
   (a) **Cotyledon**: It is leaf of embryo. It is either one or two in number. It is absent in *Cuscuta*. The largest cotyledon is seen in *Capsella*. It stores food and becomes fleshy. If it does not store food, then it becomes papery and thin.
   (b) **Tigellum**: It is main axis of embryo which is attached with cotyledon. Its micropylar end is called radicle and its opposite end is called plumule. The part of axis below the point of attachment of cotyledon or between cotyledon and radicle is called hypocotyl and the part of axis above point of attachment of cotyledon or between cotyledon and plumule is called epicotyl. The hypocotyl terminates with radicle and epicotyl terminates with plumule.
3. **Endosperm**: It is nutritive tissue formed as a result of triple fusion. In monocots, its outer covering is called aleurone layer which separates endosperm from embryo. It stores food when cotyledon cannot store food. Seed may be:
   (a) **Endospermic** (albuminous): Seed contains endosperm. It is found in monocots and few dicots such as Castor, Custard apple, Cotton, Papaya, Sunflower, Rubber).
   (b) **Nonendospermic** (exalbuminous): Seed does not contain endosperm because it was fully consumed by developing embryo. It is found in dicots and few monocots such as *Alisma*, Orchid, *Pothos, Sagittaria, Vallisneria.*

Dicot seed	Monocot seed
Seed coat is not fused with pericarp (fruit wall).	Seed coat is fused with pericarp (fruit wall).
Seed has two cotyledons.	Seed has one shield shaped cotyledon known as scutellum.
Embryo is large and centrally placed.	Embryo is small and placed on one side.
Endosperm is small and food is stored in cotyledon.	Endosperm is bulky where food is stored.
Aleurone layer is absent.	Aleurone layer is found outside endosperm.
Radicle is not covered with coleorhiza.	Radicle is covered with coleorhiza.
Plumule is not covered with coleoptile.	Plumule is covered with coleoptiles
Plumule is situated between two cotyledons.	Plumule is situated at one end near cotyledon.

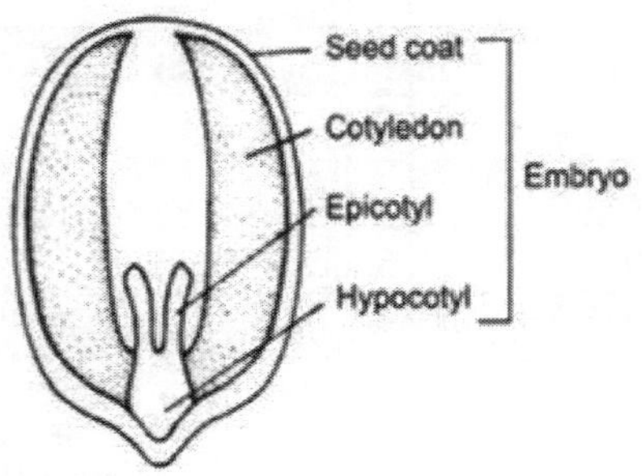

Fig: Dicot seed

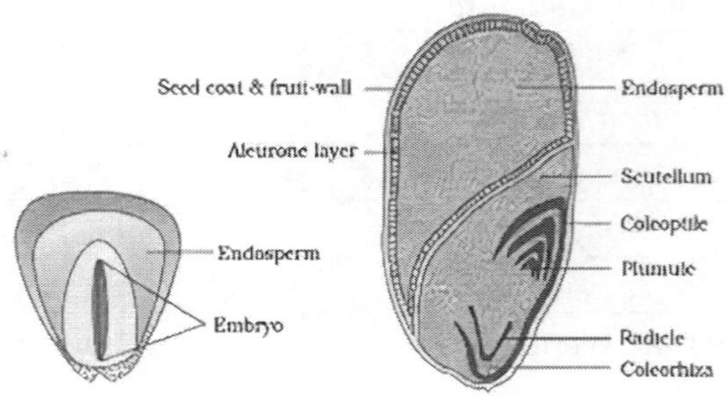

Fig: Monocot seed

Special structures: There are several structures which develop from various parts of ovule during post fertilisation development.

1. **Perisperm**: It is remnant of nucellus in seed. Nucellus is usually consumed after fertilisation but sometimes it is left. It is seen in Beet, Black pepper, *Nymphaea.*
2. **Caruncle**: It is an outgrowth of outer integument either on the side of funicle or all around micropyle. It is whitish and fleshy structure. It helps in seed germination because it is hygroscopic by absorbing water from soil and passes to embryo. It is seen in Castor.

3. Aril: It is fleshy appendage or outgrowth which arises from funicle or testa or both. It surrounds the ovule. It is regarded as third integument. It is seen in *Litchi*, Nutmeg etc.

Types of seed:

1. **Orthodox seed**: It tolerates low moisture (dehydration), low temperature (freezing) and anaerobic condition. e.g., Cereals, Legumes.
2. **Recalcitrant seed**: It gets killed under low moisture and low temperature. It is regarded as non-biodegradable. e.g., Coconut, Jackfruit, Tea.

Advantages:

(a) Seed helps the plants to colonise in other areas as aresult of dispersal.
(b) Seed nourishes the seedling till photosynthesis because it has enough food
(c) It protects young embryo due to hard seed coat.
(d) It produces new genetic combination which leads to variation.

DESCRIPTION OF FAMILIES

Largest family	Asteraceae
Second largest family	Orchidaceae
Third largest family	Fabaceae
Fourth largest family	Poaceae
Fifth largest family	Rubiaceae
Smallest family	Nympheaceae

The various symbols and their full forms are as follows:

Symbol or abbreviation		Full form
Br	–	bracteate condition
Ebr	–	ebracteate condition
Brl	–	bracteolate
Epik	–	epicalyx
0 (zero)	–	absence of a particular whorl
∞	–	indefinite number of floral parts in a whorl
$\oplus$	–	actinomorphic (regular) condition
o|o	–	zygomorphic (irregular) condition
K	–	calyx (sepals)
C	–	corolla (petals)
P	–	perianth (tepals)
A	–	androecium (stamens)
G	–	gynoecium (carpels)
♂	–	unisexual, staminate (male) flower
♀	–	unisexual, pistillate (female) flower
⚥	–	bisexual flower
( )	–	cohesion of floral parts in a whorl
( )	–	partial or slight cohesion of the parts
$\overset{\frown}{P\ A}$	–	epiphyllous stamens
$\overset{\frown}{C\ A}$	–	epipetalous stamens
$\underline{G}$	–	hypogyny (ovary superior)
G–	–	perigyny (ovary half-inferior)
$\overline{G}$	–	epigyny (ovary inferior)
G(2)	–	bicapellary syncarpous
$G\underline{(2)}$	–	bicarpellary syncarpous ovary superior
G2	–	bicarpellary free carpels
K_{2+2}	–	presence of two whorls of calyx, each of 2 sepals (Brassicaceae or Cruciferae)

Old name of family	New name of family
Graminae	Poaceae
Leguminosae	Fabaceae
Cruciferae	Brassicaceae
Labiatae	Lamiaceae
Umbelliferae	Apiaceae
Palmae	Arecaceae
Compositae	Asteraceae
Guttiferae	Clusiaceae

FABACEAE:

It is also known as bean family. Its earlier name was Papilionaceae, a subfamily of family Leguminosae.

Classification: (According to Bentham and Hooker's system)

Kingdom	Plantae
Division	Phanerogamia
Class	Dicotyledonae
Subclass	Polypetalae
Series	Calyciflorae
Order	Rosalaes
Family	Fabaceae

Vegetative features:

- Habit: Herbs, shrubs and trees
- Root: Tap root. Root nodule is mostly seen.
- Stem: Erect or climbing
- Leaf: Alternate, pinnately compound or simple, leaf base shows pulvinus, stipulate, reticulate venation.

Floral characters:

- Inflorescence: Raceme or solitary axillary
- **Flower:** Ebracteate, Pedicellate, Zygomorphic, Bisexual, Hypogynous, Complete, Pentamerous
- **Calyx:** Sepals- 5, Gamosepalous, Odd sepal anterior, Valvate aestivation
- **Corolla:** Petals-5 (unequal), Polypetalous, Largest petal upper and posterior, Descending imbricate (Vexillary) aestivation,
- **Androecium:** Stamens-10, Diadelphous, Free stamen posterior and fused stamens anterior, Anther dithecous, dorsifixed and introrse
- **Gynoecium:** Monocarpellary, Ovary superior, Marginal placentation, Ovary unilocular, numerous ovules
- **Fruit:** Mostly legume or pod
- **Seed:** Non-endospermic

Floral formula: $\% K_{(5)} C_{1+2+(2)} A_{(9)+1} \underline{G}_1$

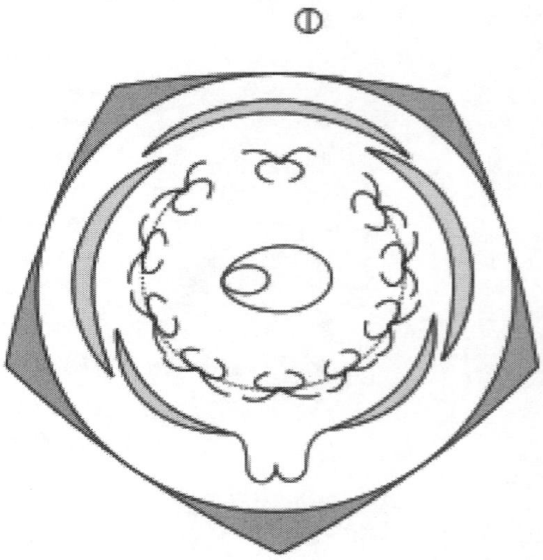

Floral diagram

Economic importance:

1. **Food:** Seeds are the source of pulses.
   - Gram/Chickpea (*Cicer arietinum*)
   - Red gram/Pigeon pea (*Cajanus cajan*)
   - Green gram (*Phaseolus radiatas=Vigna radiata*)
   - Black gram (*Phaseolus mungo*)
   - Garden pea (*Pisum sativum*)
   - Sweet pea/ Grass pea (*Lathyrus odoratus*)
   - Cow pea (*Vigna unguiculata*)
   - Soya bean (*Glycine max*)
   - French bean/Kidney bean (*Phaseolus vulgaris*)
   - Hyacinth bean (*Dolichos lablab*)
   - Broad bean (*Vicia faba*)
   - Lentil (*Lens culinaris*)
2. **Food (Spices):**
   - Fenugreek (*Trigonella foenumgraecum*)
3. **Fodder/forage crops**:
   - Alfalfa (*Medicago sativa*)
   - Cluster bean (*Cyamopsis tetragonoloba*)
   - Berseem / Egyptian clover (*Trifolium alexandrium*)
   - Senji / Indian clover (*Melilotus alba*)
4. **Oil**:
   - Groundnut/peanut (*Arachis hypogea*): Seed produces oil.
5. **Fibre**:
   - Sun hemp (*Crotolaria juncea*): The secondary phloem of the stem produces fibres.
6. **Medicine**:
   - Liquorice (*Glycyrhiza glabra*): The root is useful for cough and cold
   - Jeweller's weight / ratti (*Abrus precatorius*): The leaf is a useful part.
7. **Dye**:
   - Indigo (*Indigofera tinctoria*): Leaf produces dye.
8. **Timber**:
   - Indian redwood / shisham (*Dalbergia sisso*)
9. **Ornamental plants**:
   - Lupin (*Lupinus arboreus*)
   - Flame of forest (*Butea monosperma*)

BRASSICACEAE:

It is also called Cruciferae or mustard family.

Classification: (According to Bentham and Hooker's system)

Kingdom	Plantae
Division	Phanerogamia
Class	Dicotyledonae
Subclass	Polypetalae
Series	Thalamiflorae
Order	Parietales
Family	Brassicaceae

Vegetative characters:

- Habit: Mostly herbs, Annual, Biennial
- Root: Tap root (Fusiform root / Napiform root / Conical root)
- Stem: Herbaceous, Erect, Cylindrical
- Leaf: Simple, Sessile, Exstipulate, Reticulate venation

Floral characters:

- Inflorescence: Racemose (Raceme / corymb)
- Flower: Ebracteate, Pedicellate, Actinomorphic, Bisexual, Hypogynous, Complete, Tetramerous, Cyclic
- Calyx: Sepals-4, Arranged in two whorls of 2 each, Polysepalous, Imbricate aestivation,
- Corolla: Petals-4, Arranged in the form of a cross (cruciform), Polypetalous, Valvate aestivation,
- Androecium: Stamens-6, Arranged in two whorls, Tetradynamous, Polyandrous, Anther dithecous, basifixed and introrse, Nectaries are present
- Gynoecium: Bicarpellary, Syncarpous, Ovary superior, Parietal placentation, Ovary unilocular (becomes bilocular due to false septum called replum), Many ovules, carpels are transversely placed.
- Fruit: Siliqua
- Seed: Non-endospermic

Floral Formula: EBr ⊕ ⚥ $K_{2+2} C_4 A_{2+4} G_{(2)}$

Economic importance:

1. **Food:**
   - Cauliflower (*Brassica oleracea* var. *botrytis*)
   - Cabbage (*Brassica oleracea* var. *capitata*)
   - Turnip (*Brassica rapa*)
   - Radish (*Raphanus sativus*)
2. **Oil:**
   - Mustard (*Brassica campestris*)
3. **Spices:**
   - White mustard (*Brassica nigra*)
4. **Ornamental Plants:**
   - Candytuft (*Iberis amara*)
   - Shepherd'spurse (*Capsella bursa-pastoris*)

Floral Diagram

MALVACEAE:

It is also called mallow family.

Classification: (According to Bentham and Hooker's system)

Kingdom	Plantae
Division	Phanerogamia
Class	Dicotyledonae
Subclass	Polypetalae
Series	Thalamiflorae
Order	Malvales
Family	Malvaceae

Vegetative characters:

- Habit: Mostly herbs
- Root: Tap root
- Stem: Herbaceous or Woody, Branched, Erect
- Leaf: Simple or Compound, Alternate, Petiolate, Stipulate, Reticulate venation

Floral characters:

- Inflorescence: Solitary axillary / terminal, Raceme, Panicle
- Flower: Bracteate, Bracteolate (in the form of epicalyx), Actinomorphic, Pedicellate, Bisexual, Hypogynous, Complete, Pentamerous
- Epicalyx: 3-9
- Calyx: Sepals-5, Gamosepalous, Valvate aestivation, Persistent.
- Corolla: Petals-5, Polypetalous, Imbricate aestivation,
- Androecium: Stamens-∞, Monadelphous, Epipetalous, Anther Monothecous, basifixed and introrse
- Gynoecium: Multicarpellary, Syncarpous, Ovary superior, Axile placentation, Ovary Multillocular, 1-many ovules in each locule
- Fruit: Schizocarpic or Capsule or Berry
- Seed: Non-endospermic

Floral Formula: $\oplus$ ⚥ $\mathbf{Epi}_{3\text{-}7}\,\mathbf{K}_{(5)}\,\mathbf{C}_5\,\mathbf{A}_{\infty}\,\mathbf{G}_{(\infty)}$

Economic Importance:

1. **Food:**
   - Lady's finger (*Abelmoschus esculentus*): Fruits are used as vegetables.
2. **Fibres:**
   - Cotton (*Gossypium herbaceum*): Its seeds are source of fibres.
   - Deccan hemp (*Hibiscus cannabinus*):
3. **Ornamental plants:**
   - China rose (*Hibiscus rosa-sinensis*)

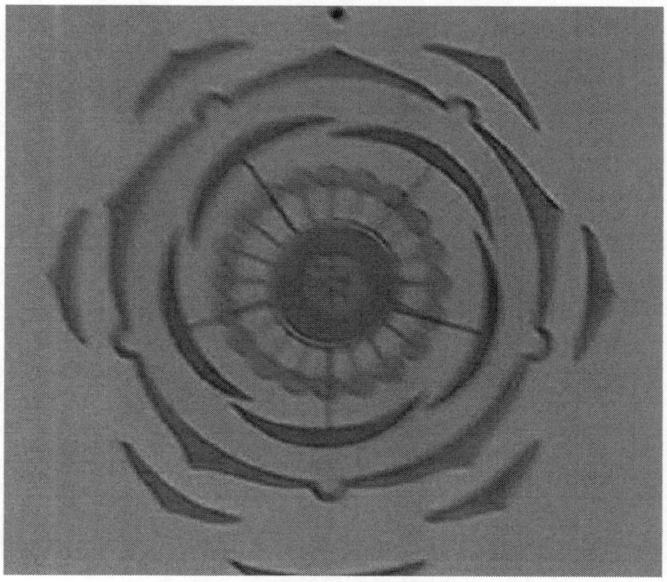

Floral Diagram

SOLANACEAE:

It is also called potato family.

Classification: (According to Bentham and Hooker's system)

Kingdom	Plantae
Division	Phanerogamia
Class	Dicotyledonae
Subclass	Gamopetalae
Series	Bicarpellatae
Order	Polemoniales
Family	Solanaceae

Vegetative characters:

- Habit: Mostly herbs
- Root: Taproot
- Stem: Mostly herbaceous
- Leaf: Alternate, Simple, Exstipulate, Reticulate venation

Floral characters:

- Inflorescence: Helicoid cyme or solitary axillary
- Flower: Ebracteate, Pedicellate, Actinomorphic, Bisexual, Hypogynous, Heterochlamydeus, Complete, Pentamerous
- Calyx: Sepals-5, Gamosepalous, Valvate aestivation, odd sepal posterior, Persistent.
- Corolla: Petals-5, Gamopetalous, Valvate aestivation,
- Androecium: Stamens-5, Polyandrous, Epipetalous, Alternipetalous, Anther dithecous, basifixed and introrse
- Gynoecium: Bicarpellary, Syncarpous, Ovary superior, Axile placentation, Ovary bilocular (multilocular due to false septum as in Tomato, Datura etc), Many ovules in each locule, Swollen placenta, Carpel (ovary) obliquely placed (Posterior carpel to the right about 45° from median and anterior to left).
- Fruit: Capsule or berry
- Seed: Endospermic

Floral formula: $\oplus$ ⚥ $K_{(5)}\ \overparen{C_{(5)}\ A_5}\ G_{(2)}$

Floral diagram

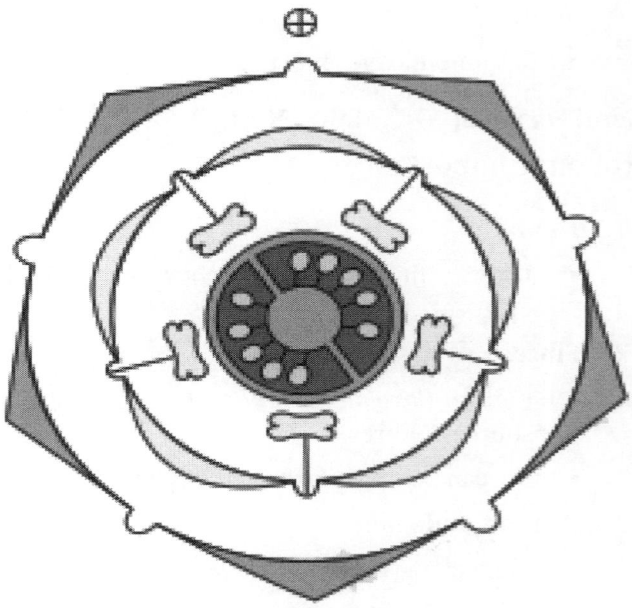

Floral Diagram

Economic importance:

1. **Food:**
   - Potato (*Solanum tuberosum*): The underground stem is used as a vegetable.
   - Brinjal/Eggplant (*Solanum melongena*): Fruit is used as a vegetable.
   - Tomato (*Lycopersicon esculentum*): Fruit is used as a vegetable.
   - Chilly (*Capsicum annum*): Fruit is used as a condiment.
   - Red pepper (*Capsicum frutescens*): Fruit is used as a vegetable.

- Gooseberry (*Physalis peruviana*):

2. **Medicine**:
   - Deadly nightshade (*Atropa belladonna*): The root produces a sedative called belladonna, and its leaf produces an alkaloid, atropine.
   - Black nightshade (*Solanum nigrum*): Fruit is given in fevers, diarrhoea and hydrophobia.
   - Jimson weed/Datura (*Datura stramonium*): It is a sedative.
   - Ashwagandha (*Withania somnifera*): It is an aphrodisiac, but root is used in rheumatism.
   - Henbane (*Hyoscyamus niger*): It produces alkaloids such as hyoscyamine and scopolamine.
3. **Narcotics**:
   - Tobacco (*Nicotiana tabacum*): Leaf produces nicotine and anabasin.
4. **Ornamental plants**:
   - Night jasmine/Queen of night (*Cestrum nocturnum*),
   - Henbane (*Hyoscyamus niger*)
   - Yesterday-today-tomorrow (*Brunfelsia hopeana*),
   - *Petunia.*

ASTERACEAE

It is also called Compositae or sunflower family.

Classification: (According to Bentham and Hooker's system)

Kingdom	Plantae
Division	Phanerogamia
Class	Dicotyledonae
Subclass	Gamopetalae
Series	Inferae
Order	Asterales
Family	Asteraceae

Vegetative characters:

- Habit: Herbs or shrubs
- Root: Taproot
- Stem: Herbaceous or woody, Erect or prostrate
- Leaves: Simple, Alternate, Exstipulate, Reticulate venation

Floral characters

- Inflorescence: Head (capitulum); A few or a large number of florets.

Disc florets:

- Flowers: Bracteate, Actinomorphic, Bisexual, Complete, sessile, Epigynous, Pentamerous
- Calyx: modified into pappus, persistent
- Corolla: Petals: 5, gamopetalous, Tubular
- Androecium: Stamens: 5, Epipetalous, syngenesious, Dithecous
- Gynoecium: Bicarpellary, Syncarpous, Ovary inferior, Basal placentation, Ovary unilocular, Single ovule, Stigma bifid

Floral formula: Br $\oplus$ ⚥ $K_{pappus}\ \overset{\frown}{C_{(5)}\ A_{(5)}}\ \bar{G}_{(2)}$

Ray florets:

- Flowers: Bracteate, zygomorphic, Unisexual (female) or neutral, Ligulate, Epigynous, Pedicellate, Incomplete
- Calyx: modified into pappus
- Corolla: Petals: 5, gamopetalous, Ligulate, Valvate aestivation
- Androecium: Absent
- Gynoecium: Bicarpellary, Syncarpous, Ovary inferior, Basal placentation, Ovary unilocular, Stigma bifid.

Floral formula: Br % ♀ $K_{pappus}\ C_{(5)}\ A_0\ \bar{G}_{(2)}$

Br % $K_{pappus}\ C_{(5)}\ A_0\ G_0$

- Fruit: Cypsela
- Seed: Non-endospermic

Economic importance

1. Food:
    - *Lactuca sativa:* Leaves are eaten as salad.
2. Oil:
    - Sunflower (*Helianthus annus*): seeds are source of oils.
    - *Artemisia:* Seeds are source of oils.
3. Medicine:
    - *Artemisia* is source of santonin which is used as vermifuge.
4. Weed:
    - Congress grass (*Parthenium hysterophorus*) is a serious menace.
5. Insecticide:
    - *Chrysanthemum:* Head or inflorescence is source of insecticide in powder form.
6. Ornamental plants:
    - *Chrysanthemum*
    - Dahlia
    - Marigold (*Tagetes indica*)
    - Sunflower (*Helianthus annus*)
    - Calendula
    - Cosmos
    - *Zinnia*

POACEAE

It is also called Graminae or grass family. It comes under monocots.

Classification: (According to Bentham and Hooker's system)

Kingdom	**Plantae**
Division	Phanerogamia
Class	Monocotyledonae
Series	Glumaceae
Order	Poales
Family	Poaceae

Vegetative characters:

- Habit: Mostly herbs, Annual or perennial
- Root: Fibrous roots, Adventitious roots
- Stem: Herbaceous, hollow (Fistular), Swollen nodes
- Leaf: Simple, Alternate, exstipulate, sessile, Leaf base forms tubular sheath, Ligulate, Parallel venation

Floral characters:

- Inflorescence: Spike of spikelets. Each spikelet is composed of 2-5 flowers which are enclosed by a pair of bracts called glumes.
- Flower: Bracteate (2 bracts called palea and lemma), Zygomorphic, Sessile, Bisexual, Incomplete, Hypogynous, Trimerous
- Perianth: Lodicules 2-3 (membranous scales)
- Androecium: Stamens-3 or 6, Polyandrous, Anther dithecous, Versatile.
- Gynoecium: Monocarpellary, Basal placentation, Ovary superior, Ovary unilocular, One ovule, Stigma feathery.
- Fruit: Caryopsis
- Seed: Endospermic

Floral formula: Br % ⚥ P2 $\widehat{\text{(lod)}}$ A6 G 1

Floral Diagram

Economic importance:

1. **Food**: All cereals and millets belong to this family.
   - Rice (*Oryza sativa*),
   - Wheat (*Triticum aestivum*),
   - Maize (*Zea mays*)
   - Barley (*Hordeum vulgare*)
   - Oat (*Avena sativa*)
   - Pearl millet=Jowar (*Sorghum vulgare*)
   - Finger millet=Bajra (*Pennisetum typhoides*)
2. **Fodder:**
   - Grasses (*Agrostris*)
3. **Sugar**:
   - Sugarcane (*Saccharum officinarum*)
4. **Essential oils**:
   - Lemon grass (*Cymbopogon citratus*): It produces aromatic oil.
   - Khus (*Vitivera zizanoides*): Its roots and rhizome produce oil of vetiver.
5. **Timber**:
   - Bamboo (*Bambusa*)
6. **Papers**: Some grasses

SHORT ANSWER TYPE QUESTIONS

1. Draw a well labelled diagram of different zones of root.
2. Define the respiratory root with examples.
3. How corm of underground modification of stem differ from tuber?
4. Distinguish between phylloclade and cladode.
5. Distinguish between thorn and spine.
6. Explain types of phyllotaxy with examples.
7. Distinguish between racemose inflorescence and cymose inflorescence.
8. Explain the different symmetries of flowers.
9. Discuss the position of floral parts with respect to ovary.
10. Discuss the several conditions of cohesion of stamens in flower.
11. What do you mean by aggregate fruits? Give examples.
12. Distinguish between dicot seed and monocot seed.
13. Draw a floral diagram of the Solanaceae family.
14. Mention the botanical name of three plants of Fabaceae, Solanaceae and Liliaceae families.

LONG ANSWER TYPE QUESTIONS

1. Describe the various modifications of a leaf with examples.
2. Discuss the special types of inflorescences.
3. What is aestivation? Describe its various types with diagrams.
4. What is placentation? Describe its several types with diagrams.
5. Mention the five distinguishing features of each Fabaceae, Solanaceae, and Liliaceae families with their floral formulae.

MULTIPLE CHOICE QUESTIONS

1. Roots are absent in

(a) *Wolffia* (b) *Victoria*
(c) *Nymphaea* (d) All of the above

2. A stem with a distinct solid node and hollow internode is called

(a) Sobole (b) Culm
(c) Scape (d) Coleoptile

3. Most reduced stems occur in

(a) Rhizome (b) Corm
(c) Bulb (d) Tuber

4. Prickles in roses are

(a) Exogenous in origin
(b) Endogenous in origin
(c) Modified stipule
(d) Modified leaf

5. The lateral roots arise from

(a) Pericycle cells (b) Endodermal cells
(c) Cortical cells (d) Hypodermal cells

6. Which of the following features would distinguish monocot from a dicot plant?

(a) Aestivation (b) Vernation
(c) Phyllotaxy (d) Venation

7. Fibrous roots develop in maize from

(a) Lower nodes (b) Upper nodes
(c) Upper internodes (d) Radicle

8. If a growing plant is decapitated

(a) Leaves become yellow
(b) Growth stops
(c) Axillary buds are inactivated
(d) Axillary buds are activated

9. **Tunicated bulb is found in:**
(a) Lily (b) Onion
(c) Garlic (d) *Oxalis*

10. **Maximum growth in root occurs**
(a) Just behind the root apex
(b) At the root apex
(c) In the light
(d) In the dark

11. **Dicotyledonous leaf showing parallel venation is seen in:**
(a) *Smilax* (b) *Ranunculus*
(c) *Calophyllum* (d) *Dioscorea*

12. **Velamen tissue is found in:**
(a) Aerial roots of Orchids
(b) Roots of Screw pine
(c) Leaves of *Ficus*
(d) Stem of Orchids

13. **The plumule in monocot seeds is covered by**
(a) Coleorhiza (b) Coleoptile
(c) Scutellum (d) Aleurone layer

14. **A fibrous root system is adapted better than a tap root system to perform the function of:**
(a) Absorption of water and minerals
(b) Transport of water and organic food
(c) Anchorage of plants into the soil
(d) Storage of food

15. **The presence of a sheathing leaf base and a ligule are characteristic of:**
(a) Fern leaf (b) Banana leaf
(c) *Cycas* leaf (d) Grass leaf

16. **Sessile flowers of only one sex are found in the inflorescence are called:**
(a) Spadix (b) Catkin
(c) Umbel (d) Panicle

17. **Which one of the following fruits is not type of composite or multiple fruits?**
(a) Pineapple (b) Mulberry
(c) Jack fruit (d) Custard apple

18. **Which of the following aestivation shows an arrangement of sepals in one direction only?**
(a) Valvate (b) Twisted
(c) Vexillary (d) Quincuncial

19. **Multiple root cap is present in:**
(a) *Pandanus* (b) *Eichhornia*
(c) *Pistia* (d) *Lemna*

20. **Largest and edible bud is found in:**
(a) Cabbage (b) Cauliflower
(c) Onion (d) *Agave*

21. **Imparipinnate leaf is the one where:**
(a) Leaflets are borne in pairs
(b) Leaflets are small
(c) Leaflets are large
(d) Rachis is terminated by an odd leaflet

22. **The main difference between thorns and prickles is in having:**
(a) Endogenous origin
(b) Vascular supply
(c) Bark
(d) All of the above

23. **Circular internodes are present in**
(a) Groundnut (b) Ginger
(c) *Amorphophallus* (d) Turmeric

24. **The entire leaf is modified into leaf tendrils in:**
(a) *Lathyrus odoratus* (b) *Lathyrus aphaca*
(c) *Gloriosa* (d) *Clematis*

25. **Persistent calyx is a feature of**
(a) Fabaceae (b) Solanaceae
(c) Liliaceae (d) Malvaceae

26. **Which of the following conditions belongs to the adhesion of stamen?**
(a) Syngenesius (b) Diadelphous
(c) Synandrous (d) Epipetalous

27. **Which of the following inflorescence shows false whorl**
(a) Cyathium (b) Spadix
(c) Hypanthodium (d) Verticillaster

28. **The number of stamens in the tetradynamous condition is**
(a) 2 (b) 4
(c) 6 (d) 8

29. **The number of leaves on each node of the opposite phyllotaxy is**
(a) One (b) Two
(c) Three (d) Many

30. **The root hairs are formed in**
(a) Zone of maturation
(b) Zone of elongation
(c) Zone of division
(d) Zone of cap

31. Which of the following inflorescence is highly evolved?

(a) Spikelet (b) Head
(c) Cyathium (d) Hypanthodium

32. Leaf apex is modified into the spine in

(a) *Gloriosa* (b) Pea
(c) *Smilax* (d) Date palm

33. Which of the following is not correctly matched?

(a) Stilt root-Maize
(b) Clinging root-Orchid
(c) Photosynthetic root-*Tinospora*
(d) Sucking root- *Jussiea*

34. Which of the following vegetative characteristics is highly conserved and used for classification of plants?

(a) Phyllotaxy (b) Venation of leaf
(c) Presence of bud (d) Node and internode

35. Fabaceae family belongs to order

(a) Rosales (b) Personales
(c) Passiflorales (d) Polemoniales

36. Monothecous anther is seen in:

(a) Cotton and Tulip
(b) *Calotropis* and lady's finger
(c) *Calotropis* and *Ficus*
(d) Lady's finger and cotton

37. The condition of tetradynamous stamen is

(a) Two outer and longer longer stamens
(b) Four outer and two shorter stamens
(c) Four outer and four shorter stamens
(d) Two inner and two longer stamens

38. Ventral suture condition is seen in

(a) Parietal placentation
(b) Marginal placentation
(c) Axilc placentation
(d) Basal placentation

39. Which of the following placentations is multilocular?

(a) Marginal (b) Axile
(c) Parietal (d) Free central

40. *Withania sominifera* produces an important drug that belongs to

(a) Liliaccae (b) Malvaceae
(c) Solanaceae (d) Fabaceae

41. The formula of petals in the Fabaceae family is

(a) 1+ (2) + (2) (b) (1) + (2) + 2
(c) 1 + 2 + (2) (d) 1 + (2) + 2

42. Ovary is obliquely placed in

(a) Fabaccae (b) Solanaceae
(c) Liliaceae (d) All of the above

43. Perigynous condition is seen in

(a) Rose, Plum, *Ficus*
(b) Plum, *Ficus*, *Morus*
(c) *Morus* Pineapple, Radish
(d) Plum, Peach, Rose

44. Unisexual flowers in dioecious condition are been in

(a) Mulberry, Castor, Rice
(b) Papaya, Date palm, Mulberry
(c) Papaya Date palm, Maize
(d) Maize, Mulberry, Papaya

45. The placentation which develops from an inner wall of ovary is called

(a) Axile (b) Basal
(c) Parietal (d) Free central

46. Syngenesius condition exhibits

(a) Filamens are free but anther are united
(b) Filaments are united but anthers are free
(c) Both filaments and anthers are free
(d) Filaments anthers are united

47. Parallel venation is found in

(a) Fabaceae (b) Solanaceae
(c) Poaaceae (d) Brassicaceae

48. The colour of sepaloid petal is:

(a) Green
(b) Coloured other than green
(c) White
(d) Mosaic coloured

49. The number of stamens in Fabaceae is

(a) 5 (b) 10
(c) 8 (d) 5

50. In which aestivation, all sepals are partly outer and partly inner?

(a) Valvate (b) Twisted
(c) Descending imbricate (d) Quincuncial

51. Composite or multiple fruits are seen in

(a) Mulberry (b) Pineapple
(c) Jackfruit (d) all of the above

52. Zygomorphic and diadelphous conditions are seen in

(a) Pulses (b) Cereals
(c) Cucurbits (d) Citrus plants

53. Perianth in the form of lodicules is found in

(a) Fabaceae (b) Solanaceae
(c) Poaceae (d) all of the above

54. Stilt roots are produced by

(a) Lower internodes (b) Upper internodes
(c) Lower nodes (d) Upper nodes

55. Pneumatophores growing in waterlogged soils are helpful in

(a) Transpiration (b) Guttation
(c) Food storage (d) Respiration

56. The leafless stem of an onion which is produced to bear flowers is called

(a) Rachis (b) Peduncle
(c) Scape (d) Torus

57. Pseudostem is found in

(a) Grass (b) Banana
(c) Bamboo (d) Cauliflower

58. Axillary buds arise

(a) Endogenously from pericycle
(b) Endogenously from the inner cortex
(c) Exogenously from pericycle
(d) Exogenously from the inner cortex

59. The lateral branch growing obliquely from the underground stem is called

(a) Bulbil (b) Sucker
(c) Offset (d) Stolon

60. Stem is reduced to disc in

(a) Rhizome (b) Corm
(c) Bulb (d) Tuber

61. Ginger is a stem, not a root because

(a) It stores food material
(b) It grows parallel to the surface
(c) It has nodes and internodes
(d) It does not have chlorophyll

62. Finely dissected leaves are commonly found in

(a) Submerged plants
(b) Free-floating plants
(c) Fixed floating plants
(d) Amphibious plants

63. The petiole is modified into a leafy green structure known as

(a) Phylloclade (b) Cladode
(c) Phyllode (d) Bract

64. In sweet peas, the tendrils are modified

(a) Stipule (b) Petiole
(c) Leaflet (d) Leaf apex

65. Veins of leaf are useful in

(a) Transport of water
(b) Transport of minerals
(c) Transport of organic solutes
(d) all of the above

66. A compound leaf looks like a simple leaf in

(a) *Calotropis* (b) Citrus
(c) Mango (d) Mustard

67. The time taken for the development of two adjacent leaves is called

(a) Leaf index (b) Plastochron index
(c) Foliar time index (d) Stomatal index

68. Obdiplastemonous arrangement of stamens shows

(a) Inner antipetalous and outer alternipetalous
(b) Outer antipetalous and inner alternipetalous
(c) Both inner and outer stamens are alternipetalous
(d) Both inner and outer stamens are antipetalous

69. Swollen placenta is a characteristic feature of

(a) Fabaceae (b) Solanaceae
(c) Liliaceae (d) Malvaceae

70. Eyes on potato are modified

(a) Axillary bud (b) Apical bud
(c) Internode (d) Scars

71. A monocarpic plant is that which

(a) produces only one seed per fruit
(b) produces only one fruit in its lifetime
(c) has only one carpel in its flower
(d) flowers only once in their life cycle

72. The term 'Epicalyx' represents

(a) Involucre
(b) An additional whorl of calyx
(c) A whorl of bracteoles
(d) A whorl of bracts

73. The coloured and attractive part of Bougainvillea is

(a) Bract (b) Sepal
(c) Petal (d) Leaf

74. The vexillum or standard in Pea flowers constitutes

(a) Calyx (b) Corolla
(c) Androecium (d) Gynoecium

75. Which of the following parts is considered equivalent to perianth?

(a) Lodicules (b) Glumes
(c) Inferior palea (d) Superior palea

76. For the study of parts of a flower, the most suitable flower would be of

(a) Mustard (b) Cucumber
(c) Sunflower (d) Champak

77. When the inflorescence axis is branched, it is called

(a) Head (b) Panicle
(c) Umbel (d) Cyme

78. Three types of flowers are found in

(a) Cyathium (b) Hypanthodium
(c) Verticillaster (d) Capitulum

79. The most highly evolved type of inflorescence is

(a) Hypanthodium (b) Cyathium
(c) Capitulum (d) Spadix

80. Which of the following plant sets is monocarpic?

(a) *Bambusa* and *Physalis*
(b) *Agave* and *Pinus*
(c) *Cycas* and *Pinus*
(d) *Bambusa* and *Agave*

81. Nectiferous glands are found in

(a) Cyathium (b) Hypanthodium
(c) Verticillaster (d) Head

82. Spike of spikelets type of inflorescence is found in

(a) Fabaceae (b) Solanaceae
(c) Poaceae (d) Brassicaceae

83. Potato is included in the Solanaceae family because

(a) It has a gamopetalous corolla
(b) It has epipetalous stamen
(c) It has pentamerous flowers
(d) It has obliquely placed ovary

84. Tegmen develops from

(a) Inner integument (b) Outer integument
(c) Nucellus (d) Micropyle

85. Caruncle is a specialised growth develops from

(a) Inner integument (b) Outer integument
(c) Nucellus (d) Micropyle

86. Replum is a false septum found in

(a) Lily (b) Pea
(c) Mustard (d) All of the above

87. Geocarpic fruit is found in

(a) Mustard (b) Potato
(c) Sweet potato (d) Groundnut

88. The Aleurone layer of maize grain is especially rich in

(a) Carbohydrate (b) Fats
(c) Oil (d) Proteins

89. Hypogeal germination occurs in

(a) Cereals (b) Beans
(c) Castor (d) Cucurbits

90. The plants such as *Lupin* and *Tulipa* belong to the families, respectively

(a) Solanaceae and Liliaceae
(b) Liliaceae and Fabaceae
(c) Fabaceae and Liliaceae
(d) Liliaceae and Solanaceae

91. Fibrous root system is found in

(a) Monocots (b) Dicots
(c) Gymnosperms (d) Bryophytes

92. The character of the flower which is represented by a floral formula but not by floral diagram is

(a) Aestivation (b) Position of gynoecium
(c) Placentation (d) Adhesion of stamens

93. Gall flowers are characteristic of inflorescence

(a) Capitulum (b) Hypanthodium
(c) Cyathium (d) Verticillaster

94. Colchicine is an alkaloid produced from *Colchicum autumnale* which belongs to

(a) Fabaceae (b) Solanaceae
(c) Liliaceae (d) Malvaceae

95. The censor mechanism works for the dispersal of seeds in

(a) Poppy (b) Bean
(c) Cucurbits (d) Berry

96. Vivipary germination of the seed is seen in

(a) *Lilium* (b) *Artocarpus*
(c) *Rhizophora* (d) *Eugenia*

97. Perisperm is a remnant of nucellus seen in

(a) Beet (b) Black pepper
(c) *Nymphaea* (d) All of the above

98. Which of the following is incorrect about monocot seed?

(a) Seed coat is fused with pericarp
(b) Scutellum is seen
(c) Aleurone layer is found
(d) Radicle is not covered with coleorhiza

99. The plants such as *Lupinus, Physalis, Petunia* and *Tulipa* belong to

(a) One family (b) Two families
(c) Three families (d) Four families

100.Which inflorescence shows single female flower which is surrounded by several male flowers?

(a) Cyathium (b) Hypanthodium
(c) Verticillaster (d) Head

101.Parietal placentation is ca haracteristic feature of

(a) Brassicaceae (b) Malvaceae
(c) Poaceae (d) Fabaceae

102. Epicalyx is found in

(a) Brassicaceae (b) Malvaceae
(c) Poaceae (d) Fabaceae

103. Calyx modified into pappus is seen in

(a) Brassicaceae (b) Malvaceae
(c) Poaceae (d) Asteraceae

104. Monadelphous condition in Malvaceae means

(a) Only filaments are united in one bundle
(b) Only anther is united in one bundle
(c) Both anther and filaments are united in one bundle
(d) None of the above

105. Feathery stigma is seen in

(a) Brassicaceae (b) Malvaceae
(c) Poaceae (d) Fabaceae

ANSWERS

1.a	16.b	31.b	46.a	61.c	76.c	91.a
2.b	17.d	32.d	47.c	62.a	77.b	92.b
3.c	18.b	33.d	48.a	63.c	78.b	93.b
4.d	19.a	34.b	49.b	64.c	79.c	94.c
5.a	20.a	35.a	50.b	65.d	80.d	95.a
6.d	21.d	36.d	51.d	66.b	81.c	96.c
7.a	22.d	37.b	52.a	67.b	82.c	97.d
8.d	23.d	38.b	53.c	68.b	83.d	98.d
9.b	24.c	39.b	54.c	69.b	84.a	99.c
10.a	25.b	40.c	55.d	70.a	85.b	100.a
11.d	26.d	41.c	56.c	71.d	86.c	101.a
12.a	27.d	42.b	57.b	72.d	87.d	102.b
13.a	28.c	43.d	58.d	73.a	88.d	103.d
14.c	29.b	44.b	59.b	74.b	89.a	104.a
15.d	30.a	45.c	60.c	75.a	90.c	105.c

CHAPTER-5

ANATOMY OF FLOWERING PLANTS

Anatomy deals with the internal organisation of plants. The plants are made up of cells and these are organised into tissues. These tissues in turn develop into organs that vary in their internal structure.

Tissue: It is a group of cells having the same origin and functions. The term tissue was coined by N. Grew.

The Father of plant anatomy is Nehemiah Grew.

The Father of Indian anatomy is K.A. Chowdhury

Plant tissue is divided into two types: **Meristematic tissue and Permanent tissue**.

1. **Meristematic tissue (Meristem):** The term 'Meristem' was coined by Nageli (1859). Meristem is a group of young, immature and undifferentiated cells that can divide. It is a localised and growing region of the plant body and contributes to the growth of plants.
2. **Permanent Tissues:** It is tissue in which cells (living or dead) have lost their capacity to divide. It develops from meristematic tissue. It is of two types on the basis of composition.
   (a) Simple permanent tissue: It is structurally similar and performs the same function. e.g., Parenchyma, Collenchyma and Sclerenchyma
   (b) Complex Permanent tissue: It is structurally dissimilar but performs the same function. e.g., Xylem, Phloem.

Meristematic tissues	Permanent tissues
These are actively and continuously dividing.	These do not divide and carry out specialised function.
Cells are undifferentiated.	Cells are differentiated.
Cells are smaller and isodiametric.	Cells are variable in shape.
Cell wall is thin.	Cell wall is thick.
Nuclei are large but vacuoles and intercellular spaces are absent.	Nuclei are small but vacuoles and intercellular spaces are present.
Metabolic rate is very high.	Metabolic rate is very low.
It does not attain a definite form or size.	It attains a definite size or form.

Tissue System: It is the group of different plant tissues irrespective of their position in the plant body performing a common function. Sachs (1875) divided tissue system into three systems:

1. **Epidermal Tissue system:** It develops from the outermost layer of the apical meristem. It consists of the following parts:
   (a) **Epidermis**: It is outermost and single-layered but multiple or multiseriate epidermis is seen in the velamen of aerial root of Orchids and leaf of *Ficus, Nerium* and *Peperomia* (on upperside) and Maize (on the upper side and lower side). It is living, thin-walled, compactly arranged and parenchymatous. Its cells have a small cytoplasm and a large vacuole. It is not uniform in structure and function due to the presence of many types of hairs and guard cells of stomata. It bears unicellular hairs in the roots and multicellular hairs in stem. It secretes cutin, a waxy substance that forms a cuticle. Cuticle is absent in roots and hydrophytes but thick in xerophytes. It contains cystolith ($CaCO_3$) hence called lithocyst. It protects plants from desiccation (drying) and infection.

(b) **Stomata**: The epidermis bears a small opening called stomata. It is absent in the root. It is present in herbaceous stems and leaves. Its number varies from 14-1038 per square mm. but the average number is around 200. It is kidney or bean-shaped in dicots and dumbbell-shaped in monocots. It has a pore called stoma bounded by two guard cells and surrounded by several subsidiary or accessory cells collectively called stomatal apparatus. The inner wall of the guard cell is thicker and non-elastic. Guard cell unlike subsidiary cells bears small chloroplast but performs no photosynthesis due to the absence of RuBisCO in spite of the presence of pigment systems. Guard cells are heterotrophic and import sugar from adjacent cells. Guard cell has large vacuoles and nuclei and is not linked by plasmodesmata to subsidiary cells. Stomata performs transpiration and exchange of gases (O_2 and CO_2).

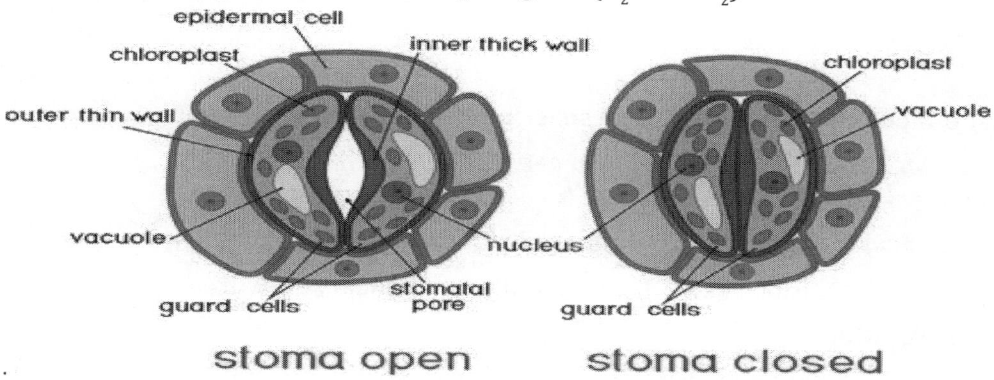

Fig: Internal structure of Stomata

Distribution of stomata:

(i) Astomatic type (Potamogeton type): Stomata are absent. e.g., *Potamogeton, Vallisneria* (submerged hydrophytes), Algae

(ii) Amphistomatic (Potato type): Stomata are present on both surfaces but more on the lower side as in mesophytes; but more on the upper side. e.g., Sciophytes.

(iii) Amphistomatic (Oat type): Stomata are present equally on both surfaces of leaf. e.g., Monocots

(iv) Epistomatic (Water lily type): Stomata are present on the upper side only. e.g., Water lily, Lotus, *Nymphaea*.

(v) Hypostomatic (Apple type): Stomata are present on the lower side only. e.g., Apple, Banyan, Oak, Peach, Walnut.

(vi) Sunken stomata: Stomata is present deep inside. e.g., xerophytes.

(c) **Appendages**: The epidermis contains the following two types of appendages:

(i) Hairs: These are unicellular in roots and multicellular in stem.

(ii) Trichomes: These are multicellular structures on the stem. These check transpiration. These are also protective and secretive.

2. **Ground Tissue system:** It is developed from ground meristem or partly from periblem and partly from plerome.

(a) **Hypodermis**: It is 2-3 layered. It is absent in roots. It is collenchymatous in the dicot stem and sclerenchymatous in the monocot stem. It is protective and photosynthetic.

(b) **Cortex:** It is several layered, thin-walled and parenchymatous. It usually contains intercellular spaces. It contains several types of cells such as resin duct, oil gland, laticifers. It stores food materials.

(c) **Endodermis:** It is single-layered, elongated or barrel-shaped and compactly arranged. It contains a thickened band called **Casparian strips** on the radial and tangential walls of roots. It is formed by a waxy substance, suberin which makes water impermeable. The Casparian strips force water and minerals through the membrane of the endodermis to protect from plasmolysis; prevent water entry into the xylem through apoplast; and regulate water and minerals in plants. It stores starch hence called starch sheath. It controls the movement of water between the cortex and xylem and hence acts as a water jacket between vascular and non-vascular regions. It regulates root pressure.

(d) **Pericycle:** It is the outermost part of the stele. It is usually one layered (but multilayered pericycle is seen in root of *Agave* and *Smilax*), thin-walled and parenchymatous. It gives rise to lateral root in monocot root but gives rise to lateral root as well as vascular cambium in dicot root. It provides mechanical support and acts as a storage organ.

(e) **Pith:** It is thin-walled, parenchymatous with or without intercellular space and is present at the centre. It stores food materials as well as excretory materials (tannin and phenol).

3. **Vascular tissue system:** It develops from the procambium. It includes xylem and phloem. These are usually associated with forming vascular bundles. Cambium is a strip of primary meristem between the xylem and phloem.

Types of cambium:

(a) Vascular cambium: It is formed inside a stele. It is of two types:

(i) **Fascicular cambium:** It is developed between the xylem and phloem of the vascular bundle during secondary growth.

(ii) **Interfascicular cambium:** It is developed between two vascular bundles during secondary growth.

(b) Cork Cambium: It is formed inside the cortex during secondary growth.

Types of vascular bundle:

(a) **Open:** Cambium is present between the xylem and phloem. e.g., Dicots

(b) **Closed:** Cambium is absent between xylem and phloem. e.g., Monocots.

Conditions of Protoxylem:

(a) **Exarch:** Protoxylem is present at the periphery and metaxylem at the centre. The primary xylem is developed from periphery to centre (centripetal development). e.g., Roots

(b) **Endarch:** The Protoxylem is at the centre and the metaxylem is at the periphery. The primary xylem is developed from the centre to the periphery (centrifugal). e.g., Stem

(c) **Mesarch:** The Protoxylem is at the centre and is surrounded by the metaxylem. e.g., Rhizome of Ferns, Roots of *Ophioglossum*.

Types of vascular bundle: It is three types on the basis of the relative position of the xylem and phloem.

(a) **Radial**: Xylem and phloem are present in separate bundles which are arranged alternately on different radii. e.g., roots.

(b) **Conjoint**: Both the xylem and phloem are present in the same bundles which are situated on the same radii. It may be of two types:

(i) **Collateral**: The xylem forms the inner part and the phloem forms the outer part. e.g., stem.

(ii) **Bicollateral**: The Xylem is surrounded by two layers of phloem from both inner and outer sides hence xylem is sandwiched.

e.g., Stem of cucurbits and *Solanum*.

(c) **Concentric**: One of two vascular tissues (xylem or phloem) occupies at centre and is surrounded by another (xylem or phloem). In this cambium is absent. It may be of two types:

(i) **Amphivasal** (Leptocentric): Xylem surrounds phloem from the peripheral side.

e.g., Stem of *Dracaena* and *Yucca*.

(ii) **Amphicribal** (Hadrocentric): Phloem surrounds xylem from peripheral side.

e.g., Stem of Pteridophytes.

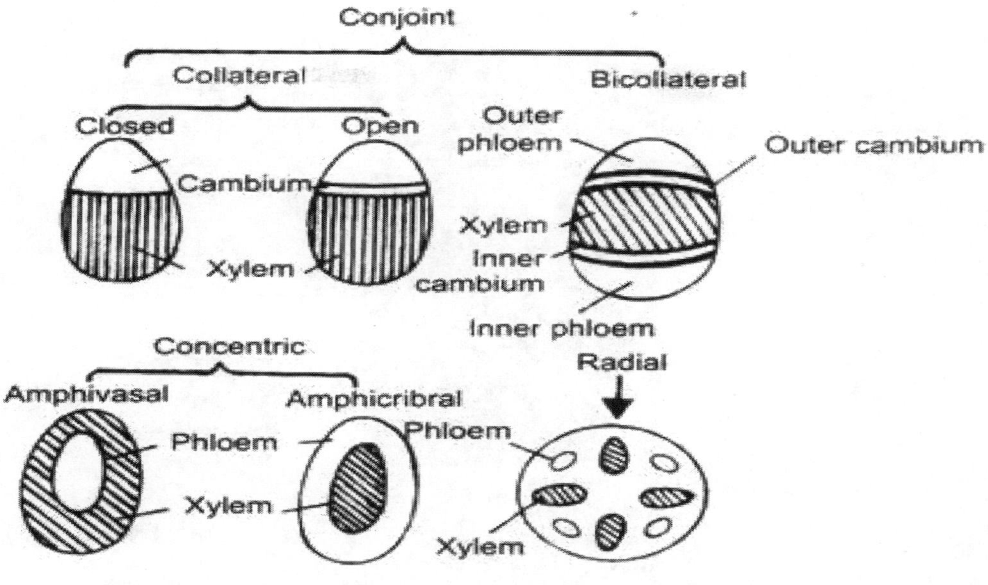

Different types of vascular bundles

ANATOMY OF ROOT

Dicot root: The internal structure on the transverse section of the root shows the following features:

(a) **Epidermis**: It is single-layered, compactly arranged and thin-walled. It is not cutinised (without a cuticle). It does not contain stomata. It bears unicellular hairs that absorb water and minerals hence epidermis is also called epiblema or rhizodermis or piliferous layer.

(b) **Cortex**: It consists of several layered, larger thin-walled parenchymatous cells with intercellular spaces. In some dicots, one to several layers of cortex become suberised below the epidermis to form exodermis which is a protective tissue. It does not contain chloroplast. It stores food materials.

(c) **Endodermis**: It is single-layered, closely packed and situated below the cortex. It contains thickened bands called **Casparian strips** in radial and transverse walls. These strips are made up of suberin. These strips prevent plasmolysis in the endodermis; permit water to pass through the symplast of the endodermis; prevent leakage of nutrients from vascular tissues; and regulate the flow of water with minerals.

(d) **Pericycle**: It is single-layered, thick-walled and parenchymatous. It gives rise to lateral root and vascular cambium during secondary growth.

(e) **Vascular bundles**: These consist of separate alternating strands of xylem and phloem (radial). These strands are usually tetrarch. These are closed due to the absence of cambium. These show exarch condition (protoxylem is situated towards the periphery and metaxylem towards the centre). Conjunctive parenchymatous cells are present between the xylem and phloem.

(f) **Pith**: It is less developed.

Monocot root:

(a) **Epidermis:** It is single layered, compactly arranged, thin-walled. It is not cutinised.

(b) **Cortex:** In many monocots, exodermis occurs.

(c) **Endodermis:** It is single-layered, closely packed and situated below the cortex. It contains thickened bands called **Casparian strips** in radial and transverse walls. These strips are made up of suberin. A few cells are thin-walled and without Casparian strips called passage cells situated opposite to the protoxylem. These passage cells help in the radial diffusion of water from the cortex into the xylem.

(d) **Pericycle:** It is single-layered, thick-walled and parenchymatous. It gives rise to lateral roots.

(e) **Vascular bundles:** These are polyarch (containing more than 6 strands of xylem and phloem).

(f) **Pith:** It is well developed and occupied at the centre. It is parenchymatous.

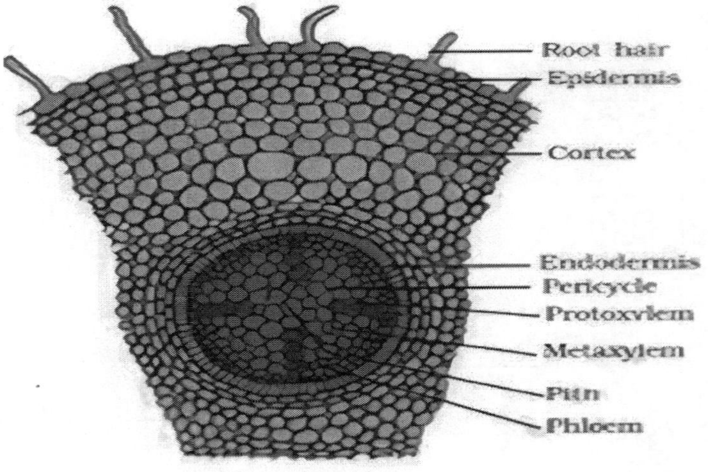

Dicot Root

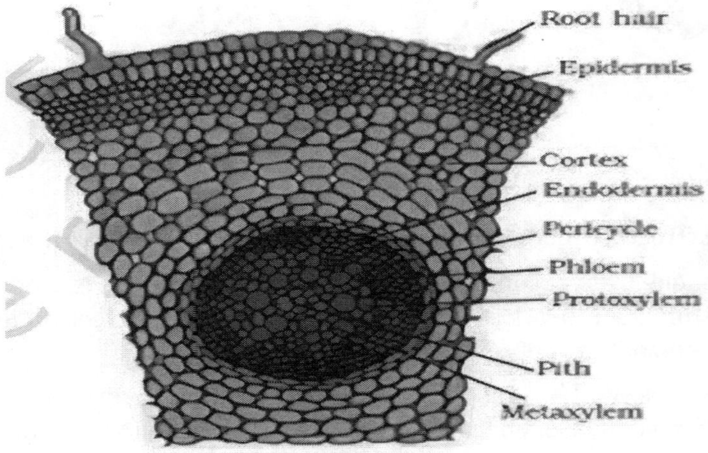

Monocot Root

Dicot root	Monocot root
Cortex is narrow.	Cortex is wide.
Passage cells are absent in endodermis.	Passage cells are present in endodermis.
Pericycle gives rise to lateral roots and vascular cambium.	Pericycle gives rise to lateral roots only.
Vascular bundles are usually tetrarch.	Vascular bundles are polyarch (≥6).
Pith is less developed.	Pith is well developed.
Cambium is present.	Cambium is absent
Secondary growth occurs.	Secondary growth does not occur.

ANATOMY OF STEM

Dicot stems: The internal structure on the transverse section of the stem shows the following common features:

(a) **Epidermis**: It is outermost, compactly arranged and single-layered. It is heavily cutinised (thick cuticle) to prevent loss of water and protects inner tissues from pathogens. It bears multicellular hairs having a few stomata.

(b) **Hypodermis**: It is present below the epidermis. It is multilayered and collenchymatous. It provides mechanical support to the young stem and also performs photosynthesis if chloroplasts are present.

(c) **Cortex:** It is made up of several layered, rounded and parenchymatous cells having intercellular spaces. It bears aerenchyma in aquatic plants. It bears resin duct, latex tubes and a reservoir of waste products.

(d) **Endodermis**: It is single-layered, compactly arranged situated just below the cortex. It separates the vascular tissues from the cortex.

(e) **Pericycle:** It is single-layered and the outermost part of the stele (central vascular cylinder).

(f) **Vascular bundles:** These are several in number. Each vascular bundle consists of xylem and phloem. The vascular bundles are ringed and conjoint collateral and open (cambium is present). The endarch condition is seen. In some plants such as cucurbits, bicollateral vascular bundles are seen.

(g) **Pith:** It is the central and innermost part which is made up of parenchymatous cells having large inter-cellular spaces. It shows the presence of medullary rays (narrow and vertically elongated bands of parenchymatous cells) extending from the pith to the periphery. These rays help in the radial conduction of food. The pith stores food materials.

Monocot stem:

(a) **Epidermis:** it is outermost, compactly arranged and single-layered. It is heavily cutinised (thick cuticle). It contains stomata.

(b) **Hypodermis:** It is present below the epidermis. It is multilayered, sclerenchymatous and lignified. It provides mechanical strength to the stem.

(c) **Ground tissue**: It constitutes the entire tissue below the hypodermis. It is made up of parenchymatous cells rich in food reserves. These cells are larger and more loosely arranged on the inner side towards the centre compared to the peripheral region. It is not differentiated into cortex, endodermis, pericycle and pith.

(d) **Vascular bundles**: These are scattered, conjoint collateral and closed (no cambium) in ground tissues. Each vascular bundle is surrounded by a 2-3 layered sclerenchymatous bundle sheath.

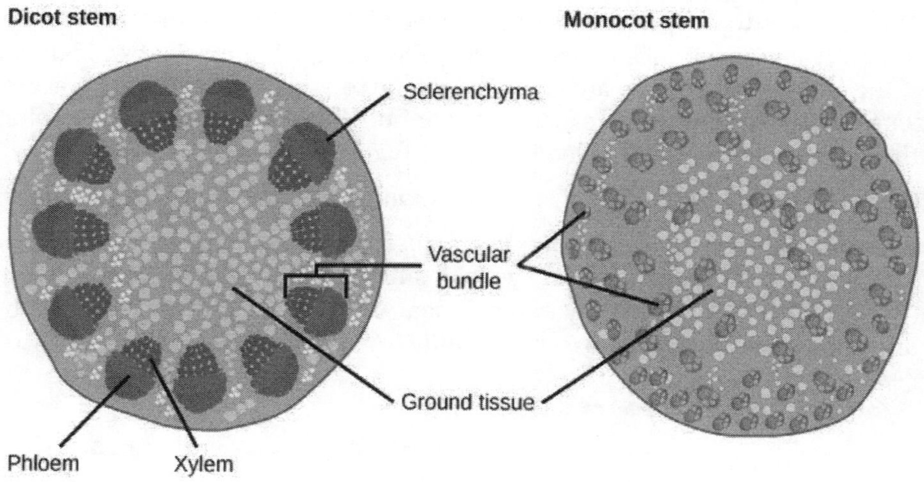

Fig: Dicot stem and Monocot stem

Dicot stem	Monocot Stem
Multicellular hairs are present.	Multicellular hairs are absent.
Stomata is bean or kidney-shaped.	Stomata is dumbbell-shaped.
Hypodermis is collenchymatous.	Hypodermis is sclerenchymatous.
Ground tissue is differentiated into hypodermis, cortex, endodermis, pericycle, and pith hence these are present.	Ground tissue is not differentiated into hypodermis, cortex, endodermis, pericycle, and pith hence these are absent.
Bundle sheath is absent.	Bundle sheath is present.
Vascular bundles are wedge-shaped, of the same size, ringed and open.	Vascular bundles are oval, of different sizes, scattered and closed.
Phloem parenchyma is present.	Phloem parenchyma is absent.
Medullary rays are present between vascular bundles.	Medullary rays are absent.
Schizo-lysigenous cavity is absent.	Schizo-lysigenous cavity is present in protoxylem.

Dicot stem	Monocot Stem
Cambium is present.	Cambium is absent.
Secondary growth occurs.	Secondary growth does not occur.

ANATOMY OF LEAF

Dicot (Dorsiventral) leaf: The internal structure on vertical section of leaf shows the following features:

(a) **Epidermis**: It consists of the upper (adaxial) and lower (abaxial) epidermis. It is compactly arranged and heavily cutinised (thick cuticle) to check the loss of water. it contains a variable number of stomata on both surfaces of the leaf. It protects the internal tissues from damage and infection.

(b) **Mesophyll**: It is present between two layers of epidermis. It is differentiated into two distinct layers-(i) the palisade cells which are upper, vertically elongated, parallel and closely packed with numerous chloroplast, and (ii) spongy cells which are inner (below palisade cells), rounded and loosely arranged with fewer chloroplast extending into the lower epidermis. The palisade cells perform photosynthesis whereas spongy cells allow gaseous exchange between the atmosphere and leaf.

(c) **Vascular bundles**: These are present in the midrib and veins. If a vein is thicker then, the size of the vascular bundle will be larger. These are conjoint, collateral and closed (no cambium). The xylem is situated towards the adaxial (upper) side and the phloem towards the abaxial (lower) side. Each vascular bundle is surrounded by thick-walled bundle sheath cells which are distinct from mesophyll cells.

Monocot (Isobilateral leaf): The internal structure on a vertical section of the leaf shows the following features:

(a) **Epidermis:** It consists of the upper (adaxial) and lower (abaxial) epidermis. It is single-layered, compactly arranged and heavily cutinised thick cuticle). It contains an equal number of stomata on both epidermis. In some grasses, the epidermal cells are arranged in long furrows, larger in size, colourless, empty, thin-walled and highly vacuolated and highly turgid called motor or **bulliform** cells which help in rolling of leaves upwards to minimise the exposed surface in dry weather (scarcity of water).

(b) **Mesophyll:** It is not differentiated into palisade and spongy cells. It is compactly arranged.

(c) **Vascular bundles**: These are conjoint, collateral and closed (no cambium). Each vascular bundle is surrounded by a chlorenchymatous bundle sheath. The xylem is situated towards the upper epidermis and the phloem towards the lower epidermis. Sclerenchyma is present on both sides of the vascular bundle. The vascular bundles are uniform in size in all veins except the midrib.

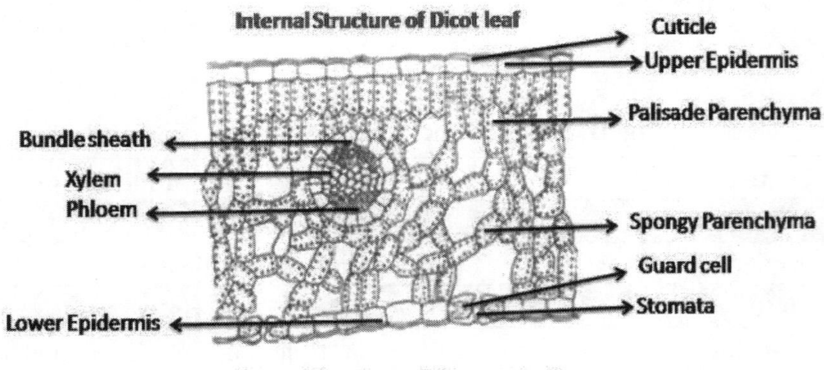

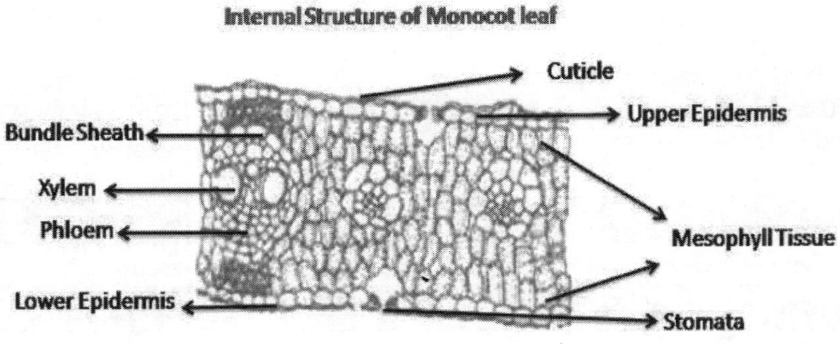

Dicot (Dorsiventral) leaf	Monocot (Isobilateral) leaf
Both surfaces of the leaf are distinct. (upper/ventral/ adaxial) and Lower/dorsal/ abaxial)	Both surfaces of the leaf are uniform.
The cuticle is thicker at the upper surface.	The cuticle is uniform in structure on both surface
Stomata are bean or kidney-shaped and more on lower side.	Stomata are dumbbell-shaped and equal in number on both sides.
Bulliform cells or Motor cells are absent.	Bulliform cells are present on the upper epidermis.
Hypodermis is collenchymatous.	Hypodermis is sclerenchymatous.
Mesophyll cells are differentiated into spongy cells and palisade cells.	Mesophyll cells are not differentiated into spongy cells and palisade cells. All are spongy cells.
Bundle sheath is collenchymatous.	Bundle sheath is sclerenchymatous.
Vascular bundles are dissimilar in size and arranged irregularly due to reticulate venation.	Vascular bundles are similar in size and are arranged in a row due to parallel venation.

SHORT ANSWER TYPE QUESTIONS:

1. Draw a well-labelled diagram of stomata.
2. Explain bulliform cells in the leaf.
3. Explain vascular cambium.
4. Define conjoint vascular bundles.
5. Write notes on stomata.

LONG ANSWER TYPE QUESTIONS:

1. Explain the distribution of stomata
2. Distinguish between the anatomy of dicot root and monocot root
3. Distinguish between the anatomy of the dicot stem and monocot stem
4. Distinguish between the anatomy of the dicot leaf and monocot leaf
5. Draw a well-labelled diagram of the anatomy of the dicot stem

MULTIPLE CHOICE QUESTIONS

1. Amphivasal vascular bundles are found in

(a) Maize and wheat
(b) *Dracaena* and *Yucca*
(c) *Cycas* and *Dryopteris*
(d) *Cucurbita* and *Helianthus*

2. Well-developed pith is found in

(a) Monocot stem and dicot root
(b) Monocot root and dicot stem
(c) Dicot root and dicot stem
(d) Monocot root and monocot stem

3. Endodermis of the dicot stem is also called

(a) Starch sheath
(b) Mesophyll
(c) Pith
(d) Bundle sheath

4. Phloem parenchyma is absent in

(a) Dicot leaf
(b) Monocot stem
(c) Dicot stem
(d) Dicot root

5. Hypodermis in a monocotyledonous stem is

(a) Parenchymatous
(b) Collenchymatous
(c) Sclerenchymatous
(d) Chlorenchymatous

6. Protoxylem in dorsiventral leaf is directed towards

(a) Adaxial side
(b) Abaxial side
(c) Surrounded by phloem
(d) Surrounds phloem

7. The correct situation of mesophyll in isobilateral grass leaf is shown by

(a) Palisade towards adaxial surface
(b) Palisade towards abaxial surface
(c) Undifferentiated mesophyll
(d) Palisade along both the surface

8. Wilting of plants takes place if

(a) Few roots are damaged
(b) Xylem is blocked
(c) Phloem is blocked
(d) Light is cut to 50%

9. Which of the following would most likely show totipotency in plants?

(a) Xylem vessels (b) Meristem
(c) Cork (d) Sieve tube

10. Passage cells occur in

(a) Epidermis (b) Cortex
(c) Pericycle (d) Endodermis

11. In a dorsiventral leaf, the protoxylem and metaxylem are located, respectively on

(a) Abaxial and adaxial sides
(b) Adaxial and abaxial sides
(c) Adaxial and adaxial sides
(d) Abaxial and abaxial sides

12. The inner wall of the stomata is

(a) Thinner and elastic
(b) thicker and elastic
(c) Thinner and inelastic
(d) Thicker and inelastic

13. Biocollateral vascular bundles have tissue arrangement

(a) Outer phloem – outer cambium – middle xylem – inner cambium – inner phloem
(b) Outer xylem – outer cambium – middle phloem – inner cambium – inner xylem
(c) Outer phloem – outer xylem – middle cambium
(d) Outer cambium – outer phloem – middle xylem – inner phloem – inner cambium

14. Casparian strips contain deposition of

(a) Cutin (b) Pectin
(c) Suberin (d) Lignin

15. Radial vascular bundles occur in

(a) Dicot root (b) Dicot stem
(c) Monocot stem (d) Monocot leaf

16. Trichomes take part in

(a) Transpiration and exchange of gases
(b) Exudation of water drops
(c) Protection and reduction
(d) Desiccation

17. In barley stem, vascular bundles are

(a) Closed and radial
(b) Open and scattered
(c) Closed and scattered
(d) Open and in a ring

18. The cork cambium, cork and secondary cortex are collectively called

(a) Phelloderm (b) Phellogen
(c) Periderm (d) Phellem

19. Which of the following is non-nucleated at maturity?

(a) Palisade cell (b) Cortical cell
(c) Periderm (d) Companion cell

20. Passage cells are found in

(a) Dicot stem (b) Monocot root
(c) Aerial root (d) Monocot stem

21. The maximum harm to a tree occurs due to

(a) Loss of half of its leaves
(b) Loss of half of its branches
(c) Loss of all of its leaves
(d) Loss of all of its bark

22. Palisade parenchyma is absent in leaves of

(a) Soyabean (b) Grain
(c) Sorghum (d) Mustard

23. The dicot root is distinguished from dicot stem by

(a) Absence of passage cells
(b) Position of protoxylem
(c) Absence of secondary xylem
(d) Presence of medullary rays

24. The girdling experiment cannot be performed in monocot plants because

(a) Vascular bundles are scattered
(b) Vascular bundles are closed

(c) Condition of xylem is exarch
(d) Protoxylem is situated outside

25. Stomata are present only on the upper side of a leaf in

(a) *Potamogeton* (b) Apple
(c) Water lily (d) *Nerium*

26. Interfascicular cambium and cork cambium are formed as a result of

(a) Differentiation (b) Dedifferentiation
(c) Undifferentiation (d) Redifferentiation

27. Which of the following thickenings is considered as advanced?

(a) Annular thickening
(b) Scalariform thickening
(c) Spiral thickening
(d) Pitted thickening

28. Which of the following cells is present in gymnosperms instead of companion cells?

(a) Complimentary cell
(b) Sister cell
(c) Albuminous cell
(d) Non-albuminous cell

29. Multiseriate epidermis is present in

(a) *Ficus* (b) *Nerium*
(c) Maize (d) All of the above

30. During the night, the stomatal apparatus contains

(a) Guard cell only
(b) Guard cell and subsidiary cell
(c) Guard cell, stoma and subsidiary cell
(d) Stoma and subsidiary cell

31. Sunken stomata is found in

(a) Hydrophytes (b) Mesophytes
(c) Halophytes (d) Xerophytes

32. Multilayered pericycle is present in

(a) Root of *Ficus* (b) Root of onion
(c) Root of *Smilax* (d) Root of orchid

33. Mesarch condition of protoxylem is found in

(a) Root of Cycas (b) Rhizome of ferns
(c) Root of orchid (d) Stem of cucurbits

34. Pith is absent in

(a) Monocot root (b) Monocot stem
(c) Dicot root (d) Dicot stem

35. The vascular bundles are wedge-shaped and ringed in

(a) Monocot root (b) Monocot stem
(c) Dicot root (d) Dicot stem

36. Which of the following is incorrect regarding the palisade cells of the dicot leaf?

(a) These are elongated
(b) These are compactly arranged
(c) These are more photosynthetic
(d) These contain less chloroplast

37. Silica is abundant in

(a) Grasses and cereals
(b) Grasses and horsetails
(c) Horse tails and ferns
(d) Mosses and ferns

38. The bundle sheath cells in Krantz anatomy are characterised by

(a) Thick wall, no intercellular space and large number of chloroplasts
(b) Thin wall, no intercellular space and large number of chloroplasts
(c) Thick wall, having intercellular space and a small number of chloroplasts
(d) Thick wall, having intercellular space and a large number of chloroplasts

39. Bicollateral vascular bundle is seen in

(a) Stem of tomato
(b) Root of orchid
(c) Root of *Ficus*
(d) Stem of cucurbits

40. Cambium is absent in

(a) Radial vascular bundle
(b) Conjoint vascular bundle
(c) Concentric vascular bundle
(d) All of the above

41. Stomata in epidermis consists of

(a) One guard cell and two subsidiary cells
(b) Two guard cells and two subsidiary cells
(c) Two guard cells and several subsidiary cells
(d) Several guard cells and several subsidiary cells

42. Which of the following features about trichomes is correct?

(a) These check transpiration
(b) These are protective in nature
(c) These are secretive in nature
(d) All of the above

43. In monocot leaves, stomata are

(a) Epistomatic (b) Hypostomatic
(c) Amphistomatic (d) Astomatic

44. Closed vascular bundles are found in

(a) Dicot root (b) Dicot stem
(c) Dicot leaf (d) Both a and b

45. Bulliform cells or motor cells are found in

(a) Upper epidermis of dicot leaf
(b) Upper epidermis of monocot leaf
(c) Mesophyll part of the leaf
(d) All of the above

ANSWERS

1.b	11.b	21.d	31.d	41.c
2.b	12.d	22.c	32.c	42.d
3.a	13.a	23.b	33.b	43.c
4.b	14.c	24.a	34.b	44.c
5.c	15.a	25.c	35.d	45.b
6.a	16.c	26.b	36.d	
7.c	17.c	27.d	37.b	
8.b	18.c	28.c	38.a	
9.b	19.c	29.d	39.d	
10.d	20.b	30.b	40.c	

CHAPTER-6

CELL: THE UNIT OF LIFE

Cell biology (formerly called cytology) is the study of the structure and function of cell and cell organelles.

Father of Cytology	Robert Hooke
Father of Modern Cytology	Swanson

DISCOVERY

Robert Hooke discovered the cell in 1665 when he examined the thin slices of bottle cork (outer dead layer of bark) of the Oak plant (*Quercus suber*) under his crude microscope. He noticed the presence of several small hexagonal boxes which together appeared as honeycomb. Externally he observed only the cell wall that surrounds the protoplast. He coined the term 'Cell' (meaning: a small room) in his book '*Micrographia*'. He considered the cell a hollow space (the empty cell wall of plant tissue). Cell is the basic structural, functional and biological unit of all known living organisms. A cell is the smallest unit of life that can replicate independently hence called the building block of life. Now it is not regarded as a hollow space but it is a separable unit having living contents called protoplast enclosed by a cell wall.

Dutrochet (1824) defined the cell is the basis of life.

Corti studied the living content of plant cells.

Rudolf Virchow (1855) proposed the 'Cell lineage theory'. It states that cells arise from pre-existing cells ('*Omnis cellula e cellula*'). His proposal established that cell division occurs in the continuity of life.

Cell theory: Matthias Schleiden (German Botanist) and Theodor Schwann (German Zoologist) proposed the 'Cell theory' in 1839 and it was later modified by Rudolf Virchow. The general unifying principles of cell theory include:

1. All living organisms are composed of one or more cells and cell products.
2. All cells arise from pre-existing cells by divisions (not derived from spontaneous generation).
3. All cells are fundamentally similar in chemical composition and metabolic activities.
4. The function of organisms as a whole is the result of activities and interaction of cells.
5. Energy flow occurs within cells.

EXCEPTIONS TO CELL THEORY

1. Viruses, viroid and prion: These do not have cellular organisation.
2. Prokaryotes: These do not have nuclei and membrane-bound cell organelles.
3. Mammalian RBCs and sieve tube: These lack nuclei.
4. *Rhizopus* and *Vaucheria*: These are multinucleated and without cell differentiation.

This theory is considered as significant because

(a) It shows that cells of different organisms are similar in structure.
(b) It explains that life exists in the form of cells.
(c) It explains that all cells are similar in metabolic activities.
(d) It confirms that life passes from one generation to another in the form of cells.
(e) All living organisms arise from a primitive cell.
(f) It shows that all living organisms arise from a primitive cell.

ORGANISATION OF CELL:

CELL

↓ Cell wall ↓ Protoplasm ↓ Vacuole

Protoplasm → ↓ Plasma membrane ↓ Cytoplasm ↓ Nucleus

Cytoplasm → ↓ Cell organelles ↓ Deutoplasm

Deutoplasm ↓

↓ Storage substances (carbohydrates, proteins, fats) ↓ Secretory substances (enzymes, pigments, nectar) ↓ Excretory substances (gums, resins, alkaloids, tannins)

Terms:

- Protoplast: It is a cell without a cell wall.
- Cytosol (Hyaloplasm or Cytoplasmic matrix): It is a liquid part of cytoplasm without cell organelles. It contains scattered ribosomes, cytoskeletons, tRNA, and enzymes of anaerobic respiration.

FEATURES OF CELL:

Number: There is one (unicellular organisms) to indefinite cells (multicellular organisms).

Shape: There is great variation. It is usually rounded but may be elongated, polygonal, columnar, discoidal, etc. It depends on the function.

Size: It varies from 0.2μm to 20μm. It depends on Nuclear-cytoplasmic ratio (N/C or N: C ratio) and metabolic activity. The N-C ratio is the ratio of the size (volume) of the nucleus of a cell to the size of the cytoplasm of that cell. It indicates the maturity of a cell. As the cell matures (increases its size), the size of the nucleus will decrease and the surface-to-volume ratio of the cell decreases. Small size is extremely important and allows for rapid uptake and intracellular distribution of nutrients and excretion of wastes due to the large surface-to-volume ratio.

Smallest cell	*Mycoplasma gallisepticum* (0.2μm)
Largest cell	Egg of Ostrich (6 inches in diameter with shell)
Longest cell	Nerve cell (90 cm)
Longest Plant cell	Fibre of Ramie (50 cm)
Longest unicellular plant	*Acetabularia*, a green alga (10 cm)

Cell size (maturity) ∞ **1/efficiency** =1/**metabolic activity** =1/**N:C ratio** =1/Surface-to-Volume ratio

- Smaller cell (immature cell) = More efficient = More metabolically active = High N:C ratio = High surface: volume ratio
- Larger cell (mature cell) = Less efficient = Less metabolically active = Low N:C ratio = Low surface: volume ratio

Chemical composition of a cell: (percent of total weight):

Water	70
Protein	18
Lipid	5
Small metabolites	3
Carbohydrates	2

RNA	1.1
Inorganic ions	1
DNA	0.25

Types of cells: Dougherty (1957) divided cells into three types based on the nucleus.

1. **Prokaryotic cell:** The nucleus is unorganized and undeveloped (primitive) and histone proteins are absent. e.g., Archaebacteria, Bacteria, and Cyanobacteria.
2. **Mesokaryotic cell:** It was coined by Dodge (1964). The nucleus is well developed but histone proteins are absent. e.g., *Peridinium balticum* (a dinoflagellate).
3. **Eukaryotic Cell:** The nucleus is well-developed, and histone proteins are present. (Advanced). e.g., Protista, Fungi, Plants and Animals.

Differences between prokaryotic cell and eukaryotic cell

Prokaryotic cell	Eukaryotic cell
Its size is small (between 0.2-5μm)	Its size is large (between 5-100μm).
Flagella do not show a 9+2 arrangement and are made up of flagellin protein. They are not enclosed in the plasma membrane and show rotatory movement.	Flagella show a 9+2 arrangement and are made up of tubulin protein. They are enclosed in the plasma membrane and show undulatory movement.
The cell has a single envelope organization (only plasma membrane is present).	The cell has a double envelope organization (plasma membrane and membrane of cell organelles).
The plasma membrane contains respiratory enzymes.	The plasma membrane does not contain respiratory enzymes.
Membrane-bound cell organelles are absent.	Membrane-bound organelles are present.
Cytoplasm does not show cyclosis or streaming movement.	Cytoplasm shows cyclosis or streaming movement.
Vacuoles are absent.	Vacuoles are present.
Plasma membrane infolds to form mesosome.	The plasma membrane does not infold to form mesosomes.
Single thylakoid lies free in the cytoplasm.	Several thylakoids are grouped in the granum of chloroplast.
Ribosomes are of 70S. These lie free in the cytoplasm or join to form polyribosomes.	Ribosomes are of 80S. These are free or attached to the nuclear membrane and endoplasmic reticulum.
Cytoskeletons are absent.	Cytoskeletons are present.
Endocytosis and exocytosis do not occur.	Endocytosis and exocytosis occur.
Cell is haploid.	Cell is generally diploid but haploid cells are also found.
The nucleus is unorganised and undeveloped.	The nucleus is well developed.
The nuclear membrane and nucleolus are absent.	Nuclear membrane and nucleolus are present.
A nucleoid is equivalent to a single chromosome.	The nucleus contains more than one chromosome.
Histone proteins are absent.	Histone proteins are present.
DNA is small, circular, and naked and occurs in the cytoplasm	DNA is large, linear (organellar DNA is circular), closed and occurs in the nucleus.
DNA does not have introns.	DNA has introns.
Plasmid is found in the cytoplasm.	Plasmid is not found in the cytoplasm.
Cell cycle is of short duration (30-60 minutes)	Cell cycle is of long duration (12-24 hours).
The mitotic spindle is not formed during cell division.	The mitotic spindle is formed.
Meiosis does not occur.	Meiosis occurs.

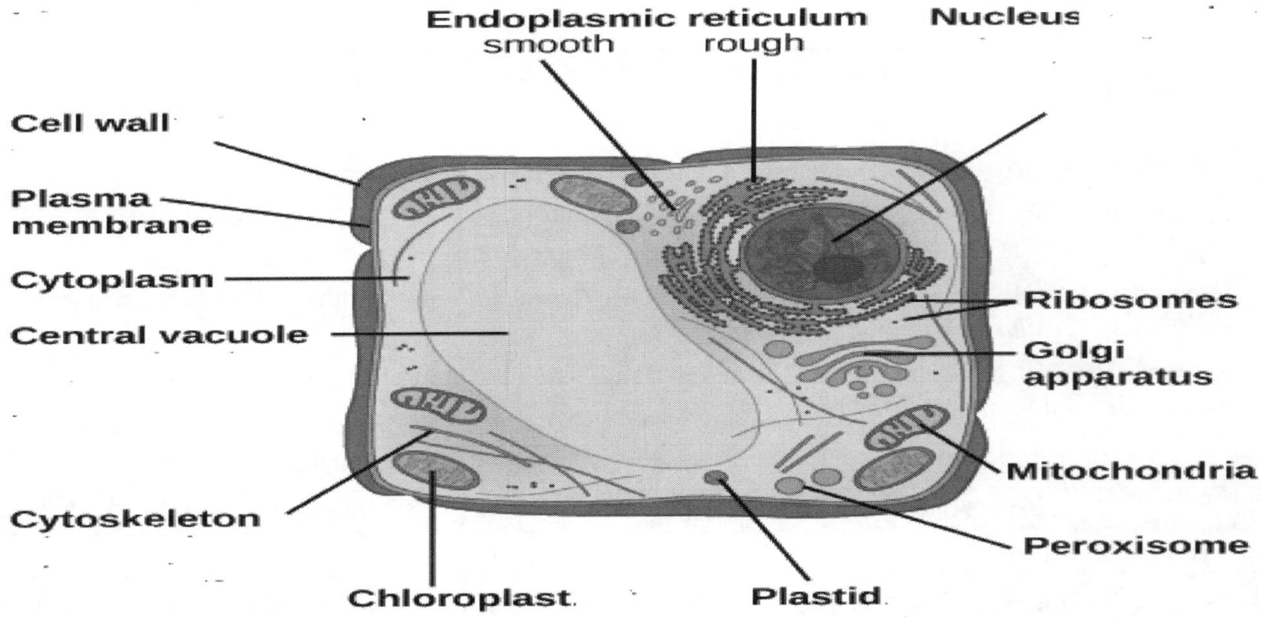

Fig: Plant cell

Differences between Plant cell and Animal cell

Plant Cell	Animal Cell
It is larger (>100μm)	It is smaller (<100μm).
The cell wall is present.	The cell wall is absent.
The vacuole is single, large, central and permanent.	Vacuoles are more, small, peripheral and temporary.
Contractile vacuoles and food vacuoles are absent.	Contractile vacuoles and food vacuoles are present.
The nucleus is peripheral in position.	The nucleus is central in position.
Mitochondria are lesser in number.	Mitochondria are larger in number.
Cristae of mitochondria are tubular.	Cristae of mitochondria are plate-like.
Plastid is present.	Plastid is absent.
The Golgi complex contains more (10-20) cisternae.	Golgi complex contains less (3-7) cisternae.
Lysosome is rarely found.	Lysosome is commonly found.
Glyoxysome is present.	Glyoxysome is absent.
The number of chromosomes is relatively small.	The number of chromosomes is relatively large
The size of the chromosome is relatively large.	The size of the chromosome is relatively small.
Reserve food is starch and fat.	Reserve food is glycogen and fat.
Mitosis occurs in both haploid and diploid cells.	Mitosis occurs in diploid cells only.
The spindle is without aster (anastral).	The spindle has an aster at each pole (amphiastral).
Cytokinesis occurs by cell plate.	Cytokinesis occurs by cell furrow (constriction).
Adjacent cells are connected through plasmodesmata.	Adjacent cells are connected through cell junctions.
Tight junctions and desmosomes are absent.	Tight junctions and desmosomes are present.
Tissue fluid is absent.	Tissue fluid is present.
It can synthesise all amino acids, coenzymes and vitamins.	It cannot synthesise amino acids, coenzymes and vitamins.
It becomes turgid if placed in a hypotonic solution.	It bursts when placed in a hypotonic solution.

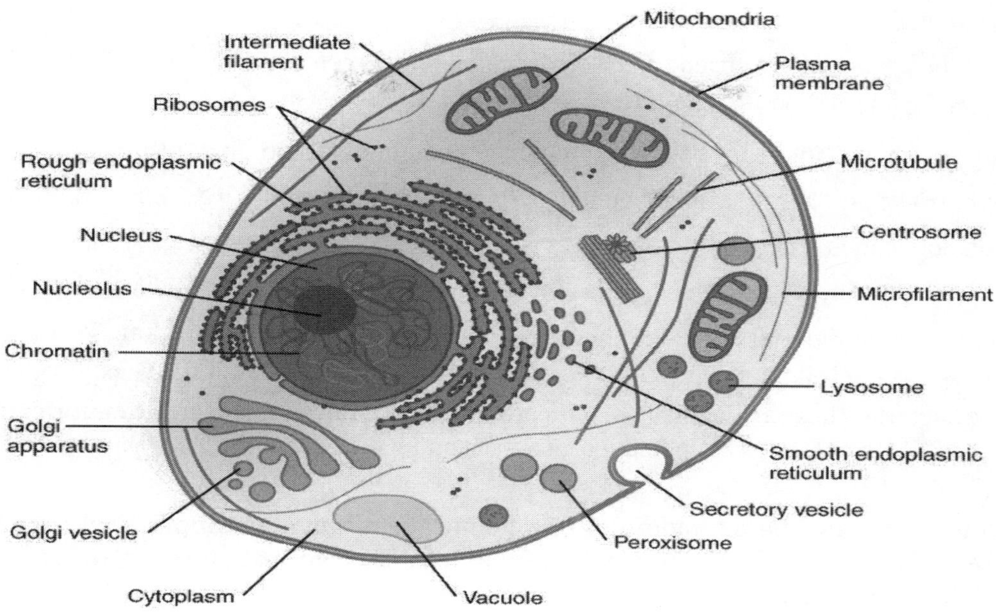

Fig: Animal cell

BACTERIAL CELL

Bacteria (sing: bacterium) are the simplest, most primitive and microscopic organisms. Antonie Van Leeuwenhoek (1875) discovered bacteria from rainwater, sewage water and saliva using a single-lens microscope of his design. Bacteria being unicellular consist of the following parts:

1. **Surface appendage:**

   (a) **Flagella:** It is long, filamentous and much thinner than eukaryotic flagella. It protrudes from the cell wall. It develops from basal granules present in the plasma membrane. It consists of a single fibril (single-stranded) surrounded by 3-8 fibrils without differentiation of axoneme and sheath. It lacks a 9+2 arrangement. It is made up of flagellin protein. It contains a basal body, hook and filament (shaft). The basal body has a central rod and rings. There are two rings (S ring in peptidoglycan layer and M ring in the plasma membrane) in Gram-positive bacteria. There are four rings (L ring associated with lipopolysaccharide; P ring associated with peptidoglycan; S ring directly attached to the plasma membrane and M ring embedded in the plasma membrane) in Gram-positive bacteria.

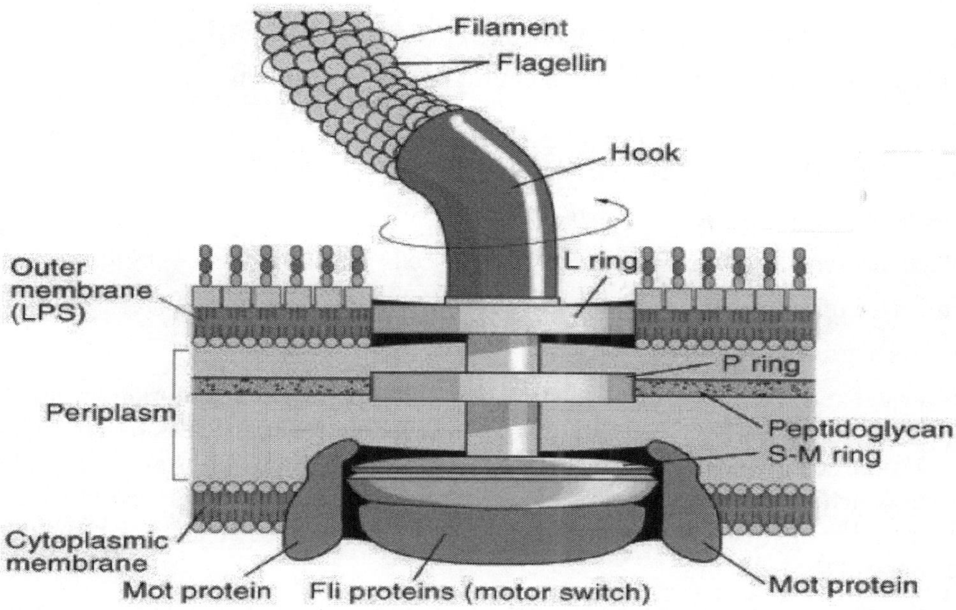

Fig: Bacterial flagella (Internal structure)

Types of flagella:

1. Atrichous: Flagella are absent. e.g., *Lactobacillus, and Streptococcus.*
2. Monotrichous: There is one flagellum. e.g., *Vibrio*
3. Amphitrichous: There are two flagella; each at two opposite ends. e.g., *Alkaligenes, Nitrosomonas*
4. Lophotrichous: There is a single bundle of flagella at one end only. e.g., *Spirillum, Pseudomonas*
5. Cephalotrichous: There are two bundles of flagella each at both ends. e.g., *Pseudomonas fluorescence.*
6. Peritrichous: There are many flagella all over the surface. e.g., *E. coli, Bacillus, Clostridium, Salmonella.*

Function: It helps in locomotion (60 cell length per sec so it takes 245 days to reach 1 km.

(b) **Pili**: These are long, a few (1-4) in number, thick and non-helical appendages. These are found in Gram-negative bacteria. These are made up of pilin protein (glucophosphoprotein). Their formation is controlled by plasmid (F-factor). These help in the process of conjugation by serving as a passage of DNA from donor to recipient cell.

(c) **Fimbriae**: These are shorter, more (300) in number, and thinner than pili. These are found in Gram-negative bacteria. These are made up of pilin protein. Their formation is controlled by nuclear genes. These help in attachment.

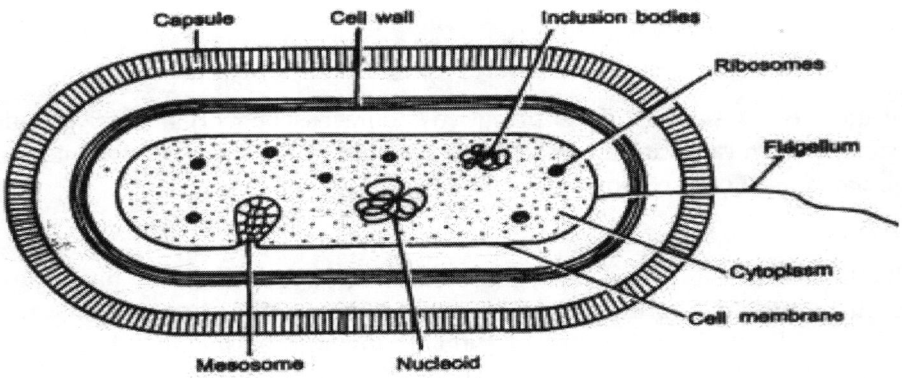

Fig: Bacterial cell

2. **Glycocalyx**: It is the outermost mucilage layer. It differs in composition and thickness in different bacteria which are as follows:

(a) **Capsule**: It is a thick and tough layer made up of polysaccharides.

(b) **Slime layer**: It is a thin and loose layer made up of polysaccharides.

(c) **S-layer**: It is a thin and loose layer made up of protein or glycoprotein.

Functions:

(i) It protects bacteria from toxic materials.

(ii) It protects bacteria from bacteriophage (virus).

(iii) It protects bacteria from engulfing by eukaryotic cells (macrophages)

(iv) It protects bacteria from desiccation or drying by binding water.

(v) It prevents aggregation of cells.

3. **Cell wall**: It is a rigid and solid covering made up of peptidoglycan or mucopeptide or murein. Its sugar part forms the backbone and is made up of alternating units of amino sugars such as N-acetyl glucosamine (NAG) and N-acetyl muramic acid (NAM) joined together by β-1,4 linkage (glycosidic bond). Its protein part contains four amino acids which are attached to NAM. These are D-alanine, L-alanine, D-glutamic acid and L- lysine (D-aminopimelic acid in the case of Gram-negative bacteria). It forms the structural framework of the cell and maintains the shape of the cell.

4. **Plasma membrane**: It is the innermost and trilaminar membrane made up of lipids and proteins. It does not contain sterol but contains pentacyclic sterol (hopanoids) in some bacteria. It has respiratory enzymes in the inner face. It is selectively permeable. It takes part in respiration by performing oxidative phosphorylation.

5. **Mesosome**: It is a convoluted structure that develops from the infolding or invagination of the plasma membrane. It increases the surface area of the plasma membrane for secretory functions. It is confined to the peripheral region. It is found in Gram-positive bacteria. It is considered analogous to mitochondria due to the presence of respiratory enzymes such as dehydrogenase and cytochromes associated with respiration. It takes part in binary fission by forming septum and separation of replicated DNA.
6. **Chromatophore**: It contains a single thylakoid which is found freely or diffused in cytoplasm. It contains bacteriochlorophyll and bacterioviridin. It is analogous to chloroplast.
7. **Ribosomes:** These are of 70S and are found freely in the cytoplasm and also attached to the plasma membrane. Their size is 20 x15 nm. These form polyribosomes when 4-6 ribosomes are attached to m-RNA.
8. **Nucleoid** (Genophore): It is fibrillar and lobular chromatin. It is either directly attached to the plasma membrane or through a mesosome. It is single (haploid), circular, double-stranded, naked (without histone proteins), supercoiled and tightly folded due to the presence of polyamines. It does not have a nuclear membrane and nucleolus thus it is considered a primitive chromosome. It occupies 10-20% of the cell. It is 1000 longer than the cell itself. It varies between species of bacteria. It can code 2000 to 3000 different proteins. It does not follow Chargaff's rule. *E. coli* has 4300 genes having 4.6 million base pairs.
9. **Plasmid:** It is extra-chromosomal, circular, double-stranded DNA found in cytoplasm. It is analogous to bacterial chromosomes. It is called episome when it is integrated with bacterial DNA. It undergoes autonomous replication. It is of several types
   (a) F-factor (fertility factor): It helps in conjugation.
   (b) R-factor: It confers resistance to various antibiotics and heavy metals.
   (c) Col-factor: It produces a chemical, colicin which kills closely related species of bacteria or sensitive *E. coli*.
   (d) Ti-plasmid: It induces the formation of crown gall (tumour).
10. **Cell inclusions:** These are nonliving, membrane-less and found freely in the cytoplasm. These are
   (a) Poly beta-hydroxybutyrate (PBH) granule: It stores carbon and energy source.
   (b) Gas vacuole: It stores gases which provides buoyancy and protects from radiations.
   (c) Volutin granule: It is a metachromatic granule that stores phosphate.
   (d) Sulphur granule: It picks up H_2S as reducing power in photosynthesis.

Gram Staining Method: It is a differential staining technique developed by Christian Gram (1884). This difference is due to differentiation in lipid content and thickness of the cell wall.

Procedure: The smear was prepared from the sample and then heat dried. There are four stains which are given one by one for about one minute, then washed gently with tap water and observed microscopically.

Stain	Action	Time	Gram-positive	Gram-negative
Crystal violet	Primary stain	1 min	Violet	Violet
Iodine (0.5%)	Mordant	1 min	Violet	Violet
Absolute alcohol	Decolourising agent	1 min	Violet	Colourless
Safranin	Counterstain	1 min	Violet	Red/pink

Mechanism for Gram-positive bacteria	Mechanism for Gram-negative bacteria
Crystal violet-iodine (CV-I) complex is formed.	Crystal violet-iodine (CV-I) complex is formed
A low amount of lipid is dissolved by alcohol.	A high amount of lipid is dissolved by alcohol.
A small pore is formed in cell wall.	A large pore is formed in the cell wall
The pore is closed due to the dehydration of proteins.	The pore is not closed due to the dehydration of proteins.
CV-I complex is trapped.	CV-I complex is leaked.
Decolourisation does not occur.	Decolourisation occurs.
Thus, it remains violet/purple	Thus, it appears red/pink

Difference between Gram-positive and Gram-negative bacteria

Gram-positive	Gram-negative
The cell wall is thick, single-layered, simple, homogeneous, smooth and rigid.	The cell wall is thin, two-layered, complex, heterogeneous, wavy and plastic.
The amount of peptidoglycan is more (90%).	The amount of peptidoglycan is less (10%).
The amount of lipids is low (3-4%).	The amount of lipids (phospholipids, lipoprotein and lipopolysaccharide) is high (25%).
L-lysine is present in peptides.	D-aminopimelic acid is present in the peptide.
Sialic acid is absent.	Sialic acid is present.
Techoic acid (acidic polymer of glucose, phosphate and alcohol) is present. It protects from high pH and temperature fluctuation.	Techoic acid is absent.
These are mostly capsulated.	These are mostly non-capsulated.
The cell wall is digested by lysozyme hence protoplast is formed.	The cell wall is resistant to lysozyme.
Porins are absent.	Porins are present
Pili are absent	Pili are present.
Mesosome is present.	Mesosome is absent.
These are more sensitive to penicillin.	These are less sensitive to penicillin.
These are less sensitive to streptomycin.	These are more sensitive to Streptomycin.
A few are pathogenic and produce exotoxins.	The majority are pathogenic and produce endotoxins.
Examples: Bacillus, Coccus, Clostridium, and Mycobacterium (lesser in number).	Azotobacter, E. coli, Rhizobium, Salmonella, Vibrio, Xanthomonas (more in number).

CELL WALL

It is the outermost, non-living, tough, flexible (sometimes rigid) and structural layer outside the plasma membrane of the cell. It was discovered (observed) and coined by Robert Hooke.

Distribution: It is found in prokaryotes, some protists (diatoms, dinoflagellates and spores of slime moulds), fungi and plants but it is absent in *Mycoplasma*, *Euglena*, the somatic phase of slime moulds. protozoans and animals.

Composition: Water (60%), cellulose (15%), hemicellulose (15%), pectin (10%). The composition differs according to cell type and functions. Cellulose microfibrils form the skeletal of the cell wall whereas hemicellulose and pectin form the matrix of the cell wall.

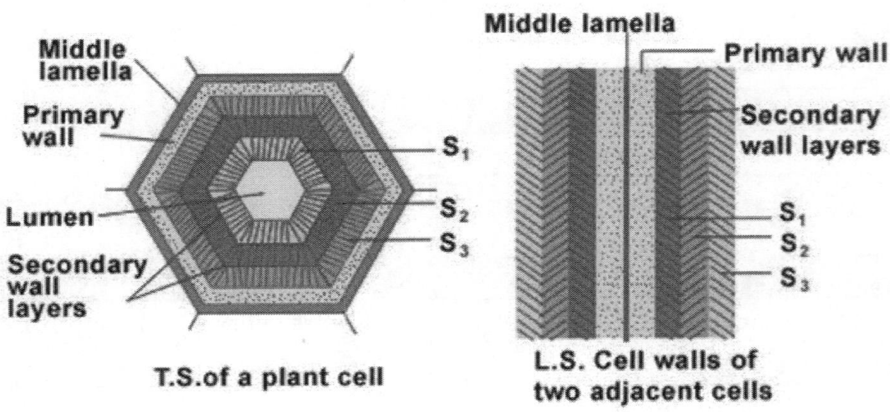

Structure: It consists of the following layers

1. **Middle lamella:**

   (a) It is the outermost, thin and sticky cementing layer between two adjacent cells.

   (b) It is made up of calcium pectate (more) and magnesium pectate.

(c) It is the first layer formed as a result of cytokinesis.

(d) It ruptures to create intercellular spaces in many tissues.

(e) It is dissolved or gelatinised by pectinase hence softening of ripened fruits occurs.

2. **Primary wall:**

   (a) It is thin, elastic and single layered present below the middle lamella.

   (b) It is made up of cellulose (25%), hemicellulose (25%) in the form of organised microfibrils and pectin (35%).

   (c) It develops by intussusceptions (deposition of wall materials in the form of grains).

   (d) It is stretchable because cell wall microfibrils are transversely oriented (anisotropic).

   (e) It is present in parenchymatous cells and expands in meristematic cells.

3. **Secondary wall:**

   (a) It is thick, non-elastic (rigid) and three-layered.

   (b) It is made up of more cellulose and less hemicellulose. It may contain a small amount of lignin, cutin or suberin.

   (c) It develops by accretion (deposition of wall materials in the form of layers).

   (d) It is present in collenchyma, sclerenchyma, xylem tracheid and xylem vessel.

   (e) It is found in hard and woody parts of plants when the cell has stopped growing.

   (f) It forms pits because the secondary wall is not of uniform thickness.

4. **Tertiary wall:**

   (a) It is present occasionally.

   (b) It is made up of cellulose.

   (c) It is found in tracheid of gymnosperms.

Functions

1. It protects the protoplasm from mechanical injury by the middle lamella.
2. It protects the protoplasm from the attack of pathogens.
3. It provides rigidity or mechanical support to the cell.
4. It maintains the shape of the cell due to the middle lamella.
5. It prevents cells from bursting (undue expansion) as a result of endosmosis.
6. It helps in the apoplast transport of water.

Intercellular spaces: These are spaces formed in mature cells only. There are three types:

1. Schizogenous cavity: It is formed by the separation of the cell wall due to the dissolution of the middle lamella. e.g., Mucilage duct, Resin duct of *Pinus*.
2. Lysigenous cavity: It is formed by the breakdown of the cell wall. e.g., Oil cavities in citrus plants and *Eucalyptus*.
3. Schizo-lysigenous cavity: It is formed by the separation and breakdown of the cell wall. e.g., Protoxylem lacunae in monocot stem (Maize).

Plasmodesmata: These are microscopic channels or pores that directly connect the cytoplasm of neighbouring plant cells. These establish living bridges between cells. A typical plant cell may have between 10^3 and 10^5 plasmodesmata.

PROTOPLASM

It is a living substance of a cell and performs all vital functions of a cell. It is the fundamental material of which living organisms are composed.

Discovery: Felix Dujardin (1835) discovered protoplasm and called it 'Sarcode'. The term 'protoplasm' was coined by Purkinje (1837).

Von Mohl (1844) stated the importance of protoplasm in cell division.

Huxley (1868) defined it as the 'Physical basis of life'.

Max Schultze (1861) proposed the 'Protoplasmic theory' in which the living matter of a cell is protoplasm which is limited by an outer membrane and contains a nucleus.

Fisher and Hardy (1894) proposed 'colloidal theory'. It states that protoplasm is a heterogeneous polyphasic colloidal system that appears different, fibrillar or granular in different phases.

Composition: Water (80%), Protein (15%), lipid (3%), carbohydrate (1%), minerals (1%). Among elements, the amount of Oxygen is very high (62%) followed by Carbon (20%), Hydrogen (10%) and Nitrogen (3%). It is difficult to determine the exact composition because it undergoes continuous change. Thus, it is a complex crystallo-colloidal mixture where some solutes form a solution and most of the solutes remain colloidal.

Properties:

(a) It is complex, granular, elastic, colourless and viscous.

(b) It is selectively permeable.

(c) Its pH is slightly alkaline but in plants it is acidic.

(d) Its specific gravity is slightly more than one.

(e) It coagulates on heating up to 60°C or treating with conc. Acid or base.

(f) It shows Brownian movement (zig-zag movement due to random collision of molecules) and Tyndall effect (scattering of light by particles).

(g) It shows intracellular movement like cyclosis or streaming. It occurs in all eukaryotic cells. It depends on viscosity and temperature. It helps in movement of cell organelles and cell inclusions.

It helps in the formation of pseudopodia in *Amoeba*. Cyclosis is of two types:

(a) **Rotation:** It moves in one direction (either clockwise or anticlockwise) around the vacuole inside the cell. e.g., Leaf cell of *Hydrilla* and *Vallisneria.*

(b) Circulation: It moves in different directions around different vacuoles inside the cell. e.g., Staminal hairs of *Tradescantia (=Rhoeo discolor).*

VACUOLE

It is a non-cytoplasmic, nonliving and fluid-filled reservoir. It was discovered by Spallanzani in a plant cell. It is occupied by 30% of the volume of the cell. It is found in cells of plants, fungi, some protists and some animals. It is very small or absent in young cells but fuses to form very large central vacuoles in mature cells such as parenchymatous and collenchymatous cells when cells increase in size. It is bounded by a single membrane called tonoplast which is lipo-proteinaceous and selectively permeable. Its structure varies according to the requirement of the cell. It varies in different tissues. It is formed in plant cells because plants cannot produce enough protein to fill their enlarging cells with protoplasm. It contains watery fluid called cell sap which is generally neutral but slightly acidic at maturity. Cell sap is hypertonic (higher concentration than cytoplasm) which generates turgor pressure in plant cells.

Vacuole is known as a storehouse of cell products because cell sap contains inorganic substances such as K^+ (more), and Na^+ and organic substances such as acetic acid (more), fumaric acid, amino acids, enzyme, sugar and secondary metabolites such as pigments, alkaloids, latex, tannins, crystals of calcium oxalates). It also contains excretory products not useful for the cell.

Origin: It develops from the invagination of the plasma membrane. It also develops from the endoplasmic reticulum.

Types:

1. Contractile vacuole: It picks up water from the surrounding cytoplasm, expands and collapses to throw water outside thus performing excretion and osmoregulation in Protozoa.
2. Food vacuole: It helps in digestion because it contains digestive enzymes.
3. Gas vacuole: It stores metabolic gases and helps in floating in prokaryotes.

Functions:

1. Vacuole performs osmoregulation by maintaining the osmotic relation of the cell. Thus, it helps in the absorption of water.
2. It provides colouration to flowers of rose and sunflower; and roots of beet due to the presence of anthocyanin (red, blue, purple) in cell sap.
3. It stores food reserves such as sucrose, salts, and inulin and digests waste products.
4. It maintains turgor pressure inside the cell.
5. It maintains an acidic internal pH.
6. It contains hydrolytic enzymes hence it acts as a lysosome in plant cells.
7. It exports unwanted substances.

PLASMA MEMBRANE

It is a thin, non-elastic, flexible and film-like membrane enclosing the cell. It was discovered and coined by Nageli and Krammer. It is also known as the cell membrane or plasmalemma. The term 'Plasmalemma' was coined by Plower. It is considered living because it is an essential part of the protoplasm that makes cells alive. It constitutes 5% of all membranes.

Composition:

(a) Lipids (32%)-Phospholipids are common (phosphatidyl serine, phosphatidylcholine (lecithin), phosphatidyl ethanol amine (cephalin). Others are glycolipid and sterol (cholesterol in animal cells and *Mycoplasma*; phytosterol in plant cells and ergosterol in bacterial cells).

(b) Proteins (42%):

(c) Carbohydrate (6%): It is an oligosaccharide associated with glycoprotein (60%) and glycolipid (10%).

(d) Water: 20% by weight.

Types of plasma membrane	Lipid	Protein	Carbohydrate
Human RBC	40%	52%	8%
Nerve cell	80%	20%	-
The outer membrane of mitochondria	47%	51%	2%
The inner membrane of mitochondria	23%	76%	1%

Models: The following models were proposed to explain the structure of the Plasma membrane:

1. **Overton** (1902) proposed that plasma membrane is composed of thin layers of lipid.
2. **Danielli and Davson** (1935) proposed the trilamellar (trilayer) model. The plasma membrane is composed of three layers in which a layer of phospholipid is sandwiched between two layers of proteins. This model had proteins present as sheets neighbouring a lipid layer, rather than incorporated into the phospholipid bilayer, thus it is not acceptable.
3. **Robertson** (1959) proposed a unit membrane model. It is called a unit membrane because of the unity in the structure of all membranes of bacteria, plants and animals. The plasma membrane is composed of one light middle layer of lipid which is sandwiched between two dark layers of proteins (trilaminar structure).

   Thickness: Protein + Lipid + Protein

   $20\ A^0 + 35\ A^0 + 20\ A^0 = 75A^0$

4. **Jonathan Singer and Garth Nicolson** (1972) proposed a fluid mosaic model. It is the most accepted model which explains the selectively permeable nature of plasma membrane.

The plasma membrane is composed of a continuous bilayer of phospholipids in which the globular proteins are embedded. It looks like an iceberg (proteins) floating in a sea of phospholipids. The plasma membrane is a two-dimensional (2-D) quasi-fluid structure where lipids and proteins are arranged in a mosaic fashion. Proteins are not arranged in continuous layers but they are present separately and asymmetrically. Proteins can diffuse freely in a lateral direction within the lipid matrix thus there is fluid movement of proteins and lipids. Thus, the lateral movement of proteins makes their nature fluid-like. Proteins cannot move from one side of the membrane to the other side (flip-flop) diffusion. Lipids are in the form of fluid and rarely show flip-flop movement (movement of lipids from inside to outside). The components of membrane (lipids, proteins and oligosaccharides are held in their positions by non-covalent interactions. Lipid molecules are in constant lateral movement and not rigidly locked hence these provide permeability, fluidity and elasticity to the membrane. The degree of fluidity depends on the percentage of unsaturated fatty acid present in the lipid bilayer and the temperature. Lipids have kinks (double bond in the tail) which prevents from close packing of molecules and make the membrane more fluid. Fluidity is due to the hydrophobic interaction of lipids and proteins.

Lipids containing phosphate and glycerol are stable and amphipathic molecules. It consists of

(a) **Phosphate heads** are hydrophilic (water soluble) and polar containing phosphate and glycerol. These face outward and lie at free surface.

(b) **Hydrocarbon tails** are hydrophobic (water insoluble) and non-polar containing long chains of fatty acids. These face inside. Tails remain in contact with each other by Vander Wal's forces hence tail is protected from an aqueous environment by its hydrophobic nature.

Lipid bilayers are generally impermeable to ions and polar molecules. The arrangement of hydrophilic heads and hydrophobic tails of the lipid bilayer prevents polar solutes (e.g., amino acids, nucleic acids, carbohydrates, proteins and ions) from diffusing across the membrane but generally allows for the passive diffusion of hydrophilic molecules.

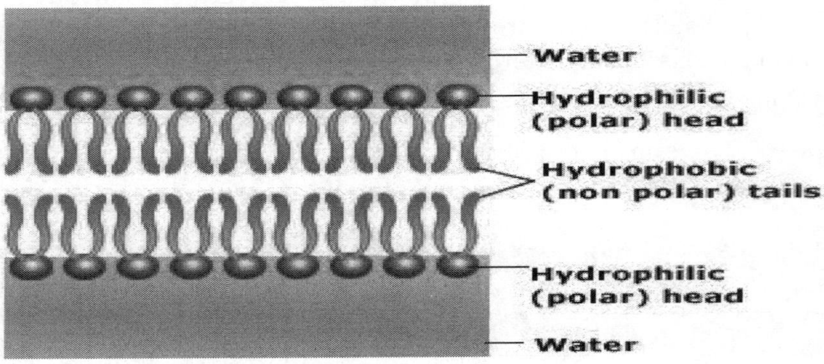

Fig: Arrangement of lipids and proteins in Plasma membrane

Proteins are of three types:

(a) **Extrinsic proteins** (peripheral proteins): These constitute 30% of total proteins. These are loosely bound (covalently attached) at the polar surface of the lipid head. These are soluble. These can be extracted or removed from the membrane. These are involved in cell interaction and cell recognition (cell signalling). e.g., ATPase in mitochondria, spectrin in RBC etc.

(b) **Intrinsic proteins** (Integral protein): These constitute 70% of total proteins. These are deeply penetrated or embedded into lipid bilayer. These are insoluble. These cannot be extracted or removed from the membrane due to strong hydrophobic interaction with a non-polar hydrophobic tail of lipids. These are involved in the transport of polar solute molecules. e.g., porin, glycophorin, and rhodopsin in rod cells of the retina.

(c) **Tunnel protein**: It is the intrinsic protein that projects on both surfaces hence known as transmembrane protein. It may be monotopic (exposed only at one surface), bitopic (exposed at both surfaces) and polytopic (exposed across surfaces). It provides a mechanical framework.

Carbohydrates are in the form of oligosaccharides which are covalently attached to the external surface of lipid (called glycolipid) and protein (called glycoprotein). Sialic acid (N-acetylglucosamine or NAG) is the terminal sugar of the oligosaccharide chain present in the glycoprotein of the plasma membrane.

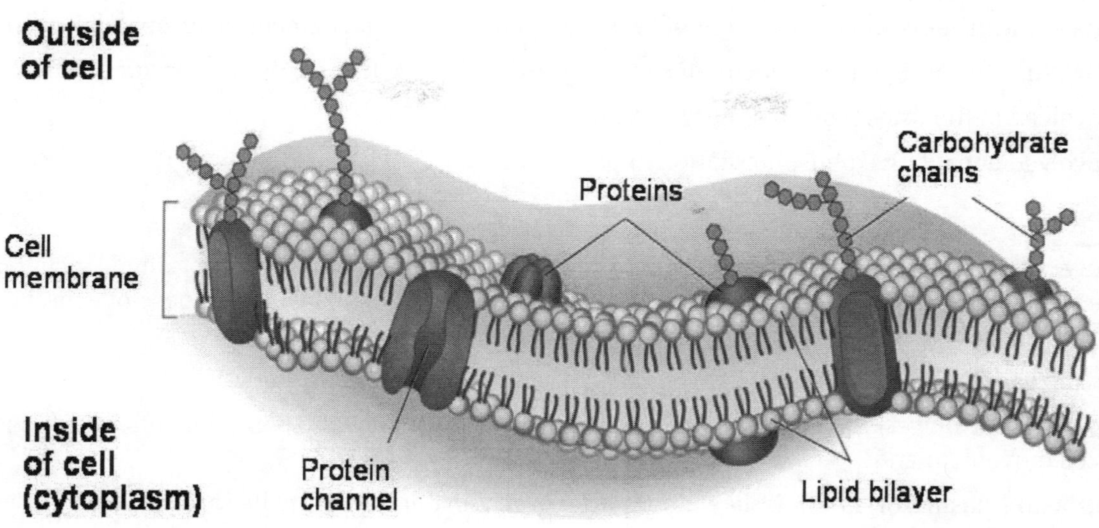

Fig: Plasma membrane

Origin: The plasma membrane is developed due to the accumulation of surface-active substances provided by cytoplasm.

In the case of animal cells, plasma membranes of neighbouring cells in tissues are linked to each other at specialised sites called cell junctions. These may be

- Adhering junction (Desmosome): It holds cells together (adhesion).
- Impermeable junction: It seals the cells together.
- Communicating junction: It mediates the passage of small molecules from one interacting cell to the other.

Functions:

1. Plasma membrane transports the materials into and outside the cell. It is selectively or differentially permeable. It takes place by
   (a) Simple diffusion: O_2, CO_2, H_2O etc.
   (b) Facilitated diffusion: Charged ions, glucose in RBC etc.
   (c) Active transport: Salts, glucose and amino acids (up to size of 1-15A^0).
   (d) Bulk transport: It occurs in eukaryotic cells. It takes place by **exocytosis** in which cholesterol is removed by carrier vesicle. It also takes place by **endocytosis** which occurs by following two methods:
      1. **Pinocytosis (cell drinking)**: It is the intake of ions, sugars and amino acids in liquid form by forming a small vesicle (100 nm in diameter). It was reported by Lewis.
      2. **Phagocytosis (cell eating)**: It is the intake of materials in solid form by forming a large vesicle (>250 nm in diameter). It was reported by Metchnikoff.
2. It protects cells from injury and provides mechanical support.
3. It maintains the shape of cells and cell organelles.
4. It helps in cell-to-cell recognition and cell-to-cell adhesion due to the presence of sugar of glycoprotein and glycolipid at the surface.
5. It delineates and separates intracellular components from the external environment (compartmentalisation).
6. It helps in chemoreception because it contains receptors that recognise specific hormones and convey the flow of information to the interior of the cell.
7. It forms the sheath of cilia and flagella.
8. It is involved in plasmolysis.

9. It helps in the pseudopodial movement of cells (e.g., WBC, fibroblast, *Amoeba*, Macrophage).
10. It determines the antigenic specificity due to the presence of glycophorin on the surface of RBC.
11. It is involved in the transmission of nerve impulses.
12. It is involved in oxidative phosphorylation in prokaryotes.

CYTOPLASM

It was discovered by Kolliker (1862). It is jelly like semi-fluid general mass of protoplasm. It includes cytosol or cytoplasmic matrix or hyaloplasm which is a clear fluid part (pH 7.2) of cytoplasm and does not contain nucleus, cell organelles and cell inclusions.

It exists in two zones:

(a) **Ectoplasm** (plasmagel) is a dense, gelatinous, outer layer just below the plasma membrane and is concerned with cell movement and

(b) **Endoplasm** (plasmasol) is a fluid-like inner part where most of the organelles are found.

It contains 90% of water. It is a crystalloid-colloidal complex where some chemicals (glucose and minerals) are present in the form of a true solution and others (proteins (maximum), enzymes, fatty acids and some sugars) are present in the form of a colloidal solution.

Functions:

1. It is the seat of synthesis of enzymes, proteins, fats and nucleotides.
2. It contains raw materials for the functioning of cell organelles.
3. It is the seat of catabolic activities such as glycolysis, anaerobic respiration, pentose phosphate pathway (PPP).
4. It performs cytoplasmic streaming or cyclosis which distributes various materials inside the cell.
5. It receives products from cell organelles.
6. It is involved in β-oxidation of fats in prokaryotes.

MITOCHONDRIA

Mitochondrion (pl: Mitochondria) is a hollow, and sac-like organelle.

Discovery: It was discovered by Kollicker (1850) in the striated muscle of insects and called sarcosome (granule). Flemming (1882) called it 'fila' (thread-like). Altman (1890) called it a 'bioblast' (a granule). Carl Benda (1898) coined the term 'Mitochondria'. Friedrich Meves (1904) first observed it in the cells of a plant, a white-water lily (*Nymphaea alba*).

Distribution: It is found in eukaryotic cells except in mammalian RBCs. It is absent in prokaryotes.

Size: Its average size is 0.5 μm in diameter. It is the largest cell organelle in animal cells and the second largest in plant cells. It approaches bacterial cells. Its smallest size is found in yeast cells and the largest size occurs in the oocyte of *Rana pipens.*

Shape: It is usually oval or sausage-shaped. It changes due to chemical or osmotic changes.

Number: It shows great variation in organism, tissue and cell types. It is one in *Chlorella* (a green alga), *Microasterias* (a green alga), and *Trypanosoma* (a protozoan) to 500,000 in flight muscle of birds. Its average number is 1000 per cell. Its number is lesser in plant cells due to the presence of chloroplast which meets the energy requirement. Its number depends on the metabolic requirements of the cell.

Activity: It remains stable in cells performing the same function. It undergoes continuous movement within the cytoplasm during interphase.

Stain: It can be stained with a vital dye, Janus green. It appears yellow due to pigment, riboflavin.

Composition: Phospholipids (30%), Proteins (70%), RNA (3%), and DNA (1%). It contains 70% of total enzymes of the cell.

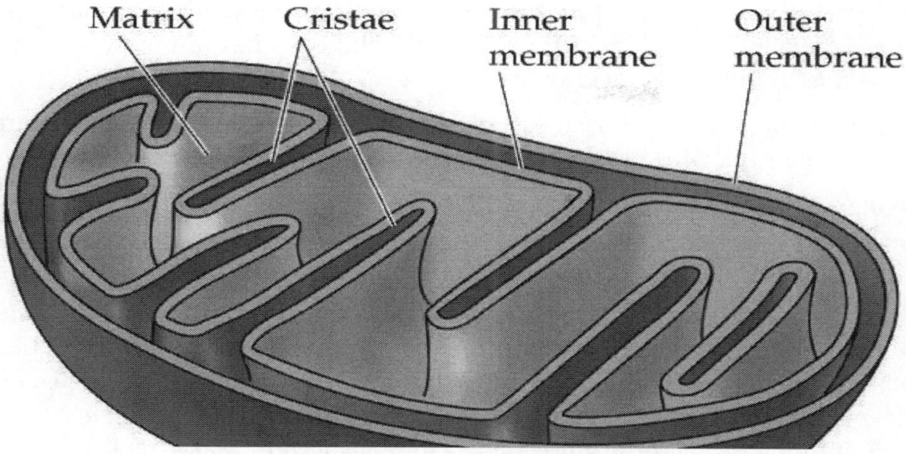

Fig: Mitochondria

Structure: It is a double membrane-bound cell organelle. The two membranes have different properties. It consists of the following parts:

(a) **Outer membrane**: Its thickness is $60A^0$. It is smooth and permeable to several metabolites because of pores having porins (intrinsic protein). It contains more lipids, and sialic acid (glycoprotein and glycolipid) but less proteins. It contains enzymes such as monoamine oxidase and fatty acid Co-A ligase.

(b) **Perichondrial space**: It is space between two membranes. It contains enzymes such as adenylate cyclase, nucleoside diphosphokinase, cytochrome c etc. It shows the lowest pH.

(c) **Inner membrane**: Its thickness is $60A^0$. It contains more proteins but less lipids and sialic acid (glycoprotein and glycolipid). It contains succinate dehydrogenase, cytochrome oxidase, ATPase etc.

(d) **Cristae**: These are several finger-like structures that develop from the inner membrane and project into the matrix. These are tubular in plant cells and plate-like in fungal cells and animal cells. The number of cristae indicates the intensity of respiration. Cristae increase the surface area of the inner membrane for reactions to occur. Its outer surface is covered with 10^4-10^5 elementary particles or oxysomes. These are also known as subunits of Parson or Racker's particles or Fo-F_1 particles. These were discovered by Fernandez Moran (1961). The oxysome is the functional unit of mitochondria. These are evenly spread over an interval of $100A^0$. Each oxysome consists of a head (F_1), stalk and base (F_0). F_0 is formed of 3 proteins whereas F_1 is formed of 4 proteins. Each oxysome with one of the respiratory chain complexes constitutes a phosphorylating unit. The whole electron transport system occurs in base and each base is a complex containing a portion of enzymes of the respiratory chain. Oxysome is lipoproteinaceous. The head of oxysome contains a coupling enzyme, ATPase which synthesises ATP in the matrix.

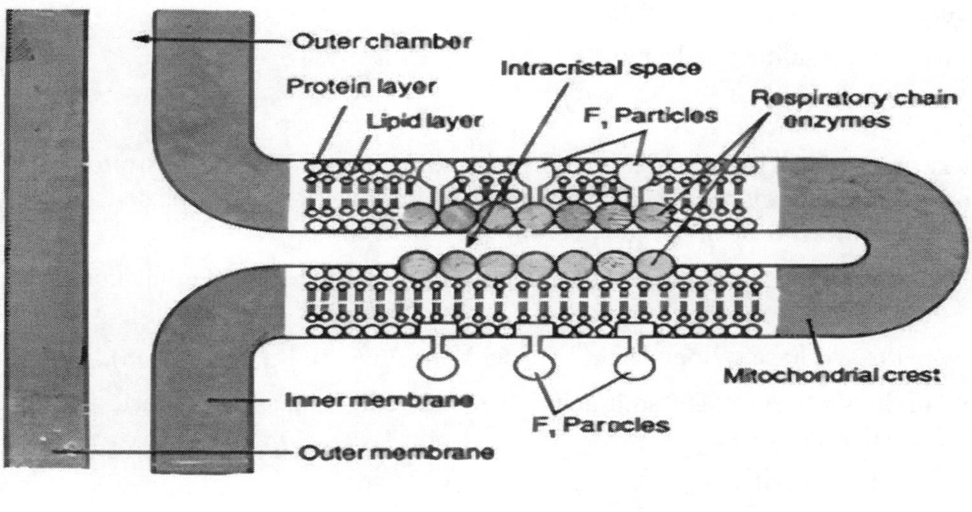

Fig: Internal structure of Mitochondria

(e) **Matrix**: It is the space enclosed by the inner membrane. It contains dense granules (inorganic salts), ribosomes (ranging from 55S to 70S), DNA, RNA and enzymes of the Krebs cycle, fatty acid oxidation, protein synthesis, and nucleic acid synthesis.

Ribosomes are smaller than bacterial ribosomes and cytoplasmic ribosomes. Its subunits are 25 S (containing 12 S r-RNA) and 35 S (containing 16S r-RNA and 5S r-RNA).

DNA is single, circular, double-stranded, and naked (without histone proteins). It was discovered by Chevremont and Nass. It is 1% of the total DNA of the cell. It is rich in G-C ratio. It has 16,569 base pairs encoding 37 genes. It does not have a repair mechanism and has a higher mutation rate than nuclear DNA.

Origin: Small free-living prokaryotes were engulfed (but not consumed) by larger prokaryotes. The two organisms developed a symbiotic relationship over time, the larger organism providing nutrients to the smaller organism and in return, they got ATP from the smaller organism. Eventually, larger organisms developed into eukaryotic cells and smaller organisms into the mitochondria. Thus, mitochondria were originally aerobic bacteria-like organisms that entered accidentally into a host cell of primitive nucleated eukaryotes and established with it for the successful endosymbiotic association. The conditions inside the host cell were favourable for prokaryotes. The bacteria provided increased capacity to produce ATP and provided the host the ability to respire aerobically. Thus, mitochondria are also called intracellular prokaryotic parasites or bacterial endosymbionts. Bacteria and mitochondria show resemblances in several respects:

(a) The size of both mitochondria and bacteria is almost the same.

(b) Both mitochondria and bacteria undergo binary fission.

(c) Porin protein is present in the plasma membrane of both mitochondria and bacteria.

(d) DNA of mitochondria and bacteria is single, circular and double-stranded.

(e) The ribosome of mitochondria and bacteria is 55S close to 70S.

(f) Respiratory enzymes in both are present on the inner surface of the inner membrane.

(g) Protein synthesis in both is sensitive to streptomycin and chloramphenicol.

The mitochondrion is considered a semiautonomous organelle (cell within the cell) because it has its DNA, RNA and enzymes but it is not fully autonomous because it is partially controlled by the nucleus and availability of materials from cytoplasm.

Terms:

- Mitoplast: Mitochondria without outer membrane.
- Chondriome: All mitochondria in a cell.
- Chondriocont: Small rod of mitochondria.
- Chondriomite: Fragmented mitochondria
- Chondriosphere: Fused mitochondria

Functions:

1. Mitochondria produces large amounts of energy by oxidation of organic substrates such as carbohydrates, proteins and fats. The energy is used in biosynthesis, membrane transport, cell division, movement etc. Hence it is known as the powerhouse of the cell (as coined by Siekevitz).
2. It forms several intermediates for the synthesis of chlorophyll, cytochromes, alkaloids, steroids, pyrimidines etc.
3. It is involved in the synthesis of long-chain fatty acids in animal cells (β-oxidation).
4. It is involved in the synthesis of glutamic acid (an amino acid) from α-ketoglutaric acid, and aspartic acid (an amino acid) from oxaloacetic acid.
5. It shows cytoplasmic inheritance (maternal influence).
6. It is involved in cell signalling, cellular differentiation and cell death.

PLASTID

The plastid is a flat and circular cell organelle found in plant cells and some Protista. The term 'plastid' was coined by Haeckel (1861).

Types of Plastids: Schimper divided plastids into three types based on types of pigments.

1. **Leucoplast**: It is the largest and colourless plastid. It does not contain chlorophylls. It stores food in roots, underground stems and seeds where light is not available. It is of three types
   (a) Amyloplast: It stores starch (carbohydrate).
   (b) Aleuroplast: It stores proteins.
   (c) Elaioplast: It stores oils and fats.
2. **Chromoplast**: It is second largest and coloured other than green. It contains carotenoids. It is found in the petals of flowers; and fruits in which colouration is provided by a pigment, anthocyanin which is dissolved in cell sap. e.g., Phaeoplast (brown algae) and rhodoplast (red algae).
3. **Chloroplast**: It is a green plastid containing a high concentration of chlorophyll.

   Discovery: It was discovered by Julius von Sachs. The term 'chloroplast' was coined by Strasburger.

   Distribution: It may change with the amount of light. When chloroplasts are exposed to sunlight, they stack along the anticlinal cell walls to minimise exposure but in the dark, they spread out in sheets along the periclinal walls to maximise light absorption.

   Size: Its size is 3-5μm in diameter. The chloroplast of shade plants (sciophytes) is larger having more grana with more thylakoid area than that of sun plants (heliophytes). It is larger in polyploid cells than in diploid cells.

Shape: It is a biconvex lens or disc or oval shaped in higher plants. Its position and orientation change with the intensity of incident light. Its shape is variable in algae.

Ribbon or spiral shaped	*Spirogyra*
Cup-shaped	*Chlamydomonas*
Collar (girdle) shaped	*Ulothrix*
Rounded	*Chlorella*
Reticulate or branch-like	*Oedogonium*
Star-shaped or stellate	*Zygnema*

Number: It is present in the green parts of a plant. It is usually 20-40 chloroplasts per cell of higher plants. It is one in *Chlamydomonas* and 100 in the palisade cell of the leaf.

Composition: Phospholipids (30%), Proteins (50%), Chlorophyll (10%), carotenoids (5%), RNA (2%) and DNA (1%).

Structure: It is a double membrane cell organelle. It consists of the following parts:

(a) **Outer membrane**: It is highly permeable because of the pores containing porins (intrinsic proteins). It is lipoproteinaceous. It is rich in lipids and porin protein.

(b) **Periplastidial space**: It is space between two membranes. Its thickness is 20 nm.

(c) **Inner membrane**: it contains a central space called stroma which is granular and selectively permeable. It is rich in proteins.

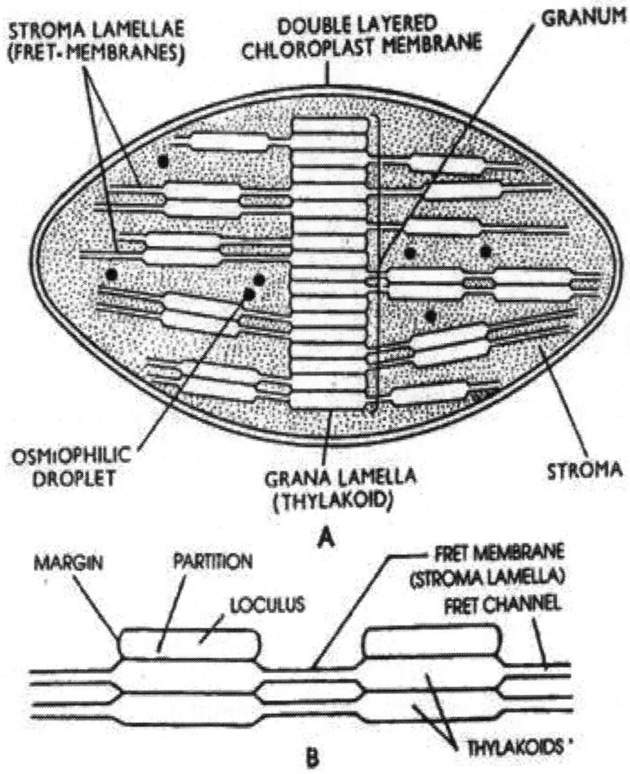

Fig: A. chloroplast, B. Thylakoid

(d) **Stroma**: It is a semi-fluid, alkaline (pH=8) and colloidal complex which contains 50% of soluble proteins. It contains starch granules (pyrenoid in algae), ribosomes, DNA, and RNA. The cpDNA is single, circular, double-stranded, naked (without histone proteins), heavily stained. It contains more genes than mitochondrial genes. It has high G+C content containing about 150,000 base pairs.

Chloroplast is considered as a semiautonomous organelle (cell within cell) because it has its DNA, RNAs, and ribosomes but depends on nuclear DNA for the synthesis of chlorophyll, carotenoids and enzymes for photosynthesis hence it is not fully autonomous.

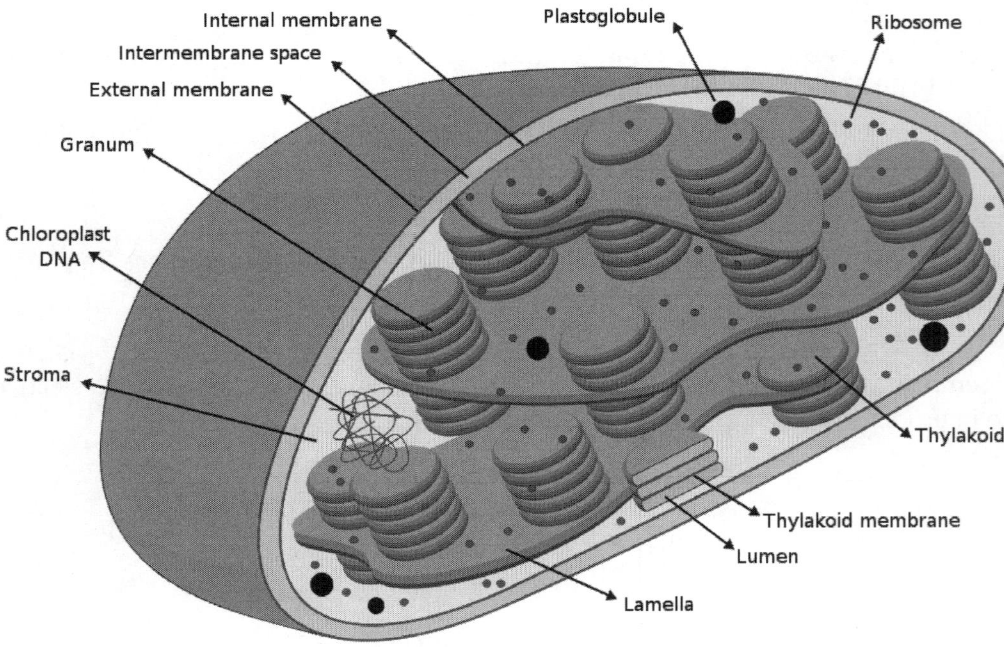

Fig: Chloroplast

5. **Grana**: It consists of membranous, flattened or circular sacs and highly organised structures called thylakoids which are arranged in a stacked manner. It develops from lamella. There are 40-60 grana per chloroplast. Grana are absent in algae and bundle sheath cells of C_4 plants. Each granum is made up of 2-100 thylakoids (usually 10-20). Thylakoid is the structural unit of chloroplast. The light reaction takes place on the thylakoid membranes. The interior of the thylakoid is acidic. Grana are interconnected by a network of tubules called stroma lamellae or intergranal lamellae or Fret. These lamellae are continuous with numerous thylakoids within the granum and also with thylakoids in different granum. These lamellae act as channels for transporting various photosynthetic substrates. The inner side of the thylakoid membrane contains a group of fine, ultrastructural particles called quantasomes (185x155x110A°) arranged in a specific manner. These were discovered by Park and Biggin but the term 'quantasome' was coined by Park and Pons. Quantasome is the smallest functional unit of chloroplast and photosynthesis. Each quantasome contains 230 chlorophyll molecules (160 chlorophyll-a and 70 chlorophyll-b) and 50 carotenoids. These photosynthetic pigments are organised into distinct units or particles of protein complexes called photosystems or Pigment systems. Quantasome has been resolved into four major complexes such as Pigment system-II, cytochrome b_6-f complex, pigment system I and ATP synthase complex (cf_o-cf_1 or coupling factor).

   There are three spaces in chloroplast such as periplastidial space, stromal space and thylakoidal space.

   Origin: Chloroplast divides by fission. Like mitochondria, it also originates from cyanobacteria through endosymbiotic relationships with eukaryotic cells. Secondary endosymbiosis occurs in which chloroplast is developed after mitochondria. Thus, chloroplast is also known as an intracellular prokaryotic parasite or bacterial endosymbiont.

   Evolution: Chromatophore (Photosynthetic bacteria) → Single thylakoid → Granal thylakoid or Chloroplast (Plants)

Terms:

- **Chromatophore:** It is the scattered thylakoids in the cytoplasm.
- **Proplastid:** It is a colourless, immature (undeveloped) and double membranous structure. It is a precursor to plastid. It can divide and differentiate into various types of plastids.
- Etioplast: It is a plastid of etiolated plant (growing in the dark) that lacks chlorophylls but contains chlorophylloid (protochlorophyll) and carotenoids.
- Plastidome: All chloroplasts in a cell.

Functions:

1. It performs photosynthesis (light reaction occurs in thylakoid of grana and dark reaction takes place in stroma).
2. It traps solar energy and converts it into chemical energy (ATP and $NADPH_2$).
3. It stores starch permanently in pyrenoid in algae and temporarily in higher plants.
4. It stores fats in the form of plastoglobulin.
5. It provides colours to flowers and fruits by converting them into chromoplast.

Photochemical conversion: All three types of plastids are photochemically interconvertible.

(a) In the case of Carrot, the leucoplast is converted into chromoplast.

(b) In the case of Tomato, leucoplast is firstly changed to chloroplast then to chromoplast.

(c) In the case of Chilly and Mango, chloroplast is converted to chromoplast.

Proplastid

⇓

Etioplast

⇓

Chromoplast ↔ **Chloroplast** ↔ Amyloplast

(Proplastid also → Chromoplast and → Amyloplast directly)

ENDOPLASMIC RETICULUM

Discovery: It was discovered by Porter, Claude and Fullam (1945). The term 'Endoplasmic reticulum' was coined by Porter (1953).

Distribution: It is absent in prokaryotes. It is present in eukaryotic cells except mammalian RBCs and sperm.

It was the first cell organelle to be revealed by an electron microscope. It constitutes 40% of the endomembrane system (having endoplasmic reticulum, Golgi body, lysosome and vacuole) because these structures remain coordinated but these have no coordination with other membrane-bound organelles). It is poorly developed during cell division. It is well developed in metabolically active or enzyme-secreting cells. It is not a stable structure because it undergoes breakdown and reconstruction. It undergoes partial fragmentation during cell division. It is basophilic due to the high content of RNA hence it is called ergastoplasm.

Endoplasmic reticulum of muscle cell	Sarcoplasmic reticulum
Endoplasmic reticulum of retinal cell	Myeloid body
Endoplasmic reticulum of nerve cell	Nissl's granule

Structure: The endoplasmic reticulum divides the intracellular space into two distinct compartments: luminal (inside ER) and extraluminal (cytoplasmic face). It is made up of a single membrane containing several channels which are interconnected systems. These channels run through the cytoplasm. It may be continuous with the nucleus and plasma membrane. It consists of the following parts:

(a) Cisternae: These are long, flattened, unbranched and sac-like structures that are parallel to each other in bundles. Their size is 400-500A^0.

(b) Tubules: These are tubular and branched or unbranched. These are connected with cisternae or vesicles to form a reticular system. Their diameter is 50-100A^0.

(c) Vesicles: These are small, and oval and remain isolated in cytoplasm. Their diameter is 25-500A^0.

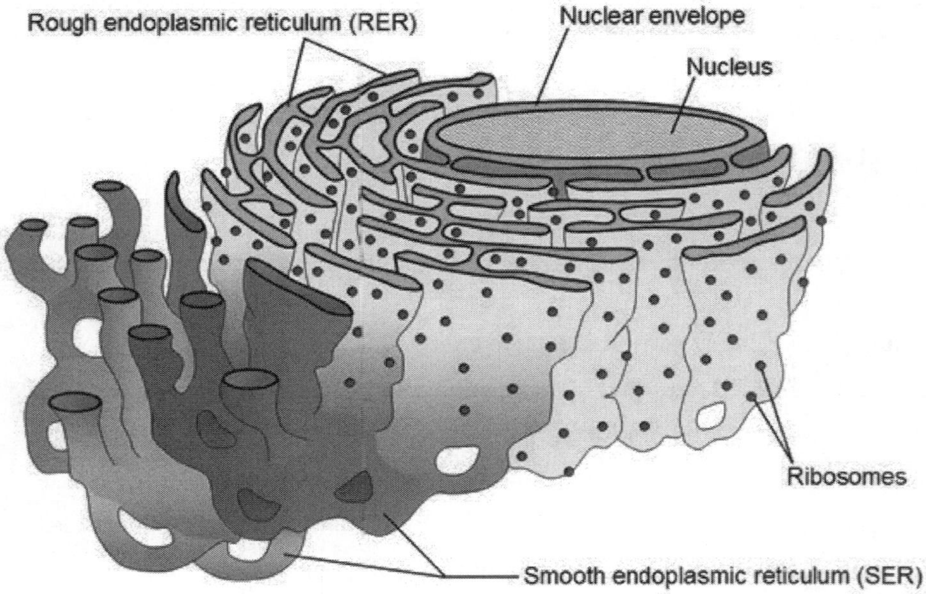

Fig: Endoplasmic reticulum

Types: ER is of two types on the basis of nature of its membrane

Rough Endoplasmic reticulum	Smooth endoplasmic reticulum
It has a rough or granular membrane due to the presence of ribosomes.	It is smooth or agranular due to the absence of ribosomes.
It is first formed.	It is later formed.
It develops from a nuclear membrane.	It develops from a rough endoplasmic reticulum.

It is attached to the nucleus.	It is attached to the plasma membrane.
It occupies two-thirds of total ER.	It occupies one-third of total ER
It is mainly made up of cisternae.	It is mainly made up of tubules and vesicles.
It is more stable and internal.	It is less stable and peripheral.
It contains ribophorins which provide attachment to ribosomes.	It does not contain ribophorins.
It contains more proteins and less lipids.	It contains less proteins and more lipids.
It develops lysosomes.	It develops sphaerosome.
Function: Protein synthesis	**Functions:** Fat synthesis except fatty acids Glycogen synthesis Synthesis of steroidal hormones (testosterone, estrogen) in animal cells. Synthesis of visual pigments from vitamin A in retinal cells. Detoxification of drugs, chemicals, and pollutants in liver cells by cytochrome P_{450}

General Functions:

1. It acts as a cytoskeleton (endoskeleton) along with microfilaments, microtubules and intermediate filaments by providing mechanical support due to the formation of a network-like structure inside the cell.
2. It keeps all cell organelles in position.
3. It provides a passage for quick intracellular transport of proteins by diffusion via nuclear membrane → nuclear pore → RER → transport vesicle → cis-Golgi → trans-Golgi → Secretory vesicle → plasma membrane and thus transports outside for extracellular use.
4. It forms plasmodesmata which join two adjacent cells.
5. It increases the surface area inside the cell for various enzymatic actions.
6. It gives rise to vacuoles.

Terms:

- **Transitional Endoplasmic reticulum**: It is a ribosome-free region of the endoplasmic reticulum. It is considered partly RER and partly SER.
- **Sarcoplasmic reticulum**: It is a modified form of smooth endoplasmic reticulum found in striated muscle. It stores Calcium ions and pumps them into sarcoplasm when muscle fibre is stimulated to permit the relaxation of myofibrils following each round of muscle contraction.
- **Microsomes**: These were discovered by Claude. These are vesicle-like artefacts reformed from pieces of ER when eukaryotic cells are subjected to homogenisation. These can be concentrated and separated from other cellular debris by differential centrifugation. These are not found in healthy and living cells. These are of two types:
  (a) Smooth microsome: It consists of SER, parts of mitochondria, Golgi apparatus and plasma membrane.
  (b) Rough microsome: It consists of RER only.

GOLGI APPARATUS

Discovery: It was discovered by Camillo Golgi (1898) in the cerebrum of a neuron of a barn owl by metallic impregnation method (staining with $AgNO_3$) and termed an 'Internal reticular apparatus'. It is also known as Lipochondria or Baker's body or Idiosome, or Dalton complex or Middleman of cell.

Distribution: It is absent in prokaryotes. It is found in eukaryotic cells except mammalian RBCs, sieve tubes of phloem, antherozoids of bryophytes and pteridophytes. It is large and more numerous in synthetic and secretory cells.

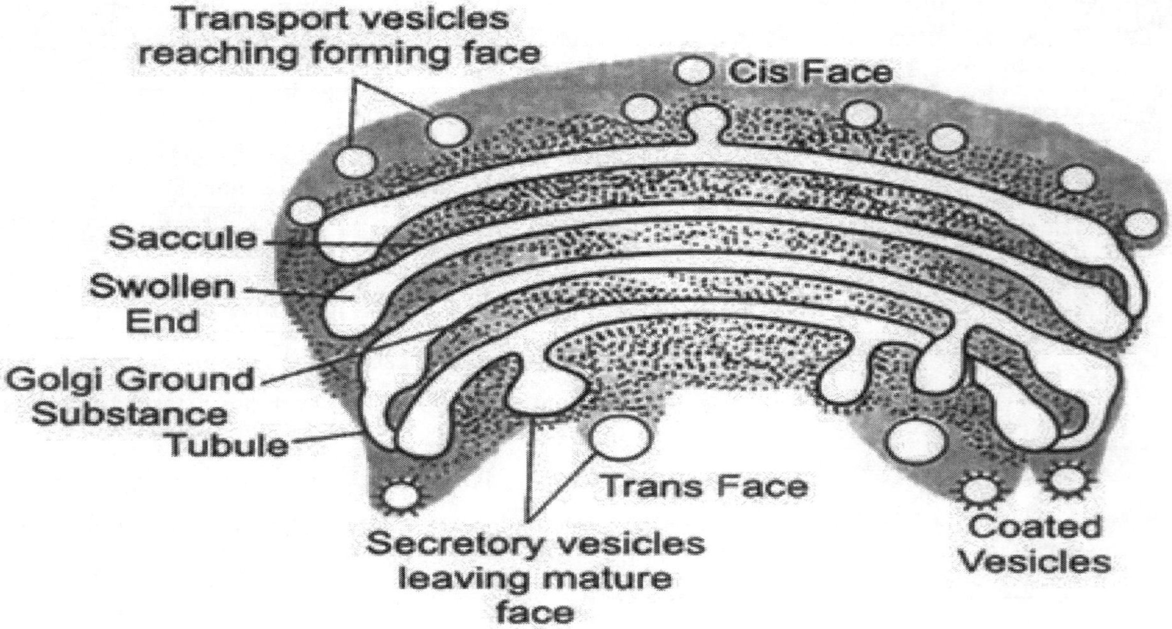

Fig: Golgi apparatus

Its shape, size and number vary because it is constantly being formed, changed, reformed or broken down. It depends on the metabolic activities of the cell. Its number increases during cell division. It is present anywhere in the cell but mostly adjacent to the ER and has a similar organisation in one type of cell.

It comes under the endomembrane system (endoplasmic reticulum, Golgi apparatus, lysosome, vacuole) which functions in close coordination with one another.

It shows a zone of exclusion because it is found in the area of cytoplasm where other cell organelles are absent.

Origin: It develops from the plasma membrane, smooth endoplasmic reticulum (SER) and nuclear membrane.

Composition: Lipids (phospholipids), proteins and enzymes (glycosyl transferase, thiamine pyrophosphatase).

Structure: It is a single membrane-bound cell organelle. It consists of the following parts:

(a) **Cisternae:** These are disc-shaped, flattened and membrane-bound intercommunicating channels and arranged in a parallel stacked manner. One stack contains 3-20 cisternae (usually 6). The cisternae vary in number, shape and organisation in different cells. Cisterna is a constant element and functional unit of the Golgi body. Cisternae are scattered and net-like and each stack of cisternae is called dictyosome. In fungi, the dictyosome is unicisternal. Each stack of cisterna is polarised structure because its margin is curved. It has two faces: cis and trans. Both faces differ in thickness, enzyme content and degree of vesicle formation.

Cis-face: It is also called convex face or proximal face or forming face. It is a collection of fused, flattened and membrane-enclosed discs. It is associated with the nuclear membrane or endoplasmic reticulum. It receives materials and vesicles from the cytosol and endoplasmic reticulum. It is also called the entry face.

Trans-face: It is also called concave face or distal face or maturing face. It is towards the plasma membrane. It gives out vesicles and Golgi vacuoles. It is also called the exit face.

(b) Tubules: These are flat and tubular structures arising from the peripheral area of cisternae.

(c) Vesicles: These are small and sac-like structures attached to tubules. These may be smooth vesicles (secretory vesicles) which are numerous and coated vesicles which are less in number.

(d) Golgian vacuoles: These are large, rounded and expanded parts of cisternae. These develop on the trans face.

Functions:

1. The Golgi body is involved in processing, sorting, packaging, transporting and release of secretory proteins

coming from ER outside the cell (for extracellular use) by exocytosis or reverse pinocytosis along the route (RER→transport vesicle→cis-Golgi→median-Golgi→trans-Golgi→secretory vesicle→Plasma membrane. The proteins are also used for intracellular use. It also performs post-translational modification of proteins.

2. It forms glycoprotein by glycosylation of protein.
3. It forms glycolipid by glycosidation of lipid (addition of oligosaccharide to phospholipid of membrane).
4. It helps in the formation of cell walls in plants during cytokinesis where cell plate is formed by secreting pectin (polysaccharide).
5. It lubricates the root tip by secreting mucilage.
6. It helps in the formation of primary lysosomes.
7. It forms an acrosome of sperm during spermiogenesis.
8. It forms insulin, melanin and yolk.

RIBOSOME

Discovery: It was discovered by Palade in animal cells and termed it 'microsome'. The term 'Ribosome' was coined by Claude.

Size: It is the smallest cell organelle. Its size is 20x15 nm in diameter.

Distribution: It is found in both prokaryotes and eukaryotic cells. It is the most abundant cell organelle. It is found floating freely in the cytoplasm of prokaryotes but in the case of eukaryotes, it is found floating freely in the cytoplasm as well as attached to nuclear membrane and RER by hydrophobic interaction. It is also found in the matrix of mitochondria and stroma of chloroplast of eukaryotic cells.

Number: It is several thousand (20,000) in prokaryotes and millions in cells of eukaryotes.

Origin: It originates from the nucleolus.

Synthesis: In prokaryotes, the ribosomes are synthesized in the cytoplasm but in eukaryotes, these are synthesized in nucleolus.

Composition: rRNA (65%) and proteins (35%). Lipid is not present.

Structure: Its structure is not stable as it is constantly being bound and released from the membrane. It is not bounded by a unit membrane. Hence it is an organelle within an organelle. The basic structure of ribosomes is due to rRNA. Proteins play a secondary role. Ribosome is strongly negative because the positive charges are not enough to neutralize the negative charges of RNA. It consists of two distinct unequal subunits- the smaller one is oval and the larger one is dome-shaped. The smaller subunit has a head, base, platform, and cleft. The larger subunit has a ridge, valley, protuberance, and stalk.

These two subunits are dependent on Mg^{+2} concentrations (0.001M). These two subunits separate if Mg^{+2} concentrations are less in the cytoplasm. These two subunits unite if Mg^{+2} concentrations are increased ten times. Both subunits are found separately in the cytoplasm but rejoin at the time of protein synthesis.

Isolation: Ribosomes are isolated from other cell components by fractionation and ultracentrifugation. These are so small and the last organelle to be sedimented in a centrifuge. It is measured in the Svedberg unit named after Svedberg. It is the rate or velocity of the sedimenting molecule per unit of gravitational field in the dense solution of sucrose. It is a measure of how rapidly a particle (organelle or macromolecule) sediments. The greater is S number, the greater the rate of sedimentation. The shape, size, and density (molecular weight) of ribosome depend on the sedimentation coefficient in an ultracentrifuge.

Types: Ribosomes are of two types based on the sedimentation coefficient

70 S	80 S
Its size is smaller.	Its size is larger.
It has a lower molecular weight.	It has a higher molecular weight.
It is found in prokaryotes.	It is found in eukaryotes.
It is synthesized and assembled in the cytoplasm.	It is synthesized and assembled in the nucleolus (5S rRNA in nucleoplasm) and final maturation occurs in the cytoplasm.

It has two subunits: 30S and 50S.	It has the 40S and 60S
Three molecules of rRNA are found →30S has 16 S rRNA (1500 bases). →50S has 23 S (3000 bases) and 5S (120 bases).	Four molecules of rRNA are found. →40S has 18S rRNA (2000bases) →60S has 28S rRNA (5000 bases), 5S rRNA (160 bases), and 5.8 S rRNA (120 bases).
rRNA component is 60% in 30 S and 50% in 50 S	rRNA component is 50% in 40 S and 65% in 60 S
Protein content is 21 in 30S and is 31 in 50S	Protein content is 33 in 40 S and 49 in 60S
The ratio of rRNA and protein is 2:1.	The ratio of rRNA and protein is 1:1.

Functions:

1. Attached ribosomes (ER bound) synthesize secretory proteins, integral membrane proteins, and lysosomal proteins for extracellular use (transport). These ribosomes form most of the proteins.
2. Free ribosomes synthesize non-secretory and structural proteins such as soluble cytoplasmic proteins, peripheral proteins, mitochondrial proteins, chlorophyll proteins, nuclear proteins, and peroxisomal proteins for intracellular use.

Microbodies: These are very small (<1μm), rounded cell organelles bounded by a single membrane. These are absent in prokaryotes. These include:

1. **Lysosome:** It was discovered by Christian de Duve (1955) in the liver cell of a rat. The term 'lysosome' was coined by Navikoff. It is also known as the pericanalicular body. It is a small, rounded, and dense body bounded by a single membrane. It is absent in prokaryotes but it is present in animal cells of eukaryotes except mammalian RBCs. It is usually absent in plant cells where the function of the lysosome is performed by vacuoles and sphaerosome. It has been reportedly found in the root tip of maize, yeast, and *Neurospora.*

   It is named a lysosome because it contains hydrolytic enzymes. It is rich in 60 types of hydrolytic enzymes which function at pH 4.5-5. Some of the enzymes are phosphatase (common and marker enzyme), protease, lipase, sulfatase, glycosidase etc. It contains a special transport protein that pumps H^+ into the cavity of enzymes. Its single membrane is impermeable to the substrate of enzymes and resistant to hydrolase. It is also called the suicidal bag of cells because it destroys cells once enzymes are released in the cell.

   It is polymorphic hence it passes through different forms in the same cell based on function. These forms are as follows:

   1. **Primary lysosome (Storage granule)**: It is very small in size and a newly formed lysosome. It is formed when an enzyme of lysosome is formed on RER and is transported to the Golgi apparatus which buds off in cytoplasm. It contains only one enzyme which fuses with larger acidic vesicles.
   2. **Secondary lysosome (Heterophagosome)**: It is formed when the primary lysosome is fused with vesicles such as phagosome or pinosome formed by plasma membrane for endocytosis. These vesicles pass digested food materials into the cytoplasm for absorption and secondary lysosome is left with undigested food materials. It contains primary lysosome and pinosome or phagosome.
   3. **Tertiary lysosome (Residual body)**: It is formed when a secondary lysosome containing undigested food materials is fused with a vesicle formed by plasma membrane for exocytosis to remove debris outside.
   4. **Autophagosome (Cytolysosome)**: It is formed by the fusion of primary lysosome with dead or worn-out cell organelles which are engulfed and later digested by autodigestion. It helps in the disposal of cell debris.

      Origin: It is formed by the joint activity of the Golgi apparatus and Endoplasmic reticulum.

      GERL= G + ER → Lysosome.

      Lysosome stabilizers: These chemicals strengthen lysosomes because these prevent enzymes from leaking into the cytoplasm and causing the destruction of cell organelles. e.g., cortisol, cortisone, cholesterol, heparin, antihistamines.

Lysosome labilizers: These chemicals weaken lysosomes because these help release enzymes outside the lysosome by rupturing the membrane and destroying cell organelles. e.g., Progesterone, testosterone.

Functions:

1. It helps in the intracellular digestion of stored substances such as proteins and fats present inside the cell during starvation or fasting and thus it provides nourishment.
2. It helps in the extracellular digestion of substances by releasing acid hydrolases outside the cell.
3. It provides immunity to the body by phagocytic activity against microbes (viruses and bacteria).
4. It helps in scavenging by removing old, dead and useless organelles.
5. It is engaged in the autolysis of old, diseased and useless cells.
6. It secretes thyroid hormones in the epithelial cells of the thyroid due to the presence of hydrolase enzymes.

2. **Peroxisome:** It was discovered by Tolbert (1960) in the liver cell but in plant cells (leaf cell of spinach) it was discovered by de Duve It is also called uricosome. It is small, rounded, and bounded by a single membrane. It is the smallest membrane-bound cell organelle. It is absent in prokaryotes but present in plant and animal cells of eukaryotes. It arises from the cisternae of the ER. It is around 100 in number and found in close association with RER, mitochondria and chloroplast.

   It contains peroxide-producing enzymes such as peroxidase, D-amino acid oxidase and urate oxidase. These oxidise the substrate and produce H_2O_2 which is a highly reactive, toxic, and oxidising agent. It also contains catalase which degrades H_2O_2 to H_2O and O_2. Catalase is the fastest-acting enzyme.

   Functions:

   1. It is involved in the process of photorespiration (glycolate cycle or C_2 metabolism) along with mitochondria and chloroplast in the cell of C_3 plants.
   2. It carries out lipid metabolism (β-oxidation of long-chain fatty acids) in animal cells.

3. **Glyoxisome:** It was discovered by Beevers (1951) in the endosperm cell of the castor. It is small rounded and bounded by a single membrane. It contains enzyme enzyme-rich matrix and crystalloid core. It is found in plant cells but absent in animal cells. It arises from RER. It occurs in seedlings (germinating seeds) of oil seed plants. It contains isocitrate lyase, malate synthetase, glycolate oxidase, and catalase hence it is also called specialized peroxisome.

   Functions:

   1. It is involved in gluconeogenesis or glyoxalate cycle (conversion of fats into glucose).
   2. It breaks down fatty acid into acetyl-CoA (β-oxidation) in cells of oil seeds.

4. **Sphaerosome (Oleosome):** It was discovered by Dangeard and Hanstein (1880). It is small, rounded and bounded by a single membrane. It is found in plant cells. It is also known as plant lysosome. It arises from SER. It is abundantly found in the endosperm and cotyledon of oil seeds. It is stained with osmium tetroxide and Sudan dye. It contains hydrolytic enzymes such as hydrolase, protease, esterase, phosphatase, ribonuclease etc., and a high percentage of lipids (95%).

   Function: It synthesizes and stores fats (lipid metabolism).

CENTRIOLE

It is a small, hollow and cylindrical structure (diameter=150 nm).

It was first observed by Von Beneden and coined by Theodor Boveri.

It is not bounded by any membrane. It is absent in prokaryotes, fungi, gymnosperms and angiosperms. It is found in motile cells of lower plants (from algae to pteridophytes) and animal cells except mammalian RBCs.

It is situated in a clear area of cytoplasm near the interphase nucleus called centrospheres or kinoplasm. The position of the centriole determines the position of the nucleus. It is present in two sets (pairs) which are at right angles to one another called diplosomes. Thus, centrioles and centrospheres constitute centrosomes. Thus, the centrosome is an organelle containing two cylindrical structures called centrioles.

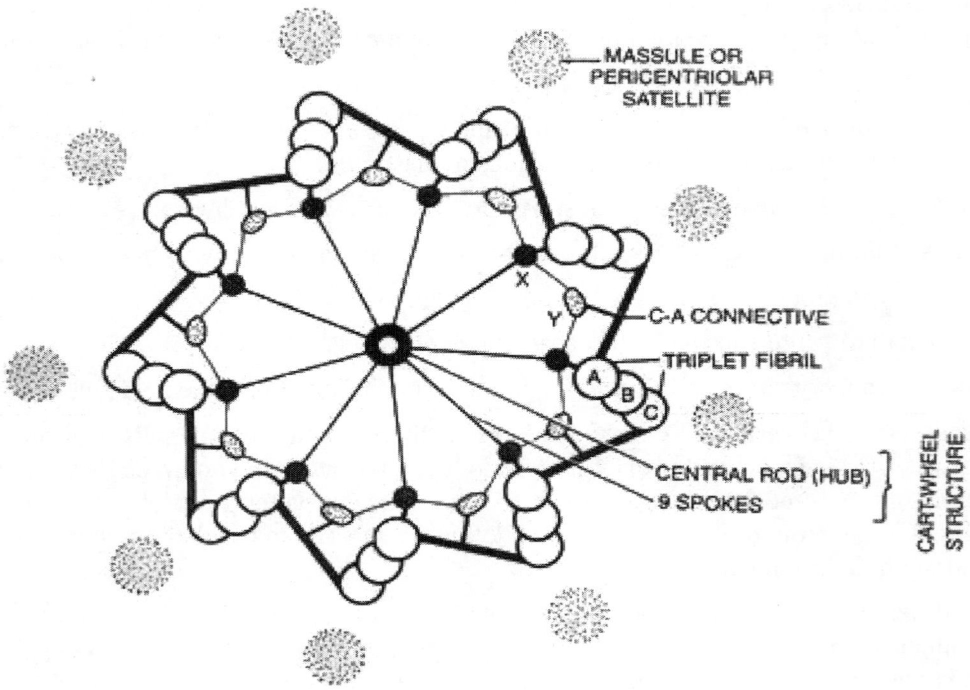

Fig: Centriole

Structure: It is a cylindrical structure. Centriole is surrounded by a dense, amorphous and protoplasmic sphere in one or more series called massules or perinuclear satellites. These form new centrioles.

Each centriole is made up of 9 sets of peripheral triplets which are arranged parallel to each other. These triplets are evenly spaced showing a 9+0 arrangement. Each triplet fibril has 3 subfibres (A-B-C). Each triplet is situated at an angle of 40° with a radius of the cylinder. The adjacent triplets are connected by A-C proteinaceous linkers. The first tubule is connected with the third tubule of the next. Each subfibre has 13 protofilaments like microtubules. At the centre, there is a proteinaceous mass called hub which is connected with 9 triplets by 9 radial spokes. Each radial spoke has a thickening called X before joining with A subfibre. Another thickening called Y is present nearby and attached both to X and A-C linkers by connectives. Thus, the whole structure looks like a cartwheel.

Origin: Centriole is synthesised during S phase but not replicated.

Functions:

1. Centriole gives rise to the basal body (9+0) which forms cilia and flagella (9+2).
2. Centriole forms spindle fibres and asters (rays radiating in all directions) during cell division.

CILIA AND FLAGELLA

These are fine hair-like and contractile appendages of cells. These are outgrowths of the plasma membrane and extend outside the cell through pores of the cell wall. These are absent in cyanobacteria, rhodophyceae, *Pinus* of gymnosperm, and angiosperm. These may be isokont (of same length) and heterokont (of different lengths). Flagella show different places of insertion.

(a) Akrokont: Flagella are inserted apically

(b) Opisthokont: Flagella are inserted posteriorly

(c) Pleurokont: Flagella are inserted laterally.

Structure: It is cylindrical and the same for both cilia and flagella. It contains a core (axoneme or shaft) which is surrounded by a plasma membrane. The axoneme is made up of 9 peripheral or radially arranged doublets of microtubules and 2 central singlets of microtubules showing a 9+2 arrangement. Each doublet gives rise to a pair of arms which is made up of dyenin (acting as ATPase to provide energy for the movement). The doublets are connected by A-B linkers that are made up of nexin. The central singlet is connected by a bridge and enclosed in a central sheath. Each singlet fibril is made up of 13 protofilaments. The central sheath is connected to each doublet by radial spoke which is made up of nexin. The movement of cilia and flagella is generated by the microtubule.

Origin: These arise from the basal body.

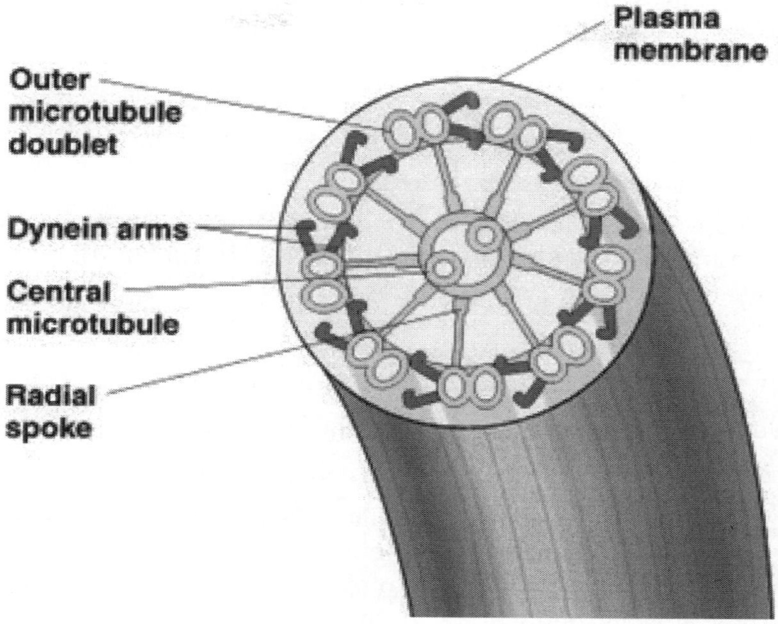

Fig: Flagella

Differences between cilia and flagella

Cilia	Flagella
These are hair-like appendages.	These are thread-like appendages.
These are present in eukaryotic cells.	These are found in prokaryotic and eukaryotic cells.
These are larger in number (300-14000).	These are lesser in number (1-4).
These are shorter in length.	These are longer.
These are present all over the surface.	These are present at the end of cell.
These beat isochronously or synchronously (simultaneously) and metachronously (successively).	These beat independently.
These move very fast.	These move very slow.
These show rotational movement like a motor.	These show wavelike or undulating movement.
These help in locomotion, feeding, circulation.	These help in locomotion.

Differences between prokaryotic flagella and eukaryotic flagella

Prokaryotic flagella	Eukaryotic flagella
These run in rotary movement.	These run in bending movement.
These use rotary motors.	These use a sliding filament system.
These are proton-driven due to the concentration of gradient set up by the cell's metabolism.	These are ATP-driven.

Cytoskeletons: These are minute and fibrous structures found in eukaryotic cells. These include microfilaments, intermediate filaments and microtubules.

Microfilaments: These are long, narrow, solid (rod-like) unbranched fibrils. These were discovered by Paleviz (1974). These are found just below the plasma membrane. These are made up of G-actin proteins. Their thickness is 7 nm. These look like beads due to the close helical arrangement of actin. These are contractile. These form extensive networks in the cytoplasm.

Functions:

1. These help in cyclosis (cytoplasmic streaming) or cell movement by which transport of nutrients occurs.
2. These help in the movement of the plasma membrane during endocytosis.
3. These form cell plates and cell furrows.
4. These form pseudopodia.
5. These help in sol-gel transformation.

Intermediate filaments: These are tubular, solid and branched filaments. These are scattered in the cytoplasm. These are made up of structural proteins such as desmin, keratin, kinesin etc. These are cross-linked with one another containing 5 protofilaments. These are non-contractile. Their thickness is 10 nm.

Functions:

1. These form a nuclear matrix (nuclear lamina) present on the surface of the inner membrane of the nucleus.
2. These form a scaffold (supporting array) for chromatin and cytoplasm.
3. These support all membrane-bound cell organelles including the nucleus

Microtubules: These are very long, tubular hollow and unbranched fibrils. These were discovered by de Robertis and Franchi (1953) in the axon of the medullated nerve fibre. These are found in centriole, cilia and flagella. Their wall is made up of 13 parallel protofilaments which are composed of alternate spirals of α and β- tubulin proteins. These are rich in ATPase. These are non-contractile. Their thickness is 25 nm. These require GTP and Ca+2 for their assembly and disassembly.

Functions:

1. These help in the formation of spindle and cell plate during cell division.
2. These maintain the shape, form and rigidity of the cell.
3. These help in the anaphasic movement of cells.
4. These help in the movement of cilia and flagella.
5. These help in the intracellular transport of materials.

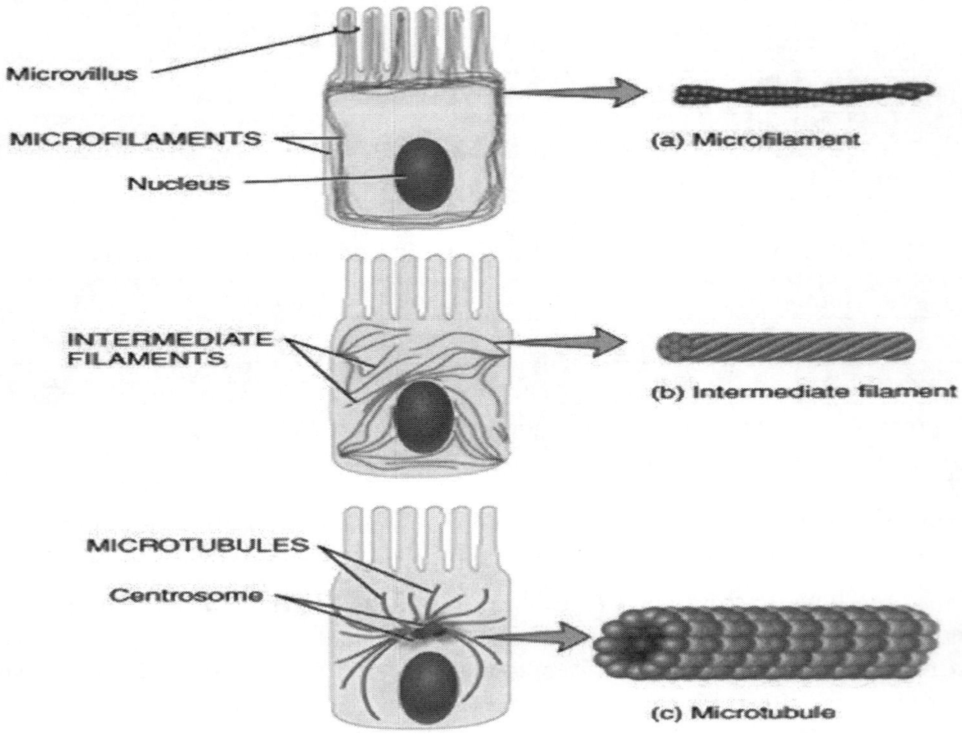

NUCLEUS

Discovery: It was discovered by Robert Brown (1833) in the cell of the root of the Orchid. It is a rounded, dense and highly specialised body. It is studied best at the interphase stage. It is considered a component because it is extracytoplasmic.

Size: Its size is 5-25µm. It is large in active cells. It depends on protein and DNA content; metabolic activity and volume of cell.

Shape: It is lens-shaped and peripheral in plant cells and rounded and central in animal cells.

Number: It is usually one per cell but there are several exceptions which are as follows.

(a) Binucleate condition (micronucleus and meganucleus) is seen in *Paramecium* but dikaryotic condition is seen in *Puccinia* (rust fungus) containing nuclei from opposite mating types.

(b) Multinucleate condition (syncytium) is seen in the epidermis of *Ascaris* where fusion of cells occurs.

(c) Multinucleate condition (coenocyte) is seen in *Rhizopus, Mucor* (fungi), plasmodium of slime moulds and liquid endosperm of coconut where nuclear division occurs without cytokinesis.

Distribution: It is absent in prokaryotes but present in eukaryotic cells except mammalian RBCs and sieve tubes of phloem. It occupies 10% of the volume of the cell.

In a cell, there is a definite nucleo-cytoplasmic ratio.

Nucleo-cytoplasmic index = Volume of Nucleus/ (Volume of Cell - Volume of a nucleus)

Composition: Its dry weight contains protein (90%), DNA (9%) and RNA (1%).

Structure: It is a double membrane cell component. It consists of

(a) **Nuclear membrane**: It is a double membrane and porous. It is simple in plants and annulated in animals. It is lipoproteinaceous. The outer membrane communicates with ER at several points. The outer membrane shows the presence of ribosomes at the outer surface. Perinuclear space (20 nm) is the space between outer and inner membranes. It is filled with fluid. It has 1000-10,000 minute circular pores (size=20-80 nm) which serve as a means of nucleo-cytoplasmic exchange of materials (mRNA, proteins, micronutrients, water, ions etc.) in both directions. It is lined with a fibrous protein network called nuclear lamina which provides shape to the nucleus, strength to the nuclear membrane and holds nuclear pores in shape. The area of the membrane where ribosomes are absent, forms vesicles (blebbing). The inner membrane is smooth.

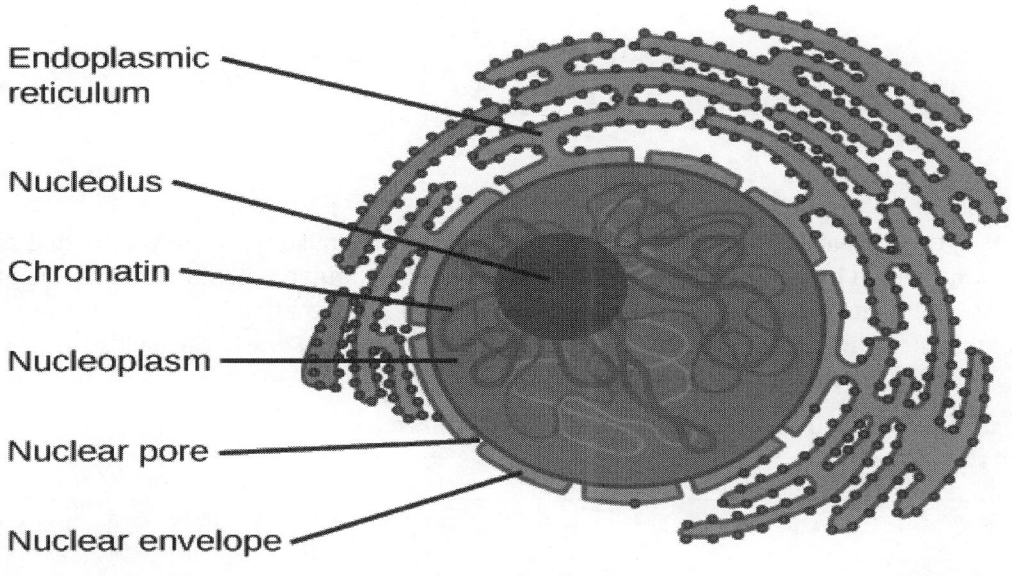

Fig: Nucleus

(b) **Nucleoplasm**: It is a clear, gel-like, and granular ground matrix (pH 7.4). It is more dense than cytoplasm. It contains ions, enzymes (polymerases), proteins, water and nucleotides. Its aqueous nature allows ions and organelles to be transported within the cell. It is a site of enzyme activities and synthesis of DNA, RNA and ribosomal subunits.

(c) **Nucleolus**: It was discovered by Fontana (1781). The term 'nucleolus' was coined by Bowman. It is the rounded and densest structure inside the nucleus. It is not bounded by any membrane. It is absent in prokaryotes. It is present in eukaryotic cells except mammalian RBCs, mammalian sperms, young embryo cells, and yeast. It is one in haploid but one to four in diploid cells but one in haploid cells. It is present anywhere (either at the centre or periphery). It is large and more in number in actively (protein) synthesising cells. It occupies a one-third area of the nucleus. It is found at a secondary constriction or nucleolar organising region (NOR) of the chromosome. This region contains repetitive DNA which forms rRNA.

Composition: Nonhistone proteins (65-70%), rRNA (30%), Ca^{+2}

Structure: It has the following parts:

(a) Nucleolar chromatin: There is perinucleolar chromatin that surrounds nucleoli and intranucleolar chromatin embedded in the matrix.

(b) Amorphous matrix: It is homogeneous containing granules and fibrils. It is rich in RNA polymerase enzymes.

(c) Granules: These are made up of ribonucleoproteins (rRNA and protein=1:2). These are precursors to ribosomes.

(d) Fibrils: These are made up of ribonucleoproteins. This part contains many copies of repetitive DNA which directs the synthesis of rRNA.

Function: It is involved in the synthesis of rRNA and ribosome. It also stores nucleoproteins.

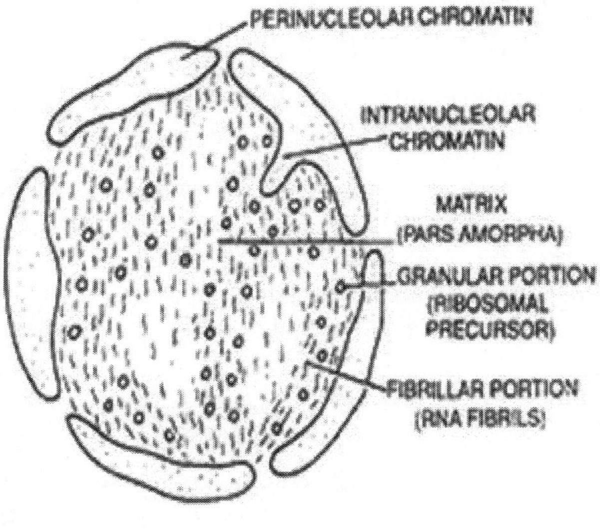

Fig: Nucleolus

(d) **Chromatin**: It is a network of nucleoprotein fibrils. It is a chromosome in fully extended form. The term 'Chromatin' was coined by Flemming (1979) because it can get stained with basic dyes such as acetocarmine or Fuelgen stain. It is found in the interphase of the cell cycle. It is condensed to more tightly coiled threads called chromosomes during the M phase of the cell cycle (cell division) due to the packaging of DNA by histone proteins. Emil Heitz developed a staining technique for chromatin.

Chromatin is differentiated into two regions during interphase.

Euchromatin	Heterochromatin
It is a loosely coiled or diffused region of chromosomes.	It is a tightly coiled or highly condensed region of the chromosome.
It gets a light stain in the interphase.	It gets a dark stain in the interphase.
It is fibrous and less labile.	It is granular and more labile.
It is present towards the centre of the nucleus.	It is present towards the nuclear membrane.
It is present in large areas.	It is present in certain areas.

Euchromatin	Heterochromatin
It contains a large amount af DNA	It contains a small amount of DNA.
It contains loosely packed DNA hence it is available for transcription and recombination (genetically active).	It contains tightly packed DNA hence it is unavailable for transcription and recombination (genetically inactive).
It is condensed during cell division only.	It remains condensed throughout the life cycle.
It replicates during the early S-phase.	It replicates later at the S-phase.
It does not show heteropycnosis.	It shows heteropycnosis.

Types of Heterochromatin

1. Constitutive Heterochromatin: The heterochromatin remains permanently in all cell types at all stages of the life cycle. It is located near the telomeres and centromere. It shows heteropycnosis (difference in thickening between different regions). It contains repetitive sequences.
2. Facultative Heterochromatin: The heterochromatin remains permanently in some special cell types or at some particular stages in the life cycle. It is sometimes expressed. e.g., Barr body.

Functions of the nucleus:

1. It contains genetic information for metabolism, development and reproduction. **Hammerling's grafting experiment:** John Hammerling performed a grafting experiment on *Acetabularia* to prove that the nucleus is the storehouse of all hereditary information. *Acetabularia* has a base, stalk and cap. The nucleus is present in the base. He selected two species: *Actabularia crenulata* and *Acaetabularia mediterranea*. The stalk of one species is grafted on the base of another species. The cap formed is similar to the species to which the base belongs. It proves that the shape of the cap is determined by the nucleus which is present in the base.
2. It passes the genetic information to daughter cells.
3. It contains directions for regulating cellular activities through information stored in mRNA.

CHROMOSOME

It is a thread-like structure found in the nucleus. It is an organised structure of DNA and protein.

Discovery: It was discovered by Hofmeister in the nucleus of the pollen mother cell of *Tradescantia*. The term 'Chromosome' was coined by Waldeyer due to its ability to take up colours in basic dyes.

Number: Its number varies. Chromosomes are present in sets (ploid). Haploid has a single set of chromosomes and diploid has a double set of chromosomes inherited from maternal and paternal parents. Animal cell contains a larger number of chromosomes than plant cells.

Among animals, the lowest number (2n=2) is found in *Ascaris microcephala* and the highest number (2n=1600) is found in *Aulocantha* (a radiolarian protist).

Among plants, the lowest number (2n=4) is found in *Haplopappus gracilis* and *Brachycome lineariloba* and the highest number (2n=1262) is found in *Ophioglossum reticulatum* (a pteridophyte).

Plants	Number of chromosomes	Animals	Number of chromosomes
Garden pea, Barley	2n=14	*Drosophila*	2n=8
Onion	2n=16	Housefly, *Coenorhabditis*	2n=12
Neurospora (a fungus)	n=7	Cockroach	2n=24(male), 2n=23 (female)
Broad bean	2n=12	Mosquito	2n=6
Rice, Tomato	2n=24	Silk worm	2n=56
Wheat	2n=42	Frog	2n=26
Maize	2n=20	Mouse	2n=20
Cotton	2n=52	Dog, Chicken	2n=78

Plants	Number of chromosomes	Animals	Number of chromosomes
Tobacco, Potato	2n=48	Guinea pig, Horse	2n=64
Cabbage, Mustard	2n=18	Rhesus monkey, Rat	2n=42
Sugarcane	2n=80	Man	2n=46
Yeast	2n=32	Gorilla, Chimpanzee	2n=48

Size: Its size varies from 0.5 to 30µm. Its average size is 3µm. It can be measured at the metaphase stage. Its smallest size is found in fungi and birds and the largest size is seen in *Trillium*. The plant cell has a larger size than an animal cell. Among plants, monocots have a larger size than dicots. Each human cell contains roughly 2 meters of DNA.

Shape: It can be observed at the anaphase stage. A chromosome shows different shapes.

(a) Metacentric (V-shaped): The Centromere is at the centre hence both arms are equal in length.

(b) Submetacentric (L-shaped): The Centromere is slightly away from the centre hence one arm (p arm) is shorter than the other (q arm).

(c) Acrocentric (J-shaped): Centromere is close to its end hence one arm (p arm) is much shorter than the other (q arm).

(d) Telocentric (I-shaped): Centromere is terminal because one arm is absent. It is rare.

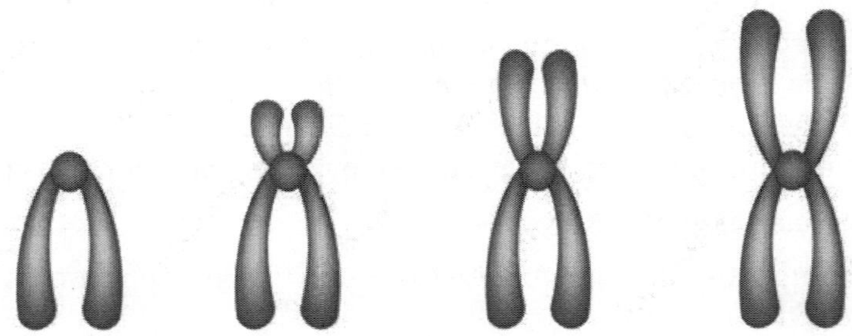

Composition: Chromosome contains

(a) **DNA**: It constitutes 40%. The amount of DNA is the same in all cells of an organism. The amount is constant in different tissues. The amount is constant within one species but varies from one species to another. The amount is related to the number of chromosomes. The amount is proportionate in haploid, diploid and tetraploid.

(b) **Proteins** (60%): These may be

(1) **Histones**: These constitute 35%. These are basic and structural proteins. These are highly heterogeneous and synthesised during S-phase. The amount of DNA and histones is constant. These lack tryptophan (amino acid). These protect DNA from the action of nuclease. These are modified by acetylation, methylation and phosphorylation. These are of different types based on differences in the sequence of amino acids.

H_1: It is the most divergent protein. It is lysine-rich. It is the largest in size. It is more easily extracted from chromatin. It is almost species-specific in its sequence.

H_2a and **H_2b**: These are less conserved than H_3 and H_4. These are slightly lysine-rich (Lysine is moderately more than arginine).

H_3 and **H_4**: These are highly conserved. These are arginine-rich.

Functions:

1. Histones stabilise the structure of chromosomes.
2. These protect DNA from the action of nuclease.
3. These neutralise the anionic charges of phosphate in DNA.
4. These repress the activity of gene.

(2) **Non-histones**: These constitute 25%. These are acidic proteins. These are synthesised during the entire life cycle. These are 15-20 in number. These differ greatly in tissues and species. These are complex and not as observed as histones.

Functions:

1. These stimulate gene activity.
2. These induce the formation of DNA in eukaryotes.

Structure: It is a cylindrical double structure that consists of two longitudinal halves called sister chromatids. A chromatid is called a half chromosome or daughter chromosome. Each chromatid has two short (p) and long (q) arms. These chromatids are held together at a single point called centromere (also known as primary constriction). The chromosome is narrower in this region. The centromere is functional during cell division while the rest part of a chromosome is generally inactive. Usually, there is a single centromere (monocentric) in a chromosome but several centromeres (polycentric) chromosome) are found in diffused conditions (holocentric) along the length of a chromosome in *Ascaris*, insects, and *Spirogyra*. The position of the centromere is fixed for a particular chromosome, but its position varies in different chromosomes. The position of the centromere determines the relative length of the arm.

$$\text{Arm ratio} = \frac{\text{Length of the longer arm}}{\text{Length of the short arm}}$$

The arm ratio is high in acrocentric chromosomes. Centromere contains highly repetitive DNA called satellite DNA.

The protein disc attached to the centromere is called a kinetochore. It is C-shaped and two in number. It is the actual site of attachment of spindle fibres to the centromere.

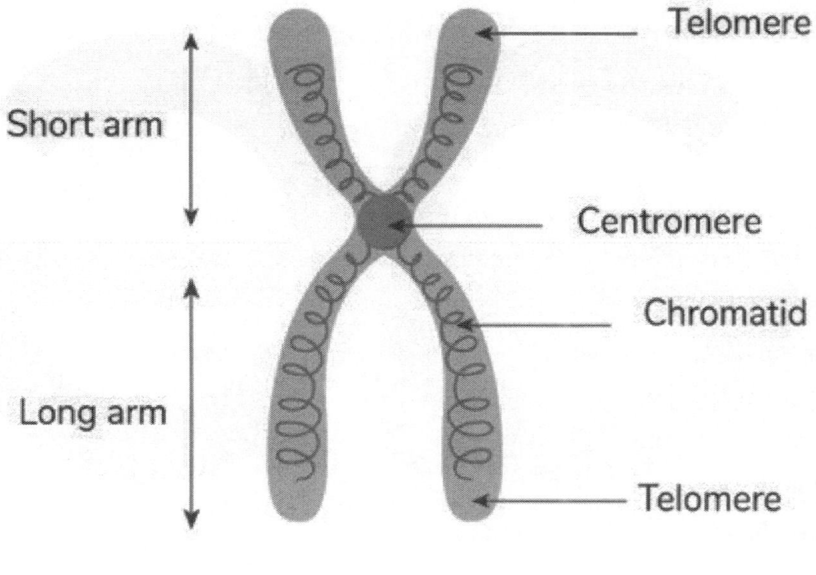

Fig: Chromosome

Secondary constriction is also observed in some chromosomes at constant positions in addition to primary constriction. Secondary constriction lacks centromere and is narrower and metabolically more active than primary constriction. It occurs in the distal end and is fixed in position hence used as a marker. Secondary constriction cuts off a small fragment of chromosome at the distal end called trabant or satellite (terminal part of a chromosome beyond secondary constriction). It remains attached to the rest of the chromosome by a thread of chromatin. The chromosome containing a satellite is called a satellite chromosome. Chromosomal satellite is different from satellite DNA. The number of SAT (Sine acid Thyminucleinico) chromosomes varies in the genome. A human somatic cell has 10 SAT chromosomes (13th pair, 14th pair, 15th pair, 21st pair and 22nd pair). Secondary constriction is the site of the formation of nucleolus hence called nucleolar organising region (NOR). The NOR of each SAT chromosome contains several hundred copies of the genes coding for rRNA. The SAT-DNA constitutes about 10% of the total DNA present in the genome of a species. The NOR is essential for the formation of nucleolus.

Each chromatid is composed of an elongated, coiled filament called chromonemata which bears several beaded or gene-bearing structures called chromomeres. It lies embedded in the matrix which is further enclosed in a proteinaceous sheath called the pellicle. It is present throughout the length of the chromosome. Each chromonemata consists of a single thread of DNA associated with histones. The number of chromonemata is two in prophase and metaphase but one in anaphase.

Telomere is the end of a chromosome. It is highly stable. It does not unite with the telomere of another chromosome hence it is not sticky and it ensures survival and polarity. It contains a short tandem repeat DNA sequence. It also contains a terminal single strand which is folded to form a hairpin loop giving it stability. The telomere facilitates the attachment of chromosomes with the nuclear membrane.

Types of giant chromosomes:

1. **Polytene chromosomes:** These were discovered by Balbiani (1881). These are found in the cells of the salivary gland of the larva of the chironomus insect (dipteran). These are also found in the gut, trachea, fat body cells and malpighian tubules. These are wide and 1000 times longer than somatic chromosomes. These contain several light bands and dark bands. The bands are enlarged to form puff or Blbiani rings involved in the synthesis of RNA.
2. **Lampbrush chromosomes:** These were discovered by Ruckert (1892). These are found in the diplotene stage of prophase 1 of meiosis in nuclei of amphibian oocytes. These contain loops which are involved in the synthesis of RNA and proteins.

SHORT ANSWER TYPE QUESTIONS

1. Write the general unifying principles of cell theory
2. Explain the differentiated cell, dedifferentiated cell and redifferentiated cell.
3. Distinguish between prokaryotic cells and eukaryotic cells.
4. Distinguish between plant cells and animal cells.
5. Write notes on vacuole.
6. Draw a well-labelled diagram of fluid mosaic model of the plasma membrane.
7. Why mitochondria are considered as semi-autonomous organelle?
8. Discuss the functions of the Golgi body.
9. Distinguish between 70 S and 80 S ribosomes
10. Discuss the functions of lysosomes.
11. Draw a well-labelled diagram of centriole.
12. Distinguish between cilia and flagella.
13. Distinguish between heterochromatin and euchromatin.
14. Write notes on nucleolus.
15. Explain the different shapes of chromosomes.

LONG ANSWER TYPE QUESTIONS

1. Describe the structure of a typical bacterial cell with a well-labelled diagram.
2. Describe the structure and functions of the chloroplast.
3. Describe the structure and functions of the Golgi body.
4. Describe the structure and functions of the nucleus.
5. Draw a well-labelled diagram of a chromosome and describe its different parts.

MULTIPLE CHOICE QUESTIONS

1. **Which of the following parts of the cell is differentially permeable?**
(a) Cell wall (b) Middle lamella
(c) Cell membrane (d) Cuticle

2. **The first layer formed in the cell wall as a result of cytokinesis is**
(a) Middle lamella (b) Primary wall
(c) Secondary wall (d) Tertiary wall

3. **The permeability of the plasma membrane is provided by**
(a) Extrinsic proteins
(b) Intrinsic proteins
(c) Lipids
(d) Sugar

4. **The absorption of dead cells is brought about by**
(a) ER (b) Lysosomes
(c) Vacuole (d) Golgi apparatus

5. **Thickness of the plasma membrane is**
(a) 20 nm (b) 7.5 nm
(c) 15 nm (d) 10 nm

6. **The examples of semiautonomous organelles are**
(a) Mitochondria and nucleus
(b) Chloroplast and Endoplasmic reticulum
(c) Mitochondria and chloroplast
(d) Endoplasmic reticulum and Golgi apparatus

7. **Hammerling performed experiments to prove that the nucleus was the physical basis of heredity while working on**
(a) *Neurospora* (b) *Drosophila*
(c) *Acetabularia* (d) *Aspergillus*

8. **A cluster of ribosomes held together on mRNA is called**
(a) Mesosome (b) Dictyosome
(c) Polysome (d) Microsome

9. **Hydrophobic ends of phospholipid molecules in the plasma membrane are**
(a) Polar (b) Neutral
(c) Non-polar (d) Bipolar

10. **Lysosome along with food content is known as**
(a) Primary lysosome
(b) Secondary lysosome
(c) Residual body
(d) Microbody

11. **The subunit of chromatids is**
(a) Chromomere (b) Chromonemata
(c) Telomere (d) Centromere

12. **Lysosomes occur mostly in:**
(a) Algae (b) Bacteria
(c) Plants (d) Animals

13. **The resin duct of *Pinus* is a:**
(a) Schizogenous cavity
(b) Lysigenous cavity
(c) Schizo-lysigenous cavity
(d) None of the above

14. **The correct sequence of the cell membrane as per the model of Singer and Nicolson is:**
(a) L – P – P – L (b) P – P – L – P
(c) P – L – L – P (d) L – P – L – P

15. **Which of the following chemical constituents is absent in ribosomes?**
(a) Protein (b) Lipid
(c) rRNA (d) None of these

16. **Rough ER is well developed in cells involved in**
(a) Lipid synthesis
(b) Protein synthesis
(c) Secretory functions
(d) Nucleotide synthesis

17. **Succinic dehydrogenase enzyme of Krebs cycle is present at**
(a) Outer membrane of mitochondria
(b) Inner membrane of mitochondria
(c) Perichondrial space of mitochondria
(d) Matrix of mitochondria

18. **Continuity of cytoplasm from cell to cell is maintained through**
(a) Middle lamella (b) Plasmodesmata
(c) Endoplasmic reticulum
(d) Cell membrane system

19. All are membrane-bounded cell organelles except

(a) Spherosomes (b) Mitochondria
(c) Ribosomes (d) Lysosomes

20. Golgi apparatus is absent in

(a) Liver cells (b) Mammalian RBCs
(c) Green algae (d) Higher plants

21. Which of the following cell organelles does not belong to microbodies?

(a) Lysosome (b) Ribosome
(c) Peroxisome (d) Glyoxysome

22. The largest membrane-bound cell organelle is

(a) Lysosome (b) Ribosome
(c) Peroxisome (d) Glyoxysome

23. The middle lamella of the cell wall is composed of

(a) Lignin (b) Suberin
(c) Calcium pectate (d) Calcium phosphate

24. Cellulose and hemicelluloses which are the constituents of the cell wall are synthesized by

(a) Lysosome (b) Microbodies
(c) Smooth ER (d) Golgi apparatus

25. Which of the following cell organelles is common to both Protista and Monera?

(a) Lysosome (b) Ribosome
(c) Chloroplast (d) Vacuole

26. Leucoplast which stores oil is called

(a) Amyloplast (b) Aleuroneplast
(c) Elaioplast (d) Proplast

27. An electron microscope has revealed the presence of

(a) Leucoplast (b) Chloroplast
(c) Chromosome (d) Ribosome

28. Endoplasmic reticulum is attached (continuous) with

(a) Ribosome (b) Golgi body
(c) Mitochondria (d) Nuclear membrane

29. The centrosome is responsible for

(a) Cell wall formation
(b) Cell plate formation
(c) Inhibition of cell division
(d) Initiation of cell division

30. Cell-to-cell recognition in the cell membrane is by

(a) Carbohydrate part (b) Lipid part
(c) Protein part (d) Nucleotides

31. Isochromosomes (chromosome with identical arms) are formed by

(a) Pairing of identical chromosomes
(b) Longitudinal splitting of centromere
(c) Transverse splitting of centromere
(d) Disjunction of bivalents

32. Maximum polymorphism is exhibited by

(a) Lysosomes (b) Ribosomes
(c) Golgi bodies (d) Endoplasmic reticulum

33. Peroxisome does not have

(a) Urate oxidase
(b) Glycolate oxidase
(c) Catalase
(d) D-amino acid oxidase

34. Pinocytosis and phagocytosis are concerned with

(a) Cell wall (b) Plasma membrane
(c) Nucleus (d) Mitochondria

35. The shape of a chromosome having a terminal centromere is called

(a) Metacentric (b) Acrocentric
(c) Telocentric (d) Acentric

36. The organelle common in both eukaryotic and prokaryotic cells is

(a) Microtubules (b) Microbodies
(c) Microfilaments (d) Microsome

37. Nucleoli are rich in

(a) RNA (b) DNA and protein
(c) Protein and RNA (d) Carbohydrates

38. The photo-respiratory carbon-oxidation cycle involves the interaction of

(a) Chloroplast (b) Peroxisome
(c) Mitochondria (d) All of these

39. Axoneme having 9 + 0 doublet microtubule arrangement is found in

(a) Cilia (b) Flagella
(c) Cilia and flagella (d) Centriole

40. The number of protofilament found in microtubules is

(a) 5 (b) 9
(c) 10 (d) 13

41. Which of the following structures is an organelle within an organelle?

(a) Ribosome (b) Peroxisome
(c) Endoplasmic Reticulum
(d) Mesosome

42. Which of the following features is not correct for bacteria?

(a) Flagella lacks 9+2 arrangement
(b) Pilus helps in conjugation
(c) Fimbriae are involved in attachment
(d) Flagella are made up of tubulin protein

43. Which of the following features is appropriate for immature cells?

(a) These are less metabolically active
(b) These are more efficient
(c) These have a low nuclear-cytoplasmic ratio
(d) All of the above

44. The number of kinetochores in the centromere of a chromosome is

(a) One (b) Two
(c) Three (d) Four

45. Which of the following is the densest and membraneless?

(a) Ribosome (b) Peroxisome
(c) Nucleolus (d) Chloroplast

46. The number of mitochondria increases in the cell of

(a) Dry seeds (b) Dormant seeds
(c) Germinating seeds (d) Dead seeds

47. Intracellular compartments are not found in

(a) Lower plants (b) Eukaryotes
(c) Prokaryotes (d) Higher plants

48. The trigger of cell division is

(a) Peroxisome (b) Ribosome
(c) Lysosome (d) Vacuole

49. The pure fractions of cellular components can be isolated by

(a) Chromatography
(b) X-ray diffraction
(c) Differential centrifugation
(d) All of the above

50. Which of the following layers is present nearest to the plasma membrane in plant cells?

(a) Middle lamella
(b) Primary wall
(c) Secondary wall
(d) Tertiary wall

51. Lomasome found in fungi is present in

(a) Cytoplasm (b) Mitochondria
(c) Vacuole
(d) Between cell wall and plasma membrane

52. Which of the following enzymes catalyses ATP in the flagellar membrane?

(a) Tubulin (b) Nexin
(c) Flagellin (d) Dynein

53. In germinating seeds, fatty acids are degraded exclusively in

(a) Peroxisome (b) Glyoxysome
(c) Mitochondria (d) Chloroplast

54. Ribosomal RNA is actively synthesised in

(a) Ribosome (b) Lysosome
(c) Peroxisome (d) Nucleolus

55. A multicellular organism possesses

(a) Differentiated cells
(b) Undifferentiated cells
(c) Dedifferentiated cells
(d) All of the above

56. Which of the following is incorrect regarding bacterial cells?

(a) The cell wall is made up of NAM and NAG
(b) Cytoskeletons are absent
(c) Ribosomes are of 70 S
(d) Nucleoid contains histone proteins.

57. The sugar component of the plasma membrane belongs to

(a) Monosaccharide (b) Oligosaccharide
(c) Polysaccharide (d) All of the above

58. The tail of the lipid of the plasma membrane is

(a) Polar and hydrophilic
(b) Polar and hydrophobic
(c) Nonpolar and hydrophilic
(d) Nonpolar and hydrophobic

59. Which of the following cell organelles does not belong to the endomembrane system?

(a) Endoplasmic reticulum
(b) Golgi complex
(c) Ribosome
(d) Lysosome

60. Triplet fibrils of a centriole are tilted at an angle of

(a) 40° (b) 45°
(c) 60° (d) 65^{0}

61. Which of the following acts as a nucleating centre for the growth of microtubules during the formation of new centrioles?

(a) Spindle (b) Massule
(c) Aster (d) Chromocentre

62. The plane of cell wall formation in dividing cell is determined by

(a) Microtubules (b) Microfilaments
(c) Microsome (d) All of the above

63. If cells are broken up and sedimented by centrifugation, the new structure formed in one of the cells was

(a) Centrosome (b) Microsome
(c) Ribosome (d) Lomasome

64. Cell theory is not obeyed by

(a) *E. coli* (b) Bacteriophage
(c) *Salmonella* (d) *Nostoc*

65. Lipo-polysaccharide (LPS) is found in

(a) All Gram-positive bacteria
(b) All Gram-negative bacteria
(c) Both Gram-positive and Gram-negative bacteria
(d) Viruses

66. Plasma membrane helps in bulk transport in case of

(a) Active transport
(b) Facilitated diffusion
(c) Pinocytosis
(d) All of the above

67. Mitochondria are stained with

(a) Acetocarmine (b) Orchill
(c) Safranin (d) Janus Green

68. Smooth endoplasmic reticulum cannot perform

(a) Glycogen synthesis
(b) Fatty acid synthesis
(c) Detoxification of drugs
(d) Synthesis of testosterone

69. The processes of glycosylation and glycosidation are performed by

(a) Mitochondria
(b) Endoplasmic reticulum
(c) Golgi complex
(d) Ribosome

70. Multinucleate condition is seen in

(a) Epidermis of *Ascaris*
(b) Mycelia of *Rhizopus*
(c) Mycelia of *Mucor*
(d) All of the above

71. Flagella of prokaryotic and eukaryotic cells differ in

(a) Type of movement and placement in the cell
(b) Location in cell and mode of functioning
(c) Microtubular organisation and type of movement
(d) Microtubular organisation and function

72. Which of the following does not occur during the maturation of a proplastid into a chloroplast?

(a) Increase in size
(b) Increase in the number of grana
(c) Bleaching of chlorophyll
(d) Synthesis of pigments

73. Telomeres of eukaryotic chromosomes consist of a short sequence of

(a) Thymine rich repeats
(b) Cytosine rich repeats
(c) Adenine rich repeats
(d) Guanine rich repeats

74. Which part of the translational modification of proteins does not occur in the lumen of ER?

(a) Glycosylation
(b) Formation of disulphide bonds
(c) Conformational folding
(d) Ubiquitination

75. Receptors for steroid hormones are found in

(a) Cell membrane (b) Cytoplasm
(c) Mitochondria (d) Ribosome

76. The maximum size of a cell is limited by

(a) The need for enough surface area for exchange with the environment
(b) The number of organelles that can be packed inside
(c) The materials needed to build it
(d) The amount of flexibility it needs to be able to move

77. Which of the following functional processes results from the presence of protein within the plasma membrane?

(a) Enzymatic activity
(b) Signal transduction
(c) Intercellular joining
(d) All of the above

78. **A karyotype would be least likely to show**
(a) An extra chromosome
(b) Part of a chromosome duplicated
(c) A missing chromosome
(d) Part of a chromosome turned around

79. **The mitochondrial membrane contains a transporter for**
(a) NADH (b) Acetyl Co-A
(c) GTP (d) ATP

80. **The portion of Golgi which contains irregular cisternae and tubules is known as**
(a) Cis-Golgi (b) Trans-Golgi
(c) Median Golgi (d) Intercisternal Golgi

81. **Which of the following does not have in common between chloroplast and mitochondria?**
(a) ATP synthase (b) ETS
(c) ATP (d) NADH

82. **Euchromatin contains**
(a) Single copy and genetically active DNA
(b) Repetitive and genetically inactive sequences
(c) Single copy and genetically inactive DNA
(d) Repetitive and genetically active sequences

83. **Which part of the ribosome gets attached to the ER?**
(a) Peptidyl site (b) Aminoacyl site
(c) Large subunit (d) Small subunit

84. **Lysosome contains hydrolytic enzymes which can digest any foreign materials except**
(a) Cellulose (b) Starch
(c) Glycogen (d) Lipids

85. **Polyribosomes are found in**
(a) Bacteria (b) Fungi
(c) Angiosperms (d) Mammals

86. **Lysosomes are abundant in**
(a) WBC and osteoblast
(b) RBC and spleen
(c) Liver and spleen
(d) WBC and spleen

87. **The middle lamella of plant cells is made up of calcium and magnesium pectate. Thus, pectic acid is the polymer of**
(a) α-1,4-D glucose
(b) β-1,6-D glucose
(c) α-1,4-D galacturonic acid
(d) β-1,4-D galacturonic acid

88. **In a cell if the acidity of the lysosome is lost, then there is a loss of**
(a) Phagocytosis of invading bacteria
(b) Elevated phosphatase level
(c) Glycogen degradation
(d) No major effect

89. **The lipid nature of the plasma membrane can be destroyed by**
(a) Hexane (b) Benzene
(c) Chloroform (d) NaOH

90. **In the plasma membrane, the functional and structural roles are played by**
(a) Proteins (b) Lipids
(c) Cholesterol (d) Oligosaccharides

91. **The most abundant intracellular cation is**
(a) K^+ (b) H^+
(c) Na^+ (d) Ca^+

92. **Which of the following is called the system of membranes?**
(a) ER (b) Golgi
(c) Lysosome (d) Chloroplast

93. **Which of the following is not considered a true cell organelle?**
(a) Lysosome
(b) Peroxisome
(c) Endoplasmic reticulum
(d) Ribosome

94. **Which of the following organelles reduces the number of other cell organelles?**
(a) Ribosome
(b) Lysosome
(c) Endoplasmic reticulum
(d) Mesosome

95. **Chromosomes are most distinct at**
(a) Interphase (b) Prophase
(c) Metaphase (d) Anaphase

96. **Mitochondria performs all of the following functions except**
(a) Nucleic acid synthesis
(b) Steroid synthesis
(c) ATP synthesis
(d) Polysaccharide degradation

97. In a chloroplast, the highest number of protons occurs in

(a) Stroma
(b) Lumen of thylakoid
(c) Intermembrane space
(d) Antenna complex

98. If a meristematic cell is placed in a medium containing active thymidine, then the radioactivity will be first observed in

(a) Euchromatin
(b) Heterochromatin
(c) Both simultaneously
(d) None of the above

99. In plants, the maximum number of chromosomes is found in

(a) *Haplopappus* (b) *Acetabularia*
(c) *Lilium* (d) *Ophioglossum*

100. The Centromere is placed at the centre in

(a) Acrocentric chromosome
(b) Metacentric chromosome
(c) Submetacentric chromosome
(d) Telocentric chromosome

101. An increase in the cell size results in

(a) Increase in surface: volume ratio
(b) Decrease in surface: volume ratio
(c) No change in surface: volume ratio
(d) All of the above

102. The exchange of information and materials through the cell surface

(a) Is increased by an increase in cell volume
(b) Is limited by increase in cell volume
(c) Is independent of change in cell volume
(d) Is decreased by an increase in cell volume

103. The surface: volume ratio sets the limit for the size of

(a) Unicellular organisms
(b) Multicellular organisms
(c) Both of the above
(d) None of the organisms

104. An undifferentiated fibrillar nucleus is seen in

(a) Prokaryotic cells
(b) Eukaryotic cells
(c) Both of the above
(d) Higher animals

105. The Singer and Nicholson model of plasma membrane differs from Robertson's model in

(a) Number of lipid layers
(b) Arrangement of the lipid layer
(c) Arrangement of proteins
(d) Absence of proteins

106. The lipid bilayer of a cell membrane is impermeable to

(a) Hydrocarbons
(b) Hydrophobic molecules
(c) Large uncharged polar molecules
(d) Small uncharged polar molecules

107. Cholesterol molecules are taken inside by the cell through

(a) Diffusion
(b) Pinocytosis
(c) Phagocytosis
(d) Receptor-mediated endocytosis

108. Cells having a large nucleus in proportion to the cytoplasm are

(a) Dividing cells (b) Dehydrated cells
(c) Dying cells (d) Active cells

109. Histone proteins found in the nuclei of eukaryotes are rich in

(a) Glycinandphenylalanine
(b) Lysine and arginine
(c) Glycine and arginine
(d) Phenylalanine and lysine

110. Nucleocytoplasmic traffic through nuclear pore is facilitated by

(a) Nucleoplasmin
(b) Nuclein
(c) Nuclear lamina
(d) Lipid bilayer of cell membrane

111. Nucleoproteins in a cell are synthesised in

(a) Nucleolus
(b) Nucleoplasm
(c) Nuclear membrane
(d) Outside the nucleus

112. Histone proteins found in the nuclei of eukaryotes are

(a) Acidic (b) Basic
(c) Neutral (d) Amphoteric

113. The function of histone proteins of nuclei is

(a) To provide amino acids for enzyme synthesis

(b) To provide structural proteins for stabilising chromosomes

(c) To neutralise anionic charges of internucleotide phosphate in DNA

(d) Both b and c

114. Nucleosomes are linked together by

(a) Histone H_1 (b) Histone H_2

(c) DNA (d) RNA

115. The core of nucleosome is made up of

(a) H_1, H_2A, H_2B, H_3

(b) H_1, H_2A, H_2B, H_4

(c) H_2A, H_2B, H_3, H_4

(d) H_1, H_2A, H_2B, H_3, H_4

116. Acetabularia, used in nucleocytoplasmic experiments by Hammerling is

(a) Unicellular, uninucleate green alga

(b) Unicellular, multinucleate green alga

(c) Unicellular, uninucleate brown alga

(d) Unicellular, uninucleate red alga

117. In the absence of histone proteins in prokaryotes, the anionic charge of each nucleotide phosphate in DNA is neutralized by

(a) Phospholipids (b) glycolipids

(c) Polyamines (d) Polysaccharides

118. Golgi bodies are maximum in

(a) Calyptrogen (b) Root tip

(c) Root cap (d) Shoot tip

119. Maximum hydrolytic enzymes are found in

(a) Peroxisomes (b) Glyoxysome

(c) Ribosome (d) Lysosome

120. The colour of rose petals is due to the presence of water-soluble pigments in

(a) Cytoplasm (b) Intercellular space

(c) Nucleus (d) Vacuole

121. A vacuole contains

(a) Only water

(b) Water and dissolved substances

(c) Only air

(d) Air and dissolved substances

122. Desmosomes are concerned with

(a) Cell division (b) Cytolysis

(c) Cell adherence (d) All of the above

123. The enzyme present in Golgi bodies is

(a) PEP carboxylase

(b) Peptidyl transferase

(c) Glycosyl transferase

(d) RuBisCO

124. Cyclosis refers to the movement of cytoplasm within

(a) Bacterial cells

(b) Prokaryotic cells

(c) Eukaryotic cells

(d) All of the above

125. Smooth ER is made up of

(a) Cisternae and tubules

(b) Cisternae and vesicles

(c) Tubules and vesicles

(d) Cisternae, tubules and vesicles

ANSWERS

1.c	11.b	21.b	31.c	41.a	51.d	61.b	71.a	81.d	91.a	101. b	111.d	121.b
2.a	12.d	22.c	32.a	42.d	52.d	62.a	72.c	82.a	92.a	102. b	112.b	122.c
3.c	13.a	23.c	33.b	43.b	53.b	63.b	73.d	83.c	93.d	103.a	113.d	123.c
4.b	14.c	24.d	34.b	44.b	54.d	64.b	74.d	84.a	94.b	104.a	114. c	124.c
5.b	15.b	25.b	35.c	45.c	55.d	65.b	75.b	85.a	95.c	105. c	115.c	125.b
6.c	16.b	26.c	36.d	46.c	56.d	66.c	76.a	86.d	96.d	106. c	116. b	
7.c	17.b	27.d	37.c	47.b	57.b	67.d	77.d	87.a	97.b	107. d	117. c	
8.c	18.b	28.d	38.b	48.c	58.d	68.b	78.d	88.c	98.a	108. d	118. c	
9.c	19.c	29.d	39.d	49.c	59.c	69.c	79.d	89.b	99.d	109.b	119.d	
10.b	20.b	30.a	40.d	50.c	60.a	70.d	80.c	90.a	100.b	110.a	120.d	

CHAPTER-7

CELL CYCLE AND CELL DIVISION

The cell cycle is the series of events that take place between the formation of a cell and its division into daughter cells. It is also called the cell division cycle. It differs in different organisms and different cell types. It is characteristic of plants and animals. It is 20 minutes in bacteria, 2-3 hours in yeast, 20 hours in onion root tip and 24 hours in human beings.

The **G_0 phase** (Quiescent phase) is a resting phase where the cell has left the cell cycle and has stopped dividing. The cell cycle starts with this phase in response to some stimulus. Non-proliferating or non-dividing cells in multicellular eukaryotes generally enter the quiescent (G_0) state from G_1 and may remain quiescent for long periods of time (as in neurons). This is very common for cells that are fully differentiated but cells are metabolically active. Many cells do not enter G_0 and continue to divide regularly and occasionally throughout an organism's life.

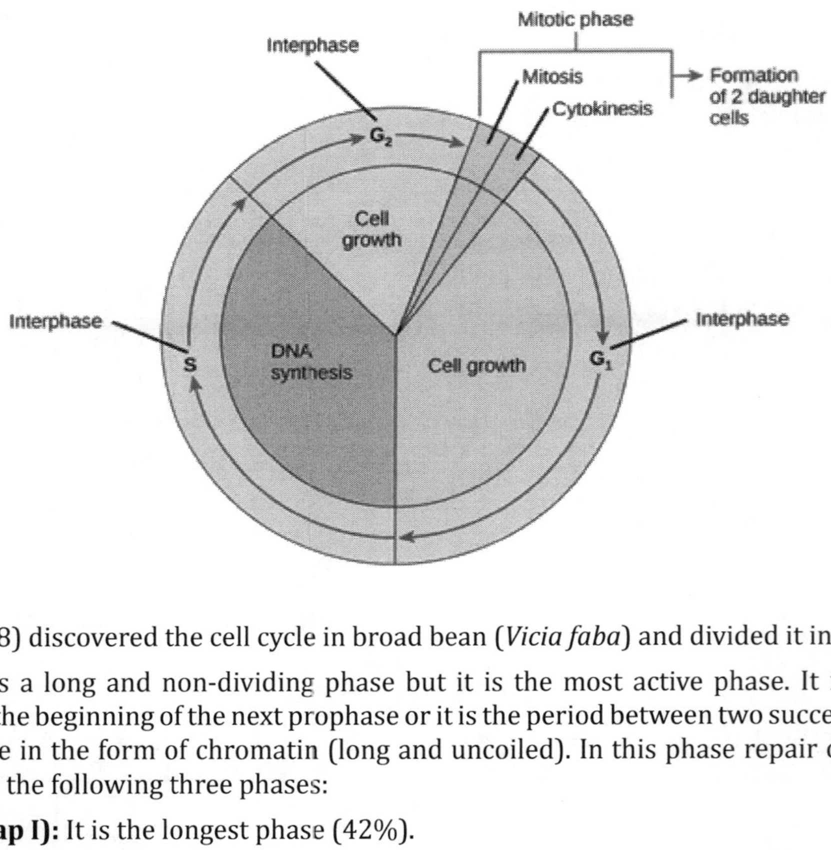

Howard and Pelc (1958) discovered the cell cycle in broad bean (*Vicia faba*) and divided it into two stages.

1. **Interphase**: It is a long and non-dividing phase but it is the most active phase. It is the period between cytokinesis and the beginning of the next prophase or it is the period between two successive mitotic divisions. Chromosomes lie in the form of chromatin (long and uncoiled). In this phase repair of damaged DNA takes place. It involves the following three phases:

   (a) **G_1 phase (Gap I):** It is the longest phase (42%).

   The size of the cell and nucleus increases. Cell shows high metabolic rate. Synthesis of structural and functional proteins (enzymes) and RNA occurs. Multiplication of cell organelles occurs. It is shorter in unicellular organisms (fast-dividing cells) and longer in multicellular organisms (slow-dividing cells). The difference in the time of the cell cycle or rate of cell division is due to variation in the length of the G_1 phase.

   G_1-specific cyclins are synthesised and these activate cyclin-dependent kinase (CDK) specific for the G_1 phase.

   First restriction (**R_1**) or major checkpoint: It occurs at the late G_1 phase (G_1 to S phase transition). It requires G_1 cyclin and cyclin-dependent kinase-2 (CDK-2).

(b) **Synthesis (S-phase):** It is the second longest phase (33%).

Replication of DNA (doubling the amount) occurs. Synthesis of histone proteins occurs. Growth of the nucleus occurs. Duplication of centrioles occurs at the late S-phase but completes in the G_2 phase.

S-specific cyclins are synthesised and G_1-specific cyclins are destroyed at the end of the G_1 phase.

(c) **G_2 phase (Gap 2)**: It is the second growth phase. It is the shortest phase in the interphase.

The size of the cell and nucleus increases. Synthesis of RNA and proteins resumes. Repair of damaged DNA occurs. Synthesis of tubulin proteins occurs that are essential for the formation of spindle fibres. Condensation of chromatin into chromosomes occurs. ATP is stored in the cell. Removal of inhibitors occurs. It prepares cells to divide.

G_2-specific cyclins are synthesised and S-specific cyclins are destroyed at the end of the S phase. It requires CDK hence called maturation promoting factor which acts as a trigger.

Second restriction point (**R_2**) or minor checkpoint: It occurs at the late G_2 phase (G_2 to M phase transition).

2. **M phase:** It is the shortest phase. It is a phase of actual division. The process of cell division consists of two sequential processes- (i) karyokinesis (a division of the nucleus) and (ii) cytokinesis (a division of cytoplasm).

   M-specific cyclins are synthesised.

   After the M phase, a cell may either enter a fresh cycle or pass into G_0 (quiescent) stage.

Regulation of cell division: It is regulated in the following manner.

Cell division cycle (cdc) gene →Cyclin protein →Cyclin dependent kinase (CDK) →Phosphorylation of histone (H_1) proteins → break down of nuclear membrane→condensation and packaging of chromatin into Chromosome in nucleus→Cell division →Inactivation and disappearance of CDK at late prophase.

Cyclin is a regulatory component whereas CDK is catalytic and acts as protein kinase. Kinase adds a phosphate group to the target substrate. Cyclin undergoes a cycle of synthesis and degradation in each cycle. Cyclins bind to CDK and control their abilities to phosphorylate particular target protein that triggers cell division.

The cell cycle depends on nutrients (growth factors), inhibitors and overcrowding.

TYPES OF CELL DIVISION

Amitosis:

It was discovered by Robert Remak (1855) in the RBC of a chick embryo. It is a direct nuclear division in which there is no formation of spindle fibres and the appearance of chromosomes. It rarely occurs. In this, the nuclear membrane does not degenerate. The nucleus elongates and constricts in the middle to form two daughter nuclei. Later centripetal constriction of cytoplasm occurs by which two daughter cells are formed. It occurs in prokaryotes, protozoans, yeast, cartilage and foetal membranes of mammals, and degenerated, damaged or old tissues.

Mitosis

It is the division of the cell in which two genetically identical daughter cells are formed but the number of chromosomes and amount of genetic material remain the same as that of the parent cell. It is a direct nuclear division.

It was discovered by Strasburger (1875) in plants but in animals, it was discovered by Flemming. The term 'mitosis' was coined by Flemming (1882).

It is a common method of cell division. It occurs in all types of cells (haploid, diploid and polyploid) in plants. It occurs in somatic cells. In plants, cell division is confined to root, stem, young leaves, floral buds and cambium but in animals, it is confined to blood cells, outer skin cells and gut epithelium.

Mitosis is best observed in the root tip of an onion in the case of plants because continuous growth of plants occurs throughout their life. The cells of bone marrow, base of nails and skin are also used to study mitosis in the case of animals. The total time taken for completion of mitosis is usually from 30 minutes to 3 hours. The duration depends on temperature. As the temperature increases, the duration of mitosis will decrease.

Mitotic apparatus = Aster + Spindle + Centriole = Achromatic figure

In lower plants and animals, mitosis is astral (amphiastral) due to the presence of centrioles but in higher plants, it is anastral because asters (rays) are not formed due to the absence of centrioles.

Types of mitosis:

1. Endomitosis: It is the duplication of chromosomes and DNA without division of the nucleus. It leads to the doubling of chromosomes (polyploidy). It is also called endoduplication or endopolyploidy. It leads to polyteny when the separation of chromatids does not occur hence there are several chromatids in a chromosome.
2. Dinomitosis: It is intranuclear (closed) mitosis in which the nuclear membrane persists (does not break down) but spindle fibres are not formed. It occurs in dinoflagellates.
3. Karyochorisis: It is intranuclear (closed) mitosis in which the nuclear membrane persists and spindle fibres are formed inside the nucleus. It occurs in fungi.
4. Eumitosis: It is extranuclear mitosis in which the nuclear membrane is lost and spindle fibres are formed outside the nuclear membrane. It commonly occurs in plants and animals.

Mitosis occurs in four stages which are as follows:

1. **Prophase:**
   (a) It is the first, longest and most active phase.
   (b) Cell becomes viscous and refractive.
   (c) Nuclear membrane and nucleolus are present but later start disappearing.
   (d) Chromatin starts condensing into chromosomes.
   (e) Chromosome becomes coiled, shortened, thickened and distinct (spireme stage) due to plectonemic coiling (spiralisation or coiling in the same direction).
   (f) Chromosome splits longitudinally into two sister chromatids attached with centromere (appearing as double-stranded).
2. **Metaphase:**
   (a) Nuclear membrane and nucleolus are absent.
   (b) Chromosome undergoes maximum contraction or thickening or condensation (dehydration).
   (c) Chromosome is visible because plectonemic coiling between sister chromatids is absent.
   (d) Chromosome is arranged at the central equatorial plate in one plane (congression) while chromatids extend freely in any direction.
   (e) The spindle fibres get attached to the chromosome by kinetochore.
   (f) The structure, size and number of chromosomes are observed.
3. **Anaphase:**
   (a) It is the shortest phase.
   (b) The centromere of chromosome divides into two sister chromatids with the division of kinetochore. Each chromatid has its centromere and is called a daughter chromosome.
   (c) Chromatids separate, start repelling each other and reach opposite poles of the spindle due to contraction of chromosomal fibres and relaxation of interzonal fibres but are connected by interzonal fibres.
   (d) Spindle fibres (made up of tubulin (95%), RNA (3.5%), and traces of actin and myosin) are of four types:
      (i) Continuous fibres: from pole to pole
      (ii) Discontinuous fibres: fibres do not reach other poles.
      (iii) Chromosomal fibre: from pole to centromere
      (iv) Interzonal fibre: from one centromere to another.
   (e) The shape of a chromosome is observed.

(f) Spindles of microtubules are polar with their polymerising '+' ends (fast-growing) facing centre and non-polymerising '-' ends (slow-growing) facing poles.

4. **Telophase:**

(a) Nuclear membrane and nucleolus reappear.

(b) Daughter chromosomes uncoil, lengthen and decondense (hydrate) hence finally indistinguishable.

(c) Spindle fibres disappear.

(d) Each pole has a nucleus.

Cytokinesis: The division of cytoplasm occurs by **cell plate** in plants. The central part of the spindle and intrazonal microtubule grows in size and traps the Golgian vacuole of the Golgi body to form the phragmoplast (barrel-shaped). It is a vesicle budded out from the Golgi complex rich in pectin and lines up at the centre to form the middle lamella which grows centrifugally towards the periphery (lateral wall) and fuses with plasma membrane. The content of the Golgian vacuole rich in pectin forms the middle lamella of cell wall and membrane of cell plate develops the plasma membrane of the new daughter cell.

In the case of animal cells, **cell or cleavage furrow** develops due to contraction of the inner surface of the plasma membrane and microfilament. This deepens and grows centripetally and finally divides the cytoplasm.

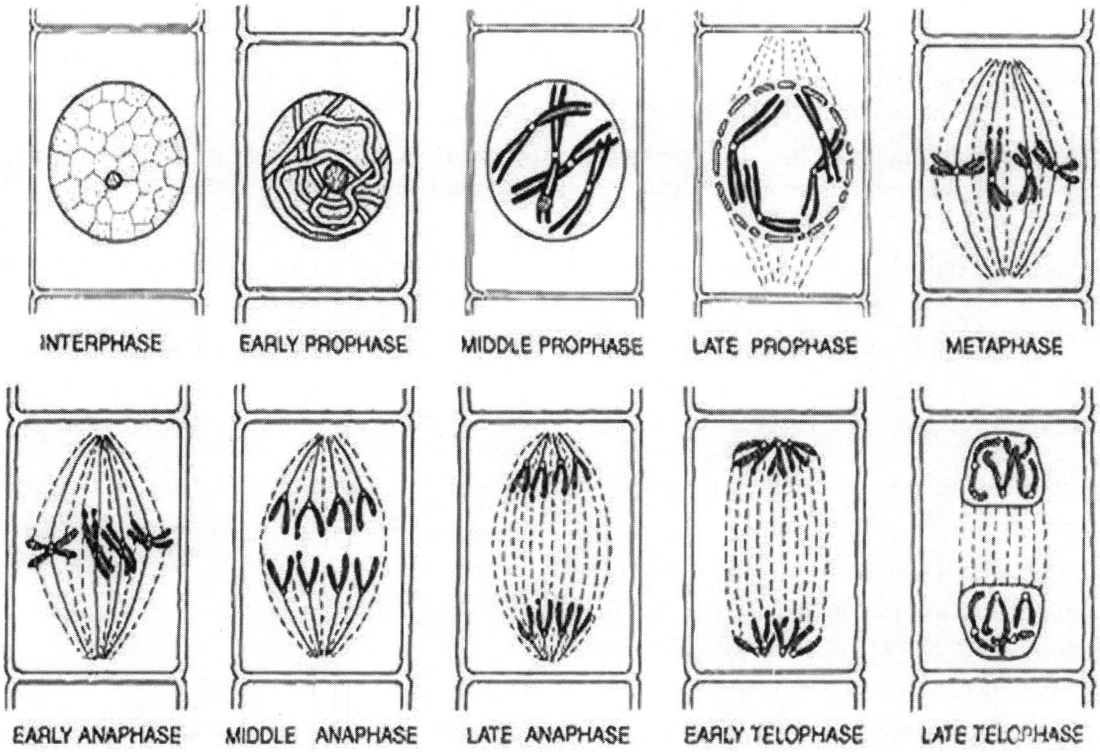

Plant cytokinesis	Animal cytokinesis
It takes place in the meristematic region.	It takes place at many places throughout body.
Centrioles are absent.	Centrioles are present.
Cytokinesis occurs by cell plate which grows centrifugally	Cytokinesis occurs by cell furrow which grows centripetally.
The new cell membrane is formed from vesicles of Golgi body.	A new cell membrane is formed from ER.

Significance:

(a) It maintains the same genetic constitution (chromosome number) qualitatively and quantitatively.

(b) It maintains the size of the cell by establishing a nuclear-cytoplasmic ratio.

(c) It is the only method of reproduction (multiplication) in unicellular organisms.

(d) It is essential for growth and development in multicellular organisms by forming tissue.

(e) It produces the condition of multicellularity.

(f) It helps in the repair, regeneration and healing of wounds by forming new cells in some animals.

(g) It leads to cancer due to uncontrolled division.

Mitogens: These are chemical substances that initiate cell division. In plant cells, temperature, auxin, cytokinin etc. are mitogens but in animal cells, the epidermal growth factor, platelet-derived growth factor, insulin, lymphokines etc are mitogens.

Mitotic poisons: These are chemical substances that check mitotic division. These include

(a) Chalone: It inhibits cell division.

(b) Mustard gas: It causes agglutination of chromosomes.

(c) Colchicine: It dissolves microtubules hence spindle fibres are not formed at metaphase.

(d) Azide, cyanides, and ribonucleases: These inhibit prophase by inhibiting phosphorylation at interphase hence chromosomes remain as chromatin.

MEIOSIS

It is two successive nuclear divisions in which four daughter cells are formed thus the number of chromosomes and amount of DNA are reduced to half of the parent cell.

It was discovered by Von Beneden. The term 'meiosis' was coined by Farmer and Moore (1905). Flemming coined the terms 'Heterotypic division' for meiosis I and 'Homotypic division' for meiosis II which take place in quick succession.

It occurs in diploid and polyploid cells (but not in odd ploidy levels such as triploid). It occurs in reproductive cells or germ cells.

It is best observed in the anther of the unopened young flower of onion and *Tradescantia*. It is antithetical or opposite to fertilisation.

Types of meiosis

1. Zygotic meiosis (Initial meiosis): It occurs in the zygospore of algae and ascus mother cell of fungi.
2. Sporic meiosis (Intermediate meiosis): It occurs in the capsule of bryophytes; sporangium of pteridophytes; spore mother cell of gymnosperms and angiosperms.
3. Gametic meiosis (Terminal meiosis): It occurs in gametocytes (primary spermatocytes and primary oocytes) of animals. It also occurs in some algae.

Meiosis occurs in two successive stages:

Meiosis-I: It is called reductional or heterotypic division. It includes:

1. **Prophase-1:** It is the longest phase. It is divided into the following five stages:

   (a) **Leptotene:**

   - Nuclear membrane and nucleolus are present.
   - Chromatin fibres condense and coil to form chromosomes.
   - Chromosome becomes thread-like and visible.
   - Chromatids are not clear. Each chromosome gets attached at both its ends to the nuclear membrane called attachment plaque or bouquet stage or synizesis.

   (b) **Zygotene**:

   - Chromosomes shorten and thicken.
   - Homologous chromosomes (one chromosome of a pair comes from the maternal side and the other from the paternal side having identical in shape and size) join and pair laterally (side by side) to form

synapsis. It was reported by Montgomery. Synapsis is usually procentric (starting from centromere). It may be proterminal or random.

- Bivalent condition is seen. Both chromosomes of pair have same length, position of centromere and number of genes. Bivalent has four chromatids and two centromeres. Bivalent is preferred term than tetrad.
- Synaptonemal complex (nucleo-proteinaceous complex) is formed which stabilises bivalent. It was discovered by Moses. It consists of two lateral elements (formed by cohesion protein) and one central element (formed by zip protein). It helps in the pairing of homologous chromosomes.
- A small amount of DNA (0.3%) of each nucleus replicates.

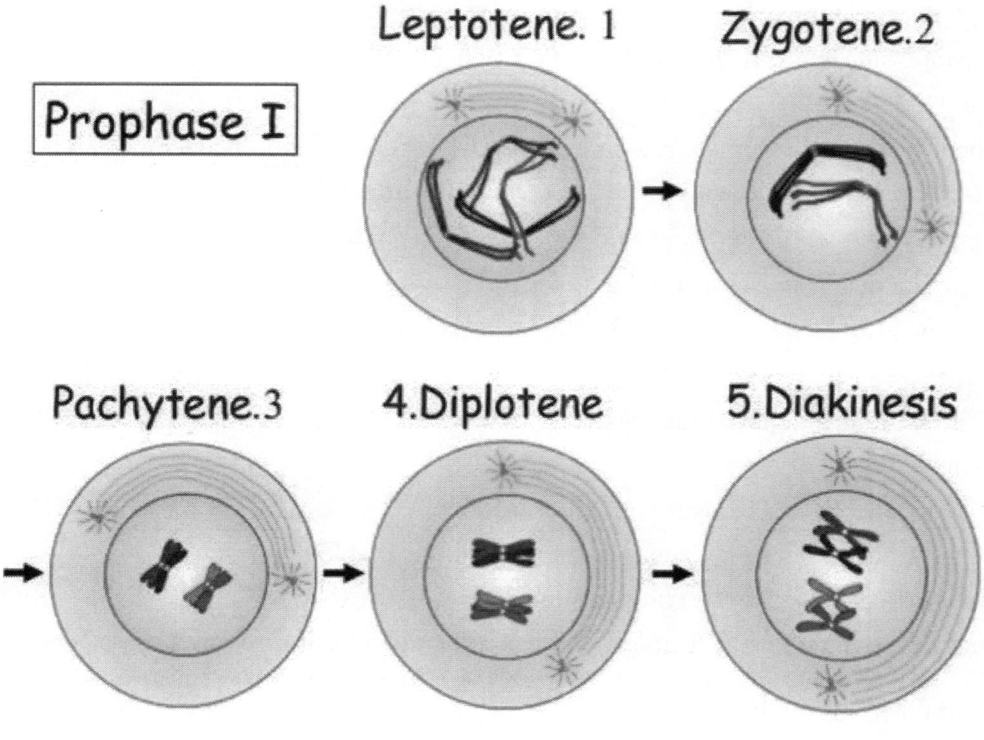

Fig: Stages in prophase I of meiosis

(c) **Pachytene:**

- Bivalent (tetrad) becomes short and thick.
- Recombination nodule appears between chromosome pairs.
- Crossing over occurs at chiasma (chiasmata is not the cause but the result of crossing over).
- Morphological details of chromosomes are best studied because chromosomes are much longer.

(d) **Diplotene:**

- It is the longest phase.
- Synaptonemal complex dissolves (desynapsis).
- Bivalent (Tetrad) is visible.
- Chiasma (a result of crossing over) is visible (the minimum number of chiasma is one). It was first observed by Janssen in amphibians.
- Chromosomes move apart.
- Chromosomes become enlarged in size and last for months and years called Lampbrush chromosomes as in amphibian oocytes.

(e) **Diakinesis:**

- Nuclear membrane and nucleolus disappear.
- Terminalisation (shifting at the end) of chiasma occurs.
- Tetrad undergoes maximum thickening, shortening or condensation.
- Repelling of bivalents starts occurring.

2. **Metaphase I:**
   - Two or more equatorial plates are formed.
   - Bivalent is arranged at the equatorial plate.
   - Each equatorial plate will have the total number of chromosomes.
   - Each chromosome gets attached to the spindle pole of its side.
3. **Anaphase I:**
   - Homologous chromosomes separate due to splitting off chiasmata (disjunction).
   - Separated chromosome (univalent) shows divergent chromatids (dyad).
   - Univalents move towards opposite spindle poles and form two groups of haploid chromosomes.
   - Real reduction in chromosome number takes place.
   - Independent assortment of paternal and maternal chromosomes occurs due to disjunction.
4. **Telophase I:**
   - Nuclear membrane and nucleolus reappear.
   - Spindle fibres disappear.
   - Chromosome uncoils and elongates (despiralisation).
   - The daughter nucleus has only one set of chromosomes (haploid).
   - Amount of DNA is halved in the daughter nucleus.

Interkinesis: It occurs in animal cells. It does not occur in plant cells, if it occurs then it is very short. It shows the synthesis of RNA and proteins. Replication of DNA does not occur because there is no S phase. Chromosomes are elongated but do not form chromatin.

It is known as equational or homotypic division. It occurs in the following phases:

1. **Prophase-II:**
   - Nuclear membrane and nucleolus start disappearing.
   - Univalent chromosome condenses, shortens or thickens.
   - Chromosomes are not plectonemically thickened (compactly twisted).
2. **Metaphase-II**
   - A single equatorial plate is formed at the right angle to metaphase-I.
   - Chromosomes are arranged on equatorial plates.
   - Sister chromatids are widely separated.
   - The centromere of each chromosome gets attached to the spindle pole of its side.
3. **Anaphase-II:**
   - The centromere of each chromosome (univalent) divides.
   - Chromatids of each chromosome (univalent) separate.
   - Chromatid now becomes daughter chromosome.
   - Each daughter chromosome is attached to the spindle pole of its side.

4. **Telophase-II:**
   - Nuclear membrane and nucleolus reappear.
   - Spindle fibres disappear.
   - Daughter chromosome uncoils or elongates or decondenses.
   - Chromosome forms a network of chromatin.
   - Amount of DNA is halved again.

Cytokinesis: A cell plate is formed and thus four daughter (haploid) cells are formed. These cells resemble each other in the number of chromosomes but differ due to the separation of maternal and paternal chromosomes as well as the formation of new types as a result of crossing over. It is of two types in plant cells.

1. **Simultaneous type:** It takes place after the completion of meiosis II resulting in the formation of tetrahedral tetrad in dicot plants. It is considered a primitive feature.
2. **Successive type**: It takes place first after meiosis I and then after the completion of meiosis II resulting in the formation of isobilateral tetrad in monocot plants. It is considered an advanced feature.

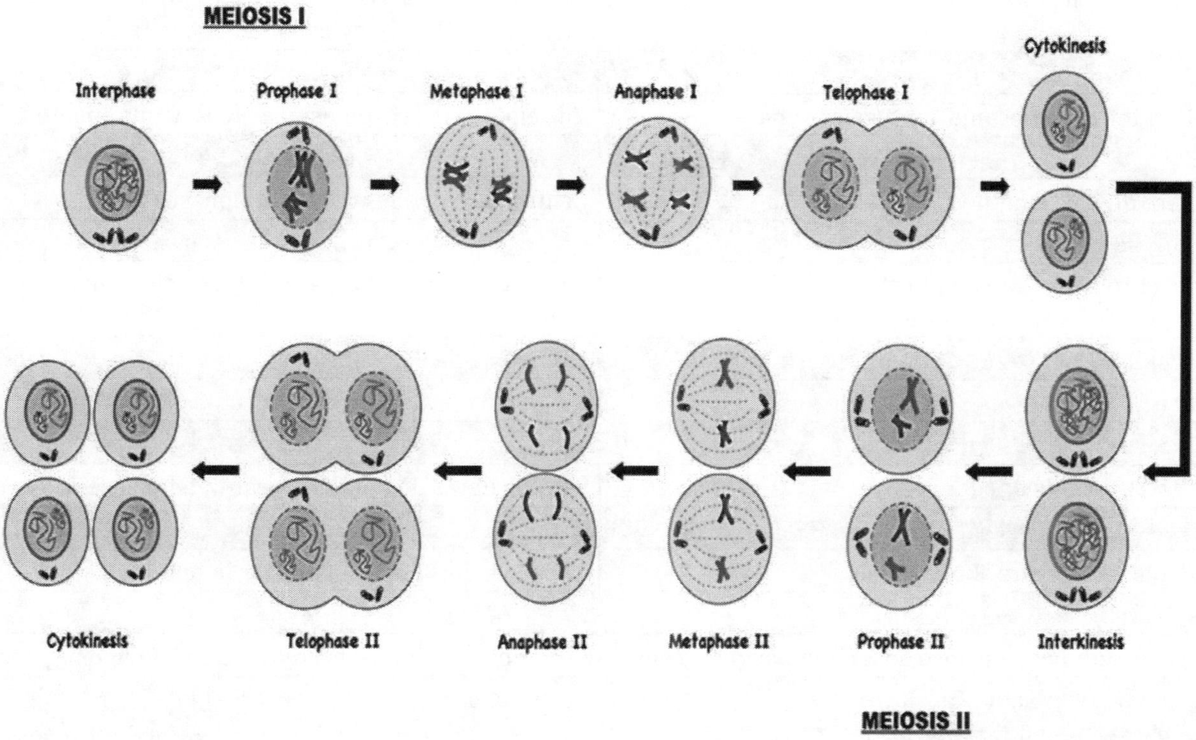

Significance:

(a) Meiosis produces haploid gametes for sexual reproduction.

(b) It brings about variation due to the formation of a new combination of chromosomes by crossing over.

(c) It causes the segregation of two alleles of a gene due to the separation of homologous chromosomes.

(d) It provides a cytological basis for Mendel's law of inheritance.

Differences between Mitosis and Meiosis

Mitosis	Meiosis
It occurs in somatic and reproductive cells.	It occurs in reproductive cells.
It is completed in one stage.	It is completed in two stages.
It is shorter in duration.	It is longer in duration.
It shows cell division only once.	It shows cell division twice.

Mitosis	Meiosis
One parent cell forms two daughter cells.	One parent cell forms four daughter cells.
Each daughter cell contains both homologous chromosomes.	Each daughter cell contains only one of each pair of homologous chromosomes.
Replication of DNA occurs in the S phase.	Replication of DNA occurs only in the S phase of interphase-I
Replication of DNA occurs once for one cell division.	Replication of DNA occurs once for two cell divisions.
Prophase	
It is simple.	It is complex.
It does not show the bouquet stage, synapsis, crossing over and chiasma.	It shows the bouquet stage, synapsis, crossing over and chiasma.
Chromosome splits off in early prophase.	Chromosome splits off in late prophase (pachytene).
Metaphase	
A single equatorial plate is formed.	Double and parallel equatorial plates are formed in metaphase-II but single in metaphase-II.
Each chromosome has two chromatids.	Paired chromosome has four chromatids.
The ends of the chromosome are towards pole.	The ends of the chromosome are towards equatorial plate.
Anaphase	
Centromere divides lengthwise.	Centromere divides only in anaphase II.
Chromatids separate.	Chromatids separate only in anaphase II.
Separated chromatids are identical.	Separated chromatids are not identical due to crossing over.
Sister chromatids move towards opposite poles.	Dissimilar chromosomes move towards opposite poles.
Telophase	
It always occurs and is longer.	Telophase-I is shorter.
Spindle fibres disappear completely.	Spindle fibres do not disappear in telophase-I.
It produces an interphase nucleus.	It does not enter into interphase.
The nucleolus reappears at telophase.	The nucleolus does not appear in telophase-I.
Results	
Chromosome number remains constant in daughter cells.	Chromosome number is reduced to half in daughter cells.
Each daughter cell is genetically identical to the parent cell qualitatively and quantitatively.	Each daughter cell is different from each other qualitatively and different from parent cell qualitatively and quantitatively.

SHORT ANSWER TYPE QUESTIONS

1. What do you mean by quiescent phase?
2. Write the functions of the S phase.
3. How cell division is regulated?
4. Mention the different types of mitogens and mitotic poisons.
5. Distinguish between metaphase of mitosis and metaphase I of meiosis.
6. Distinguish between the anaphase of mitosis and anaphase I of meiosis.
7. Distinguish between anaphase I and anaphase II of meiosis.
8. What do you mean by bivalent?
9. Discuss the importance of mitosis.

10. Distinguish between plant cytokinesis and animal cytokinesis.
11. Discuss the different types of meiosis
12. Discuss the types of cytokinesis in plants.
13. Distinguish between anastral mitosis and amphiastral mitosis.
14. Write notes on amitosis.
15. Discuss the importance of meiosis.

LONG ANSWER TYPE QUESTIONS

1. Describe the different stages of the cell cycle.
2. Describe the different types of mitosis.
3. Describe the different stages of mitosis with a well-labelled diagram.
4. Describe the prophase 1 of meiosis with a well-labelled diagram
5. Distinguish between mitosis and meiosis

MULTIPLE CHOICE QUESTIONS

1. Anastral mitosis is a feature of

(a) Higher animals
(b) All living organisms
(c) Lower animals
(d) Higher plants

2. The replication of centrioles occurs during

(a) Early prophase (b) Late prophase
(c) Interphase (d) Late telophase

3. A cell stops growing due to a shortage of nutrients at

(a) G_1 phase (b) G_2 phase
(c) S phase (d) M phase

4. The most variable length of cell cycle is noticed in

(a) G_1 phase (b) G_2 phase
(c) S phase (d) M phase

5. Nuclear membrane is broken down due to the activity of

(a) P protein (b) U protein
(c) Kinase (d) Kinetin

6. A cell contains 4n DNA at

(a) G_1 phase (b) S phase
(c) G_2 phase (d) M phase

7. Kinetin increases the rate of mitosis by reducing the duration of

(a) Interphase (b) Metaphase
(c) Anaphase (d) Telophase

8. The spindle fibres are composed of

(a) 3% RNA and 97% tubulin
(b) 10% RNA and 90% tubulin
(c) 40 % RNA and 60% tubulin
(d) 60% RNA and 40% tubulin

9. Which stage of prophase-I of meiosis is prolonged?

(a) Leptotene (b) Zygotene
(c) Pachytene (d) Diplotene

10. Replication of chromosomes without division of the nucleus is called

(a) Dinomitosis (b) Endomitosis
(c) Karyochorisis (d) C-Mitosis

11. The spindle in meiosis is oriented at

(a) Acute angle to that in meiosis-I
(b) Right angle to that in meiosis-I
(c) Obtuse angle to that in meiosis-I
(d) Same position to that in meiosis-I

12. The number of DNA in the chromosome at the G_2 stage is

(a) 1 (b) 2
(c) 4 (d) 8

13. Interzonal fibres occur in

(a) Prophase (b) Early metaphase
(c) Late metaphase (d) Anaphase

14. The angular divergence of spindle microtubules is decreased at

(a) Prophase (b) Metaphase
(c) Anaphase I (d) Anaphase II

15. The phragmoplast is organised before the cell plate at

(a) Early anaphase (b) Late phase
(c) Early telophase (d) Late telophase

16. How many mitotic divisions occur in a cell of root tip to form 128 cells?

(a) 7 (b) 127
(c) 32 (d) 16

17. How many mitotic divisions are needed for a single cell to make 128 cells?

(a) 7 (b) 8
(c) 16 (d) 127

18. How many meiotic divisions would be required to produce 101 female gametophytes in an organism?

(a) 25 (b) 26
(c) 100 (d) 101

19. Chromosomes appear as bead-like at

(a) Leptotene (b) Zygoten
(c) Pachytene (d) Diplotene

20. In meiosis, a dyad is defined as

(a) A pair of sister chromatids
(b) A pair of non-sister chromatids
(c) A pair of homologous chromosomes
(d) A pair of non-homologous chromosomes

21. During synapsis, the number of threads or chromonemata in each chromosome is

(a) 1 (b) 2
(c) 4 (d) 8

22. Chiasmata was first discovered by

(a) Morgan in *Drosophila*
(b) Beadle and Tatum in *Neurospora*
(c) Janssen in amphibians
(d) Hammerling in *Acetabularia*

23. The number of chromosome groups at the equatorial plate in metaphase-I of meiosis in a plant with 2n=50 shall be

(a) 25 (b) 50
(c) 100 (d) 200

24. During meiosis, the reduction of chromosome number to half takes place at

(a) Metaphase I
(b) Metaphase II
(c) Anaphase I
(d) Anaphase II

25. What state of the cell cycle is attained when the cell has undergone differentiation?

(a) G_1 phases (b) G_2 phase
(c) S phase (d) Go phase

26. The cells of a permanent tissue do not divide because these are arrested at

(a) G_0 stage (b) G_1 stage
(c) S stage (d) G_2 stage

27. Colchicine prevents the mitosis of the cells at

(a) Prophase (b) Metaphase
(c) Anaphase (d) Telophase

28. Chromosomes are linear in

(a) Prophase I (b) Prophase II
(c) Interphase (d) Metaphase I

29. If there are 4 chromosomes present during prophase then how many chromosomes are there in each cell at the end of anaphase II?

(a) 1 (b) 2
(c) 4 (d) 8

30. In an organism if the normal diploid number of chromosomes is 8 then how many chromatids are present in each daughter cell at the end of meiosis I?

(a) 2 (b) 4
(c) 8 (d) 16

31. Repulsion of homologous chromosomes takes place in

(a) Zygotene (b) Pachytene
(c) Diplotene (d) Diakinesis

32. In which phase of the cell cycle, cell division cannot be stopped?

(a) G_1 phase (b) S phase
(c) G_2 phase (d) G_0 phase

33. How many ATPs are required during anaphase to move chromosomes from the equator to the poles?

(a) 15 ATP (b) 30 ATP
(c) 36 ATP (d) 38 ATP

34. The nature of the coiling of chromatids is

(a) Plectonemic in both mitosis and meiosis
(b) Paranemic in both mitosis and meiosis
(c) Plectonemic in mitosis and paranemic in meiosis
(d) Paranemic in mitosis and plectonemic in meiosis

35. **A small amount of DNA is synthesised at**
(a) Leptotene (b) Zygotene
(c) Pachytene (d) Diplotene

36. **If the amount of DNA is 20 pg in the G_1 stage then how much amount of DNA is formed in each cell at the end of meiosis II?**
(a) 5 pg (b) 10 pg
(c) 20 pg (d) 40 pg

37. **If the amount of DNA in the gamete is 10 pg then how much amount of DNA is found at the G_2 stage?**
(a) 5 pg (b) 10 pg
(c) 20 pg (d) 40 pg

38. **The first restriction point occurs at**
(a) Early G_1 stage (b) Late G_1 stage
(c) Early G_2 stage (d) Late G_2 stage

39. **The stage of polyteny occurs as a result of**
(a) Dinomitosis (b) Amitosis
(c) Karyochorisis (d) Endomitosis

40. **Cytokinesis in plants occurs by cell plate which grows**
(a) Laterally (b) Centripetally
(c) Centrifugally (d) Randomly

41. **In cancer cells, mitosis stops in**
(a) Early G_1 stage (b) Late G_1 stage
(c) G_2 stage
(d) At any point in the cell cycle

42. **A cell is post-reproductive and will remain in**
(a) G_1 phase (b) S phase
(c) G_2 phase (d) M phase

43. **The number of chromatids in a cell in prophase I of meiosis of a person with Turner's syndrome**
(a) 23 (b) 25
(c) 47 (d) 90

44. **When dividing cells are examined under a light microscope, the chromosomes first become visible during**
(a) Interphase (b) Prophase
(c) Metaphase (d) Anaphase

45. **How many chromosomes will be present at the mitotic metaphase of a diploid cell if its egg has 10 chromosomes?**
(a) 10 (b) 20
(c) 40 (d) 80

46. **Cells will usually divide if they receive proper signal at**
(a) G_1 phase (b) S phase
(c) G_2 phase (d) M phase

47. **Centrioles separate during**
(a) Cytokinesis
(b) Prophase I and Prophase II
(c) Metaphase I and Metaphase II
(d) Anaphase I and Anaphase II

48. **Maize has 10 pairs of chromosomes then the number of chromosomes and chromatids at metaphase II are, respectively**
(a) 10 and 10 (b) 10 and 20
(c) 20 and 40 (d) 40 and 20

49. **Which of the following precedes the reformation of the nuclear envelope during the M phase of the cycle?**
(a) Decondensation from chromosome and reassembly of the nuclear lamina
(b) Transcription from chromosome and reassembly of nuclear lamina
(c) Formation of contractile ring and formation of phragmoplast
(d) Formation of contractile ring and transcription from chromosome

50. **Which of the following can be seen during interphase with a light microscope?**
(a) Nucleosome (b) Introns
(c) Euchromatin (d) Heterochromatin

51. **When a cell has stopped the DNA replication fork, then which of the following checkpoints should be predominantly activated?**
(a) M phase (b) G_1/ S phase
(c) G_2/ M phase
(d) Both G_2/M and M phases

52. **A somatic cell that has just completed the S phase of its cycle as compared to a gamete of the same species, has**
(a) Twice the number of chromosomes and 4 times the amount of DNA
(b) Twice the number of chromosomes and 2 times the amount of DNA
(c) Twice the number of chromosomes and 8 times the amount of DNA
(d) Same number of chromosomes and 4 times the amount of DNA

53. **Cells of certain species of animals have 6 pairs of chromosomes. How many molecules of DNA will remain in the nucleus of these animals?**
(a) 12 (b) 24
(c) 36 (d) 48

54. **Nuclear membrane is formed around the groups of daughter chromosomes during the telophase by**
(a) ER (b) Golgi
(c) Lysosome (d) Ribosome

55. **Segregation of alleles occurs at**
(a) Metaphase I (b) Anaphase I
(c) Metaphase II (d) Anaphase II

56. **Sister chromatids are**
(a) Created when DNA is replicated
(b) Separated during mitosis
(c) Attached at the centromere before division
(d) All of the above

57. **The nutritional medium was supplied to growing bacteria. If the whole nutrient gets depleted in 20 cycles, then at which stage the amount of nutrient was half used?**
(a) At the end of 10 cycles
(b) At the end of 19 cycles
(c) At the beginning of 11 cycles
(d) At the beginning of 19 cycles

58. **The shape of chromosomes is observed at**
(a) Prophase (b) Metaphase
(c) Anaphase (d) Telophase

59. **Cell plate occurs in**
(a) Plant cells
(b) Animal cells
(c) Bacterial cells
(d) Both bacterial cells and plant cells

60. **Dinomitosis occurs in the species of**
(a) Monerans (b) Protists
(c) Plants (d) Animals

61. **The synthesis of RNA stops at**
(a) G_1phase (b) S phase
(c) G_2 phase (d) M phase

62. **Chromosomes exhibit minimum coiling during**
(a) Interphase (b) Prophase
(c) Anaphase (d) Telophase

63. **Zygotic meiosis occurs in**
(a) Algae (b) Higher plants
(c) Both of the above (d) Animals

64. **Haploid number of chromosomes is found in**
(a) Root tip (b) Leaf tip
(c) Anther (d) Bulbil

65. **The number of chromatids in a chromosome at metaphase is**
(a) 2 in mitosis and 1 in meiosis
(b) 1 in mitosis and 2 in meiosis
(c) 2 each in mitosis and meiosis
(d) 2 in mitosis and 4 in meiosis

66. **Four chromatids and two homologous centromeres occur in**
(a) Zygotene (b) Pachytene
(c) Diplotene (d) Diakinesis

67. **Nuclear envelope is broken down due to the activity of**
(a) P protein (b) U protein
(c) Kinase (d) Kinetin

68. **The four daughter cells derived from a single meiosis differ from each other due to**
(a) Difference in chromosome number
(b) Crossing over only
(c) Independent assortment of chromosomes
(d) Both b and c

69. **As a result of meiosis, the four cells are produced. These are**
(a) All identical
(b) All different
(c) 50 % identical and 50 % different
(d) 75 % identical and 25 % different

70. **The thread-like structures that begin radiating from each centrosome by the end of prophase are**
(a) Spindle microtubules
(b) Aster microtubules
(c) Kinetochore microtubules
(d) Polar microtubules

ANSWERS

1.d	11.b	21.a	31.c	41.d	51. b	61.d
2.c	12.b	22.c	32.b	42.a	52.a	62.a
3.a	13.d	23.a	33.b	43.d	53.b	63.a
4.a	14.d	24.c	34.c	44.b	54.a	64.c
5.c	15.d	25.d	35.b	45.c	55.b	65.c
6.c	16.b	26.b	36.b	46.a	56.d	66.a
7.a	17.a	27.b	37.d	47.b	57.a	67.c
8.a	18.d	28.a	38.b	48.b	58.c	68.d
9.d	19.a	29.b	39.d	49.c	59.a	69.b
10.b	20.a	30.c	40.c	50.d	60.b	70.b

CHAPTER-8

PHOTOSYNTHESIS IN HIGHER PLANTS

Photosynthesis is synthesis of food (glucose) by green plants in the presence of light by utilising CO_2 and water and thus releasing O_2 into atmosphere. Photosynthesis is largely responsible for producing and maintaining the oxygen content of the Earth's atmosphere and supplies all the organic compounds and most of the energy necessary for life on Earth. It is also called carbon assimilation.

It is a physico-chemical, endothermic and anabolic process regulated by enzymes. It is ten times faster than respiration. It is a redox process because water is oxidised to form O_2, and CO_2 is reduced to form glucose.

Photosynthesis first appeared in Cyanobacteria (Blue-green algae). Most of photosynthesis (90%) is carried out by algae in the ocean and the rest (10%) by land plants. Hence these organisms are called photoautotrophs.

Generally, photosynthesis is the opposite of cellular respiration. It is the most natural process which contains life on earth. All animals and heterotrophs depend on the green plants for their organic food.

TYPES:

1. Oxygenic: Photosynthesis is mostly oxygenic in most of the organisms where oxygen is liberated. e.g., plants and some protists.
2. Anoxygenic: Oxygen is not liberated. e.g., photosynthetic bacteria such as *Chlorobium* and *chromatium.*

Objectives: Photosynthetic organisms have three objectives:

1. To capture photons with high efficiency and transfer to the reaction centre
2. To convert solar energy to chemical energy
3. To reduce CO_2 to glucose

Landmarks in Photosynthesis

1. **Stephan Hales (1717):** He is the father of plant physiology. He proposed that green plants require sunlight to obtain their nutrition from the air.
2. **Joseph Priestley (1772):** He concluded that vegetation (plants) purified the air spoilt by burning candles. He observed that a burning candle in a closed jar changed the air inside which would not allow another candle to burn or mouse to live in. But when a twig of Mint plant was placed inside a jar then air allowed the candle to burn or the mouse to live in.

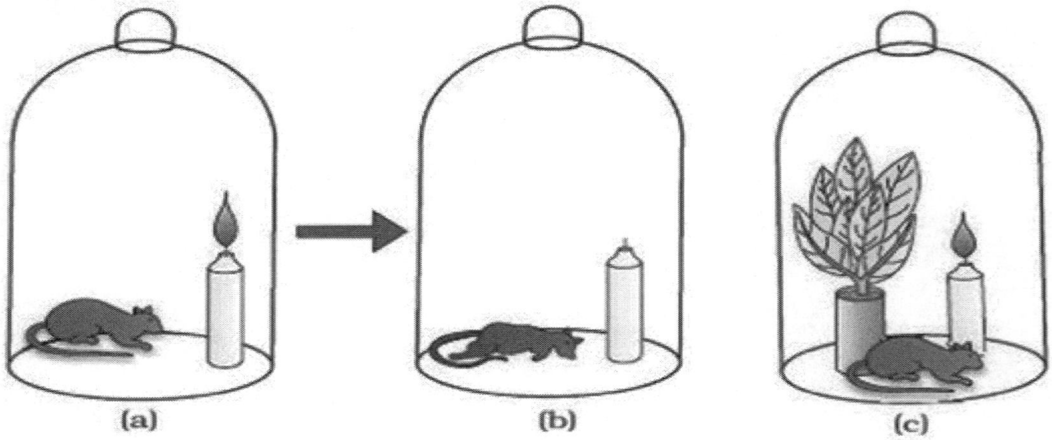

3. **Moll:** He performed the half-leaf experiment. In this, half part of the destarched leaf was inserted in an airtight bottle containing a small quantity of caustic potash (KOH) and the other half remained outside the bottle and illuminated. The part of the leaf outside the bottle turned black (positive) when the starch test (iodine) was done, and the part of the inserted leaf did not turn black (negative) because it did not show photosynthesis (formation of starch) due to lack of CO_2. All CO_2 inside the bottle was absorbed by KOH. He confirmed that CO_2 was essential for photosynthesis.

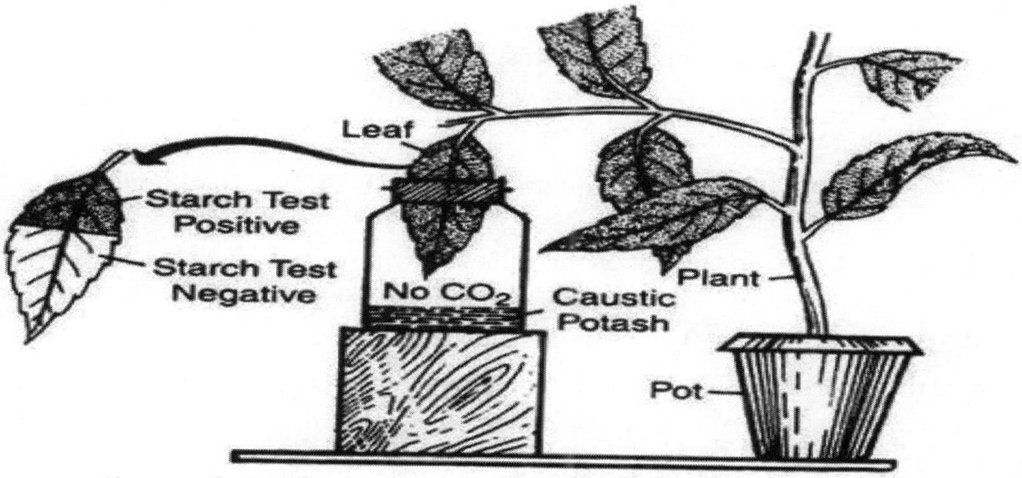

4. **Ingenhousz** (1779): He is the real discoverer of photosynthesis. He confirmed that O_2 was released from green plants when a twig of *Hydrilla* plant was inserted in a funnel covered by a test tube in a beaker and placed in bright sunlight. Small bubbles were formed which are O_2.

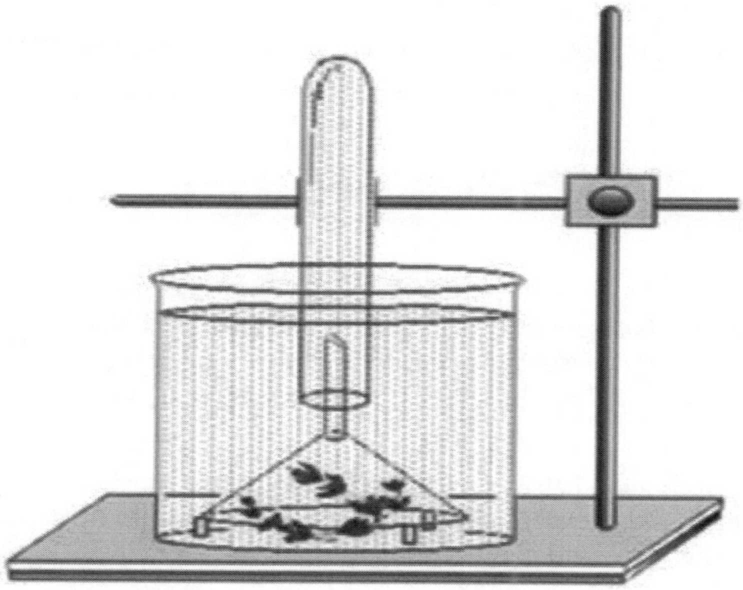

5. **Jean Senebier** (1782): He proposed that an increase in CO_2 will increase the evolution of O_2 in green plants.
6. **Theodore de Saussure** (1804): He pointed out that green plants require CO_2 water and light for photosynthesis. He further proposed that the amount of CO_2 absorbed is equal to the amount of O_2 evolved.
7. **Pelletier and Caventon** (1818): He discovered chlorophyll which could be separated from a leaf by boiling in alcohol.
8. **Dutrochet** (1837): He experimentally demonstrated that chlorophyll is necessary for photosynthesis.
9. **Liebig** (1845): He proposed that CO_2 is the main source of all organic compounds synthesised by green plants.

10. **Robert Mayer** (1845): He proposed that green plants utilise light energy and convert it into chemical energy of organic matter.
11. **Sachs** (1850): He noticed that the green part of the plant produces glucose which is stored as starch (the first stable product of photosynthesis). He discovered chloroplast in the cell where chlorophyll is located.
12. **Engelmann** (1888): He split light into its components by prism and then illuminated *Cladophora* plant placed in the suspension of aerobic bacteria. He found that bacteria accumulated in the region of red and blue light of a split spectrum. He thus discovered the effect of different wavelengths of light on photosynthesis. He concluded that chlorophyll was involved in photosynthesis.
13. **Blackman** (1905): He proposed that photosynthesis occurs in two steps (dark reaction along with light reaction). He also proposed the "Law of limiting factors".
14. **Warburg** (1920): He introduced *Chlorella* (unicellular, non-motile green alga) as an experimental material for photosynthesis.
15. **Van Niel** (1931): He demonstrated that photosynthetic bacteria fixed CO_2 in the presence of H_2S (in the case of sulphur photosynthetic bacteria) and H_2 or lactate (in the case of non-sulphur bacteria).

$$6CO_2 + 12\ H_2S \longrightarrow C_6H_{12}O_6 + 6H_2O + 12S$$

He also postulated that green plants evolved O_2 by splitting H_2O (not CO_2).

$$6CO_2 + 12H_2O \longrightarrow C_6H_{12}O_6 + 6H_2O + 6O_2$$

16. **Emerson and Lewis** (1943): They performed red drop and enhancement effects. Photosynthesis is dropped suddenly if red light of a higher wavelength (>680 nm or 700 nm) is supplied in *Chlorella*. Photosynthesis is enhanced if red light of a shorter wavelength (<680 nm) is supplied in *Chlorella*.
17. **Robert Hill** (1937): He confirmed that isolated chloroplast of green leaves of Stellaria *media* did not bring about photolysis of water and liberate O_2 when illuminated but when reducing agents or electron acceptors (Hill oxidants such as chromate, endophenol, ferricyanide, benzoquinone and ferric oxalate) were added to the illuminated suspension then chloroplast evolved oxygen.
18. **Ruben, Hassid and Kamen** (1947): They proved that O_2 evolved in photosynthesis comes from water labelled using heavier isotope (H_2O^{18}).

$$6CO_2 + 12H_2O \xrightarrow[\text{Chlorophyll}]{\text{Light}} C_6H_{12}O_6 + 6H_2O + 6O_2$$

REQUIREMENTS

1. Chloroplast: It is the main site of photosynthesis in green plants. It is located at the outer margin with its broad surface parallel to the cell wall of the mesophyll cells. This type of arrangement helps in the easy diffusion of CO_2 required for photosynthesis from the atmosphere to the inside of the chloroplast.
2. CO_2: It is taken up by the atmosphere.
3. Light: A ray of light incident upon a leaf behaves in three different ways. About 83% of the Photosynthetically Active Region (PAR) is absorbed, 12% is reflected by the leaf and 5% is transmitted. Out of total absorption, only 4%-5% is absorbed by chlorophyll for photosynthesis.
4. Water: Photosynthesis utilises less than 1% (0.2%) of total absorbed water by the plants.
5. **Photosynthetic pigments:** These are found in the thylakoid of chloroplast in plants, but these are diffused throughout the cell in the case of photosynthetic bacteria (photosynthetic bacteria and Cyanobacteria) because chloroplast is absent. They absorb light energy at specific wavelengths for photosynthesis.
    1. **Chlorophylls**: These are specialised lipid molecules and green pigments located in the lipid part of the thylakoid membrane. These were studied by Wistatter, Stoll and Fischer (1912). These are soluble in organic solvents (fat) and insoluble in water. These absorb both shorter waves (blue and violet) and longer waves (orange and red) of the visible spectrum. These transmit or reflect green light hence leaves appear to be green.

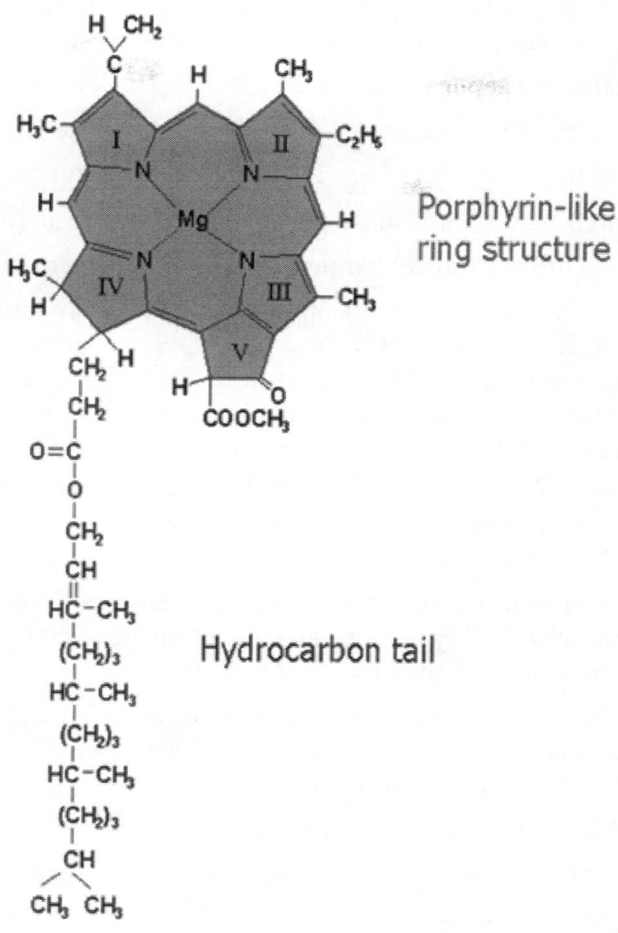

Structure: It is tadpole-shaped (square planar) and asymmetrical. It consists of

(a) **Porphyrin Head** (15x15A⁰): It is flat and square structure of alternating single and double bonds containing 4 pyrrole rings which are linked by a methine ($-CH_2$) bridge. Each pyrrole ring is made up of 5 atoms (4C and 1N). There is a non-ionic Mg atom at the centre by N atoms of pyrrole rings through two covalent and coordinate bonds. Pyrrole rings contain several side groups that change the properties of pigment. In chlorophyll-a, there is one methyl ($-CH_3$) group at the third carbon atom while in chlorophyll-b, it is replaced by an aldehyde (-CHO) group. Chlorophyll without its Magnesium core is colourless and is called phaeophytin.

(b) **Phytol tail** (20A⁰): It is a long chain of carbon and hydrogen atoms (hydrocarbon). It is attached to the 7th carbon from the head. It anchors the chlorophyll into the lipid part (hydrophobic) of the thylakoid membrane. In chlorophyll-a, there is only one double bond present in the phytol tail.

Synthesis: It is synthesised from succinyl Co-A and glycine. Several intermediates such as aminolevulinic acid, phosphobilinogen, uroporphyrinogen, protoporphyrin, Mg-Protoporphyrin, protochlorophyllide-a and chlorophyllide-a are formed. Finally, Chlorophyll-a is formed. Protochlorophyll is converted to chlorophyll by the addition of 2H atoms at carbon atoms 7 and 8 of the fourth pyrrole ring. Hence protochlorophyll lacks 2 H atoms in one of its pyrrole rings. The etiolated plants contain protochlorophyll.

Types of Chlorophyll: Aronoff and Allen (1906) recognised 8 types of chlorophyll.

(a) **Chlorophyll-a** (Blue-green): Its formula is $C_{55}H_{72}O_5N_4Mg$. It is most abundant (25%). It is three times more than chlorophyll-b. It absorbs more red light than blue and violet light. It is found in all photosynthetic organisms except bacteria.

(b) **Chlorophyll-b** (Yellow-green): Its chemical formula is $C_{55}H_{70}O_6N_4Mg$. It absorbs more blue and violet light than red light. It is found in Euglenoids, Chlorophyceae (green algae) and higher plants.

(c) **Chlorophyll-c** (Bluish green): Its chemical formula is $C_{35}H_{30}O_5N_4Mg$. It does not have a phytol chain. It is found in diatoms, dinoflagellates and phaeophyceae (brown algae).

(d) **Chlorophyll-d** (Red): Its chemical formula is $C_{54}H_{70}O_6N_4Mg$. It is found in Rhodophyceae (red algae)

(e) **Chlorophyll-e**: It is found in Xanthophyceae (yellow-green algae).

(f) **Bacteriochlorophyll-a** (Purple): $C_{55}H_{74}O_6N_4Mg$. It is found in purple sulphur bacteria (*Chromatium*).

(g) **Bacteriochlorophyll-b**: It is found in purple non-sulphur bacteria (*Rhodospirillum*).

(h) **Bacterioviridin** or **Chlorobium chlorophyll:** It is found in green sulphur bacteria (*Chlorobium*).

2. **Carotenoids**: These are formed from basic isoprene units. These are found in algae and higher plants. These are soluble in organic solvents such as chloroform, ether, fats etc. but insoluble in water. These absorb the light of shorter waves (blue and violet) and transfer it to chlorophyll for use in photosynthesis. These are called shield pigments because they protect chlorophyll from photooxidation or photobleaching. These convert nascent oxygen to molecular oxygen.

   (a) **Carotenes**: Its chemical formula is $C_{40}H_{56.}$ These are orange-coloured and hydrocarbon (tetraterpenes). These are of several types viz. α, β,γ, flavacene, and lycopene (found in tomato) but β-carotene is most common in all plants. These provide red colouration to chilly and potatoes.

   (b) **Xanthophylls**: Its chemical formula is $C_{40}H_{56}O_{2.}$ These are yellow-coloured and oxygen-containing carotene. These are of several types such as Lutein (most common), Fucoxanthin (dominant pigment in brown algae or Phaeophyceae) and Violaxanthin.

3. **Phycobillins**: These are noncyclic, linear and open tetrapyrroles attached to proteins in living cells. These are water-soluble. These lack Mg atoms and phytol tail. These absorb blue and red light. These are useful in chromatic adaptations. There are several types such as

   (a) c- Phycocyanin (blue): It is found in Cyanophyceae.

   (b) Allophycocyanin (blue): It is found in Cyanophyceae and Rhodophyceae.

   (c) r-Phycoerythrin (red): It is found in Rhodophyceae.

Ratio of pigments in chloroplast

Chlorophyll-a: Chlorophyll-b = 3:1

Chlorophyll: Carotenoids = 3:1

Carotenes: Xanthophylls = 1 :1

Photoluminescence: It is the irradiation of absorbed energy. It is of two types

1. Fluorescence: It occurs instantaneously, and it is the immediate emission of long-wave radiation. Chlorophylls show an outburst of fluorescence (Katusky effect) during the first few months of illumination. These show red fluorescence.

2. Phosphorescence: It occurs after a gap and it is the delayed emission of long-wave radiation. It continues for some time after the removal of the irradiation source.

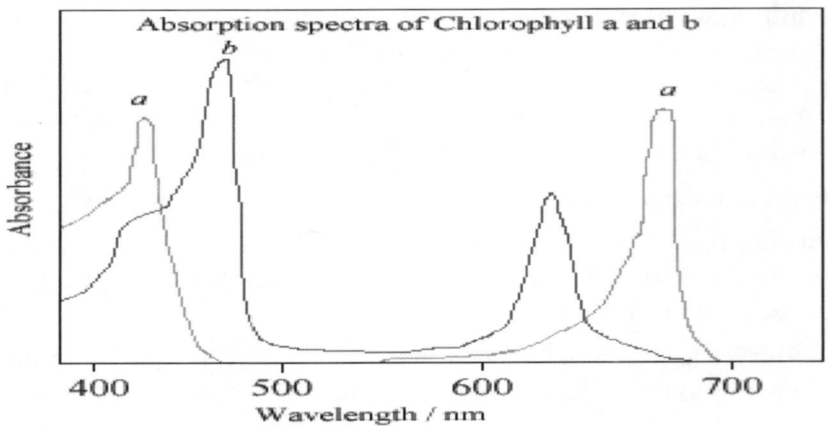

Ability to absorb light of different wavelength

1. Absorption spectrum: It is studied directly. It requires a single exposure to full light. It is a graph showing amount of light of different wavelengths absorbed by a pigment. Chlorophyll absorbs mostly blue light.
2. Action spectrum: It was studied by Engelmann (1882) in *Cladophora*. It is studied in relation to either CO_2 absorbed or O_2 evolved. It is a graph showing the rate of photosynthesis at different wavelengths of light absorbed by a pigment. The rate of photosynthesis is maximum in red light.

Anthocyanin: It is a purplish pigment and is present in leaves, flowers and fruits. It gives colouration to leaves. It is soluble in water. It is dissolved in cell sap and is not present in cell sap. It does not take part in photosynthesis.

Pigment system (Photosystem): It is a network of chlorophyll-a, accessory pigments (carotenoids and phycobilins) and associated proteins on the surface of the thylakoid membrane. It is a small group (photosynthetic unit) of 250 pigment molecules and proteins that work together for the reaction centre. It consists of

1. Light Harvesting complex (Antenna complex): It contains 200 chlorophyll molecules that absorb different wavelengths of light and transfer to the reaction centre by inductive resonance.
2. Core complex: It contains 40-60 chlorophylls, a reaction centre (P-680 / P-700), e^- donors and e^- acceptor. The reaction centre is less than 1%of chlorophyll. The core complex finally performs light reactions.

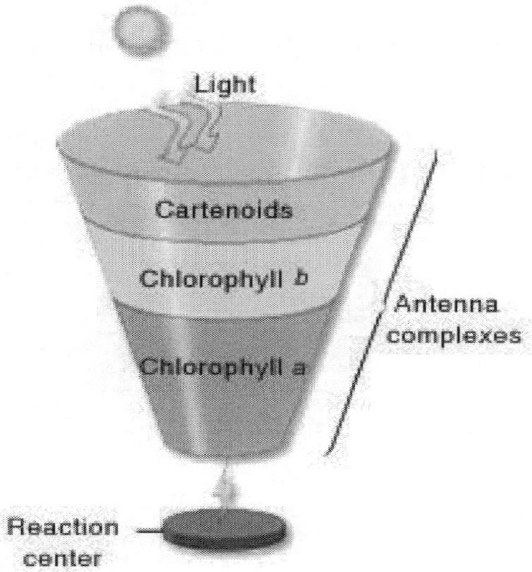

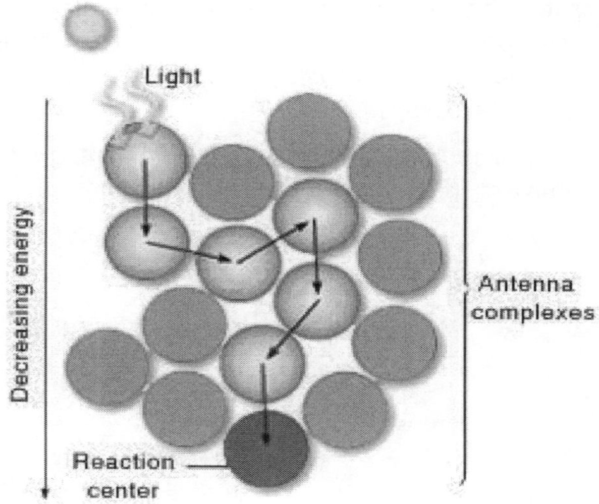

- If Mg atom is removed from the Light-harvesting complex of Chlorophyll, then the photosynthetic unit continues to harvest light for the reaction centre.
- If Mg atom is removed from the chlorophyll of the core complex, photosynthesis stops.

Types of Pigment System: Two types of pigment systems are structurally distinct but act synergistically.

Pigment system II	Pigment system I
It lies on inner surface of granal thylakoid.	It lies on outer surface of stromal thylakoid and granal thylakoid.
Its reaction centre is P-680.	Its reaction centre is P-700.
It is more fluorescent.	It is less fluorescent.
It absorbs wavelength maximum at 680 nm.	It absorbs wavelength maximum at 700nm.
It is active in red light only.	It is active in both red light and far-red light.
It contains less chlorophyll-a than chlorophyll-b.	It contains more chlorophyll-a than chlorophyll-b.
It contains less carotenes than xanthophylls.	It contains more carotenes than xanthophylls.
It performs non-cyclic photophosphorylation.	Itperformsbothcyclicandnon-cyclicphotophosphorylation.

Pigment system II	Pigment system I
It is associated with the splitting of water hence O2 is evolved.	It is not associated with the splitting of water hence O2 is not evolved
It generates ATP and $NADPH_2$.	It generates ATP.
It has O2 as a strong oxidising agent.	It has ATP as a weak oxidising agent.
It has ATP as a weak reducing agent.	It has $NADPH_2$ as a strong reducing agent.

Light reaction:

It was first proposed by Robert Hill (1939). It takes place in the thylakoid membrane of the grana of chloroplast in the presence of light. It is also called a photochemical reaction. It is a faster process where solar energy is trapped by chlorophyll and is stored in the form of chemical energy (ATP) and reducing power ($NADPH_2$). Both ATP and $NADPH_2$ are collectively called Assimilatory power. It occurs in the following steps:

1. **Absorption of solar energy**: The photosynthetic pigments such as chlorophylls, carotenoids and phycobilins absorb solar energy in different regions of visible light.
2. **Transfer of solar energy from antenna pigments to Chlorophyll-a**: Solar energy absorbed by antenna or accessory pigments (pigments other than chlorophyll-a) is transferred by resonance to chlorophyll-a which alone takes part in light reaction. Chlorophyll-a also absorbs light energy directly. In pigment system II, it is P-680 and in pigment system I, it is P-700. Light with wavelength (>680nm) is less efficient than light of wavelength (<680nm).

   Transfer of light energy: Phycoerythrin (570 nm) → Phycocyanin (670 nm) → Carotenoids → Chlorophyll-b → Chlorophyll-a →Reaction centre

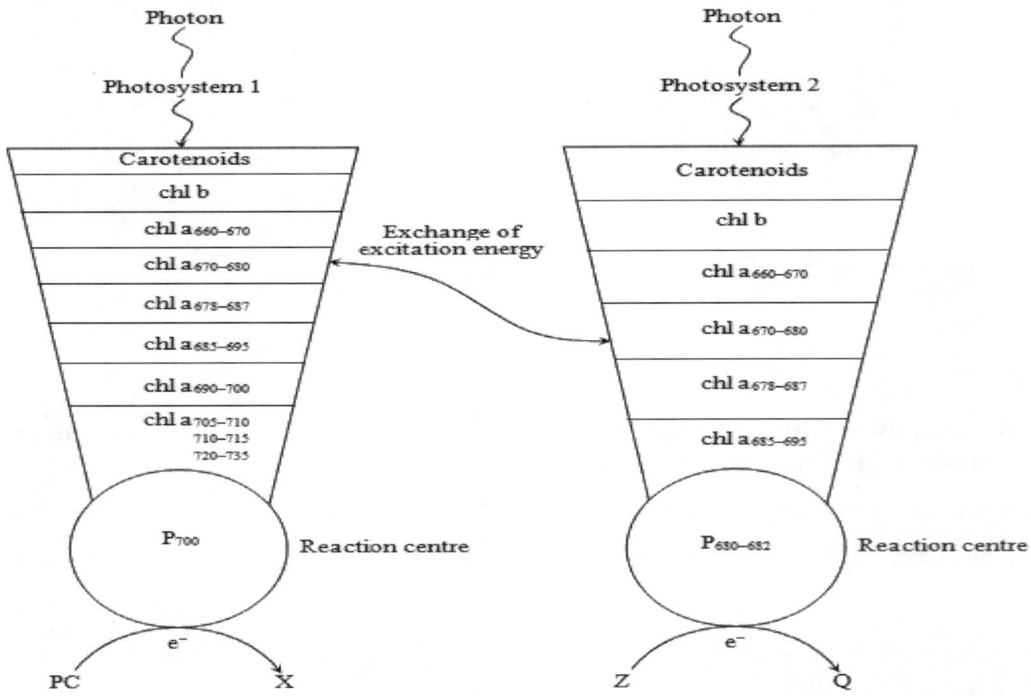

Fig. Distribution of pigments in two pigment systems

3. **Excitation of Chlorophyll-a**: Both pigment systems (Ps-II and Ps-I) after receiving a photon (quantum) of light become excited having more energy than ground state energy. After passing through the unstable second singlet state (excited state) and first singlet state, chlorophyll molecules come to the metastable triplet state. This form of chlorophyll-a takes part in light reactions. It expels its energy along with an electron and a positive charge comes on chlorophyll-a molecule. The expelled electron contains an extra amount of energy received from light. This is used in the formation of ATP during photosynthesis.

$$\text{Chlorophyll-a} \xrightarrow{\text{light}} \text{Chlorophyll-a} \longrightarrow (\text{Chlorophyll-a})^{+} + e^{-}$$

(ground state) (excited triplet state) (oxidised)

4. **Photolysis (Oxidation) of water:** It is first photochemical reaction. It is mainly concerned with Ps-II and is catalysed by Mn^{+2} and Cl^- ions. When Ps-II is active, it is photoactivated then the water molecule gets dissociated into H^+ and OH-. After that OH- releases an electron from itself and becomes a free but highly reactive form. The H^+ released in the above reaction gets associated with NADP to form $NADPH_2$. The electrons fill the vacancy in chlorophyll molecules created by the photoactivation of chlorophyll. The OH molecules now release O_2 and form H_2O. The photolysis of water maintains the supply of electrons. Hence O_2 is released during photosynthesis.

$$4H_2O \rightarrow 4H^+ + 4OH^-$$

$$4OH^- \rightarrow 4OH + 4e^-$$

$$\underline{4OH \rightarrow 2H_2O + 4e^- + O_2}$$

$$2H_2O \rightarrow 4H^+ + 4e^- + O_2$$

OR

$$12\,H_2O \longrightarrow 24H^+ + 24e^- + 6O_2$$

5. **Photophosphorylation**: It is the synthesis of ATP from ADP and Pi with the help of light. Arnon (1954) discovered photophosphorylation and observed that isolated chloroplast would synthesise ATP in the presence of light. It is a redox process (oxidation and reduction process occur simultaneously) when chlorophyll-a molecule receives a photon of light. It becomes excited and expels the extra energy along with an electron in both pigment systems. Each e-lost is accepted by another molecule, e- acceptor. Thus, chlorophyll is oxidised and the electron acceptor is reduced. Thus, instead of losing its energy, the e- is captured by the electron acceptor. This is the conversion of light energy into chemical energy. It occurs in two stages:

   (a) **Non-cyclic photophosphorylation**: Hill and Bendal proposed a Z-scheme for electron transport. In this process, the electrons are not recycled back to chlorophyll molecule and are consumed in reducing $NADPH_2$ and thus a constant supply of water molecules is required for e- and H^+. Thus, it is a unidirectional movement of electrons. It is carried out by both PS-II and PS-I. This scheme initially shows uphill transport (from PS-II to pheophytin), downhill (from pheophytin to PS-I), again uphill (from PS-I to another acceptor, FRS) and finally downhill transport (from FRS to $NADP^+$).

   Steps:

   1. When sunlight strikes PS-II, the light-harvesting complex receives the radiation first and then passes to other pigment molecules and, finally the reaction centre of PS-II (P-680) by inductive resonance. A photon of light is absorbed by P-680 of chlorophyll-a in PS-II. It gets excited and e- is emitted from it and it causes an electron deficiency which is fulfilled by e- coming from photolysis of water. Light with a wavelength smaller than 680nm is more efficient than a wavelength greater than 680 nm.
   2. The e- is trapped by a primary acceptor, pheophytin (a chlorophyll in which Mg atom is replaced by 2H atoms) then it is passed to Plastoquinone (PQ).
   3. The e- is now passed to cytochrome b_6 and cytochrome f (cytochrome complex) where e- loses sufficient energy for the synthesis of ATP. Hence, I molecule of ATP is formed by ADP and Pi.
   4. The e- is now passed to copper-containing protein, plastocyanin (PC) that acts as a shuttle between cytochrome f and PS-I. Thus e- reaches Ps-I and is accepted eventually by P-700.
   5. At the same time, PS-I also absorbs a photon of light and becomes excited to release e- so that e- deficiency occurs in the P-700 molecule. The e-deficiency is fulfilled by e-coming from PS-II.
   6. The e- is now passed to a ferredoxin-reducing substance (FRS) which is the primary acceptor.
   7. The e- is now passed to an iron-containing protein, ferredoxin which is a natural H acceptor.
   8. The e- is transferred to NADP which combines with H^+ released from photolysis of water with the help of NADP reductase to form $NADPH_2$.

$NADP^+ + 2H^+ + 2e^- \rightarrow NADPH_2$

$2NADP^+ + 4H^+ + 4e^- \rightarrow 2NADPH_2$

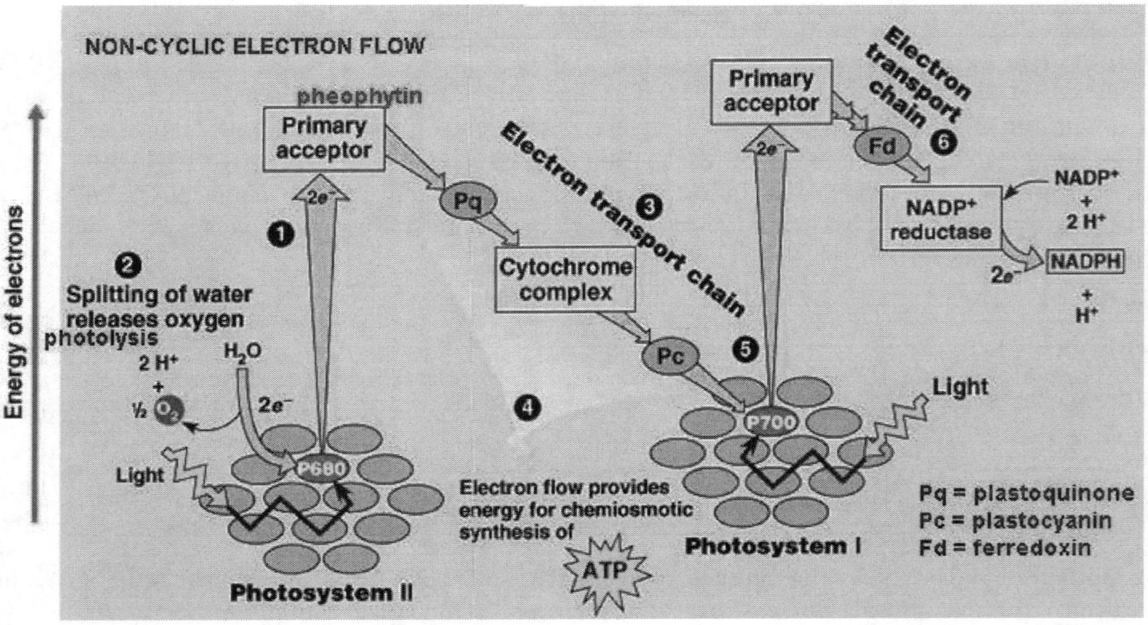

Note:

- Two e- and two H^+ are required to reduce 1 mole of NADP to $NADPH_2$.
- Two photons are required to transfer one e-
- Four e- are required to be transferred to reduce 1 mole of CO_2.
- Thus 4 moles of P-680 are excited by 4 photons. Similarly, 4 photons will release 4 electrons from 4 mole of P-680. Hence photolysis of 2 water molecules results in the production of 4e-, $4H^+$ and O_2. Now 8 photons and 2 moles of H_2O are required to reduce 2 moles of NADP+.
- Quantum yield: It is a number of oxygen molecules evolved per photon of light absorbed. Thus, it is 1/8=12.5%
- Quantum requirement: It is the inverse of quantum yield. It is a number of photons (8) of light required for each oxygen molecule to evolve.

(b) **Cyclic photophosphorylation**: In this type, electrons are recycled back to the chlorophyll molecule at low energy. It is a primitive type and the most effective anaerobic phosphorylation mechanism. The electron passes from one component of the electron transport system to the next; it progressively loses energy so there is a lowering of energy gradient. It only occurs when light of wavelength beyond 680 nm is available for excitation; and when the level of $NADP^+$ is low and NADPH is high.

Steps:

1. A photon of light is absorbed by P-700 of chlorophyll-a in Ps-I. It gets excited and an electron emitted from it is accepted by a primary acceptor, Ferredoxin reducing Substance (FRS).
2. The electron is passed from FRS to plastoquinone (PQ) cytochrome-b_6, cytochrome-f, plastocyanin (PC) and finally back to Chlorophyll-a.
3. While passing over the FRS, plastoquinone and cytochrome complex, the electron loses sufficient energy to form ATP from ADP and Pi. During this process, protons (H^+) are dragged into thylakoid building a proton gradient that forms ATP by chemiosmosis. It uses light energy to build ATP, but it cannot drive dark reactions because it does not produce sufficient ATP just like non-cyclic photophosphorylation. Hence both processes are required for this. It is the only process that occurs in certain photosynthetic bacteria.

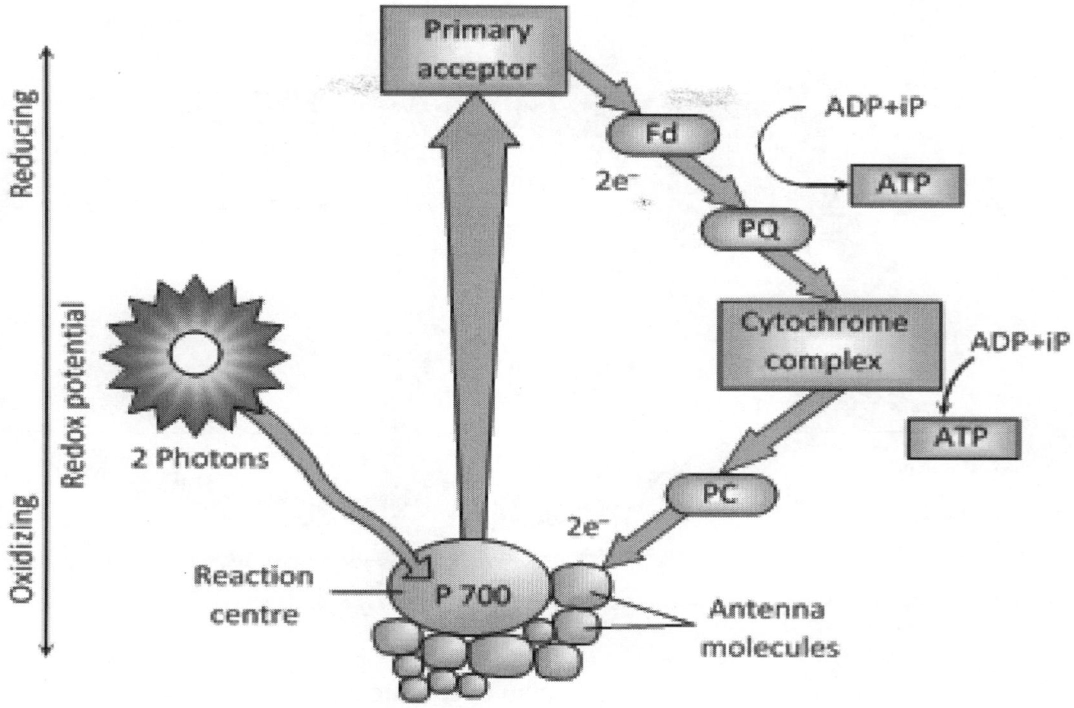

Cyclic photophosphorylation

Non-cyclic photophosphorylation	Cyclic photophosphorylation
The electron emitted from chlorophyll is not recycled back to the same chlorophyll.	The electron emitted from chlorophyll is recycled back to the same chlorophyll.
It is a unidirectional e-transfer.	It is a cyclic e-transfer.
It is carried out by both PS-I and PS-II.	It is carried out only by PS-I.
It occurs in the granal thylakoid.	It is carried out by stromal thylakoid (intergranal thylakoid).
It takes place in all plants.	It takes place in photosynthetic bacteria. It occurs in plants as a backup to provide additional ATP.
It occurs under optimum light, CO_2 and aerobic conditions.	It occurs under low light, low CO_2 and anaerobic conditions.
It synthesises 1 ATP and 2 $NADPH_2$.	It synthesises only 2ATP.
It undergoes phosphorylation at one step.	It undergoes phosphorylation at two steps.
It involves the photolysis of water and the evolution of O_2.	It does not involve the photolysis of water and evolution of O_2.
Dichlorophenyldimethyl urea (DCMU) inhibits the transport of electrons at the plastoquinone stage.	It cannot inhibit the transport of electrons.

CHEMIOSMOSIS:

It was proposed by Peter Mitchell (1961). It is the synthesis of ATP in light reaction by movement of H^+ across selectively permeable membrane along a concentration gradient. There are 4 protein complexes in the thylakoid membrane.

1. Ps-II: It is located in the stacked region of the thylakoid membrane.
2. Ps-I: It is located in stromal lamellae and unstacked region of the thylakoid.
3. Cytochrome b_6-f complex: It pumps protons.
4. ATP synthase (cf_0-cf_1): It is located in stroma lamellae and unstacked region.

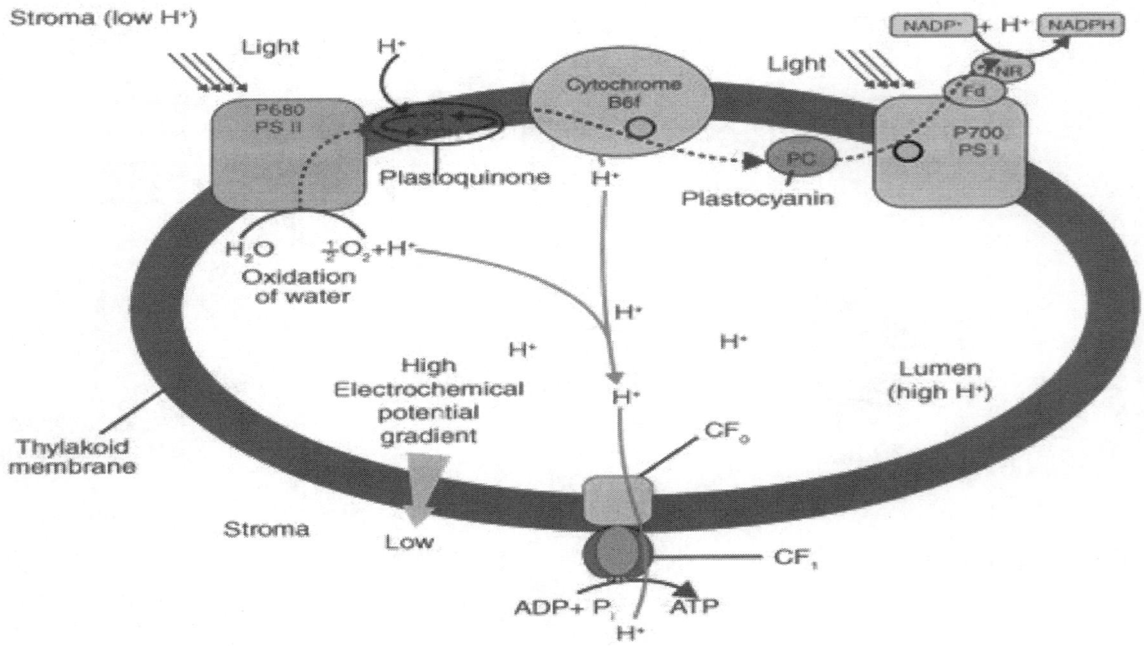

Fig: Mechanism of chemiosmosis

Chemiosmosis requires a membrane, a proton pump, a proton gradient and ATPase. It occurs in following steps:

(a) During electron transport these electrons travel down an electron transport chain through pigment systems, and some of the energy pump protons (H^+) actively across a membrane from stroma to lumen of thylakoid by cytochrome b_6-cytochrome f complex. This occurs when the primary acceptor of e- which is located towards the outer side of the membrane, transfers its e- not to the electron carrier but to H^+ carrier. Thus, this molecule removes one H^+ from stroma while transferring an e-. When this molecule passes its electrons to e- carrier on the inner side of the thylakoid membrane, then H^+ is released into the lumen of the thylakoid.

(b) Water splits on inner side of thylakoid membrane resulting in the accumulation of H^+ in the lumen that contributes to the gradient.

(c) $NADP^+$ is reduced to NADPH by NADP reductase enzyme (located on the stroma side of the thylakoid membrane) by utilising electrons that come from the acceptor of e^- of Ps-I and H^+ from the stroma.

(d) Now the number of H^+ is decreased in the stroma (alkaline condition) and increased in the lumen of thylakoid (acidic condition). This develops a proton gradient (proton motive force) across the thylakoid membrane. The gradient is broken down due to the movement of H^+ across the membrane to the stroma through the trans-membrane channel of the F_0 of ATPase.

(e) The ATPase (ATP synthetase) consists of two metallo-peptide complexes, oneisF_0 (hydrophobic) which is embedded in membrane and forms trans-membrane channel that carries out facilitated diffusion of H^+ across the membrane (from lumen to stroma). The other part is called F_1 (hydrophilic) and projects on the outer surface of thylakoid membrane (in stroma) on the side that forms stroma.

(f) The breakdown of the gradient (or diffusion of protons from lumen to stroma) provides enough energy to cause a conformational change in F_1 particle of ATPase. This energy is used to pump H^+ across the membrane to create a gradient or a high concentration of H^+ within the lumen of the thylakoid.

(g) ATPase has a channel that allows diffusion of H^+ back across the membrane. This releases enough energy to activate the ATPase enzyme that catalyses the formation of ATP.

(h) Thylakoid membrane is now impermeable to H^+. Hence H^+ can come back from lumen to the stroma through a channel provided by ATP synthase (cf_0-cf_1). The cf_0 forms a channel across a membrane and through which H^+ move from the lumen to the stroma with the help of proton motive force. This causes synthesis of ATP from ADP (phosphorylation) by cf_1 component of ATP synthase and thus ATP is released into the stroma.

$2H^+ + ADP + Pi \rightarrow 2H^+ + ATP + H_2O$

(Lumen) (Stroma)

The overall equation for the **light reaction** is:

$2H_2O + 2NADP + 3ADP + 3Pi + light \rightarrow 2NADPH + 2H^+ + 3ATP + O_2$

Anoxygenic photosynthesis	Oxygenic photosynthesis
The photosynthetic pigment is bacteriochlorophyll.	The photosynthetic pigment is chlorophyll-a.
Pigment system II is absent.	Pigment system II is present.
The electron donor is H_2S or H_2 or lactate.	The electron donor is H_2O.
Oxygen is not produced.	Oxygen is produced.
The primary product of energy conversion is ATP only.	The primary products of energy conversion are ATP and $NADPH_2$.

DARK REACTION (C_3 CYCLE):

It was first established by Blackman but later on, it was studied in detail by Calvin, Benson and Bassham (1954) who determined the sequence of reactions in *Chlorella* and *Scenedesmus* using autoradiography (C^{14}) and chromatography. It occurs in stroma of chloroplast where all enzymes of dark reaction are found. It does not require light thus it is independent of light hence it can also occur in the dark.

It is also called a biosynthetic pathway in which the assimilatory power (ATP and $NADPH_2$) produced by light reaction is utilised to reduce CO_2 to form glucose by the process of carbon fixation.

Process: The dark reaction is completed in three steps:

1. **Carboxylation:** It is the first process in which atmospheric CO_2 is accepted by a 5-carbon compound, Ribulose bisphosphate (RuBP) in the presence of Ribulose bisphosphate carboxylase oxygenase (RuBisCO) or carboxy dismutase to form a first stable 3-carbon compound, Phosphoglyceric acid (3-PGA). This process does not utilise ATP and $NADPH_2$.

   (RuBisCO is the most abundant protein which constitutes 50% of the total proteins of chloroplast. It is the slowest enzyme located in the stroma on outer surface of the thylakoid membrane. It requires Mg^{+2} as a metal activator. It was crystallised from tobacco leaves.

   $$6\,CO_2 + 5RuBP \xrightarrow{\text{RuBisCO}} 12(3\text{-PGA})$$

2. **Glycolytic reversal (Reduction):** It involves several steps:

   (a) In this process, 3-PGA undergoes a phosphorylation reaction in the presence of phosphoglyceric kinase forming 1,3-diPGA. This reaction utilises 2 ATP per molecule of CO_2 for supplying energy.

   (b) Then 1,3-diPGA undergoes a reduction in the presence of dehydrogenase enzyme forming 3-Phosphoglyceraldehyde (PGAL). This process utilises $2NADPH_2$ for reducing per molecule of CO_2 fixed. The 3-PGAL is a key product that is used in the synthesis of both carbohydrates and fats. It does not accumulate hence it is immediately converted to other products such as fructose 1,6-diphosphate, fructose 6-phosphate and then finally glucose. It seems to be a reversal of glycolysis; the only difference is that reducing power is obtained from NADPH (not NADH). Half (50%) of glucose produced is used by plants in cellular respiration to maintain itself and the rest glucose is converted to sucrose for translocation, cellulose for building cell walls and starch for storage.

3. **Regeneration of RuBP**: It is a complicated process that requires several enzymes. Simply the large amount of PGAL is converted to RuMP in the presence of kinase and later RuMP is converted to regenerate RuBP in the presence of kinase and Mg^{+2} so as to ensure the continuous running of Calvin cycle. This process utilises ATP.

   Since glucose is a 6-carbon compound thus 6 rounds of Calvin cycle are required to synthesise 1 mole of glucose because one carbon atom is reduced in each cycle.

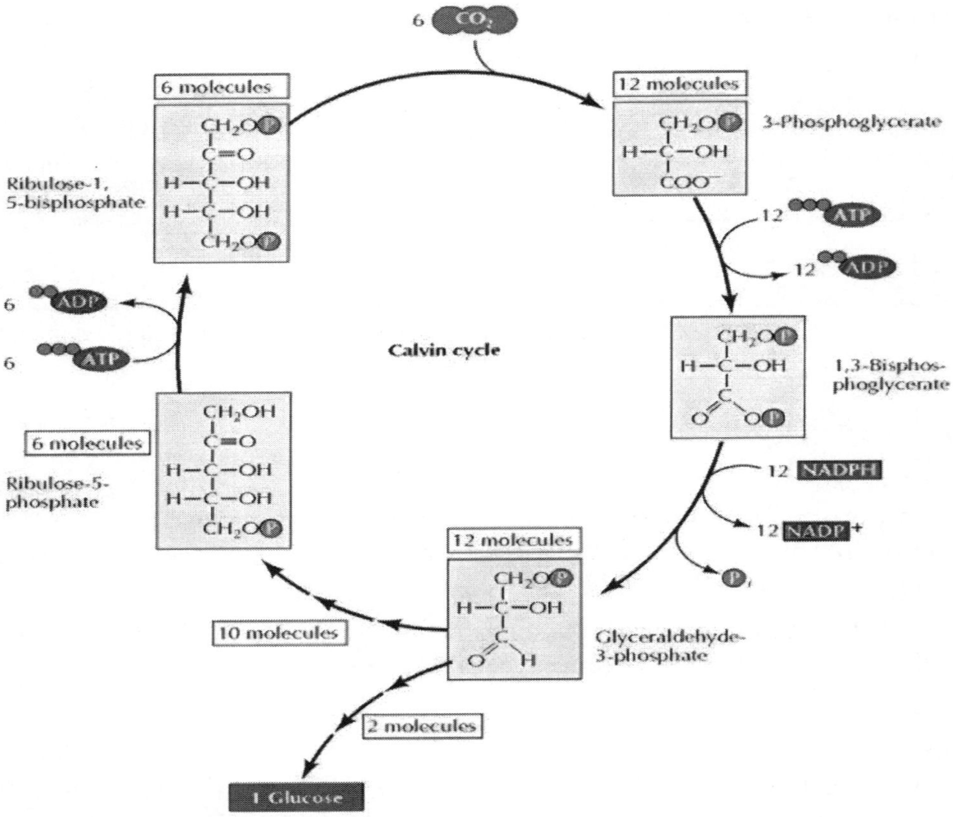

THERMODYNAMICS:

- To reduce 1 molecule of CO_2, 8 quanta of light are required.
- To synthesise 1 mole of glucose (from 6 CO_2), 48 quanta of light are required. Therefore, 42 Kcal energy is present in one quantum of red light of 680 nm.
- The minimum light energy required for reducing $6CO_2$ to synthesise glucose is 42x48=2016
- Since 1 mole of glucose is oxidised aerobically to yield 686 kcal hence 2016/686=33% of total light energy is utilised in fixing 6 molecules of CO_2.

$$1CO_2 + 2H_2O + 3ATP + 2NADPH_2 + 1RuBP \longrightarrow (CH_2O)n + H_2O + 3ADP + 3Pi + 2NADP^+ \quad + 2H^+ + 1RuBP$$

Overall reaction: $6CO_2 + 12H_2O + 18ATP + 12NADPH + 6RuBP \longrightarrow$

$$C_6H_{12}O_6 + 6H_2O + 18ADP + 18Pi + 12NADP^+ + 12\ H^+ + 6RuBP + 6O_2\uparrow$$

Conversion of energy:

1. In C_3 plants, 18 ATP and 12 $NADPH_2$ are required to synthesise I mole of glucose. 18 ATP=18 x7.3 Kcal=132 Kcal

   12 $NADPH_2$=12 x 52 kcal= 624 Kcal

   Total energy required = 132 + 624= 756 kcal

   Since I mole of glucose produces 686 kcal

   Thus 90% of the conversion of energy could be noticed in C_3 plants.

C_4 CYCLE / HATCH-SLACK PATHWAY

It was reported by Kortchak (1964) in leaves of sugarcane, but it was studied in detail by Hatch and Slack (1970).

It is a supplementary pathway because it supplies CO_2 to C_3 cycle to carry out the synthesis of sugars.

It is an alternative CO_2 fixation that occurs in tropical and subtropical grasses (monocots) and some dicots of angiosperms. It has been reported in 10 families of angiosperms such as Poaceae, Cyperaceae, Amaranthaceae, Chenopodiaceae, Asteraceae, Euphorbiaceae, Nyctaginaceae, Portulaceae, Aizoceae, and Zygophyllaceae.

The plants of these families show a very special type of anatomy called Krantz anatomy because of the presence of radially arranged cells of bundle sheath cells around a vascular bundle which looks like a ring. The cell of bundle sheath (Krantz cell) is larger in size, thick-walled and contains a large amount of chloroplast without grana. These cells do not have intercellular space but contain starch.

These plants due to high temperature and high light intensity keep their stomata partially closed due to which CO_2 concentration decreases and thus Calvin cycle cannot be operated.

It involves the following steps:

1. **Reactions in mesophyll cells:**

   (a) The low concentration of atmospheric CO_2 enters the mesophyll cell through stomata and is accepted by Phosphoenol pyruvic acid (PEP). It undergoes carboxylation in the presence of PEP carboxylase (PEPcase) to form the first stable 4C compound, oxaloacetic acid (OAA).

   (b) Later OAA is converted into malic acid in the presence of malate dehydrogenase enzyme in mesophyll cell.

2. **Reactions in bundle sheath cell:**

   Malic acid enters the bundle sheath cell through plasmodesmata where it undergoes decarboxylation to form pyruvic acid and thus CO_2 is released. As a result, RuBisCO enzyme functions as carboxylase minimising the activity of oxygenase and thus favours carboxylation. The final fixation of CO_2 (Calvin cycle) occurs in the chloroplast of bundle sheath cell where the concentration of CO_2 is high but the temperature and O_2 are low. Hence, productivity (yield) is better in C_4 plants.

3. **Reactions in mesophyll cells:**

   Finally, pyruvic acid enters into the chloroplast of mesophyll cell through plasmodesmata and reacts to form phosphoenol pyruvic acid in the presence of pyruvate phosphate dikinase enzyme.

These C_4 plants are more efficient (greater yield and productivity) because they grow fast at high temperatures, high light intensity and under low CO_2 hence these show greater affinity with phosphoenol pyruvate (PEP) at low CO_2 concentration because PEP carboxylase has much higher affinity than RuBP carboxylase and further photorespiration does not occur since it does not accept O_2.

The C_4 plants have had to adjust the low level of CO_2 by artificially raising the CO_2 concentration in bundle sheath cells at the enzyme site to prevent photorespiration. The C_4 cycle thus effectively shuttles CO_2 from the atmosphere into bundle sheath cells. As a result, the concentration of CO_2 in bundle sheath cells becomes higher than that of the atmosphere. This occurs when malic acid of the mesophyll cell undergoes decarboxylation to form pyruvic acid in the bundle sheath cell to release CO_2. These plants can survive in poor CO_2 conditions because there is an internal supply of CO_2. A high concentration of CO_2 in bundle sheath cells suppresses photorespiration which helps the C_4 plants to photosynthesise more efficiently even at high temperatures.

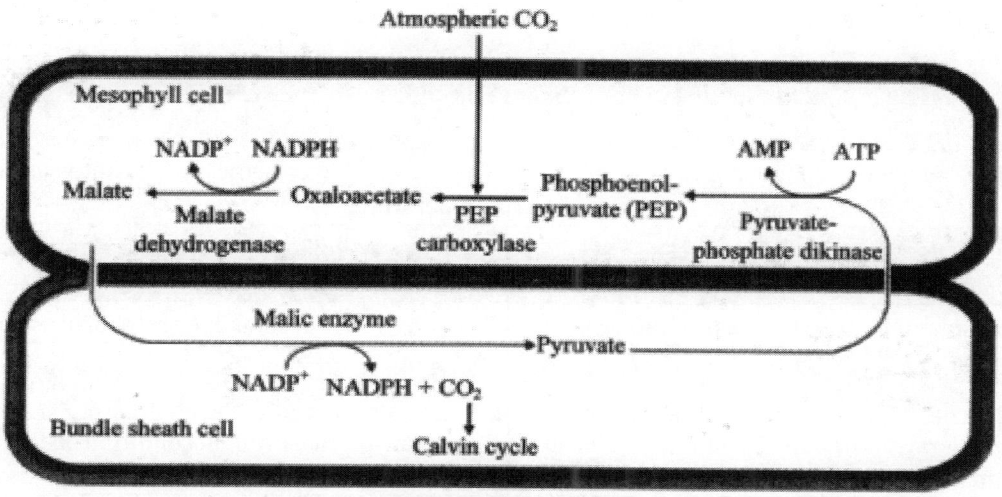

C_4 Pathway

Fig: Hatch and Slack pathway

These plants use more energy (2 more ATP per mole of CO_2 fixed) because this extra energy is required for collecting and transporting CO_2 and hydrogen from mesophyll cell to chloroplast in bundle sheath cell at the site of RuBP carboxylase activity. This is only disadvantage of C_4 plants.

Overall reaction:

$$6CO_2 + 12H_2O + 30ATP + 12NADPH + 6RuBP \longrightarrow$$
$$C_6H_{12}O_6 + 6H_2O + 6O_2 + 30ADP + 30\ Pi + 12NADP^+ + 12H^+ + 6RuBP$$

C_3 plants / C_3 cycle	C_4 plants / C_4 cycle
These are more common in dicots than monocots. e.g., rice, wheat, *Atriplex hastata, Euphorbia corallata.* These are found in 90% of plants.	These are more common in monocots than dicots. e.g., maize, sugarcane, *Sorghum, Atriplex rosea, Euphorbia maculata.*
Leaves do not show Krantz anatomy.	Leaves show Krantz anatomy
Chloroplast is monomorphic (chloroplast in mesophyll cell is granal).	Chloroplast is dimorphic (chloroplast in mesophyll cell is granal and that of bundle sheath cell is agranal).
It operates through Calvin cycle.	It operates through Calvin cycle as well as Hatch and Slack pathway.
First CO_2 acceptor is RuBP (weak CO_2 acceptor).	First CO_2 acceptor is PEP (strong CO_2 acceptor).
First product is Phosphoglyceric acid (PGA) which is 3 carbon compound	First product is oxaloacetic acid (OAA) which is 4 carbon compound
CO_2 fixation occurs in mesophyll cells.	Initial CO_2 fixation occurs in mesophyll cells and final in bundle sheath cell.
There is one carboxylation process.	There are two carboxylation processes.
CO_2 fixation is slower.	CO_2 fixation is faster.
3ATP and $2NADPH_2$ are required to fix 1 mol. of CO_2	5ATP and $2NADPH_2$ are required to fix 1 mol. of CO_2
Photorespiration occurs.	Photorespiration does not occur.
Rate of net photosynthesis is low.	Rate of net photosynthesis is high (2 to 3 times).
CO_2 compensation point (threshold value) is high (25-100 ppm).	CO_2 compensation point is low (0-10 ppm).
More CO_2 is released in light than in dark.	Less CO_2 is released in light than in dark.
Photosynthesis is not performed at very low CO_2	Photosynthesis is performed at very low CO_2.
Optimum temperature is 20°C-25°C for enzymes.	Optimum temperature is 30-45°C for enzymes.
O_2 has inhibitory effect (sensitive) on photosynthesis (Warburg effect).	O_2 has no inhibitory effect (insensitive) on photosynthesis.
These are less efficient (photosynthetic) at high temperature.	These are more efficient (photosynthetic) at high temperature.
These are more efficient (photosynthetic) at low temperature.	These are less efficient (photosynthetic) at low temperature.
It occurs at lower light intensity (1000-foot candles) which is saturation point.	It occurs at higher light intensity and even full sunlight which is not saturation point.
It requires 33% of solar energy to form assimilatory power	It requires 40% of solar energy to form assimilatory power.
These cannot tolerate saline condition.	These tolerate saline condition due to presence of organic acid.
These cannot tolerate drought condition because transpiration ratio is high.	These can tolerate drought condition because transpiration ratio is low.
These are adapted to all climates.	These are only adapted to tropical climate.

PHOTORESPIRATION:

It was discovered by Dicker and Tijo (1959) in tobacco. It is a light-dependent uptake of O_2 and release of CO_2. It is a special type of respiration occurring in many green plants (C_3 plants) when they are exposed to strong sunlight.

In hot and dry conditions, plants close their stomata to prevent water loss. Under these conditions, the CO_2 will decrease, and oxygen gas produced by the light reactions of photosynthesis will increase causing an increase in photorespiration. It was observed that photosynthesis was inhibited when there was a higher concentration of $O_{2,}$ and this inhibition was caused by the occurrence of photorespiration.

Photorespiration has no relation with normal respiration. Both processes show similarities in the uptake of O_2 and release of CO_2. It operates when the rate of photosynthesis is ten times higher than respiration due to more availability of O_2.

If the temperature is high, then oxygenation will be promoted because the solubility of CO_2 in water is decreased more rapidly than that of O_2 and further specificity of RuBisCO will be decreased. It occurs when the amount of O_2 is greater than CO_2. The competition occurs between CO_2 and O_2 for RuBisCO. The oxygenase activity increases because O_2 is a competitive inhibitor of the carboxylase activity of RuBisCO hence affinity of RuBPcase with O_2 increases and with CO_2 decreases. Now it functions as RuBP oxygenase rather than carboxylase.

It does not produce ATP, $NADH_{2,}$ and glucose. But the overall process is energy energy-consuming process and further, there is wasteful loss (or release) of around 50% of photosynthetically fixed CO_2 by utilizing ATP when glycine is converted (oxidised) into serine. The loss of carbon is due to the dual activity of RuBP carboxylase and oxygenase which can act not only as carboxylase (useful role in C_3 cycle) but also as oxygenase (enzyme catalyzing the direct insertion of molecular O_2 into the organic substrate).

It is a cyclic process that occurs in three cell organelles such as chloroplast, peroxisome, and mitochondria. It is also called glycolate metabolism or C_2 cycle because the chief metabolite and substrate is glycolate which is a two-carbon compound. Photorespiration occurs at high temperatures, bright light, high $O_{2,}$ and low CO_2.

Reactions in chloroplast:

(a) The first reaction of photorespiration is oxygenation of RuBP in the presence of O_2 and RuBP oxygenase. It forms 1 mol. of phosphoglycolate (2C) and 1 mol. of Phosphoglyceric acid (3C). While the PGA is used up in the Calvin cycle, the phosphoglycolate is dephosphorylated to form glycolate in the chloroplast.

Reactions in peroxisome:

(a) From the chloroplast, the glycolate is diffused to peroxisome where it is oxidized to glyoxalate.

(b) Glyoxalate reacts with an amino acid, glutamic acid (5C) to form another amino acid, glycine (2 C).

Reactions in Mitochondria:

(a) Glycine enters mitochondria where 2 mol. of glycine (4C) give rise to 1 mol. of another amino acid, serine (3C), and 1 mol. of CO_2.

Reactions in peroxisome:

(a) Serine is taken up by the peroxisome, where it reacts with α-ketoglutaric acid to form hydroxy pyruvic acid and glutamic acid, an amino acid.

(b) Hydroxy pyruvate is reduced to form glycerate.

Reactions in chloroplast:

(a) The glycerate leaves the peroxisome and enters the chloroplast, where it is phosphorylated to form phosphoglyceric acid (PGA). The PGA molecule re-enters the Calvin cycle to make carbohydrates, but 1 molecule of CO_2 released in mitochondria during photorespiration has to be refixed.

In other words, 75% of carbon lost by oxygenation of RuBP is recovered and 25% is lost as release of 1 molecule of CO_2. Thus, photorespiration is also called the Photosynthetic Carbon Oxidation (PCO) cycle.

Factors: Photorespiration depends on

(a) High temperature (b) Bright light
(c) High O_2 and (d) Low CO_2

Significance:

1. It protects the plants from photo-oxidative damage by dissipation of energy of excitation.
2. It also recovers some of the carbon from excess glycolate.

Reactions occurring in Mesophyll cell

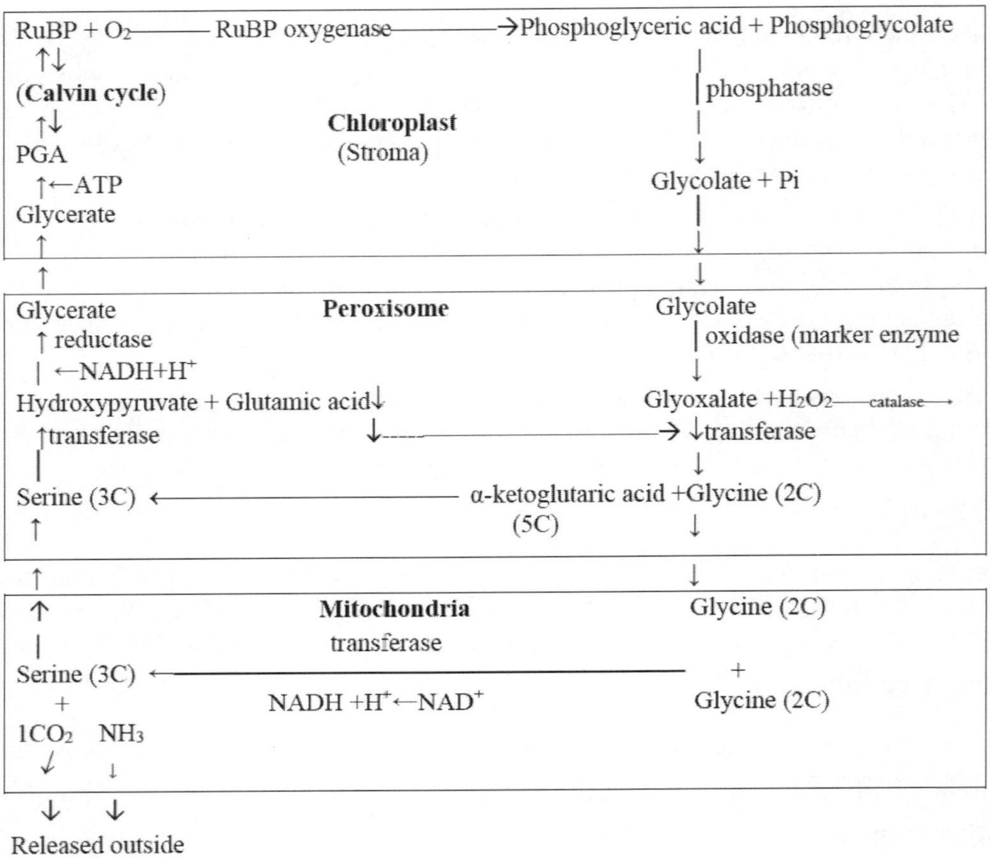

Fig: Reactions of Photorespiration

Difference between Photophosphorylation and oxidative phosphorylation

Photophosphorylation	Oxidative phosphorylation
It is synthesis of ATP in the presence of light.	It is synthesis of ATP in the presence of terminal oxidation.
It occurs in matrix of chloroplast.	It occurs in matrix of mitochondria.
It requires four complexes.	It occurs five complexes
It occurs in two ways: non-cyclic and cyclic.	It occurs in unidirectional way.
It involves two pigment systems.	It does not involve pigment systems.
It requires less number of electron carriers.	It requires more number of electron carriers.
It involves photolysis of water.	It involves formation of water.
ATP is not produced from the reduced coenzymes.	ATP is produced from the reduced coenzymes such as $NADH_2$ and $FADH_2$.
It is a part of light reaction.	It is a part of dark respiration.

Terms:

- **Compensation Point**: It is the point or threshold value of CO_2 concentration at which illuminated plant parts stop absorbing CO_2 from environment. At this point, CO_2 fixed in photosynthesis is equal to CO_2 produced in respiration and photorespiration so that there is no net exchange of gases. In other words, it is a point in which all 0_2 evolved by photosynthesis is used up in respiration and all CO_2 produced in respiration is used for photosynthesis.

  Plants cannot survive for long at the compensation point because there is a net loss of organic matter due to respiration.

  It is less in C_4 plants and more in C_3 plants.

  It is less in shady plants (sciophytes) and more in sun plants (heliophytes).

- **Law of limiting factors:** It was proposed by Blackman (1905). It states that when a chemical process is affected by several factors then its rate will be determined by the slowest factor. Hence rate of photosynthesis is limited at a given time by one and only one deficient factor. The light intensity supplied to a leaf is just sufficient to allow it to utilize 5mg of CO_2 per hour in photosynthesis.
  - A denotes no photosynthesis in the absence of CO_2. If CO_2 increases from 0 to 1 mg per hour, then the rate of photosynthesis will increase.
  - The rate of photosynthesis, however, cannot increase further because of the limiting role of CO_2. The rate will increase along line AB when the CO_2 supply is raised to 5mg per hour. The rate does not increase beyond the point B where it becomes constant along the line B-C.
  - Any further increase in CO_2 will not affect the rate of photosynthesis because light intensity has now become the limiting factor.
  - If light intensity is increased further to a medium level, the rate of photosynthesis also increases along line B-D provided there is an increased supply of CO_2.
  - Light again becomes the limiting factor at point D when the rate of photosynthesis becomes constant along line D-E.
  - Further increase in light to high level, the rate of photosynthesis increases along D-F when sufficient concentration of CO_2 is increased.
  - The rate becomes constant along F-G when light alone becomes limiting.
  - Hence B, D, and F are the points where an increase in the rate of photosynthesis stops abruptly because of the limiting nature of light.

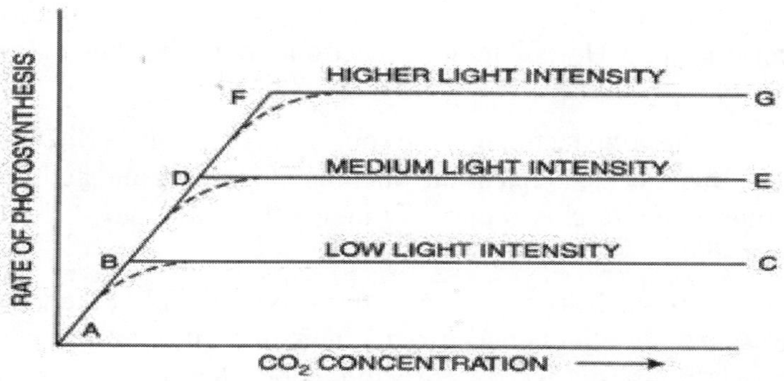

Results:

- When all the factors are present, but CO_2 concentration is low then the rate of photosynthesis is controlled by CO_2 which is limiting.
- Similarly, if light intensity is low on a cloudy day, it controls the rate of photosynthesis.
- The continued increase of the factor without a corresponding increase in the limiting factor would decrease the rate of photosynthesis.

Factors Affecting Photosynthesis:

1. **Light:**

   (a) **Intensity of light:** There are two points:

   Light compensation point: It is a point in light intensity in the evening where the rate of photosynthesis is minimum and equal to the rate of respiration. There is no net CO_2 fixation at the compensation point. The low compensation point is indicative of photosynthetic efficiency.

   Light saturation point: It is a point in light intensity where the rate of photosynthesis is maximum. Above this point, solarisation occurs where the rate of photosynthesis starts declining due to the photooxidation of chlorophyll.

   The rate of photosynthesis starts at low light intensity, starts increasing with an increase in light intensity and later decreases with high light intensity due to solarisation.

   It is more in intense light than in diffused light. Very low light intensity causes closure of stomata hence the entry of CO_2 is reduced thus the rate of photosynthesis is reduced.

   (b) **Quality of light**: It occurs in the visible light range. It is maximum in white light or sunlight (polychromatic light). It is minimum in green light, average in blue light and maximum in red light. On the other hand, red algae show maximum photosynthesis in green light and brown algae in blue light.

   (c) **Duration of light**: The rate of photosynthesis increases in long continuous light, but it is limited by the slow rate of dark reaction. Its rate is faster in intermittent light (alternate light and dark period) than in continuous light because the essential phase of photosynthesis will occur not only in light but also in the dark since assimilatory power is fully consumed in a dark reaction. Sugarcane is the most efficient converter because it utilizes 10% of light.

2. **CO_2:** The rate of photosynthesis is increased from 0.036% (360 ppm) content of the atmosphere to 0.9% but starts decreasing above 0.9%. It acts as a limiting factor under normal conditions of light and temperature (provided other factors are not limiting). CO_2 compensation point is the value of atmospheric CO_2 concentration when the rate of photosynthesis is just equivalent to the rate of respiration in the leaf so that there is no net exchange of gases. A plant cannot survive for long at the compensation point because there is a net loss of organic matter due to the respiration of non-green organs and dark respiration. The low CO_2 compensation point in C_4 plants is due to CO_2 concentration mechanism at the carboxylation site.

3. **Temperature**: The rate of photosynthesis increases with the rise in temperature (optimum 20-25°C for C_3 plants and 30-45°C for C_4 plants) under normal conditions of light and CO_2. However, the rate decreases due to denaturation of enzymes. Temperature coefficient (Q_{10}) is the ratio of the rate of reaction at a given temperature to the rate at the temperature 10°C lower. Q_{10} is 1 in a light reaction where there is no increase in the rate with temperature rise. Q_{10} is 2 in dark reaction where the rate is doubled with every 10°C rise in temperature.

4. **Water**: It acts as an indirect limiting factor. It is the source of H^+ to reduce NADP and release of O_2. It maintains the turgidity of photosynthetic cells; helps in the opening of stomata and activates enzymes. If water content of leaves is reduced, then stomata closes partially hence the entry of CO_2 is reduced by which the rate of photosynthesis is decreased.

5. **O_2:** If the amount of oxygen is less, the ratio of photosynthesis will be increased. If the amount of oxygen is more the rate will decrease (**Warburg effect**) because the high amount of oxygen causes oxidation of chlorophyll and acts as a competitive inhibitor of carboxylase activity of RuBisCO. Warburg effect is shown in C_3 plants and insensitive to C_4 plants.

6. **Chlorophyll content:** The rate of photosynthesis increases with an increase in the content of chlorophyll.

7. **Structure of leaf**: The rate of photosynthesis is affected because entry of CO_2 is dependent on size, position, and number of stomata; thickness of cuticle and epidermis; presence of epidermal hairs and amount of mesophyll cells.

8. **Age of plants**: The rate of photosynthesis is slowed down in mature plants.

9. **Accumulation of photosynthetic products**: The rate of photosynthesis is reduced in their presence.

SHORT ANSWER TYPE QUESTIONS

1. Explain Moll's experiment in photosynthesis.
2. Explain Ingenhousz's experiment in photosynthesis.
3. Write the structural formulae of chlorophyll-a and chlorophyll-b and mention their difference.
4. Write notes on carotenoids.
5. Distinguish between PS II and PS I.
6. Distinguish between the light-harvesting complex and the core complex.
7. Distinguish between non-cyclic and cyclic photophosphorylation.
8. Distinguish between anoxygenic photosynthesis and oxygenic photosynthesis.
9. What do you mean by compensation point?
10. Write notes on the law of limiting factors.
11. Discuss the role of light in photosynthesis.
12. How photolysis of water is possible in photosynthesis?
13. Write the final equation of the Calvin cycle and Hatch-Slack pathway.
14. What do you mean by Krantz anatomy?
15. Why C_4 plants are photosynthetically more efficient?

LONG ANSWER TYPE QUESTIONS

1. Discuss the mechanism of photophosphorylation in the light reaction of photosynthesis.
2. Discuss the mechanism of the chemiosmotic generation of ATP.
3. Describe the CO_2 fixation in C_3 plants.
4. Describe the CO_2 fixation in Hatch-Slack pathway and distinguish between C_3 plants and C_4 plants.
5. Describe the process of photorespiration in green plants and explain why it is considered a wasteful process.

MULTIPLE CHOICE QUESTIONS

1. Moll while working on photosynthesis confirmed that

(A) Light is essential
(B) CO_2 is essential
(C) Chlorophyll is essential
(D) Water is essential

2. Which of the plants was used by Engelmann to study action spectra?

(A) *Chlorella* (B) Mint
(C) *Chlamydomonas* (D) *Cladophora*

3. Which of the following set of pigments is called antenna pigments?

(A) All types of chlorophyll
(B) Chlorophyll-a and carotenoids
(C) Chlorophyll-a and b
(D) All pigments other than Chlorophyll-a

4. Which of the set of elements is used in the photolysis of water?

(A) Chlorine and Manganese
(B) Chlorine and Magnesium
(C) Manganese and Magnesium
(D) Iron and Chlorine

5. How much quanta of light are required to evolve an I molecule of O_2?

(A) 2 (B) 4
(C) 8 (D) 10

6. Which of the following processes occurs in photosystem II?

(A) CO_2 fixation (B) CO_2 reduction
(C) Photolysis of H_2O (D) All of these

7. **In which of the following orders, the electron carriers between PS I and PS II are correctly arranged?**
(A) Phe – PQ – Cyt b_6 - Cyt f – PC
(B) FRS –PQ – Cyt b_6 - Cyt f – PC
(C) Phe – PC – Cyt b_6 - Cyt f – PQ
(D) Phe – PC – Cyt f - Cyt b – FRS

8. **PS II is concerned with**
(A) Photolysis of water
(B) Flowering
(C) Reduction of CO_2
(D) Release of energy

9. **PS I is located on**
(A) Inner surface of thylakoid membrane
(B) Outer surface of thylakoid membrane
(C) Both surfaces of thylakoid membrane
(D) Stroma of chloroplast

10. **Ferredoxin is a component of**
(A) P680 (B) P700
(C) PS I (D) PS II

11. **Cyclic photophosphorylation produces**
(A) $NADPH_2$
(B) ATP and $NADPH_2$
(C) ATP, $NADPH_2$ and O_2
(D) ATP only

12. **ETS associated with photophosphorylation occurs in**
(A) Stroma (B) Mitochondria
(C) Thylakoid
(D) Outer membrane of chloroplast

13. **ADPis phosphorylated when three protons are shifted from**
(A) Thylakoid lumen to stroma
(B) Stroma to the cytosol
(C) Cytosol to stroma
(D) Thylakoid lumen to cytosol

14. **Most of the experimental work in photosynthesis has been done on**
(A) *Chlamydomonas* (B) *Chlorella*
(C) *Hydrilla* (D) *Cladophora*

15. **If radioactive CO_2 is given to a C_3 plant of photosynthesis, the radioactivity first appears in**
(A) 3-Phosphoglyceric acid
(B) Oxaloacetic acid
(C) Aspartic acid (D) Glucose

16. **Which of the following is not a step in the Calvin cycle?**
(A) Carboxylation
(B) Reduction
(C) Regeneration of RuBP
(D) Photophosphorylation

17. **During fixation of one molecule of CO_2 in C_3 plants**
(A) 3 ATP and 2 $NADPH_2$ are required
(B) 5 ATP and 2 $NADPH_2$ are required
(C) 12 ATP and 12 $NADPH_2$ are required
(D) 18 ATP and 12 $NADPH_2$ are required

18. **Dimorphism of chloroplast is found in**
(A) C_4 plants (B) C_3 plants
(C) CAM plants (D) All of the above

19. **Krantz anatomy is found in**
(A) Most of monocots (B) Most of dicots
(C) Algae (D) All dicots

20. **Agranal chloroplasts are found in**
(A) Mesophyll of pea leaves
(B) Bundle sheath of mango leaves
(C) Mesophyll of maize leaves
(D) Bundle sheath of sugarcane leaves

21. **The inhibitory effect of oxygen in C_3 plants on photosynthesis is called**
(A) Warburg's effect
(B) Photooxidation
(C) Solarisation
(D) Emerson's effect

22. **Chloroplasts without grana are known to occur in**
(A) Bundle sheath cells of C_4 plants
(B) Bundle sheath cells of CAM plants
(C) Bundle sheath cells of C_3 plants
(D) Mesophyll cells of C_4 plants

23. **Which of the following enzymes is not found in C_3 plants?**
(A) RUBP carboxylase
(B) PEP carboxylase
(C) NADP reductase
(D) ATP synthase

24. **Which of the following cell organelles is involved in photorespiration?**
(A) Peroxisome (B) Golgi body
(C) Endoplasmic reticulum
(D) Lysosome

25. The percentage of light energy fixed in photosynthesis is around

(A) 10 (B) 50
(C) 3 (D) 1

26. Quantasomes in chloroplast contain

(A) 200 chlorophyll molecules
(B) 250 chlorophyll molecules
(C) 100 chlorophyll molecules
(D) 300 chlorophyll molecules

27. Which of the following is incorrect regarding anthocyanin?

(A) It is soluble in water
(B) It is a purplish pigment dissolved in water
(C) It gives colouration to leaves
(D) It takes part in photosynthesis

28. During daytime, the ratio of CO_2 produced to that of CO_2 consumed is

(A) 1:1 (B) 1:5
(C) 1: 10 (D) 1:50

29. Photosynthesis is most active in

(A) Red light (B) Sunlight
(C) Blue light (D) Greenlight

30. If half of the leaves of a plant are removed, then the rate of photosynthesis will be

(A) More and the photosynthetic yield will also be more
(B) Less and the photosynthetic yield will also be less
(C) More and the photosynthetic yield will also be less
(D) Less and the photosynthetic yield will also be more

31. Photorespiration is favoured by

(A) Low O_2 and high CO_2
(B) Low light and high O_2
(C) High O_2 and low CO_2
(D) Low temperature and high CO_2

32. The chloroplast of the bundle sheath cell of C_4 plants is

(A) Large and granal (B) Large and agranal
(C) Small and granal (D) Small and agranal

33. Which of the following plants is the most efficient converter of solar energy?

(A) Wheat (B) Maize
(C) Soya bean (D) Sugarcane

34. C_4 plants are adapted to

(A) Hot and dry climate
(B) Temperate climate
(C) Cold and dry climate
(D) Hot and humid climate

35. Which of the pigment systems is inactivated in the red drop?

(A) Ps I (B) Ps II
(C) Both Ps I and Ps II (D) None of the above

36. Glycolate accumulates in chloroplast when there is

(A) Low CO_2 (B) High CO_2
(C) Bright light (D) Low temperature

37. Photorespiration shows the formation of

(A) ATP only
(B) Sugar
(C) Both ATP and sugar
(D) Neither ATP nor sugar

38. The ratio between 2-carbon and 3-carbon intermediates having $-NH_2$ group formed in photorespiration is

(A) 1:1 (B) 1:2
(C) 2:1 (D) 3:2

39. In photorespiration, the role of the peroxisome is to

(A) Reduce glycolate
(B) Oxidise glycolate
(C) Convert glycine into serine
(D) Oxidise RuBP

40. The porphyrin ring of the chlorophyll molecule is a complex ring structure of

(A) Double bonds
(B) Triple bonds
(C) Alternating single and double bonds
(D) Alternating double and triple bonds

41. Plants do not store carbohydrates in the form of glucose because

(A) Glucose dissolves in water thereby altering the osmotic balance
(B) Glucose is an unstable molecule
(C) Glucose would replace ribose
(D) Glucose attracts insect herbivores

42. Which of the following portions of chloroplast shows the lowest pH when sunlight falls on chloroplast?
(A) Stroma (B) Cytosol
(C) Periplastidial space
(D) Space enclosed by the thylakoid membrane

43. C_4 plants are found among
(A) Poaceae family
(B) Monocots only
(C) Dicots only
(D) Monocots and dicots

44. Protochlorophyll differs from chlorophyll in lacking
(A) 2 H atom in one of its pyrrole ring
(B) 4 H atom in one of its pyrrole ring
(C) 2 H atom in two of its pyrrole ring
(D) 4 H atom in two of its pyrrole ring

45. Phosphoglyceric acid (PGA) as the first CO_2 fixation product was discovered in the photosynthesis of
(A) Algae
(B) Vascular cryptogams
(C) Higher plants
(D) Bacteria

46. How would chlorophyll appear when seen in red light?
(A) Red (B) Far-red
(C) Black (D) Green

47. Bacteriochlorophyll differs from chlorophyll-a in having
(A) one pyrrole nucleus with one hydrogen
(B) one pyrrole nucleus with two hydrogens
(C) one pyrrole nucleus with three hydrogens
(D) one pyrrole nucleus with four hydrogens

48. The Emerson effect confirms the existence of
(A) Photorespiration
(B) Photophosphorylation
(C) Light and dark reactions
(D) Two distinct photochemical reactions

49. The dry weight of a green leaf will be maximum at
(A) Morning
(B) Evening
(C) Afternoon
(D) Same throughout the whole day

50. On a normal sunny day, the rate of photosynthesis is maximum during
(A) Early morning (B) Late morning
(C) Mid-day (D) Evening

51. RuBP carboxylase was crystallised from
(A) Chlorella
(B) Leaf of maize
(C) Leaf of sugarcane
(D) Leaf of Tobacco

52. Seedlings grown in the dark are
(A) Shorter than those grow in light
(B) Taller than those grow in light
(C) Similar to those grow in light
(D) They do not grow at all.

53. The process of photorespiration involves
(A) Chloroplast - Mitochondria - Peroxisome
(B) Mitochondria - Peroxisome - Chloroplast
(C) Peroxisome - Mitochondria - Chloroplast
(D) Chloroplast - Peroxisome - Mitochondria

54. Which of the following is the most limiting factor under normal conditions?
(A) CO_2 (B) Light
(C) Temperature (D) Water

55. The number of turns of the Calvin cycle required to synthesise one molecule of glucose was
(A) Two (B) Three
(C) Four (D) Six

56. Which of the following processes of the Calvin cycle does not require any energy?
(A) Carboxylation
(B) Reduction
(C) Glycolytic reversal
(D) Regeneration of RuBP

57. The first CO_2 acceptor in the Hatch-Slack pathway is
(A) RuBP (B) PEP
(C) OAA (D) PGA

58. Which of the following pigment systems is involved in bacterial photosynthesis?
(A) Ps I
(B) Ps II
(C) Both PS I and PS II
(D) None of these

59. Which of the following electron carriers contains copper?

(A) Plastoquinone (B) Plastocyanin
(C) Ferredoxin (D) Pheophytin

60. Which of the following pigments is known as shield pigment?

(A) Phycobilin (B) Chlorophyll
(C) Carotenoid (D) Anthocyanin

61. The reaction centre consists of

(A) Single-molecule of chlorophyll a
(B) Single-molecule of chlorophyll b
(C) Light-harvesting complex
(D) Both a and b

62. The electron flow from pigment system II to NADP via ETS and PS-I follows

(A) Uphill→Downhill→Uphill→Downhill
(B) Downhill→Uphill→Downhill→Uphill
(C) Both a and b
(D) None of the above

63. Which of the following materials is not recycled between dark and light reactions?

(A) NADPH +H^+ (B) ADP
(C) ATP (D) O_2 and CO_2

64. The pH of thylakoid space (lumen) in bright light becomes

(A) More acidic (B) Less acidic
(C) More alkaline (D) Neutral

65. What happens during the light reaction of photosynthesis?

(A) ADP is phosphorylated and NADP reduced
(B) ADP is hydrolysed and NADP reduced
(C) ATP is hydrolysed and NADP is reduced
(D) ATP is hydrolysed and NADP is reduced

66. The cooperation of two pigment systems of the chloroplast is required for

(A) ATP synthesis
(B) Reduction of $NADP^+$
(C) Oxidation of reaction centre of photosystem
(D) Generation of proton motive force

67. C_4 plants are mostly found in

(A) Temperate region with more humid condition
(B) Tropical region with more dry condition
(C) Temperate region with less humid condition
(D) Tropical region with more humid condition

68. In the Calvin cycle, ATP is required in

(A) Carboxylation (B) Reduction
(C) Regeneration (D) Both b and c

69. If mature leaves were enclosed for a fixed amount of time in a transparent bag that had radioactive CO_2, then maximum radioactivity was found to occur in

(A) Guard cell of leaves
(B) Mature leaves
(C) Actively growing leaves
(D) Senescent leaves

70. The productivity of crops declines when leaves begin to wilt because

(A) Stomatal closure prevents entry of CO_2
(B) Photolysis of water does not occur
(C) Mesophyll cells become flaccid
(D) Chloroplast decomposes

71. If one plant is grown in the shade of a greenhouse and another is grown under the shade of trees, then the rate of photosynthesis will be

(A) Greater in the shade of a greenhouse
(B) Lesser in the shade of a greenhouse
(C) Equal in both conditions
(D) Equally low in both conditions.

72. In an experiment on the evolution of O_2 by photosynthesising *Hydrilla* plant, a pinch of $NaHCO_3$ is added to water then the rate of photosynthesis will

(A) Decrease (B) Increase
(C) Remains constant
(D) First decrease then increase

73. How much oxygen is formed from 264 g. of CO_2 and 216 g. of H_2O?

(A) 96 g (B) 216
(C) 264 g (D) 192 g

74. Photosynthesis cannot be operated in

(A) Red light (B) Blue light
(C) Greenlight (D) Yellow light

75. C_4 plants by decreasing photorespiration and efficiency in carrying out photosynthesis can

(A) Decrease stomatal opening and thereby reduce water loss
(B) Decrease stomatal opening and thereby reduce leaf temperature
(C) Increase stomatal opening and thereby increase water loss
(D) Increase stomatal opening and thereby increase temperature rate

76. The major difference in Ps I and Ps II found in the chloroplast is
(A) Position on lamellae
(B) Chlorophyll a
(C) Position of electron carriers
(D) Energy harvesting

77. Light-harvesting complex II is located in the
(A) Thylakoid lumen
(B) Stroma
(C) Thylakoid membrane
(D) Outer membrane of chloroplast

78. If the Mg atom is removed from the core complex of the photosystem then
(A) Photosynthesis will stop
(B) Photosynthesis will initially decrease and then increase
(C) Photosynthesis will continue at the normal rate
(D) Photosynthesis will initially increase and then decrease

79. Which of the following plants is both C_3 and C_4 plant?
(A) *Atriplex* (B) *Euphorbia*
(C) Maize (D) Sugarcane

80. Photophosphorylation was discovered by
(A) Calvin (B) Blackman
(C) Robert Hill (D) Arnon

81. Which of the following plants was used by Priestley to confirm that the plant purified the air spoilt by burning a candle?
(A) Mint (B) *Hydrilla*
(C) *Cladophora* (D) *Chlorella*

82. Which of the following plants was used by Ingenhousz to confirm the release of oxygen?
(A) Mint (B) *Hydrilla*
(C) *Cladophora* (D) *Chlorella*

83. Each pyrrole ring of chlorophyll head contains
(A) 4C and 1N atoms (B) 5C and 1N atoms
(C) 3C and 2N atoms (D) 4C and 2N atoms

84. Which of the following groups is found in chlorophyll-b?
(A) $-CH_3$ (B) $-CHO$
(C) $-COOH$ (D) $=CO$

85. Chlorophyll is synthesised from
(A) Succinic acid
(B) Oxaloacetic acid
(C) Succinyl Coenzyme A
(D) Malic acid

86. Which of the following pigments does not contain a phytol chain?
(A) Chlorophyll-a (B) Chlorophyll-b
(C) Chlorophyll-c (D) Chlorophyll-d

87. Which of the following pigments is not involved in photosynthesis?
(A) Carotene (B) Xanthophylls
(C) Anthocyanin (D) Phycoerythrin

88. Which of the following pigments does not belong to antenna pigments?
(A) Chlorophyll-a (B) Chlorophyll-b
(C) Chlorophyll-c (D) Chlorophyll-d

89. Which of the following phosphorylation reactions occurs in bacteria?
(A) Non-cyclic photophosphorylation
(B) Cyclic photophosphorylation
(C) Substrate level phosphorylation
(D) Oxidative phosphorylation

90. C_4 plants perform photosynthesis under
(A) LowCO_2 (B) High temperature
(C) Bright light (D) All of the above

91. The first stable product in C_4 cycle is
(A) Phosphoglyceric acid
(B) Oxaloacetic acid
(C) Lactic acid
(D) Phosphoenol pyruvic acid

92. The first CO_2 acceptor in C_3 plants is RuBP which is
(A) 3 Carbon compound
(B) 4 Carbon compound
(C) 5 Carbon compound
(D) 6 Carbon compound

93. The first reaction in photorespiration is
(A) Carboxylation (B) Oxygenation
(C) Dehydration (D) Reduction

94. The temperature coefficient in a light reaction is
(A) 1 (B) 2
(C) 3 (D) 4

95. Photorespiration does not produce
(A) ATP (B) $NADPH_2$
(C) Glucose (D) All of the above

96. How many ATPs are required to fix 6 molecules of CO_2 in C_4 plants?

(A) 18 (B) 24
(C) 30 (D) 36

97. How much amount of glucose formed as a result of photosynthesis will undergo respiration?

(A) 10% (B) 50%
(C) 75% (D) 100%

98. The final CO_2 fixation in the Hatch-Slack pathway occurs in

(A) Mesophyll cell (B) Bundle sheath cell
(C) Companion cell (D) Xylem vessel

99. Photorespiration is also known as

(A) C_2 cycle (B) C_3 cycle
(C) C_4 cycle (D) C_5 cycle

100. Chlorophyll absorbs maximum

(A) Red light (B) Blue light
(C) Greenlight (D) Orange light

101. Chlorophyll b is present in

(A) Prokaryotes (B) Eukaryotes
(C) Both a and b (D) Bacteria

102. The electrons excited by PS II pass directly to

(A) Ferredoxin (B) Plastocyanin
(C) Plastoquinone (D) PS I

103. How many turns of the Calvin cycle produce one molecule of glucose?

(A) Two (B) Four
(C) Five (D) Six

104. C4 plants are found among

(A) Poaceae family (B) Monocots only
(C) Dicots only
(D) Monocots as well as Dicots

105. Who first demonstrated that light absorption by chloroplast releases oxyen?

(A) Robert Hill (B) Ingenhousz
(C) Engelman (D) Blackman

ANSWERS

1.B	11.D	21.A	31.C	41.A	51.D	61.A	71.A	81.A	91.B	101. b
2.D	12.C	22.A	32.B	42.B	52.B	62.A	72.B	82.B	92.C	102. c
3.D	13.A	23.B	33.D	43.D	53.D	63.D	73.D	83.A	93.B	103. d
4.A	14.B	24.A	34.D	44.A	54.A	64.A	74.C	84.B	94.A	104. d
5.C	15.A	25.D	35.B	45.A	55.D	65.A	75.C	85.A	95.D	105. a
6.C	16.D	26.B	36.A	46.C	56.A	66.B	76.D	86.C	96.C	
7.A	17.A	27.D	37.D	47.B	57.B	67.B	77.C	87.C	97.B	
8.A	18.A	28.C	38.C	48.D	58.A	68.D	78.A	88.A	98.B	
9.B	19.A	29.B	39.B	49.C	59.B	69.C	79.A	89.B	99.A	
10.C	20.D	30.C	40.C	50.B	60.C	70.A	80.D	90.D	100.B	

CHAPTER-9

RESPIRATION IN PLANTS

Respiration is an oxidative process that is mainly concerned with the production of energy that is required for all living cells to perform various vital activities.

All living organisms or processes require energy which is essential for chemical synthesis in the cell. Atoms, ions and molecules in the cell will be able to form complex chemical substances. If energy is absent, then there would be no formation of new protoplasm and hence no growth. The released energy is stored in the form of ATP in the cells.

Respiration is of three types:

1. External respiration: It is the exchange of oxygen and carbon dioxide between the environment and organisms (breathing). It occurs through stomata in land plants, entire plant surface in submerged hydrophytes, and general body surface in algae, fungi and bacteria.
2. Internal respiration: It is the exchange of O_2 and CO_2 between internal body fluids such as blood and individual cells.
3. **Cellular respiration:** It is biochemical oxidation of organic substances esp. glucose resulting in its stepwise breakdown into simpler ones with gradual and stepwise release of energy (ATP). The energy is utilised for metabolic activities in the cell. It is an exothermic process.

$$C_6H_{12}O_6 + 6O_2 \longrightarrow 6CO_2 + 6H_2O + \text{energy (686 K.cal /2870 KJ)}$$

Respiration occurs the whole day and night in all the living cells, but the rate is maximum in growing regions like root tip, stem tip, flower bud and germinating seeds. Plants take in free O_2 from the air which enters through the stomata and lenticels and diffuses through the continuous intercellular space to reach the living cell.

During daytime when the rate of photosynthesis is higher than the rate of respiration and when both processes are going on simultaneously, all the oxygen liberated in photosynthesis is not consumed in respiration, whereas all the CO_2 liberated in respiration is completely consumed in photosynthesis and some more CO_2 must enter the plant for the higher rate of photosynthesis. Thus, the effects of respiration are completely masked during the daytime. During the night, however, there is no photosynthesis and only the exchange of gases (O_2 consumed, and CO_2 liberated) between the plant and atmosphere is the effect of respiration.

Respiration occurs in the liquid medium in a cell. Respiration consumes 50% of the sugars formed in photosynthesis. The cell undergoing respiration will prefer carbohydrates (firstly monosaccharide then oligosaccharide and lastly polysaccharide), then fats and lastly proteins when both carbohydrates and fats are unavailable to them.

TYPES OF RESPIRATION

Blackman divided respiration into two types on the basis of the organic substrate used.

Floating respiration	Protoplasmic respiration
Respiratory substrate is carbohydrate.	Respiratory substrate is protein.
It is common and runs continuously throughout.	It is rare and occurs only for some periods.
It does not produce toxic products.	It produces toxic product.

Types of respiration: Based on the availability of O_2

Anaerobic respiration	Aerobic respiration
It does not require O_2 to break down the substrate.	It utilises O_2 to break down the substrate.
It occurs in the cytoplasm.	It occurs in both cytoplasm and mitochondria.
It usually occurs in microbes, yeast, parasitic worms, deep-seated tissues of plants and animals, germinating seeds, fruits and succulents.	It occurs in the living cells of plants and animals.

It involves glycolysis and incomplete breakdown of pyruvic acid.	It involves glycolysis, Krebs cycle and terminal oxidation.
It is less efficient.	It is more (15 times) efficient.
The substrate is incompletely broken and one of the end products is at least organic.	The substrate is completely oxidised, and the end products are inorganic.
Water is not formed and a lesser number (2) of CO_2 is released.	Water is formed and a larger number (6) of CO_2 is released.
One molecule of glucose produces 2 ATP on complete oxidation.	One molecule of glucose produces 36/38 ATP.
One molecule of glucose produces less amount of energy (56 Kcal).	One molecule of glucose produces a large amount of energy (686 Kcal).
It is most primitive.	It is advanced.

Anaerobic respiration can only continue for a short time in higher plants. Finally, the plants die if they are deprived of oxygen for a longer time because the energy produced in this process is not sufficient for continuing the life processes for a long time. Further accumulation of alcohol might reach the level of toxicity. Anaerobic respiration stops when the concentration of alcohol produced reaches about 13-15%.

Both anaerobic and aerobic respiration involve:

1. **Glycolysis:** It is an anaerobic breakdown of glucose into two molecules of pyruvic acid. It was discovered by Embden, Mayerhoff and Parnas (1930) hence it is known as the EMP pathway. It occurs in the cytosol of the cell because all enzymes are found. It occurs in both anaerobic and aerobic respiration. It is a reductive process. It is an oxygen-independent metabolic pathway used by both autotrophs and heterotrophs. It does not require molecular oxygen for any of its reactions, but it can take place in the presence of O_2 also hence it involves oxidation without direct use of oxygen.

ATP	Adenosine Triphosphate
NAD	Nicotinamide Adenine Dinucleotide
FAD	Flavin Adenine Dinucleotide

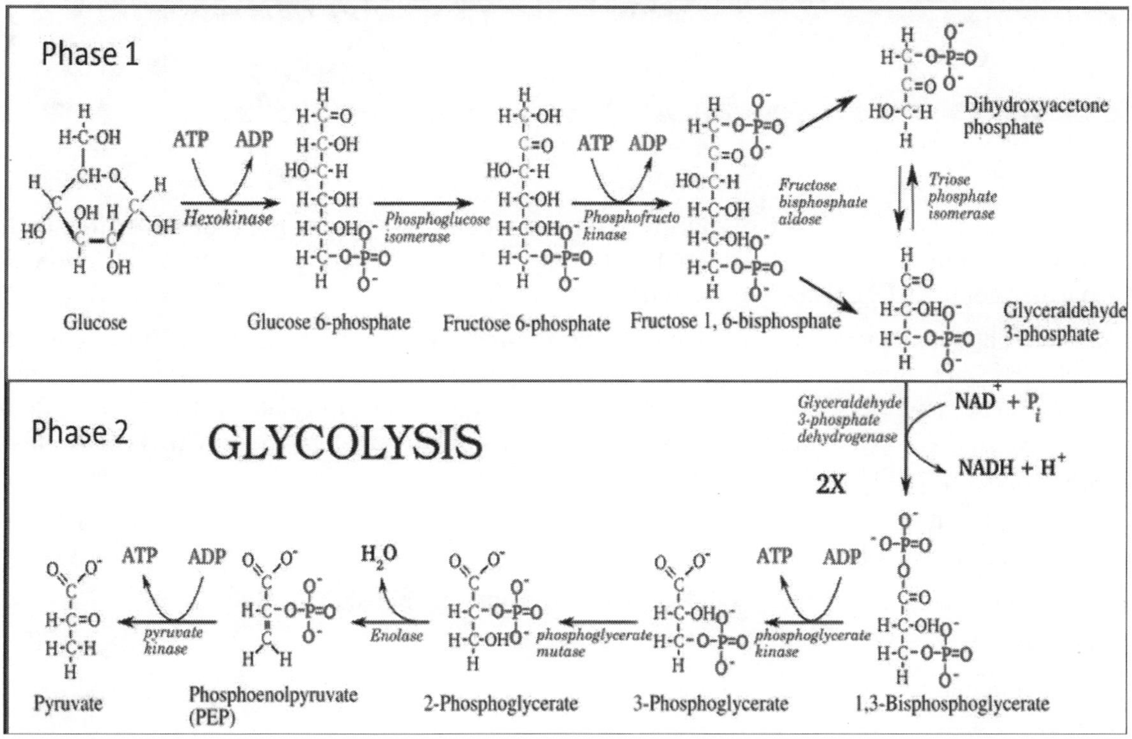

Reactions of glycolysis

Final reaction:

$$C_6H_{12}O_6 + 2ADP + 2Pi + 2NAD^+ \longrightarrow 2\ CH_3COCOOH + 2ATP + 2NADH + 2H^+ + 2H_2O$$

Steps involved in Glycolysis:

Steps	Substrate	Enzyme	Products	Reaction
1.	Glucose + **ATP**	*Hexokinase, Mg^{+2}*	Glucose-6-phosphate + ADP + Pi	Phosphorylation (irreversible)
2.	Glucose -6-phosphate	*Glucose phosphate isomerase*	Fructose-6-phosphate	Isomerisation (reversible)
3.	Fructose-6-phosphate + **ATP**	*Phosphofructokinase, Mg^{+2}*	Fructose-1,6-diphosphate + ADP + Pi	Phosphorylation (irreversible)
4.	Fructose-1,6-diphosphate	*Aldolase*	Dihydroxyacetone phosphate (ketose) + 3-Phosphoglyceraldehyde (aldolase)	Aldol splitting (reversible)
5.	Dihydroxyacetone phosphate	*Phosphotriose isomerase*	3-Phosphoglyceraldehyde.	Isomerisation (reversible)
6.	2 (3-phosphogleceraldehyde) + NAD $^+$	*Phosphoglyceric dehydrogenase*	2(1,3-diphosphoglyceric acid) **+ 2 NADH + 2H$^+$**	Oxidation (reversible)
7.	2(1,3-diphosphoglyceric acid) +2ADP + 2 Pi	*Phosphoglyceric kinase, Mg^{+2}*	2 (3-phosphoglyceric acid) + **2ATP**	Substrate level phosphorylation (reversible)
8.	2(3-phosphoglyceric acid)	*Phosphoglycero-mutase*	2(2-phosphoglyceric acid)	Isomerisation (reversible)
10.	2(2-diphosphoglyceric acid)	*Enolase*	2 (Phosphoenol pyruvic acid) + 2 H_2O	Dehydration (reversible)
11.	2 (phosphoenol pyruvic acid) + 2ADP + 2Pi	*Pyruvate kinase, Mg^{+2}*	2(Pyruvic acid) + **2 ATP**	Substrate level phosphorylation (irreversible)

Features of glycolysis:

(a) The first five reactions (first half of glycolysis) come under the preparatory or investment phase since they consume 2 molecules of ATP (one in 1st and one in 3rd step) to convert glucose into two molecules of phosphoglyceraldehyde (3 Carbon compound).

(b) The next five reactions (second half of glycolysis) are called the pay-off phase in which there is a net gain of ATP and NADH. Since glucose leads to two triose sugars (phosphoglyceraldehyde) in the preparatory phase, each reaction in the pay-off phase occurs twice per glucose molecule. This yields 2NADH molecules and 4ATP molecules (two in the 7th step and two in the 10th step), leading to a net gain of 2 NADH molecules and 2ATP molecules from the glycolytic pathway per glucose.

(c) The process of glycolysis is turned off when ATP is present in sufficient amounts because ATP is an allosteric inhibitor of phosphofructokinase enzyme.

(d) Two molecules of Pyruvic acid are formed as products from one molecule of glucose.

(e) CO_2 is not released.

The balance sheet of ATP in glycolysis

Consumption of ATP by phosphorylation of glucose at first step: – 1 ATP
Consumption of ATP by phosphorylation of glucose at third step: – 1 ATP

– 2 ATP

Production of ATP:

Substrate level phosphorylation (direct gain) at 7th step: 2 ATP
Substrate level phosphorylation (direct gain) at 10th step: 2 ATP

4 ATP

(Net gain of ATP = 4ATP - 2ATP = 2ATP)

Oxidative phosphorylation (through 2 $NADH_2$) at 6th step: **6 ATP**

Total production of ATP in glycolysis = 6ATP + 2 ATP= 8ATP

Types of phosphorylation process: In fact, phosphorylation is the synthesis of ATP by reacting with ADP and Pi. It activates sugar and prevents it from getting out of the cell. It is of 3 types:

(i) Photophosphorylation: It occurs in photosynthesis in which light is used.

(ii) Substrate level phosphorylation: It occurs in the process of respiration where there is a direct transfer of a phosphate group from one of the substrates to ADP. It does not require O_2 and Electron Transport Chain (ETC).

(iii) Oxidative phosphorylation: It occurs in the process of respiration.

Pasteur effect: It is a sudden change from anaerobic respiration to aerobic respiration due to the availability of O_2 hence there is a reduction in the consumption of respiratory substrate. Pasteur observed that when yeast was subjected to aerobic conditions then their production of alcohol and consumption of glucose declined due to the process of fermentation. He concluded that yeast consumed far more sugar when growing anaerobically rather than aerobically.

Fate of Pyruvic acid: Pyruvic acid is a key compound that stands at the junction of anaerobic respiration and aerobic respiration. It shows following three fates which are as follows:

(a) If O_2 is available then pyruvic acid will enter into the Krebs cycle via oxidative decarboxylation process.

$$2 \text{ Pyruvic acid} + 6\ O_2 \longrightarrow 6\ CO_2 + 6H_2O + \text{energy}$$

(b) If O_2 is limited as in microbes, yeast and plant cells then alcoholic fermentation occurs.

(c) If O_2 is absent as in skeletal muscle, liver and microbes then lactic acid fermentation occurs. If lactic acid accumulates in muscles, it causes fatigue and pain in muscles. However, during rest, it is again converted into pyruvic acid which undergoes aerobic respiration.

2. **Fermentation:** It is a process in which chemical changes are brought about by the action of enzymes produced by living organisms (microbes and some fungi) on an organic substrate (carbohydrate, protein, fats). It can also be defined as 'a process of converting carbohydrate to alcohol or organic acids using microorganisms (bacteria and yeast) under anaerobic conditions". The science of fermentation is called zymology. It is less efficient because it produces only 2 ATPs from glucose.

   It was discovered by Gay Lussac (1815). He described the production of alcohol from sugar.

$$C_6H_{12}O_6 \longrightarrow 2C_2H_5OH + 2CO_2$$

Pasteur (1875) first carried out alcoholic fermentation of yeast to investigate why wine sometimes turned distasteful. He showed that fermentation occurs by the action of living microorganisms and concluded that fermentation was catalysed by a vital force called 'ferment' within yeast cells. Thus, yeast's glucose consumption decreased under aerobic conditions of fermentation, in comparison to the anaerobic effect (Pasteur effect). Later Edward Buchner (1897) demonstrated that the conversion of glucose to ethyl alcohol was possible using a nonliving extract of yeast due to the action of enzymes in the extract. He coined the term 'zymase'. It requires temperature (32-36°C) and pH=4.5-5.0

Types: Fermentation is of several types:

1. **Alcoholic fermentation**: It occurs in a limited amount of oxygen. It involves two steps:

   (a) In the first step, pyruvic acid in the absence of oxygen undergoes decarboxylation forming acetaldehyde

$$2 \text{ Pyruvic acid} \xrightarrow[\text{Thiamine pyrophosphate (TPP)}]{\text{Decarboxylase, } Mg^{+2}} 2 \text{ Acetaldehyde} + 2CO_2$$

   (b) Acetaldehyde now undergoes dehydrogenation and accepts protons and electrons forming ethyl alcohol. It is used to produce beer, wine and distilled spirits.

$$2 \text{ Acetaldehyde} + 2NADH + 2H^{-} \xrightarrow{\text{Alcoholic dehydrogenase}} 2\text{Ethyl alcohol} + 2NAD^{+}$$

2. **Lactic acid fermentation**: It occurs in the complete absence of oxygen and pyruvic acid itself becomes an acceptor of protons (H^+) and electrons (e-) in the absence of oxygen. In strenuous exercise, when energy demands exceed the energy supply, the respiratory chain cannot process all of the H atoms joined by NADH. It produces lactic acid in skeletal muscle, liver and microbes. If lactic acid accumulates in muscle, it will cause fatigue. However, during rest, the same lactic acid is again converted into pyruvic acid and undergoes aerobic respiration. It is of two types:

   (a) **Homolactic**: It is carried out by *Lactobacillus, Streptococcus, Pediococcus* etc. It is used for souring milk and also for the preparation of cheese, and yoghurt. It involves a single step in which pyruvic acid is converted into lactic acid in the presence of lactic dehydrogenase enzyme (it also acts as an isoenzyme in which different molecular forms catalyse the same reaction). It occurs in one step.

$$2 \text{ Pyruvic acid} + 2 \text{ NADH} + 2H^{+} \xrightarrow[\text{Flavin Mononucleotide (FMN), Zn+2}]{\text{Lactic dehydrogenase}} 2 \text{ Lactic acid} + 2 \text{ NAD}^{+}$$

   (b) **Heterolactic**: It is carried out by *Leuconostoc*. It is used for preparing sauerkraut. It uses the Pentose phosphate (PPP) pathway. It produces lactic acid, ethyl alcohol and CO_2 in several steps.

Fermentation and anaerobic respiration are identical because

(a) Glucose is a substrate in both processes.

(b) The principal end products are CO_2 and ethyl alcohol.

(c) Zymase occurs commonly.

(d) Pyruvic acid and acetaldehyde are formed as intermediates.

Fermentation	Anaerobic respiration
It is an extracellular process.	It is an intracellular process.
It usually occurs in microbes, yeast, roots and seeds of higher plants, and muscles of animals.	It usually occurs in plants and animals.
It occurs in a small (limited) quantity of O_2.	It occurs in complete absence of O_2.

2. **Pyruvate oxidation**: Pyruvic acid, the end product of glycolysis will undergo oxidative decarboxylation in the mitochondrial matrix. It requires pyruvate dehydrogenase complex and 5 cofactors such as, Thiamine pyrophosphate (TPP), Lipoic acid, Coenzyme-A, Nicotinamide Adenine Dinucleotide (NAD) and Mg^{+2}. The function of Co. A is to activate the acetyl group. This reaction is complicated and involves several steps which are as follows:

(a) $$\text{Pyruvic acid} + \text{TPP} \xrightarrow{\text{Pyruvate decarboxylase, } Mg^{+2}} \text{TPP-acetate} + CO_2$$

Acetyl transferase

(b) TPP-acetate + lipoic acid ⟶ Acetyl hydrolipoate + TPP

Dehydrogenase

(c) Acetyl hydrolipoate + Co.A ⟶ Acetyl Co.A + Reduced lipoic acid

(d) Reduced lipoic acid +NAD^+ ⟶ Oxidised lipoic acid +$NADH^+$ + H^+

Substrates/Reactants	Enzyme	Products	Reaction
Pyruvic acid + Thiamine pyrophosphate + Lipoic acid + CoA + NAD+	Pyruvate dehydrogenase	Acetyl CoA + CO_2 + **NADH + H+**	Oxidative decarboxylation

$$2CH_3.CO.COOH + 2CoA + 2NAD^+ \xrightarrow{\text{Pyruvate dehydrogenase}} 2CH_3CO.CoA + 2CO_2 + 2NADH + 2H^+$$

Acetyl CoA (derived from the oxidation of pyruvic acid or β-oxidation of fatty acid) is the only fuel to enter the Krebs cycle. It is a connecting link between glycolysis and the Krebs cycle. With each turn of the cycle, one molecule of acetyl CoA is consumed for every molecule of oxaloacetic acid present in the mitochondrial matrix and is never generated. Actually, it is the oxidation of the acetate part of acetyl CoA that produces CO_2 and water. Two molecules of CO_2 are released since two molecules of pyruvic acid are involved in this step.

Balance sheet of ATP in pyruvate oxidation:

Oxidative phosphorylation through 2 $NADH_2$= 2 x3= **6 ATP**

3. **Krebs cycle:** It is a series of chemical reactions used by all aerobic organisms to release stored energy through the oxidation of acetyl CoA. It was investigated by Hans Krebs (1940) when he observed the increase in oxygen consumption in the breast muscle of pigeons due to the addition of certain chemicals. It occurs in the matrix of mitochondria where all enzymes are present except succinic dehydrogenase which is found in the inner membrane.

   It is called the citric acid cycle because citric acid is the first product and is constantly generated at every turn of the cycle. It is also called the tricarboxylic acid (TCA) cycle because citric acid contains three carboxylic groups. It is called the central metabolic pathway involving the oxidation of carbohydrates, proteins and fats into smaller compounds (catabolism) and synthesis of macromolecules (amino acids, chlorophylls, cytochromes etc) from intermediate products (anabolism or anapleurotic reaction). Hence it is called the amphibolic process.

   It consumes acetyl coenzyme A and water, reduces NAD^+ to NADH and produces 2 molecules of CO_2 as a waste product. At the end of the Krebs cycle, the final product is oxaloacetic acid (OAA). This is identical to the OAA that begins the cycle. Now the OAA is ready to accept another acetyl CoA to begin another turn of the cycle.

Steps	Substrate	Enzyme	Products	Types of Reaction
1.	Oxaloacetic acid + Acetyl CoA + H_2O	*Citrate synthetase*	Citric Acid + CoA	Aldol condensation (irreversible)
2	Citric acid	*Aconitase*	Cis-aconitic acid + H_2O	Dehydration (reversible)
3	Cis-aconitic acid +H_2O	*Aconitase*	Isocitric acid	Hydration (reversible)
4	Isocitric acid + NAD^+	*Isocitrate dehydrogenase*	Oxalosuccinic acid **+$NADH^+$ + H^+**	Oxidation
5	Oxalosuccinic acid	*Oxalosuccinic decarboxylase*	α-ketoglutaric acid + CO_2	Decarboxylation (irreversible)
6	α-ketoglutaric acid + NAD^+ + CoA	*α-ketoglutarate dehydrogenase*	Succinyl CoA + **$NADH^+$ + H^+** +CO_2	Oxidative decarboxylation (irreversible)

Steps	Substrate	Enzyme	Products	Types of Reaction
7	Succinyl Co A + GDP + Pi	*Thiokinase*	Succinic acid + CoA + **GTP** (animal) **/ATP** (plant)	Substrate level phosphorylation
8	Succinic acid + FAD	*Succinic dehydrogenase*	Fumaric acid + **$FADH_2$**	Oxidation
9	Fumaric acid + H_2O	*Fumarase*	Malic acid	Hydration
10	Malic acid + NAD^+	*Malate dehydrogenase*	Oxaloacetic acid + **$NADH^+$ + H^+**	Oxidation (reversible)

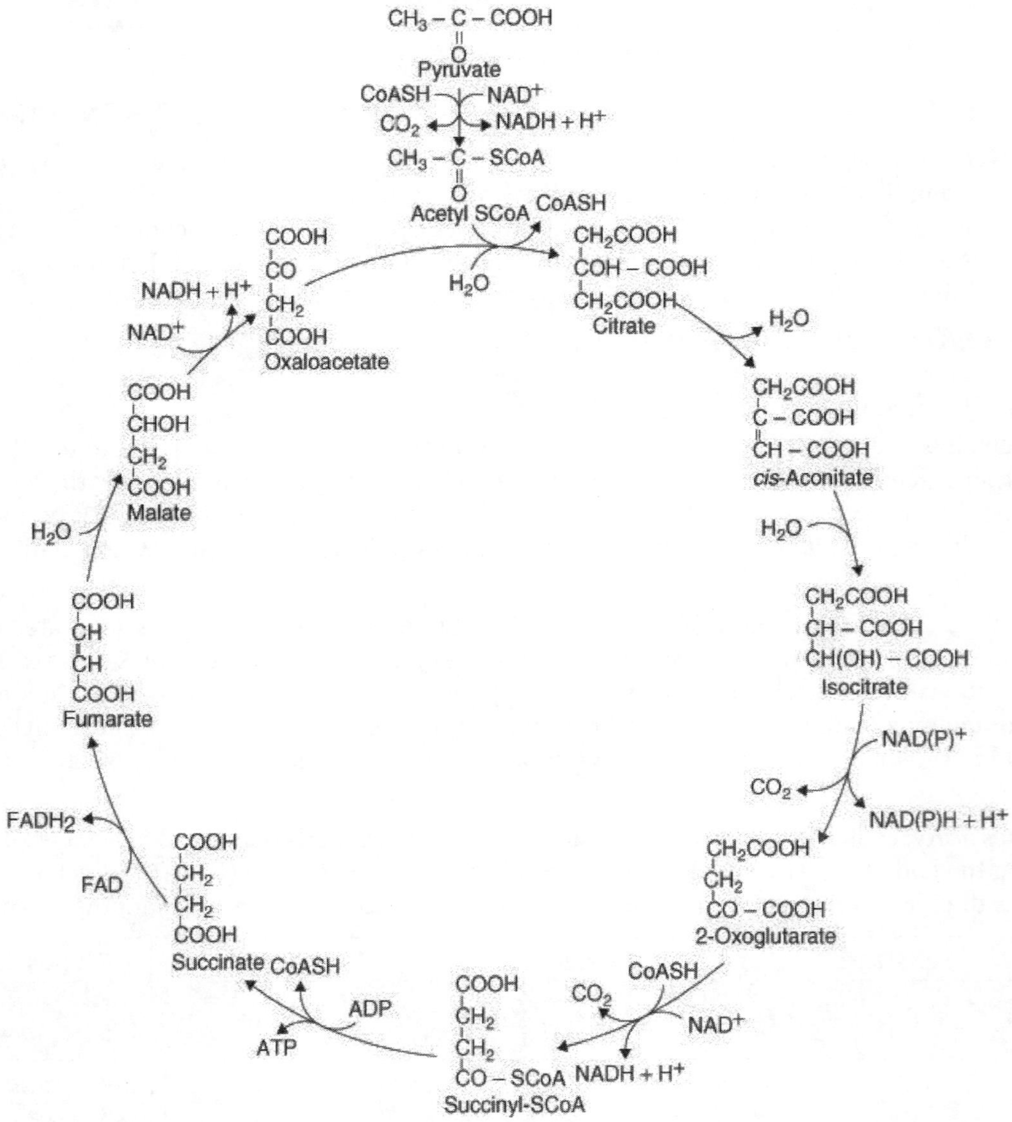

Fig: Pyruvate oxidation and Krebs cycle

The final reaction of Krebs cycle (single cycle considering one molecule of acetyl Co. A):

$CH_3COCH_2COOH + CH_3CO.CoA + 3NAD^+ + 1\ FAD + 1\ ADP + Pi + 2H_2O \longrightarrow OAA + 2CO_2 + 3NADH + 3H^+ + 1FADH_2 + 1\ ATP + H_2O$

The final reaction of Krebs cycle (two cycles considering two molecules of acetyl Co. A):

$2\ CH_3COCH_2COOH + 2\ CH_3CO.CoA + 6NAD^+ + 2FAD + 2ADP + 2Pi + 4H_2O \longrightarrow 2OAA + 4CO_2 + 6NADH + 6H^+ + 2FADH_2 + 2\ ATP + 2H_2O$

Balance sheet of ATP in Krebs cycle

Substrate level phosphorylation (Direct gain) at 7th step:	→	2ATP
Oxidative phosphorylation (Indirect gain)		
Through 2 molecules of $NADH_2$ at 4th step:	→	6 ATP
Through 2 molecules of $NADH_2$ at 6th step:	→	6 ATP
Through 2 molecules of $NADH_2$ at 10th step:	→	6 ATP
Thorough 2 molecules of $FADH_2$ at 8th step:	→	4ATP
Total production of energy:		**24** ATP

Features of Krebs cycle:

(a) Krebs cycle undergoes twice since 2 molecules of acetyl CoA are formed from two molecules of pyruvic acid. Acetyl CoA is consumed in the Krebs cycle and is never regenerated.

(b) It produces reducing power to the electron for transport of electron.

(c) There are three oxidation steps.

(d) In each cycle, 2 molecules of CO_2 (2x2 = 4 CO_2) are released outside mitochondria and lost as waste product.

(e) There are 4 major regulatory enzymes such as citrate synthetase, isocitrate dehydrogenase, α-ketoglutarate dehydrogenase and succinate dehydrogenase.

(f) The first 4-carbon compound involved was oxaloacetic acid.

(g) The first 5-carbon compound formed was α- ketoglutaric acid (first dicarboxylic acid).

(h) -ketoglutaric acid undergoes both decarboxylation and dehydrogenation reactions.

(i) The first 4-carbon compound formed was succinyl coenzyme A.

Importance:

(a) It produces a large amount of energy in a controlled manner.

(b) Pyruvic acid produces an amino acid, alanine.

(c) Acetyl Co-A is a raw material for the steroid, alkaloids, carotenoids, terpenoids, and gibberellins.

(d) Citric acid is used as a precursor for the synthesis of fatty acid.

(e) -ketoglutaric acid produces an amino acid, glutamic acid and thus helps in transamination reaction.

(f) Succinyl Co-A is involved in the synthesis of chlorophyll, phytochrome, cytochrome, haeme etc.

(g) Malic acid is used as a precursor for the synthesis of glucose.

(h) Oxaloacetic acid produces an amino acid, aspartic acid which forms alkaloids and pyrimidines.

Combined equation of glycolysis, pyruvate oxidation and Krebs cycle

$$\text{Glucose} + 10\,NAD^+ + 2FAD + 4ADP + 2\,Pi + 6\,O_2 \longrightarrow 6\,CO_2 + 10\,NADH^+ + 10\,H^+ + 2\,FADH_2 + 4\,ATP + 6\,H_2O$$

Inhibitors:

Fluoroacetate	It acts as aconitase hence synthesis of isocitric acid is inhibited.
Arsenic dehydrogenase	It acts as α-ketoglutarate dehydrogenase hence synthesis of succinyl Co-A is inhibited.
Malonate	It acts as succinic dehydrogenase hence synthesis of fumaric acid is inhibited

Entry of organic materials in different stages of aerobic respiration

1. **Starch**
   ↓amylase
   Glucose→**Glycolysis**
2. **Glycogen**
   ↓phosphorylase
   Glucose-1-phosphate
   ↓phosphoglucomutase
   Glucose-6-phosphate→**Glycolysis**
3. **Lactose**
   ↓lactase
   Glucose→**Glycolysis**
   +
   Galactose
   ↓
   UDP Galactose
   ↓
   UDP Glucose
   ↓
   Glucose-1- phosphate
   ↓phosphoglucomutase
   Glucose-6-phosphate→**Glycolysis**
4. **Sucrose**
   ↓ invertase
   Glucose→**Glycolysis**
   +
   Fructose
   ↓hexokinase
   Fructose-6-phosphate→**Glycolysis**
5. **Fats**
   ↓
   Fatty acid→Acetyl coenzyme-A→**Krebs cycle**
   +
   Glycerol →Dihydroxyacetone phosphate→3-PGAL→**Glycolysis**
6. **Proteins**→Amino acids→Different intermediates→**Krebs cycle**

Biosynthesis of amino acids:

(a) Glucose -6-phosphate→ Histidine
(b) Phosphoglycerate→Serine→ Cysteine, Glycine
(c) Phosphoenolpyruvate→ Phenylalanine, Tryptophan, Tyrosine
(d) Pyruvate→ Alanine, Valine, Leucine, Isoleucine
(e) Alpha ketoglutaric acid→ Glutamic acid→ Glutamine, Arginine, Proline
(f) Oxaloacetic acid→ Aspartic acid→ Asparagine, Methionine, Threonine, Lysine

ELECTRON TRANSPORT SYSTEM (RESPIRATORY CHAIN):

It is an assembly of electron-transporting coenzymes (carriers) to transfer electrons present in the inner membrane in a redox and sequential manner.

It is the last event of aerobic respiration and the only part of glucose metabolism that uses atmospheric oxygen and releases maximum energy. It is a unidirectional flow of electrons from NADH and $FADH_2$ to molecular oxygen. These electron carriers are arranged according to their increasing positive redox potential. Each electron donor will pass the electron to a more electronegative acceptor until e^- are passed to oxygen (the most electronegative and terminal electron acceptor in the chain). In other words, each electron carrier or acceptor has a more positive redox potential (higher electron affinity) than the previous one and with each successive transfer, the electrons lose additional free energy which is utilised to generate a proton gradient.

As the electrons travel through the chain, they move from a higher to a lower energy level because each electron carrier holds electrons at a lower energy level than the previous electron carrier. The transport of e^- releases energy to generate a proton gradient across the inner membrane by active pumping of protons into intermembrane space or intracristal space (C side).

Hence it is a process that oxidises NADH and $FADH_2$ produced in glycolysis, pyruvate oxidation and Krebs cycle. The reduced coenzymes of glycolysis enter the inner membrane of mitochondria through the C-side and those of the Krebs cycle reach the inner membrane through the M-side and finally these release energy in the form of ATP.

The main function of ETC in mitochondria is to provide energy for the synthesis of ATP during oxidative phosphorylation. Most eukaryotic cells have mitochondria which produce ATP from the products of the Krebs cycle, fatty acid oxidation and amino acid oxidation.

ETC consists of **5 complexes** of integral membrane protein.

1. Complex-I (NADH dehydrogenase complex): It contains FMN and Fe-S protein. It transfers e- from NADH to UQ and it pumps $2H^+$ across the inner membrane into C-side.
2. Complex-II (Succinate dehydrogenase complex): It contains FAD and Fe-S protein. It accepts e- from $FADH_2$ and transfers e- from succinate oxidation to UQ (Ubiquinone). It does not pump H^+.
3. Complex-III (Cytochrome b-c_1 complex / Cytochrome c reductase complex): It contains cytochrome b, Fe-S protein, ubiquinone (UQ) and hydroquinone (UQH_2) and cytochrome c_1 (located in C side). It accepts e^- from UQ and transfers to cytochrome-c. It pumps H+ into the C-side.
4. Complex-IV: (Cytochrome-c oxidase complex): It accepts e- from cytochrome-c and transfers to reduce molecular O_2 to water. It pumps H^+ into the C-side.
5. Complex-V (ATP synthase complex): It catalyses the formation of ATP from ADP and Pi. It contains a transmembrane channel through which H^+ diffuse down a concentration gradient.

Mechanism: The intermediates such as 3-PGAL, pyruvic acid, isocitric acid, α-ketoglutaric acid, succinic acid and malic acid formed as a result of the breakdown of glucose are dehydrogenated (oxidised) hence a pair of H atoms is removed from each one of them in the form of reduced coenzyme ($NADH_2$). A pair of H atoms dissociates into $2H^+$ and $2e^-$.

$$(2H \longrightarrow 2H^+ + 2\ e^-)$$

(a) NAD^+ accepts 2 H^+ and 2 e^- and is reduced to NADH and transfers 2 e- and 1 H^+ to FMN. Thus, NADH is oxidised back to NAD^+ to Krebs cycle and returns.

(b) FMN accepts 2e- and 1 H^+ from NADH and further accepts 1 H^+ extra from matrix to be reduced to $FMNH_2$. The $FMNH_2$ pumps $2H^+$ to the C-side (intracristal side or intermembrane space) and transfers 2e- to Fe-S protein and thus, $FMNH_2$ is oxidised back to FMN.

(c) Like NADH, $FADH_2$ deposits its electrons in the ETC but it does so bypassing the complex I entirely. $FADH_2$ transfers 2 e- to Fe-S protein and then to ubiquinone (UQ).

(d) Fe-S protein transfers 2e- to UQ (ubiquinone), a lipid-soluble carrier and a mobile link between complex-I and complex-III; and complex II and complex III. It is highly oxidised. It receives 2 H^+ from the matrix.

(e) UQ transfers 2e- and 2 H^+ to UQH_2 (hydroquinone) which pumps $2H^+$ to C-side from complex III.

(f) UQH_2 (hydroquinone) transfers 2e- to cytochrome b.

(g) Cytochrome- b transfers 2e- to Fe-S protein.

(h) Fe-S protein transfers 2 e- to Cytochrome-c_1.

(i) Cytochrome-c_1 transfers 2 e- to cytochrome-c (the mobile link between complex-III and complex IV of ETC).

(j) Cytochrome-c accepts 2 H^+ from matrix and transfers 2 e- and 2 H^+ to cytochrome-a.

(k) Cytochrome-a pumps 2 H^+ to C-side and transfers 2 e- to cytochrome-a_3 and it is reduced.

(l) Finally reduced cytochrome-a_3 (cytochrome oxidase), a marker enzyme of mitochondria) transfers 2e- to oxygen (final acceptor molecule) to activate it. In this activated state, oxygen will extract the free $2H^+$ from the M-side to form water. This is called terminal oxidation.

$$\frac{1}{2} O_2 + 2\ e^- + 2\ H^+ \leftrightarrow H_2O$$

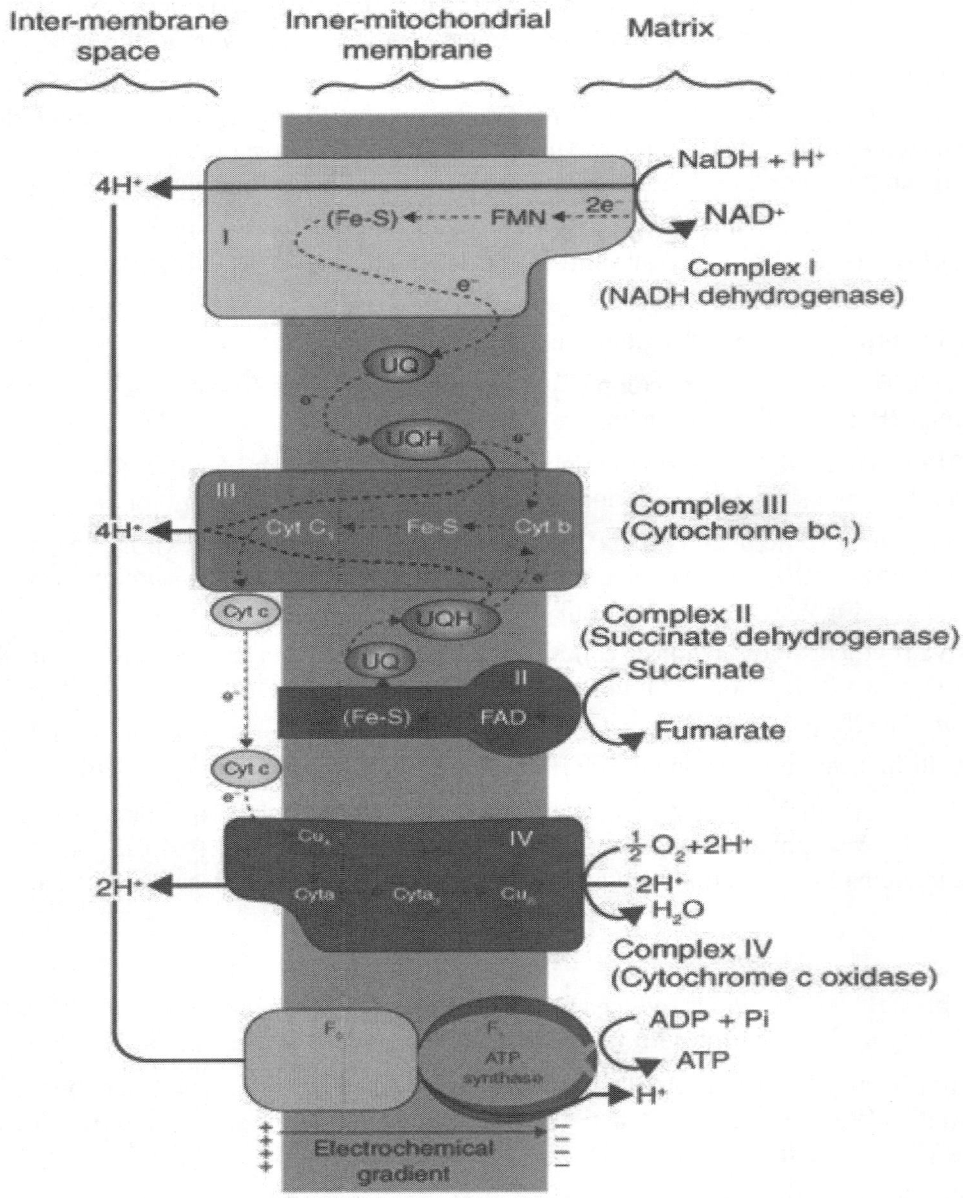

Fig: Mechanism of chemiosmosis

Thus, oxygen is a terminal acceptor of e- and most electronegative in ETS. In the Krebs cycle, $FADH_2$ is also produced and transfers e- and H^+ to ubiquinone (UQ) directly at a later stage in the chain bypassing the first stage. All cytochromes have Fe only but cytochrome a_3 has both Fe and Cu. The cytochrome a_3 accepts 2e- through its Fe part and transfers 2e- to oxygen through its Cu component.

OXIDATIVE PHOSPHORYLATION

It is the synthesis of ATP by direct oxidation of substrates or reduced coenzymes such as NADH and $FADH_2$ by e-transport in respiration. It occurs in the plasma membrane of prokaryotes but in the inner mitochondrial membrane of eukaryotes.

Peter Mitchell proposed a chemiosmotic hypothesis where electron transport and ATP synthesis are coupled by a proton gradient. Chemiosmosis is the chemical formation of ATP driven by diffusion force similar to osmosis.

The enzyme complexes such as complex-I, complex-III, and complex IV present in the inner membrane act as proton pumps that transfer protons (H^+) from the M-side to the C-side.

Most of the energy contained in food material is ultimately used to generate an H^+ gradient in the inner membrane of mitochondria. This gradient is used to drive the synthesis of ATP.

As the electrons move energetically downhill, the complexes capture the released energy and use it to pump H^+ ions from the M side to the C side. The H^+ concentration is increased on the C side and becomes acidic or positively charged and thus matrix becomes slightly negatively charged. This internal negativity attracts the positively charged H^+ and induces them to re-enter the matrix to equalise the pH. Now the inner membrane becomes impermeable to protons. This pumping across the inner membrane forms proton motive force which is the electrochemical potential difference between protons in the aqueous phases on different sides of the membrane. Proton motive force causes H^+ to move back to the matrix passing through ATP synthase which uses the energy of gradient to catalyse the synthesis of ATP from ADP.

The elementary particle or oxysome of mitochondria has three parts: Base (F_0), Stalk and Head (F_1). The base (F_0) is an integral membrane protein complex embedded in the inner membrane. It forms a proton channel through which H^+ cross the inner membrane. The head (F_1) is a peripheral membrane protein complex containing ATP synthase protruding into the matrix. It contains the site for the synthesis of ATP from ADP and Pi. The isolated F_1 cannot make ATP from ADP and Pi but it can hydrolyse ATP into ADP and Pi.

For the synthesis of one ATP, two protons (H^+) are passed through a proton channel formed by F_0 from the C-side to the M-side along high concentration to low concentration. Since 3 ATPs are produced from the complete oxidation of 1 NADH, then 6 protons (3 pairs) are passed from the C-side to the M-side but in the case of oxidation of each $FADH_2$, 2 pairs of protons are passed from the C-side to the M-side hence 2ATPs are produced because it enters electron transport chain at a later stage and transfers e- and H^+ to UQ (ubiquinone) bypassing the complex I.

The energy of the proton gradient (as H^+ flows down the gradient from the C-side to the M-side via the inner membrane) is used in attaching an inorganic phosphate (Pi) radical to ADP by high energy bond producing ATP.

$$ADP + Pi \longrightarrow ATP + H_2O$$

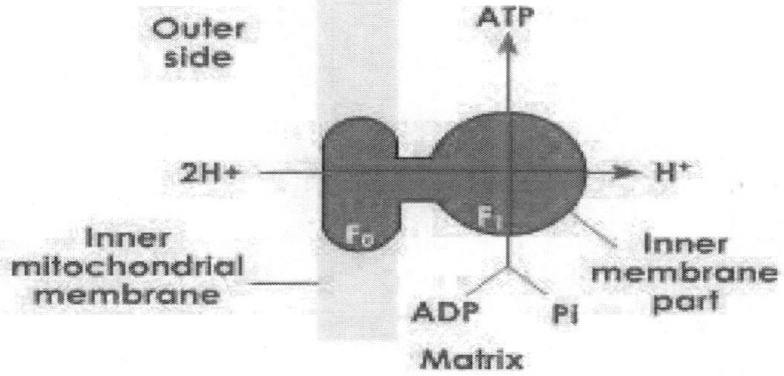

ATP (Adenosine Triphosphate): It was discovered by Lehmann (1920) but its function was discovered by Lipmann.

It is an immediate source of energy within the cell. The hydrolysis of ATP releases energy. Since ATP is found in all living cells hence it is called a universal energy carrier or energy currency of cells. It is mobile and transports chemical energy to energy-requiring processes within the cell. ATP like GTP, CTP and UTP is essential for various activities such as the synthesis of carbohydrates, proteins, and fats; active transport, translocation of food, growth and streaming of protoplasm etc.

ATP is a derivative of Adenosine monophosphate (AMP) to which two additional phosphate groups are attached through an anhydride linkage. When the first high-energy phosphate bond is hydrolysed, 8.9 Kcal energy is released;

the second phosphate bond is broken, 6.5 Kcal energy is released and if the third phosphate group (of AMP) is hydrolysed, then only 3.4 Kcal energy is released. The energy released by the breaking down of phosphate groups in ATP and ADP is much more than the energy released on the hydrolysis of most of the other covalent bonds.

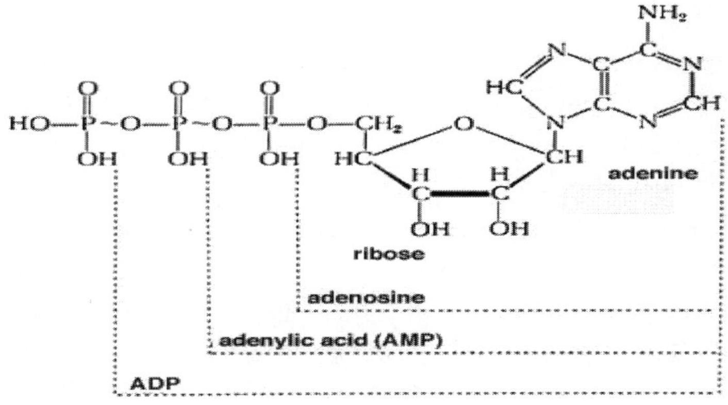

Its catabolism to ADP yields 8.15 kcal/mol of energy.

$$ATP + H_2O \longrightarrow ADP + Pi + Energy$$

ADP may be rephosphorylated to ATP by respiratory activity or another high-energy compound.

$$ADP + Pi \longrightarrow ATP + H_2O$$

The actual number of ATP produced may be lower in certain eukaryotic cells (skeletal muscles) because energy (2ATP) is consumed in transporting the NADH across the mitochondrial membrane hence there is the formation of 36 ATP instead of 38 ATP.

The **final equation of aerobic respiration** is as follows:

$$C_6H_{12}O_6 + 6\,O_2 + 38\,ADP + 38\,Pi \longrightarrow 6\,CO_2 + 6\,H_2O + 38\,ATP$$

List of reactants consumed, and products formed:

Number of glucose molecules consumed: **1**

Total number of O_2 consumed: **6**

(1 by ETS in 6th step of Glycolysis)

(1 by ETS in the step of Pyruvate oxidation)

(1 by ETS in 4th step of Krebs cycle)

(1 by ETS in 6th step of Krebs cycle)

(1 by ETS in 8th step of Krebs cycle)

(1 by ETS in 10th step of Krebs cycle)

Total number of CO_2 produced: **6**

(2 CO_2 in the step of Pyruvate oxidation)

(2 CO_2 in 5th step of Krebs cycle)

(2 CO_2 in 6th step of Krebs cycle)

Total number of H_2O consumed: **10**

(2 H_2O in 6th step of Glycolysis)

(2 H_2O in 1th step of Krebs cycle)

(2 H_2O in 3th step of Krebs cycle)

(2 H_2O in 7h step of Krebs cycle)

(2 H_2O in 9th step of Krebs cycle)

Total number of H_2O produced: **16**

($2\ H_2O$ in 9th step of Glycolysis)

($2\ H_2O$ by ETS in 6th step of Glycolysis)

($2\ H_2O$ by ETS in the step of Pyruvate oxidation)

($2\ H_2O$ in 2nd step of Krebs cycle)

($2\ H_2O$ by ETS in 4th step of Krebs cycle)

($2\ H_2O$ by ETS in 6th step of Krebs cycle)

($2\ H_2O$ by ETS in 8th step of Krebs cycle

($2\ H_2O$ by ETS in 10th step of Krebs cycle

Net H_2O produced: 16 – 10 = **$6\ H_2O$**

Process	Final hydrogen (electron) acceptor	Types of phosphorylation
Fermentation	An organic molecule	Substrate level phosphorylation
Anaerobic respiration	Usually an inorganic substance (nitrate, sulphate or carbonate)	Substrate level and oxidative phosphorylation
Aerobic respiration	Molecular oxygen	Substrate level and oxidative phosphorylation

Inhibitors:

Compound	Use	Effect	Site of action	Action
Cyanide, CO, Azide, H_2S	Poisons	Inhibitors of ETS	Complex IV	These inhibit the ETC by binding strongly with cytochrome-a_3 thus preventing the transfer of electrons from cytochrome-a_3 to oxygen.
Rotenone	Pesticide	Inhibitor of ETS	Complex I	It blocks the transfer of e- from complex I to UQ by blocking the UQ binding site.
Malonates and Oxaloacetate	Poisons	Inhibitors of ETS	Complex II	These are competitive inhibitors of succinic dehydrogenase.
Antimycin A	Piscicide	Inhibitor of ETS	Complex III	It binds to cytochrome-c reductase and thus inhibits oxidation of ubiquinone.
2,4-dinitrophenol	Poison	Uncoupler of oxidative phosphorylation	Inner membrane	It uncouples (dissociates) ETC and phosphorylation reaction. It increases the permeability of mitochondria to H^+. It helps H^+ travel down their gradient and back into the M-side without passing through ATP synthase.
Oligomycin	antibiotic	Inhibitor of oxidative phosphorylation	Complex V	It inhibits ATP synthase by blocking the flow of H^+ through F_0 subunit (proton channel).

Factors affecting Respiration:

1. O_2: The rate of respiration is decreased if atmospheric O_2 concentration is below 5% for whole tissue. The rate of respiration is increased with an increase in O_2.
2. CO_2: The rate of respiration is decreased if the amount of CO_2 in the air is more than usual (0.03%) because stomata will be closed hence O_2 will not enter.
3. Light: The rate of respiration is increased because light promotes photosynthesis hence organic substances will be formed.

4. Temperature: The rate of respiration is decreased if temperature is increased from 30°C. The rate of respiration is increased when the temperature is between 0°C and 30°C (Q_{10}=2).
5. Water: The rate of respiration increases if there is a shortage of water or slight desiccation of tissues.
6. Inorganic salts: The rate of respiration is increased if plant tissue is transferred from water to salt concentration.
7. Injury: The rate of respiration is increased if tissue is injured or wounded.
8. Age: The rate of respiration is decreased in the old stage, steady in the growing or meristematic stage and increased (climacteric) in mature fruit.

The efficiency of aerobic respiration:

1ATP produces 8.15 kcal or 34kJ /mol and thus total energy obtained from 38 ATP=38x 8.15=309 Kcal. Glucose stores energy= 686 Kcal

Efficiency $= \frac{309 \times 100}{686} = 45\%$ and the rest 55% of energy is lost as heat.

Respiratory quotient (RQ): It is the ratio of the volume of CO_2 evolved to the volume of O_2 consumed in respiration. It is measured by Ganong's Respirometer. It depends on the type of substrate.

$$RQ = \frac{\text{Volume of } CO_2 \text{ evolved}}{\text{Volume of } O_2 \text{ consumed}}$$

Case:

1. If the substrate is **carbohydrate** (as in germinating cereals; sprouting potatoes, and green leaves kept in the dark) then RQ will be one or unity.

   $C_6H_{12}O_6 + 6O_2 \longrightarrow 6\,CO_2 + 6H_2O$ + energy

   RQ = 6/6 =1 (Unity)

2. If the substrate is **protein** as in germinating pulses or **fat** as in germinating oilseeds such as castor, then RQ will be less than one (0.7 in fats and 0.9 in proteins)

   $2\,(C_{51}H_{98}O_6) + 145\,O_2 \longrightarrow 102\,CO_2 + 98\,H_2O$ + energy

   (Tripalmitin)

   RQ = 102/145 = 0.7

   Fats are poorer in oxygen hence they require more O_2 for complete oxidation.

   RQ is also less than 1 in a mixed diet, coloured petals and starved leaves.

3. If the substrate is an **organic acid (Malic acid or Oxalic Acid)** then RQ will be more than one

   $CH_2\text{-}COOH + 3O_2 \longrightarrow 4\,CO_2 + 3\,H_2O$

   CHOH-COOH

   (Malic acid)

   RQ = 4/3 = 1.3 (more than 1)

   $2\,COOH + O_2 \longrightarrow 4\,CO_2 + 2\,H_2O$

   COOH

   (Oxalic acid)

   RQ = 4/1= 4 (more than 1)

4. If **anaerobic** respiration occurs as in microbes (bacteria and yeast), deep-seated tissue of plants and animals; and fruits then RQ will be infinite

$$C_6H_{12}O_6 \xrightarrow{\text{Zymase}} 2\ C_2H_5OH + 2CO_2$$

RQ= 2/0 =

5. If the material is **succulent or CAM** plants (e.g., *Opuntia* and *Bryophyllum)*, RQ will be zero. Glucose is not completely oxidised in CO_2 and H_2O, but it is incompletely oxidised or fixed into malic acid. Hence CO_2 is not released during the night when the stomata is open.

   $2\ C_6H_{12}O_6 + 3O_2 \longrightarrow 3\ CH_2\text{-}COOH + 3\ H_2O$

   CHOH.COOH

   (Malic acid)

   RQ = 0/3 = Zero

Importance:

(a) It determines the type of respiratory substrate.

(b) It determines the type of respiration.

β-oxidation: It is a catabolic process by which a very long chain of fatty acids is broken down to generate acetyl coenzyme-A which enters the Krebs cycle. It occurs in the cytoplasm of prokaryotes and mitochondria of eukaryotes.

C_n acyl Co-A + FAD + NAD^+ + H_2O + Co-A→C_{n-2} acyl Co-A + $FADH_2$ + NADH + H^+ + Acetyl Co.A

One turn of **-oxidation** of fatty acid produces 15 ATPS.

Consumption of ATP=2ATPs and Production of ATP= 17 ATPs

Net gain = 15 ATPs

β-oxidation of Palmitic acid (16C fatty acid) produces 129 ATPs

β-oxidation of Stearic acid (18C fatty acid) produces 146 ATPs

SHORT ANSWER TYPE QUESTIONS

1. Distinguish between floating respiration and protoplasmic respiration.
2. Distinguish between anaerobic respiration and aerobic respiration.
3. What are the features of glycolysis?
4. Write notes on the Pasteur effect.
5. Distinguish between fermentation and anaerobic respiration.
6. What do you mean by pyruvate oxidation?
7. How do different carbohydrates enter the process of respiration?
8. What are the features of the Krebs cycle?
9. Discuss the importance of the Krebs cycle.
10. Write the final equations of glycolysis and Krebs cycle.
11. Write notes on the ATPase complex.
12. Write notes on ATP.
13. What happens if cyanide acts as an inhibitor?
14. What is the efficiency of aerobic respiration?
15. What will happen to RQ in case of anaerobic respiration?

LONG ANSWER TYPE QUESTIONS

1. Describe the different reactions of glycolysis and show the generation of ATP.
2. What do you mean by fermentation? Describe the alcoholic and lactic acid fermentation.
3. Describe the different reactions of the Krebs cycle and show the generation of ATP
4. Describe the transport of electrons in the inner membrane of mitochondria.
5. Discuss the mechanism of oxidative phosphorylation in mitochondria.

MULTIPLE CHOICE QUESTIONS

1. Formation of phosphoenol pyruvate from 2-phosphoglycerate takes place by

(a) Dehydration (b) Dehydrogenation
(c) Oxidation (d) Hydration

2. The number of glucose molecules required to produce 38 ATP molecules under anaerobic conditions by a yeast cell is

(A) 2 (B) 4
(C) 19 (D) 38

3. The number of water molecules produced and used in aerobic respiration, respectively, is

(A) 18 and 12 (B) 16 and 10
(C) 12 and 6 (D) 8 and 2

4. The ratio of CO_2 liberated between anaerobic and aerobic respiration is

(A) 3:1 (B) 2:4
(C) 2:2 (D) 1:3

5. If glycerol enters the Krebs cycle, then it will first convert to

(A) Acetyl CoA
(B) Phosphoglyceraldehyde
(C) Dihydroxyacetone phosphate
(D) Pyruvic acid

6. The ratio of NADH formed in glycolysis and that in the Krebs cycle is

(A) 1:2 (B) 1:3
(C) 1:4 (D) 2:3

7. The number of CO_2 released during one Krebs cycle is

(A) 2 (B) 3
(C) 4 (D) 6

8. The number of ATP molecules produced as a result of oxidative phosphorylation of glucose in respiration, is

(A) 30 (B) 34
(C) 36 (D) 38

9. The substrate-level phosphorylation reaction does not occur.

(A) 1, 3-diphosphoglyceric acid → 3-Phosphoglyceric acid
(B) Pyruvic acid → Acetyl Co-A
(C) Succinyl Co.A →Succinic acid
(D) Phosphoenol pyruvic acid → Pyruvic acid

10. The number of ATP molecules produced from the complete oxidation of one molecule of acetyl Co. A, is

(A) 12 ATP (B) 15 ATP
(C) 36 ATP (D) 38 ATP.

11. The coenzyme capable of reducing succinic acid is

(A) FAD (B) FMN
(C) NAD (D) NADP

12. The number of carbon atoms present in cis-aconitic acid, malic acid and lactic acid are, respectively

(E) 6, 3 and 3 (F) 6, 4 and 3
(G) 5, 4 and 3 (H) 6, 4 and 2

13. The number of ATP molecules synthesized by substrate-level phosphorylation during aerobic respiration of one glucose molecule is:

(A) 8 (B) 6
(C) 4 (D) 2

14. A single turn of the Krebs cycle yields

(A) 2 $FADH_2$, 2 NADH and 2 GTP
(B) 1 $FADH_2$, 3 NADH and 1 GTP
(C) 2 $FADH_2$, 3 NADH and 1 GTP
(D) 2 $FADH_2$, 5 NADH and 1 GTP

15. In the Krebs cycle, FAD participates as an electron acceptor during the conversion of

(A) Succinyl CoA to Succinic acid
(B) α-ketoglutaric acid to Succinyl CoA
(C) Succinic acid to Fumaric acid
(D) Fumaric acid to Malic acid

16. **The number of reduced NAD molecules produced in the complete breakdown of one molecule of glucose during aerobic respiration, is**
(A) 5 (B) 10
(C) 12 (D) 38

17. **In ETS of aerobic respiration, oxygen receives electrons directly from**
(A) Ubiquinone (B) Cytochrome-a
(C) Cytochrome-a_3 (D) $NADH^+$

18. **The sequence of cytochromes in ETS is**
(A) Cyt a, b, c, c_1, a (B) Cyt b, c_1, c, a, a_3
(C) Cyt b, a, a_3, c, c_1 (D) Cyt b, c, a_3, a, c_1

19. **The respiratory quotient in organic acids is**
(A) One (B) More than one
(C) Less than one (D) zero

20. **The number of ATP molecules formed from one molecule of pyruvic acid in mitochondrial respiration are**
(A) 2 (B) 4
(C) 38 (D) 15

21. **Which one of the following can respire in anaerobic conditions?**
(A) Leaves (B) Stem
(C) Flowers (D) Seeds

22. **The final electron acceptor in lactic acid fermentation is**
(A) Pyruvate (B) NAD
(C) Lactic acid (D) Oxygen

23. **The terminal electron acceptor in ETS during aerobic respiration**
(A) CO_2 (B) O_2
(C) H_2O (D) NADH

24. **The metal activator used in lactic acid fermentation is**
(A) Mn^{+2} (B) Zn^{+2}
(C) Mg^{+2} (D) Cu^{+2}

25. **The number of pairs of electrons removed from each glucose molecule is**
(A) 6 (B) 10
(C) 12 (D) 18

26. **The enzyme, pyruvate dehydrogenase does not contain**
(A) NAD
(B) Thiamine pyrophosphate
(C) Lipoic acid (D) FAD

27. **The number of substrate level phosphorylation reactions in glycolysis, is**
(A) One (B) Two
(c) Three (d) Four

28. **The enzyme released by yeast cells during fermentation is**
(A) Oxidase (B) Zymase
(c) Invertase (d) Lipase

29. **The highest rate of respiration occurs in**
(A) Root tip
(B) Growing shoot tip
(C) Germinating seed
(D) Leaf bud

30. **In glycolysis, electrons are removed by**
(A) ATP (B) NAD
(c) FAD (d) Molecular oxygen

31. **In cytochromes, the electrons are picked up and released by**
(A) Copper (B) Iron
(C) Zinc (D) Molybdenum

32. **In mitochondria, the proton gradient required for ATP synthesis develops across the**
(A) Outer membrane (B) Inner membrane
(c) C-side (d) M-side

33. **Chemiosmotic theory of ATP synthesis in mitochondria is based on**
(A) Electron gradient
(b) Accumulation of Na^+ ions
(c) Accumulation of K^+ ions
(d) Proton gradient

34. **The number of ATP molecules formed from the complete oxidation of fructose-1,6 diphosphate is**
(A) 30 (B) 34
(C) 38 (D) 40

35. **Excess of ATP inhibits**
(A) Pyruvate dehydrogenase
(B) Phosphofructokinase
(C) Isomerase (D) Enolase

36. **The formation of acetyl Co. A from pyruvic acid occurs as a result of**
(A) Substrate level phosphorylation
(B) Oxidative decarboxylation
(C) Dehydrogenation
(D) Isomerisation

37. Removal of hydrogen and CO_2 from a substrate is called

(A) Dehydrogenation
(B) Oxidative decarboxylation
(C) Dehydration
(D) Oxidation

38. In the Krebs cycle, malic acid hands over hydrogen to

(A) FMN (B) NAD^+
(C) FAD (D) TPP

39. ATP equivalents produced during the conversion of succinic acid to fumaric acid, are

(A) 2 (B) 3
(C) 4 (D) 6

40. The universal hydrogen acceptor is

(A) ATP (B) CoA
(c) TPP (d) NAD

41. GTP is formed during the conversion of

(A) Pyruvic caid to Acetyl Co. A
(B) Succinyl CoA to Succinic acid
(C) Succinic acid to Fumaric acid
(D) Fumaric acid to Malic acd

42. Krebs cycle starts with the formation of six carbon compounds by a reaction between

(A) Malic acid and Oxaloacetic acid
(B) Oxaloacetic acid and Acetyl Co. A
(C) Malic acid and Acetyl Co. A
(D) Puruvic acid and Oxaloacetic acid

43. The electron acceptors in ETS are arranged according to

(A) Decreasing positive potential
(B) Increasing positive potential
(C) Increasing negative potential
(D) Decreasing pressure potential

44. In an electron transport system, each electron carrier in the series holds an electron at

(A) Lower energy than the previous carrier
(B) Higher energy than the previous carrier
(C) Same energy level as the adjacent one
(D) None of the above

45. ATP is injected in cyanide poisoning because

(A) It is necessary for the Na^+-K^+ pump
(B) It breaks down cyanide
(C) It is necessary for cellular function
(D) It stops electron movement

46. When a molecule of pyruvic acid is subjected to anaerobic oxidation and produces lactic acid, there is

(A) Gain of 2 ATP
(B) Gain of 3 ATP
(C) Loss of 2 ATP
(D) Loss of 3 ATP

47. Fruits and vegetables are kept in refrigerator

(A) To reduce the rate of respiration
(B) To increase the rate of respiration
(C) To reduce the rate of transpiration
(D) To promote fermentation

48. CO_2 is not liberated in

(A) Alcoholic fermentation
(B) Lactic acid fermentation
(C) Aerobic respiration in plants
(D) Aerobic respiration in animals

49. A competitive inhibitor of succinic dehydrogenase is

(A) Malonate (B) Oxaloacetate
(C) Acetate (D) Sulpha drugs

50. Which one of the following cells does not respire?

(A) RBC
(B) Sieve tube cell
(C) Cork cell
(D) Epidermal cell

51. $FADH_2$ produces a lesser amount of ATP and enters into ETS at

(A) FMN (B) Ubiquinone
(C) Cytochrome-b (D) Fe-S

52. Cytochrome-a_3 contains

(A) Copper and Manganese
(B) Iron and Copper
(C) Iron and Manganese
(D) Magnesium and Zinc

53. During alcoholic fermentation of glucose, the number of ATP produced is

(A) Zero (B) 2
(C) 4 (D) 6

54. The first four carbon compounds taking part in the Krebs cycle is

(A) Oxaloacetic acid
(B) Fumaric acid
(C) Succinic acid
(D) Malic acid

55. Glutamic acid, an amino acid is produced from

(A) Oxaloacetic acid
(B) α-ketoglutaric acid
(C) Succinic acid
(D) Citric acid

56. The chemical nature of cytochrome is

(A) Metallo flavoprotein
(B) Iron-containing porphyrin pigment
(C) Glycoprotein
(D) Glycolipid

57. Cyanide-resistant respiration occurs in

(A) Bacteria (B) Viruses
(C) Plants (D) Animals

58. An aquatic plant was placed in a test tube containing distilled water and the tube was stopped. The tube was left outdoors for 24 hours, and pH value of the water was measured at regular intervals. What is the result?

(A) pH was lowest just before sunrise
(B) pH was highest just before sunrise
(C) pH was lowest at noon.
(D) pH was lowest just before sunset.

59. Cut fruits become brown due to

(A) Oxidation of tannins and phenolics by enzymes
(B) Oxidation of sugar by enzymes
(C) Dehydration of organic compound
(D) Fermentation

60. The net gain of energy from one molecule of sucrose in aerobic respiration is

(A) 8 ATP (B) 38 ATP
(C) 40 ATP (D) 76 ATP

61. The proton channel of oxysome is

(A) F_0 particle (B) F_1 particle
(C) F_5 particle (D) F_6 particle

62. The ETS in prokaryotes is located at

(A) Nucleoid (B) Ribosome
(C) Plasmid (D) Plasma membrane

63. Cyanide is an inhibitor to the electron transport chain when it prevents the transfer of electrons to

(A) FMN (B) Ubiquinone
(C) Cytochrome-a (D) Oxygen

64. The number of protons required to produce one molecule of ATP is

(A) Two (B) Three
(C) Four (D) Six

65. Oxaloacetic acid, an intermediate of the Krebs cycle, gives rise to

(A) Glutamic acid (B) Glutamine
(C) Lysine (D) Aspartic acid

66. The first dicarboxylic acid in the Krebs cycle is

(A) Citric acid
(B) Alpha ketoglutaric acid
(C) Succinic acid
(D) Oxaloacetic acid

67. Lactic acid fermentation requires a coenzyme known as

(A) FMN (B) FAD
(C) NAD (D) TPP

68. The electron transport system occurs in

(A) Cytoplasm
(B) Inner membrane of mitochondria
(C) Inter-cristal space of mitochondria
(D) Matrix of mitochondria

69. The efficiency of aerobic respiration is

(A) 25% (B) 45%
(C) 50% (D) 55%

70. The respiratory quotient in succulent plants is zero because

(A) Glucose is not completely oxidised into CO_2
(B) Glucose is incompletely oxidised into malic acid
(C) CO_2 is not released
(D) All of the above

71. The stoichiometry of anaerobic and aerobic degradation of glucose by yeast are as follows:

$$C_6H_{12}O_6 \longrightarrow 2C_2H_5OH + 2CO_2$$
$$C_6H_{12}O_6 + 6O_2 \longrightarrow 6CO_2 + 6H_2O$$

In an experiment, the complete utilisation of 0.5 mol. of glucose, partly under anaerobic and partly under aerobic conditions, yielded 1.8 moles of CO_2. What will be the Respiratory quotient for the overall combined process?

(A) 0.5 (B) 1
(C) 1.5 (D) 2

72. If temperature is increased above 35°C, then what happens?

(A) The rate of decline of respiration will be earlier than the decline of photosynthesis
(B) The rate of decline of photosynthesis will be earlier than the decline of respiration
(C) Both declines simultaneously
(D) Both runs usually

73. If a leaf is kept in a sugar solution, then the rate of respiration will

(A) Increase
(B) Decrease
(C) First decrease then increase
(D) Remains unchanged

74. In glycolysis, during oxidation electrons are removed by

(A) ATP (B) 3-PGA
(C) NAD+ (D) Molecular oxygen

75. In the absence of oxygen, cells capable of fermentation

(A) Accumulate glucose
(B) Oxidise NADH+ to NAD+
(C) Accumulate pyruvate
(D) Do not produce ATP

76. Fermentation is performed by

(A) All microbes
(B) All fungi and microbes
(C) Some fungi and some microbes
(D) Yeast only

77. Which of the following steps is associated with substrate-level phosphorylation?

(A) 1,3 diPGA→3PGA
(B) PEP→Pyruvic acid
(C) Succinyl CoA→Succinic acid
(D) All of the above

78. What percentage of ATP is formed by substrate-level phosphorylation in glycolysis?

(A) 25% (B) 100%
(C) 10% (D) 50%

79. The inner membrane of mitochondria is permeable to

(A) ATP (B) Glucose
(C) Fructose (D) Sucrose

80. How many ATPs are produced by the complete oxidation of two molecules of isocitrate by ETS only?

(A) 22 (B) 24
(C) 15 (D) 30

81. Which of the following is the first compound which is common for both glucose and fructose in glycolysis?

(A) Glucose-6-phosphate
(B) Fructose-6-phosphate
(C) Fructose-1,6-diphosphate
(D) Fructose-1,6-phosphate

82. The enzyme which occurs in all eukaryotes and shows great homology is

(A) Cytochrome b (B) Cytochrome c
(C) Cytochrome a (D) Cytochrome a_3

83. Complex III receives electrons directly from

(A) Ubiquinone (B) Cytochrome c
(C) $NADH^+$ (D) $FADH_2$

84. ATP synthase complexes are found in

(A) Outer membrane of mitochondria
(B) Inner membrane of mitochondria
(C) Attached to the cristae
(D) Floating in the matrix

85. Which of the following is most conserved in all eukaryotes and acts as respiratory pigment and accepts electrons?

(A) Ferredoxin
(B) Cytochrome c
(C) Cytochrome c oxidase
(D) NAD dehydrogenase

86. In aerobic respiration, the release of redox equivalents as H atom for the first time in the mitochondrial matrix is catalysed by

(A) PGAL dehydrogenase
(B) Isocitrate dehydrogenase
(C) Pyruvate dehydrogenase
(D) Alpha-ketoglutarate dehydrogenase

87. Energy metabolism may have evolved when early prokaryotes

(A) Used up the oxygen in their environment
(B) Began producing poisonous waste products
(C) Used up the ATP in their environment
(D) Began producing a significant amount of oxygen

88. If the organic substrate is malic acid, then RQ will be

(A) Zero (B) 1
(C) Less than 1 (D) Greater than 1

89. Flavin mononucleotide (FMN) requires

(A) Mg (B) Zn
(C) Mn (D) Fe

90. The net gain of ATP in glycolysis is

(A) 2 (B) 4
(C) 6 (D) 8

91. Succinate dehydrogenase enzyme occurs in

(A) Outer membrane of mitochondria
(B) Inner membrane of mitochondria

(C) Mitochondrial matrix
(D) Perichondrial space

92. The respiratory substrate in protoplasmic respiration is

(A) Carbohydrate
(B) Fats
(C) Both carbohydrates and fats
(D) Protein

93. How many isomerisation reactions occur in glycolysis?

(A) One (B) Two
(C) Three (D) Four

94. Which of the following acids accumulates during muscular fatigue?

(A) Pyruvic acid
(B) Oxaloacetic acid
(C) Phosphoglyceric acid
(D) Lactic acid

95. How many cofactors are required during pyruvate oxidation?

(A) Two (B) Three
(C) Four (D) Five

96. Which of the following cofactors reacts first with pyruvic acid during pyruvate oxidation?

(A) Coenzyme A
(B) Lipoic acid
(C) NAD
(D) Thiamine pyrophosphate

97. The first 4-carbon compound formed in the Krebs cycle was

(A) Citric acid
(B) Cis-aconitic acid
(C) Alpha ketoglutaric acid
(D) Succinyl coenzyme A

98. Which of the following electron carriers is a mobile link between complex III and complex IV?

(A) Cytochrome-b (B) Cytochrome-c
(C) Ubiquinone (D) Cytochrome-a_3

99. In aerobic respiration, the action of 2,4-dinitrophenol is

(A) Uncoupler of oxidative phosphorylation\
(B) Inhibitor of Electron transport system
(C) Inhibitor of Oxidative phosphorylation
(D) It does not act

100. The respiratory quotient in germinating castor seed is

(A) One (B) Less than one
(C) More than one (D) Zero

101. The presence of ATP in an adequate amount

(A) Accelerates the rate of glycolysis
(B) Reduces the rate of glycolysis
(C) Turns off the glycolysis
(D) Initiates the process of glycolysis

102. Most of the protons useful in oxidative phosphorylation enter the mitochondria as

(A) Glucose (B) CO_2
(C) O_2 (D) Pyruvic acid

103. The energy released from electrons during electron transport is used to

(A) Synthesise ATP
(B) Start the biochemical reaction
(C) Move protons out of the matrix
(D) Move electrons out of the matrix

104. The multienzyme complex needed for the conversion of pyruvic acid into acetyl CoA is larger than

(A) Small subunit of the ribosome
(B) Larger subunit of the ribosome
(C) Entire ribosome
(D) None of the above

105. The post-glycolysis phase of lactic acid fermentation aims at

(A) Substrate level phosphorylation
(B) NADH mediated phosphorylation
(C) Freeing of NAD^+
(D) Freeing of ATP

ANSWERS

1.a	11.a	21.d	31.a	41.b	51.b	61.a	71.c	81.b	91.b	101. b
2.c	12.b	22.c	32.b	42.b	52.b	62.d	72.	82.d	92.d	102.d
3.b	13.b	23.b	33.d	43.b	53.b	63.d	73.a	83.a	93.c	103.c
4.d	14.b	24.b	34.d	44.a	54.a	64.a	74.a	84.c	94.d	104.c
5.c	15.c	25.c	35.b	45.c	55.b	65.d	75.b	85.b	95.d	105.c
6.b	16.b	26.d	36.b	46.d	56.b	66.b	76.c	86.c	96.d	
7.a	17.c	27.d	37.b	47.a	57.c	67.a	77.d	87.c	97.d	
8.b	18.b	28.b	38.b	48.b	58.a	68.b	78.b	88.d	98.b	
9.b	19.b	29.c	39.c	49.a	59.a	69.b	79.a	89.b	99.a	
10.a	20.d	30.b	40.d	50.c	60.d	70.d	80.b	90.a	100.b	

CHAPTER-10

PLANT GROWTH AND DEVELOPMENT

Development: It is a multiphasic sequence of qualitative structural change in the structure and functioning of a cell, tissue, organ or organism towards a higher or more complex state or it is the conversion of one phase to the next. In plants, it starts from seed germination and ends with senescence. It follows as such:

Cell division, Plasmatic growth, Elongation, Differentiation, Maturation

Meristematic cell ——————————→ Mature cell

The development of plants involves the following stages:

(a) Juvenile stage: It starts with the seedling phase and ends in the vegetative stage.

(b) Reproductive stage: It shows the budding phase.

(c) Ripening stage: It shows the flowering phase.

(d) Senescent stage: It is the terminal phase where death occurs.

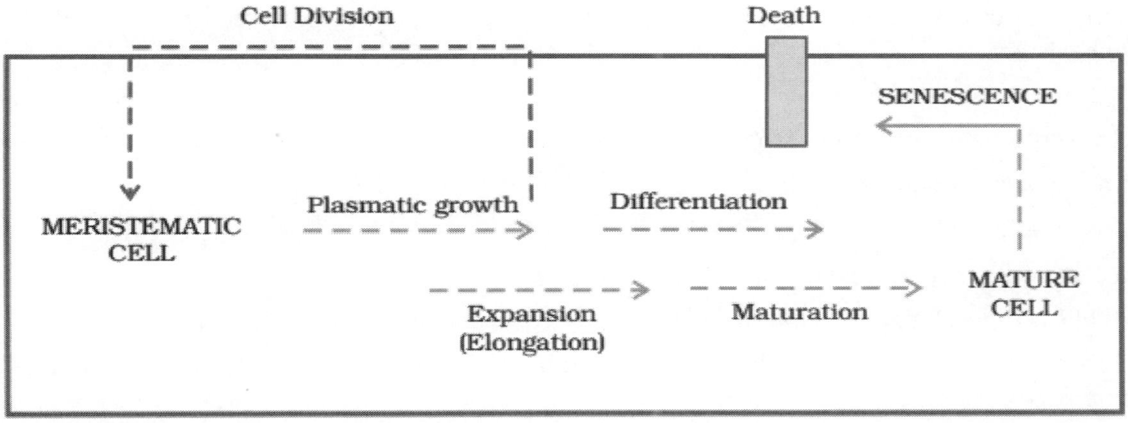

Seed dormancy: It is the condition in which seeds are unable to germinate even after getting all the necessary conditions. During this period, growth of seeds remains suspended which occurs due to internal factors and some specific environmental factors.

Causes of dormancy:

(a) Hard seed coat impermeable to gases: e.g., Apple

(b) Hard seed coat impermeable to chemicals: e.g., *Xanthium*

(c) Hard seed coat impermeable to water: e.g., Legumes, *Chenopodium, Trigonella*

(d) Mechanically resistant seed (embryo is unable to grow): e.g., *Alisma, Amaranthus, Capsella,* Mustard.

(e) Immaturity of embryo at the time of shedding: e.g., *Eranthis, Ginkgo, Ranunculus*

(f) Absence of light (Negatively photoblastic): e.g., Lettuce, Onion, Tobacco, Tomato, *Phlox* etc.

(g) Absence of after-ripening period (seeds undergo a period of rest even when the condition is favourable): e.g., Iris, Peach, Poplar, Tulip etc.

(h) Presence of excessive salts at seed shedding: e.g., *Atriplex*

(i) Presence of inhibitors (ABA, NH_3, Coumarin, Phenolics) in seeds: e.g., Cucurbits, *Xanthium, Iris*

Methods to break seed dormancy (or to start seed germination):

1. Scarification: It is a method of softening and weakening of seed coat by hot water, acids, and alcohol (chemical method) and by thrashing with a machine, rupturing with a hammer and cutting from scalpel (mechanical method). It is done in *Cassia, Melilotus*
2. Stratification: It is the treatment of seed in well-aerated and moist conditions at very low temperatures (0-5°C) just after a freezing point.
3. Alternating high and low temperature: This is done in the seed of an immature embryo and the difference of temperature should be 10°C to 20°C.
4. Exposure to light: Red light (<730 nm] breaks dormancy (seed germination) and Far red light (>730 nm) maintains dormancy (inhibits seed germination).
5. Soaking seeds in chemicals (KNO_3, H_2SO_4, thiourea, gibberellin, cytokinin, ethylene).
6. Destruction of Inhibitors:

Importance:

(a) Seeds can be stored during unfavourable periods (drought, cold, high temperature)

(b) Seeds can remain alive in soil for several years.

(c) Seed dormancy is essential for seed disposal.

SEED GERMINATION

It is the sprouting of a seed and the growth of an embryo present inside the seed into a seedling capable of independent existence.

Process:

(a) Seed absorbs water through micropyle by imbibition and osmosis.

(b) Hydration of cell organelles occurs

(c) Embryo is activated.

(d) Seed coat swells, softens and ruptures by cell elongation.

(e) Metabolic activity stars and respiratory enzymes become active.

(f) Seed respires initially anaerobically and later aerobically when O_2 enters into seed coat and releases heat energy.

(g) Seed gets requisite energy for the growth of embryo.

(h) Embryo secretes hydrolytic enzymes such as amylase, protease, lipase which convert insoluble reserve food (starch, protein, oil) into soluble substances.

(i) Embryo utilises these soluble substances

(j) Cell division starts in radicle and plumule.

(k) Radicle firstly comes out of seed coat and forms tap root.

(l) Plumule comes out rupturing seed coat and forms shoot and

(m) Finally seedling is formed.

Types of seed germination

Epigeal germination	Hypogeal germination
Cotyledon comes out of the soil surface by greater elongation of the hypocotyl (part of the axis below cotyledon). Cotyledon is flat, green and leaf-like. e.g., Bean, Castor, Cotton, Cucurbits, Mustard, Papaya, Poppy, Tamarind, Mustard, Onion	Cotyledon remains inside the soil or just on the surface but plumule comes out of the soil by greater elongation of epicotyl. In this, hypocotyl does not grow much. e.g., Cereals, Gram, Coconut, Date palm, Fan palm, Groundnut, Mango, Jack fruit, Pea.

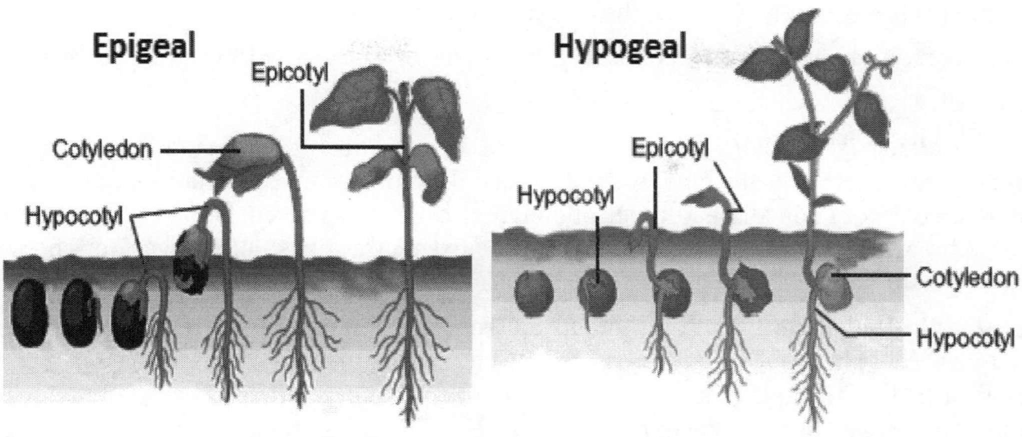

Factors:

(a) Water: It causes softening of seed coat due to seed coat.
(b) Light: Red light is stimulatory but far-red light and blue light are inhibitory to seed germination.
(c) Temperature: The required optimum temperature is 25-30°C.
(d) Oxygen: It is essential for seed germination except Rice and *Typha*.
(e) Availability of food:
(f) Viability: It is the ability to germinate. It is maximum in Lotus (*Nelumbo*).
(g) Inhibitors: These chemicals such as phenolics, tannins, alkaloids, lactone, mustard oil, scopoletin etc. inhibit seed germination.

Vivipary: It is the germination of seed while it is still attached to the parent plant or before detachment from fruit. This seed does not undergo dormancy and thus forms seedling. Due to its increasing weight, the seedling separates from the parent plant and falls into mud or soil hence the plant is established. It is a halophytic adaptation and is found in *Rhizophora* and other mangrove plants. It is not desirable in crop plants because seeds cannot be stored in this condition.

GROWTH

It is a permanent and irreversible increase in the number, length, area (size), and volume (mass) of an individual cell, organ and its parts. In this, anabolic processes dominate over the catabolic process. Growth is a fundamental property of living organisms but it is not a defining feature. In plants, limited growth occurs in leaves, flowers and fruits but unlimited growth occurs in roots and stems. In vascular plants, growth begins from a single-celled zygote which divides to form an embryo. Once the embryo germinates from its seed then it produces vegetative organs (root, stem and leaf). In plants, growth may be primary (increase in length) and secondary (increase in girth or width) due to lateral meristems (vascular cambium and cork cambium).

Parameters of Growth:

(a) Increase in Number: Bacteria, yeast and algae increase their number.
(b) Increase in Length: Growth of pollen tube
(c) Increase in Size: Watermelon
(d) Increase in the Surface area: Leaf
(e) Increase in Volume: Storage organs, Fruits.
(f) Increase in Weight (fresh or dry): Storage organs, Fruits.

Measurement of Growth:

(a) Direct method: It measures dry weight, volume and number of cells.
(b) Horizontal or Traveller's microscope: It measures growth on length.

(c) Auxanometer: It measures growth in the length of a shoot.

(d) Crescograph: It was developed by J.C. Bose. It measures primary growth very accurately in seconds.

Phases of growth: Growth involves three phases:

1. **Phase of Cell division** (Formative phase): It occurs in the root apex and shoot apex. Cells are small, isodiametric and metabolically highly active. Mitotic division occurs continuously in the apical meristem (root apex and shoot apex) hence the number of daughter cells is increased. Some of the daughter cells retain the meristematic activity, while others enter into the next phase- phase of elongation. The resulting cells have dense cytoplasm, small vacuole and a distinct centrally placed nucleus.
2. **Phase of cell elongation**: It occurs behind the root and shoot apex. It may be linear or in all directions (isodiametric cells) hence there is an irreversible increase in the size of cells due to an increase in the volume of cytoplasm. Due to the high metabolic rate, the solute concentration will be high in their cell sap hence water potential is decreased by which water enters into cells. The elongation occurs by the synthesis of protoplasm, absorption of water (hydration), photosynthesis and adding a new cell wall to make it permanent. The small vacuoles unite to form large centrally placed vacuoles. Due to endosmosis, the size of the vacuole increases, and it exerts turgor pressure on the cell wall. The cell wall being elastic expands and the cell gets elongated.
3. **Phase of Cell differentiation**: It occurs further away from the apex. Cells become mature and specialised and attain their final size due to changes in form, cell wall thickening, protoplasmic modifications by permanent changes in structure (shape and size) and physiological activities. These cells form tissues in the form of epidermis, cortex, xylem and phloem in both root and stem; and mesophyll cells in leaves. After differentiation, mature cells do not grow further and remain unchanged until death. The differentiated cells form different kinds of simple and complex tissues which form different functions.

Rates of Growth:

1. **Arithmetic growth**: It occurs at a constant rate from the very beginning and progresses arithmetically (2...3....4...6...). In this one cell divides forming two daughter cells as a result of mitotic divisions and only one daughter cell divides continuously while the other cell undergoes differentiation where the cell becomes mature, permanent and loses its capacity to divide.

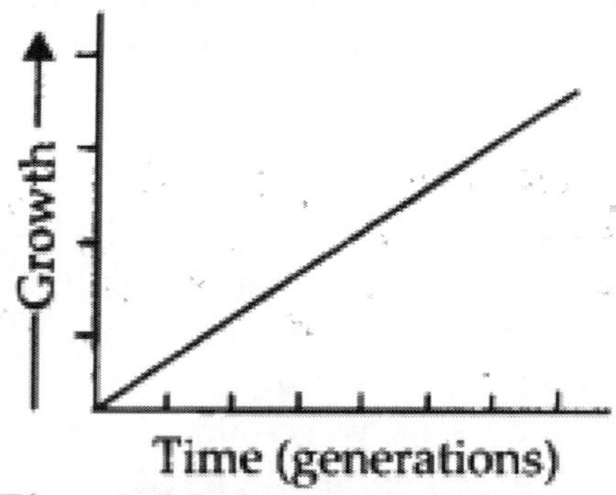

Fig.: Arithmetic growth curve

It occurs in the root and stem elongating at a constant rate. It shows a linear curve. Mathematically, it is expressed as

$$Lt = Lo + rt$$

Lt= Length (growth) of an organ after time 't'

Lo=Length (growth) of organ at the beginning

rt= Rate of growth or elongation per unit time.

2. **Geometric growth**: It occurs when every cell divides and all daughter cells grow and divide again in a geometrical manner (2....4...8...16....32....).

It occurs in many higher plants and unicellular organisms when grown in a nutrient-rich medium. It involves as such

(a) Lag phase: It is the initial phase in which slow or little growth occurs because the organism is in the process of mobilising resources.

(b) Log phase: It is the phase in which active or rapid growth occurs exponentially and reaches maximum because there is a considerable increase in the volume of cell sap along with cell wall materials.

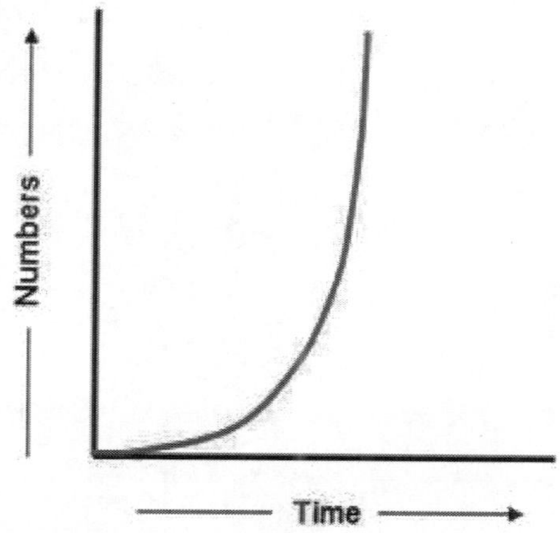

Mathematically, it is expressed as

$$\mathbf{W_1 = W_0 + e^{rt}}$$

W_1= Final size (weight), Wo= Initial size at beginning of growth

e= Base of natural log; rt= Rate of growth

3. **Sigmoid growth:** It shows when geometric growth cannot be sustained for long because some cells die due to limited nutrient availability by which growth slows down leading to a stationary phase. It is a characteristic feature of living organisms growing in their natural environment. It involves:

   (a) **Lag phase:** It corresponds to the phase of cell division. It is the initial phase in which slow or little growth occurs because an organism is in the process of mobilising resources.

   (b) **Log phase:** It corresponds to the phase of cell elongation. It is a phase in which active or rapid growth occurs exponentially and reaches maximum because there is a considerable increase in volume of cell sap along with cell wall materials.

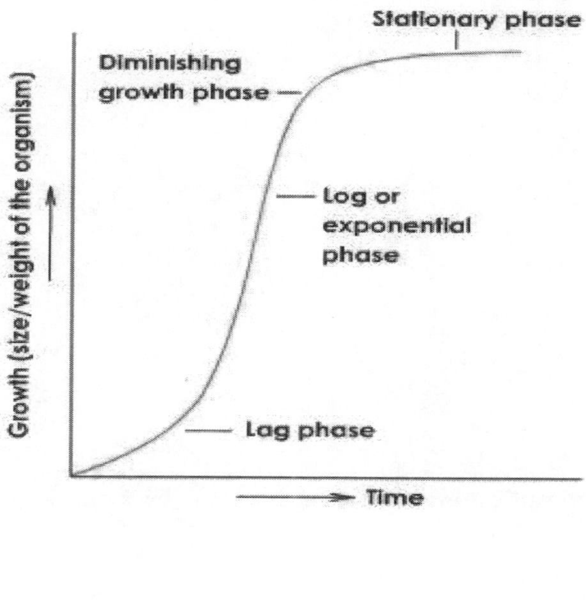

(c) **Stationary phase**: It corresponds to the phase of differentiation. It starts with the inflexion point where growth starts declining. It is a phase in which the size of body remains fairly constant because growth has stopped due to shortage of food, space and accumulation of toxins.

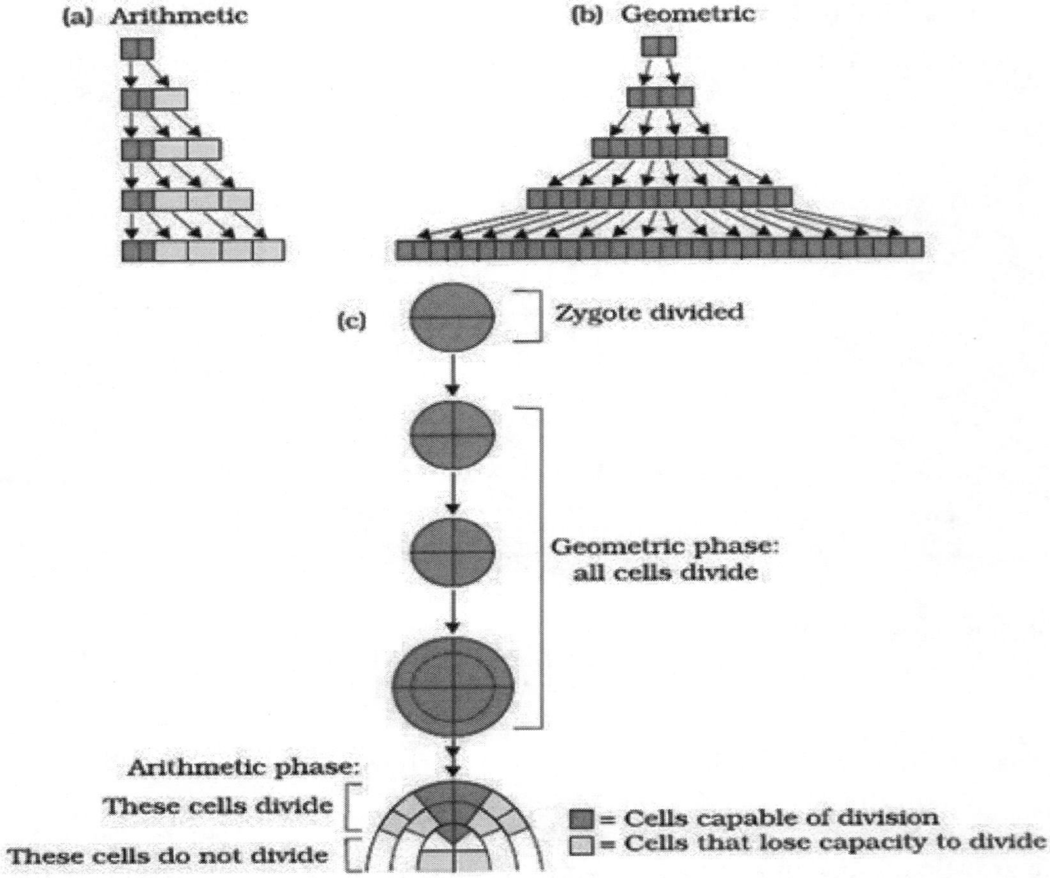

4. **Absolute growth**: It is actual or total growth per unit time.
5. **Relative growth:** It is a growth of given system per unit time. It is expressed as:
   (a) If the initial size of leaf A is 5 cm^2 and leaf grows up to 10 cm^2
   (b) If the initial size of leaf B is 50cm^2 and leaf grows up to 55 cm^2.

   Thus, in both cases, absolute growth rate is same (5cm^2) but relative growth rate in leaf A is 100% (faster / higher) and in leaf B it is 10% (slower / lower).

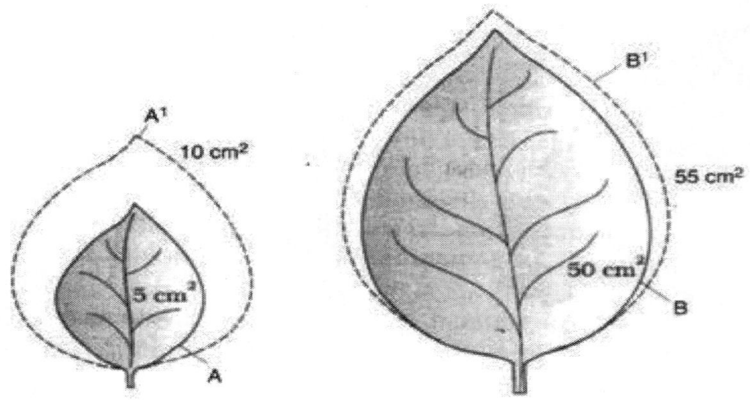

Diagrammatic comparison of absolute and relative growth rates. Both leaves A and B have increased their area by 5 cm^2 in a given time to produce A^1, B^1 leaves

Factors affecting growth:

(a) Light: Growth will be maximum in the presence of red light because it is must for photosynthesis. If light is absent then it will cause etiolation.

(b) Temperature: The optimum temperature is 20-30^{0}C. It varies from species to species. Low temperature inhibits growth due to inactivation of enzymes and high temperature also inhibits growth due to denaturation of enzymes and excess transpiration.

(c) Water: Its availability favours growth because it is essential for the activity of enzymes.

(d) Phytohormones: Growth is increased in the presence of auxin and gibberellins,

(e) Carbohydrate / Protein (C/N) ratio: The presence of more carbohydrates compared to nitrogenous compounds favours good growth.

(f) Anabolism/catabolism (A/C) ratio: Growth is greater if the A/C ratio is more.

(g) Soil nutrients: Meganutrients are essential raw materials for growth and micronutrients are essential for the activity of different enzymes responsible for growth.

(h) Oxygen: It must be present for the growth of higher plants.

Process of the growth: On the basis of the ability of cells to divide

1. Undifferentiation: It is a process in which cells usually possess the power of division hence these are unspecialised. e.g., root and shoot apex (Apical meristem), cambium, zygote., Stem cell (animal)
2. Differentiation: It is a process in which cells do not undergo division hence these become specialised and undergo maturation and qualitative permanent change with respect to structure and function. e.g., Root and shoot apex (Apical meristem) form parenchyma, collenchyma, sclerenchyma, aerenchyma, chlorenchyma, xylem, phloem, stomata, mesophyll cell.
3. Dedifferentiation): It is a process in which the cells of permanent or specialised tissues lose their specialisation, regain the power of division and form new cells. e.g., Interfascicular cambium, cork cambium.
4. Redifferentiated cells (Redifferentiation): It is a process in which the dedifferentiated cells again lose their ability to divide and become specialised and permanent cells. e.g., Secondary xylem, Secondary phloem, Phellem, phelloderm, and Callus in suspension culture.

Plasticity: It is the ability of plants to follow different pathways in response to the environment or to develop different structures. Plant growth is dependent on the environment. e.g., Heterophylly (presence of two types of leaf on the same plant). It is exhibited in the following ways:

(a) The leaves of juvenile plants of cotton, coriander, larkspur (*Delphinium*), mustard, Pea are different in shape than those of mature plants.

(b) The aerial and floating leaves of buttercup (*Ranunculus*) are generally broad and unlobed while submerged leaves are ribbon-shaped, linear and much dissected

PHYTOHORMONES:

The term 'phytohormone' was coined by Thimann (1948). These are organic compounds and are capable of influencing various physiological activities in plants. These are required in small quantities at very low concentrations. When two hormones act together and perform the same action then it is said to be a synergistic action of hormones.

AUXIN

It was the first plant hormone to be discovered.

Discovery: Charles Darwin (1880) reported the existence of auxin while demonstrating of bending of coleoptiles in canary grass (*Phalaris canariensis*) towards light. He observed

(i) If the tip was exposed towards light and then coleoptile, the tip observed bending towards light

(ii) If the coleoptile tip was covered with an opaque tin foil cap, no bending was observed.

(iii) If coleoptile was removed then bending was not noticed because the stimulus was not perceived.

Finally, it was proposed that stimulus was transmitted from tip to subapical or subterminal position where it caused differential growth resulting in bending.

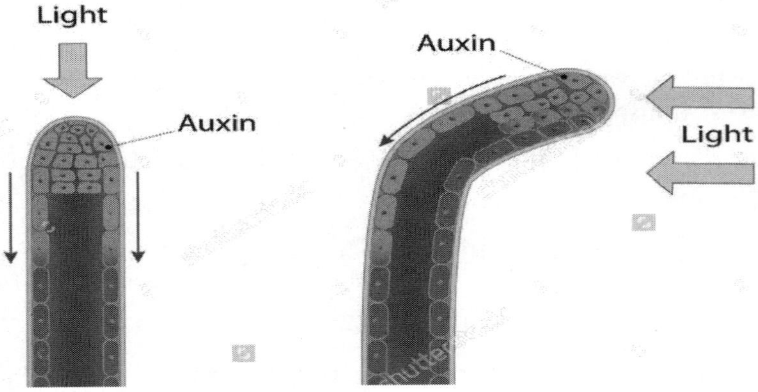

Frits Went (1928) discovered auxin while studying '***Avena* curvature test**'. He placed freshly cut coleoptile tips of oats on small0 blocks of agar for some time. The agar blocks (A and B) were then placed asymmetrically on the coleoptile stumps for 2 hours in the dark. He noticed that coleoptiles containing tips have a certain substance that could diffuse down (basipetal) into agar blocks. He termed this substance 'auxin'. He also found that the degree of curvature of coleoptiles is proportional to the amount of active substance present in agar blocks. This could be used to estimate the amount of auxin in any tissue. This test came to be known as *Avena* curvature test.

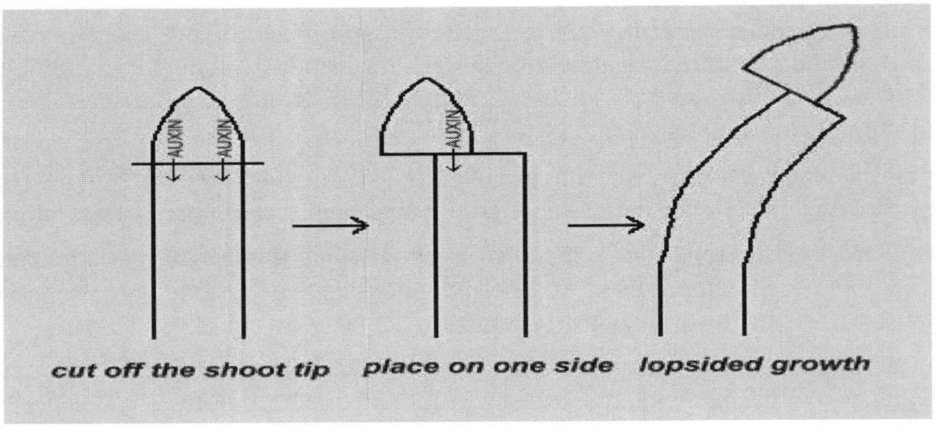

Isolation:

(i) Kogl and Hagensmit (1931) isolated auxin-a (Auxentriolic acid=$C_{18}H_{32}O_5$) from human urine of pellagra patient.

(ii) They isolated another substance and called it auxin-b (Auxenlinic acid=$C_{18}H_{30}O_4$) from corn germ oil.

(iii) They isolated another auxin and called it Auxin-c or heteroauxin (indole-3-Acetic acid=$C_{10}H_9O_2N$) from human urine.

Biosynthesis: Tryptophan amino acid is precursor of auxin. It is synthesised in the chloroplast as well as cytosol. Zn^{+2} is required for synthesis of Indole Acetic Acid (IAA).

Tryptophan

↓decarboxylase

Tyramine

↓aminooxidase

Indole-3-acetaldehyde

↓dehydrogenase

Indole Acetic Acid (IAA)

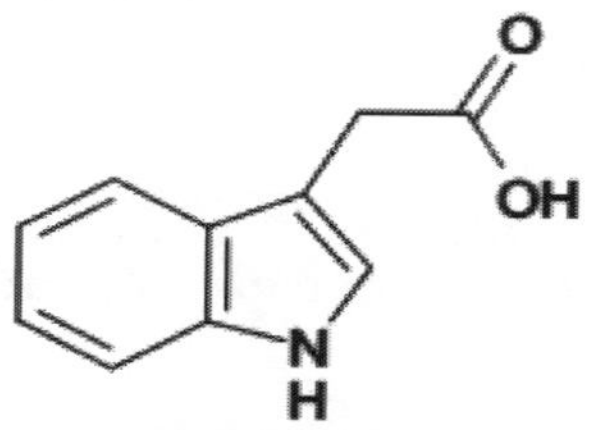

Indole-3-Acetic Acid
(an Auxin)

Nature: Auxins are weak acids and water soluble.

Structure: It shows an unsaturated planar aromatic ring structure and a carboxyl side chain. More than 100 types of auxins have been reported so far. These are classified on the basis of function only.

Distribution: Auxins are widely found throughout the plant body. These are found in the shoot apex in highest amount=10 ppm), root apex, coleoptile tip, and bud. These are found throughout day and night. These are found more on the darker side than the illuminated side.

Transport: Auxins show polar transport where they move from stem apex to root (basipetal) and also from root to shoot (acropetal). The basipetal transport is dominant because a higher concentration is found in the stem apex. Their speed is slow (6.4-26 mm/hour) and move against a concentration gradient.

Bioassay: *Avena* curvature test

Types: Auxins are of two types: Natural and Synthetic

Natural auxins: These are Indole acetic acid, Indole pyruvic acid, and Indole butyric acid.

Roles of Auxin or Indole Acetic acid (IAA):

1. Cell division: It stimulates cell division in the vascular cambium resulting in secondary growth.
2. Cell elongation: It stimulates the growth and elongation of the stem by causing the cell wall to elongate in more than one direction. It occurs in three ways: (i) with the increase in osmotic pressure; (ii) with a decrease in wall pressure, and (iii) with the increase in permeability of cytoplasm to water (endosmosis).
3. Cell differentiation: Low auxin induces phloem differentiation and high auxin induces xylem differentiation.
4. Phototropism: It promotes of bending of the shoot towards unilateral light. The differential growth occurs due to the lateral distribution of auxin when light stimulus falls (Cholodny-Went theory). There is a higher concentration of auxin on a shaded or darker side than on an illuminated side.
5. Geotropism: If a growing plant is placed horizontally then it stimulates the stem (shoot curves upwards) and inhibits the root (to become vertical again). Its low concentration stimulates the growth of the root.
6. Apical dominance: It promotes apical dominance (dominance of apical or terminal bud by inhibiting the growth of lateral buds). If the apical bud is cut or excised then the lateral bud will start growing. The apical dominance was common in plants having excurrent habits (palms, conifers, *Polyalthia*).
7. Abscission: It prevents organised shedding or fall of leaves, flowers and fruits by preventing the formation of the abscission layer at the base of the petiole or pedicel.
8. It prevents or delays senescence.
9. Parthenocarpy: It promotes the development of seedless fruits by stimulating the swelling of the ovary wall before fertilisation.
10. Respiration: It stimulates respiration indirectly through an increased supply of ADP by rapidly utilising ATP in expanding cells.
11. Secretion of Ethylene: Its higher concentration induces the synthesis of ethylene.
12. Sex expression: It induces femaleness in flowers.

Roles of Indole Pyruvic acid (IPA):

(a) It induces rooting in cut parts (cuttings)

(b) It induces the setting of natural fruits

Role of Indole Butyric Acid (IBA): It is natural as well as synthetic auxin.

(a) It induces rooting in cut parts (cuttings)

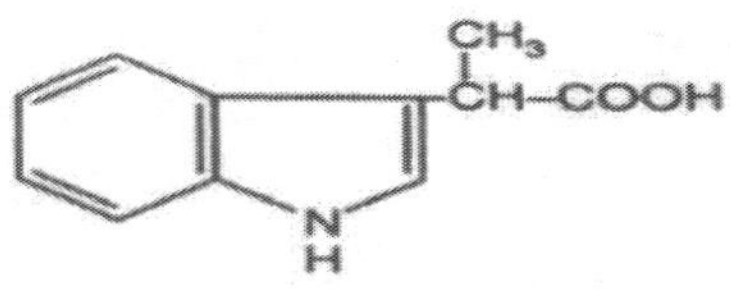

α-Indole propionic acid (IPA)

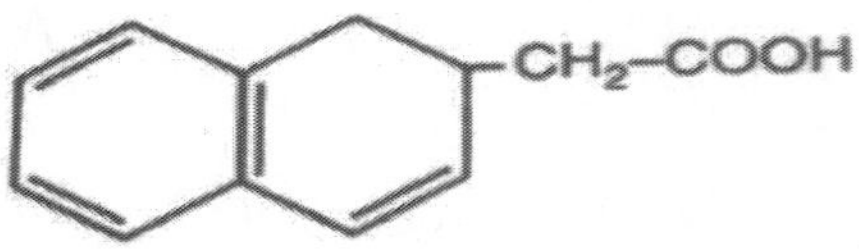

β-NAA

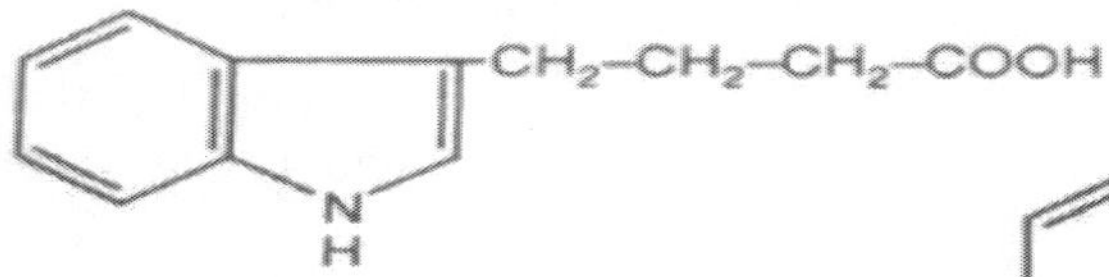

α-Indolebutyric acid (IBA)

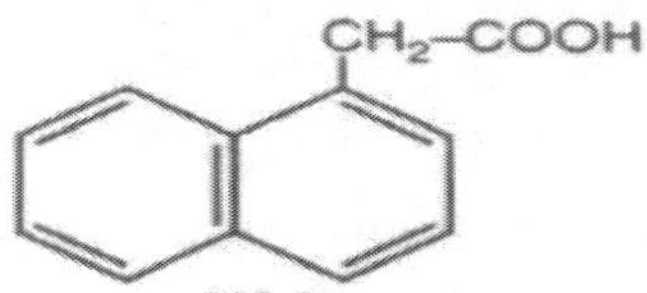

α-NAA

Synthetic auxins: These are more stable and active than natural auxins. Further these are not destroyed by enzymes present in the plants. These move in all directions.

Roles of Naphthalene acetic acid (NAA):

(a) It induces rooting in cut parts (cuttings)

(b) It induces flowering in Litchi, Pineapple

(c) It induces the setting of natural fruits

(d) It prevents lodging (falling)

(e) Its methyl ester prevents the sprouting of potatoes.

Roles of 2,4-Dichlorophenoxyacetic acid (2,4-D):

(a) It kills broad-leaved weeds (usually dicots) hence it is a powerful weedicide.

(b) It induces flowering in Litchi and pineapple.

(c) It prevents the abscission of fruits.

Roles of 2, 4, 5-Trichlorophenoxyacetic acid (2,4,5-T or Dioxin):

(a) The mixture of 2, 4-D and 2, 4, 5-T is called Agent Orange which causes the defoliation of forests (used as chemical warfare in the Vietnam war by US).

Other synthetic auxins are 2-methyl 4-chloro Phenoxy acetic acid (MCPA) and Indole-3-Propionic acid.

Synthetic antiauxins: Triiodobenzoic acid (TIBA), p-chlorophenoxy Isobutyric acid (PCIB) and Naphthylphthalamic acid (NPA).

GIBBERELLIN:

Discovery: The existence of Gibberellin was first noticed by Kurosawa (1926) while working on a rice field. He observed that some rice seedlings grew taller and pale in colour than others. He further observed that such seedlings were infected by a fungus, *Gibberella fujikuroi* (the perfect stage of an imperfect fungus, *Fusarium monoliforme*) which caused a disease called bakanae disease (foolish seedling disease) because seedlings grew foolishly so tall that they resulted into death of plants.

He observed that a sterile filtrate of fungus could stimulate unusual growth in plants and cause the symptoms of bakanae disease in normal seedlings.

Isolation:

(i) Yabuta and Hayashi (1930): They isolated the substance from the culture filtrate of *Gibberella fujikuroi* and named it 'Gibberellin'.

(ii) Yabuta and Sumiki (1938): They isolated gibberellins in crystalline form from the culture of fungus.

Biosynthesis: It occurs in plastid. It is synthesised through the isoprenoid pathway of terpene starting from acetyl CoA. It is synthesised mainly in apical tissue and young leaves.

Acetyl CoA → Acetoacetyl CoA → Mevalonic acid → Mevalonate Pyrophosphate → Isopentyl pyrophosphate → Formaacyl pyrophosphate → Transgeranyl pyrophosphate → Copalyl pyrophosphate→Kaurene → Kaurenoic acid → Hydroxykaurenoic acid → Gibberellic acid.

Nature: It is weak acid and chemically it is gibberellic acid.

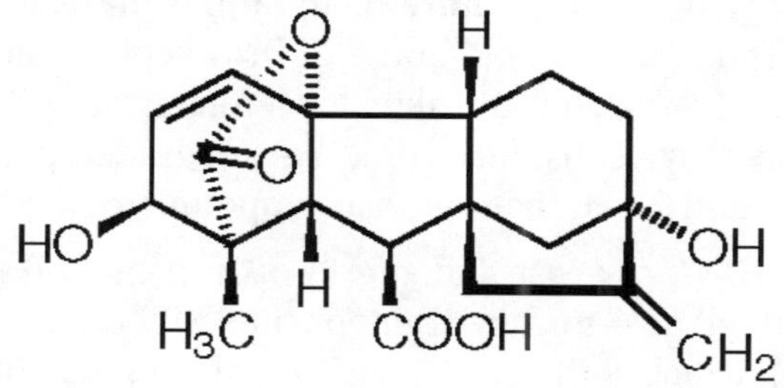

gibberellic acid

Structure: More than a hundred types of gibberellins have been reported so far. The most common and original is GA_3 which is the simplest and first to be discovered. Its chemical formula is $C_{18}H_{22}O_6$. It shows a gibbane ring and has a tetracyclic terpenoid structure (a complex of group of chemicals related to lipids) having 4 isoprene units. The individual members differ from each other in their side attachments of various groups such as –OH group and $-CH_3$ group. All gibberellins are classified on the basis of structure and functions.

Distribution: It is found everywhere in plants but more in young leaves and seeds. The reproductive parts of plants show a greater amount than the vegetative parts.

Transport: Its transport is acropetal as well as basipetal but it is not polar. It can be transported from the region of synthesis by simple diffusion, or through xylem and phloem. Gibberellins that are synthesised in the shoot are transported to the roots through the phloem. Gibberellins that are synthesised in the root are transported to the shoot through the xylem.

Bioassay: (i) Dwarf Pea test (ii) Barley endosperm test

Roles:

1. Gibberellin stimulates the elongation of the stem (internode) in genetically dwarf plants by causing cell elongation and cell division esp. in the intercalary meristem of the stem. Thus. it restores the normal size and growth of genetically dwarf plants.
2. It causes bolting (elongation of the stem into floral axis) in plants that show rosette habit (retarded growth of internode and profuse development of leaves) such as cabbage, carrot, beet etc. If the cabbage head is kept under warm night then it will return to the rosette habit. It helps in leaf expansion.
3. It induces *de novo* (new) synthesis of hydrolytic enzymes (α-amylase, protease, lipase) which are mobilised to endosperm during seed germination to break down (hydrolyse) the stored food reserve.
4. It substitutes the requirement of cold treatment (vernalisation) and induces flowering in one year bypassing the winter season in biennial plants.
5. It induces seed germination by hastening the maturity period hence it breaks seed dormancy.

6. It induces parthenocarpy in apple, and pear and also increases the size of fruit (apple) and bunch length in grapes esp. by applying Pomalin (Benzyl adenine+GA_4+GA_7).
7. It prevents or delays senescence.
8. It stimulates flowering in long-day plants and inhibits in short-day plants.
9. It stimulates fruit set in apple.
10. It stimulates sex expression by inducing maleness in flowers.
11. It induces malting process from barley in brewing industry.

CYTOKININ:

Discovery: Skoog (1954) reported a substance in coconut milk that stimulated cell division while working on Tobacco pith culture.

Isolation: Miller (1955) while working on tobacco pith culture wanted it to grow indefinitely. He added several growth substances, vitamins, nutrients etc. in the culture medium but these substances did not show any growth of tobacco. By chance, he added the contents of an old bottle of DNA kept for several years in his lab to the culture medium. He then noticed the growth of tobacco pith callus for a longer period but he did not get positive results from fresh DNA. He proposed that an active substance could be a degraded product of DNA. He then isolated this substance by autoclaving the old DNA and found that this is a compound of purine.

He obtained similar results with the DNA of yeast and sperm of Herring and named it kinetin due to its role in cell division. Later, kinetin was identified as 6-furfurylaminopurine. Kinetin is a derivative of Adenine which bears furfuryl substituent at the 9th position which shifted to 6th position of adenine during the autoclaving of DNA.

Letham (1964) isolated the first naturally occurring cytokinin from the endosperm of young or unripe maize grain (kernel) and named it 'zeatin'. Later it was identified as 'Benzyl Amino Purine (BAP)', a substituted adenine.

Nature: It is a weak base.

Biosynthesis: It is synthesised from isopentenyl adenosine monophosphate. The root tip is the usual site for the synthesis of cytokinin.

Structure: The most common cytokinins are zeatin and isopentenyl adenine ($C_{10}H_9N_5O$).

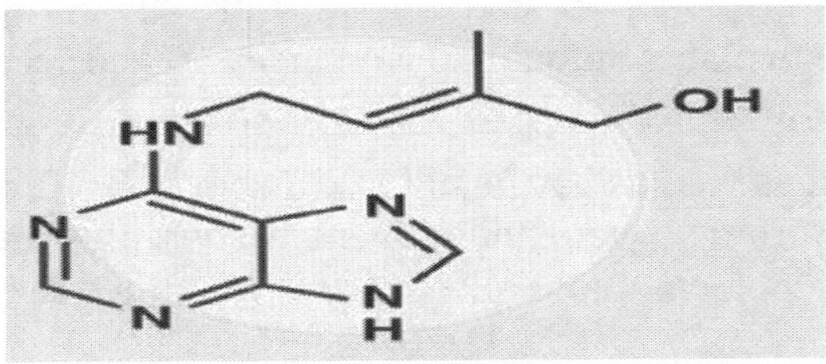

Cytokinin

Distribution: It is mostly found in meristematic and growing tissues such as embryo and endosperm of seed (more), developing roots and fruits.

Transport: It is translocated from root to shoots by xylem

Bioassay: (i) Tobacco pith culture (ii) Retardation of leaf senescence

Roles:

1. Cytokinin stimulates cell division in the presence of auxin. The ratio of cytokinin and auxin controls cell differentiation or morphogenesis. These never act alone hence no response occurs if auxin and cytokinin act alone.
   (a) If a low amount of cytokinin and a high amount of auxin are applied then the root develops.
   (b) If a high amount of cytokinin and a low amount of auxin are applied then the shoot (bud, stem and leaf) develops.
   (c) If both cytokinin and auxin are in equal amounts then the cell divides but fails to differentiate and produce a callus (undifferentiated, unorganised and irregular mass of cells).

2. It counteracts apical dominance by inducing the growth of lateral buds if cytokinin is directly applied to the axillary bud.
3. It delays the senescence of plants by promoting protein synthesis, mobilising nutrients or resources and preventing the degradation of chlorophyll and protein (Richmond-Lang effect). It promotes the conversion of the etioplast into chloroplast.
4. It increases the shelf life of vegetables by several days.
5. It induces seed germination and thus breaks seed dormancy.
6. It induces parthenocarpy.
7. It induces flowering in short-day plants.
8. It stimulates the development of chloroplast (synthesis of anthocyanin) in callus and leaves.
9. It helps in the opening of stomata.
10. It induces sex expression by inducing femaleness in flowers.
11. It promotes cell expansion in leaves and seeds.
12. It helps plants to develop resistance against pathogens.
13. It helps to produce new leaves, lateral shoots and adventitious roots.

ABSCISIC ACID (ABA):

It is a plant growth regulator (inhibitor) and antagonistic to growth promoters esp. Gibberellin.

Discovery: Addicott (1963) discovered it from old cotton fruit and named it dormin which induced dormancy when applied to bud. Wareing (1964) discovered it from young cotton fruit and named it Abscissin-II which caused abscission. Both these substances were found to be Abscisic acid (ABA).

Distribution: It is found everywhere in plants but more in fruit and seed. It is produced in some fungi, algae, bryophytes and all vascular plants.

Synthesis: It is synthesised in the chloroplast of mesophyll cells of leaves by two processes:

1. Direct synthesis: It is produced from mevalonic acid.

   Mevalonic caid→Isopentenyl pyrophosphate→Farnesyl pyrophosphate→ABA

2. Indirect synthesis: It is produced from xanthophyll (a carotenoid).

   Xanthophyll→Violaxanthin→xanthoin→ABA

Transport: It is transported through the xylem and phloem.

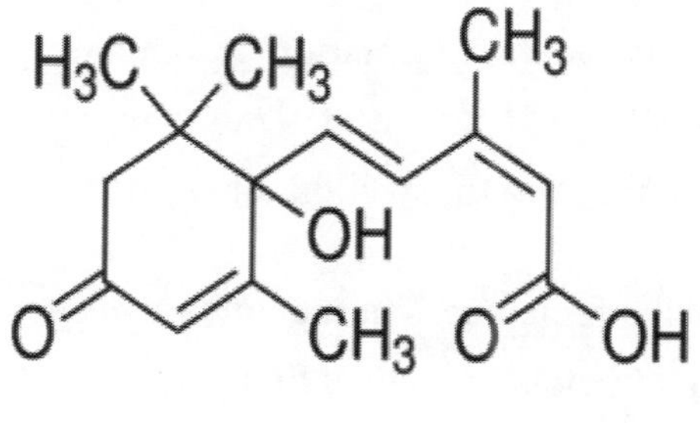

Fig: Abscisic acid

Structure: It is dextrorotatory and cis-sesquiterpene (15-carbon).

Roles:

1. It helps plants to cope with environmental stress conditions such as drought, water logging, frost (cold) and saline conditions. Thus, it is called stress hormone.
2. It causes stomatal closure by inhibiting K^+ uptake in guard cells. Thus, it is called anti-transpirant. It prevents water loss under severe drought.
3. It stimulates abscission and senescence of leaves, flowers and fruits.

4. It inhibits seed germination and induces bud and seed dormancy (to withstand desiccation and other unfavourable conditions)
5. It induces parthenocarpy.
6. It stimulates positive geotropism
7. It inhibits mitosis in vascular cambium and the synthesis of RNA.

ETHYLENE:

It is only a gaseous natural plant growth regulator (inhibitor).

Discovery: Crocker reported the release of volatile substances from ripened oranges which hastened the ripening of unripe bananas. Later it was confirmed by Cousins that the volatile substance was ethylene using the technique of Gas chromatography.

Distribution: It is found in root, stem, leaf, flower, fruit and seed of all higher plants. It is emitted from ripening fruits and wounded organs.

Synthesis: It is produced from an amino acid, methionine in all tissues. Its synthesis is induced by a high concentration of auxin but ethylene lowers the level of auxin in tissue (feedback mechanism). The shoot apex is an active site for ethylene biosynthesis.

Transport: It is a gas and hence moves by diffusion from its site of synthesis. It is not easily soluble in an aqueous medium and volatilises into intercellular space from where it is released in the outer atmosphere.

Structure: It is a simple structure. $H_2C=CH_2$

Bioassay: Triple response

Ethephon (2-chloro ethyl phosphonic acid): It releases ethylene within plants when it is sprayed in the aqueous medium.

Roles:

1. It induces the ripening of climacteric fruits such as apples, bananas, and mangoes with a rise in the rate of respiration.
2. It induces seed germination and sprouting of potatoes hence it breaks seed or bud dormancy.
3. It stimulates the senescence and abscission of leaves, flowers and fruits.
4. It stimulates horizontal (transverse) elongation of root and stem but inhibits vertical (longitudinal) elongation. Hence the root and stem look swollen.
5. It induces flowering and fruit setting in pineapple.
6. It induces the growth of adventitious root and formation of root hair thus increasing the absorption surface.
7. It induces epinasty (downward bending of leaves due to more growth on upper side than the lower side).
8. It induces triple response (horizontal growth, swelling of stem (axis) and formation of apical hook) in pea seedling.
9. It induces sex expression by inducing femaleness in the flowers of cucumber.
10. It induces sleep disease (enrolling of petals in opened flowers).

SHORT ANSWER TYPE QUESTIONS

1. What are the stages of development in plants?
2. Distinguish between arithmetic and geometric growth.
3. Explain the differentiation and dedifferentiation.
4. Explain the plasticity in plants with examples.
5. Explain the *Avena* curvature test with respect to auxin.
6. What is the full form of 2,4-D and mention its two roles?

7. What is Bakanae disease with respect to the discovery of gibberellin?
8. Explain the roles of cytokinin in morphogenesis.
9. Discuss the five roles of abscisic acid (ABA).
10. Discuss the roles of gaseous growth regulators.

LONG ANSWER TYPE QUESTIONS

1. What is the difference between growth and development? Explain the different phases of growth.
2. Explain geometrical growth in plants.
3. Explain sigmoid growth in plants.
4. Discuss the discovery of auxins.
5. Write notes on synthetic auxins.
6. Describe the roles of auxin in plants.
7. Describe the important roles of gibberellins in plants.
8. Describe the important roles of cytokinins in plants.
9. Describe the important roles of ABA in plants.
10. Describe the important roles of ethylene in plants.

MULTIPLE CHOICE QUESTIONS

1. Which of the following ions is pulled out in apoplast by auxin during growth?

(a) H^+ (b) K^+
(c) Ca^{+2} (d) Mg^{+2}

2. Which hormone is used in preserving fruits?

(a) Auxin (b) Gibberellins
(c) 2,4-D (d) Cytokinins

3. Which of the following combinations of gases is suitable for fruit ripening?

(a) 80% CO_2 and 20% C_2H_4
(b) 80% C_2H_4 and 20% CO_2
(c) 80% CH_4 and 20% CO_2
(d) 80% CO_2 and 20% O_2

4. Gibberellins were isolated from

(a) Oat coleoptile (b) Soyabean seedlings
(c) Rice (d) Fungus

5. Leaves show maximum expansion in

(a) Red light (b) Green light
(c) Blue light (d) Violet light

11. Which of the following instruments can be used to record plant growth in seconds?

(a) Arc indicator (b) Arc auxanometer
(c) Space marker disc (d) Crescograph

12. The S-shaped growth curve and grand period of growth may change with

(a) Sudden fluctuation in light intensity
(b) Change in temperature
(c) Fluctuation in humidity
(d) These remain unaffected

13. Leaf fall occurs as an abscission layer is formed when the content of

(a) Auxin decreases (b) Auxin increases
(c) ABA decreases (d) Gibberellin decreases

14. Exogenous application of gibberellins induces male flowers on genetically female plants in

(a) *Coccinia* (b) *Cucurbita*
(c) *Cucumis* (d) All of the above

15. Which of the following hormones is sprayed in sugarcane to increase the production of sugar?

(a) Auxin (b) Gibberellin
(c) Cytokinin (d) Ethylene

11. When two hormones act together and contribute to the same function hence this action is said to be

(a) Synonymous (b) Antagonistic
(c) Symbiotic (d) Synergistic

12. Which of the following hormones is not found in plants?

(a) Auxin (b) Gibberellins
(c) Cytokinin (d) Ethylene

13. If a plant is given only visible light, then its growth will

(a) Decrease
(b) Increase
(c) Firstly, decrease then increase
(d) Remain unaffected

14. Which of the gibberellins is most studied?

(a) GA_3 (b) GA_5
(c) GA_7 (d) GA_9

15. RNA formation is induced by

(a) Florigen (b) Kinetin
(c) Auxin (d) Gibberellin

16. Pruning of plants promotes branching because the axillary buds get sensitised to

(a) IAA (b) GA
(c) Kinetin (d) Ethylene

17. Which of the following is the motivative force for growth?

(a) Root pressure
(b) Turgor pressure
(c) Atmospheric pressure
(d) Osmotic pressure

18. Maximum amount of growth in root occurs

(a) At the apex
(b) Behind the apex
(c) In the presence of light
(d) In the presence of soil

19. An apple tree can be made to produce larger fruits than usual by

(a) Applying hormones
(b) Thinning of blossoms
(c) Decapitating the stem apex
(d) Defoliating

20. Which of the following is incorrect with respect to auxin?

(a) Zn is essential for its synthesis
(b) Tryptophan is the precursor to auxin
(c) It is abundantly produced in the shoot apex
(d) It cannot be synthetic

21. Phototropism in the shoot is attributed to

(a) Auxin (b) Gibberellins
(c) Cytokinin (d) ABA

22. Phototropic curvature is promoted by

(a) Red light (b) Blue light
(c) Greenlight (d) Violet light

23. If a growing plant is decapitated then

(a) Leaves become yellow
(b) Axillary buds are inactivated
(c) Axillary buds are activated
(d) Leaves become small

24. In decapitated plants, axillary buds are activated because

(a) These get more light
(b) The amount of auxin is increased
(c) Amount of food material is increased
(d) The amount of cytokinin is increased

25. Pruning makes the hedge plant dense because

(a) Injury induces dense growth
(b) Apical dominance is removed
(c) Root sprouts additional branches
(d) Pruning removes shade and allows germination of new seedlings to impart a dense growth

26. Reactivation of cambial cells in the spring season is caused by

(a) Auxin (b) Gibberellins
(c) Cytokinin (d) Ethylene

27. Which of the following hormones is concerned chiefly with root initiation?

(a) IBA (b) GA
(c) ABA (d) Kinetin

28. Parthenocarpic fruits can be produced by the application of

(a) IBA (b) 2,4-D
(c) IAA (d) All of these

29. If the seedling is grown in darkness, then

(a) It is of the same size as that in grown in light
(b) It is much healthier than that grown in light
(c) It is similar to that grown in light
(d) It is taller than that grown in light

30. A genetically dwarf plant can be converted into tall plant by the use of

(a) Kinetin (b) GA
(c) IAA (d) 2,4-D

31. **The habit of cabbage plant can be changed drastically by the application of**
(a) IAA (b) GA_3
(c) ABA (d) 2,4-D

32. **Which of the following plant materials is widely used in the preparation of culture medium?**
(a) *Pinus longifolia*
(b) *Cocos nucifera*
(c) *Colchicum autumnale*
(d) *Cycas revoluta*

33. **Richmond-Lang effect is shown by**
(a) Auxin (b) Gibberellin
(c) Cytokinin (d) ABA

34. **Ethylene is formed from the amino acid**
(a) Glycine (b) Methionine
(c) Tryptophan (d) Lysine

35. **Sleep disease of flowers is caused by**
(a) IAA (b) Ethylene
(c) ABA (d) IBA

36. **Pineapple can be made to flower in off season by**
(a) Zeatin (b) Ethylene
(c) Temperature (d) Short days

37. **Which of the following plant hormone is known as stress hormone?**
(a) Gibberellin (b) Kinetin
(c) Auxin (d) ABA

38. **Green organs are turned yellow by application of**
(a) IAA (b) IBA
(c) 2,4-D (d) ABA

39. **The movement of auxin is largely**
(a) Acropetal (b) Latera
(c) Upwards (d) Basipetal

40. **Apical dominance is due to**
(a) Cytokinin in the leaf tip
(b) Gibberellin in the lateral bud
(c) Auxin in the shoot tip
(d) ABA in the lateral bud

41. **Which of the following hormones can replace vernalization?**
(a) Auxin (b) Gibberellin
(c) Cytokinin (d) Ethylene

42. **Sprouting of potatoes can be prevented by**
(a) 2,4-D (b) ABA
(c) Maleic hydrazide (d) IBA

43. **Which of the following is not synthetic auxin?**
(a) IAA (b) 2,4-D
(c) NAA (d) NPA

44. **Indole acetic acid (IAA) is applied to the root tip of a seedling in high concentration then the growth of root would be**
(a) Inhibited (b) Increased
(c) Decreased (d) Unaffected

45. **Indole acetic acid (IAA) occurs in the human urine especially in person suffering from**
(a) Diabetes (b) AIDS
(c) Pellagra (d) Typhoid

46. **Which of the following hormones is involved in polar transport in plants?**
(a) Auxin (b) Gibberellin
(c) Cytokinin (d) ABA

47. **Gibberellins are more effective when applied to**
(a) Intact plants (b) Coleoptiles
(c) Internodes (d) Roots

48. **Gibberellic acid induced synthesis of α-amylase in barley aleurone is**
(a) Stimulated by ABA
(b) Reversibly inhibited by ABA
(c) Irreversibly inhibited by ABA
(d) Not affected by ABA

49. **Which of the following hormone shows maximum concentration in root tip?**
(a) Auxin (b) Gibberellins
(c) Cytokinin (d) ABA

50. **Which of the following hormone shows maximum concentration in lateral buds?**
(a) Auxin (b) Gibberellin
(c) Cytokinin (d) ABA

51. **Maximum concentration of gibberellins is found in**
(a) Young leaves (b) Elongating stem
(c) Mature leaves (d) Mature stem

52. **Differentiation of shoot is controlled by**
(a) High gibberellin: auxin ratio
(b) High gibberellin: cytokinin ratio
(c) High auxin: cytokinin ratio
(d) High cytokinin: auxin ratio

53. Bud dormancy can be induced by
(a) IAA (b) GA
(c) ABA (d) Ethylene

54. Which of the following hormone is not found in plants?
(a) 2,4-D (b) Ethylene
(c) IAA (d) ABA

55. Which of the following hormone is used in preserving fruits?
(a) IAA (b) 2,4-D
(c) Gibberellins (d) Cytokinin

56. Which of the following hormone causes seasonal leaf fall
(a) Auxin (b) Cytokinin
(c) Gibberellins (d) Ethylene

57. The cutting of the upper portion of the shoot tip
(a) Delays growth of plant
(b) Growth becomes faster
(c) Growth is not affected
(d) None of the above

58. ABA is synthesised from
(a) Chlorophyll (b) Terpene
(c) Xanthophyll (d) Anthocyanin

59. Seed dormancy can be broken by
(a) IAA and ABA (b) IAA and Ethylene
(c) GA_3 and IAA (d) GA_3 and Ethylene

60. The correct sequence of the developmental process in a plant cell is
(a) Plasmatic growth, Differentiation, Maturation, Senescence
(b) Differentiation, Plasmatic growth, Maturation, Senescence
(c) Differentiation, Maturation, Plasmatic growth, Senescence
(d) Plasmatic growth, Maturation, Differentiation, senescence

61. Cytokinin and ethylene have opposite effects on
(a) Elongation of stem
(b) Leaf senescence
(c) Dormancy
(d) Lateral swelling of stem

62. The removal of shoot tips (decapitation) means loss of apical dominance and it is applied in
(a) Tea plantation (b) Hedge making
(c) Bolting (d) Both a and b

63. A sigmoid growth curve is characteristic of
(a) Bacteria growing in culture medium
(b) Organisms growing in natural habitat
(c) All cell, tissues and organs
(d) All of the above

64. In plants, IAA causes cell elongation due to
(a) Increase in pH of cytoplasm
(b) Decrease in pH of cytoplasm
(c) Increase in pH of apoplast
(d) All of the above

65. Gibberellin activates α-amylase enzyme activity in
(a) Endosperm (b) Aleurone layer
(c) Cotyledon (d) Embryo

66. Plant hormones act by affecting the activities of
(a) Genes
(b) Enzymes
(c) Genes and enzymes
(d) Genes, enzymes and membranes

67. Once a flower is pollinated, changes occur that make it less attractive to insects. Its petals shrivel and fall off. Pollination must
(a) Increase the output of cytokinin in the flower
(b) Block the flow of auxin from the root
(c) Trigger the release of ethylene in the flower
(d) Increase the formation of phytochrome which sets the biological clock

68. Arithmetical growth shows
(a) Straight line curve (b) J-shaped curve
(c) S-shaped curve (d) Parabolic curve

69. *Avena* curvature test is associated with
(a) Ethylene (b) Gibberelli
(c) Cytokinin (d) Auxin

70. Triple response is shown by
(a) Auxin (b) Gibberellin
(c) Cytokinin (d) Ethylene

71. Cytokinin is chemically known as
(a) 2-furfuryl amino purine
(b) 2-furfuryl amino pyrimidine
(c) 6-furfuryl amino purine
(d) 6-furfuryl amino pyrimidine

72. The equation $W_1 = W_0 + e^{rt}$ represents
(a) Arithmetical growth
(b) Sigmoid growth
(c) Geometrical growth
(d) All of the above

73. The size of apple is increased by the application of

(a) Indole acetic acid

(b) Naphthalene acetic acid

(c) Gibberellic acid

(d) Abscisic acid

74. Which one of the following hormones is not associated with sex expression?

(a) Auxin (b) Gibberellin

(c) Cytokinin (d) ABA

75. Which of the following base of DNA is present in cytokinin?

(a) Adenine (b) Guanine

(c) Thymine (d) Cytosine

ANSWERS

1.a	11.d	21.a	31.b	41.b	51.a	61.b	71.c
2.c	12.b	22.b	32.b	42.c	52.d	62.d	72.c
3.b	13.a	23.c	33.c	43.a	53.c	63.b	73.c
4.d	14.a	24.d	34.b	44.a	54.a	64.b	74.d
5.d	15.b	25.b	35.b	45.c	55.d	65.b	75.a
6.d	16.a	26.a	36.b	46.a	56.a	66.d	
7.d	17.b	27.a	37.d	47.c	57.a	67.c	
8.a	18.b	28.c	38.d	48.b	58.c	68.a	
9.c	19.b	29.d	39.d	49.c	59.d	69.d	
10.b	20.d	30.b	40.c	50.c	60.a	70.d	

CHAPTER-11

SEXUAL REPRODUCTION IN FLOWERING PLANTS

The sexual reproduction in higher plants involves the development of male and female gametes produced from respective sex organs. In angiosperms, male gametes are produced from pollen grain (microsopore while female gametes are produced inside the embryo sac.

Flower is a specialised condensed or modified shoot because it has nodes and internodes. It is the reproductive organ of angiosperms. It is meant for sexual reproduction.

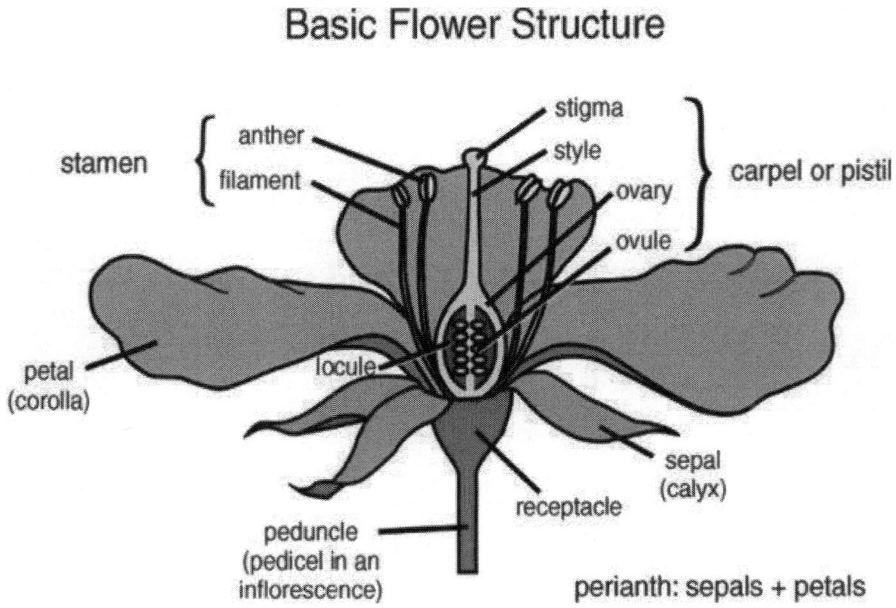

PARTS OF FLOWER

Accessory outer whorl: It does not take a direct part in sexual reproduction. It contains calyx and corolla.

1. Calyx: Its unit is sepal. It is green, lower and smaller. It protects other inner floral parts.
2. Corolla: Its unit is petal. It is coloured other than green, upper and larger. It attracts pollinators and protects inner floral parts.

Essential inner whorl: It is a fertile whorl and takes part in sexual reproduction. It contains androecium and gynoecium.

3. Androecium: Its unit is stamen. It is a male reproductive organ concerned with the production of microspores or pollens. It consists of a broad terminal anther, connective and cylindrical filament. The proximal end of the filament is attached to the thalamus or the petal of a flower.
4. Gynoecium: Its unit is carpel. It is a female reproductive organ concerned with the production of megaspores. It consists of a basal swollen ovary, elongated style and terminal stigma.

DEVELOPMENT OF MALE GAMETOPHYTE

It involves the following events in stamen.

DEVELOPMENT OF ANTHER

It is a fertile part of stamen. It is usually dithecous (two lobes having four sporangia). Both lobes of anther are attached to each other by sterile parenchymatous tissue called connective, but connective is absent in monothecous stamen. Anther is monothecous (one lobe having two sporangia) in members of Malvaceae family such as cotton, and lady's finger. Anther consists of only one microsporangium in *Arceuthobium* (smallest parasitic plant).

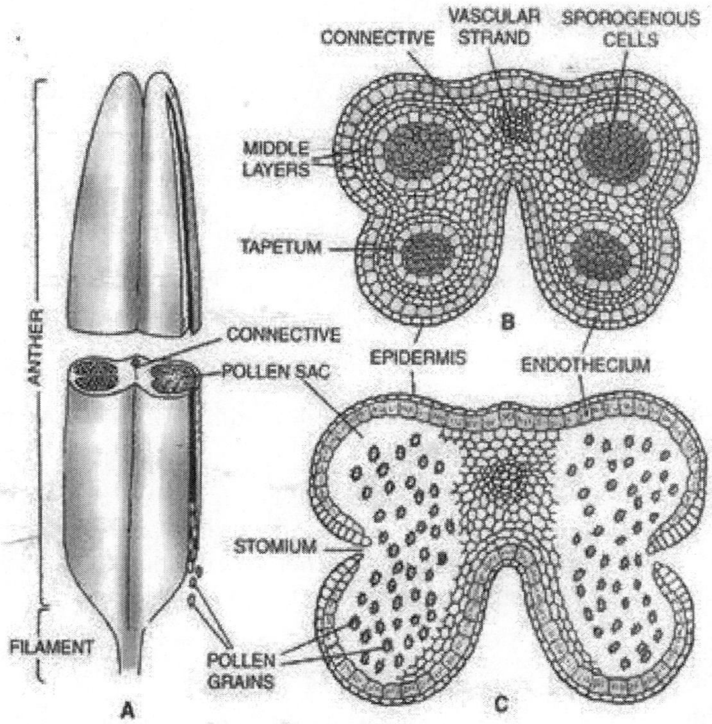

Structure of Anther. A, longitudinally dehisced anther cut transversely to show pollen sacs and connective. B, T.S. young anther; C, T.S. anther at the time of dehiscence (common or longitudinal type).

Structure: Each anther lobe has two long and cylindrical microsporangia or pollen sacs. Now four sporangia of anther are situated at four corners of anther. These run parallel to one another. The development of microsporangia is an eusporangiate type in which microsporangia develop hypodermally from a group of archesporial cells. The archesporial cells divide periclinally (parallel to outer wall of the anther lobe) forming the outer parietal cell (towards epidermis) and the mass of the inner sporogenous cell. The parietal cell undergoes a series of periclinal and anticlinal divisions forming 3-5 concentric layers of anther wall which consists of the epidermis, endothecium, middle layers and tapetum. The outer three wall layers are protective and also help in the dehiscence of anther by which pollens are released.

(a) **Epidermis**: It is firstly formed, outermost and one layered. It becomes stretched, flattened and finally shrivels off at maturity. It is a protective layer.

(b) **Endothecium:** It is sub-epidermal, one-layered and radially elongated. It develops fibrous thickening containing α-cellulose in radial and tangential walls. It develops maximum at the time of dehiscence. It is hygroscopic. The cells of endothecium become thin-walled and lack fibrous thickening at the place of dehiscence called stomium. It helps in the dehiscence of anther at maturity due to the presence of fibrous bands, differential expansion of outer and inner tangential walls and hygroscopic nature.

(c) **Middle layer**: It is short-lived (ephemeral) and 1-3 layered. It is absent in *Lemna* and *Najas*. It disintegrates at maturity to provide nourishment to the growing microspore mother cell.

(d) **Tapetum**: It is innermost, one-layered and viscous. It is rich in dense cytoplasm and more than one prominent nuclei. It possesses mitochondria, plastids, dictyosomes and a number of spherical lipid bodies called Ubisch bodies. It surrounds sporogenous cell. It attains maximum development at the pollen tetrad stage. It breaks down at the time of maturity. It either becomes multinucleate (coenocytic) without cytokinesis or its nucleus becomes polyploid due to endomitosis (duplication of chromosomes due to separation of chromatids without division of the nucleus). It is of two types:

(i) **Glandular or secretory tapetum**: It is the most common type. It remains *in situ* (original position) throughout microspore development. It secretes sporopollenin, pollen kit (the oily and sticky substance of lipid and carotenoids) and compatibility proteins. It helps in ornamentation (external thickening) of the exine of pollen grain due to Ubisch bodies.

(ii) **Amoeboid or Invasive tapetum**: It is short-lived. The inner and radial walls of tapetum break down at an early stage and their entire protoplasts move towards the centre to nourish microspores where they join to form plasmodium or periplasmodium. It is considered primitive and is found in *Alisma, Typha, Potamogeton, Tradescantia* etc.

Functions:

(a) It secretes enzymes, hormones and nutrients to provide nutrition to the developing pollen grains.

(b) It secretes callase enzyme which dissolves the callose (β-1, 3-glucan) layer of pollen tetrad to separate pollen grains.

MICROSPOROGENESIS:

It is the formation of haploid microspores (pollen grains) from diploid microspore mother cell by meiotic division inside the pollen sac (microsporangium) of anther.

Primary sporogenous cells increase their numbers due to divisions and fill the entire cavity of the microsporangium. These are transformed into microspore mother cells or pollen mother cells which are diploid. Initially, it is covered by callose and later separated. The separated microspore mother cell undergoes meiosis to produce a tetrad of microspore or pollen grains. Callose binds all four microspores into tetrad.

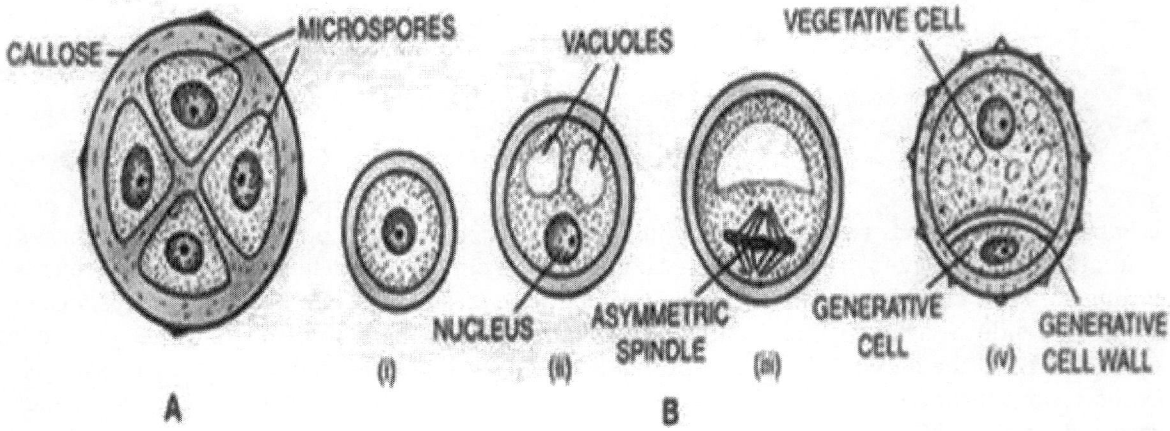

Microsporogenesis. A, a microspore tetrad. B, a microspore maturing into a pollen grain.

$$\text{Primary sporogenous cell} = \text{Microspore mother cell} \xrightarrow{\text{meiosis}} \text{Microspore tetrad}$$

Types of tetrads:

1. Tetrahedral: It shows a **simultaneous** type of cytokinesis (first and second cytokinesis occur after completion of meiosis). It is a primitive feature. It is most common and found in dicots.
2. Isobilateral: It shows **successive** type of cytokinesis (first cytokinesis occurs after meiosis I and second cytokinesis after meiosis II). It is an advanced feature. It is found in monocots.
3. Decussate: It is found in *Atriplex, Crocus, Magnolia* etc.
4. Linear: It occurs in *Halophila.*
5. T-shaped: It is found in *Aristolachia, Butomopsis.*

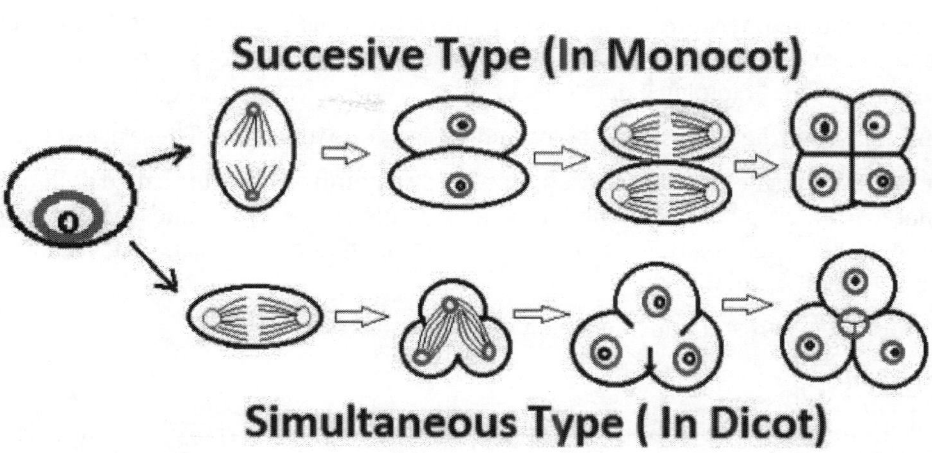

Peculiarities:

1. Out of four microspores, three degenerate and one remains functional in the case of *Cyperus*. Hence one meiotic division produces one functional pollen grain.
2. All five types of tetrads are seen in a species of *Aristolachia* and *Butomopsis*.
3. Polyspory condition (occurrence of more than 4 spores esp.11) is seen in *Cuscuta*.
4. Compound pollen grains (all the four microspores do not separate) are seen in *Juncus, Jatropha, Drimys, Dracaena, Drosera, Typha* etc.
5. Masulae (compound pollen grains are attached in small groups) is seen in Orchid.
6. Pollinia (all four pollen grains do not separate and remain united in a single sac) are seen in Orchid.
7. Translator (two pollinia of adjacent anthers are attached) is found in *Calotropis*.

Dehiscence of anther: It occurs mostly longitudinally. Pollen grains dry up and become powdery inside mature pollen sacs. Tapetum becomes absorbed. Anther dries up due to dehydration. The sterile partition wall between two pollen sacs of anther lobe gets destroyed forming a single chamber. Dead cells of endothecium contract from outer thin wall due to further drying so that anther lobe forms longitudinal concavities. Slit occurs lengthwise on anther. Anther lobe ruptures in the region of stomium. These microspores after their release from tetrad are called pollen grains. Thus, thousands of pollen grains are dispersed. The pollen grain represents the male gametophyte.

Structure of Pollen grain: Pollen grain is generally rounded having a size of 25-50μm. Its shape, size, colour and design vary from species to species. It is rich in nutrients (proteins=7-26%, carbohydrate=24-46%, and fat=1-15%) and water (7-16%). The study of pollen grains is called Palynology.

Father of Palynology	Erdtman
Father of Indian Palynology	P.K.K.Nair
Smallest pollen	Myostosis
Largest pollen	*Mirabilis* (250μm)
Longest pollen	Zostera
Largest number of pollen	Maize (25 million)

Pollen grain consists of two layers:

1. **Exine**: It is outermost, thick and cuticularised. It is made up of sporopollenin (a fatty substance). It is further divided into ectexine (sexine) and endexine (nexine). Ectexine is made up of three layers: outer discontinuous tectum, middle discontinuous baculum and inner continuous foot layer. It has a clear aperture called germ pore where sporopollenin is absent. Germ pore is developed in those areas where exine is either absent or thin. It has three germ pores (tricolpate) in dicots and one germ pore (monocolpate) in monocots.

Exine can tolerate high temperatures, acids, alkalis and hazardous environments. It cannot be degraded by any enzymes. It can be preserved for a long period during fossilisation due to the presence of sporopollenin which is resistant to physical and biological decomposition.

Exine can exhibit fascinating arrays of patterns and designs due to tectum. These are of taxonomic importance.

Exine is spiny as well as covered by a yellowish, viscous, sticky and oily coating called pollenkit in entomophilous (insect-pollinated) pollen grains. Pollenkit is chiefly made up of lipids and carotenoids. It is secreted by tapetum. Pollen kit helps in the attachment of pollens to the bodies of insects, acts as an insect attractant and protects pollens from UV radiations.

2. **Intine:** It is innermost, thin and delicate. It is made up of pecto-cellulose.

MICROGAMETOGENESIS:

It is the development of male gamete from pollen grain (microspore). Microspore (partially developed) is the first cell of male gametophyte and represents immature male gametophyte. The male gametophyte is developed inside the microsporangium (pollen sac) of the anther. It is called precocious development. It is more or less uniform in all flowering plants.

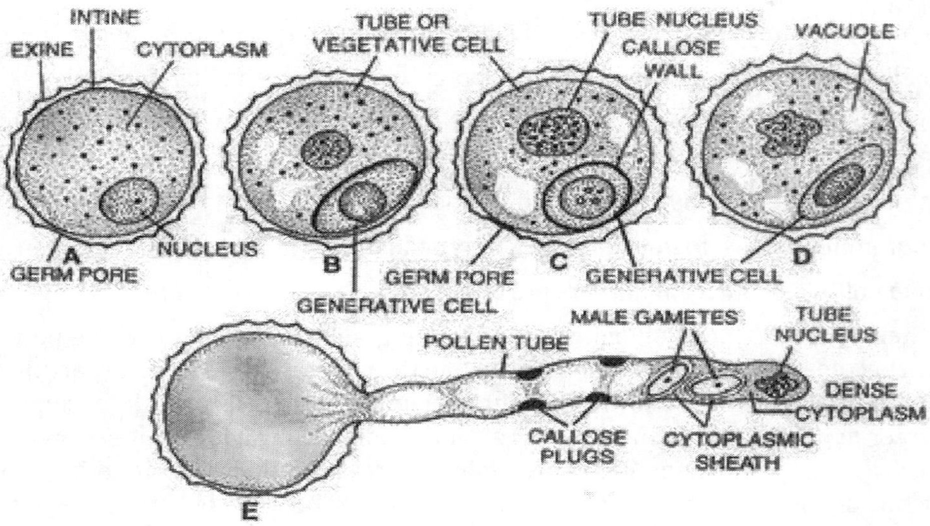

Microspore is small, unicellular, uninucleate, haploid and large in numbers. Freshly formed microspore has rich cytoplasm and centrally placed prominent nucleus. It grows in size due to flow of nutrients and the development of vacuole.

The nucleus of the microspore divides mitotically forming larger tube nucleus and a smaller generative nucleus. Both surround the cytoplasm and become larger vegetative cells and smaller and spindle-shaped generative cells. The vegetative cell has a large irregularly shaped nucleus and vacuolated cytoplasm which is rich in food reserve. The generative cell has a dense cytoplasm surrounding a prominent nucleus. A layer of callose is deposited around the generative cell in the pollen tube to separate the generative cell from pollen grain wall.

Pollen grain sheds at 2-celled stage in dicots (70% of angiosperms). This takes place after the release of pollen from anther. In this case, the generative cell divides after pollen has landed on the stigma or upon its germination.

The generative cell again divides mitotically forming two non-motile and amoeboid male gametes. The male gametes surround cytoplasm and form male cells. The pollen tube is developed from germ pore from intine. Pollen grains shed at 3-celled stage in monocots (30% of angiosperms) and *Cycas* and 4-5 celled stage in *Pinus*. The vegetative cell enlarges and descends to the tip of pollen tube where it later degenerates.

Pollen viability: It is the ability of pollen grains to germinate. Its duration is variable from 30 minutes (very small) in cereals or flowers pollinated in bud condition to several months in legumes. It is longest in *Typha* (300 days). It depends on temperature, humidity and genetic potentiality of species. It can be stored for years in liquid nitrogen (-196^{0}C) in pollen banks (cryopreservation) for future use in plant breeding programme.

Pollen products: These are in the form of tablets, creams and syrup used as food supplement by atheletes because the pollens are rich in nutrients.

Allergic Pollens: The pollen grains of *Amaranthus,* Castor, *Chenopodium,* Carrot grass (*Parthenium*), *Prosopis, Sorghum* cause allergy and respiratory trouble.

Pollen bank: Pollen grains are stored in liquid Nitrogen (-196°C).

Pollen or male sterility: It is caused by

(a) Anther suppression, abortion, phyllody.

(b) Failure of anther dehiscence

(c) Aberrant meiosis or gametophyte development

(d) Premature dissolution of callose

(e) Fungal infection

DEVELOPMENT OF FEMALE GAMETOPHYTE:

Embryo sac is known as female gametophyte. It is formed from megaspore. It involves the following events in ovary:

Development of ovule: Ovule is an integumented megasporangium present inside an ovary. It is attached with placenta. It develops into seed after fertilisation. It is generally oval in shape. There are several types of ovule based on shape and orientation.

1. **Anatropous**: It is most common. It is called an inverted ovule because the nucellus is rotated by 180^0 and the micropyle is on the lower side and comes close to the funicle due to the unilateral growth of ovule. It has the following parts:

   (a) **Nucellus**: It is the main body of ovule. It has abundant reserve food hence it provides nutrition to developing embryo. It may be

   - Crassinucellate: Nucellus is massive and well-developed. It is many layered because the parietal cell undergoes division. It is seen in *Casuarina*, members of Polypetalae subclass of Dicots; Monocots.
   - Tenuinucellate: Nucellus is thin and ill-developed. It remains single-layered because parietal cell does not divide. It is seen in members of Gamopetalae subclass of Dicots.

   (b) **Chalaza**: It is the basal part of ovule and is present just opposite to micropyle or at the junction of nucellus and integument.

   (c) **Micropyle**: It is a narrow opening at one end of the ovule. It is formed by integuments.

   (d) **Funicle**: It is a stalk of ovule. It remains attached with the placenta of ovary. It contains a vascular strand for the supply of nutrients to ovule.

   (e) **Integument**: It is a protective layer and covers the ovule except at micropyle. It arises from chalaza. It is of the following types:

   - Ategmic: Integument is absent. e.g., *Loranthus, Santalum*
   - Unitegmic: There is one integument. e.g., Gymnosperms, Gamopetalae of dicots (Solanaceae, Asteraceae).
   - Bitegmic: There are two integuments. e.g., Polypetalae of dicots (Brassicaceae, Malvaceae) and monocots.
   - Tritegmic: There are three integuments. e.g., *Asphodelus*, Litchi.

   (f) **Hilum**: It is a point of attachment of funicle with nucellus of ovule. It is junction between the ovule and the funicle.

   (g) **Raphe**: It is a ridge or bulged part when funicle gets fused with the nucellus on one side.

   (h) **Embryo sac:** It is a female gametophyte and lie at the centre inside nucellus.

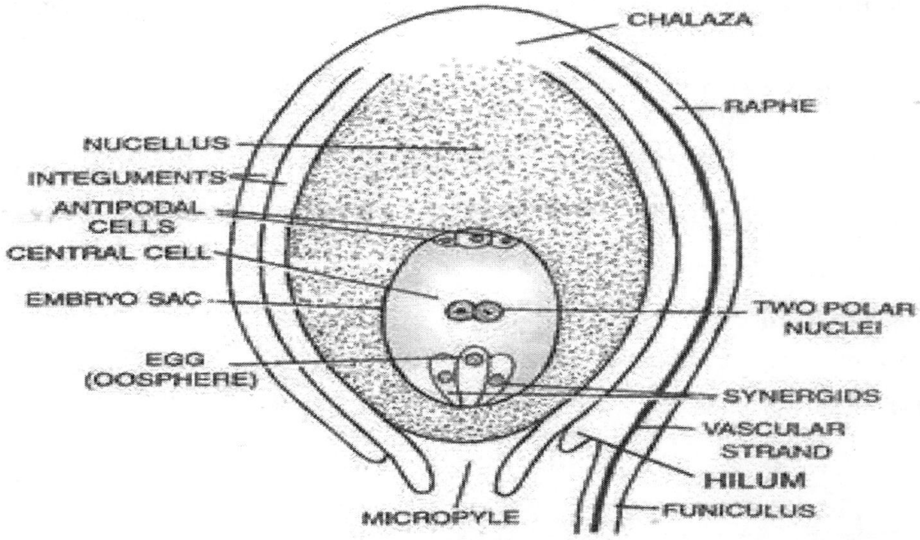

Structure of a typical ovule (anatropous ovule) prior to fertilization.

2. **Orthotropous (Atropous):** It is erect ovule in which micropyle and funicle are in one straight vertical line. It is the simplest and most primitive. e.g., Gymnosperms, Betel, *Polygonum.*
3. **Hemitropous (Hemianatropous):** It is transverse ovule in which micropyle and funicle lie at right angles. It is intermediate between anatropous and orthotropous ovule. e, g., *Primula, Ranunculus.*
4. **Amphitropous:** Both embryo sac and nucellus are horse shoe shaped. e.g., *Alisma*, *Lemna,* Poppy.
5. **Campylotropous:** Nucellus is curved hence micropyle and chalaza lie on either side of funicle. e.g., *Capparis, Chenopodium,* Members of Brassicaceae, Fabaceae.
6. **Circinotropous**: Micropyle faces upwards but funicle is large and coiled around ovule by more than 360°. It is initially orthotropous but later become anatropous due to unilateral growth of funicle. e.g., *Opuntia*

Barriers:

1. **Hypostase**: It is group of lignified (thick walled) cells present below embryo sac. It arises from cells of nucellus. It acts as barrier or boundary for growing embryo sac and thus it prevents it from pushing into base of ovule. It also maintains water balance. It is seen in *Agave*, Lily, *Aristolachia* etc.
2. **Epistase:** It is group of lignified (thick walled) cells present above embryo sac. It arises from cells of nucellus. It forms a cap like structure.

MEGASPOROGENESIS:

It is formation of haploid megaspore from diploid megaspore mother cell by meiotic division inside the ovule (megasporangium).

Archesporial cell is differentiated in hypodermis from any cell of nucellus towards micropyle. It grows and develops dense cytoplasm with prominent nucleus. Archesporial cell divides mitotically (periclinally) forming outer parietal cell and inner primary sporogenous cell. Parietal cell divides one or more times and forms parietal tissue. Primary sporogenous cell commonly functions as megaspore mother cell. It shows deposition of callose in its wall but absent in *Polygonum* type of embryo sac. Megaspore mother cell (diploid) undergoes meiotic division forming linear tetrad of haploid megaspores. At this stage, two integuments develop from base of ovule.

MEGAGAMETOGENESIS:

It is development of female gamete from megaspore. Megaspore is the first cell of female gametophyte (embryo sac). Female gametophyte like male gametophyte is reduced and dependent on nucellus (sporophytic tissue). Its development is endosporous (within megaspore).

Panchanan Maheshwari classified embryo sac into three types on the basis of (a) number of megaspores involved in the development of embryo sac, (b) number of divisions occurring in nucleus of megaspore and (c) organisation of nuclei in mature embryo sac.

1. **Monosporic Embryo sac**: It develops from a single functional megaspore and as such all the nuclei present are genetically identical. It is of two types:
   (a) ***Polygonum* type**: It was discovered by Strasburger (1879) in *Polygonum divaricatum.* It is the simplest, most common and primitive. It develops from chalazal megaspore (lowermost or fourth from micropyle).

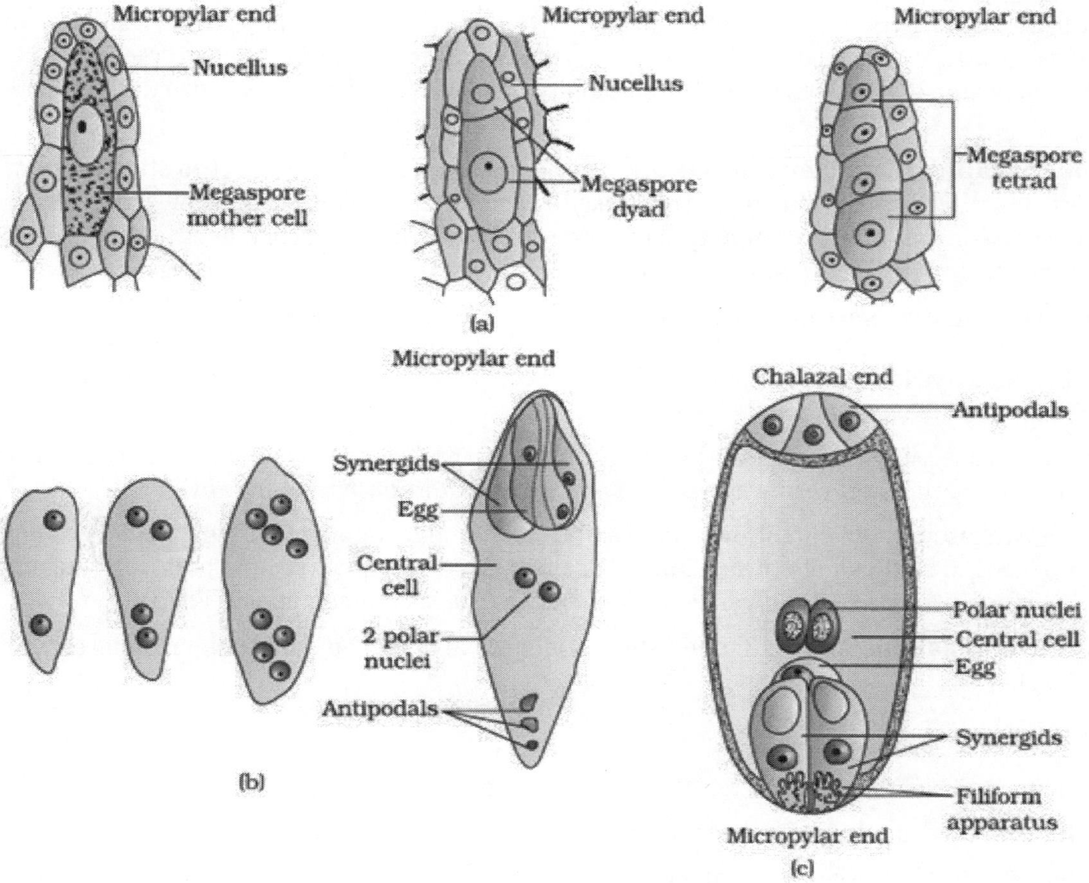

Fig: Development of *Polygonum* type of embryo sac

Functional megaspore undergoes successive three mitotic divisions without cytokinesis (free nuclear type in which karyokinesis is not followed by cytokinesis) forming 8 nucleated embryo sac in which four nuclei are present in chalazal end and four nuclei at micropylar end. Now reorganisation of nuclei occurs. One nucleus from both ends moves to middle forming two polar nuclei which unite to form secondary nucleus. The remaining three at chalazal end forms antipodal cells and rest three at micropylar end forms egg apparatus in which two are synergid cells and one is egg cell. Thus, embryo sac is 7-celled containing 8 nuclei.

Embryo sac is embedded in nucellus towards micropylar half of ovule. It is polar in organisation in which cytoplasmic elements are arranged towards chalaza and vacuolated cytoplasm towards micropyle. Embryo sac contains following type of cells.

(a) **Central cell**: It is the largest cell and mother cell of endosperm. It contains two polar nuclei (later unite to form secondary nucleus or definitive nucleus) which are present at centre. It has highly vacuolated cytoplasm.

(b) **Antipodal cells**: These are three in number and present at chalazal end. These are rich in nutrients. These provide nutrition to developing embryo sac and endosperm. These degenerate before or during or after fertilisation.

(c) **Egg**: It is one in number and present at micropylar end. It is partly embraced by synergid cells at lateral side. The nucleus is much larger than synergid cell and lies towards chalazal end and vacuole present towards micropylar end. Egg shows common walls with two synergid cells and central cell.

(d) **Synergid cells**: These are two in number and lie side by side of egg. The nucleus is present at centre. These develop several finger-like projections called filiform apparatus in the micropylar end from the cell wall to distribute nutrients in the embryo sac, to secrete chemotropic substances for attracting or directing pollen tube into synergids towards egg for fertilisation and to provide mechanical strength to synergid cells. The synergid cells usually degenerate before fertilisation and hence degenerated synergids (either one or both) form suitable space for pollen tube discharge.

POLLINATION:

It is the transfer of pollen grains from anther to stigma. It occurs before fertilisation. It is exhibited by gymnosperms and angiosperms. It is essential for fertilisation, fruit and seed formation. It has freed the angiosperms from dependence on water. It also determines the suitability of pollen for carrying out sexual reproduction. Broadly it is of types

1. **Self-pollination (Autogamy):** It is the transfer of pollen grains from anther to stigma within same flower. It is possible when anther and stigma are close to each other. It takes place when there is synchrony between pollen release and stigma receptivity. It occurs usually by wind or touch. It does not require external agents to carry pollen grains. If self-pollination is continued then it may lead to inbreeding depression.

Adaptations to ensure self-pollination or to check cross-pollination

(a) **Bisexuality**: It is a condition in which androecium and gynoecium are present in same flower. e.g., Members of Poaceae

(b) **Homogamy**: It is a condition in which both the anther and stigma of open (chasmogamous) bisexual flowers mature at the same time. e.g., Periwinkle (*Catharanthus*), *Mirabilis*

(c) **Cleistogamy**: It is a condition in which both anther and stigma of bisexual flowers remain closed (never opened) in some plants throughout their life cycle. These plants show both chasmogamous (open) and cleistogamous (closed) flowers. e.g., *Commelina, Oxalis*, Common pansy (*Viola*).

(d) **Bud pollination**: Both anther and stigma of bisexual flowers mature or open before anthesis (opening of bud).

Advantages:

(a) It maintains parental character throughout (purity of race).

(b) It maintains pure lines for the hybridisation experiment.

(c) It eliminates bad recessive characters.

(d) It saves pollen grains because it does not require large number of pollen grains.

Disadvantages:

(a) It decreases the chance of producing healthier and new varieties.

(b) It causes a loss of vigour and strength in progeny after every generation.

(c) It reduces the adaptability in a changed environment.

2. **Cross pollination (Allogamy):** It is the transfer of pollen grains from one flower to another. It is of two types:

(a) **Geitonogamy:** It is transfer of pollen grains from anther of to the stigma of another flower of the same plant. It is referred as genetically self-pollination and ecologically (functionally) cross-pollination.

(b) **Xenogamy:** It is the transfer of pollen grains from anther of one plant to stigma of another plant of same species. It is referred as genetically and ecologically cross pollination.

Adaptations to ensure cross-pollination or to check self-pollination:

(a) **Unisexuality** (Dicliny): It is the condition in which androecium and gynoecium are present on separate flowers. It is seen in some monoecious plants where male and female flowers are present separately on the same plant ensuring only geitonogamy. e.g., Maize, Castor, Coconut, Cucurbits and all dioecious plants where male and female flowers are present on two separate plants ensuring only xenogamy. e.g., Date palm, Mulberry, Papaya, *Asparagus, Cannabis, Dioscorea, Fragaria, Populus, Rumex, Smilax, Vallisneria*.

(b) **Self-sterility (Self Incompatibility)**: It is the failure of pollen grains to germinate on the stigma after falling. It ensures only xenogamy. It may be between the plants of the same species or different species. It is of two types:

(i) Sporophytic self-incompatibility: Rejection of pollen is controlled by the genotype of the sporophytic tissue of the plant from which pollen is derived. The products are proteins that are formed in the tapetum and are added to the exine of pollen grains. These products are set free on the stigma and are involved in complementary reactions.

(ii) Gametophytic self-incompatibility: Rejection of pollen is controlled by the genotype of male gametophyte. The products are formed in the pollen during its development and are stored in intine and are released on the stigma during pollination. e.g., Tea, Tobacco, *Petunia*, Potato, Radish, *Malva, Passiflora*

(iii) **Dichogamy**: It is the condition in which anther and stigma of a bisexual flower mature at different times. It may be

(c) **Protandry**: Anther matures and sheds pollens earlier than stigma. e.g., Cotton, Marigold, Rose, *Salvia*, Sunflower

(d) **Protogyny**: Stigma matures earlier (ten days before) than anther. e.g., *Aristolachia, Gloriosa, Ficus, Magnolia, Michelia, Plantago, Polyalthia.*

(e) **Heterostyly**: It is the condition in which stamen, style and stigma of bisexual flowers are of different length. Flowers may be

- **Dimorphic**: Two types of flowers are found. In one type, short stamen, and long carpel or pistil (Pin eyed) are seen and in another type, long stamen and short carpel or pistil (thrum eyed) are seen. e.g., Primrose, *Linum.*
- **Trimorphic**: Three types of flowers are found. In one type, carpel is short and stamens are long or intermediate; in the second type, the carpel is intermediate, and stamens are short or long; and in the third type, carpel is long, and stamens are short or intermediate. e.g., *Lythrum, Oxalis*

(f) **Herkogamy**: It is a condition in which a natural barrier develops between anther and stigma to check self-pollination and to ensure cross-pollination. e.g., Gynostegium disc develops in *Calotropis*; Pollinia formation occurs in Orchid; Stigma lies inside flap in Pansy (*Viola tricolor*); Anther lies in corolla basket in *Kalmia*.

Advantages of cross-pollination:

(a) It produces new varieties by bringing about genetic recombination.

(b) It produces healthy and disease-resistant plants due to hybrid vigour.

(c) It increases yield.

(d) It makes the organisms better fitted or more adaptable in changed environment.

(e) It produces large, more and viable seeds.

Disadvantages of cross-pollination:

(a) It requires a large number of pollen grains.

(b) It does not eliminate undesirable characters.

(c) It wastes lots of energy and food materials in unnecessary adaptations.

Agents of pollination:

1. **Anemophily:** It occurs by wind. It is non-directional hence it is wasteful. Flowers are small, colourless, odourless and nectarless but packed in inflorescence. Anthers are well exposed. Pollen grains are small, dry, light, non-sticky and produced in large numbers so they are blown away. Stigma is large, well exposed and feathery (hairy) to catch pollen. e.g., Gymnosperms, Cereals, Grasses, Bamboo, Sugarcane, Coconut palm, Date palm, Papaya, *Cannabis.*

2. **Hydrophily**: It occurs by water. It is not very common and occurs only in a few aquatic plants esp. monocots. Flowers are small, colourless, odourless and nectarless. Pollen grains are light, dry, unwettable due to mucilaginous covering and released on water surface. Stigma is long, sticky, unwettable due to mucilaginous covering and released on water surface. It is of two types:
   (a) **Hypohydrophily:** It occurs below the water surface. It is seen in *Ceratophyllum, Najas, Zostera*. The male flowers of *Ceratophyllum* (monoecious) breaks at base, rises upwards, reaches to water surface and undergoes dehiscence. The liberated pollen grains germinate and start sinking. They come in contact with large and sticky swaying stigma of female flowers while sinking in water. Thus pollination takes place.
   (b) **Epihydrophily:** It occurs on water surface. It is seen in *Vallisneria* (Tape grass/Ribbon weed). The male flowers of *Vallisneria* (dioecious) break from spadix and float on water surface but remain attached to female plant by long stalk. Female flowers have long sticky stigma and while floating male flowers come close to female flowers where anther bursts to release pollen. Thus pollination occurs.
3. **Entomophily:** It occurs by insects such as bees (most common), butterfly, moth, wasp, fly. The flowers are large, brightly coloured, odourful (fragrant), nectarful. The pollen grains are rough and sticky due to pollenkit. Stigma is rough, sticky and unbranched. There are several mechanisms which are as follows:
   (a) **Turn pipe /Lever mechanism:** It occurs by bees in *Salvia* plant. *Salvia* is bisexual and protandrous. It has bilipped or bilabiate corolla. A bee visits young flower and sits on lower lip in search of nectar. The pollen grains are deposited upon the back of bee. The pollen-dusted bee visits older flower and its back rubs mature stigma thus bringing about pollination.
   (b) **Trapdoor mechanism:** It occurs by a Gall wasp (*Blastophaga*) in Fig. Wasp enters inside hypanthodium (inflorescence) to lay eggs inside. As wasp comes out of hypanthodium, it pushes the pollen of male flower situated near the opening on the female flower located at the base. Thus, pollination occurs.
   (c) **Pseudocopulation:** It occurs by a hairy wasp (*Colpa aureus*) in *Ophrys speculum* (Orchid). *Ophrys* bears flowers that resemble the female wasp in colour, odour and appearance. Male wasp matures first and leave the burrow before the females come out for mating. The inexperienced males try to pseudocopulate the *Ophrys* flowers presuming them as their female partners. Wasp carries pollinia fom one flower to another and brings about pollination. Coevolution (evolutionary modification of organisms in response to other organisms) occurs because *Ophrys* flower coevolves to maintain the similarity of its petals to female wasp.
   (d) **Obligatory Symbiotic relationship:** It occurs by Pronuba moth (*Tageticula*) in *Yucca*. Coevolution occurs. Moth cannot complete the life cycle without the association of *Yucca* and in turn, yucca has no other pollinator. Female moth visits flowers at night due to fragrance. It collects pollen grains in the form of balls and it deposits the egg in ovary of flower. After that, it climbs to the top of the style and pushes the pollen ball into the hollow of stigma to bring about pollination.
4. **Ornithophily:** It occurs by birds (Crow, Parrot, Sunbird, Hummingbird). Flowers are large, leathery, brightly coloured (yellow), odourless, and nectarful. e.g., *Agave, Bignonia*, Red Silk Cotton (*Bombax*), *Butea*, Bottlebrush (*Callistemon*), Coral tree (*Erythrina*), *Sterlitzia.*
5. **Chiropterophily:** It occurs by bats. Flowers are large, dull coloured, odourful (fermenting or fruiting odour) and nectar full. Pollen grains are large in number. e.g., *Anthocephalus*, Baobab tree (*Adansonia*), *Bauhinia*, Sausage tree (*Kigella*).
6. **Malacophily**: It occurs by snail. e.g., *Colocasia, Lemna,* Lily
7. **Myrmecophily**: It occurs by ants. e.g., Guava, Litchi, Mango
8. **Ophiophily**: It occurs by snake. e.g., *Arisaemia, Michelia, Santalum*
9. **Cantherophily**: It occurs by Beetles. e.g., members of Rubiaceae

Pollen-Pistil interaction:

It is the interaction between chemicals of pollen grains and those present on the stigma. The interaction determines whether a pollen grain will germinate or not. It is a dynamic process. It is essential for fertilisation.

The exine of the pollen grain is composed of sporopollenin. The cavities present in the exine layer are filled with specific proteins (enzymes), glycoproteins, lipids and certain allergens derived from the tapetum of anther during microsporogenesis. These chemicals carry some sporophytic recognition factors since these are derived mainly from tapetum. The intine of pollen grain, on the other hand, contains carbohydrates and proteins of gametophytic origin.

The stigma of angiosperms is wet as well as dry. Wet stigma secretes free-floating amino acids, proteins and lipids, and dry stigma contains a hydrated layer of carbohydrates, proteins and lipids of sporophytic origin. Both these types of chemicals interact for the process of fertilisation.

FERTILISATION

It is the fusion of male and female gamete to form a zygote. The zygote is the first cell of the sporophyte. It was discovered by Strasburger (1894) in *Monotrapa*.

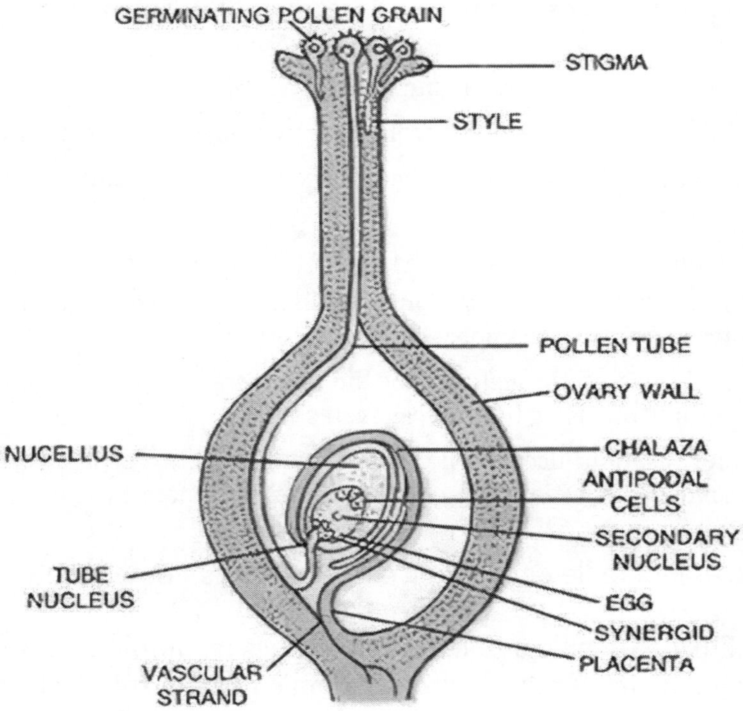

Fig: Mechanism of fertilisation

Process:

(a) Pollen grains (a 2-celled stage in most of the plants) after maturing are transferred by different agencies fall on stigma. Only right pollen belonging to the same species would be able to germinate, and wrong or incompatible pollens are rejected by the stigma so they are unable to germinate and thus they perish. Pollen grains absorb moisture from stigma and swell.

(b) Stigma secretes substances that protect pollen as well as stigma.

(c) Pollen grain germinates by rupturing exine to release pollen tube from intine through germ pore. High Relative humidity is first essential condition for pollen germination. Pollen tube was first observed by Amici in *Portulaca*. Pollen tube wall is three layered; outer pectin layer, middle pecto-cellulose layer and inner callose layer. Such fertilisation through pollen tube is called siphonogamous. Usually one pollen tube (monosiphonous) develops from a pollen grain.

(d) Growth of pollen tube is unidirectional and controlled by Ca^{+2}, a chemotropic agent present on the stigma. The generative cell divides and forms two male gametes during the growth of the pollen tube in the stigma. In some plants in which pollen grains shed at 3-celled stage, the pollen tube carries two male gametes from the very beginning. Pollen tube passes down to stigma and pollen grain remains on stigma. It grows initially at the expense of energy and food materials stored in pollen tubes.

(e) Further growth of pollen tube occurs at the expense of cells of stigma and transmitting tissue of style. Pollen tube secretes pectinase and hydrolytic enzymes to move through style. The pollen tube moves intercellularly, chemotropically and apically along the concentration gradient of Ca^{+2}-borate-inositol-sugar complex. Boron prevents the bursting of pollen. Style is either solid (dicots) or hollow.

(f) Pollen tube does not reach the ovary in a straight line. The pollen tube grows near style and curls to the bottom of the ovary and then near the receptacle. The pollen tube then breaks into the ovule through the micropyle and then the micropyle bursts into the embryo sac.

(g) Growth of pollen tube in embryo sac is now guided by obturator. It develops from funicule of ovule. It is a bridge on which pollen tube grows to reach ovule. It secretes surface exudates for nutrition of pollen tube and also provides chemical and mechanical guidance to pollen tube. It disappears after fertilisation. The duration between formation of pollen tube and its reaching to embryosac varies between a few minutes to few months.

(h) The entry of pollen tube occurs in three manners:

- **Porogamy:** Pollen tube usually enters through the micropyle. It is most common.
- **Mesogamy**: Pollen tube enters through middle part of integument (e.g., cucurbits) or funicle (e.g., *Pistachia*).
- **Chalazogamy**: Pollen tube enters through chalaza. e.g., *Casuarina*

(i) Pollen tube is now attracted by secretions of synergid cells which develop filiform apparatus. The synergid cells degenerate due to impact of pollen tube.

(j) Pollen tube usually passes through one of the degenerated synergid cells. The tip of pollen tube enlarges and ruptures releasing two male gametes.

(k) Syngamy (true fertilisation) occurs when one of male gametes fertilises with egg and thus zygote is formed. It takes much longer time because egg is in inactive state. Syngamy (generative fertilisation) is of three types.

(a) Pre-mitotic: Fusion of male and female gamete occurs immediately. e.g., Asteraceae, Poaceae

(b) Intermediate: Fusion of male and female gamete is incomplete. It is most common.

(c) Post-mitotic: The nuclei of both male and female gamete remain in contact for a while and fuse after when both nuclei enter into division. e.g., *Lilium, Fritillaria*

Post fertilisation changes:

Before fertilisation	After fertilisation
Calyx, corolla and stamen	These fall off .
Style and stigma	These shrivel and wither away.
Ovary	It forms fruit.
Ovary wall	It develops pericarp.
Ovule	It forms seed.
Outer integument	It is converted into testa of seed.
Inner integument	It is converted into tegmen of seed.
Funicle	It forms stalk of seed.
Nucellus	Remnant of nucellus is present called perisperm. e.g., beet, black pepper, *Nymphaea*
Egg	It forms Zygote.
Two polar nuclei (Secondary nuclus)	It forms Endosperm.

Double fertilisation:

It was discovered by Nawaschin (1898) in *Lilium* and *Fritillaria*. It occurs when another male gamete enters the central cell and fuses with two polar nuclei (secondary nucleus) forming the primary endosperm nucleus (3n) which later develops into endosperm. This process is also called triple fusion or pseudo-fertilisation or trophomixis. It is characteristic feature of angiosperms. It takes shorter time because central cell is in active state. Triple fusion provides a stimulus to resume growth and form nutritive tissue to provide nutrition to developing embryo. Further, it ensures that nutritive tissue is formed only when the formation of embryo has taken place by syngamy.

Thus, in the process of double fertilisation five nuclei are involved.

Syngamy= Male gamete + Egg→ Zygote (2n) → Embryo

Triple fusion= Male gamete +2 polar nuclei →Primary endosperm nucleus →Endosperm

ENDOSPERM:

It is nutritive tissue that is produced as a result of triple fusion. It develops from the central cell of the embryosac. It is a characteristic feature of angiosperms except Orchidaceae, Podostomaceae and Trapaceae. In angiosperms, it is triploid because it is formed after fertilisation. It is haploid in gymnosperms because endosperm is represented by female gametophyte which is formed before fertilisation. It is starchy in cereals; proteinaceous in legumes (pulses) and oily in Castor and Coconut.

Types of Endosperms: On the basis of development:

1. **Nuclear endosperm:** It is most common and found in 56% of angiosperms.

   The primary endosperm nucleus (PEN) undergoes first and subsequent mitotic divisions without cytokinesis (wall formation). Thus, a large number of nuclei remain free in the central cell of embryosac. Now a large vacuole appears at the centre of the central cell that pushes all the nuclei to peripheral cytoplasm. Later cytoplasm thickens hence size of the vacuole is decreased and finally, it disappears. Multinucleate cytoplasm is formed. Large-sized nuclei are aggregated at micropylar and chalazal end and small sized nuclei are present at the centre.

   Zygote undergoes resting period which is longer in nuclear endosperm.

   e.g., Coconut milk, Cereals, *Capsella*, Cotton, Sunflower

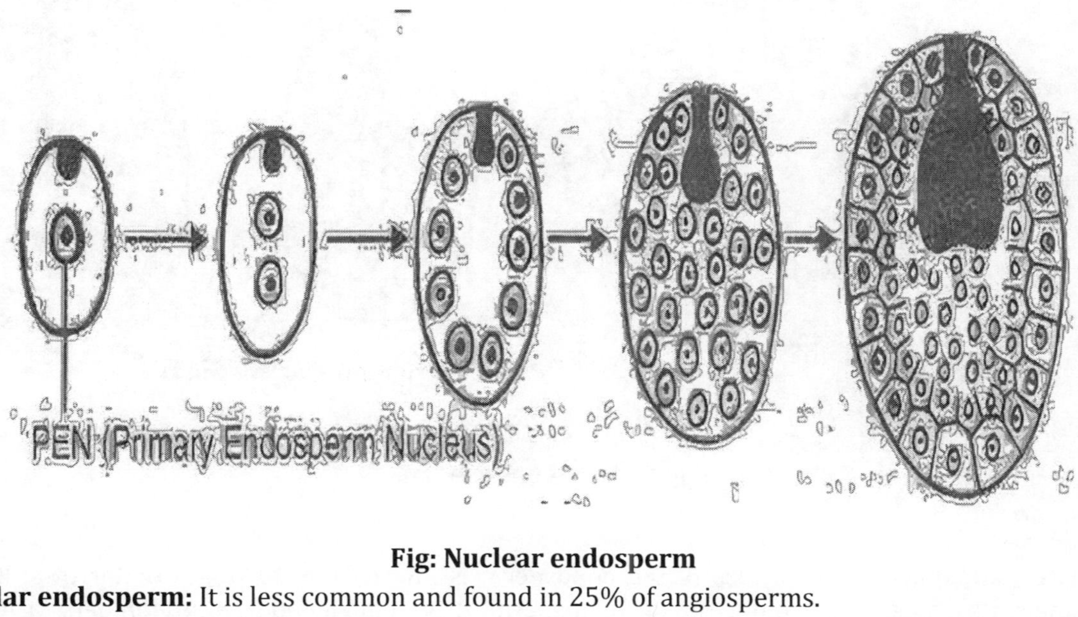

Fig: Nuclear endosperm

2. **Cellular endosperm:** It is less common and found in 25% of angiosperms.

   The primary endosperm nucleus undergoes first and subsequent mitotic divisions accompanied by cytokinesis (with wall formation). The first division is usually transverse but subsequent divisions may be transverse or longitudinal or oblique. Finally, all cells are uniformly present in central cell. It also shows the occurrence of haustoria.

Zygote undergoes resting period which is shorter in cellular endosperm.

e.g., Coconut meal (copra or white kernel), *Adoxa*, Balsam, *Datura, Magnolia, Peperomia, Petunia, Utricularia* etc.

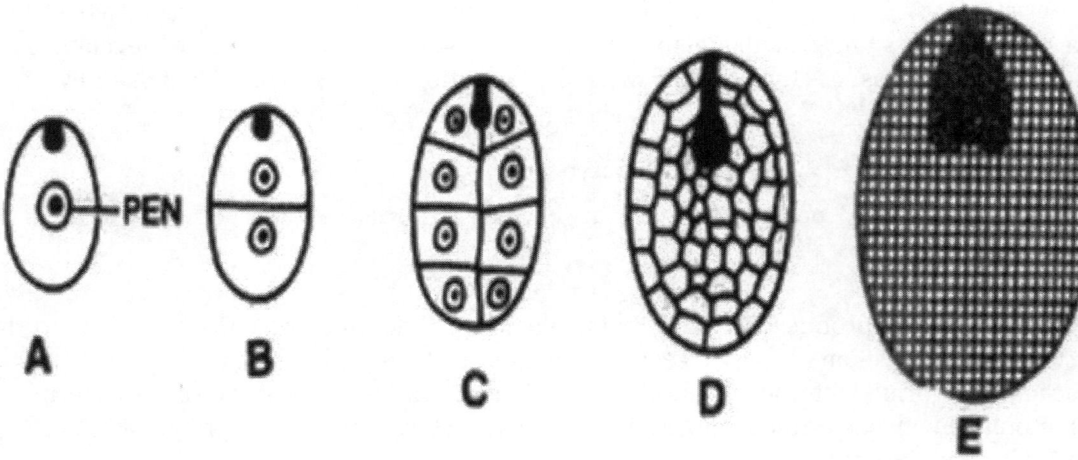

Fig: Cellular endosperm

3. **Helobial endosperm:** It is very less common (19%). It is intermediate between nuclear and cellular endosperm. The primary endosperm nucleus undergoes first division which is transverse followed by cytokinesis (wall formation) to form larger micropylar cells and smaller chalazal cell. Micropylar cell undergoes further division without cytokinesis. Chalazal cell is either remain undivided or may undergo a few divisions to form haustoria.

   e.g., *Asphodelus, Saxifraga* and Helobiales order of Monocots.

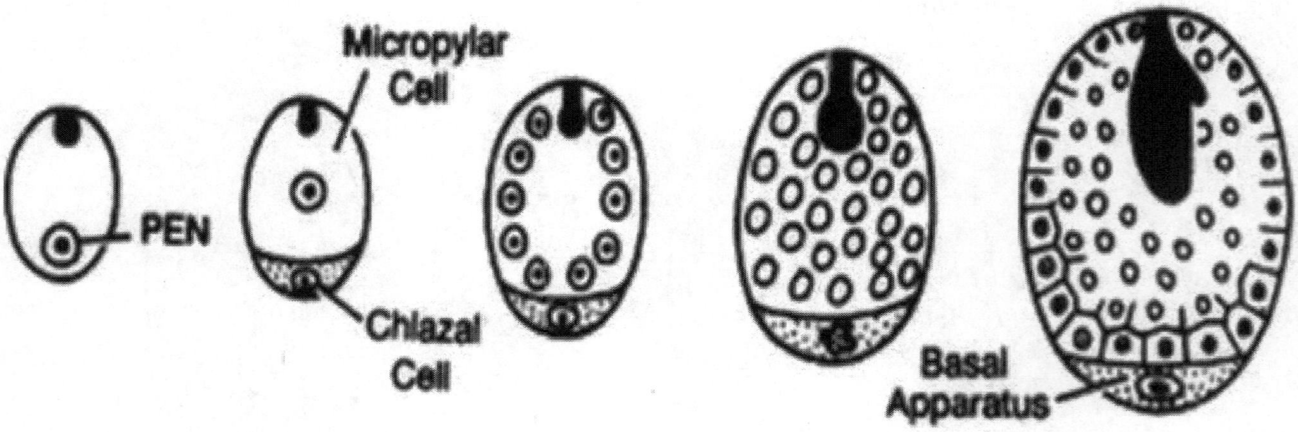

Fig: Helobial endosperm

Types of endosperms:

1. **Ruminate endosperm**: The surface of the endosperm is uneven due to ridges or furrows. It is formed due to uneven hard seed coat outer to soft endosperm. It is homogeneous. e.g., Members of Annonaceae, Aristolachiaceae, Arecaceae, Myristicaceae, Rubiaceae.
2. **Mosaic endosperm:** The surface of the endosperm contains patches of different colours. It is heterogeneous. e.g., Maize

Ploidy level of endosperm

1. Triploid: n +2n = 3n. It is most common.
2. Diploid: n + n = 2n. It is found in *Oenothera*
3. Pentaploid: n + 4n = 5n. It is found in *Penea, Plumbago, Plumbaginella, Fritillaria*
4. Nanoploid: n + 8n = 9n. It is found in *Peperomia*

Other terms

1. **Xenia**: It was coined by Focke. It is direct effect of pollen on endosperm (structure inside embryo sac except embryo). It is seen in maize.
2. **Metaxenia:** It was coined by Swingle. It is effect of pollen on tissue outside embryo sac or endosperm. It is seen in date Palm.

Fate of Endosperm:

1. It is either consumed by developing embryo before seed maturation.
2. It may persist in the seed hence seed is endospermic (albuminous).

Functions:

1. It provides nutrition to developing embryo.
2. It stores carbohydrates, protein and fats which are used in the establishment of seedlings during seed germination.

EMBRYOGENESIS:

It is the development of embryo from zygote. It occurs when endosperm development is complete. It occurs at the micropylar end of embryo sac where zygote is situated. Zygote undergoes first division which is transverse forming terminal cell and basal cell. Panchanan Maheshwari recognised five types of embryos on the basis of plane of division of terminal cell or apical cell (embryonal cell) in 2-celled embryo.

1. The terminal cell of 2-celled proembryo divides longitudinally
   (a) **Crucifer or Onagrad type**: The terminal cell takes part in the development of embryo, but the basal cell plays a minor role.
   (b) Asterad type: Both terminal cell and basal cell are involved.
2. The terminal cell of 2-celled proembryo divides transversally.
   (a) Caryophyllad type: The terminal cell is involved in the development of embryo, but basal cell does not undergo further division.
   (b) Solanad type: The terminal cell is involved in the development of embryo but basal cell forms only suspensor.
   (c) Chenopodiad type: Both terminal cell and basal cell are involved in the development of embryo.

Development of Dicot embryo (Crucifer type):

It is typically studied in shepherd's purse (*Capsella bursa-pastoris*). It involves following steps:

(a) Zygote increases its size and secretes its own cellulosic cell wall.

(b) Zygote divides transversally but unequally forming larger basal cell and smaller terminal cell towards micropylar end.

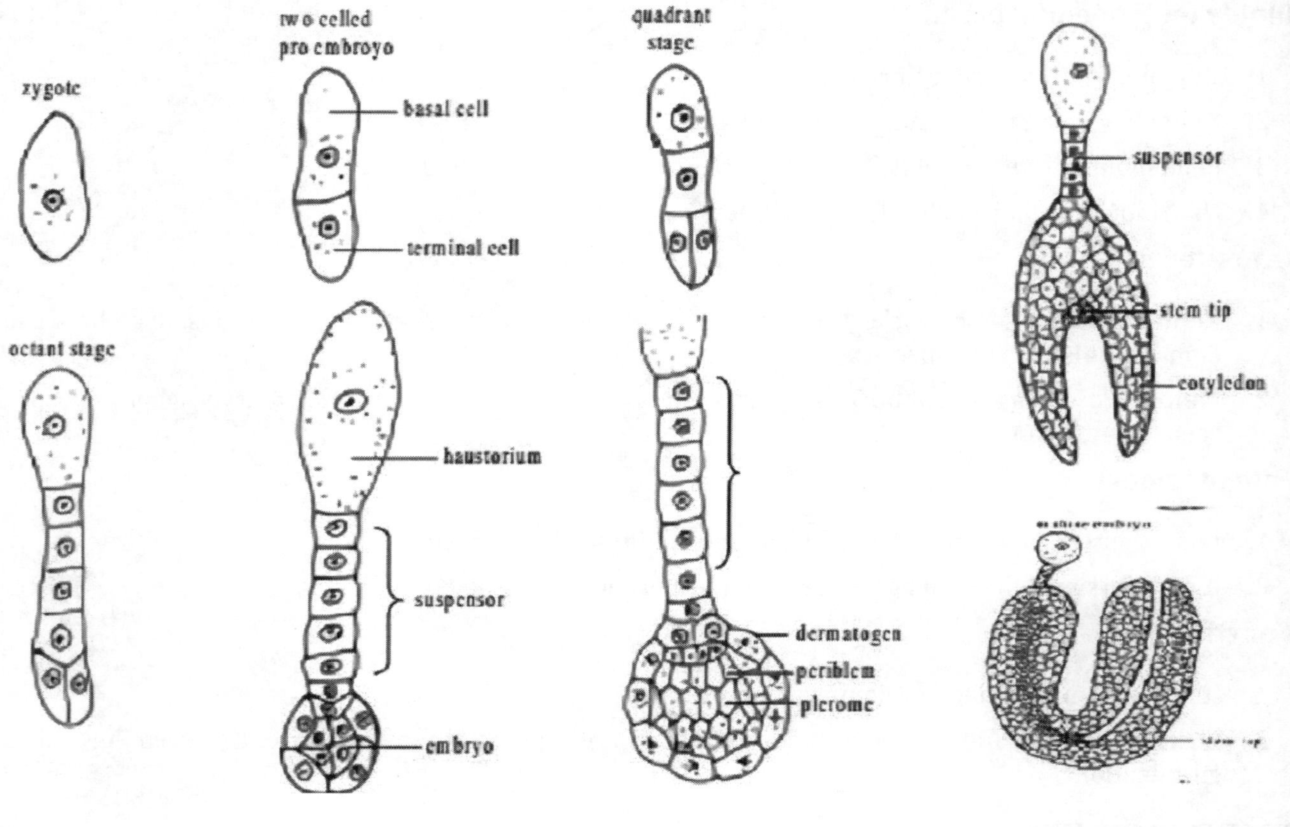

Fig: Developmental stages in dicot embryo

(c) Basal cell divides transversally three times forming an 8-celled suspensor. The first cell of suspensor towards chalaza becomes swollen and functions as a haustorium, Suspensor is short-lived structure and is concerned with the nutrition of proembryo. Suspensor pushes the embryo in the endosperm. The last cell of the suspensor towards chalaza is called hypophysis. It gives rise to root apex which later forms radicle and root cap.

(d) Terminal cell divides longitudinally twice forming a 4-celled quadrant. The quadrant again divides transversally forming an 8-celled octant. The octant divides periclinally forming two tiers of octant stage-epibasal (terminal) and hypobasal (near suspensor). The epibasal cells form two cotyledons and plumule. The hypobasal cells form hypocotyl except at its tip.

(e) Embryo is initially circular (globular) and undifferentiated. It changes to heart shaped then torpedo-shaped and finally become mature.

(f) Mature embryo consists of embryonal axis and two cotyledons. The part of axis below cotyledon is called hypocotyl which at lower end terminates into radicle (embryonic root). The part of axis above cotyledon forms epicotyl which terminates into plumule (embryonic shoot).

(g) The development of the embryo is endoscopic (embryo is pushed inside the interior of the embryo sac or female gametophyte.

Development of Monocot embryo: It involves the following steps:

(a) Zygote divides transversally forming terminal cell and basal cell.

(b) Basal cell towards the chalazal end does not divide further and gives rise to large and swollen vesicular suspensor which functions as haustoria to draw nutrition from endosperm.

(c) Terminal cell divides transversally forming two cells: the top cell and the middle cell. The top cell undergoes longitudinal division and subsequently a series of divisions forming plumule and one cotyledon. The middle cell undergoes transverse division and later several divisions forming hypocotyls and radicles. It also adds a few cells to suspensor.

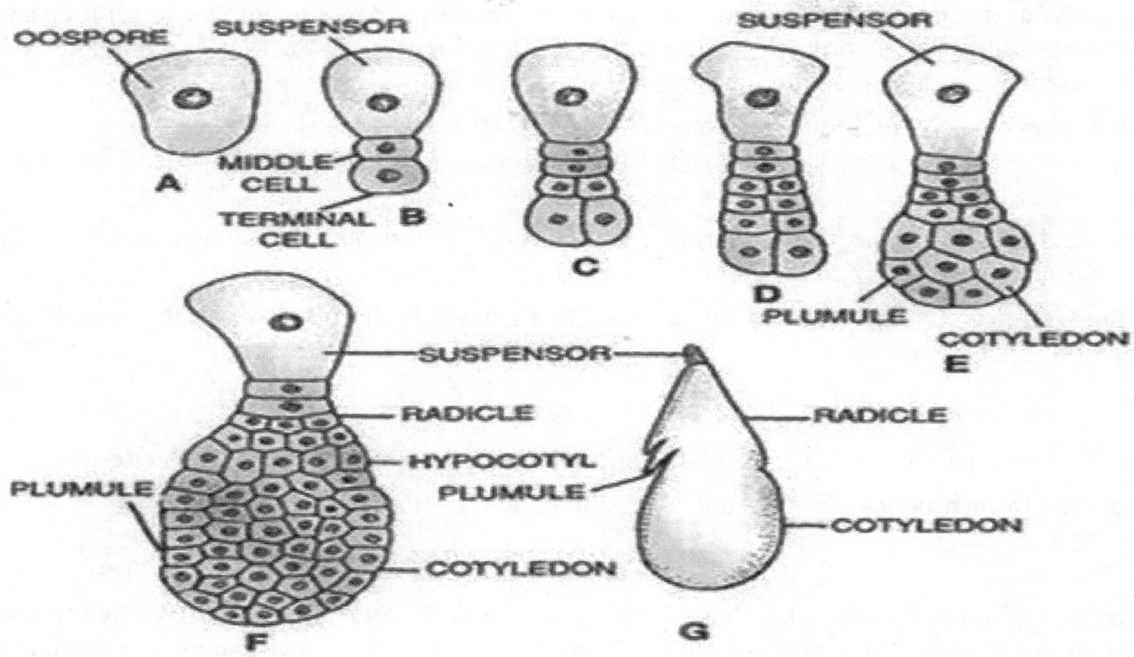

Fig: Developmental stages in monocot embryo

(d) Cotyledon called as scutellum (shield-shaped) grows rapidly and pushes plumule laterally on one side. The part of axis above the level of attachment of scutellum is called epicotyl.

(e) Radicle is covered by a sheath called coleorhiza and plumule is covered by a sheath called coleoptile.

Dicot embryo	Monocot embryo
Both terminal cell and basal cell form embryo.	Only terminal cell forms embryo.
Basal cell divides to form 8-celled suspensor.	Basal cell does not divide further and forms 1-celled suspensor.
First division of terminal cell is longitudinal.	First division of terminal cell is transverse.
It has two lateral cotyledons.	It has one terminal cotyledon.
Plumule is terminal or distal between two cotyledons.	Plumule is lateral due to excessive growth of cotyledon.
Scutellum is absent (its homologous structure is cotyledon).	Scutellum is present.
Radicle is not covered by coleorhiza (absent).	Radicle is covered by coleorhiza.
Plumule is not covered by coleoptile (absent).	Plumule is covered by coleoptile.

POLYEMBRYONY:

It is the presence of more than one embryo in one ovule/seed. It was discovered by Leeuwenhoek (1719) in Orange (Citrus) seeds. It is more common in gymnosperms. If extra embryos develop in the embryo sac, then it is called true polyembryony. If extra embryos develop outside the embryosac, then it is called false polyembryony.

Types: On the basis of origin

1. **Simple Polyembryony:** It is true polyembryony. An additional embryo develops from one or both synergid cells which become egg-like and get fertilised with a male gamete released by an additional pollen tube. e.g., *Argemone, Aristolachia, Poa, Sagittaria*.
2. **Mixed polyembryony**: It is true polyembryony. An additional embryo develops from antipodal cells which become egg-like and get fertilised with male gamete released by an additional pollen tube. e.g., *Paspalum, Ulmus.*

3. **Cleavage polyembryony:** It is true polyembryony. It is the simplest and most common. An additional embryo develops by splitting of single proembryo into two or more embryos or by cleavage of the zygote. It is more common in gymnosperms. e.g., orchids
4. **Multiple embryo sac**: An Additional embryo develops because several embryo sacs are formed by differentiation of same megaspore mother cell or two or more megaspore mother cell. e.g., Brinjal, *Casuarina, Hydrilla*.
5. **Adventive embryony**: It is false polyembryony. An additional embryo develops by differentiation of diploid cells of ovule such as nucellus and integument. e.g., Citrus plants, Mango, *Opuntia*.
6. **Induced polyembryony**: An Additional embryo develops from the cell of root, leaf, pollen grains in culture medium.

Importance:

(a) It produces uniform parental type of seedlings which are free from disease and produce better yields.

(b) It is better than those produced from cuttings (vegetative reproduction).

APOMIXIS:

It is asexual production of seed (embryo) without fertilisation. The term 'Apomixis' was coined by Hans Winkler (1907). It occurs in members of Rutaceae, Asteraceae, Rosaceae, Poaceae. It is more common in polyploids than diploids. It is also called agamospermy.

Types:

1. **Nonrecurrent apomixes**: It is also called gametophytic apomixis. Meiosis occurs hence the nucleus of an embryo is haploid. It is of two types:

   (a) **Haploid apogamy**: It is the development of a haploid embryo from any cell of the embryosac such as synergid and antipodal cells. e.g., Tobacco, Orchid.

   (b) **Haploid parthenogenesis**: It is the development of haploid embryo from an unfertilised egg. e.g., Banana, *Datura*, Maize.

2. **Recuurent apomixis**: It is also called sporophytic apomixis. Meiosis does not occur hence the nucleus of embryo is diploid. It is of two types:

   (a) **Generative apospory (Diplospory or Agamospermy)**: It is the development of a diploid embryo from megaspore mother cell (diploid). e.g., *Parthenium*

   (b) **Somatic apospory (Adventive embryony or sporophyte budding)**: It is the development of diploid embryo from the diploid cell of ovule such as nucellus (e.g., *Acalypha*, Citrus plants, Mango, *Opuntia*) or integument (e.g., *Allium, Potentilla*).

Importance:

(a) It produces genetically uniform plants.

(b) It produces multiplication of desired characters in hybrids because there is no segregation and recombination due to absence of meiosis.

(c) It produces virus-free scion materials.

(d) It bypasses the production of seeds.

(e) It provides expression of characters like hybrid vigour.

(f) It produces viable pollen.

(g) It has great scope in breeding experiment.

PARTHENOCARPY:

It is the production of seedless (sterile) fruit without pollination and fertilisation. The term 'Parthenocarpy' was coined by Noll (1902). It is commonly found in those plants which have large number of ovules per ovary such as

Citrus plants, Grapes, Pineapple, Melon, Fig etc.

Types: Nitsch (1963) recognised five categories of parthenocarpy.

1. Natural Parthenocarpy: Parthenocarpic fruit occurs naturally. e.g., Banana, Grapes, Orange, Pineapple.
2. Genetic parthenocarpy: It occurs due to mutation in axillary bud or hybridisation (crossing) e.g., Navel orange, Cucurbits, Grapes.
3. Artificial or chemical parthenocarpy: It occurs when auxin was applied after anthesis thus ovary starts maturing or ripening. e.g., Tomato. It also occurs when gibberellin was applied at anthesis hence ovary starts ripening before fertilisation. e.g., Grapes
4. Environmental parthenocarpy: It occurs when sexual reproduction is checked due to low temperature, fog or frost hence sex organs do not function. e.g., Chilly, Pear, Olive.
5. Physiological parthenocarpy: It occurs due to delayed pollination or zygotic sterility.

Advantages:

(a) Seedless fruits have an increased proportion of edible parts than normal fruits.

(b) These are widely used in the fruit processing industry for preparing jam and jelly.

Disadvantage: Parthenocarpy is undesirable in those plants where seeds are eaten such as Pomegranate, Cashew, *Pistachio*, Walnut.

SEED DORMANCY

It is the condition in which seeds are unable to germinate even after getting all the necessary conditions. During this period growth of seeds remains suspended which occurs due to internal factors and some specific environmental factors.

Causes of dormancy:

(a) Hard seed coat impermeable to gases: e.g., Apple

(b) Hard seed coat impermeable to chemicals: e.g., *Xanthium*

(c) Hard seed coat impermeable to water: e.g., Legumes, *Chenopodium, Trigonella*

(d) Mechanically resistant seed (embryo is unable to grow): e.g., *Alisma, Amaranthus, Capsella,* Mustard.

(e) Immaturity of embryo at the time of shedding: e.g., *Eranthis, Ginkgo, Ranunculus*

(f) Absence of light (Negatively photoblastic): e.g., Lettuce, Onion, Tobacco, Tomato, *Phlox* etc.

(g) Absence of after-ripening period (seeds undergo a period of rest even when condition is favourable): e.g., Iris, Peach, Poplar, Tulip etc.

(h) Presence of excessive salts at seed shedding: e.g., *Atriplex*

(i) Presence of inhibitors (ABA, NH_3, Coumarin, Phenolics) in seeds: e.g., Cucurbits, *Xanthium, Iris*

Methods to break seed dormancy (or to start seed germination)

1. Scarification: It is a method of softening and weakening of seed coat by hot water, acids, alcohol (chemical method) and by thrashing with a machine, rupturing with hammer and cutting from a scalpel (mechanical method). It is done in *Cassia, Melilotus*
2. Stratification: It is the treatment of seed in well-aerated and moist condition at very low temperature (0-5°C) just after a freezing point.
3. Alternating high and low temperature: This is done in seed of immature embryo and the difference of temperature should be 10°C to 20°C.
4. Exposure to light: Red light (<730 nm) breaks dormancy (seed germination) and Far red light (>730nm) maintains dormancy (inhibits seed germination).
5. Soaking seeds in chemicals (KNO_3, H_2SO_4, thiourea, gibberellin, cytokinin, ethylene).
6. Destruction of Inhibitors:

Importance:

(a) Seeds can be stored during unfavourable period (drought, cold, high temperature)

(b) Seeds can remain alive in soil for several years.

(c) Seed dormancy is essential for seed disposal.

SEED GERMINATION

It is sprouting of seed and growth of embryo present inside seed into a seedling capable of independent existence.

Process:

(a) Seed absorbs water through micropyle by imbibition and osmosis.

(b) Hydration of cell organelles occurs

(c) Embryo is activated.

(d) Seed coat swells, softens and ruptures by cell elongation.

(e) Metabolic activity stars and respiratory enzymes become active.

(f) Seed respires initially anaerobically and later aerobically when O_2 enters into seed coat and releases heat energy.

(g) Seed gets requisite energy for the growth of an embryo.

(h) Embryo secretes hydrolytic enzymes such as amylase, protease, and lipase which convert insoluble reserve food (starch, protein, oil) into soluble substances.

(i) Embryo utilises these soluble substances

(j) Cell division starts in radicle and plumule.

(k) Radicle firstly comes out of seed coat and forms tap root.

(l) Plumule comes out rupturing seed coat and forms shoot and

(m) Finally seedling is formed.

Types of seed germination:

Epigeal germination	Hypogeal germination
Cotyledon comes out of the soil surface by greater elongation of hypocotyl (part of axis below cotyledon). Cotyledon is flat, green and leaf-like.	Cotyledon remains inside or just on the surface but plumule comes out of the soil by greater elongation of epicotyl. In this, hypocotyl does not grow much.
e.g., Bean, Castor, Cotton, Cucurbits, Mustard, Papaya, Poppy, Tamarind, Mustard, Onion	e.g., Cereals, Gram, Coconut, Date palm, Fan palm, Groundnut, Mango, Jack fruit, peas.

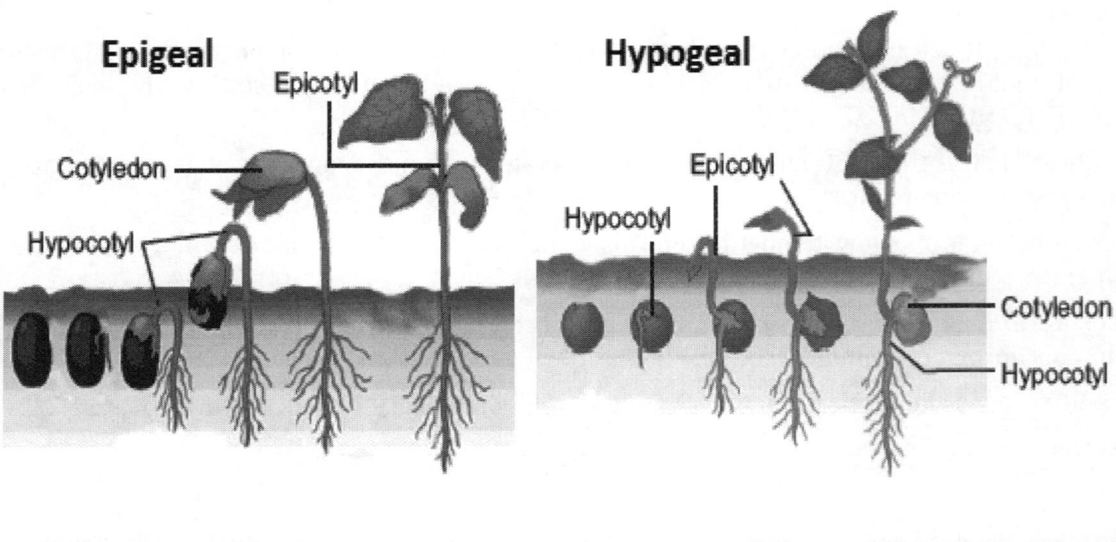

Factors:

(a) Water: It causes softening of seed coat due to seed coat.

(b) Light: Red light is stimulatory but far-red light and blue light are inhibitory to seed germination.

(c) Temperature: The required optimum temperature is 25-30°C.

(d) Oxygen: It is essential for seed germination except Rice and *Typha*.

(e) Availability of food:

(f) Viability: It is the ability to germinate. It is maximum in Lotus (*Nelumbo*).

(g) Inhibitors: These chemicals such as phenolics, tannins, alkaloids, lactone, mustard oil, scopoletin etc. inhibit seed germination.

Vivipary: It is the germination of seed while it is still attached to the parent plant or before detachment from fruit. In this seed does not undergo dormancy and thus forms a seedling. Due to its increasing weight, the seedling separates from the parent plant and falls into mud or soil hence plant is established. It is a halophytic adaptation and is found in *Rhizophora* and other mangrove plants. It is not desirable in crop plants because seeds cannot be stored in this condition.

SHORT ANSWER TYPE QUESTIONS

1. Write notes on the essential whorl of the flower.
2. Distinguish between glandular and amoeboid tapetum.
3. Distinguish between tetrahedral spore tetrad and isobilateral spore tetrad.
4. Define homogamy and cleistogamy with respect to self-pollination.
5. Write notes on pollen viability.
6. Draw a well-labelled diagram of anatropous ovule.
7. What are the advantages of cross-pollination?
8. Distinguish between epihydrophily and hypohydrophily.
9. Write notes on pseudocopulation.
10. Write notes on pollen-pistil interaction.
11. Write notes on egg apparatus.
12. Discuss the post-fertilisation changes in flower.
13. Write notes on double fertilisation.
14. Distinguish between dicot embryo and monocot embryo.
15. Distinguish between true polyembryony and false polyembryony.

LONG ANSWER TYPE QUESTIONS

1. Describe the development of anther in flower.
2. Discuss the microsporogenesis and microgametogenesis in the flower.
3. Describe the development of ovule in flower.
4. Discuss the megasporogenesis and megagametogenesis in the flower.
5. What is cross-pollination? Mention the different adaptations to ensure cross-pollination.
6. Discuss the process of fertilisation with ell labelled diagram.
7. Describe the development and types of endosperms.
8. Describe the development of the dicot embryo.
9. Mention the different types of polyembryony with examples.
10. Mention the different types of apomixis with examples.

MULTIPLE CHOICE QUESTIONS

1. **How many nuclei are involved in double fertilisation if the tetraploid male plant is crossed with a hexaploid female plant?**
(a) 7 (b) 9
(c) 11 (d) 13

2. **In anatropous ovule the micropyle is**
(a) in a straight line with funicle
(b) at right angles with funicle
(c) at 45° with funicle
(d) side by side with funicle

3. **Caruncle is derived from:**
(a) Peduncle (b) Cotyledon
(c) Integument (d) Funicle

4. **Modified shoot with highly condensed internodes is:**
(a) Leaf (b) Flower
(c) Microsporophyll (d) Megasporophyll

5. **Anthers are generally:**
(a) Monosporangiate (b) Bisporangiate
(c) Tetrasporangiate (d) Trisporangiate

6. **Pollen grains are preserved for long periods in fossils due to**
(a) Cellulose (b) Sporopollenin
(c) Carotenoids (d) Ubisch bodies

7. **The obturator in the ovule is generally derived from**
(a) Placenta (b) Funiculus
(c) Integument (d) Nucellus

8. **The constant feature of the embryo sac is**
(a) Synergids (b) Antipodals
(c) Egg (d) Polar nuclei

9. **Filiform apparatus is the feature of**
(a) Egg (b) Synergid cells
(c) Antipodal cell (d) Zygote

10. **The 2 polar nuclei of the embryo sac are derived from**
(a) Both from the micropylar end
(b) Both from chalazal end
(c) One each from the micropylar and chalazal end
(d) None of the above

11. **The flowers which expose their mature anthers and stigma to the pollinating agents are called**
(a) Cleistogamous (b) Autogamous
(c) Pseudogamous (d) Chasmogamous

12. **The number of nuclei involved in double fertilization is**
(a) 3 (b) 4
(c) 5 (d) 7

13. **The pollen-pistil rejection reaction in a plant showing gametophytic self-incompatibility usually occurs**
(a) on the stigma
(b) in the style
(c) in the ovary
(d) during gametic union

14. **Which type of spore tetrad is formed as a result of successive types of cytokinesis?**
(a) Isobilateral (b) Tetrahedral
(c) Decussate (d) T-shaped

15. **The ovule found in gymnosperms is**
(a) Anatropous (b) Orthotropous
(c) Amphitropous (d) Hemitropous

16. **Perisperm is the remaining part of**
(a) Endosperm (b) Ovule
(c) Nucellus (d) Integument

17. **How many nuclei are present in fully developed male gametophyte of flowering plants?**
(a) 1 (b) 2
(c) 3 (d) 4

18. **The development of embryosac from any cell of nucellus is called:**
(a) Apogamy (b) Apospory
(c) Parthenogenesis (d) Diplospory

19. **Tapetal cells in anther are:**
(a) Haploid (b) Polyploid
(c) Diploid (d) Triploid

20. **The stamen of the flower represents:**
(a) Microporangium (b) Megasporangium
(c) Megasorophyll (d) Microsporophyll

21. **Endosperm in coconut is:**
(a) Cellular (b) Nuclear
(c) Both cellular and nuclear
(d) Neither cellular nor nuclear

22. **In adventive embryony, embryos develop from**
(a) Endosperm (b) Synergids
(c) Nucellus (d) Egg

23. **When a diploid female plant is crossed with a tetraploid male plant, the ploidy of endosperm cells in the resulting seed is**
(a) Diploid (b) Triploid
(c) Tetraploid (d) Pentaploid

24. **At which stage, one can distinguish between the nuclear and cellular types of endosperm?**
(a) Stage following division of primary endosperm nucleus
(b) Mature stage of endosperm
(c) When the embryo has started divisions
(d) At the heart-shaped stage of the embryo

25. **Megasorangium with its protective layers called integuments is called**
(a) Ovary (b) Ovule
(c) Funicle (d) Chalaza

26. **Morphologically and physiologically specialized outer layers of endosperm of cereals constitute:**
(a) Tapetel tissue (b) Perisperm
(c) Aleurone layers (d) Chalazosperm

27. **Exine of pollen is formed by the activity of**
(a) Tapetum (b) Endothelium
(c) Middle layers (d) Endothecium

28. **Cells of tapetum are characterized by**
(a) Free cell formation
(b) Meiotic divisions
(c) Amitosis and endomitosis
(d) Mitosis and endomitosis

29. **Which is a chemotropic agent in gynoecium?**
(a) K^+ ions (b) Ca^{2+} ions
(c) Mg^{2+} ions (d) Na^+ ions

30. **The tectum, baculum and foot layer are found in**
(a) Pollen wall
(b) Integuments of ovule
(c) Pollen kitt
(d) Embryo sac

31. **If the number of chromosomes in egg cell is 8, then the number of chromosomes in the endosperm is**
(a) 8 (b) 12
(c) 16 (d) 24

32. **A close relation between flower and pollinating agent is best exhibited by**
(a) *Cocos* (b) *Salvia*
(c) *Yucca* (d) *Avena*

33. **If there are 4 cells in anther, what will be the number of pollen grains?**
(a) 8 (b) 4
(c) 16 (d) 12

34. **In flowering plants, the male gametes are formed by the:**
(a) Generative cell
(b) Vegetative cell
(c) Pollen tube
(d) Uninucleate microspore

35. **Adventive embryony in Citrus is due to**
(a) Nucellus (b) Integument
(c) Zygotic embryo (d) Fertilized egg

36. **In a mature embryo sac, the central cell is**
(a) Binucleate (b) Single nucleate
(c) Tetranucleate (d) Eight nucleate

37. **The point of attachment of the ovule to the funicle is known as:**
(a) Raphe (b) Chalaza
(c) Micropyle (d) Hilum

38. **Covering of microspores tetrad is made of:**
(a) Callose (b) Pectocellulose
(c) Sporopollenin (d) Pectin

39. **The aleurone layer in maize grain is especially rich in:**
(a) Starch (b) Lipid
(c) Hormone (d) Protein

40. **The type of pollen grains found in insect-pollinated flowers is**
(a) Hygroscopic
(b) Rough and sticky
(c) Smooth and dry
(d) Heavy and coloured

41. **Which of the following pairs of plant parts are both haploid?**
(a) Nucellus and antipodal cells
(b) Antipodal cells and megaspore mother cell
(c) Nucellus and primary endosperm nucleus
(d) Antipodal cell and egg nucleus

42. **In fully organised *Polygonum* type of embryo sac, the ratio of haploid, diploid and triploid nuclei is**
(a) 3: 1: 3 (b) 6: 0: 1
(c) 6: 1: 0 (d) 3: 2: 3

43. The direction of micropyle in anatropous ovule is

(a) Upward (b) Downward
(c) Right (d) Left

44. Cellular totipotency is demonstrated by

(a) Gymnosperms (b) All plants
(c) Eukaryotes (d) Bacteria

45. Apomictic embryos in citrus plants arise from

(a) Synergids
(b) Maternal sporophytic tissue
(c) Antipodals
(d) Egg cell

46. Even in the absence of pollinating agents, seed setting is assured in

(a) *Commelina* (b) *Zostera*
(c) *Salvia* (d) Fig

47. An organic substance that can withstand environmental extremes and cannot be degraded by any enzyme is:

(a) Cuticle (b) Sporopollenin
(c) Lignin (d) Cellulose

48. In the embryos of a typical dicot and monocot, true homologous structures are

(a) Coleorhiza and coleoptile
(b) Coleoptile and scutellum
(c) Cotyledons and scutellum
(d) Hypocotyl and radicle

49. Anthesis is a phenomenon which refers to:

(a) Reception of pollen by stigma
(b) Formation of pollen
(c) Development of anther
(d) Opening of flower bud

50. How many meiotic divisions are required for the formation of 100 seeds?

(a) 25 (b) 50
(c) 125 (d) 400

51. How many pollen grains (microspores) are produced by each head inflorescence of the Asteraceae family having 10 actinomorphic flowers if each anther produces 20 pollen grains?

(a) 500 (b) 800
(c) 1000 (d) 1200

52. A bilobed dithecous anther has 100 microspore mother cells per microsporangium, then how many male gametophytes are produced from this anther

(a) 400 (b) 800
(c) 1200 (d) 1600

53. The arrangement of megaspore tetrad in angiosperms is

(a) Linear (b) Tetrahedral
(c) Isobilateral (d) T-shaped

54. The number of female nuclei involved in double fertilisation is

(a) 1 (b) 2
(c) 3 (d) 4

55. Which of the following parts of gynoecium determines the compatible nature of pollen grain?

(a) Ovary (b) Pistil
(c) Egg (d) Ovule

56. In angiosperms, the functional megaspore develops into

(a) Ovule (b) Embryo sac
(c) Endosperm (d) Pollen sac

57. Long ribbon-like pollen grains are seen in

(a) Wind pollinated plants
(b) Bird pollinated plants
(c) Gymnosperms
(d) All of the above

58. Both autogamy and geitonogamy are prevented in

(a) Papaya (b) Castor
(c) Maize (d) Cucurbits

59. A pollen mother cell undergoes microsporogenesis and forms 4 microspores. Each of these forms 2 male gametes. How many meiosis and mitosis are required for the formation of 4 male gametophytes from a pollen mother cell?

(a) 1 meiosis and 4 mitosis
(b) 2 meiosis and 4 mitosis
(c) 1 meiosis and 8 mitosis
(d) 2 meiosis and 4 mitosis

60. One-celled suspensor is found in

(a) *Triticum* (b) *Mangifera*
(c) *Pisum* (d) *Solanum*

61. Which of the plants does not show clear-cut vegetative, reproductive and senescent phases?

(a) Annuals (b) Biennials

(c) Perennials (d) Both a and b

62. The proximal part of a filament of the stamen is attached to

(a) Thalamus or sepal

(b) Thalamus or petal

(c) Pedicel

(d) Ovary

63. In a typical embryo sac, the nuclei are arranged as

(a) 3+2+3 (b) 3+3+2

(c) 2+3+3 (d) 2+4+2

64. Long silky hairs coming out of cob of maize are meant for

(a) Fruit dispersal

(b) Attracting insects

(c) Protecting seeds

(d) Long styles and stigma

65. Emasculation is not required when flowers are

(a) Unisexual (b) Bisexual

(c) Intersexual (d) Both b and c

66. Perisperm differs from endosperm in

(a) Having no reserve food

(b) Being a diploid tissue

(c) Being a haploid tissue

(d) Formation by fusion of secondary nucleus with several sperms

67. The number of male gametes formed by 16 microspore mother cells is

(a) 16 (b) 32

(c) 64 (d) 128

68. Emasculation is not required for artificial hybridisation in

(a) Castor (b) Maize

(c) Papaya (d) All of the above

69. The aleurone layer is the outermost layer of endosperm in cereals. It helps in

(a) Protection of endosperm

(b) Growth of endosperm

(c) Mobilisation of reserve food in endosperm

(d) Accumulation of reserve food in endosperm

70. Plants growing in harsh environments such as deserts, sand dunes and arctic tundra often reproduce vegetatively because

(a) There are few animals for pollination

(b) Fruits would freeze or dry out in such condition

(c) Vegetative reproduction is not as risky as making seeds

(d) These are members of such families that reproduce asexually.

71. Egg apparatus consists of

(a) One egg cell and one synergid cell

(b) One egg cell and two synergid cells

(c) Two egg cells and one synergid cell

(d) One egg cell and several synergid cells

72. Unisexuality of flowers prevents

(a) Autogamy but not geitonogamy

(b) Autogamy and geitonogamy

(c) Allogamy and geitonogamy

(d) Geitonogamy but not xenogamy

73. Which of the following floral parts forms a pericarp after fertilisation?

(a) Outer integument

(b) Inner integument

(c) Nucellus

(d) Ovary wall

74. Male gametophyte with least number of cells is present in

(a) *Funaria* (b) *Pteris*

(c) *Pinus* (d) *Pisum*

75. Pollen grains can be stored for several years in liquid nitrogen having temperature of

(a) -50°C (b) -100°C

(c) -150°C (d) -196°C

76. A dioecious flowering plant prevents both

(a) Autogamy and xenogamy

(b) Autogamy and geitonogamy

(c) Geitonogamy and xenogamy

(d) None of the above

77. Pseudocopulation occurs between

(a) *Salvia* and Bee

(b) *Ficus* and Gall wasp

(c) *Yucca* and Pronuba moth

(d) *Ophrys* and Hairy wasp

78. If a diploid male plant and a tetraploid female plant are crossed, the ploidy level of endosperm shall be

(a) Triploid (b) Tetraploid
(c) Pentaploid (d) Hexaploid

79. Which of the following ovules shows a straight line of micropyle and nucellus?

(a) Anatropous (b) Orthotropous
(c) Hemitropous (d) Amphitropous

80. Translator condition of pollen grains is seen in

(a) Orchids (b) Cucurbits
(c) *Calotropis* (d) *Mirabilis*

81. External water is not essential for fertilization in

(a) Thallophyta (b) Bryophyta
(c) Pteridophyta (d) Spermatophyta

82. The pollen tube at the time of entering the embryo sac has

(a) 1 gamete (b) 2 gametes
(c) 4 gametes (d) 8 gametes

83. The movement of the pollen tube towards the embryo sac is

(a) Thigmotactic (b) Phototactic
(c) Thermotactic (d) Chemotactic

84. Nucellar embryo is

(a) Apomictic haploid
(b) Apomictic diploid
(c) Amphimictic haploid
(d) Amphimictic diploid

85. The best example of polyembryony is

(a) *Cycas* (b) *Citrus*
(c) *Cocos* (d) *Capsicum*

86. Meiosis takes place when there is a change in

(a) From gametophyte to sporophyte
(b) From sporophyte to gametophyte
(c) From gametophyte to gametophyte
(d) From sporophyte to sporophyte

87. Fruits have no significance if parthenocarpy is done in

(a) Grapes (b) Pomegranate
(c) Banana (d) Tomato

88. The time between germination of pollen grains and fertilization is minimum in

(a) Barley (b) Tomato
(c) Corn (d) Cabbage

89. The tapetal nuclei in anthers of many angiospermic flowers are

(a) Mostly haploid
(b) Mostly diploid
(c) Mostly aneuploid
(d) Mostly polyploid

90. A polygonum type of embryo sac is

(a) 7-celled and 8-nucleate
(b) 8-celled and 7-nucleate
(c) 7-celled and 7-nucleate
(d) 8-celled and 8-nucleate

ANSWERS

1.d	11.d	21.c	31.d	41.d	51.c	61.c	71.b	81.d
2.b	12.c	22.c	32.c	42.a	52.d	62.b	72.a	82.b
3.c	13.b	23.c	33.c	43.b	53.a	63.a	73.d	83.d
4.b	14.a	24.a	34.a	44.b	54.c	64.d	74.d	84.b
5.c	15.b	25.b	35.a	45.b	55.b	65.a	75.d	85.b
6.b	16.c	26.c	36.a	46.a	56.b	66.b	76.b	86.b
7.b	17.c	27.a	37.d	47.b	57.b	67.d	77.d	87.b
8.a	18.b	28.d	38.a	48.c	58.a	68.d	78.c	88.a
9.b	19.b	29.b	39.d	49.d	59.c	69.c	79.b	89.d
10.c	20.d	30.a	40.b	50.c	60.a	70.c	80.c	90.a

CHAPTER-12

PRINCIPLES OF INHERITANCE AND VARIATION

Genetics is the study of inheritance and variation of characters from one generation to another. The term 'Genetics' was coined by William Bateson.

Classical genetics: It is the study of phenotype leading to the isolation of genes and the study of DNA sequences.

Reverse Genetics: It is the study of DNA sequences leading to the determination of unknown phenotypic effects and their introduction in plants or animals.

Father of Classical Genetics	Gregor Johann Mendel
Father of Modern Genetics	Bateson
Father of Experimental Genetics	Thomas Hunt Morgan
Father of Cytogenetics	Muller
Father of Human and Biochemical Genetics	Archibald Garrod

Heredity is the transmission of characters from one generation (parents) to another (offspring) and fixed for a particular individual. It is based on 'like begets like'. Thus, human beings are like human beings in overall characteristics. Although the offspring differ from their parents, they resemble much and inherit the parental characters.

THEORIES OF HEREDITY:

1. **Leeuwenhoek** (1677) observed the human sperm for the first time and termed it 'animalcules'. He proposed that each sperm can develop into a new individual when introduced into the womb of a woman where it could get nourishment. This was rejected because maternal characters are also inherited in offspring.
2. **John Swammerdam** (1679) proposed **preformation theory**. The organism is already present or preformed in the sperm or egg in a miniature form called homunculus which is enlarged to form an embryo.
3. **Maupertius** proposed **particulate theory**. Particles are found in the semen of each animal and the union of particles of both parents results in the formation of a new individual.
4. **Wolff** proposed **theory of epigenesis**. The organs of the body were developed gradually from undifferentiated living materials but sperm and egg contains some definite substances.
5. **Charles Darwin** proposed **a theory of pangenesis**. Pangenes or minute gemmules are produced in every cell, tissue and organ. All the pangenes of the body collect in the gametes via blood and are passed on to the zygote where they guide the growth of different parts of the embryo to form an offspring. Hence offspring has a blend (mixture) of the characters of both parents. Several **objections** are as follows:
   (a) **Weismann** (1892) proposed the **theory of continuity of germplasm.** He observed that the tail of mice was cut and studied for up to 22 generations. He found that the complete tail structure was still inherited. (if the tail was removed then pangenes for the tail will not be available so the tail should not have developed in the next generation). He assumed that germplasm is immortal, and it alone is passed from generation to generation. The somatoplasm that forms the body is mortal and perishes when an organism dies.

   (b) Characters do not blend or mix and differ from those of parents. Some characters of a distant parent (grandparent) are expressed in the offspring. It shows that characters may remain unexpressed for one or more generations but do not blend.

   (c) The character of sex does not blend itself in unisexual organisms. Such organisms can either be male or female.

   (d) **Kolreute**r (1760) obtained fertile interspecific hybrids in Tobacco. These do not resemble either of the parents. Hybrids were self-pollinated. Some offspring resembled the hybrid while the remaining resembled one of the grandparents in different characters. This proves that traits have a particulate nature and remain discrete.

Variation: It is the degree of difference in the progeny or between progeny and parent. It is also referred to as a single difference in a trait. The difference in characters of offspring mainly depends on the unique process of crossing over which occurs during meiosis. Variation occurs commonly in sexually reproducing organisms. It does not occur in identical twins and asexually reproducing organisms. Not all organisms are similar to their parents. Certain characters appear to belong to neither of their parents. These dissimilarities are called variations. Variations are found in all characters and every direction. Variation becomes fewer if there is intensive breeding (mating between closely related organisms).

Types of variation: On the basis of cells affected:

1. **Somatic variations:** These occur in somatic or body cells of organisms. These are also called modification or acquired characters because these are got by an individual after birth during their lifetime. These are non-inheritable because they disappear with the death of an individual. These occur due to the impact of environmental factors such as **light** (If there is no light, plants become etiolated showing longer internodes, and thinner and broader leaves) and in the presence of light, plants develop more mechanical tissues and thicker smaller leaves), **temperature** (in low temperature, the root system is not well developed, metabolic activity is reduced and rate of transpiration is decreased and in high temperature, root system is well developed, metabolic activity is increased and rate of transpiration is increased.) and **water** (in scarcity of water, rate of transpiration is reduced, retentivity of water (succulence) occurs, and there is formation of spines, small leaves, sunken stomata and thick coating on leaves and in the abundance of water there is luxuriant growth in plants). The somatic variation also occurs due to the use and disuse of organs, and conscious efforts.
2. **Germinal variations:** These occur in germ cells or reproductive cells. These are heritable. These help organisms adapt to a particular environment. These are of two types:
   (a) **Continuous variations**: These are fluctuating in nature. These are already present in different organisms or races of a species. These are caused by **recombination** which occurs due to the segregation of chromosomes during gamete formation in meiosis, random fertilisation (fusion of gametes) and crossing over in meiosis.
   (b) **Discontinuous variation**: These are sudden or unpredictable in nature. These form the basis of the evolution of new species. These are caused by **mutation** which occurs due to changes in the base sequence (structure) of the gene, structural aberration in chromosomes and numerical aberration in chromosomes.

Significance of variations:

(a) Variations make organisms better fitted in the struggle for existence.
(b) These help individuals adapt to a changed environment.
(c) These provide raw materials for the evolution of species.
(d) These develop new traits.
(e) These improve the races of plants and animals with respect to resistance, yield and growth.

MENDELIAN INHERITANCE

Gregor Johann Mendel worked for 7 years from 1856 to 1863. He conducted experiments of hybridisation on 29000 pea plants. He read out his results of observations in the Natural History Society of Brunn in 1856. He published his research paper in the German language entitled "Experiments on Plant Hybridisation" in 1866 in the Annual Proceedings of the Natural History Society of Brunn. He also worked on hawkweed (*Hieracium*) and hyacinth Bean (*Dolichos lablab*) but did not get success. He died in 1884.

Mendel's experiments:

1. **Selection of pure parental plant**: Mendel selected Garden pea (*Pisum sativum*, 2n=14) because
   (a) Pea plant has distinct, well-defined, and contrasting characters.

(b) It is bisexual and self-pollinated hence it is true breeding (purity of the same trait or character is maintained from generation to generation).

(c) It is much easier to cross-pollinate (hybridise) the flowers by removing anthers (emasculation) before the maturity of pollen grains.

(d) It is annual hence it has a short life span.

(e) It is easy to grow, maintain and manage.

(f) It produces a large number of seeds in one generation.

He selected seven traits each having two forms or variants of pure or true breeding variety of pea (initially he took 34 traits, then 22 traits and finally 7 traits). All the selected traits had easily distinguishable alternate traits. The genes for 7 pairs of characters (traits) are such that some are present on the same chromosome but not linked. Thus, these seven traits are situated only on four chromosomes (1st,4th, 5th and 7th).

Trait/character	Dominant form	Recessive form	Position on chromosome
Seed colour	Yellow (Y)	Green (y)	1
Seed shape	Round (R)	Wrinkled (r)	7
Flower colour	Purple (P)	White (p)	1
Flower position	Axial (A)	Terminal (a)	4
Pod colour	Green (G)	Yellow (g)	5
Pod shape	Inflated (I)	Constricted (i)	4
Plant height	Tall (T)	Short (t)	4

2. **Hybridisation of Pure pea plants for obtaining F_1 generation**: He performed reciprocal crosses between plants having contrasting forms of a character such as tall and short. The flower on the plant (tall) was emasculated (anthers were removed) and pollens from the flower of another plant (short) were dusted over the stigma of the tall plant. Now flower was covered with a paper bag to avoid the entry of unwanted pollen. The seeds formed by crossing the plants were collected and sown. Hybrid plants (first filial or F_1) were produced the next year.
3. **Self-pollination of Hybrid plants for obtaining F_2 generation**: Self-pollination was done in F_1 hybrid plants. The flowers were covered in paper bags to keep out foreign pollen. The seeds produced were collected and sown. Later plants of the next generation (second filial or F_2) were produced and thus it was observed that seeds of tall plants produced tall plants, yellow seeds produced plants of yellow seeds and so on.

Findings of the experiments:

1. F_1 plants of reciprocal crosses were similar.
2. F_1 plants were not intermediate between the two contrasting forms of a trait. Rather they resembled one parent in having a single alternate form. There is no mixing of the two factors in the hybrid.
3. Both the parental forms of a trait are expressed in the F_2 generation.
4. One form of a trait that did not appear in the F_1 generation must be hidden or unexpressed.
5. The organism should possess two factors of each trait (**Principle of paired factors**).
6. Out of two factors (alleles) representing the particular trait, one is dominant and expresses itself in the hybrid or F_1 generation. The other factor or allele is recessive and does not show its effect (**Principle of dominance**).
7. The two factors segregate or separate at the time of gamete formation and pass into different gametes. A gamete contains one factor of a pair. Thus, Mendel predicted the occurrence of meiosis long before it was discovered. The gametes fuse randomly during fertilisation so that factors come together in the new generation and express themselves.
8. The two forms of a trait appear in the F_2 generation in the ratio of three dominant to one recessive, 3:1(monohybrid ratio) and out of dominant individuals, one is pure breeding (homozygous dominant) and two are hybrids (heterozygous dominant). This is possible only when the two factors of each trait are segregated during gamete formation (**Principle of segregation**).

9. In a dihybrid cross, four types of plants are formed in the F_2 generation, in which there are two parental and two recombinants. The ratio is 9 (both dominant): 3(one dominant second recessive): 3 (one recessive second dominant): 1 (both recessive). It is a dihybrid ratio. This shows that the factors or alleles of two traits assort independently (**Principle of independent assortment**).

Basic features of Heredity

(a) All traits have alternative or contrasting forms or variants.

(b) Traits are developed in individuals by discrete particulate entities that do not blend or change.

(c) One alternate form of a trait may be expressed more often than another.

(d) An alternate form of a trait may remain hidden for one or more generations and then reappear later in an unchanged state.

(e) Particular forms of two or more traits may occur together in one generation and separate in the next generation.

(f) One alternate form of trait present together in an individual is expressed more often than the other.

Basis of heredity:

1. **Physical basis**: Mendel proposed that heredity was controlled by factors (now called as genes) which are present in all the cells and are transferred to the next generation through gametes. These factors or genes are the physical basis of heredity.
2. **Chemical basis**: It is confirmed that heredity is controlled by genes that are situated in the chromosomes present in the nucleus of gametes. The genes are made up of DNA and protein. DNA functions as genetic material and forms the chemical basis of heredity. Genes code for proteins that develop traits of an organism.

Rediscovery of Mendel's work: His works remained unnoticed and unappreciated for 34 years due to the following reasons:

(a) He was himself not sure of his findings because he did not succeed in his earlier works.

(b) He proposed a concept of stable and unblending discrete unit factor.

(c) He failed to prove the physical existence of factors.

(d) He worked ahead of his time because at that time Biologists were impressed with the theories of Darwin (Origin of Species,1859).

(e) He introduced statistical analysis which was new at that time.

(f) He failed to project his findings due to the limited circulation of the journal and lack of communication.

However, the works of Mendel were found to be correct due to the independent findings of these three scientists who were credited with the rediscovery of Mendelism.

1. Hugo de Vries of Holland worked on evening primrose (*Oenothera lamarckiana*).
2. Carl Correns of Germany worked on maize, and
3. Erich Von Tschermak of Austria worked on different flowering plants.

Finally, Correns assigned Mendel's postulates the status of law. Bateson and Punnet found Mendel's works to be of universal application including animals.

Terms: Bateson coined the terms 'allele, F_1, F_2, homozygous, heterozygous, epistasis etc. Johannsen worked on a princess variety of red kidney beans or rajma (*Phaseolus vulgaris*) and he coined the terms 'gene', 'genotype' and 'phenotype'.

1. Unit factor: It is a gene that controls a trait.
2. Trait: It is an inherited character or feature.
3. Allele or Allelomorph: It is two contrasting forms of a factor (gene) or it is any two forms of a gene present on the same locus in a homologous pair of chromosomes and controlling the same trait (height of plant) but producing different effects (tall or short).
4. Phenotype: It is the external physical appearance of an individual.

5. Genotype: It is the genetic constitution of an individual.
6. Dominant: It is a trait that will appear in the offspring. It will appear if a dominant allele is present.
7. Recessive: It is a trait that will appear either in the absence of a dominant allele or in the presence of an identical allele.
8. Homozygous: It is a diploid condition where two alleles of a gene for a particular character are similar. It possesses dominant alleles or recessive alleles.
9. Heterozygous: It is a diploid condition where two alleles of a gene for a particular character are dissimilar. It possesses both dominant and recessive alleles.
10. Hemizygous: It is a diploid condition where one allele of a gene is missing.
11. Cross: It is mating between two individuals leading to the fusion of gametes.
12. Monohybrid cross: It is a cross involving one pair of contrasting characters.
13. Dihybrid cross: It is a cross involving two pairs of characters.
14. Hybrid: It is an offspring produced from a cross between two genetically different plants.
15. Wild allele: It is original, common or widespread in population and dominant in expression.
16. Mutant allele: It is an altered version, less common in the population and recessive in expression.
17. Epistasis: It is a type of gene interaction when one gene masks or modifies the expression of another gene at a distant locus. The former gene is called the epistatic gene, and the latter gene is known as the hypostatic gene. It is intergenic or non-allelic.
18. Punnet square: It is a checkerboard (simple box-like) developed by Punnet (1927). In this, female gametes are arranged vertically, and male gametes are arranged horizontally. It shows the result of a cross between two organisms.
19. Phenocopy: It is the different appearance (phenotypes) of plants in different seasons of the year such as the growing season, flowering season, fruiting season or season of leaf fall. It is not hereditary.

Mendel's laws of Inheritance:

1. **Law of paired factors**: Each character is controlled by a gene that has at least two alleles. It shows monogenic inheritance. These two alleles (factors) are present at the same locus of two homologous chromosomes at the same locus. These may represent the same (homozygous TT in case of pure tall plant, tt in case of pure dwarf plant) or different expression (heterozygous Tt in case of hybrid tall plant) of the same trait.

   Exceptions: (a) Multiple alleles (b) Polygenic inheritance

2. **Law of dominance**: If two organisms of contrasting characters are crossed then only one form (dominant) of character is expressed in the offspring and the other form (recessive) remains unexpressed.

Pure tall (TT) x Pure dwarf (tt)--------Parent

↓ ↓

T t --------------gamete

↓

Hybrid tall (Tt) ----------------F_1 hybrid

This law explains that individuals of F_1 express the character of one parent. This law is not universally acceptable.

Exceptions are (a) Incomplete dominance (b) Codominance

3. **Law of segregation:** The factors (two alleles of a gene) of each character segregate or separate during gamete formation hence each gamete receives only one factor (allele) for each character.

It was explained by a monohybrid cross.

TT (Pure tall) x tt (Pure dwarf)

↓ ↓

T t

↓

Tt (All hybrid tall)

Selfing of F_1 hybrids (Tt x Tt)

Female/Male gametes	T	T
T	TT	Tt
t	Tt	Tt

F_2 phenotypic ratio= 3 (tall) : 1 (short)

F_2 genotypic ratio= 1 (TT) : 2 (Tt) : 1 (tt)

(homozygous dominant) (heterozygous dominant) (homozygous recessive)

(pure tall) (hybrid tall) (pure dwarf)

This law is universally acceptable. It is applicable for diploid organisms because they form haploid gametes to reproduce sexually. It also explains that hereditary factors are discrete and do not blend when present together.

Bateson called this law as 'Principle of purity of gametes' because the segregation of two Mendelian factors of a trait results in a gamete receiving only one factor (allele) out of a pair. As a result, gametes are always pure for a character. It is also known as the law of non-mixing of alleles.

In haploid organisms such as bacteria, algae and most fungi, the characters are governed by only one allele hence law of dominance and the law of segregation do not apply to them. In these organisms, the monohybrid ratio is 1:1 instead of 3:1 and the dihybrid ratio is 1:1:1:1 instead of 9:3:3:1.

4. **Law of independent assortment**: When two individuals different from each other in two or more characters are crossed then the factors (alleles) of each character are distributed or inherited into the gametes independently of the factors of another character and anyone of a pair of characteristics may combine with anyone of another pair.

   It is explained by a dihybrid cross.

Round and Yellow seeds x Wrinkled and Green seeds

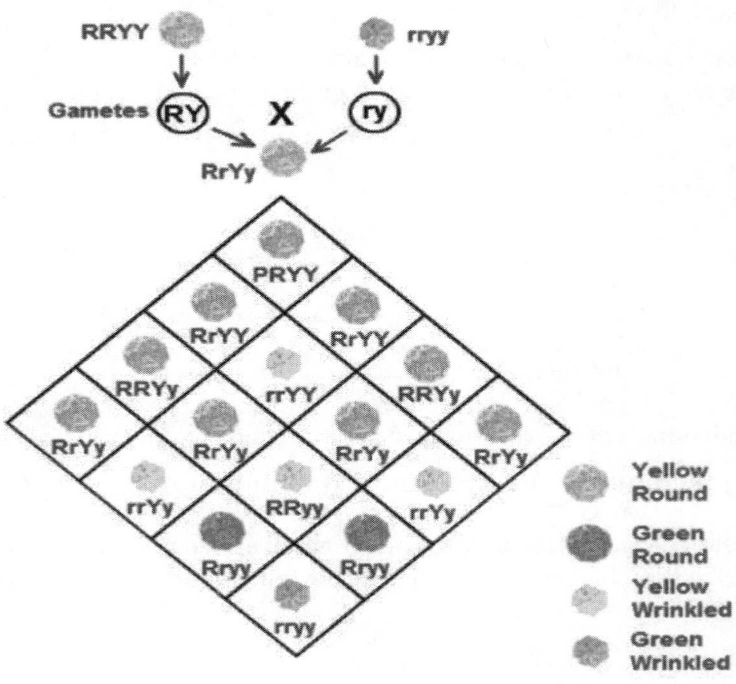

F_2 phenotypic ratio: 9 : 3 : 3 : 1

Number of Phenotypes=4 (Round yellow Round green Wrinkled yellow Wrinkled green)

F_2 genotypic ratio: 1 : 2 : 1 : 2 : 4 : 2 : 1 : 2 : 1

Number of genotypes= 9 (RRYY, RRYy, RRyy, RrYY, RrYy, Rryy, rrYY, rrYy rryy)

It produces four types of plants in F_2 generation. There are two or more parental types which show that the factors of each of the two characters assort independently of others as if the other pair of factors are not present. The outcome of the dihybrid cross makes it very clear that the segregation of seed colour is independent of seed shape and both parental and new combinations of characters appear in F_2offspring. This law clearly states that genes for one trait are not inherited together with another trait.

It is applicable for two or more than two characters. It is applied to genes present far apart on the same chromosome or unlinked genes present on different (non-homologous) chromosomes. It does not apply to genes present on the same chromosome.

This is universally unacceptable. The exception is **Linkage**.

Mendelian crosses:

1. **Reciprocal cross:** It is a cross that involves the same trait but is carried by sexes opposite to those in the original or first cross. When two parents are crossed in such a way that in one experiment, it is used as male/female and in another experiment it is female/male but the result is the same.

   TT (male) x tt (female) TT (female) x tt (male)

   ↓ ↓

   Tt Tt

   Reciprocal cross does not affect those characters which are controlled by nuclear genes. It affects those characters controlled by cytoplasmic genes.

2. **Back cross**: It is a cross between the F_1 hybrid and either of the parents (homozygous dominant or homozygous recessive). It is also called a top cross. It is of two types:

   (a) Outcross: It is a cross between F_1 hybrid and homozygous dominant parent.

   Tt x TT

Female/male gametes	T	T
T	TT	TT
T	Tt	Tt

   Results: All are tall (50% are pure tall and 50% are hybrid tall).

   It is used by breeders to bring useful characters in homozygous condition.

   (b) **Test cross:** It is a cross between F_1 hybrid and homozygous recessive parent.

   Tt x tt

Female/male gametes	t	t
T	Tt	Tt
t	tt	tt

   Results: It is a monohybrid test cross in which 50% are hybrid tall and 50% are short.

It helps to know the genotype of organisms. It is done to determine whether a dominant character is due to a homozygous or heterozygous condition.

It distinguishes (pure) homozygous dominant from (hybrid) heterozygous dominant.

Thus, all test crosses are back cross, but all back crosses are not test cross.

Monohybrid test cross ratio (Aa x aa)	1 : 1
Dihybrid test cross ratio (AaBb x aabb)	1 : 1 : 1 : 1
Trihybrid test cross ratio (AaBbCc x aabbcc)	1 : 1 : 1 : 1 : 1 : 1 : 1 : 1

Account of Mendel's work

Experiment	Number of gametes=2^n= n=number of heterozygote	Number of zygotes= $(\text{Gamete})^2$	Number of phenotype =2^n where n= number of heterozygotes	Number of genotype=3^n where n=number of heterozygote	Phenotypic ratio	Genotypic ratio
Monohybrid cross (Aa x Aa)	$2^n=2^1=2$	$2^2= 4$	$2^n=2^1=2$	$3^n=3^1= 3$	3:1	1:2:1
Dihybrid cross (AaBb x AaBb)	$2^2=4$	$4^2=16$	$2^2= 4$	$3^2= 9$	9:3:3:1	1:2:1: 2:4:2: 1:2:1
Trihybrid cross (AaBbCc x AaBbCc)	$2^3=8$	$8^2=64$	$2^3= 8$	$3^3= 27$	27:9:9: 9:3:3: 3:1	$(1:2:1)^3$

Mendel was successful due to the following reasons which are as follows:

1. He selected only pure or true breeding varieties of garden peas. He took 2 years (1857-59) to check that his experimental materials were true breeding.
2. He identified very clear contrasting characters in Pea plant.
3. He chose those traits that did not show linkage, incomplete dominance or any type of non-allelic gene interaction.
4. He concentrated on the inheritance of one particular trait with contrasting forms at one time.
5. He studied the inheritance for three or more generations.
6. He performed reciprocal crosses.
7. He took care to avoid contamination from pollen grains brought about by insects.
8. He did not explain the variation found in his results but left them as such, e.g., linkage of flower colour and seed colour.
9. He maintained a complete and accurate record of every cross, subsequent selfing and number of seeds produced.
10. He used statistical methods and the law of probability for analysing his results.

Exceptions to Mendelism or Post-Mendelian discoveries:

1. **Incomplete dominance**: It is a phenomenon in which both the alleles fail to express in an F_1 hybrid, or it is the failure of one allele to dominate the other in a heterozygous condition. It comes under intragenic interaction.

   Examples:

   (a) **Flower colour of 4 o'clock plant (*Mirabilis jalapa*) and Snapdragon or Dog flower (*Antirrhinum majus*).** It was discovered by Correns (1903).

Red flower (RR)　　　White flower (rr)

↓　　　　↓

R　　x　　r

↓

Pink flower (Rr)

Selfing of F_1 hybrid (Rr x Rr)

Female/male gametes	R	r
R	RR (red)	Rr (pink)
R	Rr (pink)	rr (white)

The genes for red and white colour do not mix in the F_1 hybrid (Pink colour) as both parental characters (red and white) reappear due to segregation in F_2 plants. Hence it is not called blending inheritance on the basis of different F_2 progeny.

Findings:

(i) Hybrid has a new phenotype (Pink) which is intermediate between those of two alleles. Hence it is called partial or semi-dominance. However, the effect of one of the alleles is more pronounced.

(ii) There is no specific gene for pink colour.

(iii) Phenotypic ratio is 1 (red) : 2 (Pink) : 1 (white) which differs from Mendelian ratio (3:1).

(iv) Genotypic ratio is 1 (RR) : 2 (Rr) : 1 (rr) which is the same as the Mendelian ratio.

(b) **Starch synthesis (size) in seed of pea.** It is not correlated with the shape of a seed. The homozygous dominant (BB) shows a round shape and large size of grain, homozygous recessive (bb) shows a wrinkled shape and small size of grain but the heterozygous dominant (Bb) shows a round shape and intermediate size of grain. Hence size of the grain shows incomplete dominance.

Round seed (large grain) x Wrinkled seed (Small grain)
BB bb
↓
Round seed (intermediate size of grain)
Bb

(c) **Size of leaf in Snapdragon** (*Antirrhinum majus*)

Narrow-leaf x broad-leaf
↓
Intermediate leaf

(d) **Feather colour of Andalusian fowl**

Black plumage x white plumage
↓
Blue plumage

2. **Codominance:** It is a phenomenon where both alleles express themselves fully, equally, independently and simultaneously in a heterozygous (F_1 hybrid) condition. It does not show a dominant-recessive relationship or Mendelian interaction. It is also called mosaic inheritance.

Examples:

(a) **Coat (skin) colour of cattle.**

Red colour (B_1B_1) x White colour (B_2B_2)
↓ ↓
B_1 B_2
↓
Roan (B_1B_2)
Inbreeding of F_1 hybrid (B_1B_2 x B_1B_2)

Female/male gametes	B_1	B_2
B_1	B_1B_1 (Red)	B_1B_2 (Roan)
B_2	B_1B_2 (Roan)	B_2B_2 (White)

Findings:

(i) Both alleles produce phenotype in F_1 hybrid.

(ii) Phenotypic ratio: 1 (Red) : 2 (Roan) : 1 (White)

(iii) Genotypic ratio: 1 (B_1B_1) : 2 (B_1B_2) : 1 (B_2B_2)

(b) **Sickle cell anaemia** of human beings: The carrier individuals (heterozygotes) show codominance because HbA and HbS alleles are present. These are more tolerant of malaria than normal homozygotes.

Normal haemoglobin (Hb^AHb^A) x Sickle celled haemoglobin (Hb^SHb^S)

↓ ↓

Hb^A Hb^S

↓

Carrier (Hb^AHb^S)

Crossing of carrier individuals (Hb^AHb^S x Hb^AHb^S

Female/male gametes	Hb^A	Hb^S
Hb^A	Hb^AHb^A (normal)	Hb^AHb^S (carrier)
Hb^S	Hb^AHb^S (carrier)	Hb^SHb^S (death)

(c) **MN** blood group individuals and **AB** blood group individuals.

3. **Lethality:** It is a condition in which the alleles created by mutations in these genes cause the death of an organism. There are several types which are as follows:
   (a) Absolute lethality: Death occurs in an embryonic state.
   (b) Sub-lethality: Death occurs at a later stage before maturity.
   (c) Delayed lethality: Death occurs after sexual maturity.
   (d) Conditional lethality: Death occurs under certain environmental conditions (temperature) only.
   (e) Semi-lethality: Death occurs in only some individuals of a population.

   Lethal alleles are of two types:

   (a) Dominant lethal alleles: These show lethal effects in homozygous conditions. A single recessive allele in a heterozygous condition can produce a normal phenotype and survive but only structural changes occur. These are rare.

   (b) Recessive lethal alleles: These show lethal effects only in homozygous conditions.

Examples:

(a) **Coat (skin) colour in mice**. It was reported by Lucien Cuenot (1905). It is an example of dominant lethal.

Inbreeding of yellow mice (Yy x Yy)

Female/male gametes	Y	y
Y	YY(death)	Yy (yellow)
y	Yy (yellow)	yy (brown)

Findings:

(i) Death occurs in an embryonic state (absolute lethality).

(ii) Lethal genes cause modification of 3:1 ratio to 2:1

(iii) Phenotypic ratio= 2 (yellow) : 1 (brown)

(b) **Leaf colour of Snapdragon** (*Antirrhinum majus*). It was reported by E. Baur (1907).

Selfing of golden leaved plant (Gg x Gg)

Female/male gametes	G	g
G	GG (death)	Gg (golden)
G	Gg (golden)	gg (green)

Findings:

(i) Death occurs before germination because it lacked the ability to make chlorophyll.

(ii) Lethal genes cause modification of 3:1 ratio to 2:1

(iii) Phenotypic ratio= 2 (golden) : 1 (green)

It is also observed in dexter gene in cattle, achondroplastic dwarfism in human being and xeroderma pigmentosum.

4. **Pleiotropy**: It is the ability of a gene to influence two or more unrelated phenotypic effects as it influences a number of characters simultaneously. It is against Mendelian interaction where a single gene controls a single trait or phenotype or character. It is not necessary that the characters are equally influenced. Sometimes, the effect of gene is more evident in the case of one trait (major effect) and less evident in the case of other traits (secondary effect). These unrelated traits are caused by the same protein produced by the same allele.

Examples:

(a) **Pea**: A gene controls the colour of the flower, and the same gene also controls the colour of the seed and red spots in the leaves.

(b) **Cotton**: A gene controls the size of the long fibre (lint), and the same gene also controls the size of the plant, the size of the boll and the number of seeds.

(c) **Drosophila**: A recessive gene controls the size of the vestigial wing, and the same gene also controls the length of the body, bristles of a wing, white colour of the eye and structure of reproductive organs.

(d) **Human beings**: A recessive gene causes **sickle cell anaemia** and the same gene also causes increased resistance to malaria; effect on the development of tissues and organs (bone, heart, liver, and lung).

A recessive gene causes **phenylketonuria,** and the same gene also causes mental retardation, reduced hair and skin pigmentation.

A recessive gene causes **Marfan syndrome**. The same gene also causes abnormally tall height; thin fingers and toes; dislocation of the lens of the eye; and heart problems.

Other examples of pleiotropy are albinism and schizophrenia.

5. **Multiple alleles:** It is the presence of more than two alleles of a gene. These alleles are located on the same gene locus of the same or homologous chromosomes. These produce different alternatives to a single trait. These alleles show dominance-recessive behaviour or codominance. These do not show crossing over because these are present on the same locus. These are found in the population.

Examples:

(a) **ABO Blood group of human beings**: There are four types of blood groups in human beings-A, B, O and AB. Landsteiner discovered A, B and O blood groups but de Castello and Steini (1902) discovered AB blood groups.

Bernstein (1925) reported the blood group inheritance. Blood groups are determined by two types of antigens (A and B) present on the plasma membrane surface of RBC. The antigens occur on the oligosaccharide-rich head region of glycophorin.

Person of blood group A has antigen A; group B has antigen B; group AB has both antigens A and B while group O does not have antigen.

There are three immunogen alleles: I^A, I^B and I^o (symbolised as I, an abbreviation of Isohaemoagglutination) occupying the same locus to determine the presence, absence and type of antigens. I^A forms antigen A; I^B forms antigen B while I^O does not form any antigen. All three alleles develop by mutation from a single ancestor. Now three alleles of a gene present on 9th chromosome are known.

First allele I^A codes for an enzyme that adds a molecule, N-acetylglucosamine to glycoprotein resulting in the 'A' blood group.

The second allele, I^B codes for an enzyme that adds a molecule, galactose to glycoprotein resulting in the 'B' blood group.

The third allele, I^O codes for a non-functional enzyme that cannot add any sugar to glycoprotein.

I^A and I^B are dominant alleles and I^O is a recessive allele. Both I^A and I^B are dominant over I^O but not over each other ($I^A = I^B > I^O$).

Case: A human being is a diploid organism, each person has two of three alleles, one from each parent.

1. If both I^A and I^B are present in an individual then both express their type of antigens in the presence of each other and the person is said to be in having AB blood group. Hence, both I^A and I^B behave as codominant alleles.
2. If both I^A and I^O are present in an individual, then only I^A will be expressed because I^O does not produce sugar. It shows a dominant-recessive pattern of Mendelian inheritance.
3. If both I^B and I^O are present in an individual, then only I^B will be expressed because I^O does not produce sugar. It shows a dominant-recessive pattern of Mendelian inheritance.

Blood group/phenotype	Genotype	antigen	Antibody
A	I^AI^A, I^AI^O	A	b
B	I^BI^B, I^BI^O	B	a
O	I^OI^O	-	ab
AB	I^AI^B	AB	-

Allele of first parent	Allele of second parent	Genotype of offspring	The blood group of offspring
I^A	I^A	I^AI^A	A
I^A	I^B	I^AI^B	AB (universal recipient)
I^A	I^O	I^AI^O	A
I^B	I^B	I^BI^B	B
I^B	I^O	I^BI^O	B
I^O	I^O	I^OI^O	O (universal donor)
		Six genotypes	**Four** phenotypes

The possible genotypes are determined by a formula

G= n (n+1)/2 where n= number of alleles.

G= 3(3+1)/2 = 6 genotypes

The possible phenotypes are determined by formula.

P= n+1 where n=number of alleles.

P= 3+1= 4 phenotypes (A, B, O and AB). The phenotypes are tested by two antisera; ant-A and anti-B.

Determination of blood groups:

Blood group of parents	Genotypes	Possible blood group of offspring	Impossible blood group of offspring
Ax A	I^AI^O x I^AI^O	A,O	B,AB
A x B	I^AI^O x I^BI^O	A,B,O,AB	--
A x O	I^AI^O x I^OI^O	A,O	B,AB
AxAB	I^AI^O x I^AI^B	A,B,AB	O
BxB	I^BI^O x I^BI^O	B,O	A,AB

Blood group of parents	Genotypes	Possible blood group of offspring	Impossible blood group of offspring
BxO	I^BI^O x I^OI^O	B,O	A,AB
BxAB	I^BI^O x I^AI^B	A,B,AB	O
OxO	I^OI^O x I^OI^O	O	A,B,AB
OxAB	I^OI^O x I^AI^B	A,B	O,AB
AB x AB	I^AI^B x I^AI^B	A,B,AB	O

6. **Polygenic inheritance**: It is the inheritance of character controlled by two or more genes being expressed at one time. It was discovered by Nilsson-Ehle who studied the inheritance of grain colour in wheat. It is also known as multiple factors inheritance because it is an additive effect of two or more genes on a single phenotype. Such genes which are non-allelic and affect the phenotype of a single quantitative trait, are called polygenes or cumulative genes.

Features:

(a) It produces a wide range of phenotypes hence called quantitative inheritance.

(b) It involves many genes at several different loci.

(c) It shows an F_1 individual which is intermediate between two parents.

(d) The intensity of character depends on the number of dominant alleles.

(e) It shows continuous variation in the phenotype.

(f) It can be easily influenced by the environment.

(g) The dominant alleles have a cumulative effect, and the full trait is shown when all dominant alleles are present.

(h) The frequency of the trait follows the shape of the bell curve.

(i) It is also expressed in metric trait expressed in length, weight and number.

(j) It controls skin colour, eye colour, hair colour, height, weight, body shape, intelligence, mechanical skills, personality and tendency to develop disease.

(k) Polygenes originate due to duplication of a part of a chromosome, increase in the number of chromosomes (polyploidy) and mutation.

Examples:

1. **Skin colour in human beings**: It was reported by Davenport (1913) in negro and white population of the US.

   (a) If skin colour is controlled by **two gene pairs** (each with 2 alleles). These genes are present at two different loci and each dominant gene is responsible for synthesis of fixed amount of pigment, melanin. The amount of melanin produced depends on the number of dominant gene.

Marriage between Negro and white

Negro (AABB) x White (aabb)

↓ ↓

AB ↓ ab

Mulatto (AaBb)

Marriage between two mulattoes (AaBb x AaBb)

Female / male gametes	AB	Ab	aB	ab
AB	AABB (Negro)	AABb (darker than Mulatto)	AaBB (darker than Mulatto)	AaBb (Mulatto)
Ab	AABb (darker than Mulatto)	AAbb (Mulatto)	AaBb (Mulatto	Aabb (lighter than Mulatto)
aB	AaBB (darker than Mulatto)	AaBb (Mulatto)	aaBB (Mulatto)	aaBb (lighter than Mulatto)
Ab	AaBb (Mulatto)	Aabb (lighter than Mulatto)	aaBb (lighter than Mulatto)	aabb (White)

Results:

(i) The number of intermediate types increases with an increase in a number of polygenes but the number of parental types remains the same.

(ii) Phenotypic ratio= **1** (negro): **4**(darker than mulatto): **6**(mulatto): **4**(lighter than mulatto) : **1**(white).

(iii) The number of dominant alleles is 4(negro), 3 (darker than mulatto), 2 (mulatto) 1(lighter than mulatto) and 0 (white).

(iv) Number of phenotypes: 2n + 1 or (2 x 2) +1= 5 where n=number of genes

(b) If skin colour is controlled by **3 pairs of genes**: Now it is considered that 3 pairs of genes are responsible for skin colour in human beings. The likely conditions are as such:

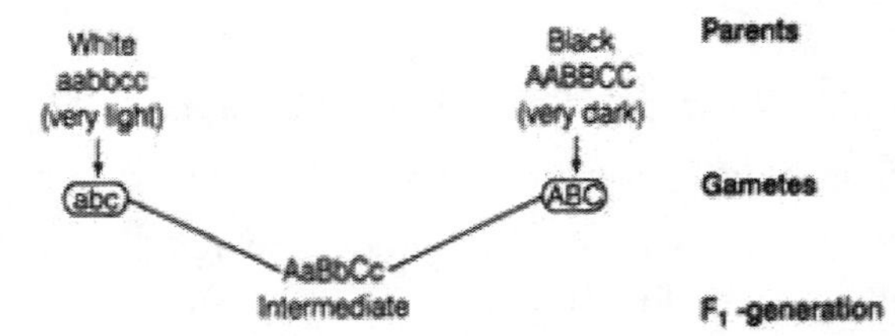

Gametes → ♂ / ♀ ↓	ABC	aBC	AbC	ABc	abC	Abc	aBc	abc
ABC	AABBCC very dark	AaBBCC dark	AABbCC dark	AABBCc dark	AaBbCC fairly dark	AaBbCc fairly dark	AaBBCc fairly dark	AaBbCc intermediate
aBC	AaBBCC dark	aaBBCC fairly dark	AaBbCC fairly dark	AaBBCc fairly dark	aaBbCC intermediate	AaBbCc intermediate	aaBBCc intermediate	aaBbCc fairly light
AbC	AABbCC dark	AaBbCC fairly dark	AAbbCC fairly dark	AABbCc fairly dark	AabbCC intermediate	AAbbCc intermediate	AaBbCc intermediate	AabbCc fairly light
ABc	AABBCc dark	AaBbCc fairly dark	AABbCc fairly dark	AABBcc fairly dark	AaBbCc intermediate	AABbcc intermediate	AaBBcc intermediate	AaBbcc fairly light
abC	AaBbCc fairly dark	aaBbCC intermediate	AabbCC intermediate	AaBbCc intermediate	aabbCC fairly light	AabbCc fairly light	aaBbCc fairly light	aabbCc light
Abc	AABbCc fairly dark	AaBbCc intermediate	AAbbCc intermediate	AABbcc intermediate	AabbCc fairly light	AAbbcc fairly light	AaBbcc fairly light	Aabbcc light
aBc	AaBBCc fairly dark	AaBBCc intermediate	AaBbCc intermediate	AaBBcc intermediate	aaBbCc fairly light	aaBbcc fairly light	aaBBcc fairly light	aaBbcc light
abc	AaBbCc intermediate	aaBbCc fairly light	AabbCc fairly light	AaBbcc fairly light	aabbCc light	Aabbcc light	AaBbcc light	aabbcc very light

Phenotypes : Very Dark (Black)–1, Dark–6, Fairly Dark–15, Intermediate–20,Fairly Light–15, Light–6, Very Light (White)–1.

Results:

(a) The number of intermediate types increases with an increase in the number of polygenes but the number of parental types remains the same.

(b) Phenotypic ratio= **1**(very dark/negro) : **6** (darker than Mulatto) : **15** (fairly darker than mulatto) :**20** (intermediate/Mulatto) :**15** (fairly lighter than mulatto) : **6** (lighter than mulatto) : 1(very light/white).

(c) Number of phenotypes= 2n +1 or (2x3) +1=7

Qualitative genetics	Quantitative genetics
It deals with the inheritance of traits of kind such as colour, form, structure etc.	It deals with the inheritance of traits of degree such as height, weight, number etc.
It shows discontinuous variation	It shows continuous variation
Each qualitative trait is governed by two or more alleles of a single gene.	Each quantitative trait is governed by many non-allelic genes or polygenes.
The phenotypic expression of a gene is not influenced by the environment.	The phenotypic expression of a gene is influenced by the environment.
It is concerned with the individual.	It is concerned with population.
Its analysis is done by counts and ratios.	Its analysis is done by statistical methods.

LINKAGE

It is the tendency of the genes situated in the same chromosome to stay together in hereditary transmission. It is an exception to Mendel's law of independent assortment.

There are so many genes present in a chromosome. The chromosomes move as units during meiosis into the gametes and all the genes present in the same chromosome will segregate as a group and travel together from generation to generation. The genes located in the same chromosome are called linked genes and those present in different chromosomes are called unlinked genes.

Linkage was explained by Morgan (1919) in *Drosophila* where red eyes and normal wings (grey body colour occur) in the same individual because the genes of red eyes and normal wings are located in the same chromosome and hence move together in the gamete and the next generation.

Types of linkage: Morgan divided linkage into two types:

1. **Complete linkage:** The genes remain linked together in the same chromosome for two or more generations and do not assort independently. It is shown by the following cross.

Red-eyed & normal-winged *Drosophila* x purple-eyed & vestigial-winged *Drosophila*

RRWW x rrww

↓ ↓

RW rw

↓

Red-eyed & Normal winged

RW/rw

Test crossing (F_1 hybrid male with homozygous double recessive female)

RW/rw (male) x rrww(female)

↓

Female/male gametes	rw	Ratio	Phenotypes
RW	RWrw	1 (50%)	Red-eyed & normal winged (parental)
rw	rwrw	1(50%)	Purple eyed & vestigial winged (parental)

Results:

(a) It shows that red eye colour and normal wings are inherited together because these genes are linked together.

(b) Purple eye colour and vestigial wing are inherited together because these genes are linked together.

(c) Only parental combinations are formed in the ratio of 1:1(50% each)

(d) Recombinants or non-parental types are not formed.

(e) There is a wide departure from the dihybrid test cross ratio (1:1:1:1) to 1:1 and the dihybrid ratio (9:3:3:1) to 3:1.

2. **Incomplete linkage:** The genes do not remain linked on the same chromosome but separate due to crossing over to form new combinations. It is shown by the following cross.

Red-eyed & normal-winged *Drosophila* x purple-eyed & vestigial-winged *Drosophila.*

RRWW   rrww

↓   ↓

RW   x   rw

↓

Red-eyed & Normal winged

(RrWw)

Test crossing (F_1 hybrid female with homozygous double recessive male)

RrWw (female)   x   rrww (male)

↓

Female/male gametes	rw	Ratio	Phenotypes
RW	RrWw	7 (Parental)	Red eyed & normal winged
Rw	Rrww	1(Recombinant)	Red eyed & vestigial winged
rW	rrWw	1(Recombinant)	Purple eyed & normal winged
rw	rrww	7(Parental)	Purpleeyed & vestigial winged

Results:

(a) The genes are incompletely or partially linked hence parental progenies are more (>50%) than recombinants (<50%).

Complete linkage	Incomplete linkage
The genes closely located in the same chromosome remain linked in hereditary transmission and do not separate.	The genes distantly located in the same chromosome remain separate.
Crossing over does not occur between closely linked genes.	Crossing over occurs.
It produces only parental progenies.	It produces more (>50%) parental progenies than less (<50%) recombinants.
It occurs rarely. (e.g., Male *Drosophila* and female silkworm).	It commonly occurs.

3. **Unlinked genes:** The two genes are not linked because these are located in different chromosomes hence assort independently. It is shown by the following cross.

Red-eyed & normal-winged *Drosophila* x Purple-eyed & vestigial-winged *Drosophila*

RRWW   x   rrww

↓   ↓

RW   rw

↓

Red eyed & Normal winged

RrWw

Test crossing (F_1 hybrid female with homozygous double recessive male)

RrWw (female) x rrww (male)

↓

Female/male gametes	rw	Ratio	Phenotypes
RW	RrWw	1(25%)	Red-eyed & normal winged (parental)
Rw	Rrww	1 (25%)	Red-eyed & vestigial winged (recombinant)
rW	rrWw	1(25%)	Purple eyed & normal winged (recombinant)
rw	rrww	1(25%)	Purpleeyed & vestigial winged (Parental)

Results:

(a) Four types of individuals are formed in equal proportions (25% each).

(b) Out of four progenies there are two parental and two recombinants.

Principles of linkage: It was discovered by Bateson and Punnet (1906) on sweet peas (*Lathyrus odoratus*) where two pairs of alleles don't show independent assortment. They got results that could not be explained by the law of independent assortment. They explained linkage on the basis of

1. **Coupling phase:** It is observed when two dominant or recessive alleles coming together from the same parent tend to enter the same gamete and inherit together. It is explained by the following cross.

Plant-1 Plant-2

Blue flower & long pollen x Red flower & round pollen

BBLL bbll

↓ ↓

BL bl

↓

Blue flower & long pollen

(BbLl)

Test crossing (Blue flower & long pollen with red flower & round pollen)

BbLl x bbll

Female/male gametes	bl	Ratio	Phenotypes
BL	BbLl	7	Blue flower & long Pollen
Bl	Bbll	1	Blue flower & round pollen
bL	bbLl	1	Red flower & long pollen
bl	bbll	7	Red flower & round pollen

Results:

(a) Both parental and recombinant progenies are found with different frequencies in the ratio of 7:1:1:7. There are 14(7+7) parental and 2 (1+1) recombinant types. Therefore, the genes are linked but undergo recombination due to crossing over in some cases.

(b) Traits were coupled and not easily assorted in an independent manner.

(c) If genes of two characters are completely linked, then only parental progenies should have a ratio of 1:1 (50% each).

(d) If genes of two characters are unlinked then four types of progenies should have a ratio of 1:1:1:1 (25% each).

2. **Repulsion phase:** It is observed when two dominant and recessive alleles coming from different parents tend to enter the different gametes to remain separate in the offspring. The separation of linked genes occurs due to the exchange of segments between a homologous pair of chromosomes during meiosis.

Plant-1 Plant-2

Blue flower & Round pollen x Red flower & Long pollen

BBll bbLL

↓ ↓

BlbL

↓

Blue flower & long pollen

(BbLl)

Test crossing (Blue flower & long pollen with red flower & round pollen)

BbLl x bbll

Female/male gametes	bl	Ratio	Phenotypes
BL	BbLl	1	Blue flower & long Pollen
Bl	Bbll	7	Blue flower & round pollen
bL	bbLl	7	Red flower & long pollen
bl	bbll	1	Red flower & round pollen

Results:

(a) Both parental and recombinant progenies are found with different frequencies in the ratio of 1:7:7:1. There are 14(7+7) parental and 2 (1+1) recombinant types. Therefore, the genes of two characters are not linked but undergo recombination due to crossing over in some cases.

Morgan suggested that two genes present on the same chromosome are in the coupling phase and when present on two different homologous chromosomes are in the repulsion phase.

Arrangement of linked genes on chromosome: If there are 2 pairs of alleles (BbLl), then the possible arrangement may be:

1. **Cis-arrangement**: It represents the coupling phase. It shows BL/bl.

   Two dominant genes are present on the same homologous chromosome.

   | **B** **L** |

   Two recessive genes are present on the other member of homologous chromosome

   | **b** **l** |

2. **Trans-arrangement**: It represents repulsion phase. It shows Bl/bL.

   One dominant and one recessive allele are located on one member of homologous pair of chromosomes.

   | **B** **l** |

   One dominant and one recessive allele are located on another member of homologous pair of chromosomes.

   | **b** **L** |

Importance of linkage:

1. Linkage maintains specific parental character because it reduces recombination.

2. It maintains the good character of the newly developed variety.
3. It provides strong proof for the linear arrangement of genes on chromosomes.

Disadvantages:

1. It dilutes the use of desirable traits if undesirable characters is also present on the same chromosome.
2. It disallows the plant and animal breeder to combine all desirable traits in a single variety.

Linkage group: It is a linearly arranged group of linked genes that are normally inherited together. It is equal to the haploid number of chromosomes found in a species and corresponds to the number of chromosomes in its genome. It was discovered by Sutton (1903). In other words, it represents all the chromosomes if an organism is haploid (fungi) or homologous pairs if an organism is diploid. All the genes present on a particular chromosome form a linkage group. The size of the linkage group is directly proportional to the size of chromosomes. If a number of chromosomes is less, then the linkage group will be small. It proves that genes are present on chromosomes.

Diploid organisms	Number of chromosome	Number of linkage group
Pea	2n=14	7
Wheat	2n=42	21
Apple	2n=34	17
Drosophila	2n=8	4
Frog	2n=26	13
Human	2n=46	22+XX=23 in female 22+X+Y=24 in male
Haploid organisms		
Neurospora	n=7	7
Mucor	n=2	2
E.coli	n=1 (nucleoid)	1

Linkage map (chromosome map or gene map): It is a linear graphic representation of the sequence, and the relative distance of various genes present in a chromosome. It was given by Sturtevant. It does not depict the physical distance among genes because of interference of one crossing over and heterochromatin.

The linkage map is easily detected in haploids (*Neurospora*) because of the following reasons:

(a) There is a single allele per locus hence products of recombination can be studied without any complication due to dominance.

(b) It is easy to analyse the products of single meiosis because these are linearly arranged in the form of an ordered tetrad, so it is possible to distinguish between first-division segregation and second-division segregation.

Significance of Linkage map:

1. It locates the different genes in specific chromosomes.
2. It shows a linear arrangement of genes.
3. It shows the strength of linkage as well as the occurrence and number of recombination.
4. It predicts the effect of the loss of chromosome segments.

Contributions of Morgan:

1. He proposed a classical theory of linkage.
   (a) Genes are located in chromosomes and arranged in linear order.
   (b) The strength of linkage is determined by the distance between linked genes in the chromosome.
   Strength of linkage = 1/distance between genes
   (c) Linked genes staying together in hereditary transmission are in the same chromosome.

(d) Genes tend to stay in parental combination except for occasional crossing over.

2. He verified the chromosomal theory of inheritance.
3. He discovered the linkage and distinguished between linked and unlinked genes.
4. He discovered the basis for variation due to sexual reproduction.
5. He proposed a chiasma-type hypothesis showing that the formation of chiasmata shows the occurrence of crossing over.
6. He along with Sturtevant proposed that the frequency of crossing over (recombination) between two linked genes is directly proportional to the distance between two genes (1% recombination= 1 Centimorgan or 1 map unit)
7. He proposed crisscross inheritance.

Problems:

1. If two heterozygous tall pea plants having genotypes (Tt) is crossed. What is the probability that two out of five plants will be dwarf?

   Ans: It is solved by binomial expression

   $$\text{Probabilty}=\frac{n!(p^{x}q^{n-x}}{x!(n\text{-}x)!}$$

   n= Total number of plants=5

   x= Number of events in one category (dwarf plant) =2

   p= Individual probability of x (dwarf plant) =1/4: obtained from cross (TtxTt)

   q= Individual probability of other category (tall plant) =3/4: obtained from cross (TtxTt)

   ! =symbol of factorial (products of integer from down to one.

   $$\frac{5!(1/4)^{2}\,x\,(3/4)^{5-2}}{2!(5-2)}$$

   $$\text{or}\quad \frac{5\times4\times3\times2\times1}{2\times1(3\times2\times1)}\times\frac{1}{16}\times\frac{3\times3\times3}{4\times4\times4}=0.26=26\%$$

2. A heterozygous pea plant that is tall with yellow seeds (TtYy) is allowed to self-fertilise then what is the probability that an offspring will be either tall with yellow seeds, tall with green seeds or dwarf with yellow seeds?

   It is solved by the sum rule.

   According to the dihybrid cross, Plants with tall and yellow seed= 9/16

   Plants with tall and green seeds= 3/16

   Plants with dwarf and yellow seeds= 3/16

   Sum rule= 9/16 +3/16 + 3/16 = **15/16**

CROSSING OVER

It is a mutual exchange of genetic material between two non-sister chromatids (adjacent paternal and maternal) chromatids of homologous chromosomes.

It was discovered by Janssen (1909).

The term 'crossing over' was coined by Morgan.

Its physical basis was first demonstrated by Creighton and McClintock.

Features:

(a) It is the separation of linked genes. If crossing over occurs, the genes separate and pass into different gametes and offspring. Thus, linkage and crossing over are antagonistic to each other.

(b) It takes place at the pachytene stage of prophase-I of meiosis.

(c) It occurs at chiasma at four stranded stages as proved by the 2:2:2:2 arrangement of ascospores in the ascus of *Neurospora*.

(d) It occurs in germinal cells.

(e) It involves breakage and reunion at a certain point.

(f) Usually, it involves two of the four chromatids of two homologous chromosomes.

(g) Its value varies from 0 to 50%.

(h) Its value can be worked out from test cross progeny.

Proof: C.B. Hutchinson (1922) took the colour and shape of the grain (seed) of maize. He performed the following crosses:

1. **Coupling cross:** He crossed a pure breeding coloured and smooth (rounded) grained variety of maize with a pure breeding colourless and shrunken variety of maize.

CCSS x ccss
↓ ↓
CS cs
↓
CcSs

Test crossing (CcSs x ccss)

Gametes	cs	Ratio	Phenotypes
CS	CS/cs	48%	Coloured and smooth (parental)
Cs	Cs/cs	2%	Coloured and shrunken (recombinant)
cS	cS/cs	2%	Colourless and smooth (recombinant)
cs	cs/cs	48%	Colourless and shrunken (parental)

Results:

(a) Four types of progenies are produced in the ratio of 48:2:2:48.

(b) The two genes are linked in the same chromosome, but 4% recombinants (2+2) occur by exchanging segments. A strong linkage exists between seed colour and seed shape.

(c) The large deviation of the observed F_2 population from the expected segregation occurred because separation in one pair of alleles is not independent of the separation in the other pair of alleles.

(d) The data shows that the two pairs of genes have not been assorted independently.

2. **Repulsion cross**: He crossed a pure breeding coloured and shrunken variety of maize with a pure breeding colourless and smooth variety of maize.

CCss x ccSS
↓ ↓
Cs cS
↓
CcSs

Crossing of F_1 hybrid with homozygous recessive parent

(CcSs x ccss)

Gametes	cs	Ratio	Phenotypes
CS	CS/cs	48%	Coloured and smooth (parental)
Cs	Cs/cs	2%	Coloured and shrunken (recombinant)
cS	cS/cs	2%	Colourless and smooth (recombinant)
cs	cs/cs	48%	Colourless and shrunken (parental)

Result: The same results were observed as those of the coupling cross.

Types of Crossing over:

1. **Single crossing over:** It occurs at one point (chiasma). It produces two parental-type (non-crossover) chromatids and two recombinant (cross-over) chromatids. Each recombinant chromatid acquires a new segment at the end.
2. **Double crossing over**: It occurs at two points (chiasma). It is two types:
   (a) Reciprocal: It involves the same non-sister chromatids. It produces two parental types (non-crossover) chromatids and two recombinant (cross-over) chromatids just like single crossing-over. Each recombinant chromatid acquires a new segment at the centre.
   (b) Complementary: It involves three or all four chromatids. It produces one or no parental (non-crossover) chromatids and three or four recombinant (cross-over) chromatids.
3. **Multiple crossing over**: It occurs at several (3 or more) points (chiasma) randomly. It is uncommon.

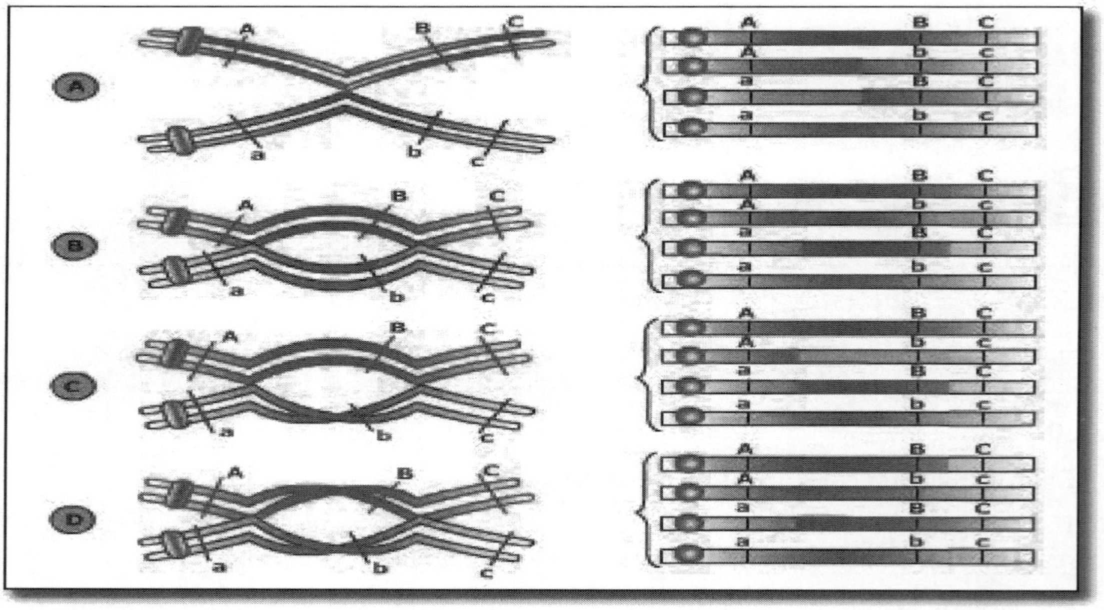

Fig: (A) Single crossing over; (B) Double crossing over (reciprocal); (C) and (D) Double crossing over (complimentary)

Mechanism:

(a) The pairing of homologous chromosomes (synapsis) takes place at the zygotene stage.

(b) Each member of the paired (synapsed) chromosome divides longitudinally into two chromatids forming tetrad.

(c) Chiasma is formed where crossing over occurs. Chiasma may be terminal (at the end of pairing chromatids) or interstitial (at the middle).

(d) Crossing over occurs between non-sister chromatids of homologous chromosomes.

(e) Non-sister chromatid breaks at a corresponding point which is facilitated by the endonuclease enzyme.

(f) The broken part of one chromatid joins the other broken part of other non-sister chromatids by ligase enzyme.

(g) Exchange of chromatids containing genes takes place.

Hybrid DNA model: It was proposed by Holliday. It is a simple model. In this only one strand in each of two DNA duplexes belonging to non-sister chromatids (from homologues) breaks. The break would occur in strands having the same polarity. The single strands released from these breaks then pair crosswise with unbroken strands by complementary base pairing. This results in the formation of hybrid DNA segments.

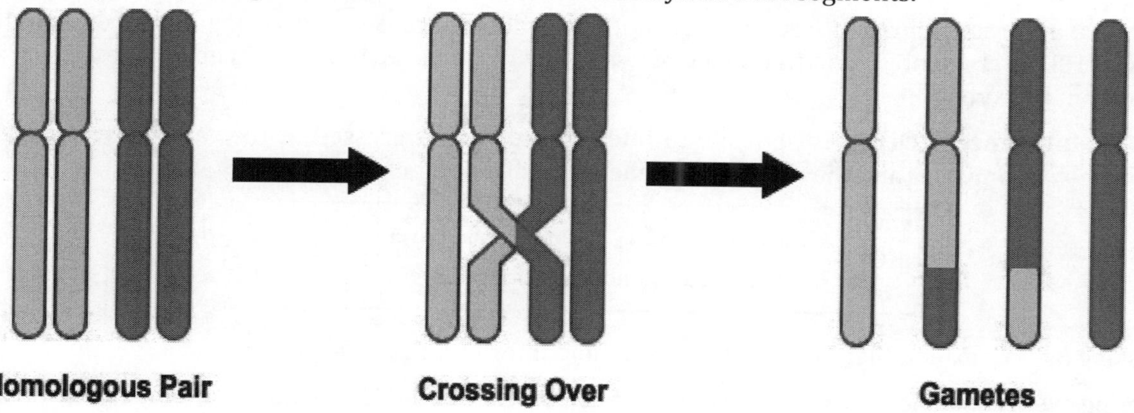

Fig: Single crossing over

Factors affecting crossing over:

1. Distance: The frequency of crossing over increases with an increase in distance between genes.
2. Temperature: The frequency of crossing over increases with a rise in temperature (from 22°C).
3. X-rays: The frequency of crossing over increases with X-ray irradiation.
4. Chemicals: The frequency of crossing over increases with the treatment of chemicals.
5. Heterochromatin: The frequency of crossing over decreases with the presence of heterochromatin.
6. Location: Frequency of crossing over decreases at the end (telomere) and centromere.
7. Age: The frequency of crossing over decreases with an increase in age.
8. Sex: The frequency of crossing over is less in male mammals. Crossing over is not seen in male *Drosophila* and female silkworms.
9. Structural changes: The frequency of crossing over reduces if translocation and inversion take place.
10. Interference: One crossing over reduces the occurrence of another in an adjacent area.

Interference: It is a phenomenon in which one crossing over reduces the occurrence of another in an adjacent area. The term 'interference' was coined by Muller (1911). It varies from chromosome to chromosome and within a chromosome from one point to another. It is maximum near the centromere and telomere (at the end of chromosome). It is affected by the distance between the genes on chromosomes. The lower the gene distance, the interference and vice-versa. It may be:

(a) Positive interference: It is most common in prokaryotes and eukaryotes.

(b) Negative interference: The presence of crossing over in one region enhances the chance of crossing over in another adjacent region. It occurs in bacteriophages, *Aspergillus* etc.

Recombination frequency: It is directly related to the distance between genes. The unit of crossing over has been termed as centiMorgan (cM) as in eukaryotes or map unit as in prokaryotes. One centimorgan is equivalent to 1% crossing over. If 1 % Crossing over means 1% centimorgan or 1 map unit, 10% crossing over means decimorgan or 10 map unit and 100 % crossing over will be Morgan or 100 map unit. A distance of 1 cM between two genes means that the genes are separated into different chromosomes on average one per 100 meiosis. When chiasma is organised in between two gene loci, only 50% of meiotic products shall be crossovers and 50% non-crossovers. Thus, the chiasma frequency is twice the frequency of cross-over products i.e., chiasma % = 2 (cross-over %) or crossover % = 1/2 (chiasma). The percentage of crossing over is calculated by test cross.

$$\text{Cross over value } (\%) = \frac{\text{Number of recombinants in test cross}}{\text{Total number of individuals}} \times 100$$

The maximum cross-over value (COV) or recombination frequency is 50% between the two genes hence these genes are linked. It is the main parameter to measure map distance. Each cross-over event results in 50% recombinant gametes and 50% parental gametes. The larger the COV, the less accurate to measure map distance. It is noted that map distance and recombination frequency do not correspond because, over a long distance, map distance will exceed even 100 but recombination frequency between any two genes will never exceed 50. It is calculated by a 2-point or 3-point test cross.

Coefficient of coincidence (COI): It is opposite to interference. It is expressed as the ratio between the actual number of double-crossing over and the expected number of such double-crossing over.

$$\text{COI} = \frac{\%\text{ actual (observed) double crossing over}}{\%\text{ expected double crossing over}} \times 100$$

If double crossing over is absent, then coincidence is zero and interference is 100%.

If double crossing over is equal to the expected value, then coincidence is 100% and interference is zero (impossible).

In practice, double-crossing over is always less than the expected value.

Significance:

(a) It produces new traits due to the formation of new chromosomes in the gametes and develops variation in newly formed individuals.

(b) It produces new and useful varieties of crop plants and animals.

(c) It gives clear proof of the linear arrangement of genes on chromosomes.

(d) It locates the point of genes on a chromosome.

(e) It helps in preparing chromosome/gene mapping.

CHROMOSOMAL THEORY OF INHERITANCE:

It was proposed by Walter Sutton and Theodor Boveri (1904). It was elaborated by Morgan, Sturtevant and Bridges. They observed a close similarity and parallel behaviour between the transmission of hereditary characters (genes) and chromosomes while passing from one generation to the next through gametes.

1. Male and female gametes play an equal role in the transmission of hereditary characters.
2. Gametes serve as a bridge between two successive generations.
3. Hereditary information is present in the nucleus because only the nucleus of sperm unites with the egg.
4. The specific number of Chromosomes in the nucleus carries the hereditary characteristics.
5. Both chromosomes and genes occur in pairs in somatic or diploid cells.
6. A gamete contains only one chromosome of a type and only one of two alleles of a character (A+X or A+Y).
7. A sort of parallelism is observed between Mendelian factors and chromosomes.
8. Numerous genes are present in chromosomes hence chromosome is carrier of Mendelian factors.
9. The paired condition of both chromosomes as well as Mendelian factors is restored during fertilisation.
10. The homologous chromosomes synapse (pair) during meiosis and then separate or segregate independently into different cells.
11. Sex is determined by sex chromosomes.
12. A wide variety of phenotypes in living organisms shows that there are far more numerous characters than chromosomes. Hence each chromosome carries several genes on it.

13. The structural and functional changes in organisms are caused due to chromosomal aberration.

SEX DETERMINATION

It is the establishment of male and female individuals or male and female organs of an individual. It differs from sex differentiation which involves the development of male and female sex organs as well as male and female secondary sexual characters. It involves:

Primary signal (environment, hormone, gene)→sex determining genes→sex differentiating genes

Discovery:

1. Tijo and Levan (1956) found that human cells have 23 pairs or 46 chromosomes where 22 homologous pairs or 44 chromosomes are called autosomes and the last or 23rd pair is known as a sex chromosome or allosome or heterosome. There are two sex chromosomes such as X and Y (XX in females and XY in males).
2. Henking discovered the X chromosome in the male bug and the Y chromosome was discovered by Nettie Stevens (1906) in beetles.
3. Henking (1891) first proposed the connection of chromosomes with sex while working on a hemipteran insect, *Pyrrhocoris apterus* and found that half of the daughter cells contained one element more than the other during anaphase-I of meiosis in spermatogenesis.
4. McClung proposed that the X chromosome is related to sex determination.

Sex chromosomes (X and Y) have two regions: (i) homologous region which carries incompletely sex-linked genes because they undergo crossing over. This region helps in pairing and (ii) non-homologous region or differential region which carries completely sex-linked genes as they do not undergo crossing over.

The sex of offspring in the animals with two types of sperms is determined at the time of fertilisation by chance event (by the kind of sperm that fuses with egg) whereas in animals with two types of eggs, sex is not determined but is already fixed during meiosis before fertilisation.

Types of sex determination:

1. Progamic: Sex is determined before fertilisation. e.g., Drones of honeybees
2. Syngamic: Sex is determined during fertilisation. e.g., Plants and animals.
3. Epigamic: Sex is determined after fertilisation. e.g., Workers and Queen of honeybees.

Different controls of sex determination:

1. **Hormonal control:**

   (a) ***Bonellia viridis*** (marine annelidian worm): Basically, all larvae are cytologically and genetically similar hence sexually undifferentiated. If a particular larva settles near the proboscis of an adult female, it will become a male individual and live in the uterus of a female. If it develops free in water, it will become a female individual. If a partly developed male is detached from proboscis then it will become intersex. It is clear that proboscis of adult females secretes a substance (hormone) that inhibits femaleness and induces larvae to differentiate into males.

   (b) The **Crew's hen** (female fowl): It is an example of sex reversal reported by Crew (1923). If a fertile hen produces offspring it changes to a fully fertile male (cock) due to a damaged ovary in a hen. The ovary in females secretes a male-suppressing hormone hence in the absence of the ovary, the testis develops. When the functional ovary is destroyed or removed, the steroid cells of the adrenal cortex become active and provoke the development of the rudimentary gonad into the testis. Both the adrenal cortex and testis secrete large amounts of male hormones which inhibit the activity of female hormone cells in the pituitary. Thus, the female sex is reversed into the male sex due to the sex reversal phenomenon.

   (c) The **Free Martin**: It is a twin of cattle in which the male is normal, and the female is sterile. It is a case of different misalignment of primary and secondary sexual differentiation. This abnormal development is only found in females who have been a member of a pair of mixed-sex twins. The embryonic blood supplies of twins in these spaces are fused in the uterus and hence XY blood cells and hormones can enter the circulation of female twins. This is sufficient to force the gonads to develop into testes and partially to block normal female development even though the animal is genetically XX. The male twin is not affected by the presence of XX cells in its blood system.

2. **Environmental control:**

   (a) *Crepidula fornicata* (slipper snail): It becomes male in the company of females and becomes female if reared alone.

   (b) *Ophryotrocha* (annelidian worm): It is male at a younger stage and becomes female at a later stage.

   (c) Turtles: Low temperature (25°C) induces maleness and high temperature (35°C) for femaleness.

   (d) Crocodiles and Snakes: Low temperature induces femaleness and high temperature for maleness.

3. **Chromosomal theory**: It was proposed by Wilson and Stevens (1906). It is most common. In diploid organisms, there are several autosomes and two sex chromosomes. In one sex, two sex chromosomes are homomorphic (XX) and in the other sex, two sex chromosomes are heteromorphic (XY). Sex is determined at the time of fertilisation. This theory proposed that one sex of any species produced two types of gametes, each responsible for the formation of one type of sex. The presence of a sex-determining region of the Y chromosome causes the development of testis in males. It shows the following cases:

   1. **Male heterogamety**: In this, a male individual produces two different types of gametes. It shows two types of sex chromosomal conditions:

      (a) **XY type** (XX-XY type): It is called lygaeus type because it was first discovered in lygaeus insects. e.g., Human beings, *Drosophila* and dioecious plants (*Coccinia, Melandrium*). The key to sex determination in humans is the SRY (sex region on Y) gene located on the short arm of the Y chromosome.

Female (AA+XX) x Male (AA+XY)

↓ ↓ ↓ ↓

A+X A+X ↓ A+X A+Y

Homomorphic/monogametic Heteromorphic/ digametic

	A+X	A+Y
A+X	AA+XX (female	AA+XY (male)
A+X	AA+XX (female)	AA+XY (male)

Sex ratio in offspring is 1:1

(b) **XO type** (XX-XO type): It is called protenor type because it was first discovered in protenor insects. e.g., *Ascaris*, bed bugs, cockroaches, grasshoppers, plants (*Dioscorea, Vallisneria*) etc.

Female (AA+XX) Male (AA+XO)

↓ ↓ ↓ ↓

A+X A+X A+X A+O

(Homomorphic/monogametic) (Heteromorphic/ digametic)

	A+X	A+O
A+X	AA+XX (female	AA+XO (male)
A+X	AA+XX (female)	AA+XO (male)

The sex ratio in offspring is 1:1.

2. **Female heterogamety:** In this, a female individual produces two different types of gametes.

   (a) **ZW type** (ZZ-ZW Type): e.g., Fish, reptiles, birds, *Fragaria* (plant).

Male (AA+ZZ) → A+Z, A+Z — Homomorphic/monogametic

Female (AA+ZW) → A+Z, A+W — Heteromorphic/ digametic

	A+Z	A+W
A+Z	AA+ZZ (male)	AA+ZW (Female)
A+Z	AA+ZZ (male)	AA+ZW (Female)

The sex ratio in offspring is 1:1

   (b) **ZO type** (ZZ-ZO Type): e.g., Moth, butterfly.

Male (AA+ZZ) → A+Z, A+Z — Homomorphic/monogametic

Female (AA+ZO) → A+Z, A+O — Heteromorphic/ digametic

	A+Z	A+O
A+Z	AA+ZZ (male)	AA+ZO (Female)
A+Z	AA+ZZ (male)	AA+ZO (Female)

The sex ratio in offspring is 1:1.

4. **Genic balance theory**: It was proposed by Calvin Bridges 1926). It was discovered on *Drosophila*. It is not universally accepted. He found a triploid female while working on *Drosophila*. The triploid female has three sets of autosomes and three sets of X chromosomes. These triploid females are normal fertile females. Bridges crossed these triploid females (3A+XXX) with diploid males (2A+XY) and noticed 8 types of individuals.

3A + XXX x 2A + XY

Female/male gametes	A+X	A+Y
2A+X	3A+XX (Triploid intersex)	3A+XY (Super male)
A+XX	2A+XXX (Super female)	2A+XXY (Diploid female)
2A+XX	3A+XXX (Triploid female)	3A+XXY (Triploid intersex)
A+X	2A+XX (Diploid female)	2A+XY (Diploid male)

The X/A ratio determines different types of individuals.

X/A ratio	Sex of Individuals
Less than 0.5	Super male or meta male (sterile)
0.5	Male (fertile)
0.5-1	Intersex •
1	Female (fertile)
More than 1	Super female or metafemale (sterile)

- **Intersex**: It is neither male nor female but a sterile intermediate between male and female. It contains XX chromosomes in some cells and XY chromosomes in some other cells. It has genitalia that seem to be between usual male and female types. A girl has a large clitoris without a vaginal opening and a boy has a small penis with a divided scrotum looking like labia.

Findings:

(i) Both autosomes and sex chromosomes are responsible for sex determination.

(ii) The sex of an individual depends on the ratio between the X chromosome and autosome.

(iii) The gene showing inclination towards maleness lies on the autosome while the gene showing towards femaleness lies on the X-chromosome.

(iv) 2A+XY produces a male because the gene for femaleness on one X chromosome cannot overcome the male tendency of the gene situated on an autosome. Similarly, 2A+XX is a female because a female character on two X chromosomes is strong enough to overcome the male tendency found in genes on 2A. Thus, one X is more powerful than one A (autosome).

(v) Autosomes show maleness. In males, autosomal material is more than in females compared to X.

(vi) X chromosome shows femaleness because of carrying the female determining gene, Sxl. So, a higher number of X produces females.

(vii) Y chromosome does not play a significant role because it does not contain sex-influencing genes.

(viii) The presence of intersex is definite proof that autosomes are also carried for sex determination.

Sex determination in Honeybee (Haplodiploidy): It occurs in Honeybees, ants, and wasps where the female is diploid (2n=32) and the male is haploid (n=16). In honeybees, there are three types of individuals.

1. Queen is diploid, fertile and female developed from fertilised egg.
2. Drone is haploid, fertile and male developed from unfertilised egg
3. Worker is diploid, sterile and female developed from a fertilised egg.

In this, sex is determined by the number of sets of chromosomes an individual receives.

Female (2n=32)   Male (n=16)

↓---meiosis---↓   ↓---mitosis

Egg (n=16)   Egg (n=16) ------ x-------- Sperm (n=16)

↓parthenogenesis   ↓

Male (n=16)   Female (2n=32)

(Drone)   **(Queen, Worker)**

- **Gynandromorph**: It is an abnormal feature (sex mosaic) where an individual has the right half of the body male and the left half female. It commonly occurs in *Drosophila* and Silkworms. It is developed due to the accidental loss of the X chromosome from a somatic cell (2A+XX) at the metaphase of the first mitotic division of the zygote after fertilisation. Hence, one of the two cells of a 2-celled proembryo will have 2A+XX with X/A=1 and another cell will have 2A+X with X/A=0.5

Sex determination in plants: Most plants are monoecious, and some are dioecious. R.P. Roy performed experiment on *Coccinia* and Warmke conducted on *Melandrium*. The findings are as follows:

(a) The ratio of X and Y determines sex as in human beings,

(b) Autosomes have no visible effect on sex determination.

(c) If Y chromosome is present, then the individual is male.

(d) If Y chromosome is absent, then the individual is female.

(e) If the ratio of X and Y is 4:1 then a hermaphrodite plant is produced.

Drosophila as experimental material for genetic research

(a) It is more suitable than pea.

(b) It is easy to be reared and bred on synthetic medium under laboratory conditions.

(c) It has a short life span (two weeks) so numerous generations are obtained.

(d) It shows several visible contrasting traits.

(e) It has a smaller number (2n=8) of chromosomes.

(f) It has a heteromorphic (XY) sex chromosome in males in and the transmission of heteromorphic chromosomes can be easily studied from one generation to another.

MUTATION:

It is a sudden change in phenotype due to qualitative or quantitative changes in the composition of genes or chromosomes. The term 'Mutation' was coined by Hugo de Vries (1900) for sudden, large and discontinuous change in evening primrose (*Oenothera lamarckiana*). He observed the sudden appearance of several new and different plants on breeding (self-pollination). Darwin called it 'sports'. Morgan studied mutation scientifically with the discovery of the White-eye mutant of *Drosophila*. Bateson called it discontinuous variation. Haploids are more suitable than diploids for the study of mutation because they allow the expression of a recessive mutation in F_1 generation.

Features:

(a) Mutation occurs in any cells or tissues at any time.

(b) It occurs in all living organisms including viruses.

(c) It may affect the structure and function of organisms.

(d) It produces mutant or changed expression.

(e) It occurs throughout the genome all the time, but some areas are more prone to mutation called hot spots.

(f) Most of the mutations are harmful.

(g) Most of the mutations are recessive.

(h) It occurs during any developmental stages of organisms.

(i) It occurs randomly in both autosomes and sex chromosomes.

(j) It shows a pleiotropic effect (a single mutant allele affects more than one character).

(k) It is recurrent (the same mutation may occur again and again).

(l) It produces visible or detectable changes (macromutation) as well as invisible or insignificant changes (micromutation).

(m) It may be forward (wild or normal allele is changed to mutant allele and backward (reverse) mutation (mutation or altered allele is changed to wild allele).

(n) It alters the morphological features of organisms such as colour, shape, size etc. and prevents the production of a biochemical in organisms.

(o) It shows lethal condition (kills all the individuals), sublethal (kills most of the individuals), vital (does not kill the individuals), subvital (kills some of the individuals) and supervital (increases the survival of the individuals).

Types: On the basis of types of cell

1. **Somatic mutation:** It occurs in the somatic cell of an organism. It is non-inheritable. It disappears with the death of organism. It affects the organisms partially. It is insignificant because it is limited to certain cells. e.g., a white spot in red eye of *Drosophila*.
2. **Germinal mutation:** It occurs in reproductive cells. It is heritable. It does not disappear with the death of an organism. It affects the whole organism. It is significant in evolution. e.g., short leg in Ancon sheep.

Types: On the basis of magnitude of effect

1. **Dominant mutation:** It occurs when the mutated gene is dominant to the wild type. It is expressed in both homozygous and heterozygous conditions. It is rare hence its frequency is low. It is easy to identify. e.g., white eye and notch wing of *Drosophila*
2. **Recessive mutation:** It occurs when the mutated gene is recessive to the wild type. It is expressed only in homozygous conditions. It is produced immediately. e.g., bar eye (narrow) of *Drosophila*

Types: On the basis of amino acid replacement

1. **Missense mutation:** It occurs when one of the bases of a codon is replaced by another base. It occurs in the first base of the codon (original sequence).

   5′-AUG CCU **AGA** GGG CAA-3′ (Original sequence)

   Met Pro ↓Arg Gly Gln

   3′-AUG CCU **GGA** GGG CAA-3′ (Mutated sequence)

   Met Pro Gly Gly Gln

   It alters the sequence of amino acids in a normal protein. It is tolerated by most of protein. It causes deleterious effects if it takes place in parts of proteins that are important for structure or function. It will produce a mutant phenotype (phenotypic effect).

2. **Nonsense mutation**: It occurs when any base of a codon is changed in a manner to form any one of the terminating codons (UAA, UAG and UGA).

   5′-AUG CCU **UGU** GGG CAA-3′ (Original sequence)

   Met Pro Cys ↓ Gly Gln

   3′-AUG CCU **UGA** GGG CAA-3′ (Mutated sequence)

   Met Pro **Stop** Gly Gln

   It terminates the elongation of the polypeptide chain hence there is the formation of a shortened protein. It has a serious effect on the activity of encoded proteins. It will produce a mutant phenotype (phenotypic effect).

3. **Same sense (Silent) mutation**: It occurs when one of the bases of a codon is replaced by another one but there is no alteration in the amino acid sequence of the protein because the changed codon specifies the same amino acid. It occurs in the third base of a codon.

   5′-AUG CCU UGU **GGG** CAA-3′ (Original sequence)

   Met Pro Cys Gly↓ Gln

   3′-AUG CCU UGU **GGA** CAA-3′ (mutated sequence)

   Met Pro Cys Gly Gln

   It is no longer functional and cannot be detected unless the nucleic acid sequencing technique is applied and the genome is screened. It has no effect on encoded protein and does not produce mutant phenotype.

Types: On the basis of origin

1. **Spontaneous mutation**: It occurs automatically without any apparent reason under natural conditions. It is also called background mutation. It is heritable. It occurs rarely. Its frequency or rate varies but it can be measured. It is higher (10^{-4}-10^{-6} per gene per generation) in eukaryotes but lower (10^{-5}-10^{-7} per gene per generation) in prokaryotes. The large gene provides a large target and tends to mutate more frequently. It is caused by the following means:

(a) UV rays of solar radiation present in the environment.

(b) Error in DNA replication due to DNA polymerase enzyme.

(c) Integration of transposons and insertion sequence at specific sites in the genome.

(d) Methylation of cytosine base of DNA at 5th position by methylase enzyme followed by spontaneous deamination.

e.g., White-eyed male *Drosophila*; and sudden occurrence of new plants and animals (sports).

2. **Induced mutation**: It occurs artificially due to the effect of either physical agents or chemicals (mutagens). It causes a change in the normal sequence of bases of DNA hence it is also known as **gene or point mutation**. It is basically intragenic. Its unit is muton and it consists of one or many nucleotide pairs.

Physical mutagens:

Ionising radiation	Particulates-α ray, β-ray, fast or thermal neutron Non-particulates- X ray, gamma ray, cosmic ray
Non-ionising radiation	UV rays

Chemical mutagens: These are more hazardous than physical mutagens. Auerbach first reported the mutagenic action of a chemical (mustard gas).

Base analogues	5-bromouracil, 5-chlorouracil, 5-iodouracil, 2-aminopurine, 2,6-diaminopurine, 5-bromodeoxyuridine
Alkylating agents	Ethyl methane sulphonate (EMS), Ethyl ethane sulphonate (EES), Methyl methane sulphonate (MMS), Dimetyl sulphonate (DMS), N-nitrosoguanidine (NTG), Mustard gas (nitrogen, sulphur).
Deaminating agents	Nitrous acid, hydroxyl amine
Intercalating agents	Acridine orange (dye)-Acriflavin (for bacteria and plants), proflavin (for viruses), Ethidium bromide
Inorganic compounds	Ferrous salts, Manganese salts
Organic compounds	Benzopyrene, Formaldehyde, organic peroxides (for fungi), Lysergic acid Dimethylamide (LSD), Aflatoxins

Induced mutation by UV ray: It is non-ionising radiation with mutagenic properties. It is present in solar radiation or produced by mercury vapour lamps or tube. Its wavelength is 1000-4000 A^0 and it is poorly penetrating. It is highly mutagenic in microbes where DNA is not covered by UV-absorbing material. It can penetrate only the surface layer of cells in multicellular organisms.

It is specifically absorbed maximum by bases of DNA at wavelength 2540 A° which is a mutagenic wavelength.

It causes a direct effect on DNA by forming a pyrimidine dimmer. It causes molecular vibration or promotion of electrons to a high energy level within thymine (target molecule). It becomes reactive and joins covalently with another thymine located nearby in the same DNA strand and forms a thymine dimer at 2600 A°. The normal condition is restored at 2400A^0.

Thymine dimer causes mis-replication due to a bend in the DNA duplex. Hence DNA polymerase cannot catalyse DNA replication in the region containing dimer. This causes strand separation leading to distortion. Finally, a mutation occurs when the cell tries to repair the damage using the error-prone repair system called SOS repair system.

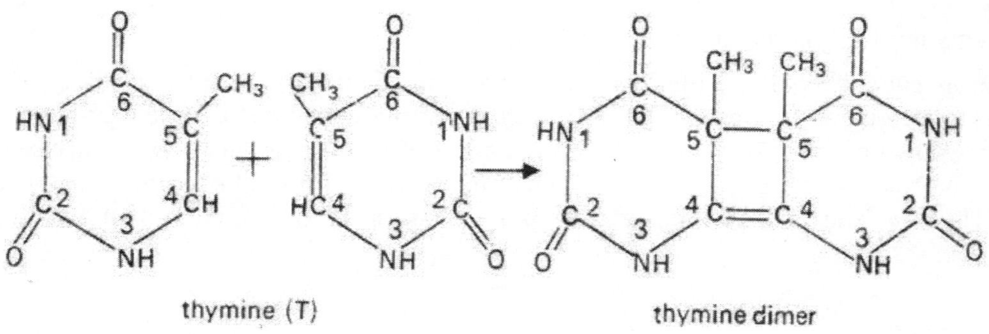

thymine (T) thymine dimer

It also causes Cytosine dimer and Thymine-cytosine dimer.

It also causes another direct effect on cytosine hydrate due to the addition of an –H atom at the 6th position and a –OH atom at the 5th position. Hence the double bond between the 5th and 6th positions is disrupted.

Induced mutation caused by X-ray: H. J. Muller (1927) discovered the mutagenic effect of X-ray on Drosophila. Stadler discovered it in Barley and maize. X-ray is produced by an X-ray tube. Its wavelength is 0.1-10 A° and highly penetrating. It can cause mutation in the seeds of plants in which the embryo is deep inside. Its frequency depends on radiation dose (measured in Roentgen unit). It causes direct major effects on DNA in several ways such as a change in a base or loss of a base disruption of the H-bond between complementary strands of DNA or break of single and double strands in DNA and cross-linkage of chromosomal proteins.

The efficiency of radiation-causing mutation depends on the stage of cell division, temperature, plant or animal species, O_2 concentration and moisture content.

Types of Gene mutation:

1. **Substitution mutation**: It occurs when one base in DNA is replaced by different base. It is most common. It is also called copy error mutation because it occurs during DNA replication due to error in base pairing. It is of two types:

   (a) **Transition mutation**: It occurs when one base of purine is replaced by another base of purine or when one base of pyrimidine is replaced by another base of pyrimidine. It is more common. There are four types of possible transitions:

   Adenine is replaced by Guanine (A→G)

   Guanine is replaced by Adenine (G→A)

   Thymine is replaced by Cytosine (T→C)

   Cytosine is replaced by Thymine (C→T)

Transition mutation caused by oxidative deamination: Nitrous acid is the best-known chemical mutagen that causes oxidative deamination by converting the amino group of a base to a keto group.

HNO_2 deaminates Adenine to hypoxanthine; cytosine to uracil; and Guanine to Xanthine. The frequency of deamination is maximum in adenine, then in cytosine and minimum in guanine.

HNO_2

Adenine ------------------→-Hypoxanthine

Deamination

Adenine is deaminated to hypoxanthine.

Hypoxanthine is guanine-like and pairs with cytosine rather than Thymine.

Cytosine pairs with Guanine in next replication.

A-T (normal)

A-T--------------→H-T---------↕ H-C

↑H-C-------------------↕

↑ **G-C** (mutant)

↑←-------------------------<-------------------------------←|

Thus, base sequence (A-T) is changed to G-C by deamination.

Transition mutation caused by Base analogue: The base analogues are unnatural bases whose molecular structure is similar to nitrogenous bases in DNA. These are derivatives of a normal base with altered base-pairing properties and cause direct mutagenesis. As a result, these are readily incorporated in DNA during replication. The higher frequency of pairing error is due to the higher electronegativity of bromine in bromouracil as compared to the $-CH_3$ group in Thymine. The analogue fits perfectly in DNA. This brings about mutation in two different ways:

Mutation by mis-incorporation or error.

Thymine----------------------------→5-Bromouracil (**5-BU**)

A-BU (normal)

A-T--------------→A-BU----------→↕ G-BU

↑ G-BU------------------↕

↑ **G-C** (mutant)

↑←---------------------------<-------------------------←|

5-bromouracil pairs with Adenine (enol form).

5-BU pairs with guanine in the next replication. Guanine will pair with cytosine in the next replication. Hence, the base sequence (A-T) is changed to G-C by base analogue.

Mutation by mis-replication:

Thymine-----------------------------------→**2-aminopurine (AP)**

A-T (normal)

A-T--------------→AP-T---------→↕ AP-C

↑ AP-C-------------→↕

↑ **G-C** (mutant)

↑←--------------------------<------------------------←|

AP pairs with thymine. AP pairs with cytosine in the next replication. Cytosine will pair with guanine in next replication. Hence, the base sequence (A-T) is changed to G-C by base analogue.

Transition mutation caused by tautomerisation: The tautomerism is the ability of a molecule to exist in more than one chemical form. The normal bases of DNA are usually in keto form but due to tautomerisation, these bases can be transformed into the rare enol forms. Thus, each base of DNA can exist in two states: normal and rare (abnormal). The rare state can exist in each base of DNA due to rearrangement or redistribution of electrons and protons in the molecule. This is a tautomeric shift. Thus, tautomerism is a phenomenon where the original keto form of base is replaced by its imino or enol form. These tautomeric shifts change the H-bonding characteristic of the base. Hence bases are converted into less stable states.

Adenine (amino form) ————→Rare adenine (imino form tautomer)

Guanine (keto form) ————→Rare guanine (tautomer)

Thymine ————————→Rare thymine (tautomer)

Cytosine ————————→Rare cytosine (tautomer)

If the base is in a rare state, it will not pair with a normal partner.

Rare adenine will pair with cytosine but not with normal thymine

Rare guanine will pair with thymine but not with normal cytosine

Rare thymine will pair with guanine but not with normal adenine.

Rare cytosine will pair with adenine but not with normal guanine

Such pairing is called forbidden base pairing. It has been found that such a situation is not stable. In the next replication, a rare state may change to a normal state. But once a forbidden base pair is formed, it will include a changed base. This will result in two kinds of DNA; one is normal DNA and the other is mutated DNA which has undergone a base pair substitution.

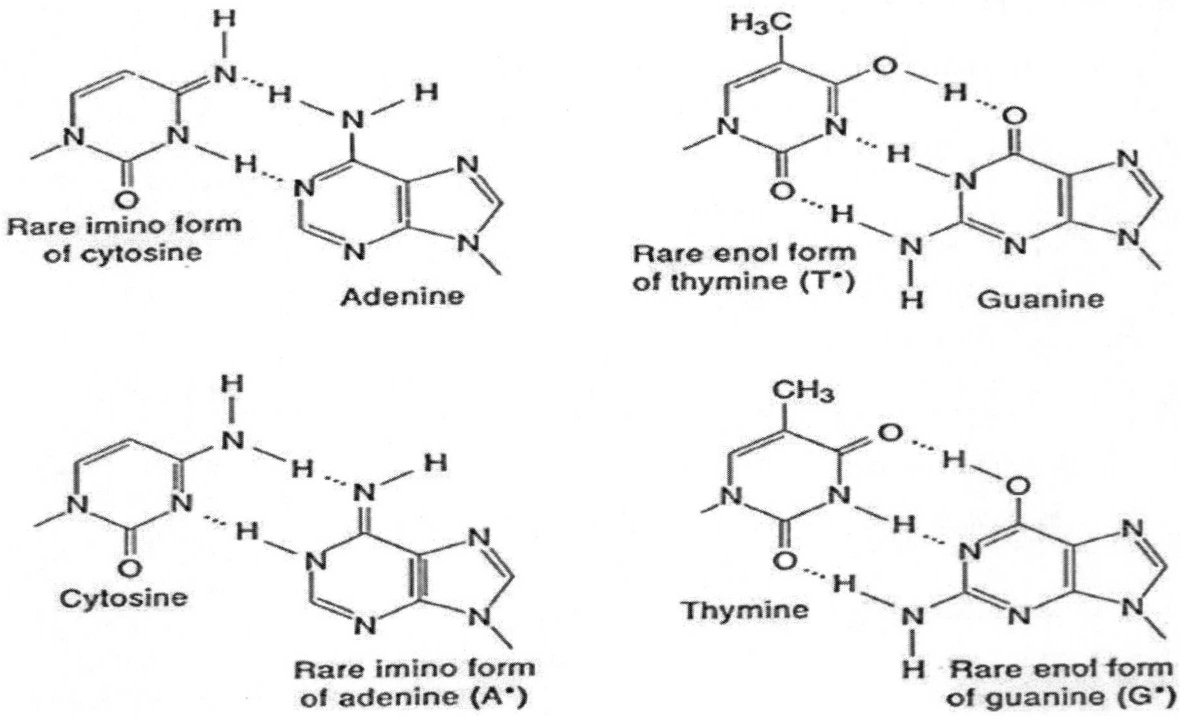

(b) **Transversion mutation**: It occurs when one base of purine is replaced by another base of pyrimidine or vice-versa. It shows more noticeable effects than transition mutation. There are eight types of possible transversions.

Adenine is replaced by Thymine (A→T)

Thymine is replaced by Adenine (T→A)

Adenine is replaced by Cytosine (A→C)

Cytosine is replaced by Adenine (C→A)

Guanine is replaced by Thymine (G→T)

Thymine is replaced by Guanine (T→G)

Guanine is replaced by Cytosine (G→C)

Cytosine is replaced by Guanine (C→G)

Transversion mutation by alkylating agents: Ethyl methane sulphonate (EMS) as an alkylating agent donates the alkyl group on the 7th position of the Nitrogen atom of Guanine in DNA. As a result, either miscoding through abnormal base pairing occurs or depurination (through gap formation) due to excision of guanine takes place. The gap may then interfere with DNA replication or may incorporate the wrong base.

EMS

Guanine------------→7-Ethyl Guanine

Guanine will become 7-Ethyl guanine (alkylated base).

The 7-Ethyl guanine will mispair with Thymine instead of pairing with Cytosine.

The 7-EG cause pairing mistake during DNA replication.

7-EG-C

G-C------------→> 7-EG------------→↕

↓ 7-EG-T

|--------------------------------------↑

2. **Frame shift mutation:** It is caused by the addition or deletion of one or more nitrogenous base pairs in DNA or mRNA because this mutation shifts the reading frame from the site of change onwards. It is called a gibberish mutation. It is much more harmful than substitution mutation. It occurs by intercalating agents (acridine dyes) which are flat, three-ringed compounds, similar in shape to base (purine) and can be intercalated between base pairs. The replication of intercalated DNA results in the addition or deletion of one to a few base pairs which produces frameshift mutation. It takes place by the following two methods:

(a) **By deletion:** When acridine dye is inserted during replication in a new chain, this does not allow any base to pair with it. Therefore, in the next replication, the base pair will be missing from DNA leading to shortening and changed organisation in DNA. The new strand will replicate with a deleted base and DNA will be formed with one base less than normal.

Original sequence: CAT **A**GT GAC GCA

↑deletion of **Adenine** base

Modified sequence: CAT GTG ACG CA

(unchanged) frameshifts towards right→

(b) **By addition (insertion):** When acridine dye is inserted during replication in a new chain, then a base is inserted opposite to acridine in the new strand. Therefore, the complementary base (X') will pair with X leading to elongation of DNA and changed organisation in DNA.

Original sequence: CATAGT GAC GCA

↑insertion of **Thymine** base

Modified sequence: CAT TAG TGA CGC CA

(Unchanged)←frame shifts towards left

Results:

1. One deletion may be neutralised by one addition or vice versa provided this occurs at the same place or very close to each other.
2. Deletion or addition of one or two bases results in a new sequence of codons which may code for entirely different amino acids. This results in a drastic change in the protein synthesised which is usually non-functional.
3. Deletion or addition of three bases may disturb only a few frames and shift by three nucleotides. The resulting protein is normal except that it lacks one amino acid or may contain extra amino acid.

Roles of mutation:

(a) It will produce better qualities of plants and animals.

(b) It will develop new species with the help of natural selection, isolation, genetic drift, and migration.

(c) It will make plants adaptable to changed environments.

(d) It will give rise to new mutant strains of microbes for better yield of antibiotics, and better fermenting ability.

(e) It will allow the genes to be identified and studied. e.g., white eye locus was identified only after the white-eyed mutant of *Drosophila*.

(f) It permits the fine structure and analysis of genes, the relationship between genes and proteins as well as some features of genetic code.

Structural chromosomal aberration (Chromosomal mutation):

These are produced due to breakage on chromosomes and later joining in chromosome segments. These aberrations occur very rarely in nature in the following ways:

1. **Deletion**: It is the loss of one or more genes due to the loss of a part of the chromosome. It was the first structural aberration detected by Bridges on the X-chromosome of *Drosophila*. It can be detected at the time of synapsis. It may be homozygous deletion if both chromosomes are involved. It may be heterozygous deletion if only one chromosome is involved. Large heterozygous deletion inhibits synapsis hence causing non-disjunction and lower fertility. It is of two types:

(a) Terminal: In this, the last segment or telomere is broken or lost due to a single break. It is rare.

(b) Intercalary: In this, any middle segment between the telomere and centromere is lost or broken due to two breaks in the G_1 period. It is most common.

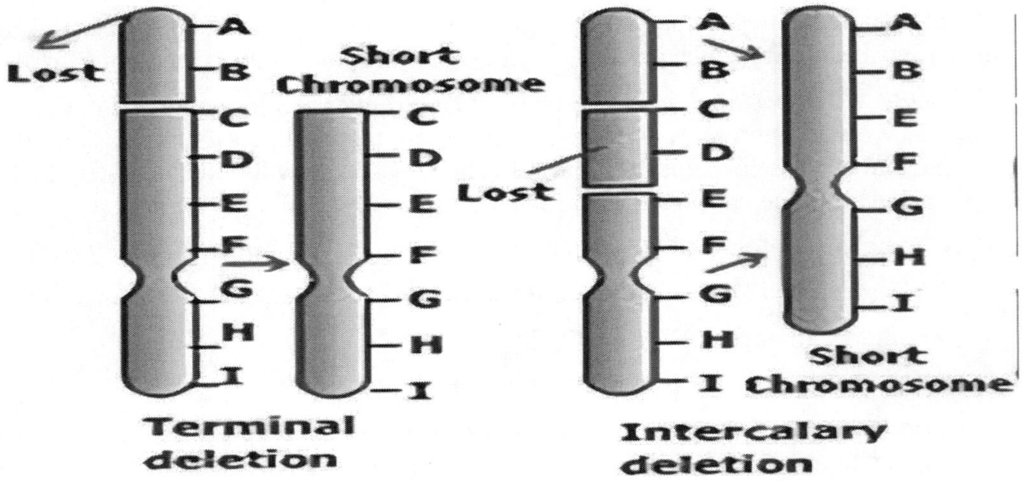

Fig: Types of deletion

Genetic effects:

(a) A deletion of small size is not harmful but of large size is always harmful.

(b) It is less harmful in polyploid species.

(c) It helps plant survive if entire chromosomes are deleted. It is observed in monosomic and nullisomic wheat and *Datura*.

(d) It forms characteristic loop.

(e) It produces sterile pollen.

(f) It produces pseudo-dominance when a recessive allele behaves like a dominant allele.

(g) Crossing over does not occur in the deleted part of a chromosome. It was confirmed by Benzer on genetic analysis of rII locus of T_4 phage of *E.coli*.

(h) Deletion of the 3c7 band of X chromosome in *Drosophila* gives rise to Notch wing.

(i) Deletion of half part of the short arm (p) of the 5th chromosome (sub-metacentric) causes cat's cry syndrome.

(j) Deletion of some part of the long arm (q) of the 22nd chromosome (acrocentric) or Philadelphia chromosome causes Chronic Myeloid Leukaemia (CML) and the deleted part is translocated to the 9th chromosome.

2. **Duplication:** It is the addition of one or more genes due to the repetition of a part of the chromosome. It involves the breakage and reunion of a chromosome segment. It occurs due to abnormal events in crossing over. It occurs in the following ways:

   (a) Tandem: It is the repetition of an adjacent part.

   (b) Reverse tandem: It is the repetition of reversal of a duplicated part.

   (c) Displaced homobrachial: It is the repetition of displaced position on the same arm of chromosome.

   (d) Displaced heterobrachial: It is the repetition of displaced position on a different arm of chromosome.

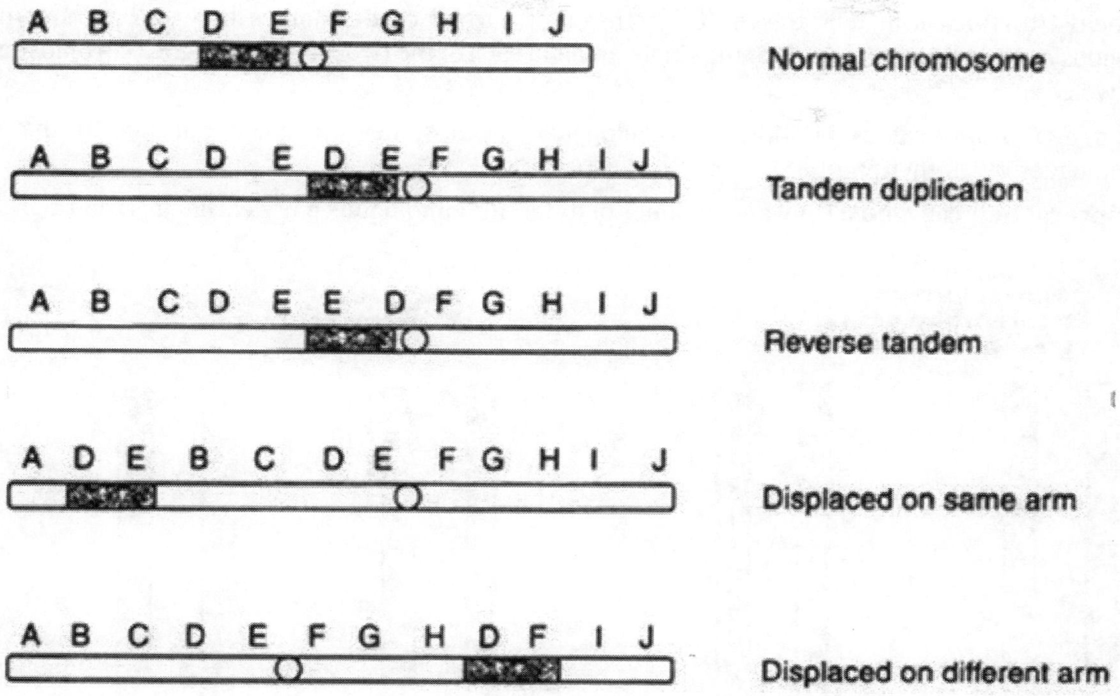

Fig: Types of duplication

Genetic effect:

(a) It is less harmful than deletion because genetic information is neither lost nor altered.

(b) It occurs frequently.

(c) It protects the organisms from lethal mutation.

(d) It is beneficial and hence the basis of the evolution of new genes because there is an increase in DNA content per cell. Hence duplication provides additional genes that mutate in new form while older genes synthesise necessary enzymes for normal activities.

(e) It can be detected if it is large due to the presence of loop.

(f) Tandem duplication of 16 A region of X chromosome in *Drosophila* produces bar eye (narrow eye) which is smaller than the wild type. This is called the position effect (phenotypic changes in the expression of genes which are due to differences in the position of genes) detected in the salivary gland chromosome.

(g) Reverse tandem duplication of 4th chromosome produces eyeless dominant.

(h) Tandem duplication of 3rd chromosome causes confluence (thickening of veins).

(i) Duplication of centromere does not survive because the two centromeres may orientate to opposite poles at division breaking the chromosome.

3. **Translocation**: It is the transfer of a part of the chromosome from one chromosome to a non-homologous chromosome. It was discovered by Bridges while working on *Drosophila*. It occurs due to irregularities during crossing over. It is of two types:

   (a) **Nonreciprocal translocation**: It is the transfer of a part of one chromosome to another non-homologous chromosome. In this, one chromosome becomes deficient and the other gains from another non-homologous chromosome. It rarely occurs. It is of two types:

   (i **Terminal:** It involves a single break in the chromosome. The broken part is directly added to another non-homologous chromosome. It is rare.

   (ii) **Intercalary:** It involves three breaks. The broken part is directly added within the break produced in the non-homologous chromosome.

(b) **Reciprocal translocation**: It is the mutual transfer of part of chromosomes between pairs of non-homologous chromosomes. It occurs at the single break in each of the two non-homologous chromosomes. It is of two types:

  (i) Reciprocal homozygous: Both the homologous chromosomes exchange parts with the two homologues of another pair.

  (ii) Reciprocal heterozygous: Only one member of different homologous pairs exchanges their parts.

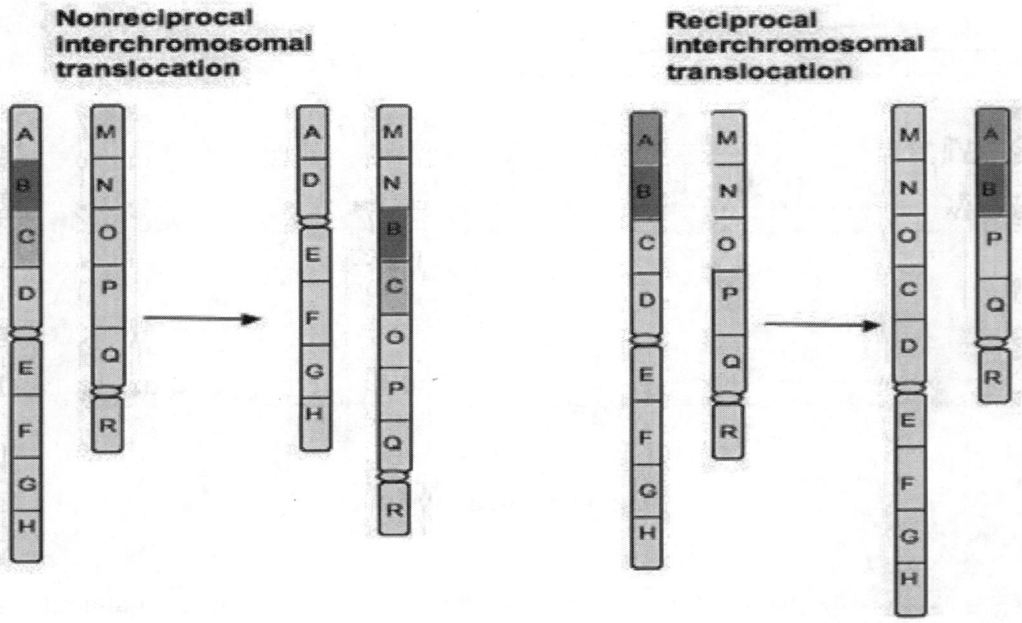

Fig: Types of Translocations

Genetic effects:

(a) Translocation gives rise to new chromosomes and new linkage groups.

(b) It forms an isochromosome.

(c) Reciprocal translocation forms multivalent rings in meiosis.

(d) Reciprocal translocation from 22nd to 9th chromosome causes chronic myeloid leukaemia (CML).

(e) Reciprocal translocation causes Down syndrome in human beings.

(f) Translocation heterozygotes in plants may produce fewer seeds than normal and such plants appear sterile.

4. **Inversion:** It is the case of reversal of an order of genes due to the reversal of a part of chromosome by 180°. It is an example of the rearrangement of genes. It was first reported by Sturtevant in *Drosophila*. It does not change the amount of DNA unlike deletion and duplication but changes the linear sequence of genetic information. It occurs due to formation of loop on a chromosome then two breaks occur at the intersection of the loop followed by the joining of intercalary part in reverse order. It is of two types:

(a) **Pericentric inversion**: It is less common. It is heterobrachial because both breaks occur in both arms. The inverted part of the chromosome includes the centromere. It may change the location of the centromere. It shows the following conditions:

- If crossing over takes place within the loop, the resulting chromatids include half of the chromatid with extra genes and deficiencies rendering it non-functional.
- If two breaks are not equidistant from the centromere, then this will change the shape of the chromosome (metacentric chromosome may become sub-metacentric or vice-versa).

(b) **Paracentric inversion**: It is more common. It is homobrachial because both breaks occur in one arm. The inverted part does not include centromere. It shows the following conditions:

- A single crossing over in inversion loop produces one monocentric fragment and one dicentric chromatid.
- A 2-strand double crossing over produces two normal and two inverted chromatids.
- A 3-strand double crossing over produces one monocentric fragment and one dicentric chromatid.
- A 4-strand double crossing over produces two monocentric fragments and two dicentric chromatids.

Thus, paracentric meiotic products include half non-functional, one-fourth functional normal and one-fourth functional inverted.

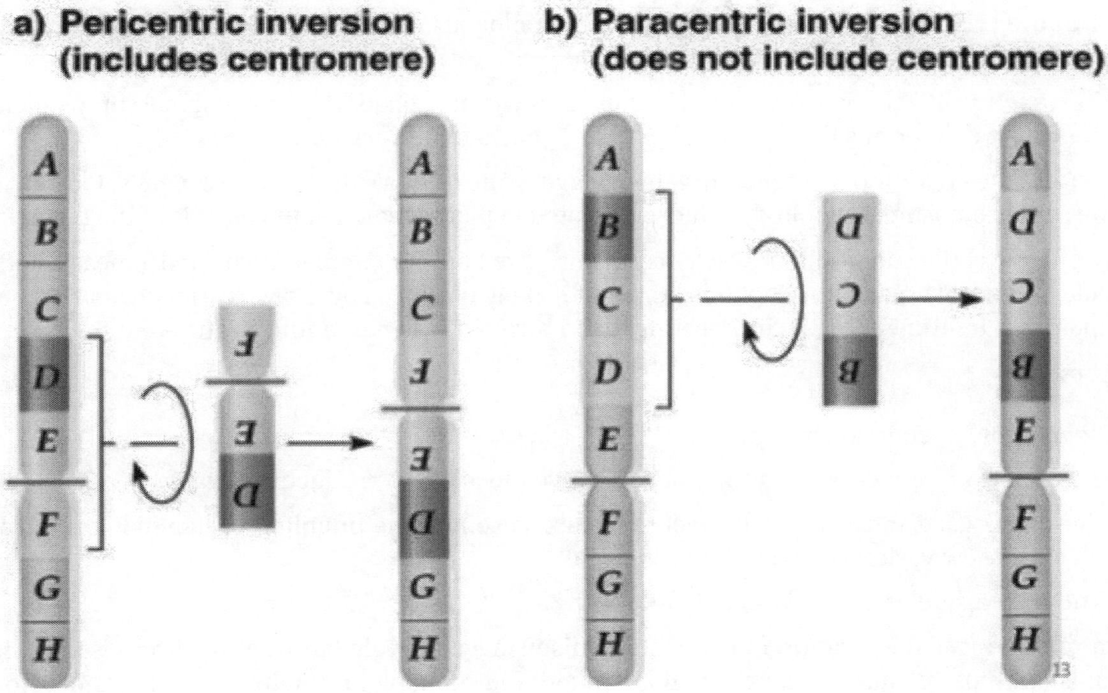

Genetic effects:

(a) It reduces the rate of crossing over and acts as crossing over suppressor because only unrecombined parental chromosomal arrangement of inverted region can be transmitted.

(b) It causes male or pollen sterility which suppresses the transcription of genes.

(c) It does not change the nature of genes.

(d) It shows position effect when genes are transferred from euchromatic segment to heterochromatic segment

(e) It maintains the heterozygosity in the species.

(f) It confirms the evidence of crossing over in which only two chromatids are inverted in crossing over.

NUMERICAL CHROMOSOMAL ABERRATION

These are produced when a number of chromosomes change from a diploid state. These are of two types:

1. **Aneuploidy**: It is the loss or gain of one or a few chromosomes in the somatic chromosome. It was discovered by Bridges. It occurs mostly due to the non-disjunction of chromosomes during mitosis and meiosis. Aneuploids rarely survive and are less viable than euploids. It is always deleterious due to genetic imbalance.

   (a) **Hypoploidy**: It is the loss of chromosomes. It is of three types:

   (i) **Monosomy**: It is the loss of one chromosome from its somatic number. Its chromosome complement is 2n-1. It causes a major genetic imbalance which is not tolerated by diploid species. It is generally weaker than normal forms. e.g., Turner syndrome (44+X).

(ii) **Double monosomy**: It is the loss of two non-homologous chromosomes (each from a different pair). Its chromosome complement is 2n-1-1.

(iii) **Nullisomy**: It is the loss of a pair of homologous chromosomes. Its chromosome complement is 2n-2. It is viable and survives in polyploids. It produces a deleterious effect. e.g., hexaploid wheat (*Triticum aestivum*).

(b) **Hyperpoloidy**: It is the gain of chromosomes. It is of three types:

(i) **Trisomy:** It is the gain of one extra chromosome. Its chromosome complement is 2n+1. It was firstly obtained by Blakeslee in jimson weed (*Datura stramonium*). It occurs spontaneously due to the production of n+1 type of gametes as a result of rare non-disjunction of a bivalent. e.g., Down syndrome, Patau syndrome, and Klinefelter syndrome.

(ii) **Double trisomy**: It is the gain of two non-homologous chromosome (each from different pairs). Its chromosomal complement is 2n+1+1.

(iii) **Tetrasomy**: It is the gain of a pair of homologous chromosomes. Its chromosomal complement is 2n+2. It is more deleterious than trisomy because it shows more variation.

2. **Euploidy:** It is the condition of a cell or organism having one or more than one complete set of chromosomes excluding sex chromosomes. Euploidy is more tolerated in plants than in animals. It is of the following types:

(a) **Haploidy**: It is the condition in which organisms have half the number of chromosomes as compared to diploid or normal organisms. Haploids can be artificially produced by X-ray treatment, delayed pollination, temperature treatment, colchicine treatment and through pollen or anther culture.

Genetic effects:

(a) Haploids show a reduction in vigour.

(i) These show a reduction in the size of vegetative as well as reproductive organs.

(ii) These are used in crop improvement because chromosome doubling of haploids produces diploid plants as new varieties) in a short period.

(iii) These are useful in the isolation of mutants.

(b) **Monoploidy**: It is the condition in which organisms have a single basic set of chromosomes. In diploid organisms (most plants and animals) the haploid and monoploid numbers are the same. However, in certain polyploidy plants such as wheat, the chromosome numbers of haploid and monoploid are different. It is observed in wheat as such:

2n (diploid) = 6X (hexaploid) = 42 (wheat)

or X (monoploid)= 42/6=7

(c) **Polyploidy:** It is the condition in which organisms have more than two sets of chromosomes. These are more common in plants than animals because most plants are monoecious. Among animals, polyploidy is very common in amphibians and reptiles. It may be:

Natural polyploidy: It occurs naturally due to non-disjunction (non-separation) of chromosomes at anaphase or due to the absence of spindle formation.

Artificial polyploidy: It is produced by

(a) Heat or cold treatment to flowers at the first division of zygote

(b) Irradiation of X-rays to vegetative and floral buds.

(c) Chemicals (colchicine) treatment to vegetative and floral buds. Colchicine is a chemical obtained from seed and corm of autumn crocus (*Colchicum autumnale*). It can be obtained either by soaking seeds and axillary bud or injecting at the seedling stage. It arrests and breaks the spindle so that cell division occurs without cytokinesis leading to the doubling of chromosome number.

Types of Polyploidy:

1. **Autopolyploidy:** It is obtained by multiplication of the same basic set of chromosomes but all chromosomes come from the same species. Natural autopolyploids are rare. e.g., doob grass (*Cynodon dactylon*) is autotriploid.

It is of two types:

(a) **Autotriploids:** These are produced from a cross between autodiploid and autotetraploid.

AA (autodiploid) x AAAA (autotetrapolod)

↓ ↓

A ↓ AA

AAA (autotriploid)

These are produced by union of haploid gamete and diploid gamete.

A x AA→**AAA** (autotriploid)

Features:

(i) These are 3 sets of homologous chromosomes hence trivalents are formed but do not disjoin normally and either disjoin 2:1 chromosome to two poles or will disjoin 1:1 leading to one chromosome as a laggard.

(ii) These are sterile because odd chromosomes do not form synapsis because there is no homologue to pair.

(iii) These are more vigorous and leafy.

(iv) These do not form seeds and hence reproduce vegetatively. These are economically important in those plants where seed is not commercially important. e.g., Orange, lemon, watermelon, grapes

(b) Autotetraploids: These are produced by doubling of autodiploid genome due to colchicine treatment.

AA (autodiploid)

↓ ←Colchicine treatment

AAAA (autotetraploid)

These are also produced by union of unreduced diploid gametes produced in theabsence of meiosis or abnormal meiosis in the diploid zygote.

AA x AA→**AAAA** (autotetraploid)

Features:

(i) These are 4 sets of homologous chromosomes hence quadrivalents are produced which disjoin in a normal 2:2 manner giving diploid gametes.

(ii) These are fertile because each gamete gets a balanced set of chromosomes.

(iii) They are vigorous and highly disease-resistant.

(iv) These have showy flowers, larger fruit and seeds and greater vitamin-c content.

(v) These form seeds. These are produced in those plants where larger leaves, flowers, fruits, and seeds are commercially important. e.g., Marigold, lily, snapdragon, primrose

2. **Allopolyploidy:** It is obtained by the multiplication of a hybrid. It is produced either spontaneously by natural hybridisation between two different species followed by spontaneous chromosome doubling or it can be artificially produced crossing two different species followed by colchicine treatment of F_1 hybrid. The allopolyploids have better adaptability than diploids. Most of the present-day crop plants are allopolyploids. It is of two types:

(a) **Segmental:** It is obtained by hybridisation of two related species hence genes of their chromosome are similar to each other.

A_1A_1 x A_2A_2

↓ ↓

A_1 ↓ A_2

A_1A_2 (sterile hybrid)

↓ Colchicine treatment

$A_1A_1A_2A_2$ (allotetraploid)

(b) **True or genomic:** It is obtained by the hybridisation of two distantly related species hence chromosomes are different (non-homologous) so the pairing does not occur in F_1 hybrid which becomes sterile.

AA x BB

↓ ↓

A B

↓

AB (sterile hybrid)

↓Colchicine treatment

AABB (allotetraploid)

Examples of new varieties:

1. **Raphanobrassica:** It is intergeneric cross between Radish (*Raphanus sativus*) and cabbage (*Brassica oleracea* var. *capitata*). It was developed by Karpechenko (1928).

Raphanus sativus (2n=18) x *Brassica oleracea* (2n=18)

↓ ↓

n=9 n=9

↓

Sterile hybrid (n+n=2n=18)

↓ Colchicine treatment

(Raphanobrassica) (2n=36=4x) or Allotetraploid or amphidiploids (fertile)

2. **Modern wheat (Common bread wheat):** It was developed by McFadden, Sears and Hitoshi Kihara (1946).

Triticum dicoccum (2n=28=4x) x *Aegilops squarrosa* (Goat grass)(2n=14=2x)

↓ ↓

n=14=2x n=7=x

↓

Sterile hybrid (2n=21=3x)

↓Colchicine treatment

Triticum aestivum (2n=42=6x)- Allohexaploid or amphitriploid (fertile)

3. **Triticale:** It is manmade cereal.

Triticum durum (2n=28=4x) x *Secale cereale* (Rye) (2n=14=2x)

↓ ↓

n=14=2x n=7=x

↓

Sterile hybrid (2n=21=3x)

↓Colchicine treatment

Triticale (2n=42=6x) - Allohexaploid or amphitriploid (fertile)

4. **New world cotton:**

Gossypium herbaceum (2n=26) x *Gossypium raimondii* (2n=26)

(Old world cotton) ↓ ↓ (new world cotton)

n=13 n=13

↓

Sterile hybrid (2n=26(13+13)

↓colchicine treatment

Gossypium hirsutum (2n=52(26+26) -allotetraploid or amphidiploid (fertile)

(Upland or new world cotton)

3. **Autoallopolypolidy:** It is obtained by union of autoplyploid and allopolyploid.

AAAA x AABB

↓ ↓

AA AB

↓

AAAB

↓colchicine treatment

AAAAAABB (autoallooctaploid)

e.g., Sun flower (*Helianthus annus*)

Advantages of polyploidy:

(a) It produces new varieties of plants which are better in quality.

(b) Polyploidy increases the size of different parts of plant giving a higher yield.

(c) It increases the size of the cell hence larger cell improves the texture of fruits and vegetables.

(d) It increases the vigour and disease resistance.

(e) It increases the vitamin-c content.

Disadvantages:

(a) Polyploids are slow to evolve because they cannot show rapid response to selection.

(b) These show a slower growth rate due to a slower rate of cell division.

(c) These show low resistance to cold and frost.

(d) These can develop sterility.

(e) These show delayed flowering.

SEX LINKED INHERITANCE (SEX LINKAGE)

It is the transmission of sex-linked characters and their determining genes borne on sex chromosomes from generation to generation.

It was discovered by Thomas Hunt Morgan (1910) on fruit fly (*Drosophila melanogaster*) when he studied the inheritance of red-white eye colour trait. He observed when a white-eyed male appeared in a culture of red-eyed males.

Types of traits:

1. **Autosomal traits**: The determining genes are found on autosomes. These show the same results in reciprocal crosses. A single recessive allele does not express its effect. These are equally affected in both sexes. There are always two alleles in both sexes. These show direct (non-criss-cross) inheritance from parents to offspring.

   (a) **Sex-limited traits:** These are autosomal inherited traits caused by a single gene and expressed only in a particular sex but their genes are also found in other sex. e.g., Development of Ovary and breast, secretion of milk; feminine musculature and voice in human female; premature baldness, beard, stout musculature and deep voice in human males; Brilliant plumage in male peacocks and antlers in male deer.

   (b) **Sex-influenced traits:** These are autosomal inherited traits observed more frequently in one sex than in another hence these are influenced differently in males and female due to sex hormones. e.g., beard, moustache, deep voice (low pitch), cleft palate (fissure in roof of mouth), forelock (area of skin without pigment), hair lip (a fissure in upper lip), stammering or stuttering (speak haltingly), pattern baldness (male show baldness if he has at least one dominant allele (Bb) and female shows baldness if she has both dominant alleles (BB). The baldness is summarised as such:

Genotypes for	Male	Female
Baldness	BB, Bb	BB
Normal	bb	Bb, bb

2. **Sex linked traits:** The determining genes are found on sex chromosomes. All sex-linked traits are inherited together. These are of two types:

   (a) **X-linked traits:** These traits are determined by the homologous region of the X chromosome. These do not have alleles on the Y chromosome. These occur more in males than females because there is no homologues of this gene on the Y chromosome. e.g., Haemophilia and colour blindness.

   Males suffer due to a single recessive allele, but females will suffer when two recessive alleles are situated.

   A man cannot be a carrier. He is always diseased. Thus, he is hemizygous having one copy of the X chromosome for the trait because his Y chromosome carries no defective allele.

   A woman is homozygous for recessive disorders because her both X chromosomes carry defective alleles hence, she is diseased and passes the allele of a sex-linked trait to both her sons and daughters.

   A woman is also heterozygous for recessive disorders because her only one X chromosome carries a defective allele hence, she is a carrier (appears normal).

   (b) **Y-linked traits:** These traits are determined by holandric genes present on the non-homologous region of the Y-chromosome only. These do not have alleles on the X chromosome. These are expressed only in males but not in females. These are rare because Y chromosome is very small and does not contain many genes. These show non-criss-cross inheritance.

   Holandric: These are inherited straight from father to son.

   Father→ Son→ Grandson. (male to male).

   e.g., Hypertrichosis (excessive growth of hairs on ears), Testis determining factor (TDF), Keratoderma dissipatum (thickened skin on extremities), Webbed toes in humans, and porcupine skin.

   (c) **X-Y linked traits:** These are determined by the genes that are localised in homologous region (the part which synapses) of the X and Y chromosome. e.g., Xeroderma pigmentosum (patches of pigments on skin and skin cancer).

Features:

(a) Most of the sex-linked traits are recessive.

(b) These traits show different results in reciprocal crosses.

(c) These traits are more common in males than in females.

(d) These are mostly harmful.

(e) These show mostly crisscross inheritance. It is of two types:

Diagynic: Father passes X-linked allele of trait to daughter who passes to son.

Father——→daughter——→Son

Diandric: Mother passes X-linked allele of a trait to son who passes to daughter.

Mother——→son——→Daughter

Proof: It was proved by Morgan while working on *Drosophila.*

The chromosome complement in *Drosophila* is XX (female) and XY (male). Its eye colour is sex-linked. The Allele of the eye colour gene is located on the X chromosome and there is no allele in the Y chromosome. Female expresses sex-linked recessive trait when it has two genes for it while male expresses even they have a single gene. The normal eye colour is red and is dominant over the mutant white eye colour which is recessive. The inheritance of X-linked eye colour in *Drosophila* is shown in the following two crosses:

1. **First cross: Red-eyed female x White-eyed male**

   If a homozygous red-eyed female is mated with a hemizygous (having a single allele for a trait) white-eyed male, then all F_1 individuals are red-eyed irrespective of sex. It shows that the gene for white colour is recessive to red colour.

   X^RX^R x X^rY (hemizygous/1-allelic)

	X^r	Y
X^R	X^RX^r (red-eyed female)	X^RY (red-eyed male)
X^R	X^RX^r (red-eyed female)	X^RY (red-eyed male)

Now red eyed females and red-eyed male of F_1 are inter-crossed.

X^RX^r x X^RY

	X^R	Y
X^R	X^RX^R (red eyed female)	X^RY (red eyed male)
X^r	X^RX^r (red eyed female)	X^rY (white eyed male)

Results:

Red-eyed individuals: White-eyed individuals

3 1

Red-eyed female: Red-eyed male: White-eyed male

2 1 1

In this cross, white-eyed females were not observed. Morgan concluded that the gene for eye colour is located on the X chromosome. *Drosophila* will have white eye colour only when both X-chromosomes contain the gene for white eye (homozygous recessive).

The above cross shows that a recessive X-linked trait follows criss-cross inheritance (the trait is transmitted from father to grandson through daughters). The latter are carriers because they have a trait but do not express it.

2. **Second cross: White-eyed female x Red-eyed male**

   If a homozygous white-eyed female is mated with a hemizygous (having a single allele for a trait) red-eyed male, then all F_1 individuals show that all females are red-eyed and all males are white-eyed. This cross showed that eye colour and sex were linked.

   X^rX^r x X^RY (hemizygous/1-allelic)

	X^R	Y
X^r	X^RX^r (red eyed female)	X^rY (whit eyed male)
X^r	X^RX^r (red eyed female)	X^rY (white eyed male)

Red-eyed female : White-eyed male

2 2

Now red-eyed female and white-eyed male of F_1 are intercrossed.

X^RX^r x X^rY

	X^r	Y
X^R	X^RX^r (red-eyed female)	X^RY (red-eyed male)
X^r	X^rX^r (white-eyed female)	X^rY (white-eyed male)

Result:

Red-eyed female	: White-eyed male	: Red-eyed male	: White-eyed male
1	1	1	1

The results of this cross are entirely different from the previous cross. In this cross, both red eye and white eye colour are in equal proportions in both the sexes of the F_2 generation.

Morgan finally concluded that the eye colour gene is linked to sex and is situated on the X chromosome.

The above cross clearly shows criss-cross inheritance (the father transmits his dominant sex-linked trait (red-eyed) to his grandsons through his daughter).

Thus, both crosses show that the pattern of inheritance of eye colour traits exactly parallels the behaviour of X-chromosomes carrying the alleles of the eye colour gene. This proves that eye colour gene (mutant white-eyed allele or normal red-eyed allele) lies on the X chromosome.

Examples: It is explained in the following five cases:

1. **Marriage between a normal man and a carrier woman**

XY x X^cX

X^c X

X	X^cX (carrier daughter)	XX(normal daughter)
Y	X^cY (diseased son)	XY (normal son)

Results: 50% of daughters are carriers and 50% are normal.

50% of sons are diseased and 50% are normal.

2. **Marriage between a normal man and a diseased woman**

XY x X^cX^c

X^c X^c

X	X^cX (carrier daughter)	X^cX(carrier daughter)
Y	X^cY (diseased son)	X^cY (diseased son)

Results: 100% daughters are carriers.

100% sons are diseased.

3. **Marriage between a diseased man and a normal woman**

X^cY x XX

X X

X^c	X^cX (carrier daughter)	X^cX(carrier daughter)
Y	XY (normal son)	XY (normal son)

Results: 100% of daughters are carriers.

100% of sons are normal.

4. **Marriage between a diseased man and a carrier woman**

X^cY x X^cX

X^c X

X^c	X^cX^c(diseased daughter)	X^cX(carrier daughter)
Y	X^cY (diseased son)	XY (normal son)

Results: 50% of daughters are diseased and 50% are carriers.

50% of sons are diseased and 50% are normal.

5. **Marriage between a diseased man and a diseased woman**

X^cY x X^cX^c

X^c X^c

X^c	X^cX^c (diseased daughter)	X^cX^c (diseased daughter)
Y	X^cY (diseased son)	X^cY (diseased son)

Results: 100% of daughters are diseased.

100% of sons are diseased.

GENETIC DISORDERS

These occur in both sex chromosomes and autosomes.

Sex-linked disorders: These occur due to gene mutation in sex chromosomes. These are of two types:

1. **Sex-linked (X-linked) recessive disorders**: The mutated gene is recessive. These are caused by a defective gene on the X chromosome. These are found more in males than females. These include:
   (a) **Haemophilia**: It was discovered by John Cotto. The affected individual is unable to clot blood properly from a minor cut hence such an individual may die of blood loss hence called **Bleeder's disease**. It is quite common in royal families of Europe. The disease spread to them through the children of Queen Victoria (1819-1901). The ancestors of queen did not possess the disease. It appears that gene for haemophilia developed either in germ cell of her father or herself through mutation. It is of two types:

   Haemophilia A: It occurs due to mutation of the Anti-Haemophilic Globulin (AHG) or Factor VIII gene hence factor VIII is not formed. It is more common (one in every ten thousand in men and one in every ten million in women).

   Haemophilia B: It occurs due to mutation of the Plasma Thromboplastin Factor or Factor IX gene hence factor IX is not formed. It is called Christmas disease. It is less common.

   (b) **Colour blindness**: It was reported by Dalton and described by Horner (1876). The affected individual is unable to distinguish between red and green colours due to the absence of some colour-sensitive cells of the retina. Red colour blindness is called protanopia and green colour blindness is called deuteranopia.

   (c) **Duchenne Muscular Dystrophy**: The mutated gene on the X-chromosome does not produce a protein, dystrophin. The affected individual suffers from muscular degeneration due to the absence of dystrophin protein. The muscular contraction is inhibited because muscle cell fails to release Calcium in the absence of dystrophin. The muscular degeneration leads to loss of skeletal muscle control.

   (d) **Glucose-6-Phosphate dehydrogenase deficiency(G6PD) Syndrome**: The mutated gene on X-chromosome does not produce Glucose-6-phosphate dehydrogenase which is essential for carrying out Hexose Monophosphate Shunt (HMS) in RBC. The haemoglobin crystallises and RBC ruptures (haemolysis) in the absence of G-6-PD which is used to form glutathione which protects Hb and RBC. The affected individual suffers from haemolytic anaemia, yellowing of skin and white eye (jaundice). It is also known as favism. Such individuals cannot suffer from malaria because they cannot have plasmodium in their RBC.

2. **Sex-linked (X-linked) dominant disorders**: The mutated gene is dominant and is present on the X-chromosome so only one copy of the allele is enough to cause disorder when inherited from a parent. These occur more in females than in males. Males are diseased when a defective allele is present only on the X chromosome. Males cannot be diseased if a mother is not diseased. Females are diseased when a defective allele is present on one X chromosome. These include:
   (a) Chondrodysplasia punctata
   (b) Hypophosphatemic ricket
   (c) Retinitis pigmentosa

Autosomal disorders: These occur due to gene mutation in autosome. These are of two types:

1. **Autosomśal recessive disorders:** The mutated gene is recessive. These occur in both males and female. Homozygous dominant individuals are normal (AA) and heterozygous dominant individuals are carriers (Aa) but homozygous recessive individuals are diseased (aa). These include

   (a) **Alkaptonuria (AKU):** It was the first recessive disorder to be discovered by Garrod. It is a defect in amino acid metabolism. It occurs due to a mutation of a gene present on 11th chromosome by which homogentisic acid oxidase is not produced. The affected individuals cannot convert homogentisic acid into acetoacetic acid due to a lack of homogentisic acid oxidase enzyme. Thus, homogentisic acid (alkapton) is accumulated in blood and is excreted in large amounts in urine. The urine turns black after coming in contact with air due to the oxidation of homogentisic acid.

   (b) **Albinism**: It is a defect in amino acid metabolism. It occurs due to a mutation of a gene present on the 11th chromosome by which tyrosinase enzyme is not produced. The affected individuals cannot convert tyrosine into melanin due to a lack of tyrosinase enzyme. Hence melanin (black pigment) is not formed in the skin, hairs and iris of an eye. The albinos have poor vision but lead a normal life.

   (c) **Phenylketonuria (PKU):** It is a defect in amino acid metabolism. It occurs due to a substitution mutation of gene present on the 12th chromosome by which phenylalanine hydroxylase enzyme is not produced. The affected individuals cannot convert phenylalanine into tyrosine due to a lack of phenylalanine hydroxylase enzyme. Thus, phenylalanine is accumulated in blood and some of it is passed from urine. The affected child is normal at birth but within a few weeks, the level of phenylalanine starts rising in plasma and by the age of six months, pale skin, walking disability and severe mental retardation occur. It occurs in one in five thousand births.

   (d) **Sickle cell anaemia:** It is a hereditary blood disorder. It occurs due to the transversion type of substitution mutation of a gene present on the 11th chromosome. The mutated gene (Hb^SHB^S) causes a single base change in the amino acid sequence of the β-chain hence glutamic acid (hydrophilic) is changed to valine (hydrophobic) at the 6th position of the β-chain. Valine being hydrophobic forms hydrophobic bonds with complimentary sites of other globin molecules due to polymerisation. It causes Hb molecules to aggregate, form fibres after collapsing and distorts the shape of RBC (biconcave) into sickle-shaped.

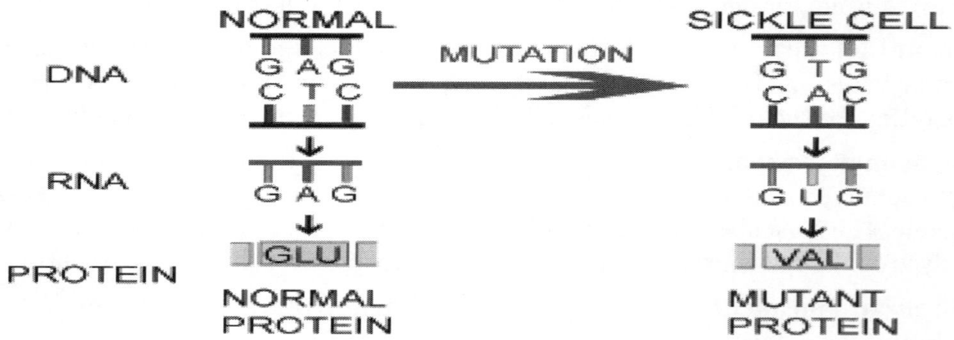

   It occurs under low oxygen tension (at high altitudes and during strenuous exercise), temperature changes and stress. It is found in a population of equatorial regions (tropical Africa) and also in American blacks whose ancestors belonged to Africa.

   Sickle-shaped RBC causes secondary effects such as severe anaemia (because it has the tendency to clot and rupture), reduction in blood circulation and damage to the spleen, liver and brain.

   It is controlled by two alleles. There are following three genotypes:

   - Hb^AHb^A: It is homozygous dominant and normal.
   - Hb^AHb^S: It is heterozygous dominant and normal but carrier. It causes periodic discomfort.
   - Hb^SHb^S: It is homozygous recessive and diseased. The affected individuals die before maturity. Natural selection keeps the lethal sickle allele (HbS) at a slow frequency in the population because affected individuals die before attaining the age of 20 without producing offspring.

   (e) **Thalassemia**: It was reported by Cooley (1925). It is an inherited blood disorder. It occurs due to either mutation or deletion of a gene controlling the formation of the globin chain of haemoglobin (made up of 4 chains-2 α and 2β) by which rate synthesis of one of the globin chains is reduced causing anaemia.

Types: On the basis of types of globin chain affected

A: Alpha thalassemia: In this production of the alpha chain is reduced and that of the beta chain is increased. It is controlled by two genes (HBA-1 and HBA-2) having four alleles on the 16th chromosome. There are four cases:

(a) If a mutation occurs and there is one defective allele, then it causes thalassemia silent. The genotype is ●ααα.

(b) If a mutation occurs and there are two defective alleles, then it causes thalassemia minor (large number of microcytic RBC and lesser number of Hb). The genotype is ●●αα.

(c) If a mutation occurs and there are three defective alleles, then it causes Haemoglobin Bart (no Hb) in infants and haemoglobin H (unstable Hb) in adults. The genotype is ●●●α.

(d) If a mutation occurs and there are all four defective alleles then it causes thalassemia major (severe haemolytic anaemia, pale skin, fussiness). It is also called Cooley's anaemia. The genotype is ●●●●.

B: Beta thalassemia: In this, the production of the beta chain is decreased. It is controlled by a single gene (HBB) with two alleles present on the 11th chromosome.

(a) If a mutation occurs and there is a defective allele then it causes thalassemia minor. The genotype is ●β.

(b) If a mutation occurs and there are two defective alleles then it causes thalassemia major. The genotype is ●●.

(f) **Cystic fibrosis**: It is a hereditary disease. It occurs due to a mutation in a gene present on 7th chromosome. This causes the inability to transport salts hence mucous thickens on the lungs, pancreas, liver, kidney, and intestine. The affected individuals contain high levels of Na^+ and Cl- in their sweat.

(g) **Galactosemia**: It is a rare metabolic disorder. It occurs due to a mutation of gene present on 9th chromosome by which galactose-1-phosphate uridyltransferase enzyme is not produced. The affected individual fails to convert or metabolise galactose into glucose due to the absence of this enzyme. Hence the level of galactose-1-phosphate is increased in various tissues and becomes toxic. It causes hepatomegaly (enlarged liver), cirrhosis, and renal failure in child and becomes fatal at the age of 3 years.

(h) **Gaucher's disease**: It was reported by Phillipe Gaucher (1882). It occurs due to mutation in a gene present on 1st chromosome by which glucocerebrosidase enzyme is not produced. The affected individual shows accumulation of glucocerebroside (Sphingolipid) in WBC of blood, spleen, liver, kidney and brain. It causes anaemia, low platelet count, enlarged spleen and limb bones.

(i) **Tay-Sach disease** (Infantile amourotic idiocy): It was reported by Warren Tay (1881). It occurs due to a mutation in a gene present on the 15th chromosome by which hexosaminidase is not produced. The affected individual shows abnormal accumulation of gangliosides in the nerve cells of the brain and spinal cord. It causes degeneration of Central nervous system (brain and spinal cord) and becomes fatal at the age of 4 years.

2. **Autosomal dominant disorders:** The mutated gene is dominant. Only one mutated gene is sufficient for disorders. These occur in both males and females. Homozygous dominant individuals are diseased (AA) and heterozygous dominant individuals are diseased (Aa) but homozygous recessive individuals are normal (aa). These include:

(a) **Adactyly**: In this, either fingers or toes or both are absent.

(b) **Brachydactyly**: In this, fingers or toes are abnormally short.

(c) **Polydactyly**: In this, extra fingers or toes are found.

(d) **Achondroplasia**: In this, bones do not grow much causing dwarfism.

(e) **Huntington chorea**: It occurs due to a mutation of a gene present on 4th chromosome. The affected individual shows gradual deterioration of muscles and brain causing uncontrollable shaking (dance-like movement) and loss of cognition (thinking ability).

(f) **Myotonic muscular dystrophy**: It occurs due to a mutation of a gene present on 19th chromosome. The affected individual shows weakness and wasting of muscles. It occurs in adulthood.

(g) **Marfan syndrome**: It occurs due to a mutation of a gene present on the 15th chromosome.

(h) **Neurofibromatosis**: It occurs due to a mutation of a gene present on the 22nd chromosome. It causes a non-malignant brain tumour.

Disorders due to gene incompatibility:

1. **Rh-factor incompatibility**: Landsteiner and Weiner (1940) discovered an antigen or protein called Rh factor in the blood of Rhesus monkeys. The formation of Rh antigen is controlled by the dominant gene (R) and its absence by the recessive gene (r). The genotypes of Rh+ individuals are RR and Rr and those of Rh- individuals are rr.

   It causes Erythroblastosis foetalis in which the death of the foetus occurs within the womb or just after birth.

   Generally, human blood is devoid of Rh antibodies. However, it has been noticed that on transfusion of blood of a Rh^+ person to Rh- person. The recipient develops Rh antibodies in its blood plasma. If Rh^+ blood is transfused for the second time it causes agglutination and leads to the death of Rh person.

   The father of Rh affected foetus is Rh^+ and the mother is Rh-. The child inherits the Rh^+ trait from the father. A few Rh^+ red blood corpuscles of the foetus in the womb enter in the blood of the mother where they develop Rh antibodies. As the mother's blood is Rh-, it causes no harm to her. These Rh antibodies along with the mother's blood on reaching the foetal circulation cause clamping of foetal RBCs or agglutination reaction. The first child is somehow born normal because by that time the number of antibodies in the mother's blood remains lesser but they increase with successive pregnancies. So, a marriage between an Rh^+ man and an Rh- woman should be avoided.

Man	Woman	Types of biological marriage
Rh^+	Rh^+	Compatible marriage
Rh-	Rh-	Compatible marriage
Rh-	Rh^+	Compatible marriage
Rh^+	Rh-	Incompatible marriage

CHROMOSOMAL ABNORMALITIES

These occur in chromosomes. These are of two types:

1. **Autosomal abnormalities:** These occur in autosomes. These occur in both males and females. These can be detected by karyotyping.

   (a) **Down syndrome** (Mongolism or Mongolian idiocy): It was reported by Langdon Down (1866). It was the first syndrome to be reported. It is caused by the presence of extra chromosome in the 21st chromosome due to trisomy. It also occurs due to Robertsonian translocation or mosaicism. Trisomy is developed due to non-disjunction hence n+1 male or n+1 female gametes are produced. Non-disjunction is a meiotic error. It occurs when a pair of homologous chromosomes fail to separate or segregate during anaphase-I of meiosis-I. This is called primary disjunction which is more common. It occurs when sister chromatids fail to separate during anaphase-II of meiosis-II. This is called secondary disjunction. As a result, one daughter cell receives both the chromosomes of a pair and the other without a chromosome of that pair. It also occurs in mitosis after fertilisation forming mosaicism.

Male gametes	Female gametes	Offspring
22 + X	**23** + X	**45** + XX (female having 47 chromosomes)
23+ Y	22 + X	**45** + XY (male having 47 chromosomes)

   It is more common in females. Its frequency of occurrence is one in 750.

   Symptoms: Rounded face, broad forehead, flat hand or broad palm (palm crease), asymmetrical skull, open mouth, slant eye, epicanthus (folded eyelid), protruding tongue, short neck and fingers, defective heart, undeveloped genitalia and gonads, mental retardation.

   (b) **Edward syndrome:** It was reported by Edward (1960). It is caused by the presence of extra chromosome number in 18th chromosome due to trisomy.

Male gametes	Female gametes	Offspring
22 + X	**23** + X	**45 +** XX (female having 47 chromosomes)
23+ Y	22 + X	**45 +** XY (male having 47 chromosomes)

It is more common in females. It is fatal at the age of 3-4 months. Its frequency of occurrence is one in 5000.

Symptoms: Smallmouth, jaws and nose; short stature, syndactyly (fingers tightly clenched against the palm of hand), defective heart, mental retardation etc.

(c) **Patau syndrome** (D-syndrome): It was reported by Patau (1960). It is caused by the presence of extra chromosome number in 13th chromosome due to trisomy.

Male gametes	Female gametes	Offspring
22 + X	**23** + X	**45 +** XX (female having 47 chromosomes)
23+ Y	22 + X	**45 +** XY (male having 47 chromosomes)

It is more common in females. It is fatal at the age of 3-4 months. Its frequency of occurrence is one in 5000.

Symptoms: Small head and chin; Deformed ears, defective eyes, cleft palate, syndactyly, mental retardation.

(d) **Cri-Du-chat syndrome**: It was reported by Lejeune (1963) in France. It is caused by the deletion of half part in the short arm of the 5th chromosome (submetacentric). It is very rare.

Symptoms: Small head, receding chin, widely spaced eyes, children cry like cats.

(e) **Chronic myeloid leukaemia** (Myelogenous leukaemia): It was reported in Philadelphia in 1959. It is caused by the deletion of some part in the distal region of the long arm of 22nd chromosome or Philadelphia chromosome (acrocentric).

Symptoms: Excess formation of granular WBC in blood, reduced formation of RBC, severe anaemia.

2. **Sex chromosomal abnormalities:** These occur in the sex chromosome.

(a) **Turner syndrome**: It was reported by Turner (1988). It is caused by the absence of one sex chromosome in females due to monosomy. It occurs due to the fusion of a normal egg (22+X) with abnormal sperm (22+O) without a Y chromosome or the fusion of an abnormal egg (22+O) with normal sperm (22+X).

Egg	Sperm	Offspring
22 + X	22 + **O (abnormal)**	44 + **XO** (sterile female having 45 chromosome)
22 + O (abnormal)	22 + X	44 + **XO** (sterile female having 45 chromosome)

It is sex chromatin negative. Its frequency of occurrence is one in 10,000.

Symptoms: Webbed skin of neck; heavy neck muscles and narrow hip (masculine feature); short stature; undeveloped breast; small uterus; reduced ovary; no egg and menstruation; hearing impairment.

(b) **Klinefelter syndrome:** It was reported by Harry Klinefelter (1942). It is caused by the presence of one sex chromosome in males due to trisomy. It occurs due to the fusion of a normal egg (22+X) with abnormal sperm (22+XY) having an extra Y chromosome or the fusion of abnormal egg with an extra X chromosome with normal sperm.

Egg	Sperm	Offspring
22 + X (normal)	22 + **XY (abnormal sperm)**	44 **+ XXY** (sterile male having 47 chromosomes)
22+**XX (abnormal)**	22+ Y (normal)	44 **+ XXY (**sterile male having 47 chromosomes)

It is sex chromatin positive. Its frequency of occurrence is one in 500.

Symptoms: Feminine features such as less muscle, body hairs and gynecomastia (enlarged breast tissue), long leg, broad hip, small penis, small testis, no sperm and reduced sex drive.

(c) **Super female (Metafemale)**: It is caused by the presence of one or two extra X chromosome in females due to trisomy. It occurs due to the fusion of an abnormal egg (22+XX) having extra X chromosome with normal sperm.

Egg	Sperm	Offspring
22 + XX (**abnormal)**	22 + X (normal)	44 + **XXX** (super female having 47 chromosomes

Symptoms: Abnormal height, abnormal sexual development, mental retardation.

(d) **Super male (meta male):** It is caused by the presence of one or two extra Y chromosomes in males due to trisomy. It occurs due to the fusion of a normal egg (22+X) with abnormal sperm (22+YY) having an extra Y chromosome. It is also known as Jacob's syndrome or criminal syndrome.

Egg	Sperm	Offspring
22+X (normal)	22 + YY (**abnormal)**	44 + **XYY** (super male having 47 chromosomes

Symptoms: Abnormal height, aggressive attitude, criminal bent of mind, mental retardation.

APPROACHES TO STUDY HUMAN GENETICS:

In human beings, several traits follow the same genetic principles of plants and animals. Around 200 traits are hereditary in human beings in which there are dominant and recessive traits. Most of the traits are transmitted in a simple Mendelian manner, others show post-Mendelian principles of genetics such as codominance, polygenes, sex-linked genes etc.

Character(Trait)	Dominant	Recessive
Hair	Dark, curly and abundant	Light, straight and less
Skin colour	Normal	Albinism
Eyes	Brown and large	Blue/green/hazel and small
Colour vision	Normal	Colour blind
Ears	Long,wide and free	Short, narrow and fused
Nose	Broad	Narrow
Nostril	Wide	Narrow
Cheeks	Dimpled	Normal
Chin	Cleft chin	Smooth chin (no cleft)
Lips	Broad and thick	Thin
Handedness	Right-handed	Left-handed
Length of toes	2nd toe longer	1st toe longer
Hearing and speech	Normal	Deaf
Vision	Astigmatism	Normal
Stature	Short (dwarf)	Tall
Baldness	Baldness in male	No baldness in female
Polydactyly	Six fingers	Five fingers (normal)
Tongue rolling	Ability to roll	Inability to roll
PTC tasting	Ability to taste	Inability to taste
Blood groups	A,B and AB	O
Blood clotting	Normal	Haemophilia
Rh factor	Positive	Negative

Inheritance in human beings is very complicated because

(a) It is controlled by two or more than two pairs of genes (not controlled by a single pair of genes).

(b) A person can observe three or four generations.

(c) It has mixed ancestry due to marriage which combines the entirely different genetic characteristics of two families in children. Hence few traits are pure.

(d) The generation time is long (often 20 years or more).

(e) The number of progenies is small.

(f) Human traits are easily influenced by the environment.

(g) Human has large number of chromosomes (46).

(h) Humans cannot be subjected to controlled mating.

There are different approaches to studying human genetics:

Population genetics: It involves the study of the distribution of traits and the estimation of gene frequency in a population of organisms. It is explained by

1. **Hardy and Weinberg law**: The alleles and genotype frequencies in a population will remain constant from generation to generation in the absence of other evolutionary influences. They proposed an equation:

$$p^2+2pq+q^2=1$$

or $p+q=1$

or $(p+q)^2=1$

p= dominant allele

p^2=frequency of homozygous dominant individual

q=recessive allele,

q^2=frequency of homozygous recessive individual

$2pq$=frequency of a heterozygous individual

The seven assumptions underlying Hardy-Weinberg equilibrium are as follows:

(a) Organisms are diploid.

(b) Only sexual reproduction occurs.

(c) Generations are non-overlapping.

(d) Mating is random.

(e) Population is large.

(f) Allele frequencies are equal in the sexes.

(g) There is no migration, mutation or selection.

2. **Pedigree analysis:** It is a diagram showing the occurrence and appearance of inheritance of certain genetic traits (autosomal or sex chromosomal or dominant or recessive or X-linked or Y-linked) of the organism and its ancestors. It is a type of genetic analysis in which trait is traced through several generations of a family to determine how a trait is inherited. It includes law of probability, the chance of difference in realised ratio based on the smallness of progeny, elimination of alternatives and inheritance of genes that follow Mendelian features.

Symbols: Several symbols show different conditions in a family.

Tips to solving the pedigree problems:

(a) Firstly, check cytoplasmic inheritance

(b) Then check Y-linked inheritance

(c) Then check X-linked dominant/autosomal dominant inheritance

(d) Finally check X-linked recessive/autosomal recessive inheritance

PEDIGREE SYMBOLS

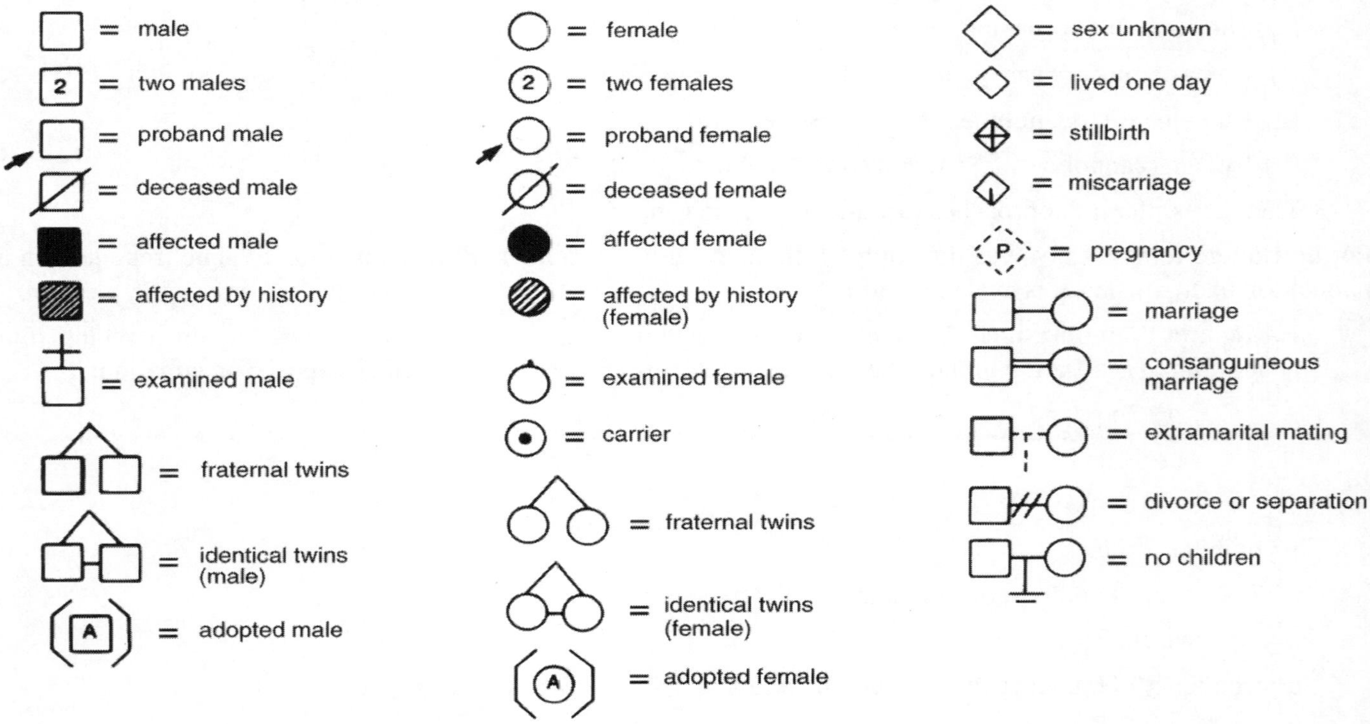

Basic facts:

(a) When one or both the parents are affected (diseased): It is often a dominant

(b) Both normal parents producing normal offspring: It is always dominant.

(c) Both normal parents producing diseased offspring: It is always a recessive.

Diseased= Affected; Normal=Unaffected

Types of traits:

1. **Autosomal recessive traits:** (AA=Normal, Aa=Carrier and aa =Diseased or affected)

   (Alkaptonuria, Albinism, Phenylketonuria, Sickle cell anaemia, Thalassaemia, Cystic fibrosis).

   (a) These occur in both sexes with almost equal frequency.

   (b) Disease does not occur in any one generation.

   (c) A diseased mother can produce normal children.

   (d) The diseased individual has no affected father.

   (e) Normal parents (must have carriers) can produce diseased children.

   (f) If both parents are heterozygous (carriers), then ¼ of the offspring will be diseased and ¾ of the offspring will remain normal.

   (g) Both the diseased parents cannot have normal offspring.

2. **Autosomal dominant traits:** (AA=Diseased, Aa =Diseased and aa=Normal)

   (Polydactyly, Huntington chorea, Myotonic dystrophy)

   (a) These occur in both sexes with almost equal frequency.

   (b) Both sexes transmit the disease to their offspring.

(c) Disease occurs in every generation (affected males and females).

(d) The diseased father transmits disease to his sons.

(e) Diseased individual has diseased parent.

(f) Normal parents do not transmit the disease.

(g) Diseased parents can transmit the disease in ¾ offspring but ¼ offspring will remain normal.

(h) When one parent is diseased (heterozygous) and the other parent is normal, ½ of the offspring will be diseased.

3. **Sex-linked (X-linked) recessive traits:**

(Colour blindness, Haemophilia, Duchenne muscular dystrophy)

(a) Disease may not occur in any one generation.

(b) These are more common in males than females.

(c) A diseased mother will transmit disease to all her sons.

(d) A diseased father will transmit disease to all his daughters.

(e) Disease is never passed from father to son.

(f) Diseased sons are usually born to normal mothers.

(g) Diseased father has carrier daughters.

(h) If a diseased mother has one son normal and another son diseased then it cannot be X-linked recessive.

4. **Sex-linked (X-linked) dominant traits:**

(Hypophosphatemic rickets, Retinitis pigmentosa)

(a) Disease occurs in every generation.

(b) These are more common in females than in males.

(c) A diseased father transmits disease to all daughters. If any daughter is normal, then it cannot be X-linked dominant.

(d) A diseased father never transmits disease to sons.

(e) Diseased mother transmits disease to sons.

(f) Diseased daughters must have either a diseased mother or a diseased father.

5. **Y-linked traits:**

(Hypertrichosis)

(a) Disease is passed from the father to all his sons only.

(a) All females will remain normal.

(b) Disease occurs in every generation.

6. **Maternal or cytoplasmic influence**:

(a) A diseased mother transmits disease to all children (sons and daughters).

(b) A normal mother has normal children.

(c) If a normal mother has at least one offspring diseased then it cannot be cytoplasmic inheritance.

(d) If a diseased mother has normal offspring, then it cannot be cytoplasmic inheritance.

Examples:

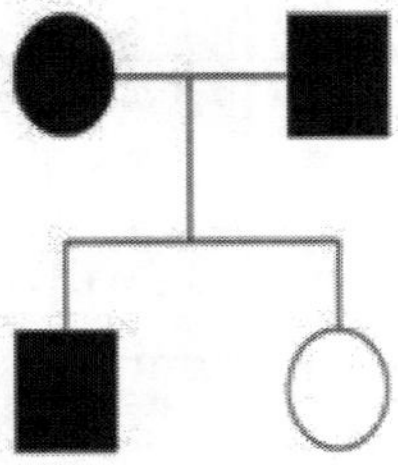

Autosomal Dominant

Cannot be recessive as affected parents could not have an unaffected offspring

Parents MUST be heterozygous

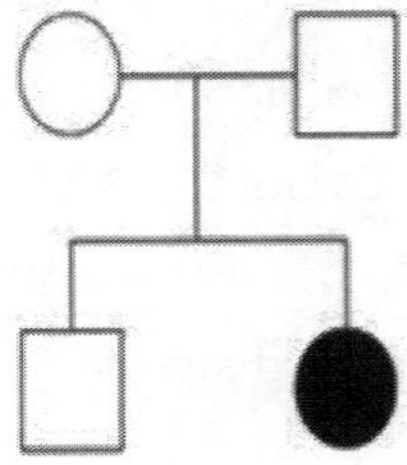

Autosomal Recessive

Cannot be dominant as unaffected parents could not have an affected offspring

Parents MUST be heterozygous

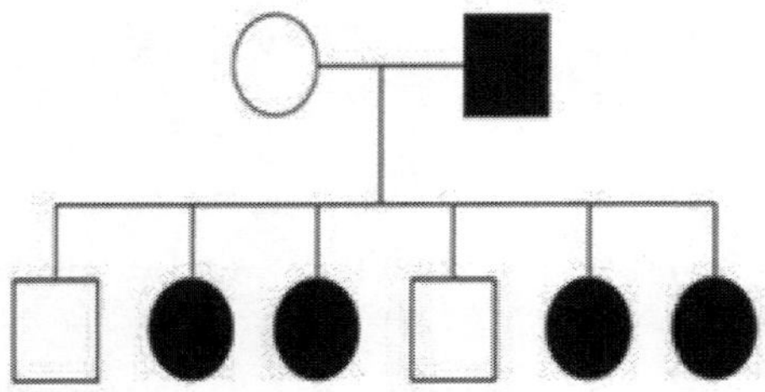

X-Linked Dominant

Mode of inheritance cannot be confirmed

However, 100% incidence of affected daughters from an affected father suggests X-linked dominant inheritance

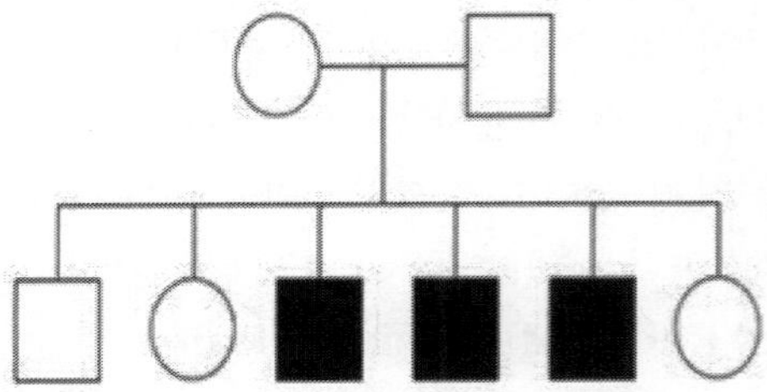

X-Linked Recessive

Sex linkage cannot be confirmed

However, high incidence of affected sons suggests X-linked recessive inheritance

Importance:

(a) It can locate the origin of a trait in the ancestor.

(b) It indicates the possibility of children with genetic defects.

(c) It can indicate the harm in the marriage of close relatives.

(d) It is used for genetic counselling.

(e) It is used in medical research.

Examples:

1. **Solve the following pedigree problem.**

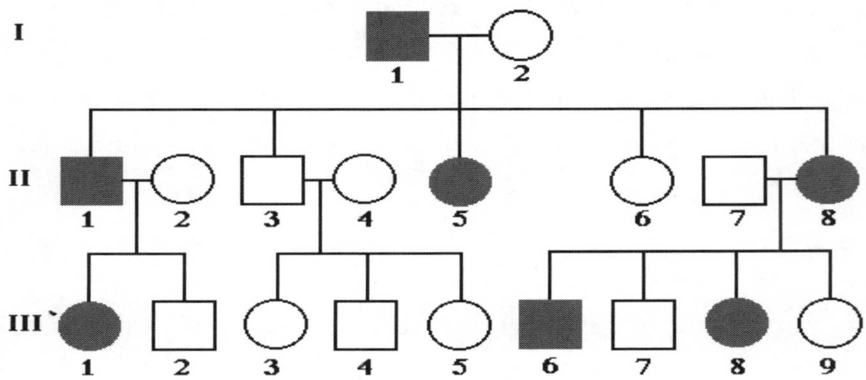

2. **Solve the following pedigree problem.**

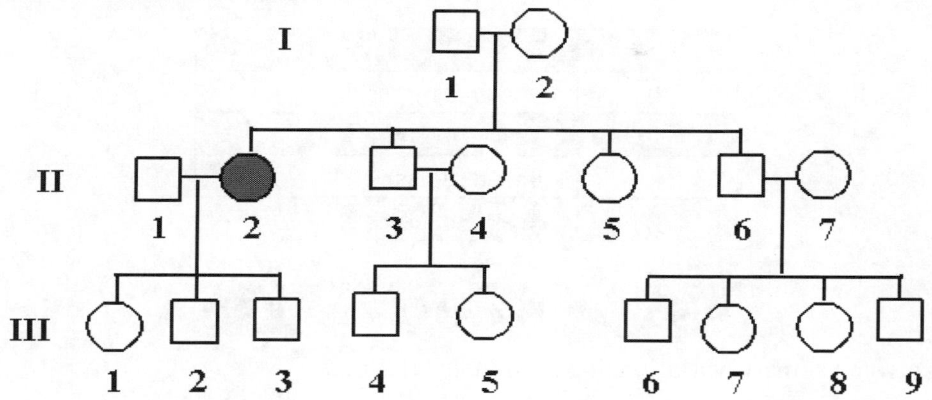

3. **Solve the following pedigree problem.**

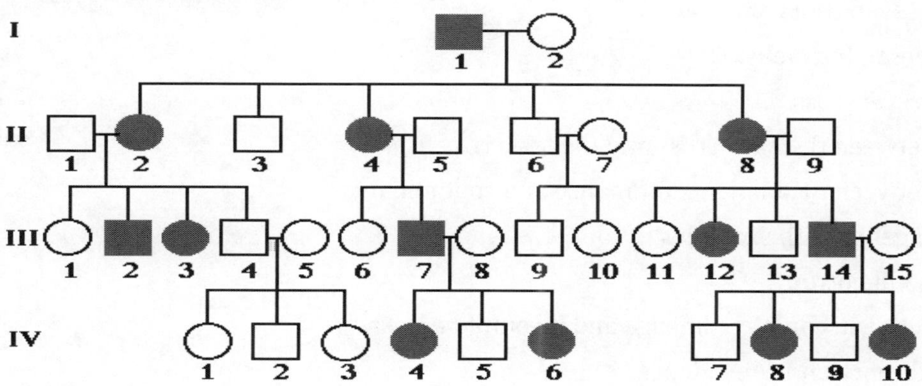

4. **Solve the following pedigree problem.**

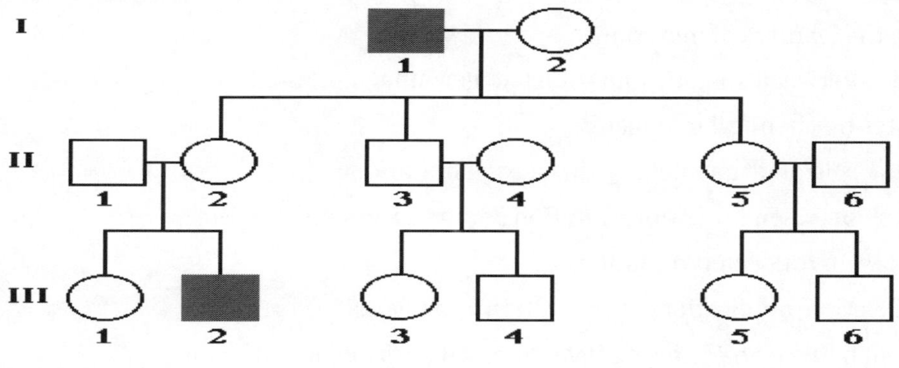

5. **Solve the following pedigree problem**

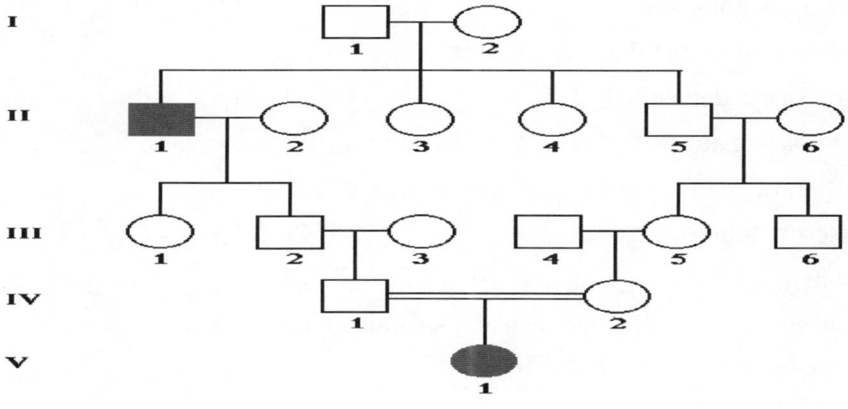

Answers:

1.	Autosomal dominant
2.	Autosomal recessive
3.	X-linked dominant
4.	X-linked recessive
5.	Autosomal recessive

SHORT ANSWER TYPE QUESTIONS

1. Distinguish between somatic variation and germinal variation.
2. Discuss the significance of variation.
3. Why Mendel selected garden peas?
4. What are the seven traits of Mendel?
5. Who rediscovered Mendelism?
6. Explain reciprocal cross.
7. Distinguish between the back cross and the test cross.
8. Distinguish between incomplete dominance and codominance.
9. Explain lethal genes with examples.
10. Write notes on pleiotropy.
11. Distinguish between complete linkage and incomplete linkage.
12. Explain interference and coincidence.
13. Explain the environmental control of sex determination with reference to *Bonellia.*
14. What are the contributions of Morgan in genetics?
15. What are the features of mutation?
16. Distinguish between somatic mutation and germinal mutation.
17. Write notes on chemical mutagens.
18. Discuss the role of ultraviolet rays in causing mutation.
19. Distinguish between transition mutation and transversion mutation.
20. Write notes on frameshift mutation.
21. Explain deletion, duplication, translocation and inversion.
22. Distinguish between paracentric inversion and pericentric inversion.
23. Explain monosomy and trisomy.
24. How colchicine causes polyploidy?
25. Distinguish between autotriploid and autotetraploid.
26. How *Raphanobrassica* is developed?
27. Mention the examples of autosomal recessive disorders.
28. Explain sex-limited traits and sex-influenced traits with examples.
29. Explain the crisscross inheritance.
30. Write notes on haemophilia.
31. Distinguish between autosomal inheritance and sex-linked inheritance.
32. How Down syndrome is developed?

33. How is Turner syndrome developed?
34. Distinguish between super female and super male.
35. How autosomal recessive and sex linked disorders are located in a family on the basis of pedigree analysis.

LONG ANSWER TYPE QUESTIONS

1. Discuss Mendel's laws of inheritance and explain the dihybrid cross.
2. Discuss the multiple alleles with reference to blood group inheritance.
3. Explain the polygenic inheritance with reference to the skin colour of human beings.
4. Discuss the chromosomal theory of inheritance.
5. Explain the linkage in sweet peas with reference to the coupling phase and repulsion phase.
6. What is crossing over? Describe its types, mechanism and significance.
7. Discuss the chromosomal theory of sex determination.
8. Describe the molecular mechanism of gene mutation.
9. Discuss the proof of sex-linked inheritance.
10. Discuss the occurrence of sickle cell anaemia and thalassemia.

MULTIPLE CHOICE QUESTIONS

1. **Mendel worked on plants for seven years from**
   (a) 1850 to 1857 (b) 1854 to 1861
   (c) 1856 to 1863 (d) 1863 to 1870
2. **A cell is heterozygous at 3 gene loci. How many different types of gametes can it form?**
   (a) 2 (b) 3
   (c) 6 (d) 8
3. **Who were three biologists who independently rediscovered Mendel's principles?**
   (a) Correns, Sutton and Tschermak
   (b) Bateson, De Vries and Punnet
   (c) Punnet, Sutton and Bateson
   (d) Correns, De Vries and Tschermak
4. **A double heterozygous tall plant with a yellow colour of seed is selfed. The ratio of dwarf plants with green cotyledons is**
   (a) 1/4 (b) 1/8
   (c) 1/16 (d) 3/16
5. **The recombination frequency between the genes A and B; A and C; B and C are 9%, 17% and 26% respectively. The sequence of genes A, B, C on the chromosome is**
   (a) ABC (b) BAC
   (c) ACB
   (d) information is insufficient
6. **A and B genes are linked. What should be the genotype of progeny in a cross between AB/ab and ab/ab?**
   (a) AAbb and aabb
   (b) AaBb and aabb
   (c) AABB and aabb
   (d) aaBB and aabb
7. **If a haploid number of chromosomes is 10, what would be the number in monosomic and tetrasomic?**
   (a) 10 and 40 (b) 19 and 22
   (c) 9 and 22 (d) 19 and 40
8. **If all the sons of a couple are colour-blind then**
   (a) Mother is colour blind
   (b) Mother is carrier and father is colour blind
   (c) Mother is normal and father is colour blind
   (d) Mother is carrier and father is normal
9. **In Drosophila, white eye colour is a recessive X-linked trait while red eye colour is dominant. A while eyed female is crossed with a red-eyed male. The female offspring with red eye colour would be:**
   (a) 0% (b) 25%
   (c) 50% (d) 100%

10. How many phenotypes are formed in the skin colour inheritance of a human being if it is controlled by 3 pairs of genes?

(a) 5 (b) 7
(c) 9 (d) 10

11. A trihybrid cross is made between two yeasts, both having genotypes AaBbCc. The proportion of offspring having genotype aaBbcc will be:

(a) 1/64 (b) 1/32
(c) 1/16 (d) 3/64

12. From a cross AABb × aaBb, the genotypes AaBB: AaBb: Aabb: aabb will be obtained in the ratio of

(a) 2 : 1 : 1 : 0 (b) 1 : 2 : 1 : 0
(c) 1 : 2 : 1 : 1 (d) 1 : 2 : 0 : 1

13. Two genes 'a' and 'b' present on the same chromosome show 12% recombination. The distance between them is

(a) 0.12 map units (b) 1.2 map units
(c) 12 map units (d) 120 map units

14. Pure homozygous offspring in a dihybrid cross in F_2 generation will be

(a) 1/4 (b) 1/8
(c) 3/8 (d) 1/16

15. A hemophilic man marries a normal woman. Their offspring will be:

(a) All normal
(b) All haemophilic
(c) All boys haemophilic
(d) All girls haemophilic

16. The phenotypic ratio obtained by crossing the F_1 dihybrid with a recessive parent is:

(a) 9 : 3: 3 : 1
(b) 3 : 1
(c) 1 : 1 : 1 : 1
(d) 1 : 1 : 1 : 1 : 1 : 1: 1 : 1

17. How many types of gametes are obtained from a plant of genotype AABbccDd?

(a) 2 (b) 4
(c) 4 (d) 8

18. The ratio of parental to recombinants in dihybrid ratio phenotypically is

(a) 6 : 10 (b) 10 : 6
(c) 12 : 4 (d) 9 : 7

19. The number of hybrids dominant of both traits in the dihybrid ratio is

(a) 4 (b) 6
(c) 12 (d)16

20. The possible phenotypes and genotypes for ABO blood groups are

(a) 4 and 6 (b) 6 and 4
(c) 9 and 6 (d) 16 and 10

21. Wheat plant is 6x = 42, then the number of chromosomes in monosomic, haploid and monoploid conditions is, respectively

(a) 43, 21 and 7 (b) 41, 21 and 7
(c) 15, 7 and 7 (d) 13, 7 and 7

22. The probability of a male child of a haemophilic father and a normal mother becoming haemophilic is

(a) 0% (b) 25%
(c) 50% (d) 100%

23. For MN blood group system, the frequencies of M and N alleles and 0.7 and 0.3, respectively. The expected frequency of MN blood group bearing organism is likely to be:

(a) 40 % (b) 21 %
(c) 42 % (d) 49 %

24. From a cross AaBB X aaBB the genotypic ratio in F_1 generation will be obtained in terms of

(a) 3 AaBB : 1aaBB
(b) All AaBB : No aaBB
(c) 1AaBB : 1 aaBB
(d) 1 AaBB : 3 aaBB

25. What proportion of the offspring of a cross AABBcc x AaBbCc will be completely heterozygous if genes segregate individually?

(a) ¼ (b) 1/16
(c) ½ (d) 1/8

26. A man with blood group B marries a woman with blood group A and the impossible blood group in children will be

(a) A and B only (b) B and AB only
(c) A and AB only (d) None

27. In a cross between genotype AB and ++, 650 out of 1000 individuals were of parental type. The distance between A and B is

(a) 35 map units (b) 45 map units
(c) 30 map units (d) 15 map units

28. Sickle cell anaemia is an example of

(a) Codominance
(b) Pleiotropy
(c) Sublethality
(d) All of the above

29. The four daughter cells derived from single meiosis differ from each other due to

(a) Difference in chromosome number
(b) Crossing over only
(c) Independent assortment of chromosomes only
(d) Crossing over as well as independent assortment of chromosomes

30. If a boy's father has haemophilia and his mother has one gene for haemophilia; what is the chance that the boy will inherit the disease?

(a) 25% (b) 50%
(c) 75% (d) 100%

31. Mr. X is heterozygous for one autosomal gene pair Bb and he carries a recessive X-linked gene d. What proportion of his sperm will be bd?

(a) 0 (b) 1/2
(c) 1/4 (d) 1/8

32. Which chromosome set is found in male grasshoppers?

(a) XY (b) X
(c) YY (d) XX

33. The amino acid substituted in sickle cell anaemia is

(a) Valine for glutamic acid in the beta chain
(b) Glutamic acid for valine in the alpha chain
(c) Glutamic acid for valine in the beta chain
(d) Valine for glutamic acid in the alpha chain

34. Heterozygous tall (Tt) is crossed with homozygous tall (TT). The percentage of heterozygous tall in the progeny would be

(a) 25% (b) 50%
(c) 75% (d) 100%

35. A human male produces sperms with the genotypes AB, Ab, aB, and ab about two diallelic characters in equal proportions. What is the corresponding genotype of this person?

(a) Aabb (b) AABb
(c) AABB (d) AaBb

36. The genes controlling the seven pea characters studied by Mendel are now known to be located on how many different chromosomes?

(a) Seven (b) Six
(c) Five (d) Four

37. To find out the different types of gametes produced by a pea plant having the genotype AaBb it should be crossed to a plant with the genotype

(a) AABB (b) AaBb
(c) aabb (d) aaBB

38. In pea plants, yellow seeds are dominant to green. If a heterozygous yellow-seeded plant is crossed with a green-seeded plant, what ratio of yellow and green-seeded plants would you expect in F_1 generation?

(a) 9 : 1 (b) 1 : 3
(c) 3 : 1 (d) 1 : 1

39. In a dihybrid cross between AABB and aabb, the ratio of AABB, AABb, aaBb, aabb in F_2 generation is:

(a) 9 : 3 : 3 : 1 (b) 1 : 1 : 1 : 1
(c) 1 : 2 : 2 : 1 (d) 1 : 1 : 2 : 2

40. What is the ratio of one pair of contrasting characters in F_2 of a dihybrid cross?

(a) 1 : 2 : 1 (b) 3 : 1
(c) 9 : 3 : 3 : 1 (d) 1 : 4 : 6 : 4 : 1

41. If two genes 'a' and 'b' are linked and show 20% recombination, the proportion of gametes produced in F_1 by a dihybrid ++/ab derived from a cross between ++/++ and ab/ab will be

(a) ++ 40, ab 40, +a 10, +b 10
(b) ++ 20, ab 20, +a 40, +b 40
(c) ++ 40, ab 10, +a 10, +b 40
(d) ++ 10, ab 10, +a 40, +b 40

42. If a-b is 10% and b-c is 20% crossing over, then the frequency of double-crossing over in a – c will be

(a) 1% (b) 2%
(c) 4% (d) 8%

43. Sexually reproducing organisms contribute to their offsprings

(a) All the genes
(b) One forth of their genes
(c) Double the number of their genes
(d) Half of their genes

44. A plant of F_1 generation has genotype AABbCC. On selfing of this plant, the phenotypic ratio in F_2 generation will be
(a) 3 : 1 (b) 1 : 1 : 1 : 1
(c) 1 : 2 : 1 (d) 9 : 3 : 3: 1

45. In a population of 1000 individuals, the gene frequency of dwarf plants is 0.4, and then the number of heterozygotes will be
(a) 480 (b) 360
(c) 160 (d) 600

46. Two alternative alleles, the dominant 'A' and recessive 'a' are in equilibrium is population the frequency of A is 0.6. The percentage of individuals showing the dominant trait in the population should be:
(a) 36 % (b) 60 %
(c) 75 % (d) 84 %

47. If one gamete in 5 carries a recessive allele then the frequency of homozygous recessive genotype in a population at Hardy-Weinberg equilibrium is:
(a) 4 % (b) 10%
(c) 16 % (d) 20 %

48. The coupling ratio in incomplete linkage is:
(a) 1 : 7 : 7 : 1 (b) 7 : 1 : 1 : 7
(c) 1 : 1 : 1 : 1 (d) 9 : 3 : 3 : 1

49. In autosomal recessive traits, the genotype of normal but carrier individual is:
(a) AA (b) Aa
(c) aa (d) Aa and AA

50. If skin colour is controlled by 2 pairs of genes and the number of individuals in a population is 960 then the number of mulattoes in the population is:
(a) 160 (b) 360
(c) 720 (d) 420

51. The probability of the offspring of genotype aaBbCCdd in a cross between two plants of genotypes AabbCcDd and AaBbCcdd is
(a) 1/8 (b) 1/16
(c) 1/32 (d) 1/64

52. A heterozygous Pea plant having genotype TtYy (tall with yellow seeds) is allowed to self-fertilise. What is the probability that an offspring will be either tall with yellow seeds, tall with green seeds or dwarf with yellow seeds?
(a) 64% (b) 74%
(c) 84% (d) 94%

53. If two heterozygous plants are crossed and genes are linked then the distribution of phenotypes in F_1 generation for a dihybrid cross will be
(a) 1:1 (b) 3:1
(c) 1:2:1 (d) 1:1:1:1

54. Starch synthesis in Pea seed is controlled by one gene having 2 alleles, B and b. BB produces large-sized starch grains while bb produces small-sized starch grains. BB seeds are round and bb seeds are wrinkled. Its heterozygote produces
(a) Large-sized starch grain and wrinkled seeds
(b) Small-sized starch grain and wrinkled seeds
(c) Medium-sized starch grain and wrinkled seed
(d) Medium-sized starch grain and round seed

55. If two heterozygous tall pea plants (Tt) are crossed then what is the probability that out of three plants picked up at random, two would be dwarf and one would be tall?
(a) 3/8 (b) 3/16
(c) 9/16 (d) 9/64

56. If two heterozygous tall pea plants (Tt) are crossed then what is the probability that out of five plants will be dwarf?
(a) 16% (b) 26%
(c) 36% (d) 46%

57. When a red-grain variety of wheat is crossed with another white-grain variety of wheat, an F_1 hybrid is produced. On selfing F_1 hybrid, F_2 generation is produced. How many offspring of F_2 resemble phenotypically their parents (grain colour) if it is controlled by 3 gene pairs?
(a) 1/16 (b) 2/64
(c) 9/64 (d) 3/64

58. An individual with 'cd' genes is crossed with wild type '++'. On test crossing the progenies were found to be '+c' 105, '+d' 115 , 'cd' 880 and '++' 900. The distance between two 'cd' genes is
(a) 11 map unit (b) 22 map unit
(c) 20 map unit (d) 50 map unit.

59. A species of plant A with 20 chromosomes is crossed with another species B with 16 chromosomes. When chromosome number in zygote is doubled then it produces a fertile hybrid C. The number of chromosomes in each cell of endosperm is
(a) 36 (b) 54
(c) 90 (d) 108

60. Colour blindness is due to sex linked recessive gene. A survey of 500 men from a local population revealed that 20 were colour-blind. What percentage of females in this population would be expected to be normal?

(a) 36 (b) 48
(c) 96 (d) 99.84

61. Albinism is known to be due to autosomal recessive mutation. The first child of a couple with normal skin pigmentation was an albino. What is the probability of their second child being albino?

(a) 100% (b) 50%
(c) 25% (d) 75%

62. One map unit corresponds to the physical length of the chromosome in which one cross-over occurs in every

(a) One meiosis (b) Two meiosis
(c) Ten meiosis (d) Hundred meiosis

63. Controlled sex mechanism occurs in

(a) *Equisetum* (b) *Luffa*
(c) *Sphaerocarpus* (d) *Melandrium*

64. The epistatic effect in which the dihybrid cross ratio 9:3:3:1 between AaBb x AaBb is modified, is the

(a) Interaction between two alleles of the same loci
(b) Interaction between two alleles of different loci
(c) Dominance of one allele on another allele of the same loci
(d) Dominance of one allele on another allele of both loci

65. If Mendel had studied the 7 traits using a plant with 12 chromosomes instead of 14 then what would have been different?

(a) He could have mapped the chromosomes
(b) He would have discovered incomplete dominance
(c) He would have discovered sex linkage
(d) He would not have discovered the law of independent assortment

66. Mental retardation in men increases with

(a) Decrease in X chromosome
(b) Increase in X chromosome
(c) Decrease in Y chromosome
(d) Increase in Y chromosome

67. *Drosophila melanogaster* has

(a) 2 pairs of autosomes and 1 pair of allosome
(b) 3 pairs of autosomes and 1 pair of allosome
(c) 4 pairs of autosomes and 1 pair of allosome
(d) 5 pairs of autosomes and 1 pair of allosome

68. The formation of multivalent at meiosis in a diploid organism is due to

(a) Reciprocal translocation
(b) Non-reciprocal translocation
(c) Inversion
(d) Deletion

69. One crossing-over results in the formation of

(a) 2 parental gametes and 2 recombinants
(b) 4 parental gametes
(c) 4 recombinants
(d) 2 recombinants

70. Which of the following bases is responsible for hot spots for spontaneous point mutation?

(a) Guanine (b) Adenine
(c) 5-Bromouracil (d) 5-Methylcytosine

71. A F_1 hybrid having the genotype Ab/aB produces 8% of cross-over gametes AB and ab, then the distance between A and B is

(a) 8 cM (b) 16 cM
(c) 24 cM (d) 4 cM

72. In a family, a father has a blood group A and a mother has B. Their children show a 50% probability of a blood group AB indicating that

(a) Father is heterozygous
(b) Father is homozygous
(c) Mother is heterozygous
(d) Either of parent is heterozygous

73. Deletion of half part of the 5th chromosome causes

(a) Down syndrome
(b) Turner syndrome
(c) Klinefelter syndrome
(d) Cat-like cry syndrome

74. The rate of crossing over is reduced by

(a) Deletion (b) Duplication
(c) Translocation (d) Inversion

75. Raphanobrassica is a classical example of

(a) Autotriploid (b) Autotetraploid
(c) Allotriploid (d) Allotetraploid

76. Triticale is manmade cereal developed from

(a) *Triticum durum* and *Aegilops squarrosa*
(b) *Triticum aestivum* and *Secale cereale*
(c) *Triticum turgidum* and *Aegilops squarrosa*
(d) *Triticum durum* and *Secale cereale*

77. Haemophilia A occurs due to lack of

(a) Factor VI (b) Factor VI
(c) Factor VIII (d) Factor 1X

78. Albinism is an autosomal recessive metabolic disorder that occurs due to lack of

(a) Arginase
(b) Tyrosinase
(c) Homogentisic acid oxidase
(d) Phenylalanine hydroxylase

79. If the red-flowered and wrinkled-seeded pea plant is crossed with white flowered and round-seeded pea plant, then F_1 generation plants are obtained. If selfing of F_1 generation then F_2 generation plants are obtained. How many plants will be with intermediate size of starch grain in F_2 generation?

(a) 9/16 (b) 6/16
(c) 8/16 (d) 1/16

80. In a population of 1000 individuals, 360 belong to genotype AA, 480 to Aa and the remaining 160 to aa. Based on this data, the frequency of allele 'A' in the population is

(a) 200 (b) 400
(c) 600 (d) 800

81. A man is heterozygous for autosomal genes A and B and is hemizygous for haemophilic gene C. What proportion of his sperm will be abc?

(a) ½ (b) ¼
(c) 1/8 (d) 1/16

82. A couple has 8 children (all boys). What is the chance that their 9th child will be another boy?

(a) ½ (b) ¼
(c) 1/6 (d) 1/8

83. A couple went to a doctor and reported that both of them are 'carriers' for a particular disorder, their first child is suffering from this disorder and they are expecting their second child. What is the probability that the new child would be affected by the same disorder?

(a) 25% (b) 50%
(c) 75% (d) 100%

84. Multiple alleles are present at

(a) different loci in the same chromosome
(b) same loci in the different chromosome
(c) same locus within the homologous chromosome
(d) different loci in the different chromosome

85. A man and woman who both appear normal have a child together who has sickle cell anaemia. The woman becomes pregnant again and is told that she is carrying fraternal twins. What is the probability that both of the couple's twins will develop sickle cell anaemia?

(a) 1/2 (b) 1/4
(c) 1/16 (d) 9/16

86. The genotype of baldness in females is

(a) BB (b) Bb
(c) bb (d) Both BB and Bb

87. Which of the following is incorrect regarding chromosomal abnormalities?

(a) Epicanthus is seen in Down syndrome
(b) Klinefelter syndrome occurs due to trisomy of the sex chromosome
(c) Cat-like cry syndrome occurs due to trisomy of the 5th chromosome
(d) Turner syndrome is sex chromatin negative

88. A woman heterozygous for colour blindness marries a colour-blind man. What is the ratio of carrier daughter, colour-blind daughter, normal sons and colour-blind sons?

(a) 1:2:1:2 (b) 1:1:1:2
(c) 1:1:1:1 (d) 1:2:2:1

89. Sickle cell anaemia is an example of

(a) Codominance
(b) Sub-lethal gene
(c) Pleiotropy
(d) All of the above

90. The total number of chromosomes in individuals suffering from Turner's syndrome is

(a) 44 (b) 45
(c) 47 (d) 48

91. The diseased offspring cannot be found in the marriage of

(a) Normal man and diseased woman
(b) Diseased man and normal woman
(c) Diseased man and carrier woman
(d) Diseased man and diseased woman

92. Which of the following laws of Mendel is not applicable for linkage?

(a) Law of unit factor
(b) Law of dominance
(c) Law of segregation
(d) Law of independent assortment

93. A red-flowered snapdragon plant is crossed with pink flowered plant then 200 plants are obtained. How many of them are red-flowered?

(a) 50 (b) 100
(c) 150 (d) 200

94. A father has AB blood group and mother's blood group is untraced. They have three children. One has 'A' blood group while second has 'B' blood group. What is the possibility of occurrence of 'O' blood group in third offspring?

(a) 0% (b) 25%
(c) 50% (d) 75%

95. If the parents has free ear lobes and the offspring has attached ear lobes, then the parents must be

(a) Homozygous (b) Heterozygous
(c) Hemizygous (d) Codominant

96. Children suffering from PKU are given food low in phenylalanine and supplemented with tyrosine because

(a) They are unable to utilise phenylalanine
(b) They do not require phenylalanine
(c) They have increased tyrosine anabolism
(d) They have increased tyrosine catabolism

97. Which of the following Mendel's laws is universally acceptable?

(a) Law of unit factor
(b) Law of dominance
(c) Law of segregation
(d) Law of independent assortment

98. Incomplete dominance was observed in

(a) *Mirabilis* and *Solanum*
(b) *Solanam* and *Lilium*
(c) *Lilium* and *Antirrhinum*
(d) *Antirrhinum* and *Mirabilis*

99. Complete linkage is observed in

(a) Male *Drosophila* and male silkworm
(b) Male *Drosophila* and female silkworm
(c) Female *Drosophila* and male silkworm
(d) Female *Drosophila* and female silkworm

100. Hormonal control of sex determination is observed in

(a) *Bonellia* (b) The Crew's hen
(c) Free martin (d) All of the above

101. Haplodiploidy condition is observed in

(a) Honeybees (b) *Drosophila*
(c) Human beings (d) Silkworm

102. Which of the following is incorrectly matched?

(a) Non-ionising radiation-UV rays
(b) Deaminating agent- Nitrous acid
(c) Alkylating agent- Benzopyrene
(d) Intercalating agent-Acridine dye

103. The number of linkage groups in *Pisum sativum* is

(a) 7 (b) 14
(c) 10 (d) 20

104. Thymine dimer formed in induced mutation is due to

(a) X-ray (b) UV ray
(c) Acridine dye (d) Nitrous acid

105. The number of transversion mutations is

(a) 2 (b) 4
(c) 6 (d) 8

106. Frameshift mutation is caused by

(a) Deletion and modification of base pair
(b) Deletion and insertion of base pair
(c) Insertion and replacement of base pairs
(d) Duplication and inversion of base pairs

107. Gene dosage depends on

(a) Total number of genes
(b) Ploidy level of the species
(c) Frequency of induced mutations
(d) Number of chromosomes involved in reciprocal translocation

108. Which one of the following alterations in DNA is likely to be lethal?

(a) Substitution of adenine for cytosine
(b) Substitution of methylcytosine for cytosine
(c) Deletion of three nucleotides
(d) Insertion of one nucleotide

109. Which one of the following phenomena leads to aneuploidy?

(a) Improper division of centromere
(b) Desynapsis
(c) Non-disjunction
(d) Disjunction

110. The spontaneous mutations in nature are

(a) Recessive and lethal
(b) Dominant and beneficial
(c) Dominant and lethal
(d) Recessive and beneficial

111. Mendel formulated the law of purity of gametes on the basis of

(a) Dihybrid cross
(b) Monohybrid cross
(c) Back cross
(d) Test cross

112. Which of the following genotypes is an example of true dihybrid condition?

(a) Ttrr (b) Ttrr
(c) TtRr (d) TTRr

113. If 120 plants are produced by crossing pure red and pure white flowered pea plants, then the ratio of offspring will be

(a) 90 red: 30 white
(b) 30 red: 90 white
(c) 60 red: 60 white
(d) All red

114. How many types of gametes are produced from the organism with genotype AABBCC?

(a) 1 (b) 2
(c) 4 (d) 8

115. Two allelic genes are located on the

(a) Same chromosome
(b) Two homologous chromosomes
(c) Two non-homologous chromosomes
(d) Any two chromosome

116. Mrs. X has an autosomal gene pair 'Bb' and she contains X-linked gene 'd' on her both X chromosomes. The percentage of gamete containing 'bd' genes is

(a) 50% (b) 25%
(c) 75% (d) 100%

117. A dihybrid plant on self-pollination produced 400 phenotypes with 9 types of genotypes. The number of seeds having genotype TtRr is

(a) 50 (b) 150
(c) 200 (d) 100

118. In a cross, Aabb x aaBb, the genotypic ratio is

(a) 1 :2: 1 (b) 1:1:1:1
(c) 3:1 (d) 9:3:3:1

119. The somatic cell of a human male contains a single Barr body then the genetic composition of the person would be

(a) XY (b) XO
(c) XXY (d) XYY

120. In which animals, sex is determined by the female individual?

(a) Human (b) *Drosophila*
(c) Bird (d) Grasshopper

121. If a normal woman marries an albino man and their children are half albino, half normal then the woman is

(a) Homozygous normal
(b) Heterozygous normal
(c) Homozygous recessive
(d) Homozygous dominant

122. The probability of four sons to a couple is

(a) 1/4 (b) 1/8
(c) 1/16 (d) 1/32

123. The probability of three daughters to a couple in three children is

(a) 1/4 (b) 1/8
(c) 1/16 (d) 3/8

124. A tobacco plant heterozygous for albinism is self-pollinated and 1200 seeds are subsequently germinated. How many seedlings would have the parental genotypes?

(a) 300 (b) 600
(c) 900 (d) 1200

125. Down syndrome is caused by an extra copy of chromosome number 21. The percentage of affected offspring produced by the affected mother and normal father is

(a) 25% (b) 50%
(c) 75% (d) 100%

126. Albinism in human beings is controlled by a recessive gene 'a'. If both parents were known to carrier (Aa) for albinism then what will be the chance of 1 normal and 3 albino in a family of four?

(a) 12/256 (b) 81/256
(c) 54/256 (d) 100/256

127. Which of the following does not bring variation at the individual level?

(a) Gene mutation
(b) Fertilisation
(c) Meiosis and crossing over
(d) Chromosomal aberration and hybridisation

128. If a girl is born with sickle cell anaemia, what must be true?

(a) Only father has sickle cell anaemia
(b) Only mother carried sickle cell anaemia
(c) Only mother had sickle cell anaemia
(d) Both parents carried sickle cell anaemia

129. Linkage decreases the frequency of

(a) Hybrid (b) Dominant allele
(c) Recessive allele (d) Both b and c

130. Hypophosphatemia is caused by an X-linked dominant gene in human beings. If a hypophosphatemic man marries a normal woman then which of their children will have hypophosphatemia?

(a) All of the daughters
(b) Half of their daughters
(c) All of the sons
(d) Half of their sons

131. Physically 1 map unit on linkage maps can be defined as the length of chromosomes in which the average crossover formed during 50 cells undergoing meiosis is

(a) 1 (b) 25
(c) 50 (d) 100

132. If two recessive mutations are alleles of different genes, then F_1 progeny are wild-type. This is due to

(a) Incomplete dominance
(b) Complementation
(c) Supplementation
(d) Codominance

133. Which of the following chromosomal alterations causes most drastic consequences?

(a) Deletion (b) Duplication
(c) Translocation (d) Inversion

134. On a pedigree having the inheritance of PKU, a horizontal line joins a black square and a half-black circle. What fraction of this couple's children would you expect to suffer from PKU?

(a) None (b) 1/2
(c) 1/4 (d) 3/4

135. Mr. X and Mrs. Z are normal, but their daughter was born with Alkaptonuria. The probability of their next child being born with alkaptonuria is

(a) 0 (b) 1/2
(c) 1/4 (d) 2/3

136. Duchenne Muscular Dystrophy is caused by a sex-linked recessive allele. Its victims are almost boys who die before the age of 20. This disorder is never seen in girls because

(a) Sex-linked traits are never seen in girls
(b) The allele is carried on the Y-chromosome
(c) Non-disjunction occurs in males but not in females
(d) A female must have two copies of genes to express it

137. Disorders involving an unusual number of sex chromosomes show that maleness is caused by the

(a) Presence of x chromosome
(b) Presence of Y chromosome
(c) Absence of X chromosome
(d) Absence of Y chromosome

138. In heterozygous cells, the gene of a recessive allele

(a) Is always turned off
(b) May be expressed but the product is quickly degraded
(c) May be expressed without measurable effect
(d) Is deleted

139. When a plant of chromosomal type 'aa' pollinates a plant of 'AA'. What will be the chromosomal constitution of the embryo and endosperm in the resulting seeds?

(a) Diploid zygote of type 'Aa' and triploid endosperm of type 'AAa'
(b) Diploid zygote of type 'aa' and triploid endosperm of type 'Aaa'
(c) Diploid zygote of type 'AA' and triploid endosperm of type 'AAa'
(d) Diploid zygote of type 'Aa' and triploid endosperm of type 'aaa'

140. Reciprocal crosses do not yield equal results if characters are

(a) X-linked
(b) Autosomal
(c) Extrachromosomal
(d) None

141. Dosage compensation in the case of human beings is achieved by

(a) Hyperactivation of X chromosome
(b) Hyperactivation of Y chromosome
(c) Heterochromatisation of X chromosome
(d) Heterochromatisation of Y chromosome

142. Which of the following plants is used as a model to study the developmental genetics of zygotic embryogenesis?

(a) *Daucus carota*
(b) *Arabidopsis thaliana*
(c) *Zea mays*
(d) *Nicotiana tabacum*

143. Which of the following genotypes would produce the greatest variety of gametes of the alleles assorted independently?

(a) aaBBCcDd (b) aabbCCDD
(c) AaBbCCDd (d) AABBCCDD

144. If bacteria possess two copies of certain genes then they are referred as

(a) Merodiploid (b) Euploid
(c) Polyploid (d) Amphidiploid

145. Mendelian inheritance is obeyed by

(a) Quantitative traits
(b) Transposons
(c) Organelles
(d) Gene for vertical transfer of disease

146. A woman with an albino father marries an albino man. The proportion of her progeny is

(a) All normal
(b) All albino
(c) One normal and one albino
(d) One normal and two albino

147. If a man of blood group 'A' marries a woman of blood group 'AB' then which type of progeny would indicate that man is heterozygous 'A'?

(a) AB (b) A
(c) O (d) B

148. The weight of fruits in a plant is determined by the number of dominant alleles of a certain number of genes. If seven categories of weight of fruit were found, then how many gene sites would be involved?

(a) 2 (b) 3
(c) 4 (d) 5

149. The ratio of heterozygous, homozygous and colour-blind hemizygous offspring of a colour-blind man and carrier woman is

(a) 1:1:2 (b) 1:1:1
(c) 2:1:1 (d) 1:2:1

150. If Mendel had studied 7 pairs of characters in a plant with 12 chromosomes instead of 14 then he could not have discovered

(a) Dominance
(b) Independent assortment
(c) Linkage
(d) Crossing over

151. The number of plants produced from a dihybrid plant with incomplete linkage on test cross is

(a) 1 (b) 2
(c) 4 (d) 8

152. The number of linkage groups in nucleoids of bacteria is

(a) 1 (b) 2
(c) 3 (d) 4

153. Test cross cannot be performed in

(a) Pea (b) Man
(c) Cat (d) *Neurospora*

154. The most suitable medium for culturing *Drosophila* is

(a) Agar agar (b) Moist bread
(c) Ripe banana (d) Dung

155. In a random mating population, the frequency of disease-causing recessive alleles is 80%. What would be the frequency of carrier individuals in the population?

(a) 32% (b) 64%
(c) 16% (d) 100%

156. A plant is heterozygous (Bb) and produces 2 gametes such as B and b. The probability of 'b' gamete fertilising B or b is

(a) Zero (b) 1
(c) 1/2 (d) 1/4

157. Albinism is determined by a recessive gene in man. The presence of albinism in 50% of children born to a couple proves that

(a) Both parents are heterozygous for albinism
(b) Father is homozygous normal and mother is heterozygous
(c) Father is homozygous for albinism but mother is heterozygous
(d) Both parents are homozygous

158. In a population where the frequency of a homozygous recessive disease is 16% then the frequency of the dominant allele would be

(a) 0.16 (b) 0.6
(c) 0.4 (d) 0.84

159. Both transition and transversion types of mutation are induced by

(a) Methyl methane sulphonate

(b) Nitrous acid

(c) Base analogue

(d) Acridine dye

160. There are two genes A and B which are linked. What shall be the genotype of progeny in a cross between AB/ab and ab/ab?

(a) AAbb and aabb

(b) AaBb and aabb

(c) AABB and aabb

(d) aaBB and aabb

161. A colour-blind woman with 'A' blood group marries a man with having 'AB' blood group and a normal parent. The couple has 2 daughters and 2 sons. The possibility of normal visioned daughter with 'B' blood group will be

(a) 25% (b) 50%

(c) 75% (d) 100%

162. In sickle cell anaemia, substitution mutation occurs between

(a) A neutral amino acid and an acidic amino acid

(b) An acidic amino acid and a neutral amino acid

(c) A neutral amino acid and a basic amino acid

(d) A neutral amino acid and an aromatic amino acid

163. Mr. Z is suffering from hypertrichosis and PKU. His father is heterozygous for PKU. The probability of Mr. Z's sperm having one recessive autosomal allele and holandric gene is

(a) 1/2 (b) 1/4

(c) 3/8 (d) 3/4

164. A diseased man marries a normal woman. They have 3 diseased daughters and 5 normal sons. The gene for this disease is

(a) Sex-linked recessive

(b) Sex-linked dominant

(c) Sex-limited trait

(d) Autosomal dominant

165. How many different genotypes can be expected in F_2 of the cross AABB x aabb when genes are completely linked (coupled) and inherited independently, respectively?

(a) 2 and 4 (b) 3 and 9

(c) 4 and 8 (d) 5 and 9

166. In a graft, the stock has 40 chromosomes and the scion has 20 chromosomes. The number of chromosomes in the root and egg, respectively

(a) 20 and 10 (b) 40 and 10

(c) 10 and 20 (d) 10 and 40

167. Which of the following is a set of discontinuous variations?

(a) Blood group and sex

(b) Hair colour and intelligence

(c) Height and excellence in maths

(d) Skin colour and mechanical skill

168. If a diploid cell is treated with colchicine, it becomes

(a) Haploid (b) Triploid

(c) Tetraploid (d) Hexaploid

169. In a population of 6000 individuals, an autosomal recessive disorder is present in 60 individuals. What is the number of homozygous normal (DD) offspring in the population?

(a) 4860 (b) 5469

(c) 1000 (d) 2000

170. In a plant, a quantitative character is controlled by 2 polygenes and each dominant allele produces 10 cm height and the basic height of the plant is 30 cm. What is the height of a plant having all dominant alleles?

(a) 40 (b) 50

(c) 60 (d) 70

171. In humans, XX males and XY females are rare but such cases are due to

(a) Deletion of the X chromosome

(b) Deletion of the Y chromosome

(c) XY translocation

(d) Duplication of the X chromosome

172. A single base pair mutation occurs in strand CATTACCG, its complimentary strand had a sequence GTTATGGC. It means mutation has changed the base pair

(a) A→T (b) T→T

(c) T→A (d) A→A

173. In human females, there is inactivation of one X chromosome for dosage compensation due to

(a) Methylation (b) Acetylation

(c) Phosphorylation (d) Formylation

174. Which of the following shows a diseased male in pedigree analysis?

(a) Black square (b) Black circle
(c) Clear circle (d) Clear square

175. Edward syndrome occurs in human beings due to

(a) Monosomy of 21st chromosome
(b) Trisomy of 21st chromosome
(c) Monosomy of 18th chromosome
(d) Trisomy of 18th chromosome

176. Two pink flowered four o'clock plants are self-fertilised then what will be the probability of the first 3 plants appearing to be red?

(a) 1/4 (b) 1/8
(c) 1/16 (d) 1/64

177. If a diseased mother transfers the disease to all children then it is termed as

(a) Autosomal recessive
(b) Autosomal dominant
(c) Sex-linked recessive
(d) Maternal inheritance

178. Turner syndrome is a case of

(a) Monosomy (b) Trisomy
(c) Nullisomy (d) Tetrasomy

179. Alpha thalassemia is caused by mutation of genes present on

(a) 9th chromosome (b) 11th chromosome
(c) 16th chromosome (d) 22nd chromosome

180. Folded eyelid or epicanthus is seen in

(a) Edward syndrome (b) Down syndrome
(c) Turner syndrome (d) Klinefelter syndrome

181. Who first proposed that the X chromosome was involved in sex determination?

(a) Morgan (b) Henking
(c) McClung (d) Sturtevant

182. The chromosomal theory of inheritance was proposed by

(a) Morgan (b) Sutton and Boveri
(c) Janssen (d) Wilson and Stevens

183. Cytoplasmic or plastid inheritance was reported in

(a) *Mirabilis jalapa*
(b) *Oenothera lamarckiana*
(c) *Pisum sativum*
(d) *Drosophila melanogaster*

184. Which of the following is incorrect about crossing over?

(a) It increases with an increase in the distance between genes
(b) It increases with irradiation of X-ray
(c) It increases due to interference
(d) It decreases at telomere

185. If the blood groups of your parents are O and AB then the impossible blood group in offspring will be

(a) O and AB (b) O and A
(c) O and B (d) A and AB

186. Which of the following blood groups does not have antigen?

(a) A (b) B
(c) AB (d) O

187. Codominance was observed in

(a) Flower colour in *Mirabilis jalapa*
(b) Skin colour in cattle
(c) Leaf size in *Antirrhinum ajacis*
(d) Skin colour in mice

188. Which of the following is sex chromatin negative?

(a) Turner syndrome
(b) Klinefelter syndrome
(c) Down syndrome
(d) Super male

189. The offspring showing chromosome complement of 44+XXY suffers from

(a) Turner syndrome
(b) Klinefelter syndrome
(c) Down syndrome
(d) Super male

190. Which of the following is Y-linked traits?

(a) Hypertrichosis
(b) Testis determining factor
(c) Porcupine skin
(d) All of the above

191. If a diseased man marries a carrier woman, then

(a) 50% of sons will be diseased
(b) 25% of sons will be diseased
(c) 75% of sons will be diseased
(d) 100% of sons will be diseased

192. The chromosome complement of nullisomy condition is

(a) 2n-1 (b) 2n-2
(c) 2n+1 (d) 2n+2

193. A frameshift mutation is caused by

(a) Nitrous acid (b) Bromouracil
(c) Acridine dye (d) All of the above

194. 5-bromouracil is a base analog that causes gene mutation by

(a) Mis-replication
(b) Mis-incorporation
(c) Tautomerisation
(d) Dimer formation

195. The frequency of oxidative deamination for causing transition mutation is maximum in

(a) Adenine (b) Guanine
(c) Thymine (d) Cytosine

196. Who discovered the mutagenic effect of X-ray on *Drosophila*?

(a) Morgan (b) Muller
(c) Sturtevant (d) Hugo de Vries

197. Which of the following is incorrect with respect to mutation?

(a) It occurs in higher plants and animals
(b) It is recessive
(c) It shows hot spots
(d) It is mostly harmful

198. The large, sudden and discontinuous change was first noticed by Hugo de Vries in

(a) *Drosophila melanogaster*
(b) *Mirabilis jalapa*
(c) *Oenothera lamarckiana*
(d) *Pisum sativum*

199. If the X/A ratio is 0.5 then it will be

(a) Super female (b) Fertile female
(c) Intersex (d) Fertile male

200. Which of the following combinations of chromosomes is shown by the moth and butterfly?

(a) XX-XY type (b) XX-XO type
(c) ZZ-ZW type (d) ZZ-ZO type

ANSWERS

1.c	21.b	41.a	61.c	81.c	101.a	121.b	141.c	161.a	181.c
2.d	22.a	42.b	62.a	82.a	102.c	122.c	142.b	162.a	182.b
3.d	23.c	43.d	63.d	83.a	103.a	123.b	143.c	163.a	183.c
4.c	24.c	44.a	64.d	84.c	104.b	124.b	144.d	164..b	184.c
5.b	25.d	45.a	65.d	85.c	105.d	125.b	145.a	165..b	185.a
6.b	26.d	46.d	66.b	86.a	106.b	126.a	146.c	166.b	186.d
7.b	27.a	47.a	67.b	87.c	107.d	127.d	147.d	167.a	187.b
8.a	28.d	48.b	68.a	88.c	108.d	128.d	148.b	168.c	188.a
9.d	29.d	49.b	69.a	89.d	109.c	129.a	149.b	169.a	189.b
10.b	30.c	50.b	70.d	90.b	110.a	130.a	150.b	170.d	190.d
11.b	31.c	51.d	71.b	91.b	111.a	131.c	151.c	171.c	191.a
12.b	32.b	52.d	72.b	92.d	112.c	132.c	152.a	172.a	192.b
13.c	33.a	53.b	73.d	93.b	113.d	133.a	153.d	173.a	193.c
14.a	34.b	54.d	74.d	94.a	114.a	134.b	154.c	174.a	194.b
15.a	35.d	55.d	75.d	95.b	115.b	135.c	155.a	175.d	195.a
16.c	36.d	56.b	76.d	96.a	116.a	136.d	156.c	176.d	196.b
17.c	37.c	57.b	77.c	97.c	117.d	137.b	157.c	177.d	197.a
18.b	38.d	58.a	78.b	98.d	118.b	138.b	158.d	178.a	198.c
19.a	39.c	59.b	79.c	99.	119.c	139.a	159.a	179.c	199.b
20.a	40.b	60.d	80.c	100.d	120.c	140.c	160.b	180.b	200.d

CHAPTER-13

MOLECULAR BASIS OF INHERITANCE

Genetic material is a chemical substance which stores biological information in a coded form, and transmits it to next generation where it is expressed in offspring.

Features of a molecule to fulfil the criteria of genetic material:

1. The genetic material should be present in the nucleus of every cell and its amount should be the same in all somatic cells of a given species.
2. It should be structurally and chemically stable.
3. It should be able to transmit Mendelian characters to the next germination.
4. It should undergo replication (doubling the amount) and
5. It should undergo mutation.

THE SEARCH FOR GENETIC MATERIAL

In fact, the works for the principle of inheritance by Mendel and the discovery of nuclein by Miescher were almost at the same time but to discover and prove DNA as genetic material took a very long time. Further, the works of Sutton and Morgan confined it (genetic material) to chromosomes located in the nucleus of most cells. Chromosome contains DNA as well as protein and was considered a structure for hereditary transmission. It was not clear whether protein or DNA was genetic material. Until 1949, proteins were considered to be playing an informational role. In the meantime, Mirsky and Ris (1949) reported that all cells of an organism have a constant amount of DNA and different amounts of proteins. This favoured DNA as genetic material.

Evidence in favour of DNA as genetic material:

(a) It is present in nucleus of every cell and its amount is same in all somatic cells of a given species.

(b) It undergoes replication in S-phase of interphase just before cell division.

(c) It passes its copies to next generation where it expresses its genetic information.

(d) It shows mutation occasionally.

(e) It stores biological information in coded form as in triplet code.

(f) It controls various life processes.

Experimental evidence in favour of DNA:

1. **Transformation experiment:** It was discovered by Frederick Griffith (1928) on *Streptococcus pneumoniae* (earlier known as *Diplococcus pneumonia),* a bacterium that causes pneumonia in mammals. There are two different strains of this bacterium.

   (a) S-strain: It is a virulent strain that is enclosed by a capsule (coat) and forms a smooth and shiny colony when grown on agar medium. It is harmful (pathogenic) and causes pneumonia.

   (b) R-strain: It is a non-virulent strain that is not enclosed by a capsule (coat) and forms a rough colony when grown on an agar medium. It is harmless (non-pathogenic) and does not cause pneumonia.

   Both S and R strains have several subtypes such as S-I, S-II, S-III and R-I, R-II, and R-III, respectively. Griffith tested the virulence of two strains such as S-III and R-II in the following four ways:

   (a) Strain S-III (Live, virulent) $\xrightarrow{\text{Injected into mice}}$ mice died of pneumonia

Injected into mice

(b) Strain S-III (heat-killed at 80oC, virulent)⟶mice alive

Injected into mice

(c) StrainR-II (Live, non-virulent)⟶mice alive

(d) Strain S-III (heat-killed, virulent)+Strain R II (live, non-virulent)⟶mice died of pneumonia

Griffith found a live S-strain of a bacterium in the blood of dead mice. He now concluded that a substance (**transforming principle**) had entered the living RII strain from dead strain III which changed (transformed) the RII strain into the S-III strain. This was a permanent genetic change as the S III strain continued to produce similar cells. It was not possible to revert back to the RII strain.

He had no idea about what component (carbohydrate or protein or DNA) caused virulence.

Oswald Avery, Colin Macleod and Maclyn McCarty (1944) conducted the experiment of Griffith in vitro (test tube) to identify the component responsible for this transformation. They separated the filtrate (extract) of the S III strain of bacterium into carbohydrate, protein and DNA fractions and added each fraction into a separate culture tube containing the R-II strain and observed.

Mixture	Result	Observations
Carbohydrate fraction of S-III strain + RII strain	R-II strain	No transformation
Protein fraction of S-III strain + RII strain	RII strain	No transformation
DNA fraction of S-III strain + RII strain	SIII strain	Transformation occurs
DNA fraction of S-III strain + DNase + RII strain	RII strain	No transformation

It was noticed that only the culture containing the DNA fraction of strain SIII transformed the RII strain to the S III strain. This confirmed that DNA was the transforming agent. When DNA was treated with deoxyribonuclease (an enzyme that digests DNA) then, transformation of the RII strain did not occur.

2. **Blender Experiment:** It was performed by Alfred Hershey and Martha Chase (1952) using T_2 bacteriophage (virus) to prove that DNA was the genetic material in the viruses because phage DNA enters into host cells and works as genetic material.

   They labelled protein and DNA with different isotopes. They developed the two cultures of *E. coli*. In one culture, a radioisotope S^{35} was labelled into sulphur containing amino acids (cysteine and methionine) of protein and in another culture, a radioisotope P^{32} was labelled into nucleotides of DNA.

   Process:

   (a) Infection: The phages (viruses) were used separately to infect *E. coli* (bacterium). They introduced T_2 bacteriophages in both cultures and infection was allowed to run for 10 minutes. Multiplication started.

   (b) Blending: Both cultures were gently shaken in a blender at 10,000 rpm to separate ghost or capsid (empty phages) sticking to the surface of *E. coli* or to break contact between viruses and bacteria.

   (c) Centrifugation: Both the cultures were centrifuged and the level of radioactivity in the supernatant (suspension) and pellet was observed.

In one culture (protein of bacteriophage labelled with S^{35}), the supernatant contained radioactive empty bacteriophage and the pellet settled down in a centrifuge tube containing non-radioactive heavier bacteria. It showed that the protein coat of bacteriophage did not enter the bacterial cell during infection.

In another culture (DNA of bacteriophage labelled with P^{32}), the supernatant contained non-radioactive empty bacteriophage and the pellet settled down in a centrifuge tube containing radioactive heavier bacteria. It showed that the DNA of bacteriophage entered the bacterial cell during infection.

Now it was confirmed that sulphur did not form a part of DNA and phosphorus only formed part of DNA hence P^{32} radioactive was found in *E. coli* infected with P^{32} phage. This proved that protein did not enter bacteria from bacteriophage and only DNA passed from phage to bacteria.

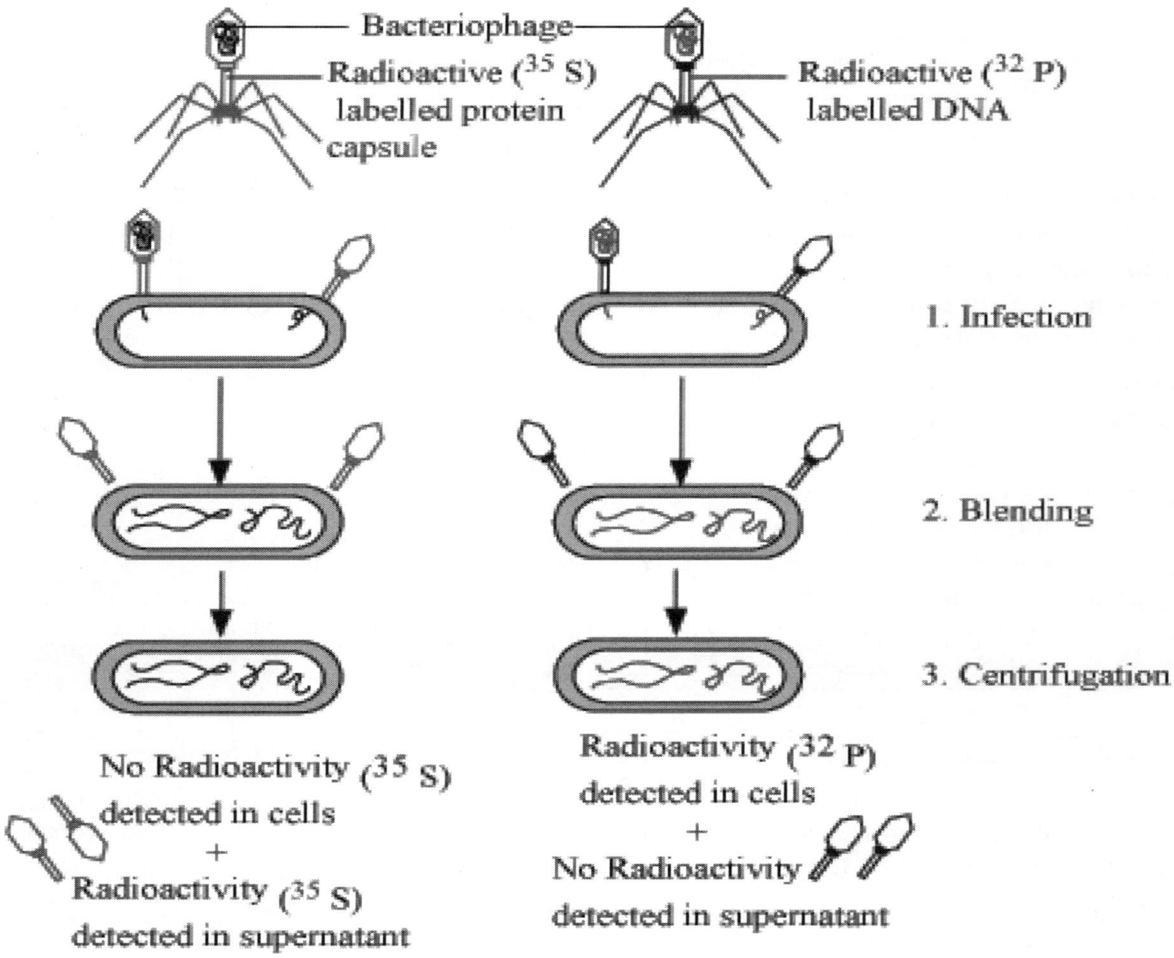

Fig: Blender experiment

STRUCTURE OF DNA

Nucleic acids are highly complex, negatively charged, acids, and the largest macromolecule. There are two types of nucleic acids found in living organisms. DNA was first isolated by Friedrich Miescher (1869) from the nucleus of a pus cell and the sperm of Salmon and termed it 'nuclein'.

Altman (1899) coined the term 'Nucleinic acid' and then called it 'Nucleic acid' because it was found in the nucleus and showed acidic nature.

Fischer (1880) discovered the presence of two types of bases-purine and pyrimidine.

Levene (1910) discovered deoxyribose sugar and phosphoric acid in it. He characterised four types of nucleotides present in DNA.

Astbury proposed through his first X-ray diffraction that DNA was a polynucleotide in which nucleotide was arranged perpendicular to the long axis of a molecule.

Maurice Wilkins and Rosalind Franklin (1953) proposed through X-ray photographs (x-ray crystallography) that DNA was a helix with a width of 2 nm and one turn of helix is 3.4 nm having ten layers of stacked bases.

The DNA molecule is a polymer of several thousand nucleotides. Each nucleotide consists of sugar, base and phosphate.

Sugar: It is deoxyribose (pentose). At the second carbon atom, oxygen is absent hence OH group is replaced by the H atom but in RNA it is ribose sugar. The 5th and 3rd carbon atoms participate in a phosphodiester bond while the 1st carbon atom is always occupied by an organic base.

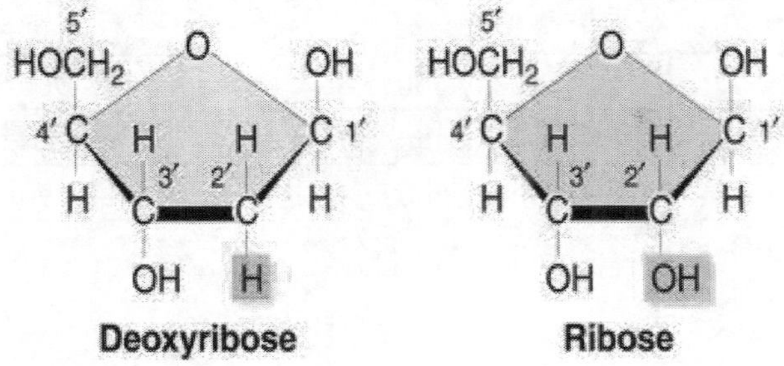

Nitrogenous bases: These are hydrophobic. There are two types of bases:

Purines: These are double-ringed and 9-membered (5 Carbon and 4 Nitrogen atoms) compounds. These have high molecular weight. In these, the N atom is located at the 1st, 3rd, 7th and 9th positions of carbon. These are of two types:

1.	Adenine	6-aminopurine
2.	Guanine	2-amino,6-oxypurine

Pyrimidines: These are single-ringed and 6-membered (4 carbon and 2 Nitrogen atoms) compounds. These have low molecular weight. In these, the N atom is present at the 1st and 3rd positions of carbon. There are of three types:

1.	Thymine	5-methyl 2,4-dioxypyrimidine
2.	Uracil	2, 4-dioxypyrimidine: It is present only in RNA in place of Thymine
3.	Cytosine	4-amino, 2-oxypyrimidine

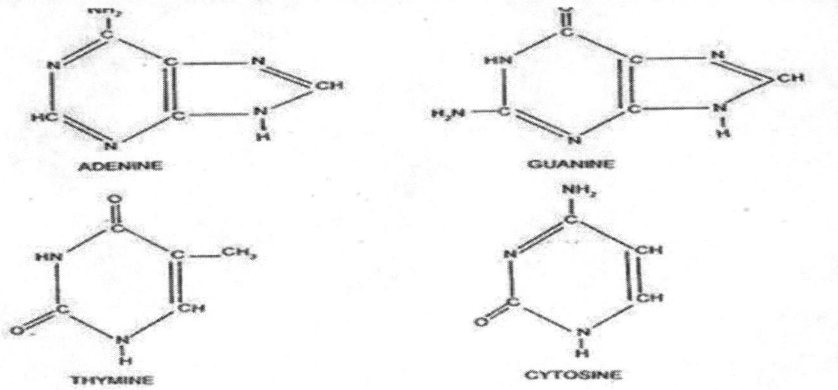

Phosphate: It is present as phosphate group of orthophosphoric acid (H_3PO_4).

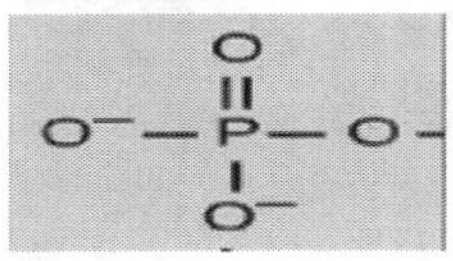

It alternates with sugar. It has three reactive OH groups that are involved in the sugar-phosphate backbone with the help of a phosphodiester bond. Each phosphate group is joined to the OH group of the 3rd carbon of one sugar at one side and joined to the OH group of the 5th carbon of another sugar of the next nucleotide by a phosphodiester bond (5′ C-O-P-O-C 3′). It gives an acidic nature to DNA. It is used as a linker in strands due to the multiple bond facility.

Nucleosides: These are formed from sugar and base through an N-glycosidic bond (C-N covalent bond). It is a condensation reaction and water is released.

Base + Sugar———————àDeoxyriboside + H_2O

A glycosidic bond (or glycosylic bond) develops between 1st carbon of sugar and the N atom of the 9th position in the case of purines and between 1st carbon of sugar and the N atom of the 1st position in the case of pyrimidines. There are four nucleosides in DNA:

Bases	Sugar	Nucleosides
Adenine +	Deoxyribose →	Deoxyadenosine
Guanine +	Deoxyribose →	Deoxyguanosine
Thymine +	Deoxyribose →	Deoxythymidine
Cytosine +	Deoxyribose →	Deoxycytidine

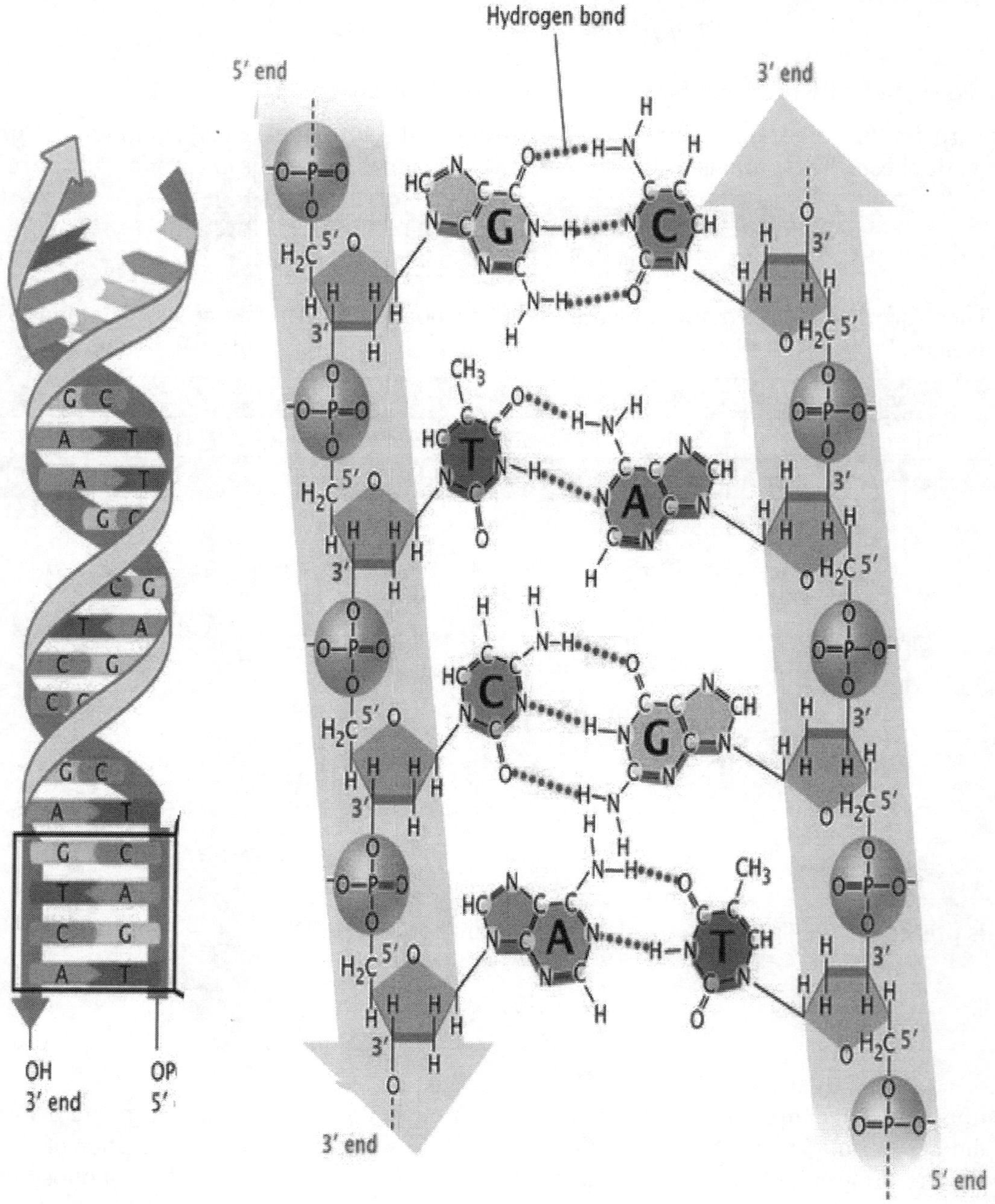

Fig: Antiparallel nature of DNA

Nucleotides: These are formed from sugar, base and phosphate through phosphodiester bonds. It is also a condensation reaction and water is released.

Phosphodiester bond develops between the –OH group of the 3rd carbon of sugar of one nucleotide and the 5th carbon of another sugar linked with the phosphate group. Thus, nucleotides of a strand are linked to each other through a phosphodiester bond (3′-C-O-P-O-C-5′) between sugar (3′) of one nucleotide and phosphate (5′) of another nucleotide. At the 3′-end of a strand, the 3rd carbon of sugar is present and usually, it has a free –OH group. At the 5′-end of a strand, the 5th carbon of sugar is present and usually, it has a phosphate group.

There are four nucleotides in DNA:

Nucleosides	Phosphate group	Nucleotides
Deoxyadenosine +	Phosphate →	Deoxyadenylate (Deoxyadenosine-5- monophosphate)
Deoxyguanosone +	Phosphate →	Deoxyguanylate (Deoxyguanosine -5-monophosphate)
Deoxythymidine +	Phosphate →	Deoxythymidylate (Deoxythymidine-5-monophosphate)
Deoxycytidine +	Phosphate →	Deoxycytidylate (Deoxycytidine-5-monophosphate)

Features:

1. **Polarity:** The polynucleotide chains show polarity (direction). One end of each DNA strand is called the 5′ end. The last deoxyribose unit at this end has a carbon atom at the 5th position. The other end of the strand is called the 3′ end. The last deoxyribose unit at this end has a carbon atom at 3rd position. There are two ends:

   (a) **5′:** The phosphorus group of the 5th carbon of sugar is free but its 3rd carbon is not free because it is attached to the P atom forming a phosphodiester bond.

   (b) **3′:** The phosphorus group of the 5th carbon of sugar is not free because it forms a phosphodiester bond, but its 3rd carbon is free because its OH group does not form a phosphodiester bond.

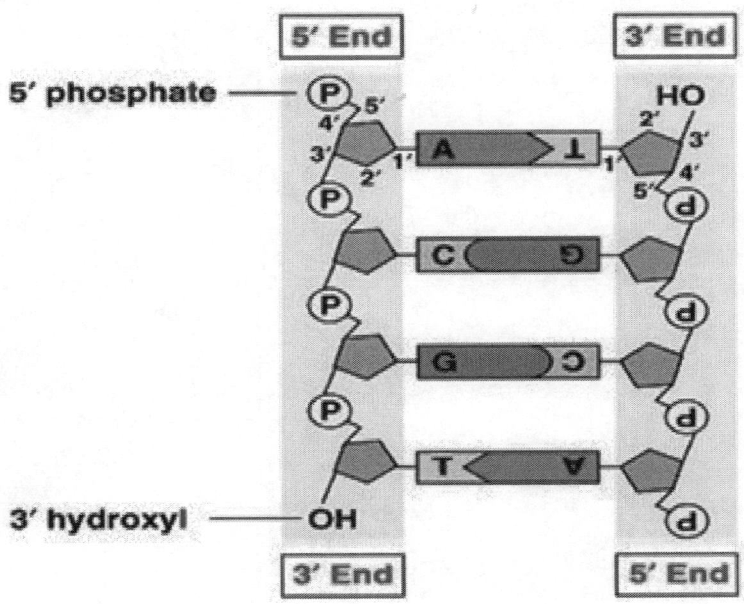

2. **Variety:** Only four types of base pairs are involved in the formation of DNA molecules. These base pairs may occur in any sequence, and these may be any number of sequences in a molecule. This gives an infinite variety to DNA. Thus, each individual has a specific sequence of base pairs in its DNA.

3. **Complementarity of base pairs:** The two polynucleotide chains of DNA are not identical but complementary to each other in base pairs. This is due to restrictions on base pairings. A base pair must consist of a purine and a pyrimidine. There are two reasons for these limitations:

   (a) The space available between the two sugar-phosphate chains of DNA (2nm) can accommodate one purine and one pyrimidine. If there are two purines then the space will be too wide. If there are two pyrimidines then the space will be too narrow. In these cases, proper hydrogen bonding will not occur.

   (b) A specific purine base pairs with a specific pyrimidine base because of a perfect match between H-donor and acceptor sites on two bases. Only adenine and thymine; and also guanine and cytosine have the proper spatial arrangements to form correct H-bonding. Adenine and Cytosine and also guanine and thymine do not form proper H-bonding and cannot align correctly thus cannot pair because these combinations are too large and too small to fit inside the helix.

Double helical model: James Watson and Francis Crick (1953) proposed the first three-dimensional double helical model of DNA from an X-ray photograph of Maurice Wilkins and Rosalind Franklin and Edwin Chargaff's analytical data. It could explain the storage of genetic information and DNA replication. There are several features which are as follows:

1. DNA consists of two polynucleotide chains.
2. Two chains spiral plectonemically around a central axis to form a pair of right-handed helixes.
3. A right-handed helix turns in a downward clockwise direction.

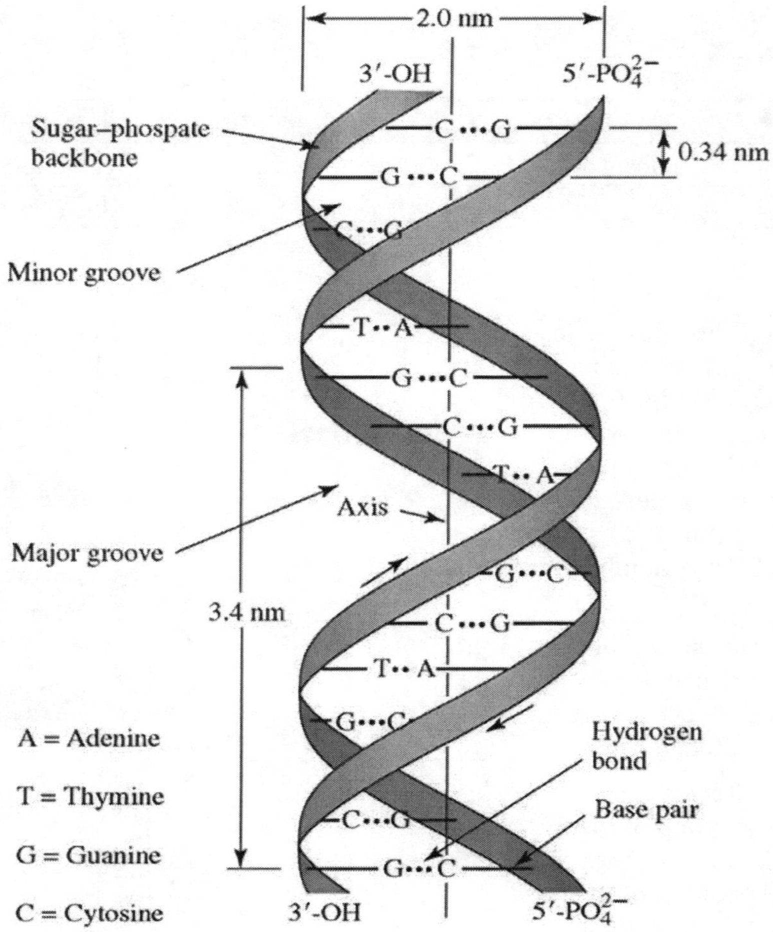

Fig: DNA double helix

4. Two chains run in opposite (antiparallel) directions. One chain is aligned in 5′→3′ direction, its partner must be aligned in 3′→5′ direction. This antiparallel direction is essential for hydrogen bond formation between base pairs.
5. Sugar phosphate makes the backbone of a double helix. Each chain has an alternate sugar-phosphate backbone. Phosphates are on the outside of the helix and bases inside.
6. The bases project at right angles to the axis of a double helix.
7. Hydrophobic interaction and Vander Wall's forces between the planar bases provide stability to DNA.
8. Two chains are held together by H-bonds between each base of one chain and an associate base of another chain. The length of the H-bond is 3A⁰. Adenine and thymine base pairs are joined by two H bonds whereas guanine and cytosine base pairs are joined by three H-bonds. Thus, there are four possible base pairs in DNA: A-T, T-A, G-C and C-G. There is an average of 25 H-bonds within each complete turn of the double helix.
9. The distance from the P atom of the backbone to the centre of the axis is 10A⁰. Thus the width (diameter) of the double helix is 20A⁰. The width between two backbones is constant and equal to the width of base pairs (width of a purine + width of a pyrimidine).
10. The association of two purines would extend beyond the specified width and the association of two pyrimidines would not be sufficiently wide.
11. The sequence of one chain determines the sequence in the other. Thus, the two chains are said to be complimentary.
12. Chargaff (1950) proposed a base equivalence rule where the amount of purine (adenine + guanine) is equal to that of pyrimidine (thymine + cytosine).

A + G = T + C or A + G / T + C = 1 for double-stranded DNA.

or A+ G + T + C=100%.

Hence, A=T or A/T=1; G=C or G/C=1.

The deoxyribose sugar and phosphate residues occur in equal proportion.

13. The base pairs are spaced at intervals of 3.4 A^0. Thus, one complete turn of a double helix containing 10 base pairs has 34 A^0.
14. The space between adjacent turns of the helix forms two grooves of different widths alternatively- a wide major groove (22 A^0) and a narrow minor groove (12 A^0). The major groove is generated by a larger angular distance between sugars and the minor groove is formed by a smaller angular distance between sugars. In major grooves, the edges of the bases are more accessible than in minor grooves. In these grooves, specific proteins interact with DNA molecules.

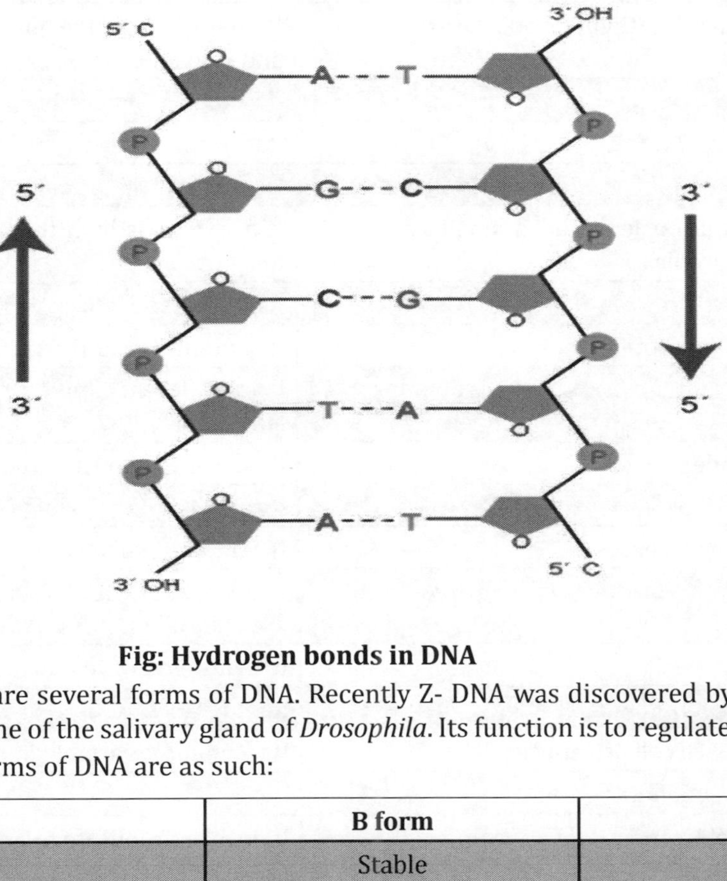

Fig: Hydrogen bonds in DNA

Polymorphism in DNA: There are several forms of DNA. Recently Z- DNA was discovered by Alexander Rich. It is found in the polytene chromosome of the salivary gland of *Drosophila*. Its function is to regulate gene expression. The differences between different forms of DNA are as such:

Features	B form	Z form
Stability	Stable	Unstable
Diameter of helix	20A^0	18 A^0
Number of base pairs per turn of helix	10	12
Distance between two base pairs	3.4 A^0	3.7 A^0
Pitch (turn) of helix	34 A^0	45 A^0
Vertical rise per base pair	3.4 A^0	5.7 A^0
Base pair tilt to axis of helix	6.3^0	7^0
Rotation per base pair (angle of helical twist)	36^0	60^0
Direction of helix coiling	Right-handed	Left-handed
Glycoside torsion angle	Anti	Syn (purine) and anti (pyrimidine)
Sugar puckering	3′-exo	3′-endo

Features	B form	Z form
Nature of phosphate backbone	Helical	Zig-zag
Distance of phosphate from the axis	10 A^0	9 A^0
Repeating unit	One base pair (mononucleotide)	Two base pairs (Dinucleotide)

Base ratio: It is A+T/G+C. It is variable (between 0.4 and 1.9) and constant for a species. It is low (less than 1) in primitive organisms (prokaryotes) and high (more than 1) in advanced organisms (eukaryotes). It can be used to identify the source of DNA and also helps in the classification of organisms.

DNA (Deoxyribonucleic acid)	RNA (Ribonucleic acid)
It contains deoxyribose sugar.	It contains ribose sugar.
It contains adenine, guanine, thymine and cytosine.	It contains adenine, guanine, uracil (in place of thymine) and cytosine.
It is usually double-stranded.	It is usually single-stranded.
It is of one type.	It is of three types.
It is usually genetic material.	It is genetic material in some viruses only.
It occurs mostly in the nucleus and a small amount in mitochondria and chloroplast.	It occurs mostly in the cytoplasm and a small amount in nucleus.
It originates later (from RNA).	It originates first (primitive).
It contains millions of nucleotides	It contains fewer (70-12000) nucleotides.
It has the same amount of adenine and thymine; and guanine and cytosine.	It has different amounts of adenine and uracil; and guanine and cytosine
It follows Chargaff's rule.	It does not follow Chargaff's rule.
It is Fuelgen positive.	It is Fuelgen negative.
It does not contain unusual bases.	It contains usual bases.
It is more stable due to the presence of thymine.	It is less stable due to the presence of uracil which causes degradation. It also contains 2′-OH group
It is less reactive.	It is more reactive due to the presence of the 2′-OH group.
It undergoes mutation at a slow rate.	It undergoes mutation at a fast rate.
Its amount is fixed in each cell of a species.	Its amount is variable in different cells.
It has base-pairing throughout the length.	It has base pairing only in the hairpin and helical region.
It replicates to form DNA.	It does not replicate to form DNA.
It transcribes to form RNA.	It does not transcribe.
Its renaturation after melting is slow.	Its renaturation after melting is fast.
It has a longer span.	It has a shorter span.

Prokaryotic DNA	Eukaryotic DNA
It occurs in the cytoplasm.	It occurs in the nucleus, mitochondria and chloroplast.
It is circular.	Nuclear DNA is linear, but the DNA of mitochondria and chloroplast is circular.
Its amount is low.	Its amount is more.
It has a smaller part of DNA which is non-functional.	It has a greater part of DNA which is non-functional.
It has a larger part of DNA which is functional.	It has a smaller part of DNA which is functional.

It does not contain introns.	It contains introns.
It can code for a few proteins.	It can code for more proteins.

Circular DNA	Linear DNA
It is found in prokaryotes; and organelles of eukaryotes (mitochondria and chloroplast).	It is found in eukaryotes.
It exists as a single chromosome.	It exists in several chromosomes.
It does not have histone proteins.	It contains histone proteins

PACKAGING OF DNA DOUBLE HELIX

It is the accommodation of long-sized DNA in small areas through compaction or packaging. It occurs in both prokaryotes and eukaryotes. A single molecule of DNA within a chromosome may be as long as 8.5 cm (3.3 inches) hence it has to be supercoiled into a very complex shape.

DNA packaging in prokaryotes: The length of the DNA helix in *E. coli* (haploid organism) is very long as compared to body size. It is calculated as

Length of DNA= Distance between base pairs x Total number of base pairs

0.34 nm x 4.6 million base pairs

$0.34 \times 10^{-9} \times 4.6 \times 10^{6} = 1.56$ mm

Prokaryotic DNA is not scattered throughout the cell in spite of the absence of a nucleus instead it is present in the cytoplasm of a cell having a size of 1µm in a supercoiled stage. This compaction in the form of a loop is called nucleoid or prochromosome or genophore. The long-sized DNA is supercoiled (coiled and recoiled) with the help of RNAs and basic nonhistone proteins like polyamines. These proteins are positively charged thus hold firmly the negatively charged DNA.

DNA packaging in eukaryotes: It is very complicated in eukaryotes because the length of DNA is greater than the dimension of a nucleus (10^{-6} m). The length of DNA varies not only in different organisms but also in different tissues or even in different regions within the same cell of the organism. The length of the DNA helix in a human being (diploid organism) is very long compared to body size. It is calculated as

Length of DNA= distance between base pairs x Total number of base pairs

0.34 nm x 3.3 billion base pairs (diploid)

$0.34 \times 10^{-9} \times 2\ (3.3 \times 10^{9}) = 2.2$ m

A human diploid cell having 46 chromosomes has 5.6 picogram of DNA hence it contains 174 cm because one pg is equivalent to 31 cm of DNA helix. All DNA in our cells is almost 60 billion km long.

Nucleosome or solenoid model: It was proposed by Kornberg and Thomas (1974). It is a widely accepted model.

Chromatin is composed of a repeating unit called nucleosome (nu particle). A nucleosome is an ellipsoid or oblate particle about the size of 6 x 11nm. It is the first order of packaging. It contains an average of 200 base pairs of DNA. The length of DNA associated with the nucleosome core is affected by H_1 histone and non-histone proteins. The length of DNA suggests that DNA should be coiled around the surface of nucleosome core particles. It further suggests that histones do not cover DNA, but DNA is wrapped around histone core particles. The nucleosome consists of the following components:

(a) **Core**: It contains a histone octamer (8) composed of two molecules of each of histones H_2a, H_2b, H_3, and H_4 stacked one on top of the other. The histone octamer forms a wedge-shaped (barrel-shaped) disc around which 140 bp long DNA is wrapped outside in 1¾ (1.8) turns in a left-hand direction. The presence of histones is necessary for the formation of the inner core of the nucleosome. The positively charged end of histone proteins are towards the outside to attract and hold firmly the negatively charged strand of DNA. This segment of DNA in chromatin is nuclease resistant. It forms beads on a string connected by sequences of linker DNA.

(b) **Linker DNA**: It contains 8 to 114 base pairs (average 60). Its size is 14nm. It connects with core particles. It is nuclease susceptible. It is additional DNA required to make up the 200 bp nucleosome.

(c) **H_1 histone**: It contains one molecule in the nucleosome. It is mostly located outside linker DNA between the nucleosomes or at the central hole of a solenoid. It is large which is due to the presence of additional C-terminal tail stabilizing DNA between the cores of nucleosome. It seals the DNA in the nucleosome by binding at the site where DNA enters and leaves the nucleosome. Thus, H_1 helps in the grouping of nucleosomes. It is most readily removed at moderate salt concentration from chromatin which yields a beads-on-a-string fiber due to the expansion of 10 nm nucleosome fiber. If it is removed completely then it does not affect the structure of a chromosome. At low salt concentration, H_1 remains bound hence beads are joined together into a more compact structure.

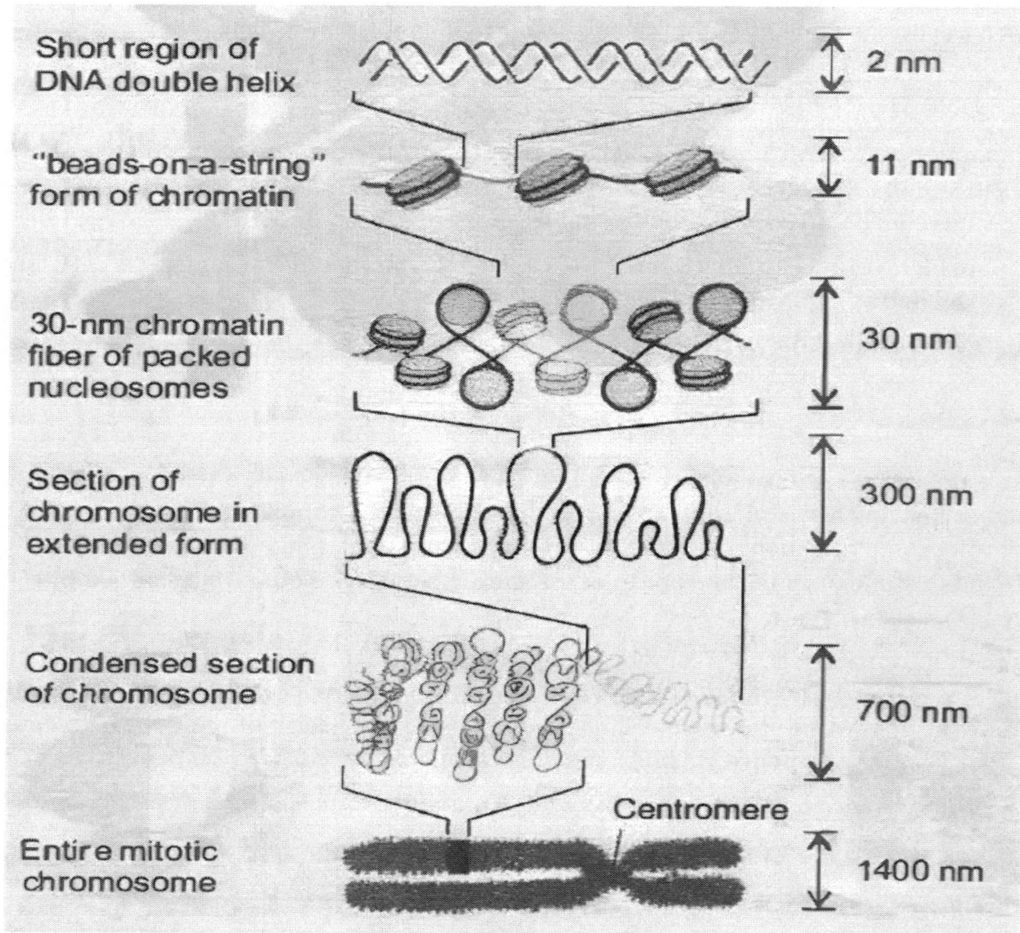

Fig: Sequence of packaging

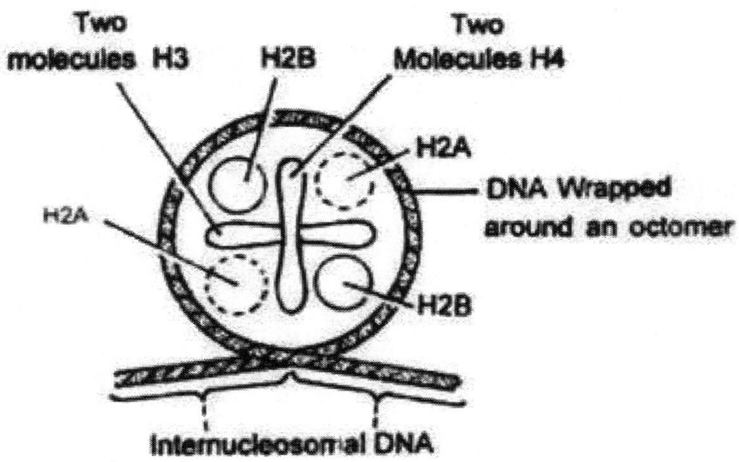

Fig: DNA packaging in nucleosome

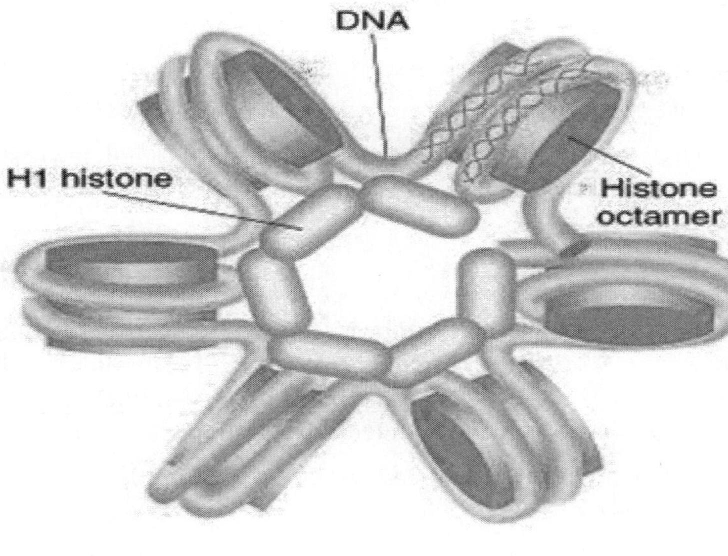

Fig: A single solenoid

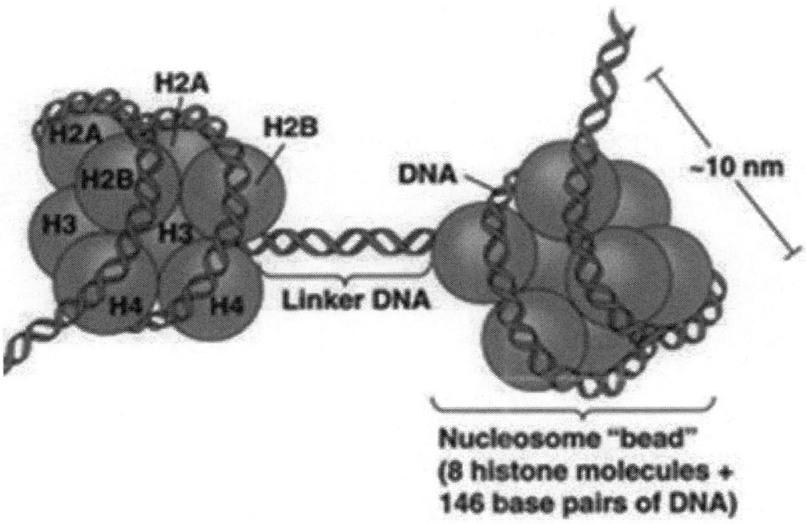

Fig: Arrangement of histone proteins in nucleosome

Nucleosome core + H_1 + linker DNA = chromatosome

Nucleosomes are arranged edge to edge with their flat faces parallel to an axis of fiber. Six nucleosomes undergo supercoiling or condensation to form a solenoid (30nm diameter) per turn. Solenoid is a second order of packaging. H_1 histone is required for its organization. Solenoid is the basic constituent of both interphase and mitotic prophase. It undergoes further coiling to produce chromatin fiber which shows euchromatin (loosely packed structure) and heterochromatin (densely packed structure). If chromatin is digested with endonuclease enzyme, DNA breaks into fragments of 200 bp in which the core of the nucleosome has 140 bp and the linker part has 60 bp. This 140 bp core is common to all types and represents the basic level of organization of chromatin. The only difference lies in that of linker DNA. Now chromatin undergoes further coiling to form chromatids and finally to chromosomes.

DNA →	Nucleosome →	Solenoid →	Chromatin fiber →	Chromatid →	Chromosome
(2 nm)	(10 nm)	(30 nm)	(300 nm)	(700 nm)	(1400 nm)

REPLICATION OF DNA

It is a process in which the amount of DNA gets doubled, or DNA makes its exact copies. It takes place continuously in the cytoplasm in prokaryotes and during the S-phase of the interphase nucleus of eukaryotes. It is bidirectional (usually) as well as unidirectional. There are three modes of replication:

1. Conservative replication: One daughter DNA contains both parental (old) strands and another daughter DNA has both newly synthesized strands.
2. Semi-conservative replication: Each daughter DNA contains one parental (old) strand and one newly synthesized strand. It is universally accepted.
3. Dispersive replication: Each daughter DNA contains both parental (old) and newly synthesized strands when parental DNA is broken into several fragments, each fragment replicates, and the old and new fragments would recombine randomly.

Proof of semi-conservative replication of DNA

1. **Messelson and Stahl's experiment:** It was proved by Mathew Messelson and Franklin Stahl (1958) in *E. coli* in the presence of N^{15}. They grew *E. coli* for many generations (up to 14) in a medium having a heavy isotope of Nitrogen (N^{15}) in the presence of NH_4Cl till the bacterial DNA became completely labeled with a heavy isotope. Now they placed heavily labeled bacteria in a fresh medium containing a normal isotope of nitrogen (N^{14}).

DNA was extracted, purified, and centrifuged at high speed (50,000 rpm) for 40-72 hours. They found that all DNA of the first generation (replication) was hybrid (N^{15}-N^{14}).

The daughter cells were allowed to divide again in the N^{14} medium. They found two hybrid DNA (N^{15}-N^{14}) and two light DNA (N^{14}-N^{14}).

Subsequent generations showed an increase in the number of light DNA and a decrease in the number of hybrid DNA. There was no heavy DNA.

Heavy and light DNA can be separated by equilibrium density gradient centrifugation using CsCl. The same CsCl is centrifuged at high speed (50,000 rpm) for 40-72 hours. This leads to the formation of a linear gradient of increasing density from the top to the bottom of the centrifuge tube.

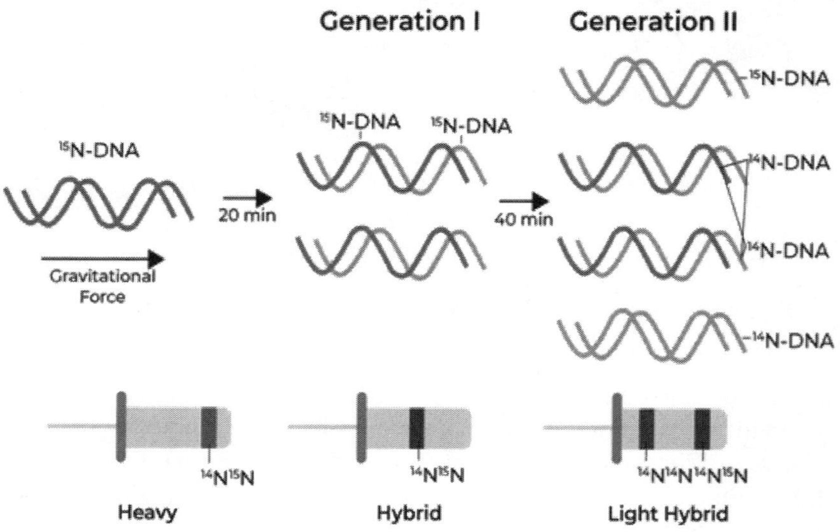

Fig: Experiment to prove semiconservative replication

1. **Cairn's autoradiography experiment:** In this, a heavy isotope of hydrogen (H^3) was used to replace thymine with radioactive (tritiated) thymidine in a DNA molecule.
2. **Taylor's experiment:** He proved a semi-conservative mode of chromosome replication in eukaryotes using radioactive (tritiated) thymidine in the root tip cell of a flat bean (*Vicia faba*). He treated the root tip with radioactive thymidine to label the DNA. The root tips were grown in the normal medium. In the first generation, both chromatids were labelled and in the second generation, one chromatid of each chromosome was labelled, and the other was normal.

Mechanism of semi-conservative replication: It is a complex and multistep process. It is fast (750-1000 bp/sec) in prokaryotes and slow (50-100 bp /sec) in eukaryotes, but the human genome can be copied only in a few hours because there are multiple initiation sites (many replication forks). *E. coli* takes 38 minutes to replicate at the rate of 2000 bp/sec. The chance of error is very low because DNA is required to preserve the meaning of the genetic message from generation to generation.

Requirements:

(a) DNA template: It uses single-stranded or double-stranded, but ssDNA is more efficient because it does not wind.

(b) DNA-dependent DNA polymerase enzyme: It was discovered by Kornberg (1956). It is a highly efficient enzyme because it can polymerize some 200 bp/sec. It is fast and catalyzes the reactions with full accuracy. There are three DNA polymerase enzymes in prokaryotes and five types in eukaryotes.

(c) Primer: It is an RNA chain with a free 3′-OH group.

Properties	DNA polymerase I	DNA polymerase II	DNA polymerase III
Molecular weight	High (109,000 daltons)	Higher (120,000 daltons)	Highest (>250,000 daltons)
Nucleotides polymerized at 37°/min	Up to 1000	50	15000
Affinity for nucleotides	Low	Low	High
5′→3′ polymerase activity in Klenow fragment (large)	occurs	Occurs	Occurs
5′→3′ exonuclease activity in smaller fragment	Occurs	Does not occur	Does not occur
3′→5′ exonuclease activity in the larger fragment (Klenow fragment)	Occurs	Occurs	Occurs
Function	Repair of DNA and excision of RNA primer	Repair of DNA	Replication of DNA, proofreading and editing.

(d) Deoxyribonucleotides: These are d ATP, d GTP, d TTP, d CTP. These act as substrates and provide energy for polymerization reactions.

(e) Helicase: It is ATP dependent enzyme.

(f) Single-strand binding (SSB) proteins

(g) Superhelix relaxing protein

(h) Ligase: It forms a covalent phosphate bond between 3′-OH and 5′-phosphate group.

Process:

(a) DNA replication starts at a specific point called origin. In prokaryotes, there is only one origin (*Ori*-c) hence it occurs in both directions. The origin is 245 bp long and has generally A-T rich weak pairs. But in eukaryotes, there are several origins due to giant molecules hence it is unidirectional.

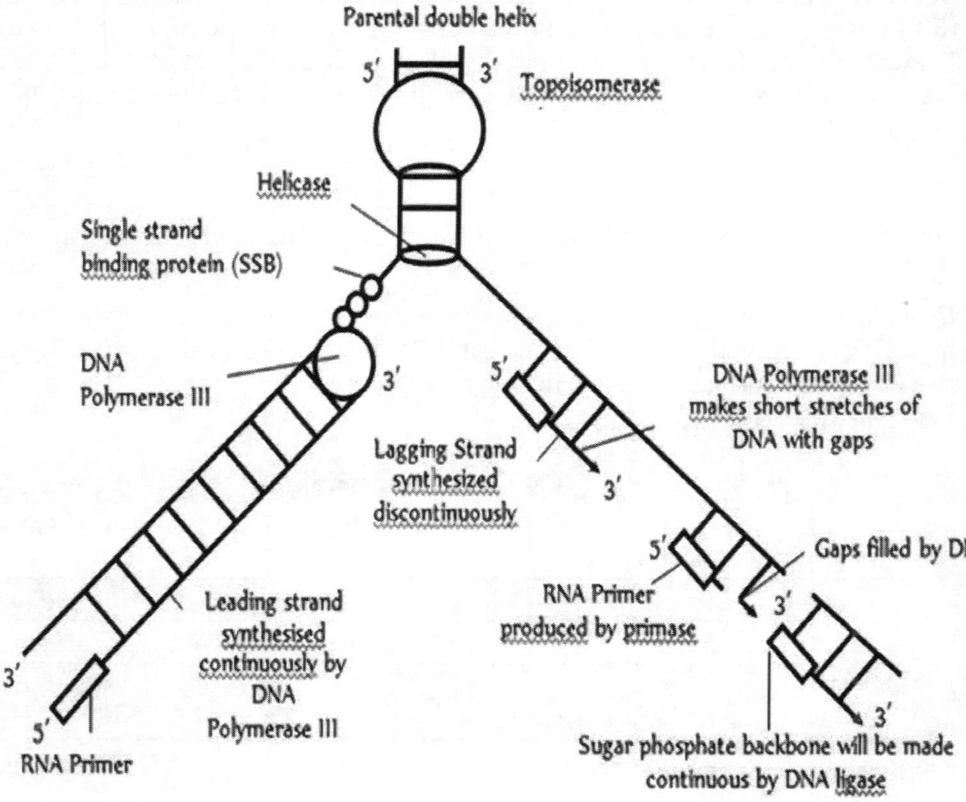

Fig: Mechanism of DNA replication

(b) Replication starts with a nick or break or incision on one of the two parental strands by endonuclease enzyme.

(c) Unwinding of DNA double helix occurs by helicase enzyme which breaks H-bonds between base pairs using ATP as a source of energy.

(d) Unwinding of helix is promoted by single-strand binding (SSB) protein which binds preferentially to a single strand of DNA so that the normal base of helix is available for replicating enzyme. It also stabilizes ssDNA and further prevents from reannealing of strands.

(e) Replication fork (Y-shaped) moves down due to unwinding of the helix.

(f) A tension or strain or torque is developed at a distal end of DNA. This tension is relieved by superhelix relaxing protein (gyrase) in bacteria and Swivelling protein (topoisomerase) in eukaryotes. This protein has nick and joining activity. Initially, it causes a nick on one strand of DNA which rotates over the other to relax the strain or coiling. This strand is joined by a ligase enzyme.

(g) RNA primer is formed which starts the synthesis of DNA. Primer is a short fragment (5-10 nucleotides) of RNA that is synthesized by template DNA at 5′-end (fork end of other strands) by primase enzyme (a modified RNA polymerase). Primer is formed because DNA polymerase cannot start the synthesis of a new strand of DNA. It is required to which polynucleotides are added.

(h) Synthesis of DNA occurs from 3′→5′ end of template DNA or 5′→3′ end on newly synthesized strand by primer. Thus synthesis starts from 3′-end of DNA.

(i) Elongation of new strand occurs by DNA polymerase III which adds deoxyribonucleotides (dATP, dGTP, dTTP, dCTP) to 3′-OH group of last nucleotides of RNA primer. The synthesis occurs in 5′→3′ direction. This is called the leading strand which grows continuously as a single piece. Now primer attached with 5′-end is removed by 5′→3′ exonuclease activity of DNA polymerase I. The gap left by the removal of primer is filled by complimentary nucleotides by 5′→3′ polymerase activity of DNA polymerase-I.

(j) Okazaki fragments are short DNA fragments are formed on the parental strand in 5′→3′ direction starting from a primer. A separate primer is formed for this fragment. This is called a lagging strand which grows discontinuously. Each Okazaki fragment has its primer. This fragment consists of 1000-2000 nucleotides in prokaryotes and 100-200 in eukaryotes. Later primer is removed by DNA polymerase I. The gap left by the removal of primer is filled by complimentary nucleotides by 5′→3′ polymerase activity of DNA polymerase I.

(k) Joining of Okazaki fragments occurs by ligase enzyme to form a complete strand by binding covalently by replacing primers with deoxyribonucleotides. Ligase requires NAD activity.

Thus, DNA replication is said to be semi-discontinuous because the leading strand is synthesized continuously, and the lagging strand is formed discontinuously.

The above replication is unidirectional in which one strand is stationary and the other strand will move with replication. In this, a single replication fork is formed.

The bidirectional replication also occurs when both strands are moving with replication and none of the strands is stationary. In this, two replication forks are formed that move in opposite directions.

Terms:

Proofreading: It is the removal of mismatched nucleotides by 3′→5′ exonuclease activity of DNA polymerase I and replacement by correct nucleotides.

Nick translation: It is a method in which DNA is nicked *in vitro* by a Klenow fragment (large) of DNA polymerase which is used to label DNA with molecules with radioactive nucleotide.

RNA

Ribonucleic acid is a long, unbranched polymer of nucleotides. It is usually single-stranded (double-stranded in Reo viruses) and laevorotatory. It consists of nitrogenous bases (adenine, guanine, uracil, and cytosine), ribose sugar, and a phosphate molecule. RNA may form secondary structures such as stem-loop and hairpin. RNA has four nucleosides which are as follows:

Base	Sugar	Nucleoside
Adenine +	Ribose →	Adenosine
Guanine +	Ribose →	Guanosine
Uracil +	Ribose →	Uridine
Cytosine +	Ribose →	Cytidine

There are four nucleotides in RNA which are as follows:

Nucleoside	Phosphate	Nucleotide
Adenosine +	Phosphate →	Adenylate (Adenosine-5-monophosphate)
Guanosine +	Phosphate →	Guanylate (Guanosine -5-monophosphate)
Uridine +	Phosphate →	Uridylate (Uridine-5-monophosphate)
Cytidine +	Phosphate →	Cytidylate (Cytidine-5-monophosphate)

It is present in the cytoplasm and nucleus of all living cells. It is unstable, more reactive, and labile than DNA and easily degradable by enzymes or chemicals due to its ribofuranose ring structure. Its instability is due to the presence of the 2′-OH group of ribose sugar, uracil, and enzymatic properties. It occurs as single stranded hence it does not show a long regular helical structure like DNA. Inside RNA forms a relatively globular conformation in which local regions of helical structure are formed by intramolecular H-bonding and base stacking within a single nucleotide chain. It acts as an adaptor and plays a catalytic role.

RNA world hypothesis: It was proposed by Alexander Rich (1962) but the term was coined by Walter Gilbert (1986). It was supported by Carl Woese (1969). RNA was the first form of genetic material to originate. It places RNA at center stage when life originated. Later life evolved to use DNA and proteins due to the relative instability and poor catalytic properties of RNA. There are a few evidence which are as follows:

(a) RNA is the genetic material in some viruses. It stores and transmits genetic information.

(b) It can duplicate itself by ribozyme.

(c) Primer RNA is required for initiation of DNA synthesis (replication).

(d) RNA of *Tetrahymena heterophila* (a ciliate) acts as a ribozyme which has catalytic ability and plays a pivotal role in peptide bond formation. It is also involved in splicing (removal of introns).

(e) The different types of RNA (mRNA, rRNA, and tRNA) are required for the synthesis of protein.

(f) RNA can catalyze the synthesis of a nucleotide, 4-thiouridine from 4-thiouracil and activated ribose.

(g) Ribonucleotide is present in ATP in all living cells.

(h) Ribonucleotides give rise to deoxyribonucleotides to make DNA.

(i) RNA can form double helices similar to DNA and tertiary structure similar to those of proteins.

Therefore, DNA has evolved from RNA with certain chemical modifications because RNA was very unstable due to its high reactivity. DNA is stable and being double-stranded with complimentary strands, it resists changes as it evolves its repair process.

RNA world	--------	**Transition world**	--------	**Present world**
(Nucleotides→RNA)		(RNA→DNA)		(DNA→RNA→Protein)

RNA as genetic material: Fraenkel-Conrat and Singer (1957) demonstrated that RNA is genetic material in some viruses. They separated TMV particles into RNA and protein. They developed a technique for forming reconstructed viruses containing protein from one mutant strain of TMV and RNA from another or vice-versa. Such hybrid viruses or chimeras were then allowed to infect tobacco leaves and the progeny was examined. In all cases, the progeny was phenotypically and genotypically identical to parent RNA from where RNA had been obtained, not the parental protein type.

It is evident from these findings that only RNA (not protein) of TMV has the ability to cause disease and the type of protein present in the virus particle is determined by RNA.

In organisms, where genetic information is contained in and transmitted through DNA, RNA does not serve as genetic material in spite of a good amount. This non-genetic RNA is synthesized on a DNA template.

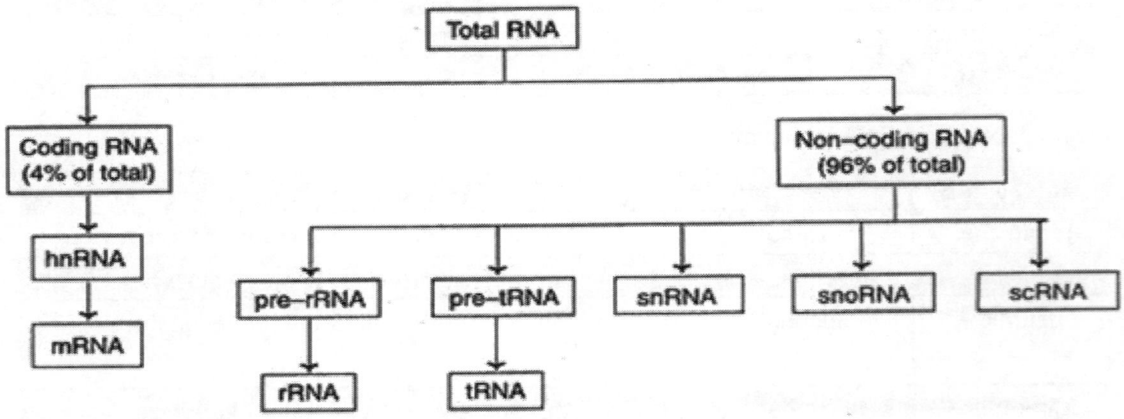

Types of non-genetic RNA: There are three types which are as follows:

1. **Messenger RNA**: The term 'mRNA' was coined by Jacob and Monod (1961).
   (a) It is the longest, linear, single-stranded, and of high molecular weight (500,0000).
   (b) It is synthesized in the nucleus on template DNA.
   (c) It is unstable, and short-lived (half-life=2 min) in bacteria.
   (d) It is metabolically stable and long-lived (half-life=6hr) in eukaryotes.
   (e) It constitutes 5% of total cellular RNA because it constantly undergoes breakdown by ribonuclease enzyme.
   (f) It is most heterogeneous in size and stability.

(g) Its sedimentation coefficient is 8S.

(h) It is complementary to the sense strand of DNA.

(i) It combines with several ribosomes to form a polysome or polyribosome.

(j) It passes from the nucleus to the cytoplasm for protein synthesis.

Types:

(a) Polycistronic mRNA: It is coded by several adjacent cistrons. It occurs in prokaryotes. It is metabolically unstable and is destroyed while the metabolic process is on.

(b) Monocistronic mRNA: It is coded by one gene (cistron). It occurs in eukaryotes. It is metabolically stable as it normally waits for the completion process.

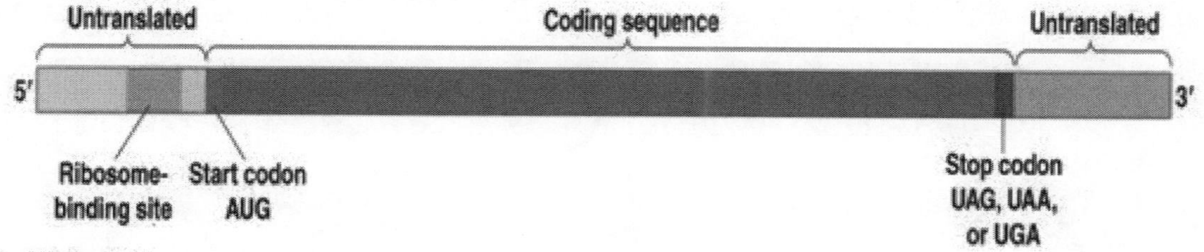

(a) Bacterial mRNA

(b) Eukaryotic mRNA

It consists of several components:

(a) **Cap**: It is a blocked 7-methylated guanylate sequence (unusual base). It is found at 5′- end of mRNA. It helps in binding of m-RNA with ribosome very firmly and protects mRNA from the degradative action of exonuclease enzyme at 5′-end. If it is removed, then there is a loss of translational process hence it controls protein synthesis.

(b) Non-coding region: It is rich in adenine and uracil residues and does not translate proteins hence called untranslated region (UTR).

(c) Initial codon: It is AUG/GUG.

(d) Coding region: It translates proteins. Its length depends on the length of a polypeptide chain.

(e) Terminating codon: It is UAA / UAG / UGA.

(f) Non-coding region: It does not translate proteins.

(g) Poly A tail: It is a polyadenylate sequence found at the 3′-end of mRNA. It is added to the nucleus before m-RNA reaches the cytoplasm. It protects m-RNA from cellular enzymes and helps in detachment from ribosomes.

Function: It conveys genetic information (message) from DNA to ribosomes for the synthesis of particular types of polypeptides.

2. **Ribosomal RNA**: The term 'rRNA' was coined by Kurland.

(a) It is greatly coiled, heaviest, and of the highest molecular weight (1.1×10^6).

(b) It is most stable.

(c) It constitutes 80% of total cellular RNA hence most abundant in cytoplasm.

(d) It is synthesized in nucleolus from nucleolar DNA (NOR of chromosome) in eukaryotes.

(e) It appears either in a compact rod (coil) or in the form of an extended region. Its length is variable.

(f) It is found in the ribosome as its basic constituent and occurs in association with proteins.

(g) It is highly conserved among organisms and presumably present in the earliest form of life.

(h) Its sedimentation coefficient is variable.

(i) The sedimentation coefficient in chloroplast rRNA is 16 S and that of mitochondrial rRNA is 12 S.

Types	Size	Large subunit rRNA	Small subunit rRNA
Prokaryotic	70S	50S has 5 S rRNA (120 nucleotides) and 23 S (2906 nucleotides).	30 S has 16S rRNA (1542 nucleotides).
Eukaryotic	80S	60S has 5S rRNA (121 nucleotides), 5.8 S rRNA (156 nucleotides), and 28 S rRNA (5070 nucleotides)	40 S has 18S rRNA (1869 nucleotides)

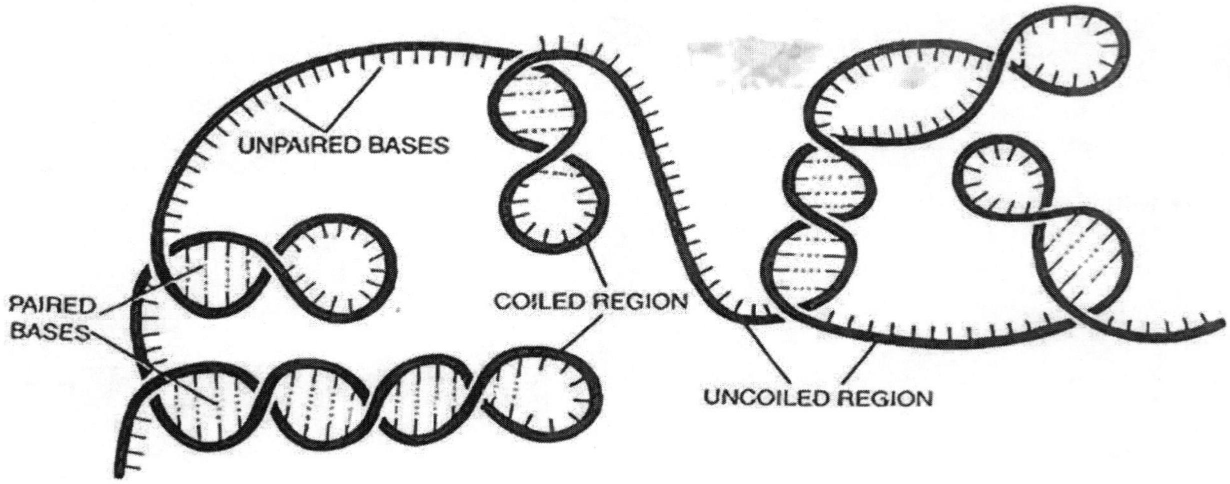

Fig: rRNA

Functions:

(a) It provides stability to ribosomal subunits.

(b) It provides a large surface area for spreading mRNA over ribosomes in translation.

3. **Transfer RNA**: The term 'tRNA' was coined by Hoagland.

(a) It is the smallest, shortest, lightest and lowest molecular weight.

(b) It is made up of 77 nucleotides (74-95).

(c) It is synthesised in the nucleus on a DNA template in the direction of $5' \rightarrow 3'$.

(d) It constitutes 15% of the total cellular RNA in the cell.

(e) It is found in cytoplasm.

(f) Its sedimentation coefficient is 3.8 S.

(g) Its number is 20 in the cell carrying specific amino acids.

(h) It is quite stable in prokaryotes and less stable in eukaryotes.

(i) It is soluble in 10 M-HCl. Its life span is short.

(j) It contains several unusual bases like pseudouridine, dihydrouridine, methyl guanine, methyl cytosine, inosine, ribothymine etc which protect tRNA from degrading action of ribonuclease (RNase). These unusual bases are produced after transcription by chemical modifications such as deamination (e.g., inosine), methylation (e.g., methylcytosine, methylguanine etc.) and double bond saturation (e.g., dihydrouridine).

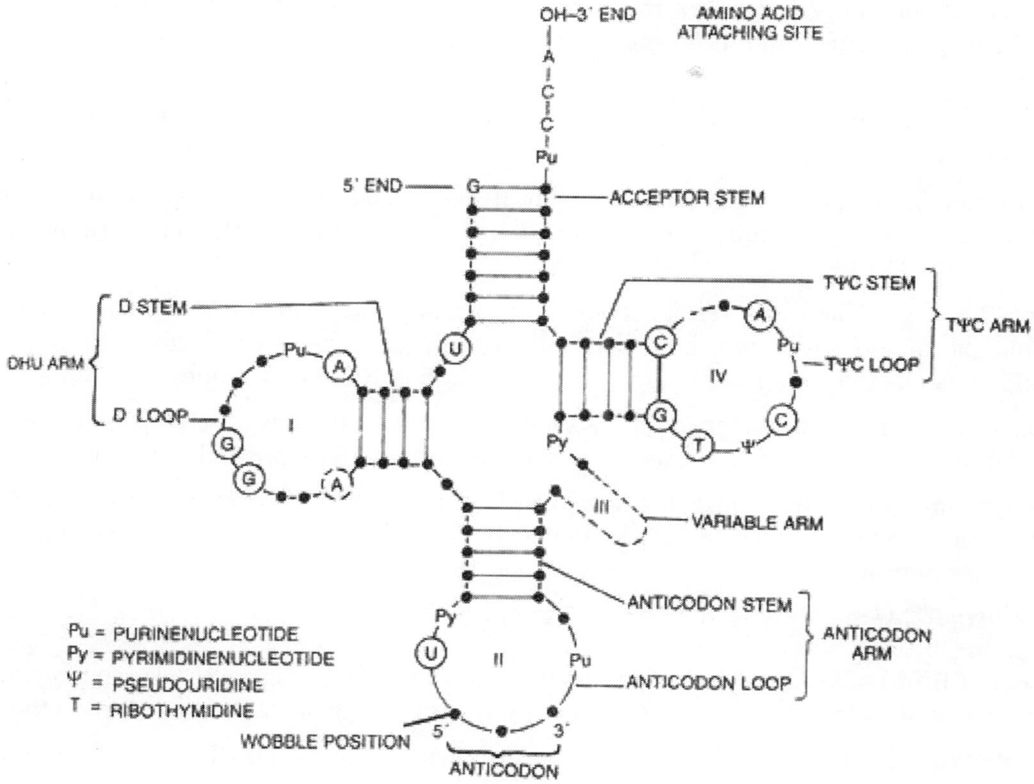

Fig: Clover leaf model of tRNA

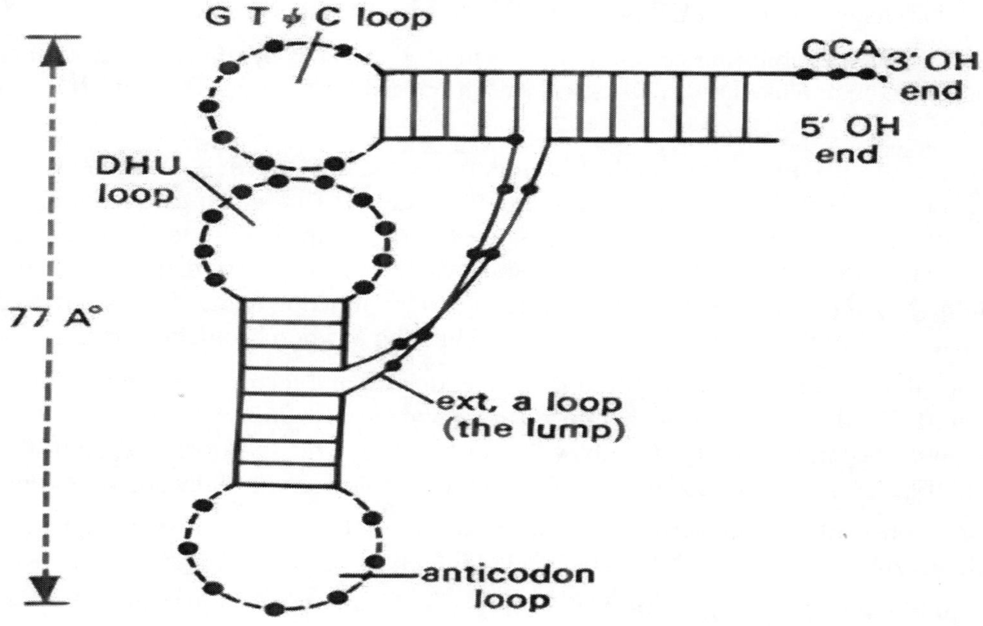

Fig: L-shaped model of tRNA

Types:

(a) Monomeric: It arises from a single gene. It is found in prokaryotes. It is a single molecule with a leader sequence of nucleotides at 5′ end and a trailer sequence at 3′-end.

(b) Multimeric: It arises from a cluster of genes.

Structure: Its tertiary structure (three-dimensional) is L-shaped as proposed by Kim and Klug (1973). Its secondary structure (2-dimensional) of yeast alanyl t-RNA is clover leaf-like as proposed by Robert Holley. It consists of several arms forming a stem and loop.

(a) Acceptor arm: It contains 7 base pairs and 4 unpaired bases (nucleotides). Its 5′-end bears Guanine having a terminal phosphate group and its 3′-end has a CCA- OH group. It lacks a loop. It is the site of amino acid binding.

(b) Pseudouridine (TψC) arm: It contains 4 base pairs in the stem and 7 unpaired bases (nucleotides) in the loop. It has pseudouracil as an unusual base. It is the site for attachment with ribosomes.

(c) Dihydrouridine (DHU) arm: It is the largest loop. It has 4 base pairs in the stem and 8 unpaired bases (nucleotides) in the loop. It has dihydrouracil as an unusual base. It is the site for binding aminoacyl synthetase enzyme.

(d) Extra arm: It is a variable arm present between the pseudouridine arm and the anticodon arm. It has a variable number of nucleotides which causes a difference in size of t-RNA. It forms two types of tRNA: Type I containing 4-5 nucleotides in the extra arm and Type II having 10-24 nucleotides.

(e) Anticodon arm: It contains 5 base pairs in the stem and 7 unpaired bases (nucleotides) in the loop including three nucleotides called anticodons. It recognises and binds codons on mRNA by H-bonds.

Function: It collects amino acids from the cytoplasm, transfers to the ribosome and aligns to the mRNA template by recognising the codon. Hence it is called an adaptor molecule. It acts as the physical link between the mRNA and the amino acid sequences of protein.

Other non-genetic RNA:

(a) **Small nuclear RNA** (snRNA): It is found in the nucleus. There are U_1, U_2, U_4, U_5, U_6 used for the formation of spliceosome leading to splicing of introns. It helps in the processing of m-RNA and rRNA in the nucleus.

(b) **Small cytoplasmic RNA** (scRNA): It is found in the cytoplasm. It helps in the binding of ribosomes with ER in the cytoplasm for producing secretory proteins.

(c) **Heterogeneous Nuclear RNA** (hnRNA): It is found in the nucleus. It is the precursor to mRNA. It is larger and higher molecular weight than mRNA. It is synthesised from split genes in eukaryotes.

(d) **Antisense RNA (**mRNA inhibiting complimentary (mic) RNA): It is synthesised sometimes on the strand complementary to the one used for mRNA synthesis. It is used for the regulation of DNA synthesis and gene expression.

Effect of temperature and pH on Nucleic acid: DNA undergoes denaturation at high temperatures (82-90°C) and high pH in which H-bonds are broken between bases of both strands. A-T base pair can undergo easy denaturation (melting) since A-T base pairs have two H-bonds which denature comparatively at low temperatures but G-C base pair can undergo denaturation (melting) at high temperatures because G-C base pairs have three H-bonds hence G-C base pair is stable and dense. The melting temperature (Tm) is a function of the G+C content of DNA. Its range is 80-100°C. The denatured strand absorbs more light energy showing a hyperchromatic effect.

Renaturation or annealing occurs when denatured DNA reassociates and forms a duplex when it is cooled at 65°C or when pH is lowered. Rapid cooling allows only the formation of local regions of dsDNA formed by base pairing or joining of short regions of complementary strands within or between strands, but slow cooling allows time for the wholly complementary DNA strands to find each other and the sample becomes fully double-stranded.

Both DNA and RNA are hydrolysed completely to their constituents (bases, sugar and phosphate) when these are subjected to very high temperature (>100°C) or strong acid (perchloric acid) or urea.

RNA is more susceptible to hydrolysis than DNA due to the presence of 2-OH in RNA. It is perfectly positioned to participate in the cleavage of the RNA backbone by an intramolecular attack on the phosphate of the phosphodiester bond.

Chromicity of DNA: Both DNA and RNA absorb UV light due to the conjugated aromatic nature of bases. The sugar-phosphate backbone does not help in absorption. Both DNA and RNA absorb maximum UV light at a wavelength of 260 nm which is quite distinct from protein which absorbs maximum UV light at a wavelength of 280nm. The absorption at 260nm is maximum for isolated nucleotides, intermediate for ssDNA and RNA (hyperchromic) and minimum for dsDNA (hypochromic).

Concept of gene:

1. **Classical concept:** 'Gene is a unit if inheritance composed of a segment of DNA situated at a specific locus which

carries coded information associated with a specific physiological function and can undergo recombination and mutation'.

2. **Modern concept:** The classical concept has changed after the discovery of DNA in 1953. Now it is confirmed that hereditary material is DNA and a gene is a linear segment of DNA called cistron. Thus cistron (synonym with gene) is the largest element within a gene. It is a unit of function. It defines 'Gene as a unit of genetic information and a segment of DNA that codes for a single polypeptide or one RNA molecule and includes the 5′ (upstream) and 3′ (downstream) regulatory sequence and noncoding sequence (intron)'.

Types of genes: Based on nucleotide sequence

1. **Normal genes:** These have a continuous sequence of nucleotides that code for a single polypeptide chain.
2. **Pseudogenes:** These resemble other genes but do not contain useful genetic information. Hence these do not direct protein synthesis. These are useless to organisms and are considered defective copies of functional genes. These have been reported in alpha and beta globulin genes of mice; *Drosophila*, human beings etc.
3. **Split genes:** These were discovered by Sharp and Roberts (1977). These are found in eukaryotes which contain non-essential introns (non-coding) interspersed with essential exons (coding region). These introns are required to be removed before entering into translation. Split gene is an ancient feature of the genome. The presence of introns represents the dominance of the RNA world. These were first reported in the ovalbumin gene of chicken.
4. **Jumping genes (Transposons):** These were discovered by Barbara Mc Clintock (1951) in maize. These move from one place to another in the chromosome and also between the chromosomes of the same genome.
5. **Overlapping genes:** These can code for two or even more proteins. These are found in tumour-producing viruses such as Φ x 174, and SV 40.

GENETIC CODE

It is the relationship or correspondence between sequences of bases in DNA / mRNA and the sequence of amino acids in a polypeptide chain. It is a set of rules by which information encoded in genetic material (DNA or RNA sequences) is translated into proteins by living cells. It is a collection of codons. It allows DNA and RNA sequences to be 'decoded' into the amino acids of a protein.

Meaning of genetic code: The uniqueness of every cell, individual or species lies in the uniqueness of its proteins. The cells are enabled to synthesise their specific proteins by the information flowing from DNA.

Function of Genetic code: It determines the sequence of amino acids in polypeptides by the sequence of nitrogenous bases in DNA or mRNA.

Nature of genetic code: There are 20 types of amino acids and only four types of nucleotides.

If one base codes for one amino acid then only four amino acids would be specified, hence protein could contain only four amino acids. This singlet code is thus inadequate.

A
G
C
U

If a sequence of two bases codes for one amino acid, the four bases could specify 16 (4x4) amino acids. This is doublet code which is also insufficient.

AA	AG	AC	AU
GA	GG	GC	GU
CA	CG	CC	CU
UA	UG	UC	UU

If a sequence of three bases codes for one amino acid, then four bases could specify 64 (4x4x4) amino acids. This is triplet code which is sufficient for 20 amino acids.

George Gamow (1954) suggested the triplet code. He postulated that sets of three bases must be employed to encode the 20 standard amino acids used by living cells to build protein.

Francis Crick (1961) demonstrated that a codon consists of three DNA bases. by producing mutations involving the addition or deletion of bases in T_4 bacteriophage.

Assignment of codons with unknown sequence (Polyuridylic acid-Poly UUU):

1. **Marshall Nirenberg** and **Mathaei** (1961) discovered and proved experimentally. They synthesised RNA (mRNA) using only uracil, so that all along the length of mRNA. There was no other base. The mRNA consisted of the same triplet UUU repeated many times. It was called polyuridylic acid. A series of 20 test tubes was prepared each containing cell-free extracts of *E. coli* including ribosome, tRNA, ATP, enzymes and different radioactive labelled amino acids. Poly U was added to each test tube and left to allow *in vitro* synthesis of polypeptides to occur. Analysis of the contents of the test tube showed that a polypeptide had been formed only in the test tube containing the amino acid, Phenylalanine. Thus, Nirenberg showed that the base triplet of mRNA or codon UUU determines the position of Phenylalanine in a polypeptide chain.

   Later on, Poly A produced polylysine: Poly C produced polyproline. It means UUU coded for Phenylalanine, AAA for lysine and CCC for proline. Using various polymers most of the remaining codons were then determined. Thus, Nirenberg deciphered the first codon, UUU which coded for phenylalanine.

2. **Khorana** synthesised the first artificial (non-functional) gene, the Alanine tRNA gene with 77 base pairs with a defined combination of bases (homopolymers and copolymers).
3. **Khorana** synthesised a second artificial (functional) gene, the tyrosine tRNA gene with 207 base pairs with a defined combination of bases (homopolymers and copolymers).

Properties of Genetic Code:

(a) **Triplet**: A genetic code is a triplet because three adjacent bases called codons specify one amino acid in a polypeptide.

(b) **Degenerate**: It is degenerate or redundant because more than one codon (2-6) is required to specify an amino acid. The different codons for a given amino acid are said to be synonymous. The degeneracy or redundancy of code was discovered by Nirenberg. The degenerate code protects the organisms from many harmful mutations because if one base of a codon is mutated, other codons will code for the same amino acid and there will be no alteration in the polypeptide chain.

Partial degeneracy: It occurs when the first two bases are identical, and the third base of degenerate codons differs. e.g., CUU and CUC code for Leucine.

Complete degeneracy: It occurs when any of the four bases can take a third position and still code for the same amino acid. e.g., UCU, UCC, UCA, UCG code for serine.

Exceptions: Single codon (AUG) codes for methionine and another single codon UGG codes for tryptophan.

(c) **Non-ambiguous**: It is non-ambiguous because one codon specifies only one amino acid hence same codon does not code for two different amino acids.

Exceptions: A few codons are ambiguous where one codon codes for more than one amino acid.

UUU codes for phenylalanine in streptomycin sensitive strain of *E. coli* but it also codes for leucine, isoleucine and serine when ribosome is treated with Streptomycin. This ambiguity is enhanced at high Mg^{+2} concentration, low temperature and in the presence of ethyl alcohol.

GUG codes for methionine initially but later codes for valine.

(d) **Non-overlapping**: It is non-overlapping because a single letter (ribonucleotide) at a specific location in mRNA is not used for two different codons hence adjacent codons do not share any base. Thus, each letter is read only once. e.g., AUGCCGUCA is read only as AUG CGU and CAG. It is not as AUG GCC CGU UCA or AUG UGC GCC CGU GUC UCA

Exception: Overlapping codon is seen in bacteriophage ϕ x174.

(e) **Commaless**: It is commaless because there is no punctuation or gap between the codons. Once translation starts, the codons read one after another with no breaks or demarcating signals in between them.

(f) **Universal**: It is almost the same for all organisms from bacteria to human beings because all living organisms contain some 20 common amino acids and the same 5 bases. It is quite evident that all organisms on earth evolved from a common ancestor and code must have been established shortly after life originated.

Exceptions: There are variations in genetic code in the mitochondrial genome of *Mycoplasma capricolum* (a ciliate protozoan), yeast and mammals (mice, cattle and human beings).

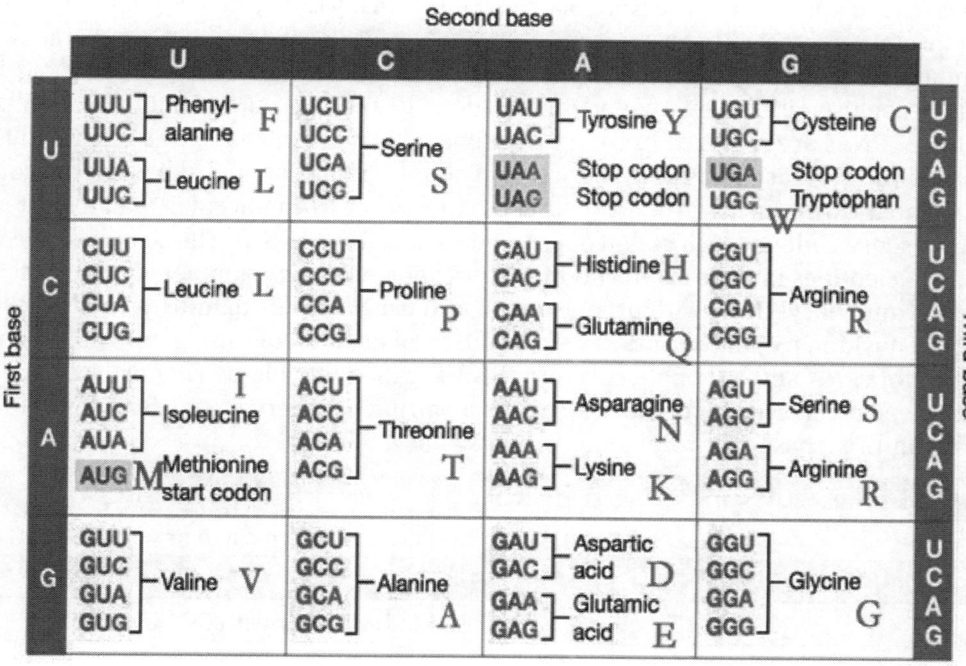

Fig: Genetic code

Universal / standard code	Non-universal code
There are 55 anticodons excluding terminal codons.	There are 22 anticodons.
AUG and GUG are initial codons.	(AUN=AUU, AUC, AUA and AUG) are initial codons for all.
UAA, UAG and UGA are terminating codons.	UGA codes for tryptophan amino acid.
AGA and AGG code for arginine.	These are terminating codons.
AUA codes for isoleuine.	AUA codes for methionine.
CUU, CUC, CUA and CUG code for leucine.	These code for threonine.

(g) **Polarity**: It has fixed polarity (direction) from left to right. It has fixed starting and termination ends. The messenger in mRNA reads in $5' \rightarrow 3'$ direction and the polypeptide chain is synthesised in $N \rightarrow C$ direction (from amino group to carboxyl group) hence polarity is from $5'$-end to $3'$-end.

(h) **Initial codon**: Genetic code has AUG which plays dual roles (acting as a start signal for initiation of the polypeptide chain and also coding formyl methionine in prokaryotes and only methionine in eukaryotes). Sometimes, GUG also acts as a start signal in prokaryotes but it is less efficient since it has lesser affinity for f-met-RNA. It initiates protein synthesis when normal AUG is lost by addition.

(i) **Terminating codons**: Genetic code has three codons such as UAG (amber), UAA (ochre) and UGA (opal) which act as a full stop to protein synthesis thus terminating the coding message. These codons are also known as nonsense codons because they do not code for any of the amino acids. UAG was the first terminating codon to be discovered.

(j) **Wobbling**: Francis Crick proposed a wobble hypothesis to explain the partial degeneracy (redundancy) of code which occurs at 3rd position of a codon. The degeneracy of code is not random. Mostly different codons for a particular amino acid have the same first two bases. The degeneracy of code accounts for the unequal distribution of different amino acids within a protein. According to this, the specificity of a codon is particularly determined by the first two bases and the third base of a codon is insignificant. For example, CGU, CGC, CGA and CGG all code for arginine. It appears that CG specifies arginine and the third letter is not important so major degeneracy occurs at the third position of a codon while the first two bases of a codon do not change. This 3rd base is called the wobble (abnormal) base which lacks specificity. This hypothesis states that the 3rd base in an mRNA can undergo non-Watson–Crick base pairing with the first base of a tRNA anticodon. These abnormal bases such as inosine (deaminated guanine), pseudouridine, tyrosine etc. can pair with more than one nitrogenous base at the same position.

Thus, organisms need not have the pressure of producing so many tRNAs hence wobbling reduces the number of tRNAs for protein synthesis thus overcoming the effect of degeneracy since several codons meant for the same amino acid are recognised by the same tRNA or it helps one t-RNA to read more than one codon. The tRNA can recognise either one, two or three codons depending upon wobble base (5′-anticodon base) hence at least 32 tRNAs are required to code 61 codons. This phenomenon gives rise to multiple codes for a given amino acid (degeneracy), possible suppression of point mutation in 3rd base of the codon, more easily removed deacylated tRNA during protein synthesis and fewer tRNA molecules than expected. There are not 61 tRNAs for 61 sense codons (each codon does not have a distinct tRNA). The 5′-nucleotide in the anticodon can vary. If the nucleotides in the second and third anticodon positions complement the first two bases in the mRNA codon, an aminoacyl tRNA with proper amino acid usually binds to mRNA. The loose base pairing will occur in the first position (5′) of the anticodon and the third position of the codon. Thus, the pairing between an unusual base of tRNA and a wobble base of mRNA is called wobble pairing or wobbling. Three types of wobbling have been proposed. A wobble base pair is a pairing between two nucleotides in RNA that does not follow the Watson-Crick base pair rule.

tRNA 5′-anticodon base	mRNA 3′-codon base	Pairing
Inosine pairs with	A, C or U	CCI anticodon can pair with GGU, GGA or GGC
Guanine pairs with	C or U	CCG anticodon can pair with GGC or GGU
Uracil	A or G	CCU can pair with GGA or GGG

Wobble hypothesis has the following attributes:

(a) The first two bases in the codon show the coding specificity. These bases form strong Watson-crick base pairs and bind strongly to the anticodon of the tRNA.

(b) If the first nucleotide in the anticodon is a C or A then pairing is specific and acknowledges original Watson-crick pairing, that is only one specific codon can be paired to that tRNA. If the first nucleotide is U or G, the pairing is less specific ant two bases can be interchangeably recognised by the tRNA. Inosine displays the true qualities of wobble where it is first nucleotide in the anticodon then any of the bases in the original codon can be matched with the tRNA.

(c) Due to specificity inherent in the first two nucleotides of the codon, if one amino acid is coded for by multiple anticodons and those anticodons differ in either the second or third position (first or second position in the codon) then a different tRNA is required for that anticodon.

(d) The minimum requirement to satisfy all possible codons (61 excluding three stop codons) is 32 tRNAs (31 tRNAs for the amino acids and one initial codon).

PROTEIN SYNTHESIS

Gene expression is the process by which information contained in the base sequences of genes is decoded to produce proteins that determine the phenotypic traits of organisms and are responsible for cellular functions.

The proteins constitute the major part of the dry weight of an actively growing cell. These proteins are built up from 20 amino acids which constitute the basic building blocks. In proteins, these amino acids are linked up by peptide bonds to form long chains called polypeptides.

The flow of information:

The sequence of bases in a particular segment of DNA will determine the sequence of amino acids in a particular polypeptide. This is central dogma. It explains how protein synthesis is controlled by nucleic acids. Protein synthesis occurs in the G_1 and G_2 phases of the cell cycle. It takes place in the cytoplasm where it involves ribosomes. Therefore, genetic information has to be carried from the nucleus to the cytoplasm. There are two modes of flow of information.

1. **Unidirectional flow of information**: It was proposed by Francis Crick (1953).

   DNA →RNA →Protein

2. **Reverse flow of information**: It was proposed by Howard Temin and David Baltimore (1975) in retroviruses (HIV, Rous sarcoma virus). It occurs in the presence of reverse transcriptase enzyme. It is RNA-dependent DNA polymerase (RNA-directed DNA polymerase). It is found in both normal and tumour cells. It is used in the synthesis of complementary or copy DNA (cDNA) from mRNA leading to the preparation of cDNA library for the study of genes expressed in specific tissues.

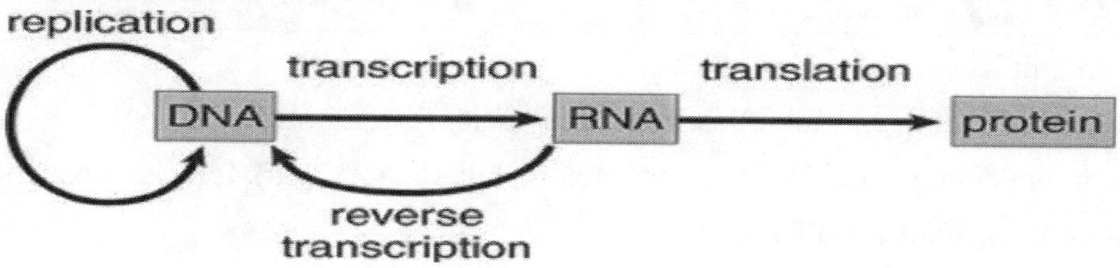

Protein synthesis occurs in the following two steps:

Transcription: It is the synthesis of mRNA from DNA in the nucleus. It requires the following components:

(a) Template strand of double-stranded DNA or single-stranded DNA.

(b) Four activated ribonucleotides (ATP, GTP, UTP, CTP)

(c) Mg^{+2}

(d) RNA polymerase: It was discovered by Severo Ochoa. It is a DNA-dependent or DNA-directed enzyme. It does not require primer and exonuclease activity. A single RNA polymerase enzyme synthesises all types of RNA in prokaryotes but in eukaryotes, three types of RNA polymerase enzymes are required to synthesise all types of RNA.

Features	RNA polymerase I	RNA polymerase II	RNA polymerase III
Location	Nucleolus	Nucleoplasm	Nucleoplasm
Sensitivity to α-aminitin	Insensitive	Strongly sensitive	Strongly sensitive
Promoters	Upstream of the start point	Upstream of the start point	Downstream of the start point
Level of activity	50-70%	20-40%	1-10%
Carboxy terminal domain	Absent	Present	Absent
Function	Synthesises 18 S, 28S, 5.8 S rRNA	Synthesises heterogeneous nuclear RNA and mRNA	Synthesises 5S rRNA, tRNA and snRNA.

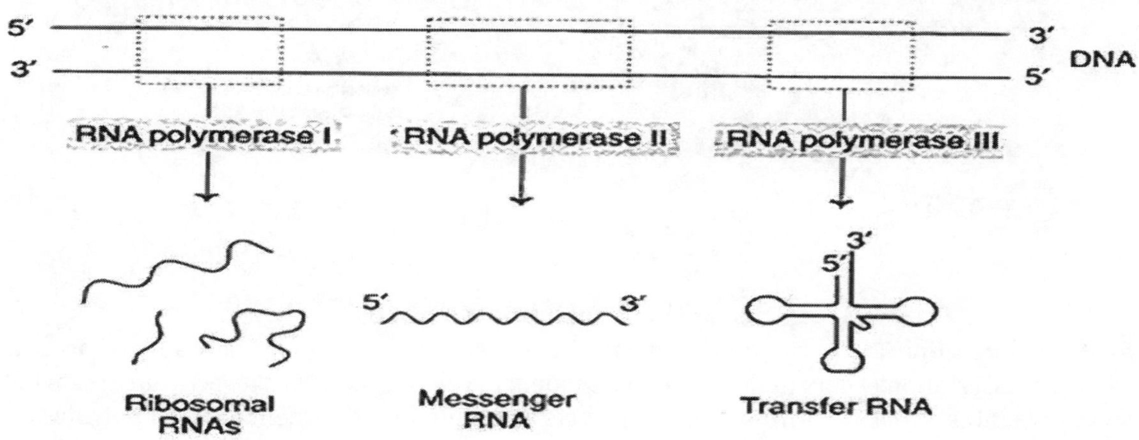

Each RNA polymerase enzyme (holozyme) consists of a core complex having four polypeptide chains and a sigma factor.

α-subunit	It recognises promoter.
α′-subunit	It recognises promoter.
β-subunit	It binds sigma factor.
β′-subunit	It binds DNA template.
Sigma factor	It recognises promoter signal on DNA template and directs binding of RNA polymerase to promoter region. It thus increases the specificity of core enzyme binding to DNA.

A typical transcription unit has

(a) Promoter at 5′-end of coding strand: It has 10 and 35 nucleotide pairs such as

- A-10 sequences Pribnow box (TATAAT) or Hogness box (eukaryotes): It is the most conserved sequence.
- A-35 recognition sequences (TTGACA)

(b) Start point: It is mostly purine on the template strand.

(c) Coding region:

(d) Terminator site:

35 sequences 10 sequences Promoter site Downstream→

Recognition sequence (TTGACA)	17 bp (spacer)	Pribnow box (TATAAT)	-7bp-spacer ▌ Coding strand ▌ Start point Terminator site Template strand

←Upstream

Types of transcription: It is of two types:

(a) Asymmetrical Transcription: A given gene is transcribed from one strand of double-stranded DNA.

(b) Symmetrical Transcription: A given gene is transcribed from both strands of double-stranded DNA.

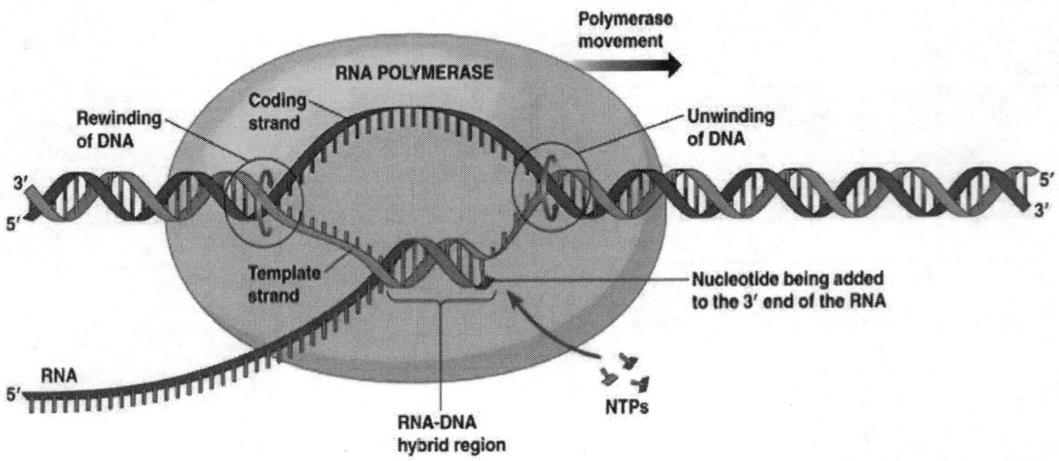

Fig: Mechanism of transcription

Mechanism of transcription: Transcription cannot start randomly but starts specifically at the start of a gene. The signal for starting transcription occurs in the promoter sequence. The promoter lies directly upstream of the coding strand. Promoter contains a specific DNA sequence that acts as a point of attachment for RNA polymerase.

(a) Sigma factor is attached to the core enzyme by beta subunit forming holozyme. The sigma factor recognises the start point at -35 sequences on the template strand.

(b) Transcription starts from left to right (5′→3′) on the coding strand when the holoenzyme binds to the promoter region of DNA. Holozyme binds to about 70-80 bp extending from -58 to +20 to form a closed binary complex on the template strand.

(c) Holozyme now creates a bubble that separates the two strands of DNA at -10 box which is rich in weak A-T bonds forming an open binary complex.

(d) mRNA starts forming on the coding strand because correct ribonucleotides are aligned opposite to the deoxyribonucleotides of the template strand according to the base pair rule. The first base is guanine transcribed for mRNA. When the first two ribonucleotides were aligned the holoenzyme formed a phosphodiester bond between them.

(e) Holozyme with template strand containing two ribonucleotides forms a ternary complex.

(f) Sigma factor now detaches from core enzymes of holoenzyme at promoter region for reuse because it is no longer required or its presence may disturb the process.

(g) The newly formed RNA chain has a triphosphate group at 5′ ends and OH at 3′ ends.

(h) Core complex binds tightly to the template strand and sequentially adds the correct nucleotides to the 3′-end of RNA and copies the base sequence of the coding strand into RNA copy.

(i) As RNA elongates, its 5′-end progressively separates from the template strand.

(j) These single strands (coding and template) progressively reassociate to form a duplex. Hence elongation involves progressive separation of duplex to form transient single strands.

(k) Core complex continues RNA synthesis at the speed of 40 nucleotides per sec till it reaches the terminator site.

(l) Termination occurs non-randomly when the core complex reaches the terminator sire. In *E. coli*, it occurs at palindrome. It requires the rho (ρ) factor which binds RNA transcript at specific sites. It involves the following events:

 (a) It stops adding ribonucleotides to the RNA chain,

 (b) It dissociates the RNA chain from the template strand by disrupting the non-covalent bond that holds newly formed RNA to template DNA.

 (c) It reassociates the separated strands to form a DNA duplex,

 (d) It causes the disappearance of bubbles and

 (e) Finally, it releases core enzymes from DNA.

Double strands of DNA

5′—GTCATGATCACGA—3′ : Non-template / coding/sense strand

3′—CAGTACTAGTGCT—5′ : Template / noncoding / antisense strand

↓

5′—GUCAUGAUCACGA—3′ : **mRNA** transcript

Prokaryotic transcription	Eukaryotic transcription
It occurs in the cytoplasm.	It occurs in the nucleus.
There is no definite phase for its occurrence.	It occurs in G_1 and G_2 phase.
It requires a single RNA polymerase for synthesising all RNAs.	It requires 3 RNA polymerases for synthesising all RNAs.
RNA is released and processed in cytoplasm.	RNA is released and processed in the nucleus.
mRNA transcript has few surplus nucleotides.	mRNA transcript has a large surplus of nucleotides.
Transcription and translation are coupled.	Transcription and translation occur separately.

Post-transcriptional processing: The released RNA is called primary transcript.

Prokaryotic mRNA undergoes little processing to become active because transcription and translation occur in the same region. It results in the beginning of translation even before mRNA is fully formed.

Eukaryotic mRNA usually undergoes chemical modification called RNA processing in the nucleus of eukaryotic cells to form functional RNA. In eukaryotes, a gene is made up of several pieces of the base sequence coding for an amino acid called exon separated by stretches of noncoding sequence called intron. Thus, the information on eukaryotic genes for assembling a protein is not continuous but split. During processing, elements that are no longer of any use are removed and functionally effective elements are added. In eukaryotes, transcribed mRNA is larger than processed or mature mRNA. The larger RNA containing pre-mRNA transcript is known as heterogenous nuclear RNA (hn-RNA). It contains both unwanted base sequences (transcribed from intron) alternated with useful base sequences (transcribed from exon). This primary transcript is converted into functional RNA after post-transcriptional processing.

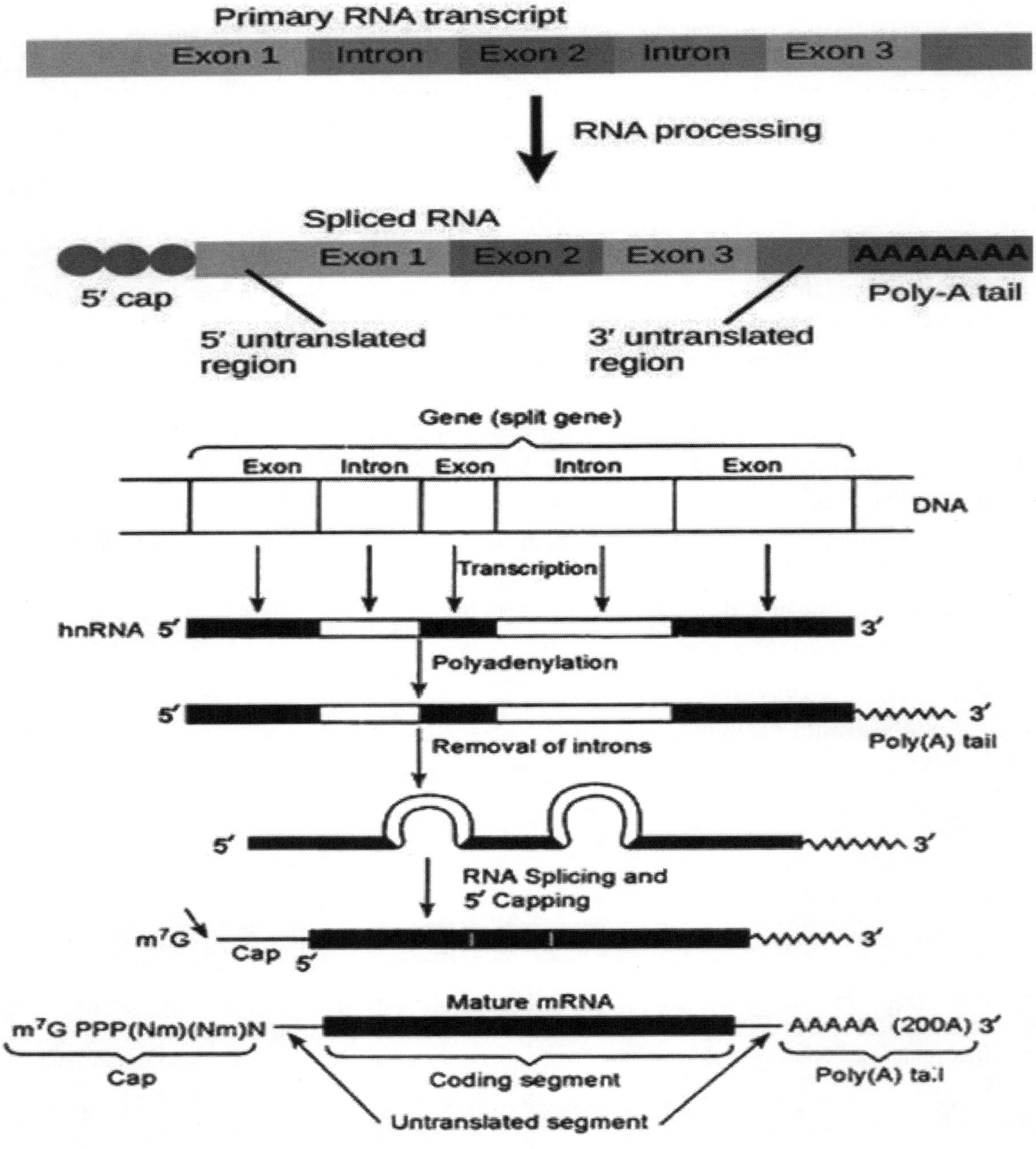

Fig: Process of post-transcriptional processing

Post-transcriptional processing: It involves several events which are as follows:

1. **Methylation (Capping)**: It is the addition of the $-CH_3$ group to the guanylate sequence at the 5′-end of nascent m-RNA by RNA polymerase II. It protects mRNA from degradation. The cap is essential for the formation of the mRNA-ribosome complex. Translation is not possible if a cap is absent.

2. **Polyadenylation (Tailing):** It is the terminal addition of a polyadenylate sequence of 200-300 bases at the 3′-end of mRNA (hnRNA) by a polyadenylate polymerase enzyme. It occurs immediately after transcription. It stabilises mRNA and protects it from degradation of coding sequence at 3′-end by exonuclease enzyme. It transports mature mRNA from the nucleus to the cytoplasm.
3. **Splicing (Tailoring):** It is the removal of introns from the primary transcript by the spliceosome. Each intron starts with dinucleotide GU and ends with dinucleotide AG (GU-AG rule). These are recognised by components of the splicing apparatus of small nuclear ribonucleoprotein (snurp) complex such as U_1, U_2, U_4, U_5, and U_6 called spliceosome which is formed between 5′-end (GU) and 3′-end (AG) of an intron. Energy is obtained from ATP and it removes ATP. It occurs in two steps:
   (a) **Phosphodiester cleavage**: In this, the 3′-hydroxyl group of Adenine and branch point sequence attacks phosphodiester bond 5′ to the G of GU (5′-splice site). The bond is broken releasing the 5 end of intron and attacking it to branch point sequence. The intron now forms a tailed loop structure called a lariat. It does not require ATP.
   (b) **Ligation reaction**: In this, the 3′-end of the intron is cleaved after G of AG (3′- splice site). The intron is released, and two exon sequences are joined together by the RNA ligase enzyme. It requires ATP.

   Splicing is of two types:
   (a) Constitutive splicing: It is common and occurs in all cells without any variation. It produces only one mature mRNA.
   (b) Alternative splicing: It occurs rarely. It produces more than one mRNA.

TRANSLATION

It is a process in which an RNA molecule (a particular nucleotide sequence present in mRNA) is translated into a particular amino acid sequence forming a polypeptide chain with the help of a ribosome. It is a very complex process involving all three types of RNA. It occurs in the following steps:

1. **Activation of amino acids**: It requires the following components:
   (a) Amino acids: There are 20 amino acids (monomers) that form the raw materials for protein synthesis. These are available in the cytoplasm as an amino acid pool. The polypeptide chain is brought about by an amino acid, methionine in eukaryotes and formyl methionine in prokaryotes.

S.No.	Amino Acid	Abbreviation	S.No.	Amino Acid	Abbreviation
1.	Phenylalanine	Phe	11.	Histidine	His
2.	Leucine	Leu	12.	Glutamine	Gla
3.	Isoleucine	Ile	13.	Asparagine	Asn
4.	Methionine	Met	14.	Lysine	Lys
5.	Valine	Val	15.	Aspartic acid	Asp
6.	Serine	Ser	16.	Glutamic acid	Glu
7.	Proline	Pro	17.	Cysteine	Cys
8.	Threonine	Thr	18.	Tryptophan	Try
9.	Alanine	Ala	19.	Arginine	Arg
10.	Tyrosine	Tyr	20.	Glycine	Gly

   (b) tRNA:
   (c) Aminoacyl tRNA: It recognises the specific codon of mRNA due to anticodon that base pairs with mRNA codon.
   (d) ATP,
   (e) Mg^{+2}

All amino acids except glycine show D and L forms (stereoisomerism or optical isomerism). Only L forms are involved in protein synthesis. Each of the 20 amino acids occurs in the cytoplasm in an inactive state. Hence each amino acid is activated by a charging enzyme, aminoacyl synthetase in the presence of ATP. This enzyme is substrate-specific.

$$\text{Amino acid + ATP} \xrightarrow{\text{Aminoacyl synthetase, } Mg^{+2}} \text{Aminoacyl adenylate (AA-AMP) + PPi}$$

It is the first energy-requiring process in the entire process. The hydrolysis of inorganic pyrophosphate with the help of pyrophosphatase enzyme provides energy for driving the initial reaction.

2. **Charging of tRNA**: Now the activated amino acid is transferred to 3′- end of tRNA. It occurs with the formation of the high-energy ester bond between the –COOH group of amino acid and the 3′-OH group of terminal adenosine of tRNA with the help of the aminoacyl tRNA synthetase enzyme. It has two active sites: One for recognising the anticodon loop and acceptor arm of tRNA and another for recognising specific amino acids.

$$\text{AA-AMP + tRNA} \xrightarrow{\text{Aminoacyl tRNA synthetase}} \text{Aminoacyl tRNA + AMP}$$

3. **Chain initiation**: It is a slow process. It requires the following components:
   (a) Formylated methionyl tRNA (f Met tRNA):
   (b) mRNA: It acts as a template. It conveys genetic information from DNA to ribosomes.
   (c) Initial codon (AUG) of mRNA:
   (d) Ribosome: It has small and larger subunits. These units are separate when not involved in protein synthesis. Several ribosomes line up on the mRNA chain during protein synthesis called polysome or polyribosome. It has been established that proteins are synthesised on polyribosomes, not at single ribosomes. This occurs in both prokaryotes and eukaryotes as well as for the cell organelles such as mitochondria and chloroplast.

   The eukaryotic ribosome has a groove at the junction of two subunits (40S and 60S). A tunnel extends from the groove through a large subunit and opens into the canal of the ER. The polypeptides are synthesised in grooves between these subunits and pass through the tunnel of large subunits into ER. The developing polypeptides are protected from the action of cellular enzymes. The small subunit forms a cap over the large subunit. The large subunit attaches to ER by two glycoproteins (ribophorin-I and ribophorin-II). The ribosome holds mRNA, tRNA and associated enzymes control until the formation of peptide bonds between adjacent amino acids.

   Prokaryotic ribosome consists of two subunits-30S and 50S in which 30S stabilises mRNA while 50S stabilises aminoacyl tRNA.

   Thus, the ribosome has three sites: A site (aminoacyl tRNA site) receives all the incoming charged tRNA while the P site (peptidyl tRNA site) holds tRNA carrying the growing polypeptide chain and E site (exit site) where discharged tRNA leaves the ribosome.

   It is advantageous for DNA (information centre) and ribosome (manufacturing centre) to be present at different sites in the cell. If both were in the nucleus, the ribosomes would be far away from energy sources and raw materials. Likewise, if both were in cytoplasm, the DNA would be exposed to respiratory breakdown.
   (e) GTP: It is an energy source.
   (f) Mg^{+2}
   (g) Initiation factors: There are 3 initiation factors in prokaryotes present in the 30S subunit of the ribosome.

IF_1	It binds over the A site of the 30S ribosome to prevent initiator tRNA from binding to the A site and thus helps bind f-met tRNA with the initial codon of mRNA at the P site. It increases the affinity of the 30S subunit for other factors. It also stimulates the activity of IF_2 and IF_3.
IF_2	It positions f-met-tRNA at the P site in the initiation complex. It binds with GTP hence it has ribosome-dependent GTPase activity.
IF_3	It binds the free 30S subunit of the ribosome with mRNA.

Chain initiation occurs in the following steps:

(a) **Formation of formylated methionyl tRNA**: Methionine carried by tRNA is formylated in the presence of transformylase enzyme and formyl tetrahydrofolate (formate) to form f-met tRNA.

$$\text{Met-tRNA + formate} \xrightarrow{\text{Transformylase}} \text{f-Met-tRNA}$$

(b) **Binding of 30 S ribosome with mRNA**: The 3′- end of 16 S rRNA in the 30 S subunit of ribosome binds with the 5′-end of Shine Delgarno sequence or purine-rich leader sequence (5′-AGGAGGU-3′) of mRNA carrying initial codon (AUG) in the presence of IF_3. This attachment helps AUG to combine with initiator anticodon (UAC) on tRNA.

5′———AGGAGGU———AUG———3′

(Shine Delgarno sequence) (Initial codon)

$$\text{30S + mRNA} \xrightarrow{IF_3} \text{30S-mRNA-}IF_3$$

(c) **Binding of 30S-mRNA-IF_3 with f-met-tRNA:** The anticodon (UAC) of f-met tRNA binds with codon (AUG) of mRNA with the help of IF_2 and GTP.

$$\text{30S-mRNA-}IF_3\text{ + f-met-tRNA} \xrightarrow{IF_2\text{, GTP}} IF_3\text{-30S-mRNA-f-met tRNA-}IF_2$$

(Initiation complex)

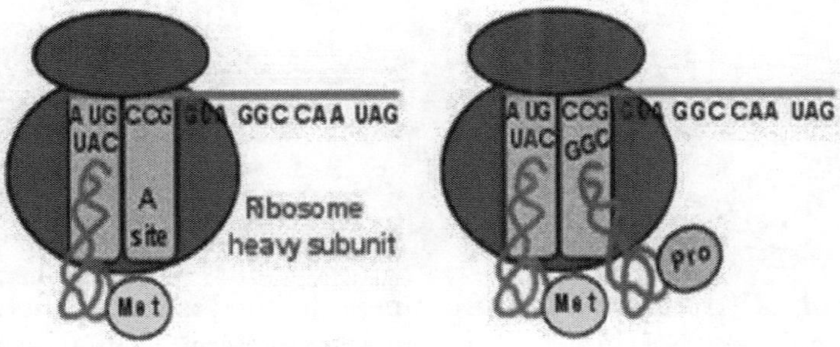

(d) **Association of ribosomal unit:** The initiation complex combines with a large subunit (50S) of ribosome hydrolysing GTP into GDP and Pi and releasing both initiation factors (IF_2 and IF_3). The initiation complex is now complete with the active ribosome in which A site is vacant and P site is occupied by f-met tRNA.

$$IF_3\text{-30S-mRNA-f-met tRNA-}IF_2\text{ + 50S} \longrightarrow \text{70S-mRNA-f-met-tRNA +}IF_3\text{+}IF_2$$

4. **Chain elongation:** It requires the following components:

(a) 70S ribosome,

(b) Codons of mRNA (other than initial codon and terminating codons),

(c) Aminoacyl tRNA,

(d) Elongation factors: There are three elongation factors required.

EF-Tu (Temperature unstable)	It directs the next t-RNA to its correct position in the ribosome.
EF-Ts (Temperature stable)	It is most abundant. It regenerates EF-Tu after hydrolysis of the attached GTP.
EF-G (Translocase or GTPase)	It translocates mRNA

GTP

(f) Mg^{+2}

Chain elongation is most rapid and cyclic process. It undergoes at the rate of 15 amino acids per sec at 37°C in prokaryotes and 2 amino acids per sec in eukaryotes. It runs by regular addition of amino acids. It involves all the reactions from the formation of the first peptide bond to that of the last peptide bond of polypeptide chain. It occurs in following steps:

(a) **Binding of aminoacyl tRNA at A site of ribosome (decoding):** The P site is occupied by f-met tRNA paired to AUG and the A site is positioned over the second codon.

Any charged aminoacyl tRNA except f-met tRNA enters the A site of ribosome depending mainly on the codon of mRNA present at the A site. Its anticodon locates (reads) and binds to the complementary codon of the mRNA chain by H-bond at A site in the presence of GTP and two elongation factors (EFTu and EFTs). EFTu mediates the entry of aminoacyl tRNA at A site by firstly binding to GTP.

Aminoacyl t-RNA + EF-Tu +GTP ⟶ Aminoacyl tRNA-EF-Tu-GDP

EF-Tu-GDP + EF-Ts ⟶ EF-Tu-EF-Ts + GDP

EF-Tu-EF-Ts + GTP ⟶ EF-Tu-GTP + EF-Ts

The activated EF-TU-GTP complex binds to tRNA. Then after, the GTP of the complex is hydrolysed to GDP and derives the binding of aminoacyl tRNA at the A site. Then, EF-Tu is released leaving new tRNA at the A site. EF-Ts mediate the release of EF-Tu-GDP from the ribosome and the regeneration of EF-Tu-GTP.

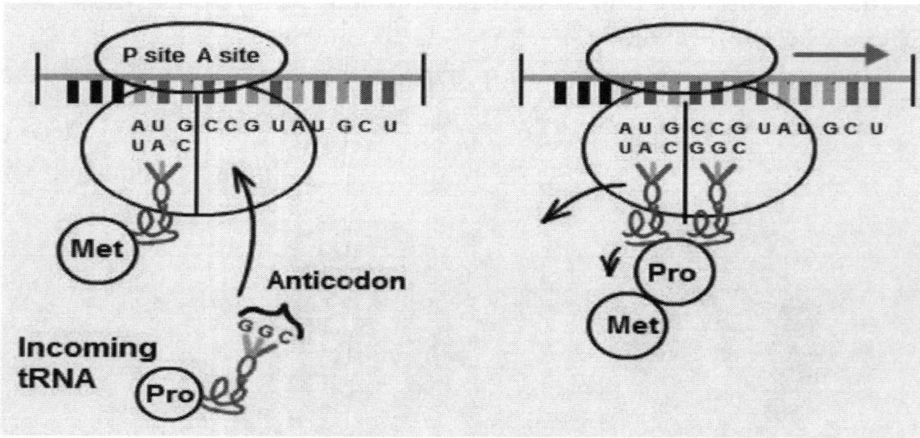

(b) **Formation of Peptide bond (transpeptidation)**: A peptide bond is formed between free –COOH group of peptidyl tRNA at P site and free $-NH_2$ group of an aminoacyl tRNA at A site which is just arrived at this site. Bond formation is catalysed by peptidyl transferase enzyme (a component of 23 S rRNA of large subunit of ribosome) and it is simultaneous with the transfer of amino acid from P site to A site. Peptide bond formation depends on release of EF-Tu from ribosome but not on the hydrolysis of GTP.

In peptide bond formation, the linkage between first amino acid and its tRNA is broken and the –COOH group now forms a peptide bond with free $-NH_2$ group of second amino acid. Thus, the second tRNA carries a dipeptide and first tRNA is without amino acid.

The reaction is accomplished by transfer of f-methione on f-met tRNA of P site to amino acid attached to tRNA bound to second codon at A site. After the formation of peptide bond, P site has a free tRNA (deacylated) while A site has peptidyl tRNA.

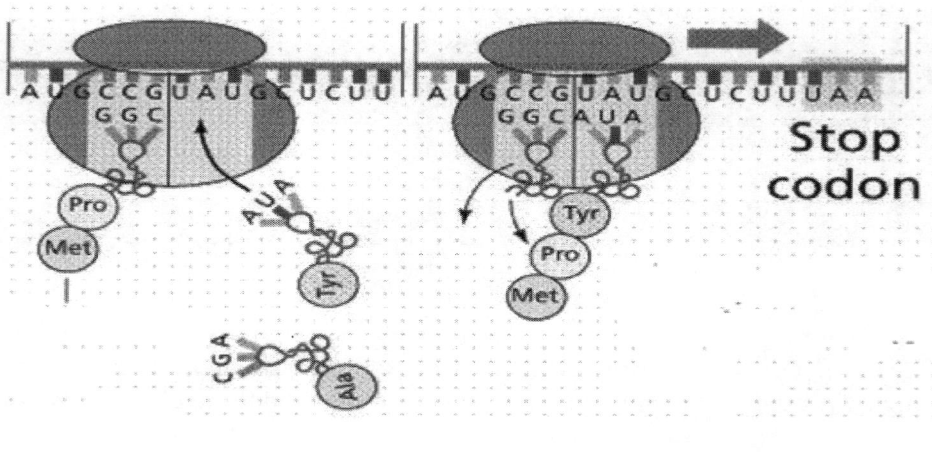

(c) **Translocation of peptidyl tRNA from A site to P site:** Ribosome now moves 3 bases along mRNA, this ejects free tRNA from P site and shifts it to Exit site and from here it leaves ribosome and returns to mixed pool of tRNA in cytoplasm.

Now dipeptidyl tRNA moves from the A site to the new free P site. It requires EF-G and GTP.

EF-G-GTP complex binds to the EF-G site of the 50S subunit. Then GTP is hydrolysed and EF-G is released which is essential for binding EF-Tu (present in ternary complex to A site).

In this manner, the ribosome moves 3 bases at a time and one amino acid is added after every activity of translocation. The first tRNA, now uncharged, separates from the mRNA chain at the P site of the ribosome and returns to the mixed pool of tRNA in the cytoplasm. Likewise, the second tRNA leaves the mRNA chain vacating the P site.

Thus, the cycle repeats as each amino acid is added to the polypeptide.

As the ribosome moves over mRNA, all the codons of mRNA arrive at the A site one after another and the peptide chain grows. Thus, the amino acids are linked up in a polypeptide in a sequence communicated by DNA through mRNA. This is called nascent polypeptide.

The growing polypeptide chain always remains attached to its original ribosome and is not transferred from one ribosome to another. Only one polypeptide chain can be synthesised at a time on one ribosome.

Energy used for protein synthesis

- One ATP is hydrolysed to AMP during activation of amino acid.
- One GTP is hydrolysed to GDP as each successive amino acyl-tRNA complex attaches to the A site of the ribosome.
- A second GTP is hydrolysed to GDP as the ribosome moves to each new codon on mRNA.

Thus 2 GTP and 1ATP are required for the formation of each peptide chain.

$$1ATP + 2GTP \xrightarrow{\text{Formation of peptide bond}} AMP + 2GDP + 4iP$$

5. **Chain Termination**: It requires the following components:

(a) 70S ribosome:

(b) Terminating codons (UAA, UAG and UGA):

(c) Release factors (RF): There are three release factors in prokaryotes:

RF_1	It recognises UAG and UAA and then activates ribosome to hydrolyse the bond between tRNA and polypeptide.
RF_2	It recognises UAA and UGA and then activates ribosome to hydrolyse the bond between tRNA and polypeptide.
RF_3	It dissociates RF_1 and RF_2 from the ribosome after termination and thus releases either RF_1 or RF_2.

The chain elongation is continued until any one terminating codon (UAG, UAA and UGA) is reached at the A site of mRNA. When it reaches A site then it does not attract any aminoacyl tRNA hence it is not read and joined by anticodon. Thus, there will be no further addition of amino acid to the polypeptide chain. The terminating codon gives the signal to the ribosome for the attachment of a protein, release factors (RF) at the A site. Thus, release factor brings about the termination of chain formation by the following events:

(a) The polypeptide chain separates from tRNA at the P site.

(b) tRNA is released from the P site of the ribosome.

(c) Ribosome is dissociated to subunits.

(d) mRNA is released.

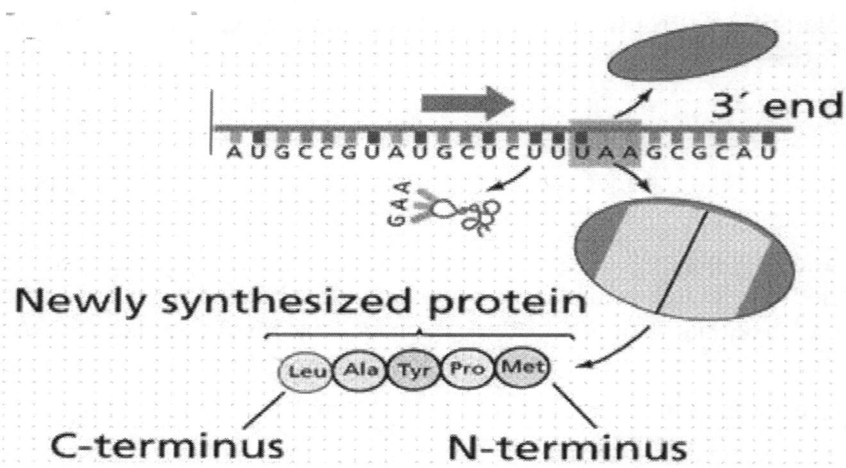

Fig: Chain termination

Prokaryotic translation	Eukaryotic translation
It is carried out by 70S ribosome.	It is carried out by 80 S ribosome.
It starts while transcription is on.	It starts after the completion of transcription.
It has polycistronic mRNA	It has monocistronic mRNA.
It starts with formyl methionine (the first amino acid).	It starts with methionine (the first amino acid). No formylation is required.
It performs chain initiation when the ribosome enters mRNA at AUG or the nearby Shine Delgarno sequence.	It performs chain initiation when the ribosome enters at capped region of mRNA at 5′-end then to AUG.
It requires 3 initiation factors (IF_1, IF_2 and IF_3).	It requires 10 initiation factors.
It requires 3 elongation factors (EFTs, EFTu and EFG).	It requires 2 elongation factors (EF_1 and EF_2).
It requires 3 release factors (RF_1, RF_2 and RF_3).	It requires 1 release factor.

Post-translational processing: The released polypeptide chain is modified in bacteria in several ways:

(a) Deformylation**:** The formyl group of f-met is removed by deformylase enzyme by hydrolysing the entire f-methionine residues.

(b) Modification of released polypeptide: The released polypeptide being linear (primary structure) loses some amino acids from the end by the activity of exopeptidase enzyme and then coil and folds on itself to acquire secondary or tertiary structure. It may combine with other polypeptides to assume a quaternary structure.

(c) Cleavage of specific region: Proezyme is converted to enzyme by trimming of amino acids from the end or removal of internal peptides or removal of aminoterminal signal peptide sequence from secreted protein.

(d) Glycosylation: It is done by attachment of sugar.

(e) Phosphorylation: It is done by attachment of phosphoryl group.

(f) Hydroxylation: It is done by attachment of hydroxyl group.

(g) Acetylation: It is done by attachment of acetyl group.

Role of antibiotics in protein synthesis: The antibiotics inhibit the process of protein synthesis at both level of transcription and translation.

Antibiotics	Action	Functions
Actinomycin D	Transcription	It blocks the progression of RNA polymerase
Rifampicin	Transcription	It inhibits RNA polymerase by binding its beta subunit.
Streptomycin	Translation	It binds 30 S subunit and causes misreading.
Tetracycline	Translation	It inhibits the binding of aminoacyl tRNA to the 30S ribosomal subunit.
Neomycin	Translation	It inhibits interaction between mRNA and tRNA.

Antibiotics	Action	Functions
Chloramphenicol	Translation	It inhibits peptidyl transferase or peptide bond formation
Erythromycin	Translation	It inhibits translocation of mRNA
	Translation	It inhibits peptidyl transferase of the 60S subunit of ribosome in eukaryotes.
Puromycin	Translation	It causes premature release of growing polypeptides.

REGULATION OF GENE EXPRESSION

It is a process by which the rate of gene expression is controlled in different tissues of organisms at different times because genes do not operate to produce enzymes all the time. Each cell of a living organism contains thousands of genes. But all genes do not function at a time..Genes function according to the requirements of the cell. Genes control the phenotypic expression through the production of enzymes. The synthesis of enzymes is not constant hence sometimes high and sometimes low depending upon the requirements of cells. The organisms regulate the activity of their genes to conserve energy and resources so that only those products essential for the functioning of cells are produced. It is essential for the organisms as it increases their versatility and adaptability by allowing the cell to express proteins when needed. The genes will be turned off or on as per the requirement.

There are two types of genes:

1. **Constitutive genes:** These are expressed continuously and remain active in a cell because their products are required for normal cellular activities. These are also called housekeeping genes. e.g., genes of glycolysis, Krebs cycle, ATPase etc.
2. **Non-constitutive genes**: These are not always expressed themselves in a cell. These are switched off or on according to the requirement of cellular activities. These are also called luxury genes. e.g., Structural genes in the lactose system in *E. coli*. There are two types:
   (a) **Inducible genes**: These are switched on in the presence of an inducer.
   (b) **Repressible genes**: These constitute to express themselves till a repressor stops their activity.

$$DNA \xrightarrow{\text{Transcription}} mRNA \xrightarrow{\text{Translation}} Protein \xrightarrow{\text{Biosynthetic pathway}} Phenotype$$

Levels of regulation: The genes regulate at following levels during protein synthesis.

Process	Organisms	Examples
Transcription (during formation of transcript)	Prokaryotes and eukaryotes	*Lac* operon, *Trp* operon, *Gal* operon
Post transcription (during splicing)	Eukaryotes	
Translation (during transport of mRNA from nucleus to cytoplasm)	Eukaryotes	*Arg* operon
Post-translation	Eukaryotes	

Operon Model: Jacob, Monod and Lwoff (1961) gave the concept of the operon model to explain the regulation of gene expression (action). Operon is a transcriptional unit (unit of gene expression) in which several genes usually encoding proteins with related functions are regulated together. Thus, an operon consists of a group of structural genes whose transcription is regulated by a set of genes; regulator, promoter and operator genes. These genes encode enzymes for the same biosynthetic or metabolic pathway. Operon is of two types:

1. **Inducible operon (*Lac* operon system):** In prokaryotes, it contains genes that encode enzymes in the presence of an inducer. These enzymes are used in the catabolism of lactose hence this operon occurs in the catabolic pathway (positively induced). It generally remains switched off but works only in the presence of an inducer. The message produced by an operon is polycistronic. Lactose is an apparent natural inducer but allolactose (isomer of lactose) is an actual inducer. Isopropyl thiogalactoside (IPTG) is an artificial inducer.

Several genes involved in this system are as follows:

(a) Structural genes: These are three genes in *E. coli* (5 in the *Trp* operon of *E. coli* and 10 in *His* operon of *Salmonella typhimurium*) adjacent to one another and each gene is controlled separately depending upon the substrate to be utilised. These genes are expressed together because in *E. coli* mRNA is polycistronic containing more than one coding region under the same regulatory control. The enzymes produced by these genes are normally present at a very low level but in the presence of lactose, their level rises rapidly. These genes are as follows:

***Lac* Z**: It is the largest gene having 3063 bp. It codes for β-galactosidase enzyme which hydrolyses lactose into glucose and galactose. It also converts lactose into allolactose.

***Lac* Y**: It contains 800bp. It codes for the permease enzyme which transports lactose into the cell through the cell wall.

***Lac* A**: It contains 800bp. It codes for trans-acetylase enzyme which transfers acetyl group from acetyl CoA to β-galactosidase to hydrolyse lactose. It is not essential.

(b) **Operator gene (*Lac* O)**: It is a small gene having 35 bp. It is situated between base -5 and +21 at 5'end of the *Lac* promoter region. It serves as a binding site for repressor proteins. It directs and regulates structural genes to transcribe. If a mutation occurs in the operator gene, then the repressor cannot bind.

(c) **Promoter gene (*Lac* P)**: It is the smallest gene containing 30 bp. It serves as a binding site for RNA polymerase at the recognition point. It has three sites: a recognition site, a binding site and an mRNA initiation site.

(d) **Regulator gene (*Lac* I)**: It is a moderately large gene having 1290 bp. It lies upstream of the operon. It codes a repressor protein which is synthesised all the time constitutively. Repressor protein is allosteric and exists in two forms; in one form it is paratactic to the operator and binds with it and in another form it is paratactic to the inducer (lactose).

The *Lac* operon system shows two types of control:

1. **Negative control**: It produces an **active repressor** protein whose binding will stop transcription. It operates in a biosynthetic pathway.

   (a) **If lactose is absent**: The regulator gene produces an active repressor protein. Repressor binds with operator gene. It stops the binding of RNA polymerase on the recognition point of the promoter gene. Hence transcription of structural genes will not occur **(Switch off).**

   (b) **If lactose is present:** The regulator gene produces an active repressor protein. Cell encounters lactose, a few molecules of *Lac* enzymes present in the cell allow lactose to be taken up and metabolised. Lactose (inducer) binds to the active repressor which becomes the inactive repressor by changing its conformation. Now inactive repressor does not bind to the operator gene. Thus, RNA polymerase binds to promoter gene. Transcription of structural genes will occur (**Switch on**).

2. **Positive control**: It produces an **inactive activator** (Catabolite activator protein or CAP) whose binding will start transcription.

   (a) **If lactose is absent**: The regulator gene produces an inactive activator protein. It cannot bind RNA polymerase on the recognition point of the promoter gene. Hence transcription of structural genes will not occur **(Switch off).**

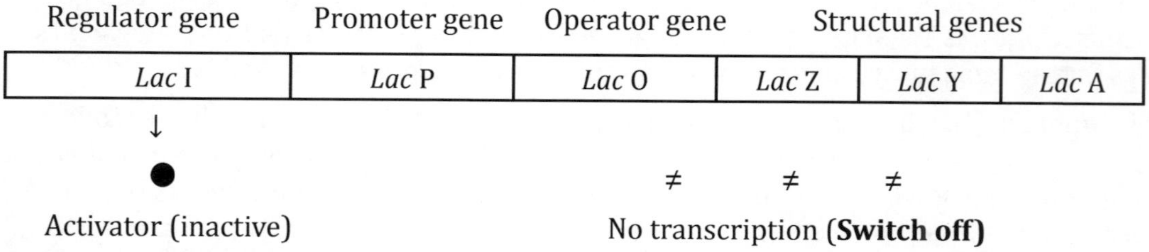

   (b) **If lactose is present:** The regulator gene produces an inactive activator protein. Cell encounters lactose, a few molecules of *Lac* enzymes present in the cell allow lactose to be taken up and metabolised. Lactose (inducer) binds to the inactive activator which becomes active by changing its conformation.

Now active activator binds to the operator gene. Thus, RNA polymerase binds to the promoter gene. Transcription of structural genes will occur (**Switch on**).

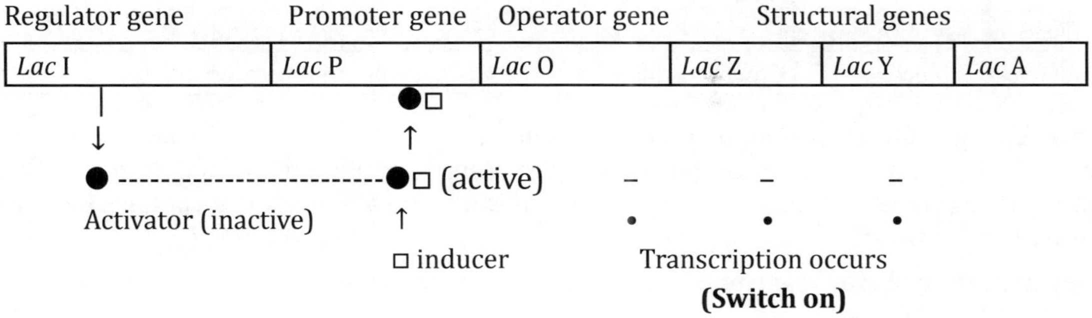

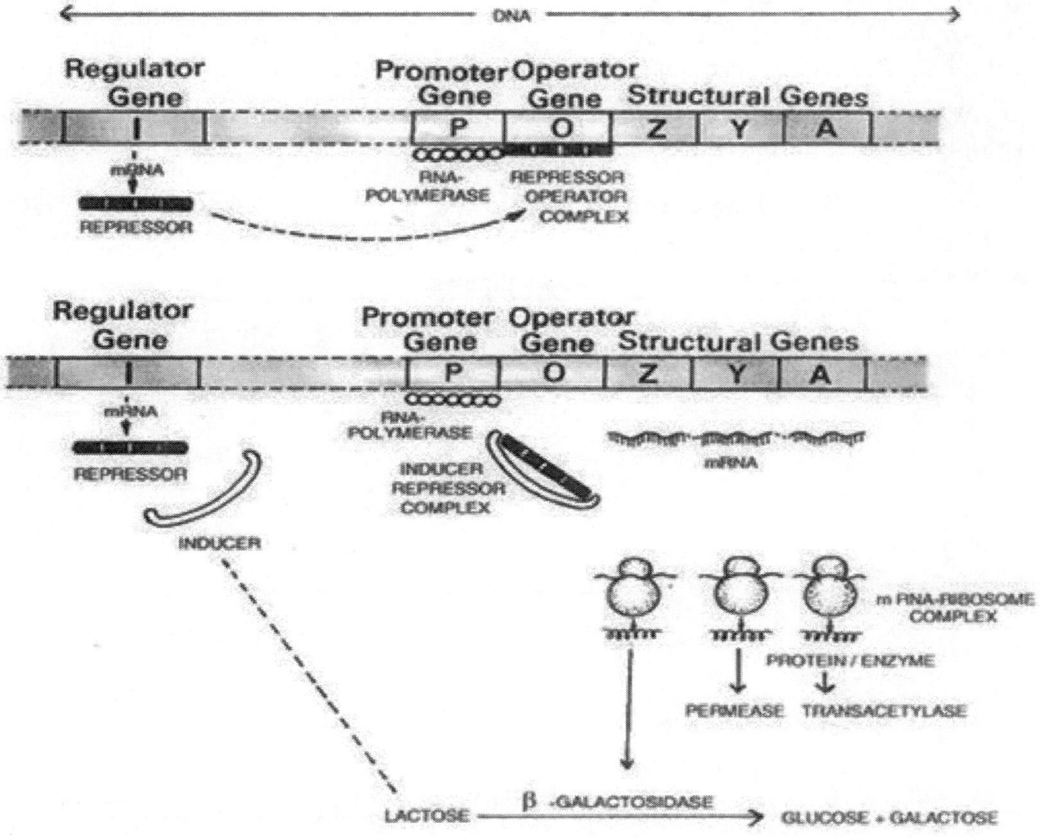

Fig: Negative control (Switch off and Switch on)

Case:

1. **If glucose is present**: The *Lac* operon will not operate indefinitely in the presence of lactose in the medium. If glucose accumulates in the medium then the *Lac* operon is switched off and *E. coli* uses glucose instead of lactose. Glucose inhibits operon by decreasing the level of cAMP because glucose inhibits the adenylate cyclase enzyme. Under this condition, catabolite activator protein (CAP) does not bind upstream of the promoter and *lac* operon is transcribed at a very low level. To switch on the *Lac* operon, a high level of cAMP and CAP are required to bind to the promoter region of DNA.
2. **If glucose is low or absent**: The level of cAMP will increase and thus

   cAMP binds CAP forming the cAMP-CAP complex which binds the promoter gene of *Lac* operon. This enhances RNA polymerase binding to promoter thus transcription is increased by 50-fold.
3. **If both glucose and lactose are present together**: In this case cell will use the glucose first hence transcription of the *Lac* operon will occur at a low level. However, when glucose is fully used up then catabolite repression will end and transcription of *Lac* operon increases allowing available lactose to be utilised.

Lactose is low	Glucose is low	The *Lac* operon is not expressed.
Lactose is low	Glucose is high	The *Lac* operon is not expressed.
Lactose is high	Glucose is low	The *Lac* operon is expressed.
Lactose is high	Glucose is high	The *Lac* operon is expressed.

2. **Repressible operon/Tryptophan operon system**: It contains genes that encode enzymes involved in the anabolic pathway. The *Trp* operon works as a corepressor and inhibits its synthesis through feedback inhibition (end product) or by an alternative mechanism called attenuation. It contains five structural genes. There are two controls:

   1. **Negative control**: The regulator gene produces an aporepressor (inactive repressor).

      It occurs in two modes:

      (a) **If tryptophan is present:** The aporepressor becomes an active repressor in the presence of corepressor, tryptophan. The active repressor binds to the operator gene. Transcription will not occur

      (b) **If tryptophan is absent**: The aporepressor will not be active due to the absence of the corepressor, tryptophan. Thus aporepressors cannot bind operator gene. Hence transcription will occur.

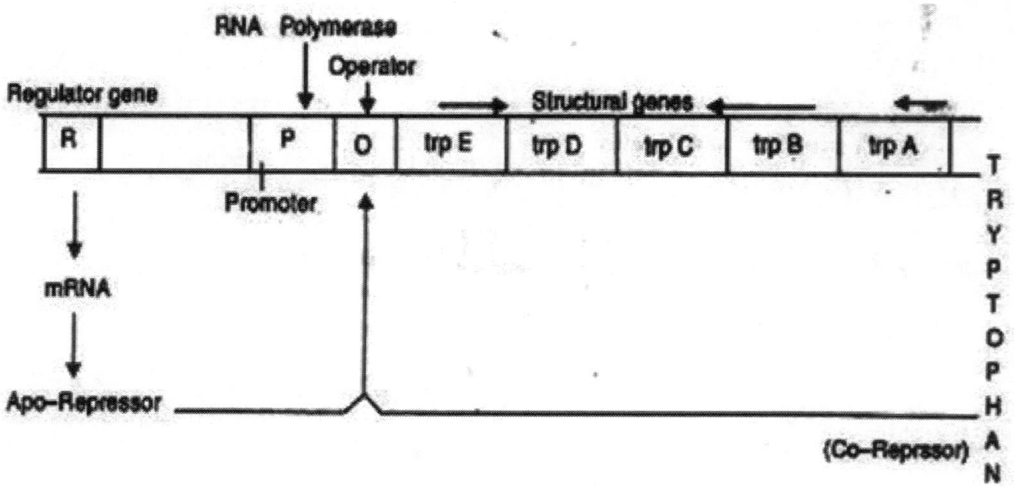

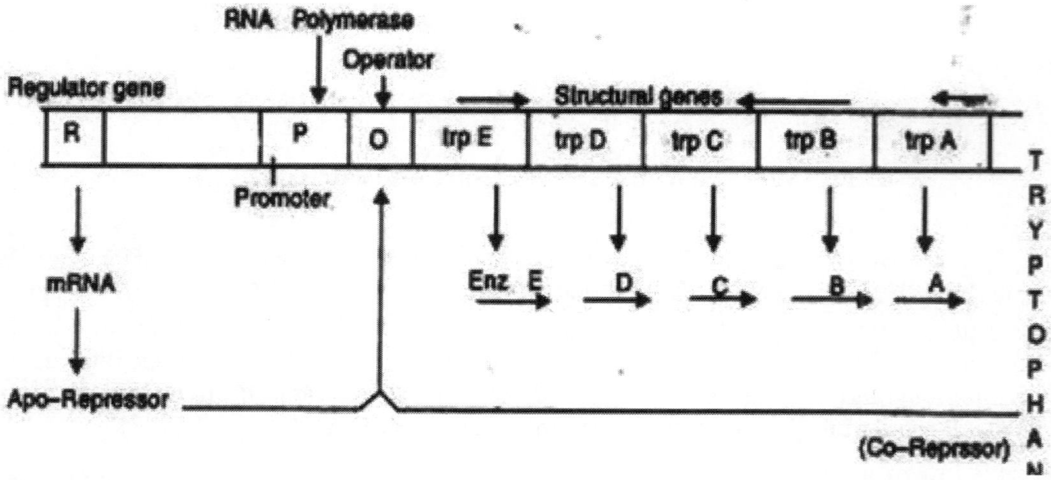

   2. **Positive control**: The regulator gene produces an active activator. It occurs in two modes:

      (a) **If tryptophan is present**: The active activator will become inactive due to the presence of tryptophan. Thus, transcription will not occur (Switch off).

      (b) **If tryptophan is absent**: The active activator will bind to the operator gene due to the absence of tryptophan. Thus, transcription will occur (switch on).

Inducible operon	Repressible operon
It operates by three structural genes.	It operates by five structural genes.
It operates in catabolic pathway.	It operates in anabolic pathway.
It turns operon on.	It turns operon off.
It starts transcription and translation.	It stops transcription and translation.
Its regulator gene produces an active repressor.	Its regulator gene produces an inactive aporepressor.
It uses lactose which changes the active repressor to inactive to bring about transcription.	It uses tryptophan which changes inactive aporepressor to active corepressor to stop transcription.

HUMAN GENOME PROJECT

The term 'Genome' was coined by Winkler. A genome is total DNA (complete set of genes) in a single chromosome (haploid set) and inherits as a unit from one parent. Thus, a single genome is found in haploid cells, a double genome in diploid cells and more than two genomes in polyploid cells. The genome includes both nuclear DNA and extranuclear DNA (of mitochondria and chloroplast. The genome of prokaryotes consists mostly of unique DNA sequences, but the genome of eukaryotes contains repetitive, interspersed and clustered DNA sequences.

The term 'Genomic' was coined by Roderick (1986). Genomics is the study of the genome through analysis, sequencing and mapping of genes and non-cistronic areas.

The human genome project is an international collaborative research programme to complete the mapping and understanding of the genome of human beings at the level of the nucleotide sequence. The genome sequenced by the project is a composite one based on the individual genomes of men and women from a variety of races and ethnic groups.

It was launched on 1st October, 1990. It was launched by the National Human Genome Research Institute (NHGRI) and coordinated by the US Department of Energy and the National Institute of Health involving 14 countries (Australia, Brazil, Canada, Denmark, France, Germany, Israel, Italy, Japan, Netherlands, Russia, Sweden, United Kingdom and US). It is called a mega project because it involves lots of money, the most advanced technologies, many scientists, numerous computers and a long time. If data is to be stored in books, with each book having 1000 pages and each page having 1000 letters then some 3300 books will be required.

It was scheduled to be completed in 2005 but completed on April 14, 2003 (only sequence of chromosome I was completed in 2006). Its total cost was US 9 billion dollars.

First genome to be sequenced	1999	*Haemophilus influenzae* (a bacterium)
First eukaryotic genome to be sequenced	1999	Yeast, *Coenorhabditis elegans* (a nematode)

HGP also includes mapping of genome of other species

Organisms	Number of genes	Number of base pairs (Genome size)
Bacteriophage	100	10,000 bp
E. coli	4600	4.6 million bp
Yeast	6000	12 million bp
Arabidopsis	25,000	130 million bp
Rice	50,000	407 million bp
Coenorhabditis	19,000	97 million bp
Drosophila	16,000	180 million bp
Human being	30,000	3164.7 million bp or 3.2 billion bp

Classification of the human genome

1. **Extragenic DNA** (70-80%): It contains
   (a) Low copy number (70-80%):

(b) Highly repetitive (20-30%): It contains Clustered repeats (60%) which have microsatellite DNA, minisatellite DNA and satellite DNA; and Dispersed repeats (40%) which have short interspersed elements (SINES) and Long interspersed elements (LINES).

2. **Genes and related sequences** (20-30%): These contain

(a) Coding sequences (<10%)

(b) Non-coding sequences (>90%): These contain gene fragments, pseudogenes, Introns, promoters, leaders and trailers.

Objectives of HGP:

1. To identify all genes in human DNA.
2. To determine the structure and function of genes.
3. To determine the sequences of genes of all 23 pairs of chromosomes.
4. To construct detailed genetic and physical maps of all base pairs.
5. To point out specific genes that cause genetic disorders.
6. To develop tools for data analysis.
7. To store the information from DNA sequences in the database.
8. To address ethical, legal and social issues arising from this project.
9. To establish bioinformatics

Methods:

1. **Sequence tagged site (STS)**: It is a short DNA segment that occurs only once in a genome and whose exact location and order of bases is known. STS serves as a landmark on the physical map of a genome. It is also known as Expressed sequence tag (EST)-the genes that are expressed as RNA.
2. **Sequence annotation (SA):** It is the sequencing of the whole genome (coding and noncoding region) and later assigning functions to different regions in the sequence. This method is followed to sequence the human genome.

Process:

(a) The total DNA is isolated from a cell.

(b) DNA is digested into small fragments. Each piece of DNA fragment is amplified.

(c) The small fragment of DNA is ligated into a vector (Yeast artificial chromosome or Bacterial artificial chromosome) to form recombinant DNA.

(d) The r-DNA is introduced into yeast or bacteria.

(e) The fragment is sequenced from two ends by an automated sequencer (developed by F. Sanger).

(f) The sequences are aligned based on overlapping segments by a specialised computer-based programme.

(g) The sequences are annotated and assigned to each chromosome.

Celera Genomics, USA used a strategy of sequencing from both ends of human DNA fragments.

Findings:

1. The human body is made up of 100 trillion cells grouped into over 260 types.
2. Every cell contains the same number of chromosomes and the same level of genome.
3. The size of the genome is not associated with the complexity of body organisation.
4. The human genome contains 3164.7 million base pairs (or almost 3.2 billion base pairs) in the 23 pairs of chromosomes within the nucleus of our cells.
5. The 99.9% of total base pairs are the same in all humans.
6. The total number of genes is 30,000.

7. Chromosome I has the highest (2,968) number of genes.
8. Y chromosome has the lowest (231) number of genes.
9. Among somatic chromosomes, the 22nd chromosome has the lowest (272) number of genes.
10. X chromosome is larger than the Y chromosome in size and has more (2,000) genes than the Y (231) chromosome.
11. Gene has an average size of 3,000 base pairs.
12. The largest gene is dystrophin containing 2.4 million base pairs long situated on the X chromosome.
13. The smallest gene is the testis determining factor (TDF), a holandric gene containing 14 base pairs long situated on the Y chromosome.
14. Genes occupy only 1.1% of chromosomal space and the rest is junk DNA or spacer.
15. Genes work in networks rather than in isolation.
16. Less than 2% (around 1.5%) of the genome is exon which codes for protein and the remaining is intron (either pseudogene or repeat sequence) which does not code for protein.
17. Repeat sequences or junk DNA repeat 100 to 1000 times.
18. Three chromosomes such as 1st, 6th and 10th are killer chromosomes which are responsible for genetic disorders. The Ist chromosome is responsible for 350 illnesses.
19. Single nucleotide polymorphism (SNP) has been identified at 1.4 million locations. **Single nucleotide polymorphism (SNP)**: It is a variation in a single nucleotide that occurs at a specific position in the genome where each variation is present to some appreciable degree within a population with a frequency higher than 0.01. On average, SNP occurs once in 300 nucleotides. Hence, out of three billion (3 x 10^9) base pairs, there will be 10^7 SNP. Thus, its number will be 1 % or more. For example, at a specific base position in the human genome, the base Cytosine may appear in most individuals but in some individuals, the position is occupied by base Adenine so there is a SNP at this specific base position and the two possible nucleotide variations- Cytosine and Adenine are said to be alleles for this base position. SNP is a single base pair position in genomic DNA at which different nucleotides occur in different individuals of a population. Each nucleotide at such a position denotes an allele of SNP. The SNP acts as a biological marker. It is caused by mutation.

Significance:

(a) It is used to study the interaction between various genes, and proteins formed in tissues and organs.

(b) It helps to diagnose the disease by the revelation of genes present on a particular chromosome.

(c) It is useful to treat incurable genetic disorders through gene therapy.

(d) It helps understand human genetics.

(e) It has developed forensic science.

(f) It is used to develop designer drugs and genetically modified diets.

(g) It is useful to achieve healthier living by studying changes in different developmental stages.

DNA FINGERPRINTING

It was developed by Alec Jeffrey (1985) while working on the sequence of myoglobin gene. It facilitates the identification at the gene level by matching DNA fragments using DNA probes.

This technology was refined by developing a more sensitive DNA profiling technique focusing mainly on highly variable minisatellites in the human genome.

The genome (DNA of a single set of chromosomes) of a person contains about 3 billion base pairs in which 99.9% of base pairs are the same in all human beings and only 0.1% of total base pairs vary and provide individuality to each human being. The chemical structure of everyone's DNA is the same. The only difference between people is the sequence of base pairs. Everyone has a different sequence hence every person would be identified solely by the sequence of base pairs.

DNA molecules can be divided into coding sequences (exons) which carry information required to synthesise proteins and non-coding sequences (introns) which do not supply any relevant information. The changes usually occur and accumulate within these non-coding sequences. These (about 10-100 base pairs) are responsible for variation among individuals and can be employed as tools to create a genetic fingerprint of an individual. Some of these non-coding sequences contain highly repetitive sequences of base pairs. These are tandemly repeated many times and are found at many places (centromere, telomere and heterochromatic area) throughout the length of DNA. The number of repeats is very specific in each individual and it is inherited. It is unlikely that the repeats of one individual are found in DNA are considered whether two samples of DNA are from same person, related persons or unrelated persons.

A DNA fingerprint is the loci (region) where test DNA hybridises with the probe. When DNA is subjected to density gradient centrifugation, these repeat sequences are called satellites because these float over the rest of the DNA.

Types of satellite DNA: Based on length of repeats and size of clusters.

1. **Microsatellite DNA** (Simple sequence repeat): It is a short repeat sequence of 4-8 (less than 10) base pairs that repeat 10 to 100 times. It is random or dispersed and frequently distributed throughout all chromosomes hence it is abundant. It shows a large stable polymorphism due to variation in the number of repeat units. It is an ideal molecular marker for gene mapping in the Human Genome Project.
2. **Minisatellite DNA**: It is a small repeat sequence of 10-60 base pairs that repeat 1000 times. It varies in number (hypervariable) from person to person hence called variable number of tandem repeats (VNTR). It is very specific to each individual and varies in number from person to person. It shows a high degree of polymorphism and is inherited from parents. It is found in many places but frequently at the telomere and centromere. It is used as a genetic marker and acts as a probe that recognises complementary sequences in DNA. Every individual receives 50% of DNA from the father and 50% from the mother thus the number of VNTRs in a particular area of the two homologous chromosome or DNA molecules will be different. Hence, an individual will have a distinct combination of VNTR.

   Genetic fingerprinting exploits highly variable repeating sequences called minisatellites. Two unrelated persons will be unlikely to have the same number of satellites at a given locus. It is used to distinguish one person from another except monozygotic twins. The probability of two unrelated persons having the same pattern of location and repeat number of minisatellites is one in ten million. For instance, one person has the unit ATG repeated 60 times at one gene locus, 120 times at the second locus and so on whereas another person is likely to have a different number of repeats at these loci. Such polymorphic gene loci are called simple tandem repeats.

Procedures:

1. **Collection** of any sample (WBC of blood stain, sperm of semen, egg of vaginal fluid, cell of hair root follicle, buccal cell of saliva, cell of skin, cell of bone marrow, cell present in urine and cell of any body tissue).
2. **Isolation (Extraction)** of DNA from any samples. It is done chemically using detergent to wash extra material from DNA. It is also done mechanically applying a large amount of pressure to squeeze DNA. If DNA is in low amounts, then it can be amplified by PCR.
3. **Digestion (Fragmentation)** of DNA into several fragments containing VNTR of different sizes by a restriction endonuclease enzyme (e.g., *Alu* I or *Hae* III).
4. **Separation** of DNA fragments (VNTR) based on size by agarose gel electrophoresis or polyacrylamide gel electrophoresis (PAGE). The VNTRs are denatured into single-stranded DNA in the presence of an alkaline chemical (NaOH).
5. **Transfer** of separated ssDNA to synthetic nylon membrane or nitrocellulose paper on the gel by Southern blotting.
6. **Preparation** of radioactive single-stranded DNA probe containing repeat sequence of base pairs complementary to VNTR to analyse Southern blot.
7. **Hybridisation (Binding)** of radiolabelled VNTR probe to specific DNA repeat sequence of base pairs complementary to VNTR sequence on paper.
8. **Removal (Washing)** of the unhybridised DNA using a buffer.

9. **Detection** of hybridised DNA fragments by autoradiography with the help of colour-inducing (chromogenic) or fluorescent dye (exposure of X-ray film on a nylon sheet to mark the place where radioactive DNA probes have bound to DNA fragments). These places on X-ray film are dark bands that represent the DNA fingerprint or DNA profile.
10. **Identification** of a particular person's DNA by the occurrence and frequency of the particular gene sequence contained in the probe.

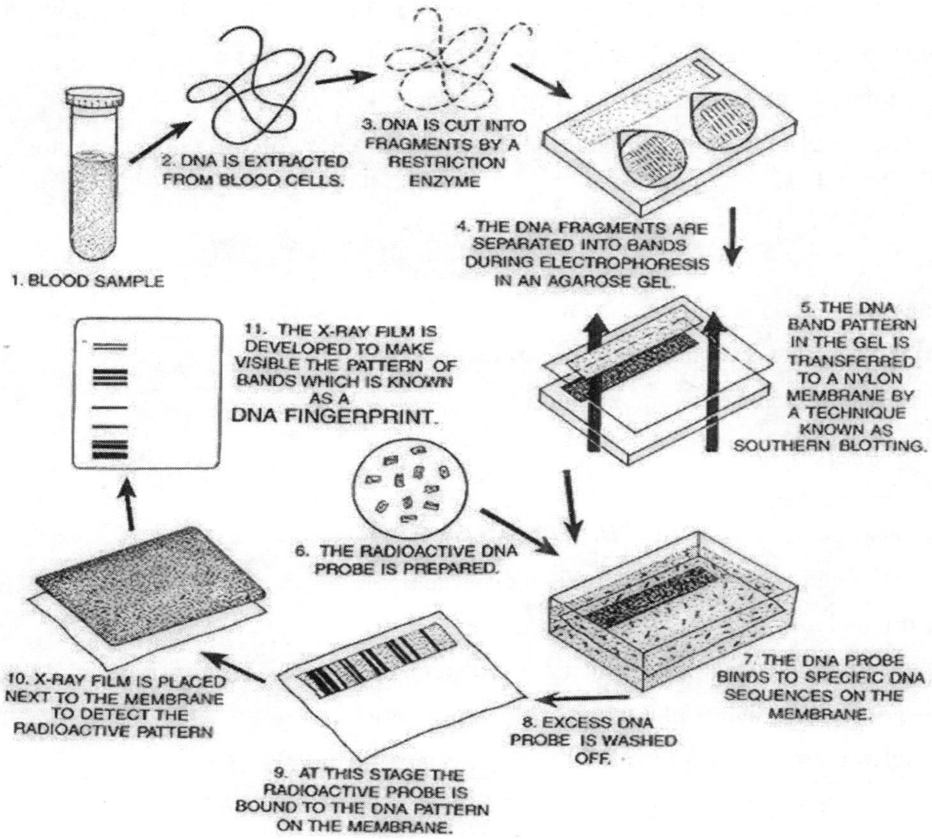

Fig: Steps involved in DNA fingerprinting

The first DNA fingerprint was performed in England in 1987.

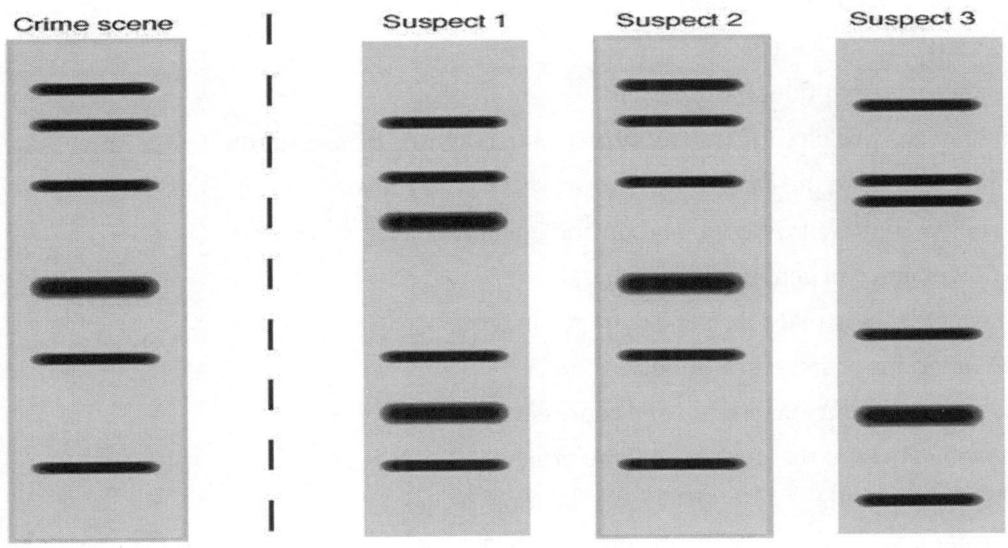

Fig: Matching of DNA fingerprints

Applications:

(a) DNA fingerprinting identifies the culprit or criminal or real victim by matching the DNA of the suspect with DNA collected from the site.

(b) It solves cases of paternity or maternity because a person inherits his or her VNTR from his or her parents (every son or daughter receives half of the chromosome from the father and half from the mother).

(c) It identifies the victims of disasters like plane crashes, building collapse, fire breakouts, floods or earthquakes.

(d) It determines the evolutionary relationship or link among organisms or races.

(e) It identifies human remains and matches organ donors.

(f) It solves the immigration dispute by establishing whether a hopeful immigrant is a close relative of already an established resident.

SHORT ANSWER TYPE QUESTIONS

1. What are the criteria of a genetic material?
2. Distinguish between the R strain and the S strain.
3. Distinguish between nucleoside and nucleotide.
4. Explain Edwin Chargaff's rule.
5. Explain the base ratio.
6. Distinguish between the B form of DNA and the Z form of DNA
7. Distinguish between prokaryotic DNA and eukaryotic DNA.
8. Write notes on nucleosomes.
9. Distinguish between conservative replication and semiconservative replication of DNA.
10. Mention the evidence in favour of RNA as primitive genetic material.
11. Distinguish between monocistronic mRNA and polycistronic mRNA.
12. Draw a well-labelled diagram of tRNA.
13. Mention the functions of mRNA, rRNA and tRNA.
14. Write notes on split genes and overlapping genes.
15. Define the modern concept of genes.
16. Explain the linear flow of information.
17. Write notes on RNA polymerase.
18. Discuss the post-transcriptional processing.
19. Distinguish between prokaryotic transcription and eukaryotic transcription.
20. What are the reactions in the chain initiation process of the translation process?
21. Distinguish between prokaryotic translation and eukaryotic translation.
22. Explain the degeneracy of genetic code.
23. Discuss the events taking place during chain termination.
24. Mention the functions of structural genes.
25. Distinguish between inducible operon and repressible operon.
26. What are the objectives of the Human genome project?
27. What are the important findings of the human genome project?
28. What is the significance of the human genome project?
29. Distinguish between microsatellite DNA and minisatellite DNA.

LONG ANSWER TYPE QUESTIONS

1. Describe the transformation experiment to prove DNA genetic material.
2. Describe the blender experiment to prove DNA is a genetic material.
3. Describe the double helical structure of DNA and show how it differs from RNA.
4. Describe the mechanism of DNA replication.
5. Describe the different types of RNA.
6. Describe the mechanism of transcription in protein synthesis.
7. Describe the mechanism of translation in protein synthesis in brief.
8. Discuss the properties of genetic code.
9. Describe the regulation of gene expression with respect to the concept of Lac operon.
10. Describe the procedure of DNA fingerprinting.

MULTIPLE CHOICE QUESTIONS

1. **The base ratio is less than one in**
   (a) *E. coli* (b) Plants
   (c) Fungi (d) Animals
2. **The length of DNA in prokaryotes is**
   (a) 1.56 mm (b) 1.56 cm
   (c) 1.56 m (d) 15.6 cm
3. **The core of nucleoplasm contains**
   (a) Dimer of histones
   (b) Tetramer of histones
   (c) Hexamer of histones
   (d) Octamer of histones
4. **The nature of DNA replication is**
   (a) Semi-conservative and continuous
   (b) Semi-conservative and semi-discontinuous
   (c) Conservative and continuous
   (d) Conservative and discontinuous
5. **The expected frequency of introduction of the wrong base during DNA replication Is**
   (a) 1:100 (b) 1:1000
   (c) 1:1 million (d) 1:1 billion
6. **A DNA molecule in which both strands have radioactive thymidine is allowed to duplicate in an environment containing non-radioactive thymidine. The exact number of DNA containing some radioactive thymidine after 3 duplications is**
   (a) 1 (b) 2
   (c) 4 (d) 8
7. **A DNA molecule of *E. coli* is heavy (fully labelled with N^{15}) and is allowed to replicate in a medium containing N^{14}. After one generation, the two daughter strands of molecules have**
   (a) Same density but do not resemble their parent DNA
   (b) Different densities but resemble their parent DNA
   (c) Different densities but do not resemble their parent DNA
   (e) Both the strands differ in density but resemble their parent DNA
8. **RNA and DNA were artificially synthesised *in vitro*, respectively**
   (a) Ochoa and Nirenberg
   (b) Ochoa and Kornberg
   (c) Kornberg and Nirenberg
   (d) Nirenberg and Khorana
9. **Nirenberg synthesised a mRNA containing 34 polyadenine (A-A-A-A...) and found a polypeptide formed of 11 polylysine. It proved that the genetic code for lysine is**
   (a) A singlet (b) A-A doublet
   (c) A-A-A triplet (d) Many adenines
10. **The 5′-end of tRNA always ends in the base**
    (a) Adenine (b) Guanine
    (c) Thymine (d) Uracil

11. The minimum length of cistron in a base pair that synthesises a polypeptide of 50 amino acid codons is

(a) 50 bp (b) 100 bp
(c) 150 bp (d) 200 bp

12. Considering that we have four nucleotides A, G, T and C, the number of base substitutions that can occur in the amino acid codons are

(a) 261 (b) 264
(c) 549 (d) 535

13. In a codon, wobbling or degeneracy of genetic code is generally restricted to

(a) First N base (b) Second N base
(c) Third N base (d) Aromatic amino acid

14. The formation of polyribosomes from ribosomes is done in the presence of

(a) Ca^{+2} (b) Mg^{+2}
(c) Na^{+} (d) Cu^{+2}

15. Nitrogenous bases are linked to sugar by

(a) Hydrogen bond
(b) Glycosidic bond
(c) Phosphodiester bond
(d) All of the above

16. Positively charged basic proteins that are found in eukaryotes are called

(a) Histones (b) Non-histones
(c) Polyamines (d) All of the above

17. Which of the following enzyme catalyses the transcription of all types of RNAs in eukaryotes?

(a) DNA-dependent RNA polymerase
(b) DNA-dependent DNA polymerase
(c) RNA-dependent DNA polymerase
(d) RNA-dependent RNA polymerase

18. A cell-free system for protein synthesis was used by

(a) Khorana (b) Nirenberg
(c) Francis Crick (d) Richard Gamow

19. In protein synthesis, tRNA carrying the amino acid enters

(a) A site of ribosome
(b) P site of the ribosome
(c) E site of the ribosome
(d) Anywhere

20. Transfer RNA is attached to mRNA by its

(a) I loop (b) II loop
(c) III loop (d) Variable arm

21. Amino acid binding site in tRNA is

(a) Anticodon loop (b) DHU loop
(c) CCA end (d) 5′end

22. Which of the following rRNA acts as structural RNA as well as ribozyme in bacteria?

(a) 23 S rRNA (b) 16 S rRNA
(c) 5 S rRNA (d) 5.8 S rRNA

23. Positively regulatory proteins are called

(a) Repressor (b) Activator
(c) Promoter (d) Enhancer

24. The Southern blotting technique involves the transfer of DNA from

(a) Membrane to gel
(b) Gel to membrane
(c) Solution to gel
(d) Gel to solution

25. The sensitivity of DNA fingerprinting can be increased by using

(a) Intron sequence (b) Exon sequence
(c) PAGE (d) PCR

26. If a base sequence of a codon in mRNA is 5′-AUG-3′, the sequence of tRNA pairing with it must be

(a) 3′-CAU-5′ (b) 5′-CAU-3′
(c) 5′-UAC-3′ (d) 3′-UAC-5′

27. The number of nucleosomes in a human cell with 6 billion base pairs is

(a) $3x\ 10^5$ (b) $3x\ 10^7$
(c) $2x\ 10^5$ (d) $2\ x10^7$

28. RNA polymerase was discovered by

(a) Kornberg (b) Ochoa
(c) Khorana (d) Nirenberg

29. The total number of nitrogenous bases in one helical turn of DNA is

(a) 5 (b) 10
(c) 12 (d) 20

30. The enormous diversity of protein molecules is mainly due to the diversity of

(a) Peptide bonds
(b) Amino group on the amino acids
(c) R groups on the amino acids
(d) Amino acid sequences

31. In a double-stranded DNA molecule, the percentage of cytosine is 18. The percentage of adenine will be

(a) 18 % (b) 32 %
(c) 36 % (d) 50 %

32. The number of genome types present in a typical green plant cell is

(a) One (b) Two
(c) Three (d) Four

33. The step of translation which does not consume a high energy phosphate bond is

(a) Activation of amino acid
(b) Formation of peptide bond
(c) Binding of aminoacyl tRNA to ribosome
(d) Translocation of aminoacyl tRNA

34. A polypeptide just released by a ribosome from mRNA has 75 amino acids. What would be the number of nucleotides in the coding region of mRNA including the initiation codon and terminating codon?

(a) 225 (b) 228
(c) 220 (d) 230

35. How many H-bonds are present in a segment of DNA if one strand contains an A-T-G-C nucleotide sequence?

(a) 4 (b) 8
(c) 5 (d) 10

36. If DNA exclusively consists of only 6 base pairs then what would be the minimum coding ratio to code only 20 amino acids?

(a) 1 (b) 2
(c) 3 (d) 4

37. If a polynucleotide contains 5 bases, then what will be the number of different sequences?

(a) 125 (b) 625
(c) 512 (d) 1024

38. How many signal codons are there?

(a) 1 (b) 2
(c) 3 (d) 4

39. Considering the average molecular mass of a base to be 500 Da, what is the molecular mass of a dsDNA of 10 base pairs?

(a) 5 KDa (b) 10 KDa
(c) 100 KDa (d) 500 KDa

40. If the molecular weight of a polypeptide is 15.3 KDa, what would be the minimum number of nucleotides in the m RNA that codes for this polypeptide if molecular weight of each amino acid is 90 KDa?

(a) 123 (b) 170
(c) 340 (d) 510

41. Which of the following sets of bacterial strains was selected by Griffith for his transformation experiment?

(a) S II and R II (b) S II and R III
(c) S III and R II (d) S III and R III

42. Which of the following sequences in the blender experiment occurred?

(a) Infection- Blending- Centrifugation
(b) Blending-Infection-Centrifugation
(c) Centrifugation-Infection-Blending
(d) Blending –Centrifugation-Infection

43. In the case of pyrimidines, N atoms are present on

(a) 1st and 7th positions
(b) 1st and 5th positions
(c) 1st and 3rd positions
(d) 1st and 9th positions

44. The number of nucleosomes in the solenoid is

(a) 2 (b) 4
(c) 6 (d) 8

45. The core of the nucleosome contains

(a) 60 bp (b) 100 bp
(c) 140 bp (d) 200 bp

46. Taylor proved the semi-conservative mode of replication in the roots of

(a) *Pisum sativum*
(b) *Vicia faba*
(c) *Lathyrus odoratus*
(d) *Oenothera lamarckiana*

47. DNA polymerase III lacks

(a) $5' \rightarrow 3'$ exonuclease activity
(b) $5' \rightarrow 3'$ endonuclease activity
(c) $3' \rightarrow 5'$ exonuclease activity
(d) $3' \rightarrow 5'$ endonuclease activity

48. Which of the following enzymes relieves tension during DNA replication?

(a) Helicase (b) Gyrase
(c) DNA polymerase I (d) DNA polymerase III

49. The gap left by the removal of primer is filled by

(a) 5′→3′ polymerase activity of DNA polymerase I
(b) 3′→5′ polymerase activity of DNA polymerase I
(c) 5′→3′ polymerase activity of DNA polymerase III
(d) 3′→5′ polymerase activity of DNA polymerase III

50. The removal of mismatched bases (proofreading) is done by

(a) 3′→5′ exonuclease activity of DNA polymerase I
(b) 5′→3′ exonuclease activity of DNA polymerase I
(c) 3′→5′ exonuclease activity of DNA polymerase II
(d) 3′→5′ exonuclease activity of DNA polymerase III

51. Which of the following enzymes catalyses the formation of RNA from a DNA template?

(a) RNA polymerase
(b) DNA polymerase
(c) Reverse transcriptase
(d) *Taq* polymerase

52. The first codon to be discovered by Nirenberg and Mathaei was

(a) AUG (b) UUU
(c) UAG (d) GUG

53. Which is not true according to Chargaff's rule?

(a) A=T (b) G=C
(c) A+G=T+C (d) A+T=G+C

54. In the sequence of the processed mRNA ready for translation

5′-AUG CUA UAC CUC CUU UAU CUG UGA-3′. The number of different tRNA molecules required to translate this mRNA is

(a) 4 (b) 6
(c) 7 (d) 8

55. The telomere of the eukaryotic chromosome contains a short segment of

(a) A rich (b) G rich
(c) T rich (d) C rich

56. A gene encoding polypeptides of 50 amino acids get mutated at 25th codon, UAU becoming UAG. The result would be

(a) Polypeptide of 24 amino acids
(b) Polypeptide of 50 amino acids
(c) Polypeptide of 25 amino acids
(d) Polypeptide of 49 amino acids

57. In the DNA segment having 6 coils, there are 22 base pairs linked by two H-bonds. How many cytosine bases are found in that segment?

(a) 22 (b) 38
(c) 44 (d) 76

58. The removal of RNA polymerase III from nucleoplasm will affect the synthesis of

(a) mRNA (b) rRNA
(c) tRNA (d) All of the above

59. The DNA fingerprinting has proved to be useful in forensic science. It involves the use of

(a) Microsatellite DNA
(b) Minisatellite DNA
(c) cDNA
(d) rRNA

60. The length of the DNA molecule consisting of 5000 nucleotide pairs will be

(a) 0.17 μm (b) 1.7 μm
(c) 17 μm (d) 170 μm

61. A DNA molecule having A+G/T+C =0.71 shows that the molecule is

(a) Linear (b) Circular
(c) Single-stranded
(d) Double-stranded

62. Feulgen staining for DNA is based on the reaction of Schiff's reagent with

(a) Free aldehyde group
(b) Purines at N-7 positions
(c) Deoxyribose at C-2 positions
(d) Pyrimidines at N-1 position

63. RNA polymerase II which synthesises mRNA in eukaryotes is generally

(a) Resistant to α-amanitin
(b) Sensitive to α-amanitin
(c) Resistant to actinomycin-D
(d) Sensitive to rifampicin

64. Genomes of prokaryotes consist mostly of

(a) Repetitive DNA sequences
(b) Unique DNA sequences
(c) Interspersed DNA sequences
(d) Clustered repeated DNA sequences

65. **RFLP (restriction fragment length polymorphism) study is a technique for**
(a) Transferring genes from unrelated species
(b) Isolating single genes
(c) Isolating single gene products
(d) Identifying genetic (DNA) homologies

66. **Which one of the following is known to act as a biological mutagen?**
(a) Acridine orange (b) Nitrosoguanidine
(c) Streptomycin (d) Transposon

67. **Which of the following is incorrectly matched?**
(a) Southern blot- DNA hybridisation
(b) Northern blot- RNA sequencing
(c) Western blot- Protein sequencing
(d) Eastern blot- Amino acid sequencing

68. **Which one of the following combinations possesses the integrated protein synthesising machinery?**
(a) Ribosome, tRNA, amino acid
(b) mRNA, tRNA, amino acid
(c) Ribosome, t RNA, mRNA, amino acid
(d) tRNA, ribosome, nucleus, mRNA

69. **Which one of the following is not a PCR-based molecular marker?**
(a) RAPD (b) RFLP
(c) AP-PCR (d) DNF

70. **Which one of the following is not correct for tRNAs?**
(a) The 5′- end is phosphorylated
(b) They have unusual bases
(c) The anticodon loop is identical
(d) The 3′-end base sequence is CCA

71. **An increased melting temperature for duplex DNA results from a high content of**
(a) Adenine + Guanine
(b) Cytosine + Thymine
(c) Guanine + Cytosine
(d) Thymine + Cytosine

72. **What will happen to transcription in the absence of the sigma factor of RNA polymerase?**
(a) Transcription will not take place
(b) Transcription will initiate from the initial codon
(c) Transcription will initiate from many random sites
(d) Transcription will initiate from the other strand

73. **The number of ribosome binding sites used by tRNA during protein synthesis is**
(a) 2 (b) 3
(c) 4 (d) 6

74. **The replication of single-stranded viral RNA is brought about by**
(a) Reverse transcriptase
(b) RNA polymerase
(c) DNA primer
(d) DNA ligase

75. **Which one of the following sequences is located upstream of transcription initiation sites in *E. coli*?**
(a) Hogness box (b) Pribnow box
(c) CAAT box (d) MADS-box

76. **During initiation in the process of translation, a GTP molecule is required in**
(a) Association of 30S-mRNA with formyl-met tRNA
(b) Association of 50 S subunit of the ribosome with the initiation complex
(c) Formation of formyl-met- tRNA
(d) Binding of 30S subunit of the ribosome with mRNA

77. **Which of the following is incorrect with respect to *lac* operon concept?**
(a) It is inducible operon?
(b) Absence of glucose is a must for transcription
(c) It applies to all prokaryotes
(d) Repressor is a part of *lac* operon

78. **How much energy is required for the formation of each peptide chain?**
(a) 1 GTP and 1 ATP
(b) 2 GTP and 1ATP
(c) 1GTP and 2 ATP
(d) 3GTP and 2 ATP

79. **Single nucleotide polymorphisms (SNP) have been identified at**
(a) 1.4 million locations
(b) 2.4 million locations
(c) 1.4 billion locations
(d) 2.4 billion locations

80. **Minisatellite DNA is a small repeat sequence of**
(a) 4-8 bp (b) 10-60 bp
(c) 80-100 bp (d) 100-120

81. Which of the following is correct about operons?

(a) These are uniform in size
(b) These are found in all eukaryotic genes
(c) These do not bind to protein
(d) These are shorter in lower eukaryotes than in higher forms

82. Penicillin inhibits bacterial multiplication because

(a) It checks spindle formation
(b) It destroys chromatin
(c) It inhibits RNA formation
(d) It inhibits cell wall formation

83. All of the following are potential control mechanisms for the regulation of gene expression in eukaryotes except

(a) Gene amplification
(b) Degradation of mRNA
(c) Lactose operon
(d) Transcription

84. DNA fragments are transferred from a gel to a membrane via a process called Southern blotting. Its purpose is to

(a) Analyse the RFLPs in the DNA
(b) Separate the PCR
(c) Attach the DNA fragments permanently to a substrate
(d) Separate the two complementary DNA strands

85. DNA and RNA synthesis takes place in

(a) $5' \rightarrow 3'$ direction
(b) $3' \rightarrow 5'$ direction
(c) Both a and b
(d) DNA in $5' \rightarrow 3'$ and RNA in $3' \rightarrow 5'$ direction

86. Which of the following amino acids is entirely incorporated in the nucleotide bases?

(a) Glutamic acid (b) Alanine
(c) Glycine (d) Serine

87. Modern molecular taxonomy involves the comparison of

(a) Chromosome morphology
(b) r DNA
(c) C-DNA (d) tRNA

88. Which of the following techniques is used for biodiversity characterisation?

(a) RAPD (b) RFLP
(c) SSR(d) SNP

89. The nucleic acid present in viruses is

(a) Either RNA or DNA
(b) RNA and DNA
(c) Only RNA
(d) Only DNA

90. Which of the following is not an example of genomic control of a gene?

(a) Alternative DNA splicing
(b) DNA rearrangement
(c) Gene amplification
(d) Gene deletion

91. Which of the following is an allosteric protein that is inactive until it binds to cAMP to activate transcription?

(a) RNA polymerase
(b) Catabolite activator protein
(c) ATP
(d) All of the above

92. Histone structure can be altered by

(a) DNA rearrangement
(b) Acetylation
(c) Sigma factor
(d) Methylation

93. In eukaryotes, DNA packing affects gene expression by

(a) Controlling access to DNA
(b) Positioning related structural genes near each other
(c) Protecting DNA from mutation
(d) Enhancing recombination of genes

94. Which of the following is arranged in correct order by size (from largest to smallest)?

(a) Codon-chromosome-gene-nucleotide
(b) Gene-chromosome-codon-nucleotide
(c) Nucleotide-chromosome-gene-codon
(d) Chromosome-gene-codon-nucleotide

95. Multigene families arise as a result of

(a) Nucleosome
(b) Introns
(c) Heterochromatin
(d) Euchromatin

96. DNA is destabilised by

(a) Polyglycine (b) PEG
(c) Sucrose (d) Formamide

97. Which of the following is most conserved amongst organisms?

(a) Mitochondrial DNA
(b) Chloroplast DNA
(c) Recombinant DNA
(d) DNA that codes for tRNA

98. Jacob and Monod proposed the lac operon concept that is applicable for

(a) All prokaryotes
(b) All eukaryotes
(c) All prokaryotes and some eukaryotes
(d) All organisms

99. Which of the following conditions is not essential for nucleic acid hybridisation?

(a) High salt concentration
(b) Primer
(c) High temperature
(d) Complimentary strands

100. Which of the following groups is present at the extreme 3′ end of a single polynucleotide strand?

(a) Phosphate group
(b) Hydroxyl group
(c) Oxo group
(d) Alcoholic group

101. In bacteria and viruses, the intertwined DNA starts separating during replication at

(a) One point
(b) Many points
(c) One and Many points, respectively
(d) Many and one points, respectively

102. The first step in the initiation of protein synthesis requires

(a) 30S ribosomal subunit
(b) ATP
(c) Ribosomal A site
(d) Translational factor Tu

103. If the molecular mass of an amino acid is 150 Dalton, then the molecular weight of its tripeptide will be

(a) 450 (b) 486
(c) 414 (d) 504

104. An open reading frame has

(a) No start and stop codon
(b) A start and stop codon
(c) No start but stop codon
(d) A start but stop codon

105. The consensus sequence of 5′ and 3′ splice junctions in eukaryotic m RNA contains

(a) GU-GA (b) GU-AG
(c) AG-GU (d) CG-AG

106. The uncoiled DNA in differentiating cells is generally involved in the synthesis of

(a) DNA (b) mRNA
(c) Protein (d) Glucose

107. The functioning of structural genes is controlled by

(a) Operator (b) Regulator
(c) Promoter (d) Enhancer

108. Which of the following will be produced in a cell if there is a nonsense mutation in the *Lac* Y gene?

(a) Beta galactosidase
(b) Permease
(c) Transacetylase
(d) All of the above

109. *E. coli* has 4.6 x 10^6 base pairs and completes the process of replication within 38 minutes. What will be the average rate of replication?

(a) 100 bp/sec (b) 200 bp/sec
(c) 100 bp/sec (d) 2000 bp/sec

110. If a polynucleotide contains 6 bases, then what will be the number of different sequences?

(a) 2064 (b) 2460
(c) 3456 (d) 4096

111. How many H-bonds are present in a segment of DNA if one strand contains an A-T-G-C nucleotide sequence?

(a) 5 (b) 10
(c) 12 (d) 16

112. The coding ratio in all organisms is 3 (three nucleotides code one amino acid). If DNA exclusively consists of only 6 base pairs, then what would be the minimum coding ratio assuming that these are only 20 amino acids to be coded?

(a) 1 (b) 2
(c) 3 (d) 4

113. The total number of thymine bases present in ds DNA if it transcribes mRNA which reads as follows: 5′ AUG CAU CAU GCA-3′

(a) 5 (b) 6
(c) 7 (d) 8

114. If there is a double bond between A and T and a triple bond between G and C, then which sequence will be most stable at higher temperatures?

(a) ATTGTACCAAA (b) AATTATATATA
(c) AGCAGAGAGTT (d) AGGCCGGCCCT

115. The relative amount (mole %) of nitrogenous bases in a genomic DNA sample is as such: A=25, G=25, T=32, C=18. The most likely source of DNA is

(a) Virus (b) Yeast
(c) Bacteria (d) Salmon sperm

116. Hershey and Chase performed a blender experiment to prove DNA as genetic material on

(a) T_2 bacteriophage
(b) T_3 bacteriophage
(c) T_4 bacteriophage
(d) T_5 bacteriophage

117. Which of the following enzymes is used in the replication of DNA?

(a) DNA polymerase I
(b) DNA polymerase II
(c) DNA polymerase III
(d) All of the above

118. The diameter of a solenoid is

(a) 10 nm (b) 20 nm
(c) 30 nm (d) 300 nm

119. Which of the following types of RNA is the smallest and shortest?

(a) mRNA (b) rRNA
(c) tRNA (d) None of the above

120. Testis determining factor (TDF) is the smallest gene containing

(a) 10 base pairs (b) 14 base pairs
(c) 20 base pairs (d) 24 base pairs

121. Codon-anticodon pairing occurs by

(a) Hydrophobic interaction
(b) Electrostatic interaction
(c) Hydrogen bond
(d) Covalent bond

122. The nature of *Trp* operon is

(a) Negatively repressed
(b) Negatively induced
(c) Positively repressed
(d) Positively induced

123. Griffith performed a transformation experiment to prove DNA as genetic material on

(a) *Mycobacterium tuberculosis*
(b) *Salmonella typhimurium*
(c) *E. Coli*
(d) *Streptococcus pneumonia*

124. Which of the sets of isotopes of elements was used by Hershey and Chase to prove DNA as genetic material?

(a) S^{35} and P^{32} (b) S^{35} and C^{14}
(c) P^{32} and C^{14} (d) C^{14} and N^{15}

125. The nitrogen atoms in purines are present on

(a) Two positions (b) Three positions
(c) Four positions (d) Six positions

126. The IUPAC name of adenine is

(a) 6-aminopurine (b) 5-aminopurine
(c) 2-aminopurine (d) 4-aminopurine

127. Which of the following positions of the carbon atom of sugar is occupied by the base?

(a) 1st (b) 2nd
(c) 3rd (d) 5th

128. The phosphate group present in DNA is

(a) PO_4^{-3} (b) PO_4^{-2}
(c) PO_4^{-} (d) PO_3

129. How many H-bonds (on average) are present within each complete turn of DNA double helix?

(a) 18 (b) 20
(c) 25 (d) 30

130. The distance of phosphate from the axis in DNA is

(a) 8A° (b) 10A°
(c) 12A° (d) 20A°

131. RNA is less stable than DNA due to

(a) Presence of uracil
(b) Presence of 2-OH group
(c) Both of the above
(d) None of the above

132. Which of the following has circular DNA?

(a) Prokaryotic DNA
(b) Mitochondrial DNA
(c) Chloroplast DNA
(d) All of the above

133. The length of DNA in eukaryotes is

(a) 2.2 mm (b) 2.2 cm
(c) 2.2 m (d) 2 m

134. Which of the histone proteins is not present in the core of the nucleosome?

(a) H_1 (b) H_2a
(c) H_2b (d) H_3

135. The average number of base pairs in linker DNA of nucleosomes is

(a) 40 (b) 60
(c) 100 (d) 160

136. RNA as genetic material was proved by Fraenkel-Conrat and Singer on

(a) Bacteriophage
(b) Tobacco mosaic virus
(c) Pox virus
(d) HIV

137. Ribozyme is RNA catalysing as an enzyme. It was discovered in

(a) *Mycoplasma* (b) *Tetrahymena*
(c) *Oenothera* (d) *E. coli*

138. Which of the following RNA is most abundant?

(a) mRNA (b) rRNA
(c) tRNA (d) None of the above

139. Which of the following RNA is soluble?

(a) mRNA (b) rRNA
(c) tRNA (d) None of the above

140. The cap of eukaryotic mRNA contains methylated residue of

(a) Adenine (b) Guanine
(c) Thymine (d) Cytosine

141. The 3-D structure of tRNA is

(a) Clover leaf-shaped
(b) L-shaped
(c) J-shaped
(d) V-shaped

142. How many anticodons are present in the anticodon loop of tRNA?

(a) Two (b) Three
(c) Five (d) Seven

143. Renaturation or annealing of DNA occurs at

(a) 40°C (b) 65°C
(c) 80°C (d) 94°C

144. DNA absorbs maximum UV light at

(a) 200 nm (b) 260 nm
(c) 280 nm (d) 300 nm

145. Which of the following sets of codons is considered an exception to degenerate codons?

(a) AUG and UGG
(b) UGG and UAG
(c) UGG and UGA
(d) AUG and UGA

146. The first terminating codon discovered was

(a) UAA
(b) UGA
(c) UAG
(d) All were discovered simultaneously

147. Reverse flow of information for protein synthesis occurs in

(a) Reoviruses (b) Adenoviruses
(c) Toga viruses (d) Retroviruses

148. tRNA in eukaryotes is synthesised by

(a) RNA polymerase I
(b) RNA polymerase II
(c) RNA polymerase III
(d) None

149. Which of the following subunits of RNA polymerase binds with sigma factor in transcription?

(a) α'subunit (b) α subunit
(c) β' subunit (d) β subunit

150. Which of the Initiation factors binds the free 30S subunit of the ribosome with mRNA?

(a) IF_1 (b) IF_2
(c) IF_3 (d) IF_4

151. How many release factors are in the protein synthesis of prokaryotes?

(a) One (b) Two
(c) Three (d) Four

152. Which of the following is the first event when chain termination occurs at the end of the translation process?

(a) Separation of polypeptide
(b) Separation of tRNA
(c) Dissociation of ribosome
(d) Release of mRNA

153. Which of the following antibiotics inhibits protein synthesis at the transcription stage?

(a) Streptomycin (b) Erythromycin
(c) Tetracycline (d) Rifampicin

154. Which of the following genes is the longest in *Lac* operon?

(a) *Lac* Z (b) *Lac* Y
(c) *Lac* A (d) Regulator gene

155. The *Lac* Y of structural genes of operon codes for

(a) β-galactosidase enzyme
(b) Permease
(c) Transacetylase
(d) Lactase

156. The human genome project was launched on

(a) 1st October, 1988 (b) 1st October, 1990
(c) 1st October, 1995 (d) Ist October, 2000

157. Dystrophin is the largest gene-containing

(a) 1 million base pairs
(b) 1.4 million base pairs
(c) 2 million base pairs
(d) 2.4 million base pairs

158. The smallest gene in the *Lac* operon is

(a) Regulator gene
(b) Operator gene
(c) Lac A of structural gene
(d) Promoter gene

159. The first step in chain elongation of protein synthesis is

(a) Binding of aminoacyl tRNA at the A site of the ribosome
(b) Binding of aminoacyl tRNA at the P site of the ribosome
(c) Translocation of peptidyl tRNA
(d) Formation of peptide bond

160. The Shine Delgarno sequence in the transcription process of protein synthesis is

(a) AUGAGGU (b) GUGAUGG
(c) AGGAUGG (d) AGGAGGU

161. Which of the following properties does not belong to the genetic code?

(a) It is degenerate
(b) It is comma less
(c) It is overlapping
(d) It is non-ambiguous

162. The inducible operon follows

(a) Anabolic pathway
(b) Catabolic pathway
(c) Inhibitory pathway
(d) Stimulatory pathway

163. Which of the following countries did not take part in the Human genome project?

(a) Japan (b) Italy
(c) China (d) Germany

164. Which of the following samples is used while doing DNA fingerprinting?

(a) RBC (b) WBC
(c) Plasma (d) Anyone

165. Which of the following processes is not involved in DNA fingerprinting?

(a) Agarose gel electrophoresis
(b) Southern blotting
(c) Autoradiography
(d) Homogenisation

166. Which of the following codons does not justify the degeneracy rule of genetic code?

(a) UGG (b) CUG
(c) GGG (d) GAG

167. The enzyme required for the post-transcriptional event and represents the dominance of the RNA world is

(a) SNRNPs
(b) Guanyl transferase
(c) Peptidyl transferase
(d) Methylase

168. Lac operon is regulated by

(a) Lac repressor alone
(b) CAP-cAMP complex
(c) Initiation complex
(d) Both a and b

169. If there are 81 bases in RNA that code for a protein with 27 amino acids and the base at the 61st position is deleted such that the length of RNA becomes 80 bases, how many codons will be altered?

(a) 7 (b) 9
(c) 11 (d) 15

170. A sequential expression of a set of human genes occurs when a steroid molecule binds to the.

(a) Ribosome (b) mRNA
(c) tRNA (d) DNA sequence

171. Which of the following is used to know the phylogeny?

(a) DNA
(b) mRNA
(c) rRNA
(d) tRNA

172. Which of the following is an ambiguous codon?

(a) GUG
(b) CUG
(c) GGG
(d) GAG

173. Which of the following rRNAs acts as structural RNA as well as ribozyme in bacteria?

(a) 23 S rRNA
(b) 5 S rRNA
(c) 18 S rRNA
(d) 5.8 S rRNA

174. Initiation of a polypeptide chain in protein synthesis is induced by

(a) Methionine
(b) Tryptophan
(c) Glycine
(d) Lysine

175. Telomeric repetitive DNA sequence controls the function of eukaryotic chromosomes because they

(a) prevent chromosome loss
(b) act as replicons
(c) are RNA transcription factor
(d) help chromosome pairing

176. The angle of helical twist in the B form of DNA is

(a) 6.3°
(b) 6°
(c) 60°
(d) 90°

177. Ribozyme, a catalytic RNA was discovered in

(a) *Saccharomyces cerevisiae*
(b) *Tetrahymena thermophila*
(c) *Caenorhabditis elegans*
(d) *Drosophila melanogaster*

178. The DHU loop in transfer RNA is the site of

(a) Amino acid binding
(b) Aminoacyl synthetase enzyme binding
(c) Codon recognising
(d) Codon binding on mRNA

179. The β-subunit of RNA polymerase

(a) recognises promoter
(b) binds sigma factor
(c) binds DNA template
(d) terminates transcription

180. The structural genes such as Lac Z, Lac Y and Lac A of the operon model are not involved in the synthesis of

(a) β-galactosidase enzyme
(b) Permease enzyme
(c) Trans-acetylase enzyme
(d) Glucose isomerase enzyme

ANSWERS

1.a	16.a	31.b	46.b	61.c	76.a	91.b	106.b	121.c	136.b	151.c	166.a
2.a	17.a	32.c	47.a	62.a	77.d	92.b	107.b	122.a	137.b	152.a	167.a
3.d	18.b	33.b	48.b	63.b	78.b	93.a	108.a	123.d	138.b	153.d	168.d
4.b	19.a	34.b	49.a	64.b	79.a	94.d	109.d	124.a	139.c	154.a	169.a
5.d	20.b	35.d	50.a	65.d	80.b	95.c	110.d	125.c	140.b	155.b	170.d
6.b	21.c	36.b	51.c	66.d	81.a	96.d	111.b	126.a	141.b	156.b	171.c
7.a	22.a	37.d	52.b	67.d	82.d	97.a	112.b	127.a	142.b	157.d	172.a
8.b	23.b	38.d	53.d	68.c	83.c	98.a	113.c	128.a	143.b	158.d	173.a
9.c	24.b	39.b	54.c	69.d	84.a	99.b	114.d	129.c	144.b	159.a	174.a
10.b	25.d	40.d	55.b	70.c	85.a	100.b	115.d	130.b	145.a	160.d	175.a
11.c	26.d	41.c	56.a	71.c	86.c	101.a	116.a	131.c	146.c	161.c	176.b
12.c	27.b	42.a	57.b	72.c	87.b	102.a	117.c	132.d	147.d	162.b	177.b
13.c	28.b	43.c	58.c	73.a	88.a	103.c	118.c	133.c	148.c	163.c	178.b
14.b	29.d	44.c	59.b	74.a	89.a	104.b	119.c	134.a	149.d	164.c	179.b
15.b	30.d	45.c	60.b	75.b	90.a	105.b	120.b	135.b	150.c	165.d	180.d

CHAPTER-14

BIOTECHNOLOGY PRINCIPLES AND PROCESSES

Biotechnology is the scientific and wilful manipulation of living organisms at molecular and gene levels for the benefit of human beings. The term Biotechnology was coined by Karl Erky, a Hungarian Engineer in 1917 to describe a process for large-scale production of useful substances.

Branches of Biotechnology

1. Traditional or Old Biotechnology: Biotechnology has been practised since the beginning of history. It has been used to bake bread, produce dairy products, brew alcoholic beverages, and breed food crops and domestic animals.
2. Non-gene Biotechnology: It is concerned with whole cells, tissues, or even individual organisms.
3. Modern or Gene Biotechnology: It is concerned with the transfer or manipulation of genes from one organism to another based on genetic engineering. It includes *in vitro* fertilisation, the development of antibiotics, vaccines, recombinant proteins, hormones, gene therapy etc.

Revolutions in Biotechnology

1. Fermentation or gelatinisation or solubilisation of starch into butanol and acetone by *Clostridium acetobutyricum* in 1920 as reported by Chaim Weismann.
2. Discovery of Penicillin antibiotic from *Penicillium notatum* in 1929 by Alexander Fleming.
3. Establishment of Recombinant DNA technology in 1972.

Initiatives by Govt. of India: The National Biotechnology Board (NBTB) was set up in 1982 to encourage and coordinate various research activities in biotechnology. Now it was replaced by the Department of Biotechnology (DBT) within the Ministry of Science and Technology in 1986.

Genetic Engineering: It is a technique of altering the genetic constitution of cells or individuals by the selective removal, insertion or modification of individual genes or gene sets. The term 'genetic engineering' was coined by Jack Williamson.

It is also called genetic modification. It is a technique to alter the nature of genetic material to introduce it into the host organism and then change the phenotype of the host organism. It is done by linking the foreign genes to bacterial control sequences then it is possible to achieve expression of these genes and thereby produce bacterial strains capable of producing new commercial products.

It is also called DNA cloning where there is production of a large number of identical DNA molecules from a single ancestral DNA molecule.

Precursor: Pasteur took the opportunity to unravel the mystery of why wine turned sour. He utilised microbes for fermentation.

Facilitator: Watson and Crick elucidated the double-helical model of DNA in 1953.

The first company to focus on genetic engineering, Genetech, was founded in 1976 and started the production of human proteins.

Achievements

1. Paul Berg (1972) is regarded as the father of genetic engineering. He developed the first recombinant DNA by combining the DNA (gene) of a monkey or Simian Virus-40 with lambda phage (virus).
2. Stanley Cohen and Herbert Boyer (1973) constructed the first recombinant DNA and created the first transgenic organism with the help of restriction endonuclease and ligase enzyme by inserting antibiotic resistance gene of *Salmonella typhimurium* (a bacterium) into a plasmid of *E. coli* (a bacterium).

TOOLS OF GENETIC ENGINEERING

1. Enzymes: Restriction endonuclease enzyme

It was discovered by Arber, Nathans and Smith (1970). It comes under the category of nuclease enzyme which is of three types such as exonuclease (cutting only one strand of ds DNA at the external position either 5′ or 3′ ends), endonuclease (cutting only one strand of dsDNA only at any internal position) and **restriction endonuclease.** It is an enzyme that cuts both strands of DNA duplex at a specific site (recognition sequence or palindromic sequence) by breaking the phosphodiester bond between the sugar and phosphate component.

It is also known as a molecular scalpel or scissor. It generally occurs in bacteria as a chemical weapon against viruses. Its number is more than 3000 which have been isolated from 230 strains of bacteria and were fully characterised. All enzymes require Mg^{+2}. It does not act on the DNA of its cell because of the presence of a parallel enzyme produced by the cell which keeps on adding the $-CH_3$ group to specific Adenine and cytosine residues within recognition sequences.

The first restriction enzyme discovered was Hind II from *Haemophilus influenzae* strain Rd. Later Eco R1 was isolated from *Escherichia coli* strain RY13.

Nomenclature of Restriction endonuclease (as in Eco RI)

(i) The first letter of the restriction enzyme denotes the name of the genus (**E**: *Escherichia*)

(ii) The second and third letters denote the epithet of species (**co**: coli)

(iii) The fourth letter (in Roman) indicates the strain of species (**R**: strain)

(iv) The fifth letter (in Roman numeral) indicates the order of identification of different enzymes from the same bacterium it produces (**I**: first identified).

Types of restriction endonuclease enzyme:

1. **Type I:** It has three different subunits**.** It is not used in RDT because it cuts DNA at sites other than the recognition site forming random cleavage patterns. It requires ATP.
2. **Type II:** It has two identical subunits. It is used in RDT because it cuts DNA at a specific site (palindromic site) and produces restriction fragments. It does not require ATP. It requires Mg+2. It is most stable. Most restriction endonuclease enzymes belong to this type. It forms palindromes by acting on recognition sequences (4-6 base pairs which are predominantly G-C rich) with rotational symmetry. The recognition sequences are different and specific for the different restriction endonuclease enzymes. Thus, in palindrome, the base sequence in both strands of DNA duplex reads the same when read from the same end (5′- or 3′- end of both strands). The enzyme that recognises 4 bases will cut DNA more frequently (called Tetra cutter) than those atch recognise 6 bases (called Hexa cutter).

   Cleavage pattern: Restriction endonuclease enzymes produce two types of cuts:

   (a) **Staggered**: The enzymes cut both strands of DNA at different locations generating protruding ends or sticky ends or cohesive ends (one strand of duplex extends some bases beyond the other but these ends readily pair with each other). This is required in RDT

   (b) **Even**: The enzymes cut both strands of DNA at the same locations generating blunt or flush ends.
3. **Type III**: It has two different subunits. It is not used in RDT because it cuts DNA at sites other than recognition sites forming random cleavage patterns. It also requires ATP.

List of restriction enzymes with their source, cleavage site and products

Enzymes	Source (Organism and strain)	Cleavage site	**Products**
Hind II	*Haemophilus influenzae* Rd	GTC↓GAG CAG↑CTC	Blunt ends
Hin d III	*Haemophilus influenzae* Rd	A↓AGCTT T TCGA↑A	Sticky ends

Hae III	Haemophilus aegypticus	GG↓CC CC↑GG	Blunt ends
Hpa II	Haemophilus parainfluenzae	C↓CGG GGC↑C	Sticky ends
Eco R I	*Escherichia coli* Ry13	G↓AATTC C TTAA↑G	Sticky ends
Eco R II	*Escherichia coli* R24	C↓CTGG GGAC↑C	Sticky ends
Eco R V	Escherichia coli	GAT↓ATC CTA↑TAG	Blunt ends
Alu I	Arthrobacter luteus	AG↓CT TC↑GA	Blunt ends
Bam H I	*Bacillus amyloliquefaciens* H	G↓GATCC CCTAG↑G	Sticky ends
Sal I	*Streptomyces albus* G	G↓TCGAC CAGCT↑G	Sticky ends
Taq I	*Thermus aquaticus* YT1	T↓CGA AGC↑T	Sticky ends
Pst I	*Providencia stuartii*	CTGCA↓G G↑ACGTA	Sticky ends
Pvu II	*Proteus vulgaris*	CAG↓CTG GTC↑GAC	Blunt ends
Sca I	*Streptomyces caespitosus*	AGT↓ACT TCA↑TGA	Blunt ends
Stu I	*Streptomyces tubercidicus*	AGG↓CCT TCC↑GGA	Blunt ends
Sma I	*Serratia marcescens*	CCC↓GGG GGG↑CCC	Blunt ends
Bgl II	*Bacillus globigli*	A↓GATCT TCTAG↑A	Sticky ends
Kpn I	*Klebsiella pneumoniae*	GGTAC↓C C↑CATGG	Sticky ends
Xba I	*Xanthomonas badrii*	T↓CTAGA AGATC↑T	Sticky ends
Sau 3AI	*Staphylococcus aureus* 3A	G↓ATC CTA↑G	Sticky ends
Nla III	*Neisseria lactamica*	↓CATG GTAC↑	Sticky ends
Ava II	*Anabaena variabilis*	G↓GACC CCTG↑G	Sticky ends
Not I	*Nocardia otitidis*	GC↓GGCCGC CGCCGG↑CG	Sticky ends

DNA Ligase: It was discovered by Mertz and Davis (1972). It seals the nicks between adjacent nucleotides in the DNA duplex by forming a new sugar-phosphate bond with 5′-phosphate and a 3′-OH. It is also known as molecular glue. It is present in both prokaryotes and eukaryotes. There are two **types**:

(a) T_4 DNA ligase (Viral ligase): It ligates both sticky and blunt ends. It requires ATP as a cofactor.

(b) *E. coli* DNA ligase (Bacterial ligase): It ligates sticky ends only. It requires NAD^+ as a cofactor.

Both enzymes perform annealing when the cofactor undergoes splitting and forms an Enzyme-AMP complex. The complex binds to the nick and makes a phosphodiester bond which seals the nick. Then the enzyme and AMP are finally released.

Alkaline phosphatase: It removes the phosphate group from the 5′-ends of DNA/RNA. It is used to prevent sticky ends from being ligated from one another (recircularisation).

2. Vector

It is DNA that carries foreign inserted DNA fragments into a host for cloning and expression. It is also known as vehicle DNA.

Features of Cloning Vectors:

(a) A cloning vector should possess an origin of replication so that it can self-replicate inside the host cell.

(b) It should have a unique recognition or restriction site for the insertion of the target DNA.

(c) It should have one or two selectable markers with an antibiotic resistance gene that facilitates screening (identification and selection) of the recombinant organisms or positively transformed cells. The marker gene may be selectable which encodes a product that allows the transformed cell to survive and grow under the condition that kills or restricts the growth of non-transformed cells or screenable which gives a product that can be detected easily.

(d) It should be small in size so that it can easily integrate into the host cell.

(e) It should be capable of inserting a large segment of DNA.

(f) It should possess multiple cloning sites.

(g) It should be capable of working under the prokaryotic and eukaryotic systems.

(h) It should be easy to isolate and purify from the cells.

(i) It should replicate autonomously independently of the host chromosome and should have an origin of replication (*Ori*). The *Ori* is a specific sequence in DNA from where replication starts. It is single in prokaryotes but several in eukaryotes. It also controls the copy number of linked DNA as per the requirement.

(j) It should show normal activity in the host cell.

Types of vectors:

(a) Cloning vector: It is used for cloning

(b) Expression vector: It is the production of protein by translation and transcription by specified DNA insert.

Examples:

(a) **Plasmid**: It was discovered by Lederberg and William Hayes. It is extrachromosomal, small (1000 times smaller than nucleoid), circular, double-stranded, closed and supercoiled DNA. It can be easily isolated.

It occurs naturally in bacteria, but it can be produced artificially. Its size is 2 kb to 200 kb, and it contains 2-250 genes. It can undergo autonomous replication independently of the nucleoid. It may exist independently or become integrated into bacterial chromosome-forming episome. It contains about 2% of genetic information which is sufficient for its replication and additional information for host properties such as antibiotic resistance. It frequently carries the genes that are useful to cells and is not essential for normal growth and division.

It may be either a **Stringent** type when it has a low copy number (1 or 2 copies) or a **Relaxed** type when it has high copy numbers (10 or more).

It may be conjugative when it mediates DNA transfer through conjugation (e.g., F-plasmid, R-plasmid) or non-conjugative (e.g., Col plasmid).

It can clone DNA insert up to the size of 10 Kb.

An ideal plasmid must have the following features:

(a) It has a small size (2-4 Kb) which makes it easy to purify from bacterial culture.

(b) It has only one recognition site for at least one restriction endonuclease.

(c) It has at least two selectable marker genes.

(d) It undergoes more than one replication for each replication of the host genome.

Artificial plasmids:

(i) **pBR322:** It is a plasmid developed by Bolivier and Rodriguez (1977). It is most popular, widely studied and widely used. It is constructed from plasmids of *E. coli* (pBR318 and pBR320). Its size is 4,363 bp. It undergoes much faster replication than bacterial DNA. It possesses two genes that confer resistance to antibiotics- *amp*^{r} and *tet*$^{r.}$ It has unique recognition sites for 20 restriction endonucleases. There are two sites such as Pst I and Pvu I are located within the amp^r gene and *Bam* H I and *Sal* I are within the tet^r gene. Some other recognition sites such as *Eco* R I, *Cla* I, *Hind* III also exist. The 'rop' codes for the proteins involved in the replication of plasmid. The 'ori' is meant for the origin of replication.

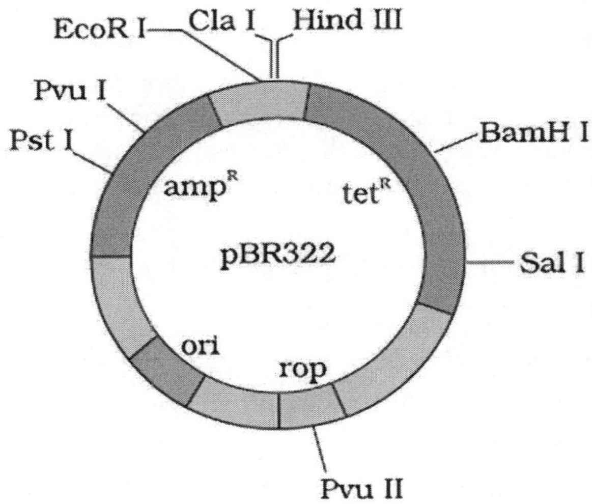

(ii) **pUC 18:** It is a plasmid developed at University of California by Messing and Vieria. It is much smaller than pBR322. Its size is 2,688 bp. It is a derivative of pBR322 and possesses all essential parts of pBR322. It has the following advantages over pBR322.

(a) Each bacterial cell produces up to 700 copies without any treatment.

(b) Cells containing r-DNA are selected in a single step.

(c) It also allows sequencing of DNA insert.

(iii) **pUC 8**: It is the smallest plasmid. Its size is 2,100 bp.

(iv) **pSC 101:** It is a plasmid developed by Stanley Cohen and Boyer (1973). It was a recombinant plasmid capable of being replicated in the host (*E. coli*).

(b) **Lambda phage:** It is the best-studied and genetically modified strain of bacteriophage. Its size is 48,502 bp. It contains the origin of replication, genes for Head and Tail protein, enzymes for DNA replication, lysis and lysogeny. It also contains single-stranded protruding cohesive ends of 12 bases. It has double-stranded and linear DNA in the Head region but in *E. coli*, the two cohesive ends join to form a circular molecule that is necessary for replication. The sealed cohesive ends are called ***cos*** site which has 12 unpaired and complimentary bases. This *cos* site is a site of cleavage and is essential for packaging of mature phage DNA into phage Head.

It can clone DNA insert up to the size of 23 kb

(c) **M-13**: It is commonly called Phagemid. Its size is 6,407 bp. It has single-stranded and circular DNA. When it infects E. coli, its single-stranded DNA replicates to produce a double-stranded DNA known as Replication Form (RF). Hence RF molecule can be regarded as being similar to a plasmid. It can clone DNA insert up to size of 1-4 kb.

(d) **Cosmid**: It was developed by Collins and Hons (1978). It is a hybrid and artificial vector constructed by the *cos* site of lambda phage and plasmid. Its size is 5400 bp. It contains a minimum of 250 bp of λ-DNA which includes a *cos* site of 12 bases of lambda phage and sequence needed for binding of and cleavage by terminase. It can clone DNA insert up to a size of 45 kb which makes it more efficient. It is also used for gene mapping.

(e) **Ti-Plasmid**: It is the largest plasmid. Its size is 213 kb. It is produced from *Agrobacterium tumefaciens* which infects mostly in dicots (broad-leaved plants). This bacterium invades plant tissue and causes a tumour called crown gall. During infection, part of Ti plasmid called T-DNA integrates into plant chromosomal DNA. It does not infect monocot plants (cereals).

(f) **Bacterial Artificial chromosome (BAC):** It contains an F-factor gene for replication, a unique restriction site and a selectable marker gene. It is better than Yeast Artificial Chromosome (YAC) because it is more stable, simpler and performs easier transformation. It can clone large DNA insert up to a size of 300 kb. It is also used for analysis of large genomes (genome sequencing project).

(g) **Yeast Artificial chromosome (YAC):** It is a linear vector that behaves like a yeast chromosome. It is capable of replicating in two different hosts (prokaryotes and eukaryotes). It contains ARS (*ori*) sequence for replication; CEN-Y sequence for centromeric function; TEL sequence (telomeric) present at two ends for protection from exonuclease; two selectable marker genes (TRP-1 and URA-3); one selectable marker gene (SUP-4) into which DNA insert is integrated and necessary sequence from *E. coli* plasmid for selection and propagation in *E. coli.*

It can clone large DNA insert of size between 200 kb and 1000 kb. It is also used in the analysis of large genomes (Human Genome Project).

(h) Shuttle Vector: It is constructed for use in both prokaryotes and eukaryotes. It is a vector that can replicate two different organisms. e.g., Yeast Episomal Plasmid (YEP).

(i) Plant Viruses: Several plant viruses such as Cauliflower mosaic virus, Tobacco mosaic Virus, Gemini virus etc are also used as vectors in eukaryotes esp. in plant cells

(j) Animal Viruses: Several animal viruses such as Adenovirus, Bovine Papillomavirus, Retrovirus, Simian Virus-40 etc have also been used as vectors in eukaryotes esp. animal cells.

3. Host organism:

The features of a good host (competent cell) are as follows:

(a) It should grow fast and in large quantity under ordinary cultural method

(b) It should be non-pathogenic

(c) It should support the replication of recombinant DNA.

(d) It should be easy to transform.

(e) It should not be any undesirable elements that interfere with the replication of recombinant DNA.

(f) It should not contain active restriction enzymes.

(g) It should not have methylase enzyme because it could methylate the r-DNA hence it can become resistant to restriction enzyme.

(h) It should have a well-developed genome.

(i) It should secrete recombinant protein from expressed foreign genes.

Usually, *E. coli*, yeast, plant cells and mammalian cells are used as host cells depending upon the aim of the cloning experiment. However, *E. coli* is mostly preferred for eukaryotic genes because

(i) It has a genome size of 4.6 million base pairs.

(ii) It has 2923 genes.

(iii) It is easy to grow in ordinary culture medium at 37°C.

(iv) It shows a rapid rate of multiplication.

(v) It is generally non-pathogenic.

(vi) It is usually sensitive to antibiotics.

4. Bioreactor

It is also called fermentor. It is a large vessel made up of glass, stainless steel or polypropylene used for culturing the host (plant/animal) on a large scale to extract the desired biotechnological product under optimum conditions of temperature, pH, substrate, salt, vitamins, O_2 etc. Its capacity is 100 to 1000 litres. A stirred tank reactor is simple and commonly used. It is a large vessel in which cell culture is stirred mechanically. It is suitable for bacterial and animal cell cultures. It is well suited for the production of primary metabolites (ethanol, acetone, butanol, amino acids) and secondary metabolites (antibiotics, plant hormones).

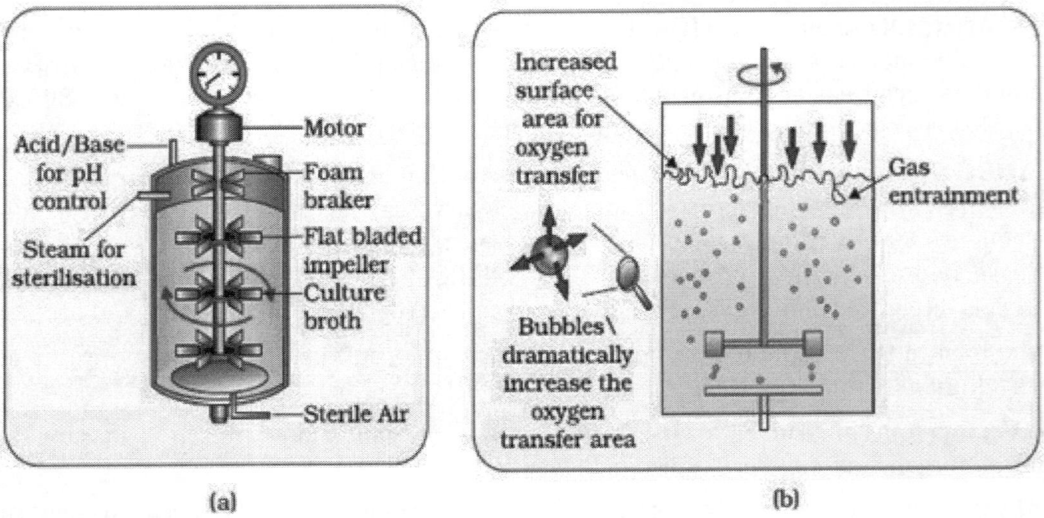

(a) Simple stirred-tank bioreactor; (b) Sparged stirred-tank bioreactor through which sterile air bubbles are sparged

PROCEDURES OF GENETIC ENGINEERING:

1. **Isolation of DNA:** It can be done by the following methods.

(a) **Conventional method:** If the source of DNA is a bacterial cell, then it is treated with a lysosome to dissolve the cell wall. If the source of DNA is a fungal cell, then it is treated with chitinase to dissolve the cell wall. If the source of DNA is a plant cell, then it is treated with cellulase to dissolve the cell wall. If the source of DNA is an animal cell, then it is treated with Sodium Dodecyl Sulphate (SDS) to lyse the cell. The naked protoplast so obtained is purified. In the case of eukaryotes, DNA is enclosed within a membrane, the cell is broken open to release DNA along with RNA and protein. Then RNA is removed by RNase and protein is removed by protease. Now pure DNA is precipitated by adding chilled alcohol. The precipitate appears as fine threads in the suspension which can be removed by spooling. Hence the gene of interest is now isolated. Now centrifugation is done by density gradient centrifugation. The localisation of DNA in the gradient can be done by instrumental method (absorption at 260 nm or visually by fluorescence). Finally, DNA is isolated by inserting a syringe into the rod of a soft plastic centrifuge tube and withdrawing the bands visible in UV radiation.

(b) **cDNA method**: A copy or complementary DNA is used when the eukaryotic gene is to be expressed in prokaryotes. In this process, m-RNA of a desirable gene is isolated from the cell and then the complementary DNA strand of RNA is produced by the reverse transcriptase enzyme. Later DNA: RNA hybrid molecule is formed. Now RNA strand is digested by RNase H or alkaline hydrolysis. The 3′-end of the DNA strand serves as a template for the synthesis of its complementary strand by reverse transcriptase enzyme. Hairpin is opened by Si nuclease. Finally, the C-DNA duplex so produced is used for cloning.

2. **Fragmentation of DNA**: The purified DNA is cut into fragments by restriction endonuclease enzyme under optimum conditions. The different restriction endonuclease enzymes produce different numbers of fragments and one of these contains the gene of interest.
3. **Isolation of desired fragments of DNA**: The desired fragment of DNA is separated by Gel Electrophoresis.

Gel Electrophoresis: It was developed by Tiselius (1937). It is a technique used to separate a mixture of molecules (charged particles) by their differential migration through a stationary phase (gel) in an electric field. It is an efficient tool to separate, identify and purify the DNA and other biomolecules such as amino acids and proteins.

Requirements:

(a) Electrophoresis box or chamber: It contains anode (positive electrode), and a cathode (negative electrode).

(b) Gel: It physically separates the molecules.

(c) A shelf: It is designed to hold gel.

(d) A comb: It is used to form well within gel.

(e) A power supply: It provides electricity that carries the molecules through the gel.

Procedure: It is done in an electrophoresis box which contains an anode (positive electrode), cathode (negative electrode), a shelf designed to hold gel, a comb used to form well within gel and a power supply. It involves the following steps:

1. **Preparation of agarose gel of appropriate concentration:** Agarose is a linear polymer of D-galactose and anhydro L-galactose. It is a purified powder isolated from agar. It has a large pore size which acts as a molecular sieve. This allows the larger molecules to move more slowly than smaller molecules. It forms a gel when dissolved in water and boiled. Agarose gel is cast by boiling agarose in the presence of buffer. A desired concentration of agarose gel (0.8%) is added to electrophoresis buffer TAE (Tris Acetate EDTA). A gel comb is inserted to make a well where DNA will be loaded.
2. **Loading of DNA sample into the well to achieve separation**: The appropriate quantity of DNA is loaded in the well and an electric field (150V DC) is applied across the gel to start electrophoresis. DNA, a negatively charged particle at neutral pH, migrates towards anode at rates determined by its molecular size and conformation. The smaller DNA fragments will travel faster. The speed of migration (electrophoretic mobility) of a molecule depends on its charge and molecular mass. All DNA fragments migrate nearly the same distance into gel and form bands. Gel electrophoresis is carried out in buffers ranging from pH 3-10.

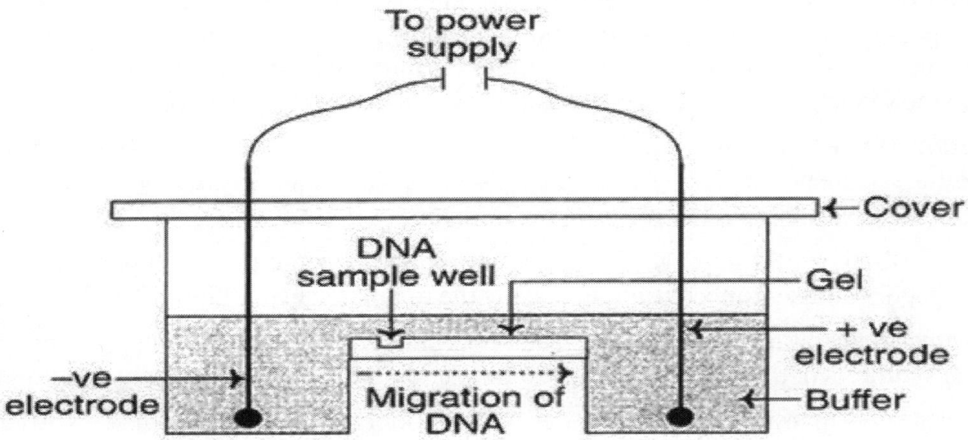

3. **Staining the gel**: The gel is now removed from the chamber and immersed in a solution of sucrose and loading dye (ethidium bromide/methylene blue) for 45 minutes for staining. Sucrose increases the density of DNA preparation and dye increases the visibility of preparation. The tracking dye such as bromophenol is used. The orange bands appear due to ethidium bromide and the blue bands due to methylene blue at room temperature. The separated bands of DNA were cut from agarose gel and extracted from gel piece. This is called elution. Hence DNA fragments are purified for constructing recombinant DNA by joining with a vector.

Advantages:

1. It contains a non-toxic gel medium.
2. Gels are quick and easy to cast.
3. It is good for separating large DNA molecules.
4. It can recover samples by melting the gel.

Polymerase Chain Reaction (PCR): It was developed by Kary Mullis (1985). It is a method of formation of multiple copies of desired DNA fragments in vitro with the help of RNA primer and DNA polymerase. It results in selective amplification of a specific region (target sequence) of DNA. It generates billions of copies of desired DNA in a few hours.

Requirements:

(a) DNA template: It is taken from any source which contains one or more target DNA. The RNA can also be used by first making a DNA copy using reverse transcriptase enzyme.

(b) Two nucleotide primers: Primers are short and single-stranded DNA (20 bp long) that hybridise to target DNA region, one to each strand of duplex is required. These primers should be complimentary and are oriented with their ends facing each other allowing the synthesis of DNA towards one another.

(c) DNA polymerase enzyme: *Taq* polymerase is required. It is isolated from *Thermus aquaticus*, a bacterium. It is DNA DNA-dependent DNA polymerase enzyme. It is stable at 92°C. Other thermostable enzymes such as *Pfu* polymerase (isolated from *Pyrococcus furiosus*) and Vent polymerase (isolated from *Thermococcus litoralis*) are considered more efficient than *Taq* polymerase because they have proofreading function which produces high-fidelity amplification.

(d) Four deoxyribonucleotides (dATP, dGTP, dTTP and dCTP).

(e) Mg^{+2}

Procedure: It is a chain reaction because newly synthesised DNA strands will act as templates for subsequent DNA synthesis. The reaction mixture was prepared containing all four components and Mg^{+2}. It involves the following steps:

1. **Denaturation:** The reaction mixture containing target DNA is heated at high temperature (94°C) resulting in the separation of two strands. Each single strand of target DNA acts as a template for DNA synthesis.
2. **Primer annealing**: The mixture is now cooled to a temperature from 50 to 70°C (depending upon the melting temperature of the expected DNA) allowing the annealing of primer to complementary sequences in DNA. Two sets of small chemically synthesised nucleotides. Primer hybridises to each of single-stranded template DNA because the sequence of primer is complimentary to 3- end of template DNA.
3. **Polymerisation (Primer extension):** *Taq* polymerase synthesises DNA region between the primers using deoxynucleotide triphosphate (dNTP) and Mg^{+2} at 74°C. Hence primer is extended towards each other so that the DNA segment lying between the two primers is copied.

   In the first cycle, DNA is synthesised which extends beyond the target sequence. This is because there is nothing to prevent DNA polymerase from continuing to copy the template beyond the end of the target sequence.

   However, in subsequent cycles, newly synthesised DNA which ends with the primer sequence acts as a template and limits the synthesis of the target sequence so that amplified DNA contains only the target sequence.

   The cycle is repeated many times until the desired level of amplification is achieved. After each cycle, the number of templates is doubled. This number is calculated by this formula: $M_f = M_i \times 2^n$ where M_f is the final number of DNA molecules produced by PCR., M_i is the initial number of molecules or template, and n is the number of cycles performed.

   PCR products have much higher mutations (base substitution) than DNA obtained from in vivo replication because *Taq* polymerase does not have a proofreading ability.

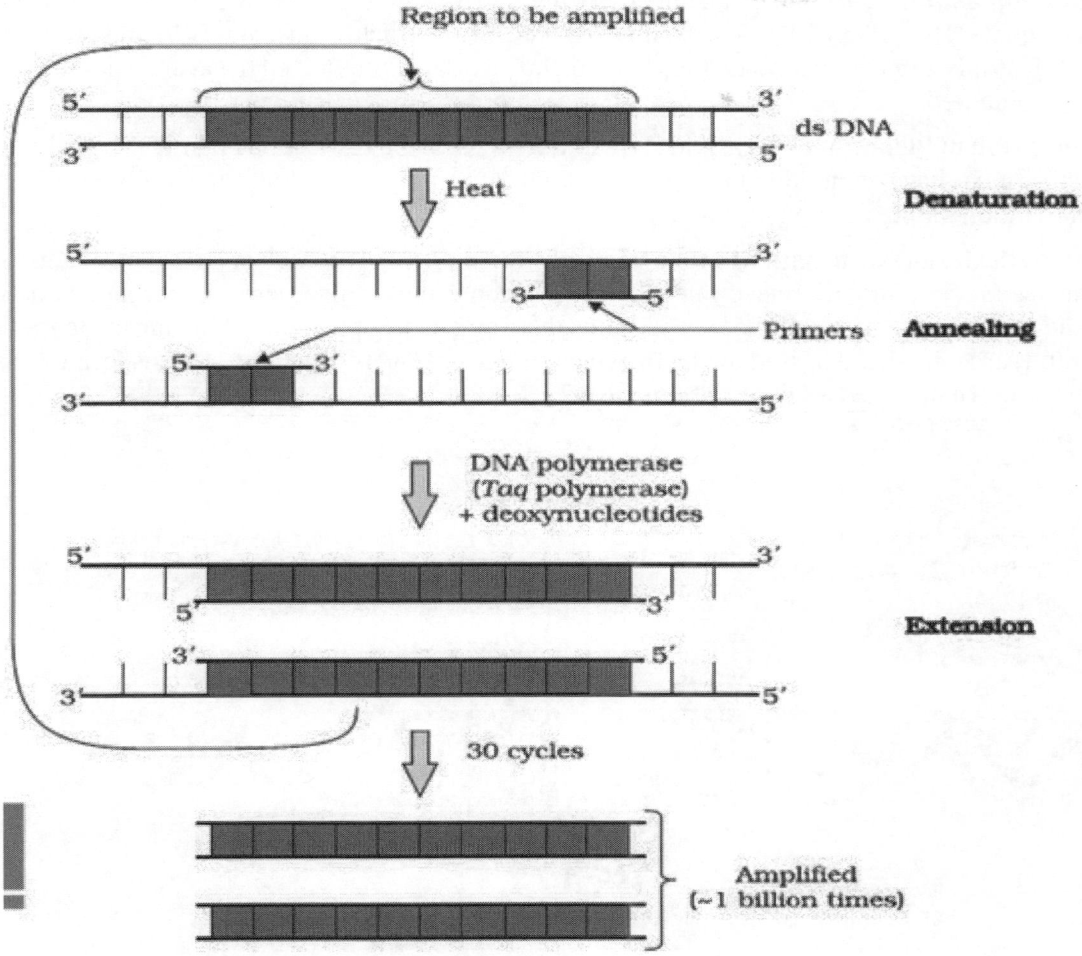

Applications of PCR:

(a) It is used for gene cloning (generation of clones).

(b) It helps in the study of DNA polymorphism as RAPD.

(c) It is useful in DNA fingerprinting, gene therapy, site directed mutagenesis.

(d) It detects bacterial and viral diseases including HIV.

(e) It is useful in antenatal (prenatal) diagnosis of genetic disease or disorders and determination of sex embryos.

(f) It is used in DNA sequencing

(g) It prepares probes in Northern and Southern blotting.

PCR	Gene cloning
It is *in vitro* manipulation.	It is both *in vitro* and *in vivo* manipulation.
It requires DNA.	It requires foreign DNA, restriction endonuclease, vector and host cell.
Its cost is less.	Its cost is more.
It is more efficient.	It is less efficient.
It requires less skill.	It requires more skill.
It takes a few hours.	It takes 2-4 days.
Its error probability is less.	Its error probability is more.
It has more application.	It has less application.

4. **Isolation of plasmid**: It is done by alkaline lysis method. Plasmid is isolated when a bacterial cell is treated with detergents. The cell membrane is solubilised. Plasmid, extrachromosomal DNA and other molecules are released. This mixture is called lysate (supernatant). The lysate is subjected to several chemicals and finally, a plasmid is obtained.
5. **Fragmentation of Plasmid**: The plasmid is treated with the same restriction endonuclease to generate sticky ends which must have complimentary residues similar to foreign DNA but opposite in direction to prevent recircularization of the digested plasmid.
6. **Ligation of desired DNA fragment into a vector to form rDNA**: The ends of both plasmid and foreign DNA are cohesive (sticky) and are mixed under annealing conditions which permits pairing between them. The nicks (the broken covalent bonds between 5-phosphate and 3-OH of two neighbouring nucleotides within the DNA strand) are joined and ligated by the DNA ligase enzyme. The recombinant DNA is formed that combines the genetic information of two different organisms.

Action of Restriction enzyme

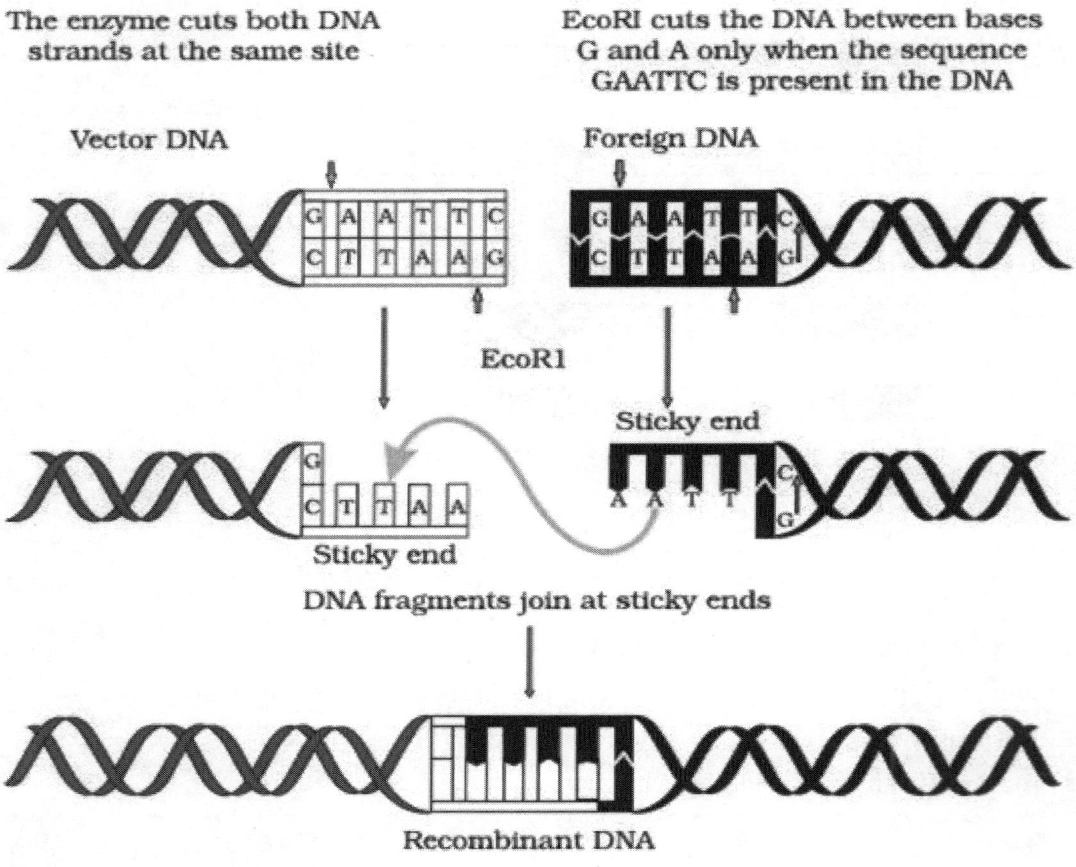

Direct or Vector-less Gene transfer:

(a) Electroporation: It is the most common and efficient method of delivering genes in bacteria, plant, yeast and animal cells. It does not require carrier DNA and is very quick. In this, both DNA and host cell are mixed in a suspension and subjected to an electric field of an optimum voltage (350 V) for a very short time. Thus, small pores develop into the cell wall and plasma membrane of the host cell. Then r-DNA enters into the host cell through pores.

(b) Microinjection: It is a technical, time-consuming and more efficient method. It is easier in animal cells (egg, embryo) due to the lack of a cell wall which interferes with the process of transformation. It is most successful in animal cells (non-vacuolated embryonic). In this, foreign DNA is directly injected inside the nucleus or cytoplasm of plant or animal cells with the help of a micropipette or microneedle under a powerful microscope. In plants, protoplast is preferred for transformation due to the absence of a cell wall. In this, 40-50 protoplasts can only be injected in an hour. It is better to take cells with dense cytoplasm with smaller vacuoles because large vacuoles hinder the delivery of DNA into the nucleus.

(c) **Short Gun/Gene gun/Microprojectile/Biolistic/Ballistic method:** It is a costly method. It is used in the leaf cells of plant cells. It does not require the removal of the cell wall. It is also used to transform shoot apex, embryos, and mature pollen. It is much easier to transform in meristematic cells than in non-dividing cells. In this, DNA is coated with very fine gold or tungsten (1-2μm) particles and shot at the target cell at very high speed. It is used more effectively in producing transgenic plants.

Technique	Developer	Objective
Southern blotting	Edwin Southern	To identify DNA segments
Northern blotting	Alwine	To identify RNA
Western blotting	Burnette	To identify protein

7. **Introduction of recombinant DNA in the host cell (Transformation):** The recombinant DNA is introduced into *E. coli* to select the recombinant DNA from the unchanged vector or obtain many copies of recombinant DNA (Gene cloning) or to synthesise a desired protein (recombinant protein).

   E. coli is not easily transformable and hence poorly accessible to r-DNA because r-DNA is a hydrophilic molecule hence it cannot pass through the cell membrane due to lipid bilayer. Thus *E. coli* is treated with cold $CaCl_2$ under hypotonic conditions at low temperature initially and then the temperature is raised to 42°C for 1-2 minutes so that the size of pores of the cell wall will increase. Now *E. coli* is suspended in 50 mM $CaCl_2$ to take up foreign DNA at a concentration of 10^8-10^{10} cells/ml and these cells are incubated for 12-24 hours to increase the efficiency of transformation. The efficiency of transformation depends on the type of vector used and the type of host cell. The transformation efficiency of 10^5 μg would be adequate for a simple cloning experiment. There is about one transformation per 10,000 plasmids. The rDNA is added, and it enters into *E. coli* hence transformation takes place within a few minutes.

   Transfection: It is the introduction of recombinant DNA in eukaryotic cells esp. animal cells. It is more efficient than transformation. The term transfection was originally coined to describe the introduction of phage DNA into bacterial cells.

8. **Screening (Identification and selection) of Transformed cells:** The following types of bacterial cells are obtained after transformation.

   (a) The majority are non-transformed cells without vectors.

   (b) A proportion of transformed cells are non-recombinant cells with unaltered vectors (without rDNA).

   (c) Rest transformed cells are recombinant cells with recombinant vectors (with rDNA).

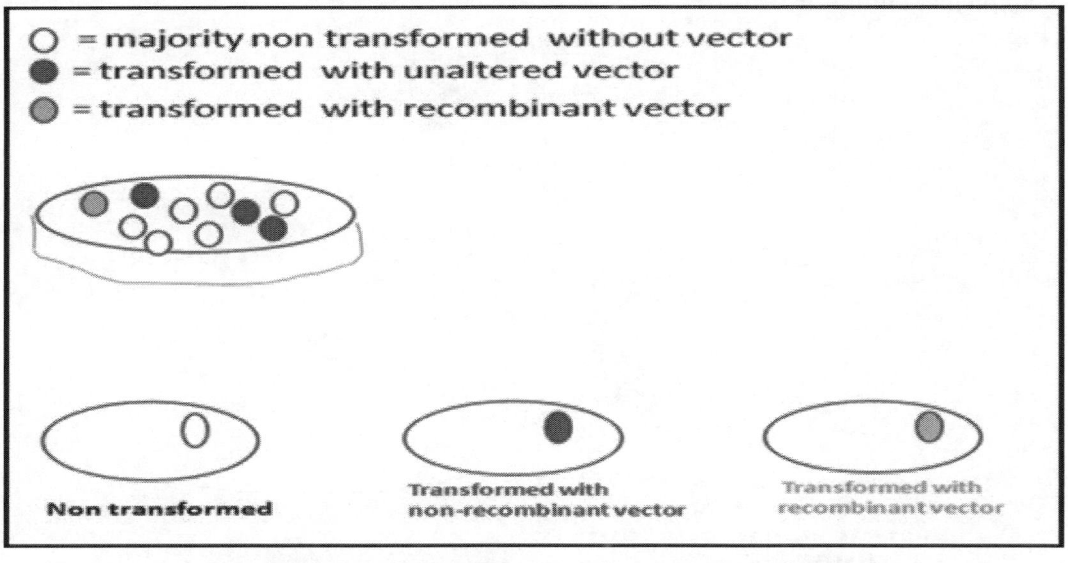

Fig: Master plate showing three types of cells

Direct selection of recombinants:

(a) **Antibiotic resistance screening:** Identification of transformed bacterial cells requires the presence of suitable marker genes on the vector molecule. Marker genes may be either selectable markers or screenable markers. The selectable markers are antibiotic-resistance genes. These are used to differentiate transformed from untransformed cells. A good selectable marker gene provides resistance to the antibiotic which arrests the growth of a particular cell. A host cell strain is selected that is sensitive to a particular antibiotic, ampicillin, tetracycline, or kanamycin. The non-transformed cells are eliminated by integrating a suitable marker gene (reporter gene) into the vector. A marker gene produces a phenotype that helps in the quick selection of transformed cells. When a population of bacterial cells is placed on a medium containing ampicillin, then those cells survive and form colonies that have an ampicillin resistance gene. The reporter gene such as *npt* II (neomycin phosphotransferase II) is a marker gene whose phenotypic expression can be monitored and thus report the activity or change in advance of the effect of modification.

But in the case of vector (pBR322) there are two selectable marker genes in which the DNA insert may be placed within one of the markers, amp^r gene. The other marker tet^r gene is used for the elimination of non-transformed cells.

Insertional inactivation of antibiotic resistance gene: It is the inactivation of a gene upon insertion of another gene in its place or within its coding sequence. It is used to detect the presence of foreign DNA within the plasmid in a transformed cell.

Case:

(a) Non-transformed colonies: These cannot grow in the ampicillin or tetracycline medium.

(b) Transformed colonies: These grow in an ampicillin or tetracycline medium.

(i) Transformed with a non-recombinant or unaltered vector can grow in both ampicillin and tetracycline-containing medium.

(ii) Transformed with a recombinant vector carrying a gene of interest can grow only in ampicillin medium but cannot grow in tetracycline medium due to insertional inactivation.

It occurs by following two methods:

1. Restriction enzyme (*Pst* 1 or *Pvu* 1) is used. If a foreign DNA insert is placed at *Pvu* 1 site of the **ampicillin-resistant** gene, the amp^r gene becomes non-functional. Antibiotic resistance by amp^r gene is lost. Cells do not grow on ampicillin. But recombinant DNA is formed. Cells will grow in the presence of tetracycline antibiotic.

2. Restriction enzyme (*Bam* H1 or *Sal* 1) is used. If a foreign DNA insert is placed at *Bam* H1 site of the **tetracycline-resistant** gene, the tet^r gene becomes non-functional. Antibiotic resistance by tet^r gene is lost. Cells do not grow on tetracycline. But recombinant DNA is formed. Cells will grow in the presence of ampicillin antibiotic.

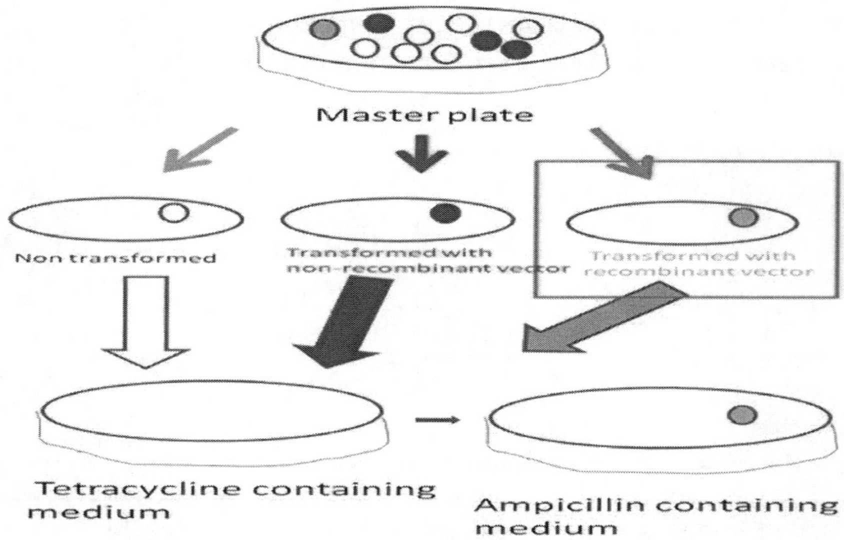

(b) **Blue-white screening**: It is a rapid and efficient technique for the identification of recombinant bacteria. It relies on the activity of β-galactosidase, an enzyme occurring in *E. coli*, which cleaves lactose into glucose and galactose. The presence of lactose in the surrounding medium triggers the *Lac Z* operon in *E. coli*. The operon activity results in the production of a β-galactosidase enzyme that metabolises lactose. This mutant form of protein however may return fully to its active tetrameric state in the presence of an N-terminal fragment of the mutant β-galactosidase by the α-peptide is called α-complementation.

The blue/white screening method works by disrupting this α-complementation process. The recombinants are screened from non-recombinants based on their ability to produce colour in the presence of chromogenic substrate. A screenable marker gene produces an enzyme that gives a specific colour in the presence of a particular substrate.

The presence of active β-galactosidase can be detected by X-gal, a colourless analog of lactose that may be cleaved by β-galactosidase to form 5-bromo-4-chloro-indoxyl which then spontaneously dimerizes and oxidises to form a bright blue insoluble pigment 5,5-dibromo-4,4-dichloro-indigo. This results in a characteristic blue colour in cells containing a functional β-galactosidase. Blue colonies therefore show that they may contain a vector with an uninterrupted *Lac Z* (therefore no insert) while white colonies, where X-gal is not hydrolyzed, indicate the presence of an insert in Lac Zα which disrupts the formation of an active β-galactosidase.

If rDNA is inserted within the coding sequence of the β-galactosidase enzyme, then the enzyme gets inactivated (non-functional). The presence of a chromogenic substrate gives blue-coloured colonies if the plasmid does not have a foreign DNA insert (non-recombinants). If there is no blue colour, then the vector has a foreign DNA insert and such colonies are called recombinants.

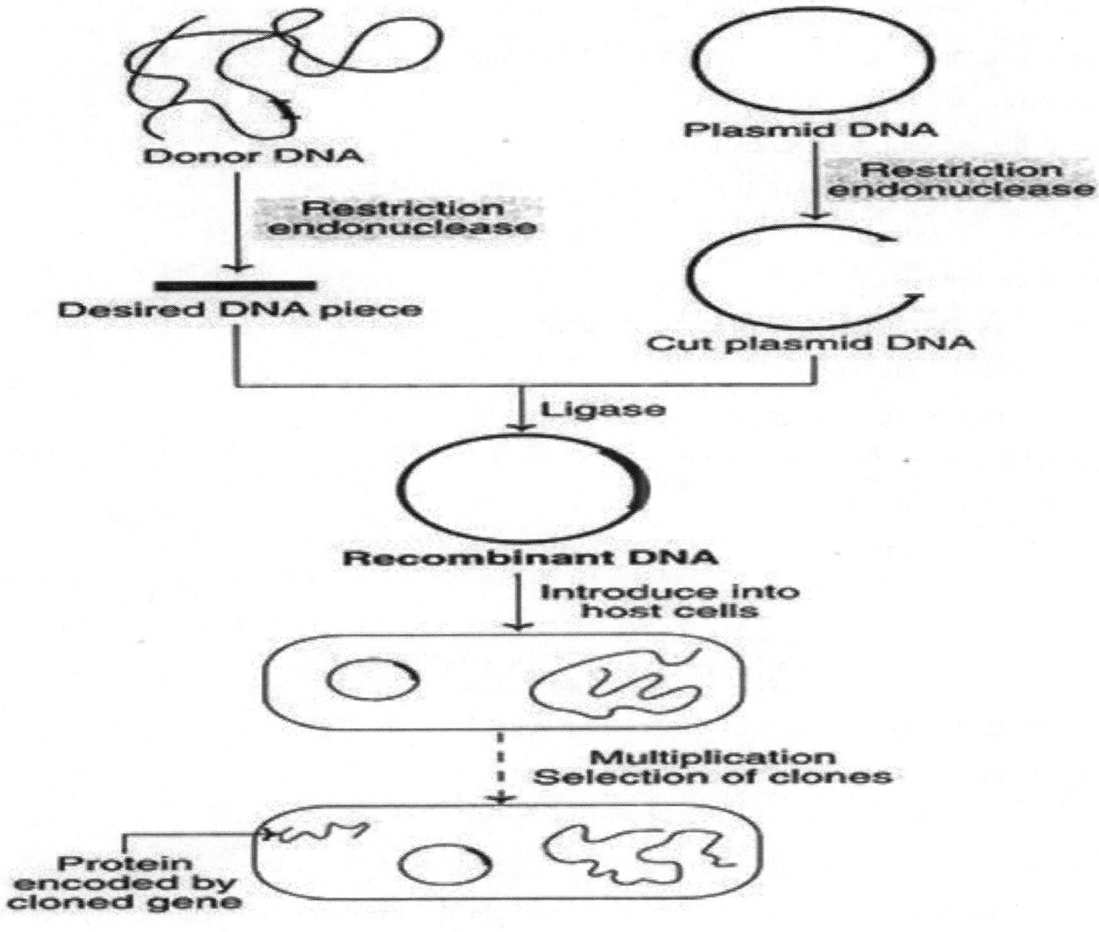

Fig: Process of genetic engineering

9. **Multiplication of Transformed Cell**: The transformed cell is multiplied in continuous culture in a bioreactor containing a suitable medium. The cells reach in physiologically most active phase-log phase where multiplication is rapid. This may be introduced into bacterium for the production of protein encoded by insert. If any protein-encoding gene is expressed in a heterologous host, it is called a recombinant protein. This produces a larger biomass to get a higher yield of desired proteins.

10. **Downstream processing:** After the formation of the product in a bioreactor, it undergoes some processes before being ready for marketing. It involves
    (a) Separation or removal of particles (insoluble): It is the separation of cells, cell debris and other particles from fermentation broth. It can be done by filtration, centrifugation, sedimentation, flocculation, electro-precipitation or gravity settling.
    (b) Isolation of product: It can be done by solvent extraction, adsorption, ultrafiltration, precipitation etc.
    (c) Purification of product: It separates contaminants that resemble the product in physical and chemical properties. It is done by chromatography, crystallisation and fractional precipitation.
    (d) Packaging of Products: It is done by crystallisation, desiccation, lyophilisation, spray drying.

    The product is subjected to quality control testing and kept in suitable preservatives. If a drug is to be manufactured, then it has to undergo clinical trials. The downstream processing and quality control testing are different from product to product.

SHORT ANSWER TYPE QUESTIONS

1. What are the contributions of Paul Berg and Stanley Cohen in genetic engineering?
2. Write notes on pBR322.
3. Distinguish between endonuclease and restriction endonuclease enzymes.
4. Mention five examples of restriction endonuclease enzymes.
5. Write notes on DNA ligase.
6. Distinguish between sticky ends and blunt ends.
7. What are the features of a good vector?
8. Write notes on cosmid.
9. Distinguish between BAC and YAC.
10. What are the features of a good host for the transformation process?
11. Write notes on bioreactor.
12. Write notes on downstream processing.
13. Distinguish between PCR and gene cloning.
14. Write notes on the gene gun.
15. Explain insertional inactivation.

LONG ANSWER TYPE QUESTIONS

1. What are the features of plasmid? Discuss its role in genetic engineering.
2. Describe the different steps in recombinant DNA technology.
3. Discuss the procedure of PCR and mention its significance.
4. Discuss the procedure of gel electrophoresis.
5. Describe the types of restriction endonuclease enzymes and discuss their role in genetic engineering.

MULTIPLE CHOICE QUESTIONS

1. **Lysozyme is used to remove cell walls in case of**
   (a) Fungus (b) Plant
   (c) Bacterium (d) All of the above
2. **The most widely used plasmid is**
   (a) pTi (b) pBR 322
   (c) pRi (d) pUC
3. **A vector specially designed for cloning 45 kb DNA fragments is**
   (a) Plasmid (b) Cosmid
   (c) Phagemid (d) pBR322
4. **The vector accommodating the largest size of DNA is**
   (a) Plasmid (b) YAC
   (c) Cosmid (d) BAC
5. **Which of the following is a palindrome?**
   (a) CCGTA/GGCAT
   (b) GAATTC/CTTAAG
   (c) CCCC/GGGG
   (d) CATTG/GTAAC
6. **Which one of the following is not a PCR-based molecular marker?**
   (a) RAPD (b) RFLP
   (c) AP-PCR (d) DNF
7. **Restriction enzymes cut DNA at specific sites known as**
   (a) Telomeric sequences
   (b) Palindromic sequences
   (c) Terminator sequences
   (d) Attenuator sequences
8. **The enzyme involved in the self-splicing of introns in *Tetrahymena thermophila* was**
   (a) Primase
   (b) Reverse transcriptase
   (c) Ribozyme
   (d) RNA polymerase
9. **The most commonly used method of choice of gene transfer using immature embryos as explants is**
   (a) Microprojectile
   (b) Electroporation
   (c) Liposome mediated
   (d) Chemically stimulated
10. **The frequency of *Agrobacterium-mediated* genetic transformation is increased by**
    (a) Ascorbic acid (b) Aflatoxin
    (c) Acetosyringone (d) Abscisic acid
11. **A shuttle vector is defined as**
    (a) A plasmid capable of replicating in one organism
    (b) A plasmid capable of accepting only large DNA insert
    (c) A plasmid incapable of replicating in cells
    (d) A plasmid capable of replicating in two different organisms.
12. **The selectable marker that provides resistance to the antibiotic Kanamycin is**
    (a) Gentamicin acetyltransferase (gat)
    (b) Hygromycin phosphotransferase (hpt)
    (c) Streptomycin phosphotransferase (spt)
    (d) Neomycin phosphotransferase (npt II)
13. **The pBR322, a frequently used vector for cloning genes in *E. coli* is**
    (a) An original bacterial plasmid
    (b) A modified bacterial plasmid
    (c) A viral genome
    (d) A transposon
14. **The gene promoter sequence commonly used in gene cloning experiments are derived from**
    (a) TMV (Tobacco Mosaic Virus)
    (b) TSV (Tobacco Satellite Virus)
    (c) CMV (Cucumber Mosaic Virus)
    (d) CaMV (Cauliflower Mosaic Virus)
15. **Ti plasmid used in genetic engineering is obtained from**
    (a) *Bacillus thuringiensis*
    (b) *Agrobacterium rhizogenes*
    (c) *Agrobacterium tumefaciens*
    (d) *Pseudomonas putida*
16. **Genetically engineered male-sterile crop plants have been produced by inserting**
    (a) Opaque 2 gene
    (b) Virus coat-protein gene
    (c) Barnase gene
    (d) Chitinase gene

17. **Transgenic plants can act as living bioreactors for the production of chemicals and pharmaceuticals. This is known as**
(a) Organic farming
(b) Molecular farming
(c) Pharmacotechnology
(d) Hybridoma technology

18. **DNA coated with tungsten or gold particles is used in direct or vector less transformation in**
(a) Biolistic method (b) Electroporation
(c) Microinjection (d) Lipofection

19. **The enzyme used in the PCR technique is**
(a) RNA polymerase
(b) DNA polymerase
(c) Reverse transcriptase
(d) *Taq* polymerase

20. **Which one of the following is not a thermostable enzyme?**
(a) *Taq* DNA polymerase
(b) RNA polymerase
(c) *Pfu* DNA polymerase
(d) Vent polymerase

21. **Crown gall disease was caused by**
(a) Bacteria (b) Pi plasmid
(c) Ti plasmid (d) Virus

22. **Gel electrophoresis is used for**
(a) Isolation of DNA
(b) Fragmentation of DNA
(c) Separation of DNA
(d) Ligation of DNA

23. **Restriction enzyme *Eco*R1 cleaves DNA at the sequence**
(a) AAGCTT (b) GAATTC
(c) TATAGC (d) CGTACG

24. **The blunt ends are formed by**
(a) *Alu* I (b) *Taq* I
(c) *Hind* II (d) All of the above

25. **Plasmid can clone DNA insert up to the size of**
(a) 10 kb (b) 25 kb
(c) 45 kb (d) 200 kb

26. **Which of the following vectors is hybrid?**
(a) Plasmid (b) Cosmid
(c) BAC (d) YAC

27. **Who first used a gene of SV-40 in the experiment of genetic engineering?**
(a) Arber, Nathans and Smith
(b) Stanley Cohen
(c) Boliver and Rodriguez
(d) Paul Berg

28. **The vector used for ligation of foreign DNA in eukaryotes is**
(a) Cosmid (b) BAC
(c) Plant virus (d) pBR[322]

29. **The first restriction endonuclease enzyme to be discovered was**
(a) Eco RI (b) Hind II
(c) Hind III (d) Alu I

30. **The loading dye used in the agarose gel electrophoresis is**
(a) Acridine dye (b) Safranin
(c) Acetocarmine (d) Ethidium bromide

31. **Which of the following processes is not associated with PCR?**
(a) Denaturation
(b) Autoradiography
(c) Annealing
(d) Extension

32. **During transformation, *E. coli* was treated with**
(a) $CuSO_4$ (b) $HgCl_2$
(c) $NaNO_3$ (d) $CaCl_2$

33. **Molecular glue used in the experiment of genetic engineering is**
(a) Restriction endonuclease
(b) DNA polymerase
(c) Helicase
(d) DNA ligase

34. **The process in which separated bands of DNA are cut out from agarose gel and extracted from the gel piece is called**
(a) Eluviation (b) Alluviation
(c) Illuviation (d) Elution

35. **Genomic DNA is digested with *Alu* I, a restriction enzyme that is a four-base cutter. What is the frequency with which it will cut the DNA assuming a random distribution of bases in the genome**
(a) 1/16 (b) 1/32
(c) 1/64 (d) 1/256

36. When the DNA fragments are observed under UV light, they are seen as

(a) Orange-coloured bands
(b) Yellow-coloured bands
(c) Blue-coloured bands
(d) All of the above

37. In gel electrophoresis, the restriction enzyme digested DNA is loaded in a well near

(a) Anode
(b) Cathode
(c) Centre of gel
(d) Anywhere in the gel

38. In insertional inactivation, the recombinant DNA is inserted with the coding sequence of

(a) Restriction enzyme
(b) Tetracycline resistant gene
(c) Ampicillin resistant gene
(d) Beta-galactosidase enzyme

39. Recombinant colonies in inserted inactivation are differentiated based on showing

(a) Blue colour (b) Red colour
(c) Orange colour (d) No colour

40. A selectable marker gene is incorporated into a plasmid to

(a) Increase its copy number
(b) Increase its transformation efficiency
(c) Eliminate the non-transformants
(d) Increase the expression of the gene of interest

41. Foreign DNA cannot pass through cell membrane because it is

(a) Hydrophilic
(b) Hydrophobic
(c) Rich in protein
(d) Heavy

42. A single strand of nucleic acid tagged with a radioactive molecule is called

(a) Plasmid (b) Vector
(c) Prion (d) Probe

43. PCR involves the use of *Taq* polymerase enzyme which is

(a) Thermostable (b) Thermolabile
(c) Psychrophilic (d) All of the above

44. A method used to distinguish the DNA of one individual from another is

(a) DNA sequencing (b) Hybridisation
(c) RFLP (d) RAPD

45. T_4 polynucleotide kinase is used for

(a) Labelling 3′ ends of DNA
(b) Labelling 5 ends of DNA
(c) creating blunt ends of DNA
(d) Dephosphorylation of DNA

46. A plasmid cloning vector shall contain all of the following except

(a) Origin of replication
(b) Inducible promoter
(c) Selectable marker gene
(d) Multiple cloning sites

47. Which one of the following sequences is most likely to be a restriction enzyme recognition site?

(a) CGGC (b) CGC
(c) GTAATG (d) GTCGAC

48. If bacterial genome and plasmid are allowed to replicate in the same manner, then

(a) Plasmid genome will replicate first
(b) Bacterial genome will replicate first
(c) Both will replicate at the same time
(d) Depends on GC content

49. Which of the following restriction endonuclease enzymes is used in genetic engineering?

(a) Type I (b) Type II
(c) Type III (d) All of the above

50. The polymerase chain reaction was discovered by

(a) Stanley Cohen
(b) Paul Berg
(c) Arber, Nathans and Smith
(d) Kary Mullis

ANSWERS

1.c	11.d	21.c	31.b	41.a
2.b	12.d	22.c	32.d	42.d
3.b	13.b	23.b	33.d	43.a
4.b	14.d	24.d	34.d	44.c
5.b	15.c	25.a	35.d	45.b
6.d	16.c	26.b	36.c	46.b
7.b	17.b	27.d	37.c	47.d
8.c	18.a	28.c	38.d	48.a
9.b	19.d	29.b	39.d	49.b
10.c	20.b	30.d	40.c	50.d

CHAPTER-15

BIOTECHNOLOGY AND ITS APPLICATIONS

The methods of genetic modifications are used to develop new plant, animal, and microbial strains for use by human beings.

Types of genetic modification methods for crops:

1. Traditional methods like selective breeding and crossbreeding
2. Recombinant DNA technology or Genetic engineering
3. Genome editing: It is a new method that involves the removal of unwanted genes.

ROLE OF BIOTECHNOLOGY IN AGRICULTURE

Production of transgenic plants: Transgenic plants are those plants that have been genetically modified utilizing recombinant DNA technology (rDT) to express exogenous genes or modify endogenous genes. These plants contain foreign genes of different species inserted artificially at the level of protoplast, gamete, or zygote. Biotechnology has produced more transgenic crops in a shorter period. These have been produced in more than 60 plants of angiosperms esp. dicots. The transgenic plant has been produced in only one gymnosperm species (e.g., *Picea*). Tobacco was the first transgenic plant (antibiotic-resistant) produced in 1982. Bt-brinjal was the first transgenic plant developed in India by Maharashtra Hybrid seed company. Bt-cotton is the only transgenic plant that has been approved for commercial cultivation in 2002 in India. A genetically modified organism (GMO) contains any foreign gene altered by any method. The first genetically modified organism was bacteria in 1973.

Objectives of transgenic plants / GMOs:

(a) To introduce a new trait to the plant (not naturally occurring)
(b) To bring together useful genes from a wide range of living species in one plant
(c) To increase yield or productivity
(d) To develop features like insect resistance, disease (virus) resistance, herbicide resistance and nematode resistance
(e) To tolerate heat, cold, salt and drought conditions
(f) To increase the efficiency of minerals uptake by plants
(g) To reduce post-harvest loss

Transgenic plants are useful in several ways:

1. **Insect resistant**: The crop plants such as cotton, corn, rice, wheat, and soya bean have been engineered to express the insecticidal toxin gene of *Bacillus thuringiensis* so that insects attempting to eat these plants get killed. *Bacillus thuringiensis* is a Gram-positive, aerobic, motile, peritrichous, endospore-forming soil bacterium. It is species-specific and does not harm other organisms. It contains about 40 *cry* genes in its large plasmid. These genes have been cloned from bacteria and introduced in plants. It forms a solid, intracellular, insoluble and toxic crystal protein (α, β, γ-exotoxins and δ-endotoxin in crystalline form) in the vegetative cell during endospore formation. This Bt toxin does not kill *Bacillus* because the toxin is present in the form of inactive protoxin. The transgenic plants having *cry* genes act as bioinsecticides. The insecticide protein is secreted as an inactive protoxin. The insects ingest inactive protoxin and reach their midgut where alkaline pH due to protease enzyme solubilises it and converts it to active toxin. The toxic function occurs in the N-terminal of protein and the C-terminal is involved in forming crystal protein. The active toxin binds to the epithelial cell of the midgut of the insect. Then lysis of the cell occurs due to osmotic imbalance. The insects die within 30 minutes to 3 days after ingestion. Bt- toxin does not harm fish, birds, humans and other mammals.

Procedure: It involves the following steps:

(a) Bt-gene that produces the desired lethal gene is isolated.

(b) Bt-gene is ligated to a marker gene to form a recombinant gene for insect resistance.

(c) The recombinant gene is inserted into plant cells.

(d) Plant cells are grown in the presence of antibiotics.

(e) Plant cells that carry the recombinant gene survive and are grown into plants.

e.g., Bollgard for cotton

Types of gene	Target
Cry I A(a)	Butterfly (lepidopteran)
Cry I A(b)	Corn borer (lepidopteran)
Cry I A(c)	Boll worm (lepidopteran)
Cry I B	Moth and butterfly (lepidopteran)
Cry I C	Moth and butterfly (lepidopteran)
Cry I D	Moth and butterfly (lepidopteran)
Cry I E	Moth and butterfly (lepidopteran)
Cry II A(b)	Bollworm (lepidopteran)
Cry II B	Flies and mosquitoes (Dipteran
Cry III A, B, C	Beetle and boll weevil (Coleopteran)
Cry IV A, B, C, D	Black flies and mosquitoes

2. **Nematode resistant**: It was discovered by Fire and Melo. Transgenic plants such as tomato, tobacco, brinjal etc. have been developed which are resistant to a nematode (*Meloidogyne incognita*) which causes root-knot disease in which plants cannot absorb water and minerals and thus roots develop cracks causing secondary infections. Nematode resistance in tobacco is based on RNA interference (RNAi) technique. All eukaryotic organisms use RNAi as a method of cellular defence.

   Procedure: It involves the following steps:

   (a) Nematode-specific gene is isolated.

   (b) It is ligated to *Agrobacterium* vector to form a recombinant gene.

   (c) The recombinant gene is introduced into the host tobacco.

   (d) Nematode-specific gene is triggered to produce single-stranded RNA which is complimentary to RNA produced by nematode to infect the host tobacco plant.

   (e) This complimentary RNA binds to the mRNA of the nematode to form double-stranded RNA. The source of ds-RNA may be infection of ribovirus or transposon.

   (f) The ds-RNA is fragmented into short interference RNA (siRNA).

   (g) The siRNA develops a silencing complex which prevents specific mRNA of nematode from performing translation.

   (h) The m-RNA becomes toxic. The nematode cannot survive in transgenic Tobacco host expressing specific interfering RNA (RNAi). Thus, the nematode is unable to infect the transgenic tobacco plant, and it gets protected from the parasite (nematode).

3. **Virus resistant**: The first transgenic virus-resistant plant was Tobacco produced in 1986.

   (a) Transgenic plants such as tobacco, tomato, and potatoes resistant to Tobacco Mosaic Virus (TMV) were developed due to the introduction of a gene for viral coat protein into plants. When these plants were inoculated with TMV then symptoms failed to develop. The resistance occurs due to blocking of the process of uncoating of virus particles which is essential for replication and expression of the viral genome.

(b) Transgenic plants such as tobacco and tomatoes resistant to the Tobacco Rattle Virus were developed due to the introduction of a gene for nucleoprotein into plants.

(c) Transgenic plants such as tobacco and tomatoes resistant to the Tobacco Satellite Virus were developed due to the introduction of a gene for viral satellite RNA into plants.

4. **Herbicide resistant**: These plants tolerate herbicides which kill the surrounding flora but leave cultivated crop plants intact.

(a) Transgenic *Petunia* resistant to herbicide, **Glyphosate** was developed due to the insertion of a gene for Enol pyruvyl shikimate Phosphate synthetase (EPSPS) of *Salmonella typhimurium* into vector, *Agrobacterium*. Its herbicide action is broad spectrum. It inhibits EPSPS and kills plants by interfering with the shikimate pathway in plants.

(b) Transgenic tobacco resistant to **atrazine** was developed due to the insertion of gene for the QB protein of *Arabidopsis* into a vector, *Agrobacterium*. Its herbicide action is broad spectrum.

(c) Transgenic tobacco, corn, wheat and soybean resistant to **phosphinothricin** were developed due to the insertion of the *bar* gene of *Streptomyces* into a vector, *Agrobacterium.* Its herbicide action is broad spectrum.

(d) Transgenic tobacco resistant to **Salphonyl urea** was developed due to the insertion of the acetolacetate synthase gene of *Arabidopsis*. Its herbicide action is selective.

5. **Nutrient-rich (Golden rice)**: It was developed by Ingo Potrykus (1990). It is yellow or orange coloured due to the presence of beta carotene. It is genetically enriched with β-carotene to combat vitamin A deficiency.

Procedure:

(a) The three genes (*psy* gene coding for the enzyme, phytoene synthase; *lyc* gene for lycopene cyclase from daffodil (*Narcissus pseudonarcissus*) and *crt*1 gene coding for carotene desaturase from a bacterium (*Erwinia uredovora*) are isolated.

(b) These genes are ligated into the Ti plasmid of *Agrobacterium* to form rDNA.

(c) The rDNA is introduced into *Agrobacterium.*

(d) Transformed bacteria are cultured and transferred to Petri dish containing embryo.

(e) These bacteria infect rice embryos resulting in the transfer of desirable gene necessary for the synthesis of β-carotene.

(f) Thus, golden rice shows golden-coloured grains.

Chemicals involved: Isoprenoid unit→Geranyl diphosphate (Carotenoid) → Phytoene → Lycopene→ β-carotene → Retinal → Retinol (Vitamin A)

6. **Delayed ripening**: Flavr Savr is a transgenic tomato which was first introduced in the USA in 1994. Ripening or softening of fruit occurs due to the breakdown of polygalacturonic acid, a component of the cell wall of the pericarp of fruit, by an enzyme polygalacturonase. The softening makes it more susceptible to being damaged by fungal infections. Ripening in tomatoes is delayed due to a lack of a sufficient amount of polygalacturonase enzyme. Tomato stays firm when fully ripened.

It is based on antisense transcription (normally m-RNA is synthesised on the sense strand of DNA. If it occurs on a complementary strand of the sense strand, then it is called antisense transcription). It is the first commercially grown genetically modified food crop and it was made more resistant to rotting by Californian company, Calgene.

Procedure:

(a) Firstly, an antisense copy of the gene encoding the polygalacturonase enzyme of tomato is isolated.

(b) It is ligated to *Agrobacterium tumefaciens* to form rDNA.

(c) Now rDNA is introduced in fresh tomato.

(d) Antisense copy of gene forms antisense RNA (opposite to sense RNA).

(e) Complimentary sense and antisense RNA bind together in cells forming dsRNA.

(f) It blocks translation.

(g) Then the usual amount of polygalacturonase enzyme is not formed in tomato.

Thus, it will delay the ripening or softening of fruit hence it will show increased shelf life, increased fungal resistance and more nutrients that add flavour.

7. **Biodegradable plastic**: Transgenic *Arabidopsis* is developed due to the insertion of the Polyhydroxybutyrate (PHB) gene of *Alcaligenes eutrophus* for the production of biodegradable plastic.
8. **Molecular farming**: Transgenic potato, tomato and tobacco plants are used as bioreactors or factories for manufacturing pharmaceuticals such as sugar, fatty acid, starch, cellulose, wax etc.
9. **Pest resistant**: The transgenic tobacco plant is developed due to the insertion of the Cow Pea Trypsin inhibitor (CpTI) gene.
10. **Hybrid seed**: Transgenic *Brassica napus* was produced for the production of hybrid seed. The male sterile plant was produced due to the insertion of a *barnase* gene (encoding a ribonuclease) derived from *Bacillus amyloliquefaciens.* Likewise, a fertility restorer plant was produced due to the insertion of the *barstar* gene (encoding ribonuclease inhibitor) derived from *Bacillus amyloliquefaciens.*

ROLE OF BIOTECHNOLOGY IN MEDICINE

Production of Therapeutic Drugs: The recombinant DNA technology is used to produce around 30 recombinant drugs (therapeutic drugs) for human use. These drugs do not induce immunological reactions like some other drugs obtained from non-human sources. These are produced from genetically engineered microbes esp. *E. coli* and yeast). The genes are expressed in animal cell culture.

Recombinant proteins	Application
Humulin	Teatment of diabetes mellitus
Blood clotting factor-VIII	Treatment of haemophilia-A
Blood clotting factor-IX	Treatment of haemophilia-B
Reo pro	Prevention of blood clot
Tissue plasminogen activator (tPA)	Treatment of heart disease by thrombolysis (it dissolves blood clot)
Urokinase	Treatment of heart disease by thrombolysis (it dissolves blood clot)
OKT-3	Reversal of acute kidney transplantation
Asparaginase	Treatment of cancer
Tumour necrosis factor	Treatment of cancer
Interleukin	Treatment of cancer
DNase	Treatment of cystic fibrosis
Monoclonal antibodies	Diagnosis and treatment of diseases
Interferon	Treatment of Hepatitis-C, cancer, AIDS, viral infection
Hepatitis-B surface antigen (HBSAg)	Treatment of hepatitis-B
Hirudin	Anticoagulant
Erythropoietin	Treatment of anaemia
Human growth hormone	Treatment of dwarfism
Calcitonin	Treatment of rickets
Chorionic gonadotrophin	Treatment of infertility
Epidermal growth factor	Treatment in wound and burn healing
Platelet derived growth factor	Treatment of Diabetic ulcers
Antitrypsin	Treatment of emphysema

1. **Production of Humulin (Human insulin)**: Insulin is produced by β-cells of the islet of Langerhans of the pancreas which controls blood sugar levels. In the human body, two genes A and B genes produce polypeptides independently which undergo association in the cytoplasm. Humulin was synthesised by Tsan in 1978.

Procedure:

(a) m-RNA synthesising pro-insulin is extracted from pancreatic cells.

(b) cDNA (gene) is synthesised by reverse transcriptase enzyme using mRNA.

(c) cDNA is ligated to pBR322 (a vector) to form rDNA.

(d) Now recombinant DNA is introduced into *E. coli* (transformation). Transformed bacterial cell divides rapidly every 20 minutes. Hence numerous transformed cells containing r-DNA are produced within a short period.

(e) Transformed bacterial cells for the presence of rDNA are selected using a marker gene (ampicillin).

(f) Transformed cells are grown in a large bioreactor. Lactose is used as an inducer to switch on the *Lac* Operon linked to the gene of interest in *E. coli.*

(g) The hybrid protein containing β-galactosidase and proinsulin is isolated from this culture and purified. The hybrid protein is then treated with cyanogen bromide to separate β-galactosidase and pro-insulin. Finally, pure human insulin (humulin) is formed.

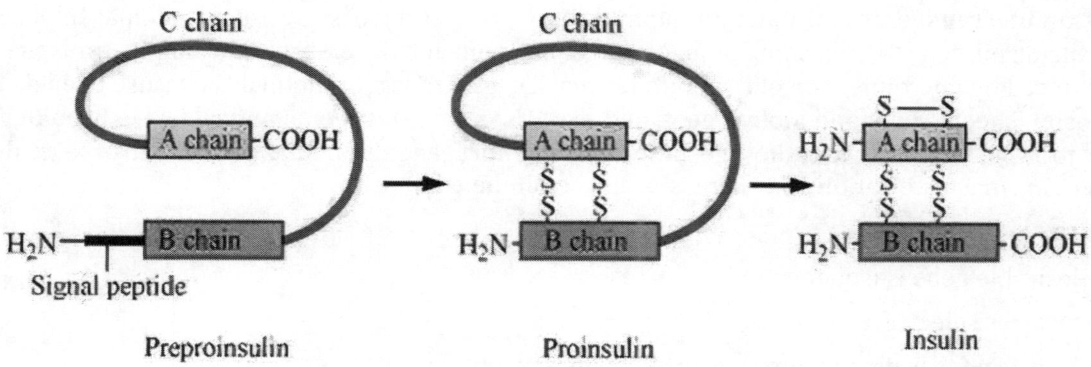

The freshly synthesised polypeptide is pre-proinsulin. It is composed of 109 amino acids in a chain. In the first step, pre-proinsulin undergoes processing, loses about 23 amino acids and becomes proinsulin. In the second step, it undergoes denaturation using trypsin and carboxypeptidase hence losing 35 amino acids to become functional insulin. Functional insulin is made up of 51 amino acids arranged in two polypeptide chains (A chain contains 21 amino acids and B chain contains 30 amino acids) held by three disulphide bonds.

N-terminal-------------B-chain------------C-chain------------A-chain---------C-terminal

Preproinsulin (109 amino acids)

↓ -23 amino acids

Proinsulin (86 amino acids)

↓ -35 amino acids

Insulin (51 amino acids)

2. **Gene therapy**: It is the treaccctment of disease by inserting a normal, new and functional extra gene into a cell or tissue of an individual with the intention that the gene product will play a therapeutic role. Michaele Blease and French Anderson performed the first clinical gene therapy successfully in 1989 on a human patient (a 4-year-old girl) suffering from Severe Combined immuno-deficiency (SCID). This disease is caused by the absence of adenosine deaminase (ADA) an enzyme involved in purine metabolism and is essential for the immune system to function. Its deficiency destroys selectively T- and B-cells of the immune systems leading to SCID (Bubble Boy disease) in which patients die of ordinary disease due to lack of immunity. It is caused by a defect in the gene for expressing the enzyme, Adenosine deaminase.

Procedure:

(a) A defective gene is identified.

(b) Normal healthy gene encoding Adenosine Deaminase (ADA) is either isolated or synthesised.

(c) It is ligated to a vector (retrovirus) forming recombinant DNA.

(d) A functional recombinant ADA gene is introduced in lymphocytes isolated from the blood or bone marrow of the patient at the embryonic stage to function. The difficulties of gene therapy include the choice of vector being used.

Use:

Pati**ent therapy**: Cells with healthy genes are introduced in affected tissue so that healthy gene overcomes the effect without affecting the inheritance of the patient.

ROLE OF BIOTECHNOLOGY IN INDUSTRIES

Microbes have been genetically engineered for the production of enzymes, hormones, organic acids, antibiotics, Single cell protein, alcoholic beverages, and biofuels such as ethanol and butanol. These microbes are also useful in the extraction of minerals by bacterial leaching.

ROLE OF BIOTECHNOLOGY IN ANIMAL BIOTECHNOLOGY:

1. **Production of Transgenic Animals:** Rudolph Jaenisch created the first transgenic animal (mouse) in 1980 by introducing foreign DNA encoding human growth hormone into its embryo. Most of the transgenic animals were produced by microinjection of DNA into fertilised eggs. Transgenic animals have also been developed in cows (Rosie), sheep (Polly and Molly), goat, and fish. Maximum transgenic animals have been developed in mice due to a short estrous cycle, short gestation period, short generation time, production of several offspring, convenient *in vitro* fertilisation and successful culture of the embryo.

   Procedure:

   (a) The desirable gene is isolated.

   (b) The vector is selected.

   (c) The desired gene is ligated into a vector to form recombinant DNA.

   (d) The adult female is injected with hormone for superovulation (collection of eggs). The eggs are mated to form fertilised egg.

   (e) The recombinant DNA is injected into a fertilised egg (transformation).

   (f) The transformed fertilised egg is allowed to grow for some time.

   (g) The selected transformed egg is implanted into the uterus of a surrogate mother.

   Importance: The transgenic animals are used

   (a) To study the expression of normal genes on normal body function and development.

   (b) To test the toxicity of drugs

   (c) To test the vaccines before use on human beings.

   (d) To serve as a model to study the role of genes for various human diseases such as cystic fibrosis, cancer, Alzheimer's disease, and rheumatoid arthritis.

   (e) To serve as a model to exhibit the disease symptoms to treat the diseases.

   (f) To produce a desired substance (recombinant protein) in large quantities (bioreactors)

   (g) To Produce organs for organ transplantation.

2. **Cloning:** A clone (identical organism) is produced by the fusion of the nucleus of an adult somatic cell from one sheep with an unfertilised enucleated egg from another followed by implantation into a surrogate mother. Dolly (clone of sheep) was successfully created by Ian Wilmut in February 1996 after failures of 277 times. This lamb was exactly similar to her mother. Other examples are George and Charlie (cloned calves) and ANDI

(the World's first genetically altered primate, monkey) produced by inserting a jellyfish gene into the embryo of a rhesus monkey.

Procedure:

(a) Mammary cell was isolated from donor sheep (6 years old) and cultured *in vitro.*

(b) The egg cell was isolated from another donor sheep and its nucleus was removed.

(c) Both cells were fused and this fused cell contains the nucleus of the mammary cell.

(d) This fused cell was grown (cultured *in vitro*) into an early embryo up to the blastula or morula stage.

(e) The embryo was implanted into the uterus of a third sheep (surrogate mother) who has been treated hormonally to stimulate pregnancy. Finally surrogate mother gave birth to a normal lamb which grew into a normal adult sheep (Dolly) that is genetically identical to the mammary cell donor.

Importance: The cloned animals are used in

(a) Multiplication of desirable genotype without the risk of variation

(b) Understanding of curing genetic disease

(c) Genetic research.

ROLE OF BIOTECHNOLOGY IN THE ENVIRONMENT

Genetically engineered microbes are used for

1. Pollution control: The Oil spill in a water body is cleared by digesting hydrocarbons of crude oil. It is done by developing a superbug or oil-eating bug (as developed by Anand Mohan Chakrabarty) due to the introduction of plasmid from different strains into a single cell of *Pseudomonas putida*. The superbug degrades oil (hydrocarbon) into non-polluting and harmless chemicals.
2. Detoxification of waste and effluents (waste treatment)
3. Removal of xenobiotics (non-biological wastes)
4. Restoration of degraded land.

Hazards of genetic engineering (Disadvantages of genetic engineering):

(a) The genetically modified organisms for herbicide and pest resistance can transfer the genes into sexually compatible wild relatives or weeds that acquire resistance to become superweeds and cause more harm.

(b) These may cause allergic reactions because most allergens are protein.

(c) These may disturb the local ecosystem by inhibiting or promoting the growth of indigenous plants.

(d) The transgenic plants for pest resistance may be toxic to human beings because these produce toxic substances.

(e) These plants may produce harmful pathogens in the environment.

(f) These plants may also carry undesirable features such as necrosis, sterility and reduced growth along with desirable features.

(g) These show the expression of many transgenes at low levels which usually limit their usefulness.

Safeguards of Genetic Engineering and Bioethics:

(a) Care should be taken to avoid accidental escape of dangerous genetically modified microbes.

(b) Strict laboratory procedures should be followed to protect people from infection by genetically modified microbes

(c) Extreme precautions should be taken while cloning harmful genes.

(d) Prior approval should be required from the concerned Govt. authority while doing research work.

Regulatory agencies of India

(a) Recombinant DNA Advisory Committee (**RDAC**): To evolve guidelines for research involving RDT.

(b) Review Committee on Genetic Manipulation (**RCGM**): To recommend special suitable conditions for experiments and trials.

(c) Genetic Engineering Approval Committee (**GEAC**): To monitor large-scale manufacturing and release of genetically modified organisms into the environment; import and export of genetically modified organisms.

Biosafety issues: Modern Biotechnology has resulted in a legal characterisation and treatment of trade-related biotechnological process and products commonly called Intellectual Property. Simply it is product of mind. The Intellectual property right (IPR) is the right to protect this property. The IPR includes right related scientific works, inventions in all fields, industrial design, trademarks, patent, copyright, trademarks, recombinant DNA technology (RDT).

Biopatent: It is the protection or right given by the Government to the inventor of biological material to prevent others from the commercial use of his invention. It prevents others from making, using or selling the invention. A patent is a personal property which can be licensed or sold like any other property. Different countries have different patent laws, and these can be modified from time to time. The infringement of patents is decided by the court. The protection is usually given for a finite term (typically 20 years in the case of patents). The laws related to patents in India are governed by Indian patents act, 1970 as amended by Patent amendment, act 1999 and Patents amendment act, 2002 which came into force with effect from May 2, 2003.

A patent application can be filed at any of the four patent offices in India such as Delhi, Mumbai, Kolkata and Chennai. Its head office is in Kolkata.

The biopatent is the patent of life forms. The biopatents are granted by the Government for

(a) Discovery of new strain of microbes

(b) Transgenic or genetically modified plants and animals

(c) New biological or biotechnological method/process

(d) Discovery of DNA sequence

(e) Proteins encoded by DNA sequences

(f) Production process

(g) Product application

(h) New varieties of plants developed through breeding techniques.

Biopiracy: It is the unlawful use of wild plants, biological compounds or genetic sequences by advanced countries without obtaining consent from the countries or people concerned. Several examples of biopiracy are as follows:

1. The government of India challenged the patent of Indian seed oil of **neem** granted to the US Dept. of Agriculture in **1992**. This patent was got cancelled.
2. The government of India challenged the patent of Indian **turmeric** powder granted to the University of Mississippi Medical Centre of the USA in **1995.** This patent was got cancelled.
3. The government of India challenged the patent of Indian **basmati rice** granted to Rice Tech of the USA in **2001**. This patent was got cancelled.

SHORT ANSWER TYPE QUESTIONS

1. What are the objectives of transgenic plants?
2. Write notes on golden rice.
3. Discuss the procedure of nematode-resistant plants.
4. Explain antisense technology in producing Flavr Savr.
5. Mention five examples of therapeutic proteins.
6. Discuss the role of genetically engineered microbes in the environment.
7. Discuss the procedure of producing clones.

8. Write notes on gene therapy.
9. Write notes on biopatent.
10. Write notes on superbugs.

LONG ANSWER TYPE QUESTIONS

1. How insect-resistant transgenic plant is produced?
2. How nematode resistant plants are produced?
3. Discuss the role of recombinant DNA technology in medicines regarding Humulin.
4. Discuss the procedure of producing transgenic animals.
5. What are the hazards of genetically modified organisms?

MULTIPLE CHOICE QUESTIONS

1. **The microbe provides a natural system for gene transfer to produce transgenic plants is**
   (a) *E. coli*
   (b) *Pseudomonas putida*
   (c) *Bacillus thuringiensis*
   (d) *Agrobacterium tumefaciens*

2. **Transgenic plants are produced by introducing new genes by using**
   (a) Mutation (b) Crossing over
   (c) Radiation (d) Gene gun

1. **Silencing of mRNA has been used in producing transgenic plants resistant to**
   (a) Boll worms (b) Nematodes
   (c) Viruses (d) Bacteria

2. **Human insulin (Humulin) is being commercially produced from transgenic species of**
   (a) *Rhizobium* (b) Yeast
   (c) *E. coli* (d) *Agrobacterium*

3. **The first clinical gene therapy was given to a SCID patient who suffered from a deficiency of**
   (a) Restriction endonuclease
   (b) Polygalacturonase
   (c) Adenosine deaminase
   (d) Neuroamine deaminase

4. **Insect-resistant transgenic cotton has been produced by inserting a piece of DNA from**
   (a) an insect (b) a bacterium
   (c) a virus (d) a yeast

5. **In transgenics, the expression of a transgene in the target tissue is determined by**
   (a) Enhancer (b) Promoter
   (c) Reporter (d) Attenuator

6. **The number of amino acids contained in Proinsulin is**
   (a) 106 (b) 86
   (c) 51 (d) 21

7. **Cry 1 A (b) gene is resistant to**
   (a) Bollworm (b) Corn borer
   (c) Mosquitoes (d) Honeybees

8. **A therapeutic drug, Tissue plasminogen activator (tPA) is used in the treatment of heart disease because**
   (a) It does not form clots in blood vessel
   (b) It dissolves blood clots in blood vessel
   (c) It decreases blood pressure
   (d) It provides oxygen.

9. **The RDAC stands for**
   (a) Recombinant DNA Approval Committee
   (b) Recombinant DNA Advisory Committee
   (c) Recommendation of DNA in Advisory Committee
   (d) Report of DNA on Approval Committee

10. **The first introduced transgenic crop in India (2002) is**
    (a) Bt-cotton (b) Bt-brinjal
    (c) Bt-rice (d) Bt-turmeric

11. A transgenic plant Golden rice was discovered by
(a) Tsan (b) Ingo Potrykus
(c) Paul Berg (d) Bayliss

12. Root-knot disease was caused by a
(a) Fungus (b) Virus
(c) Bacterium (d) Nematode

13. FLAVR SAVR is a transgenic plant of tomato based on
(a) Same sense RNA technology
(b) Antisense RNA technology
(c) Terminator technology
(d) Nonsense technology

14. Which country has decided to 'label' those Genetically modified (GM) foods that are found to be safe?
(a) US (b) England
(c) Japan (d) Mexico

15. The most common trait that was improved in commercialised crops was
(a) Nutritional quality
(b) Disease resistance
(c) Drought resistance
(d) Insect resistance

16. The crystals of Bt-toxin produced by some bacteria do not kill the bacteria themselves because
(a) Bacteria are resistant to the toxin
(b) Toxin is inactive
(c) Toxin is immature
(d) Bacteria enclose toxins in a special sac

17. In RNAi, genes are silenced using
(a) ssRNA (b) dsRNA
(c) ssDNA (d) dsDNA

18. In which of the following plants, resistance against a nematode was introduced by implying RNAi?
(a) Tomato (b) Tobacco
(c) Golden rice (d) Bt-cotton

19. A functional ADA cDNA can be introduced into the stem cells of the patients receiving gene therapy by using vector
(a) Retrovirus
(b) *E. coli*
(c) *Bacillus thuringiensis*
(d) *Agrobacterium*

20. Which of the following genes was introduced in the first transgenic cow?
(a) *Cry* I Ac
(b) α-1-antitrypsin
(c) β- 1-antitrypsin
(d) Human α-lactalbumin

21. Which of the following has been covered under the broad patent category?
(a) *Brassica* (b) *Triticum*
(c) *Pisum* (d) *Oryza*

22. The new strain of bacteria produced by biotechnology in the alcohol industry is
(a) *E. coli*
(b) *Saccharomyces cerevisiae*
(c) *Pseudomonas putida*
(d) *Bacillus subtilis*

23. Which of the following microbes was used to clear the oil spill in the water body?
(a) *Bacillus megatherium*
(b) *Anabaena azollae*
(c) *Pseudomonas denitrificans*
(d) *Pseudomonas putida*

24. Dolly is a clone of sheep that was successfully created by
(a) Ingo Potrykus (b) Ian Wilmot
(c) Paul Berg (d) Stanley Cohen

25. In transgenics, the expression of transgene in the target tissue is determined by
(a) Reporter (b) Promoter
(c) Enhancer (d) All of the above

26. The first transgenic plant produced was
(a) Tomato (b) Tobacco
(c) Potato (d) Cotton

27. Bollgard is a transgenic plant of
(a) Tomato (b) Tobacco
(c) Potato (d) Cotton

28. A transgenic plant was found to be resistant to a nematode known as
(a) *Caenorhabditis elegans*
(b) *Meloidogyne incognita*
(c) *Ascaris lumbricoides*
(d) *Wuchereria bancrofti*

29. The first transgenic virus-resistant plant produced was
(a) Tomato (b) Tobacco
(c) Potato (d) Cotton

30. Transgenic *Petunia* was found to be resistant to a herbicide known as

(a) Glyphosate (b) Atrazine
(c) Phosphinothricin (d) Sulphonyl urea

31. Which of the following therapeutic recombinant proteins is used in the treatment of heart disease?

(a) Humulin
(b) Urokinase
(c) Tissue plasminogen factor
(d) Hirudin

32. Humulin contains

(a) 21 amino acids (b) 51 amino acids
(c) 86 amino acids (d) 109 amino acids

33. Which of the following diseases is treated by Gene therapy?

(a) Tuberculosis
(b) Huntington chorea
(c) SCID
(d) AIDS

34. Rossie is a transgenic animal of

(a) Mice (b) Cow
(c) Sheep (d) Monkey

35. Maximum transgenic animal has been developed in

(a) Mice (b) Cow
(c) Sheep (d) Monkey

36. ANDI was the world's first genetically altered

(a) Mice (b) Cow
(c) Sheep (d) Monkey

37. Which of the following regulatory agencies monitors the import and export of genetically modified organisms?

(a) GEAC (b) RDAC
(c) RCGM (d) CPCB

38. The Government of India in 2001 challenged the patent of a certain hybrid of

(a) Cotton (b) Rice
(c) Turmeric (d) Brinjal

ANSWERS

1.d	11.b	21.a	31.b
2.d	12.b	22.d	32.a
3.b	13.b	23.a	33.c
4.c	14.d	24.d	34.b
5.c	15.b	25.b	35.b
6.b	16.c	26.b	36.b
7.c	17.d	27.a	37.a
8.b	18.b	28.b	38.d
9.b	19.b	29.d	39.a
10.b	20.b	30.b	40.b

CHAPTER-16

MICROBES IN HUMAN WELFARE

Microbes are microscopic organisms that are too small to be seen with the unaided eye. Microbes include bacteria, viruses and subviral agents (viroids and prions), some fungi, algae and protists. Louis Pasteur was the first person who demonstrate the role of microbes in food. He reported that buttermilk and Beer are the products of fermentation carried out by yeast. Microbes play an important role in the processing of many foods, improving their nature, digestibility, nutritive value, flavour and appearance of raw material and the modified product is called fermented food.

ROLES OF MICROBES IN HOUSEHOLD PRODUCTS

Microbes are helpful in the preparation of household products such as

1. Dairy products

Milk is an aqueous solution (pH=7.0) of proteins (casein and lactalbumin), fats and carbohydrates (lactose) with numerous vitamins and minerals. *Lactobacillus* form of bacteria grows in milk and converts the lactose sugar of milk into lactic acid. It causes coagulation and partial digestion of casein into calcium paracaseinate.

(a) **Curd**: It is prepared by inoculating milk with *Lactobacillus acidophilus* at 30-40°C. Curd is more nutritious than milk as it contains a number of organic acids and vitamins including vitamin B_{12}.

(b) **Yoghurt**: It is prepared by inoculating milk with *Lactobacillus bulgaricus* and *Streptococcus thermophilus* at 45°C.

(c) **Buttermilk**: It is prepared by inoculating skimmed milk with a combination of *Lactobacillus acidophilus, Streptococcus cremoris* and *Streptococcus lactis* at 22°C for 18 hours. Buttermilk (milk remains after cream) is acidulated liquid after churning butter from curd.

(d) **Sour cream**: It is prepared by inoculating churned milk with *Streptococcus lactis* to produce lactic acid.

(e) **Cheese**: It is the oldest food item. It is prepared by separating curd from liquid part or whey by *Propionibacterium shermanii*. The large holes in cheese occur due to the production of large amounts of CO_2 released by bacteria. Cheese may be soft (>50% water), semisolid (40-50%) and hard (<40% water).

Roquefort cheese is ripened by growing *Penicillium roqueforti* for flavour.

Camembert cheese is ripened by growing *Penicillium camemberti* for flavour.

2. Bakery products:

(a) **Bread:** Bread is a staple food that is prepared by baking, steaming or frying dough. It mainly consists of wheat flour, water, salt and a leavening agent, Baker's yeast (*Saccharomyces cerevisiae*). Firstly, selected strains of yeast are grown on molasses, and these are harvested after sufficient growth. These are converted into powder or cakes. A small quantity of yeast is added to wheat flour and kneaded. The kneaded flour is kept at a warm temperature for a few hours. Finally, it swells up called leavening which is caused by the secretion of three enzymes by yeast.

Firstly, amylase breaks down starch into maltose; maltase converts maltose into glucose and finally, zymase brings about the fermentation of glucose forming ethyl alcohol and CO_2. These two compounds cause swelling or leavening of the dough. The leavened dough is baked and finally, ethyl alcohol and CO_2 evaporate making the bread porous and spongy.

ROLES OF MICROBES IN INDUSTRIAL PRODUCTS

Microbes are suitable for industrial fermentation processes because

(a) They grow very rapidly after inoculation.

(b) They grow very easily on simple nutrient requirements.

(c) Their growth conditions such as temperature, and pH can be controlled very precisely.

(d) They require temperature for the processes lower than conventional processes.

(e) They can be easily manipulated through genetic engineering.

(f) They produce a wide variety of chemicals (products).

1. **Brewery**: It is the production of alcohol from sugar with the help of microbial enzymes as a result of fermentation. It is an anaerobic process carried out by yeast. It was demonstrated for the first time by Louis Pasteur. A variety of alcoholic beverages are made by using the yeast-*Saccharomyces* acting on different substrates. There are several species of yeast which are involved in brewery.

Baker's yeast or Brewer's yeast	Saccharomyces cerevisiae
Wine yeast	*Saccharomyces ellipsoideus*
Sake yeast	Saccharomyces sake
Food yeast	Torulopsis utilis

Process:

(a) Selection of strain: The microbes used by fermentation industries are mutant strains of fungi or bacteria that selectively synthesise large amounts of various metabolic intermediates or metabolites.

(b) Preparation of medium: Alcoholic beverages require a specific substrate that provides flavour, colour, aroma etc. These differ with respect to the production process, but microbe is the same in all cases.

Fermentation of barley malt	Beer and Whisky
Fermentation of grape juice	Wine, Champagne and Brandy
Fermentation of sugarcane molasses	Rum
Fermentation of rice starch	Sake
Fermentation of rye malt	Gin
Fermentation of potato starch	Vodka

(c) Sterilisation of medium and bioreactor (fermentor).

(d) Inoculation of selected strains of yeast in a bioreactor: Fermentation is carried out in three ways:

(i) Batch process: It is a discontinuous process in which the entire culture medium with the organism is removed from the bioreactor after every batch.

(ii) Continuous process: A fraction of the medium with a fermented product is drawn out at regular intervals and fresh sterilised medium and inoculums are fed into the bioreactor continuously.

(iii) Fed-batch process: A small amount of medium is regularly fed in the bioreactor to optimise the working of fermenting microorganisms without inhibition.

(e) Filtration: The culture fluid is filtered when the growth of microorganisms is over. Later yeast cells (biomass) are separated from the supernatant.

(f) Recovery of products: The products are recovered from supernatant. It depends on the culture medium, size of the fermentor, culture condition, potential microorganism and design for separation of solid and liquid. There are two types of beverages which are as follows:

Fermented beverages such as beer (5% alcohol), wine (10% alcohol) and sake (15% alcohol) are produced without distillation.

Distilled beverages such as gin (40% alcohol), rum (40% alcohol), vodka (40% alcohol), whisky (50% alcohol) and brandy (70% alcohol) are produced by distillation of fermented medium. The process of distillation is required for hard or strong liquors containing high concentrations of alcohol. Actually, when the fermenting mixture reaches an alcoholic concentration between 14% and 18% then yeast cells die. As a result, the fermenting mixture must be distilled heating it above the boiling temperature of ethyl alcohol (78%) and below the boiling temperature of water (100%).

2. **Production of enzymes**: Microbial enzymes may be intracellular (produced inside the cell and start functioning in the same cell) and extracellular (produced inside the cell, later secreted out the cell and start functioning outside the cell). Microbial enzymes have advantages over plant and animal enzymes because these are produced on a large scale within limited space and time and further these are easily purified. Microbes produce a wide variety of enzymes.

Enzymes	Sources	Application	Industry
Amylase	*Bacillus subtilis*	Starch coating	Paper
	B.amyloliquefaciens	Removal of stains	Detergent
	Aspergillus niger	Digestive aid	Pharma
	Aspergillus oryzae, Rhizopus oryzae	Softening of bread	Baking
Protease	*Bacillus subtilis*	Removal of stain	Detergent
	B. licheniformis	Tenderisation of meat	Meat
	Aspergillus oryzae	Bread	Baking
Lipase	*Aspergillus niger*	Removal of stain	Detergent
	Candida lipolytica	Hydrolysis of butter fat	Dairy
Cellulase	*Streptomyces spp.*	Clarification of wine	Brewery
	Aspergillus niger	Clarification of wine	Brewery
	Trichoderma reesei	Softening of fruit or hydrolysis of fruit juices	Fruit processing
Pectinase	*Bacillus subtilis*	Pectin hydrolysis in fruit juices	Fruit processing
	Aspergillus niger	Clarification of wine	Brewery
	Rhizopus oryzae	Retting of fibres	Textile
Lactase	*Saccharomyces fragilis*	Coagulation of milk protein	Dairy
	Torula cremoris	Hydrolysis of lactose in milk and whey	Dairy
Invertase (Sucrase)	*Saccharomyces cerevisiae*	Hydrolysis of sucrose	Confectionary
Glucanase	*Bacillus subtilis* *Aspergillus niger*	Hydrolysis of fruit juices	Fruit processing
Glucose isomerase	*Bacillus coagulans*	Conversion of corn starch into high fructose corn syrup	Soft drink
Glucose oxidase	*Aspergillus niger*	Removal of glucose or oxygen	Food
Streptokinase (Tissue Plasminogen Activator)	*Streptococcus*	Dissolution of blood clots. Plasminogen→plasmin	Medicine
Cyclosporin-A	*Trichoderma polysporum*	Inhibition of T-cells and prevents rejection reaction in organ transplantation as immunosuppressant	Medicine
Statin	*Monoascus purpureus*	Lowering of blood cholesterol	Medicine

3. **Production of organic acids:**

   (a) **Acetic acid**: It is the earliest microbial product used by human beings. It is produced in two steps through anaerobic fermentation of sugarcane juice or molasses by yeast and *Acetobacter aceti*. It is used in vinegar, sauce, pickle, and pharma industry and also used as a preservative of meat and vegetables.

   (b) **Lactic Acid**: It is the first acid to be produced through the fermentation of corn starch by *Lactobacillus delbrueckii*. It is used in confectionery, soft drinks, fruit juices, leather and the textile industry. It is also used as a preservative for fruits and vegetables.

   (c) **Citric acid**: It is produced through fermentation of beet molasse (sucrose), glucose by *Aspergillus niger*. It is used in the dye, and pharma industry and also used as a flavouring agent or preservative in jam, jelly, and confectionery.

   (d) **Propionic acid**: It is produced through fermentation of sugar by *propionibacterium.*

   (e) **Butyric acid**: It is produced through fermentation of sugar by *Clostridium acetobutyricum*. It is used in the preparation of solvents.

   (f) **Fumaric acid**: It is produced through fermentation of sugar by *Rhizopus nigricans, Rhizopus stolonifer*. It is used as a wetting agent and in the manufacture of resins.

   (g) **Gluconic acid**: It is produced through fermentation of glucose by *Aspergillus niger, Penicillium purpurogenum*. It is used in the pharma industry as it forms calcium gluconate and ferrous gluconate to provide calcium and iron in the human body.

   (h) **Oxalic acid**: It is produced through fermentation of sugar by *Aspergillus niger*. It is used as an organic solvent.

   (i) **Gallic acid**: It is produced through fermentation of sugar by *Aspergillus niger*. It is used in manufacturing ink and dye.

   (j) **Itaconic acid**: It is produced through fermentation of sugar by *Aspergillus terreus*. It is used in manufacturing plastics.

   (k) **Kojic acid**: It is produced through fermentation of sugar by *Aspergillus oryzae*. It is used in manufacturing insecticides and fungicides.

4. **Production of vitamins:** Vitamins are small organic compounds provided to the body in small amounts. These usually make up all or part of the enzyme cofactors. These help in the normal maintenance and optimal growth of an organism.

Vitamin	Discoverer
A (Retinol)	McCollum
B_1 (Thiamine)	Casimir Funk
C (Ascorbic acid)	Szent Gyorgy
D (Calciferol)	Mellanby

   (a) **Vitamin C** (Ascorbic acid): It was the first vitamin to be produced from the fermentation of sugar by *Acetobacter, Corynebacterium* etc.

   (b) **Vitamin B_2** (Riboflavin): It is produced through fermentation of maize steep liquor by *Ashbya gossypii, Clostridium acetobutyricum, Mycobacterium smegmatis* etc.

   (c) **Vitamin B_6** (Pyridoxine): It is produced from *E. coli*.

   (d) **Vitamin B_{12}** (Cyanocobalamin): It is produced through the fermentation of sugar by *Propionibacterium shermanii, Pseudomonas denitrificans* and *Bacillus megatherium*.

   (e) **Vitamin K**: It is produced from fermentation by *E. coli*.

5. **Production of vitamins:** Vitamins are small organic compounds provided to the body in small amounts. These usually make up all or part of the enzyme cofactors. These help in the normal maintenance and optimal growth of an organism.

Vitamin	Discoverer
A (Retinol)	McCollum
B_1 (Thiamine)	Casimir Funk
C (Ascorbic acid)	Szent Gyorgy
D (Calciferol)	Mellanby

(f) **Vitamin C** (Ascorbic acid): It was the first vitamin to be produced from the fermentation of sugar by *Acetobacter, Corynebacterium* etc.

(g) **Vitamin B_2** (Riboflavin): It is produced through fermentation of maize steep liquor by *Ashbya gossypii, Clostridium acetobutyricum, Mycobacterium smegmatis* etc.

(h) **Vitamin B_6** (Pyridoxine): It is produced from *E. coli.*

(i) **Vitamin B_{12}** (Cyanocobalamin): It is produced through the fermentation of sugar by *Propionibacterium shermanii, Pseudomonas denitrificans* and *Bacillus megatherium.*

(j) **Vitamin K**: It is produced from fermentation by *E. coli.*

6. **Production of Antibiotics**: These are chemical substances produced by certain microorganisms at very low concentrations that inhibit or kill other microorganisms without affecting the host. The term 'antibiotic' was coined by Paul Vuillemin. Babes (1855) demonstrated the phenomenon of antibiosis.

Alexander Fleming (1928) discovered the first antibiotic, Penicillin (wonder drug) from *Penicillium notatum* which inhibits the growth of *Staphylococcus aureus*. This antibiotic was widely used to treat wounded American soldiers in World War II. It is the first natural compound to show antimicrobial properties. More than 7000 antibiotics have been isolated so far, and only 100 or so are being used to treat human, animal and plant diseases. The antibiotic was, however, commercially extracted by Chain and Florey. They got the Nobel Prize in 1945.

Sources:

(a) Bacteria: *Streptomyces* (mostly) *Bacillus, Pseudomonas, Noccardia, Micromonospora, Actinoplanes* etc.

(b) Fungi: *Penicillium, Cephalosporium* etc.

Features of a good antibiotic:

(a) It is nontoxic to host cells (humans and animals).

(b) It does not affect normal microflora of the alimentary canal of the host.

(c) It does not interfere with the normal host defence mechanisms.

(d) It shows quick action.

(e) It does not cause allergies to the host.

(f) It remains stable and active in body fluids.

Procedures for Commercial production of penicillin:

(a) Selection of high-yielding microbial strain.

(b) Preparation of inoculum in the form of suspension.

(c) Preparation of the suitable medium (broth).

(d) Inoculation of the medium in the fermenter.

(e) Incubation with forced aeration during fermentation.

(f) Separation of fungal biomass.

(g) Extraction, purification and recovery of the penicillin

Chemical nature:

(a) Aminoglycosides: Streptomycin, Neomycin, Kanamycin, Gentamicin

(b) Polypeptides: Polymixin-B, Bacitracin.

(c) Beta-lactams: Penicillin, Ampicillin, Amoxycillin, Cephalosporin.

(d) Tetracyclines: Aureomycin (Chlorotetracycline), Terramycin (Oxytetracycline).

(e) Macrolides: Erythromycin, Azithromycin, Carbomycin, Novobiocin.

Range of effects:

(a) Narrow spectrum: Bacitracin, Polymixin, Penicillin

(b) Broad spectrum: Streptomycin, Cyclohexamide, Tetracycline, Terramycin, Gentamicin, Rifampicin

Mode of action: Antibiotics function as bactericides (killing bacteria) or bacteriostatic (inhibiting the growth of bacteria). These perform their action by the following ways:

(a) Disruption of cell wall synthesis: Penicillin, Bacitracin, Cephalosporin

(b) Destruction of plasma membrane synthesis: Amphotericin, Polymixin, Nystatin etc.

(c) Inhibition of 30S ribosome function in protein synthesis: Streptomycin, Tetracycline, Kanamycin etc.

(d) Inhibition of 50S ribosome function in protein synthesis: Erythromycin, Chloramphenicol.

(e) Inhibition of 30 S ribosome function in protein synthesis: Streptomycin

(f) Inhibition of chain initiation in protein synthesis: Tertracycline

(g) Inhibition of DNA synthesis: Rifampicin

Reasons for resistance to antibiotics: Bacteria show resistance to antibiotics by undergoing the following processes:

(a) Bacteria modify the structure of their cell wall to reduce the ability of antibiotics to penetrate.

(b) Bacteria modify the antibiotics enzymatically. e.g., Penicillinase or β-lactamase deactivates Penicillin, Ampicillin, Cephalosporin etc.

(c) Bacteria methylate their ribosomes obscuring the target of antibiotics.

(d) Bacteria undergo mutation to alter the target of antibiotics.

List of antibiotics and their sources

Antibiotics	*Source*
Subtilin	*Bacillus subtilis*
Bacitracin	*Bacillus licheniformis*
Polymixin-B	*Bacillus polymyxa*
Polymixin-E	*Bacillus colistinus*
Tyrothricin	*Bacillus brevis*
Nisin	*Streptococcus cremoris*
Streptomycin	*Streptomyces griseus*
Grisin	*-do*
Cyclohexamide	*-do*
Aureomycin (Chlorotetracycline)	*Streptomyces aureofaciens*
Tetracycline	*-do*
Rimocidin	*Streptomyces rimosus*
Terramycin (Oxytetracycline)	*-do*
Chloramphenicol (Chloromycetin)	*Streptomyces venezuelae*
Erythromycin	*Steptomyces erythreus*
Kanamycin	*Streptomyces kanamyceticus*
Neomycin	*Streptomyces fradiae*
Streptothricin	*Streptomyces lavendulae*

Antibiotics	Source
Carbomycin	*Streptomyces halstedii*
Cycloserine	*Streptomyces orchidaceous*
Novobiocin	*Streptomyces niveus*
Nystatin	*Streptomyces noursae*
Vancomycin	*Streptomyces orientalis*
Viomycin	*Streptomyces puniceus*
Oleandomycin	*Streptomyces antibioticus*
Lincomycin	*Streptomyces lincolensis*
Gentamicin	*Micromonospora purpurea*
Rifampicin (Rifamycin)	*Noccardia mediterranei*
Thermorubin	*Thermoactinomyces antibioticus*

ROLE OF MICROBES IN SEWAGE TREATMENT:

Sewage is the wastewater of domestic, commercial and industrial units. It is rich in organic wastes and microbes many of which are pathogens. Industrial sewage is called effluent, and it carries huge amounts of organic residues. The effluents of sugar industries (factories), distilleries, and paper and cloth factories are heavily loaded with organic wastes and thus require proper treatment before draining them into any freshwater bodies. When these wastes are drained in the untreated state to the river, they are unable to recycle them and then water becomes heavily polluted.

Sewage treatment is the process of removing contaminants from wastewater or sewage through physical, biological and chemical methods to produce environmentally safe water.

Sewage treatment is also known as wastewater treatment.

The pollution level of water is measured by the level of Biochemical Oxygen Demand (BOD). It is the amount of oxygen consumed by various microbes in the aerobic decomposition of organic wastes present in one litre of sewage water. If organic wastes are more in water, then the BOD of water will be increased hence the Dissolved oxygen (DO) will be decreased due to more turbidity of water.

The sewage water needs to be treated to decrease the BOD and increase the DO before draining in water bodies.

Treatment of water occurs through the following **three** stages once the level of pollutants in wastewater is determined. These stages are as follows:

1. **Primary treatment (Physical or mechanical method):** It removes 50% of suspended solids. It involves the following steps:
   (a) The large, suspended and floating solid matters are physically removed by sequential filtration.
   (b) The filtrate is now kept in a large, open and circular settling tank where grit containing soil, sand, silt and pebbles settle down.
   (c) Grit is now removed by sedimentation.
   (d) All sediments or solids settle down to form primary sludge and supernatant forms primary effluent which is industrial sewage containing huge amount of organic residues.
   (e) Effluent (settled sewage) is now moved to secondary treatment.
2. **Secondary treatment (Biological method):** It involves microbial degradation of organic matter. It removes 85% of impurities. It employs the following three methods: Activated sludge system, Oxidation tank and Trickling filter system.

   Activated sludge system: It involves the following steps
   (a) The primary effluent is taken to a large and rectangular aeration tank.
   (b) Liquid is regularly agitated mechanically.

(c) Air is pumped into water (primary effluent) to encourage the growth of a large number of aerobic and heterotrophic bacteria (*Bacillus, Arthrobacter, Beijerinckia, Mycobacterium, Pseudomonas, Enterobacter, Corynebacterium* and Actinomycetes), protozoans and yeast.

(d) These saprotrophs form flocs which are bacterial colonies entangled with sewage fungal mycelia floating over the surface of effluent.

(e) These organisms decompose (digest) a lot of organic matter in effluent, convert it into microbial biomass and release a lot of minerals. This action reduces BOD.

(f) Effluent is now passed into the settling tank.

(g) The flocs (bacterial colonies) are allowed for sedimentation and the sediment of the settling tank is called activation sludge. The effluent or supernatant is generally passed into the river, or it is treated chemically to purify further.

(h) A large amount of activation sludge is passed into a large tank called an anaerobic sludge digester and the remaining small part is used as inoculum in an aeration tank.

(i) The aerobic microbes get killed in it.

(j) The anaerobic heterotrophic bacteria (Methanogens) digest the organic mass as well as aerobic microbes of activated sludge to produce biogas (CH_4, CO_2, H_2S etc).

(k) Now effluents from secondary treatment are generally released into natural water bodies such as rivers and streams. If the wastewater is to be used for domestic purposes then it will be subjected to tertiary treatment.

Primary sludge	Secondary sludge
It is produced from primary treatment.	It is produced from secondary treatment.
It does not form flocs.	It forms flocs.
It does not require aeration.	It requires aeration.
It requires little decomposition.	It requires large decomposition.

3. **Tertiary treatment (Chemical method):** It removes 99% of all impurities from sewage. It involves the removal of inorganic nutrients and the killing of pathogenic microbes by the following processes:

   (a) UV filtration: It kills pathogenic microbes (viruses and bacteria)

   (b) Precipitation: It removes phosphates

   (c) Chlorination: It removes NH_3 and kills pathogenic microbes.

   (d) Carbon filter: It removes residual toxins.

   (e) Reverse Osmosis: It removes Total dissolved solids (TDS) in the form of salts.

ROLE OF MICROBES IN BIOGAS PRODUCTION

Biogas is a fuel gas produced by anaerobic breakdown of biomass with the help of methanogens. It is also called as gobar gas, sewage gas, landfill gas, and biofuel. It is a colourless and odourless gas. It contains average 60% CH_4, 40% CO_2, 4% N_2, 1% H_2S and traces of H_2O, H_2, and O_2. The energy released from biogas depends on the proportion of methane present in it. It is a renewable and alternative source of energy. China is the largest producer of biogas. The biogas production was first started in India by the joint efforts of Indian Agricultural Research Institute (IARI) and Khadi and Village Industries Commission (KVIC).

It is mainly produced from agriculture waste, plant materials, manure, municipal waste, animal wastes esp. dung because these contain cellulosic materials and methanogens. Cattle dung is the most important substrate for biogas production as it contains cellulose as well as methanogens.

Production Unit:

1. Biogas plant consists of a 10-15 feet deep and air-tight cylindrical concrete tank (digester) in which slurry of dung is fed

2. A floating cover is placed over the slurry which keeps on rising as the gas is produced in the tank by microbial activity.
3. It has an outlet which is connected to pipe for the removal of biogas.
4. It has another outlet to remove spent slurry which can be used as fertiliser.

Process: Plant and animal wastes (cattle dung and manure from poultry) containing polymers (cellulose, hemicellulose, starch, lignin, protein, fats) are the raw materials for biogas. Biogas is produced in three stages:

1. **Solubilisation of polymers** (enzymatic breakdown) by decomposer microbes (facultative anaerobes) into simpler and soluble compounds (monomers).
2. **Acidogenesis of monomers** by fermenting microbes (facultative anaerobes esp. Bacteroides and *Clostridium*) into organic acid (acetic acid).
3. **Methanogenesis of organic acid** by methanogens (obligate anaerobes esp. *Methanobacterium* and *Methanococcus*) into **biogas** (Methane and Carbon dioxide).

Factors useful for biogas production

(a) Type of waste: Cow dung is preferable. If human excreta are added then biogas production will increase because it is rich in nitrogen content.

(b) Temperature: The optimum temperature should be 35 to 38^{0}C. Low temperatures will reduce its production.

(c) Optimum pH should be 7.

(d) Slurry composition where water and solid should be in the ratio of 1:1

(e) Anaerobic condition

(f) Presence of inhibitors: There should not be ammonia, sulphates and antibiotics. Agricultural waste should be avoided because it is rich in ammonia. Likewise, paper residues should be avoided because it is rich in sulphates.

Advantages:

(a) It provides fertilisers or manure with its leftover

(b) It is cheaper than fuel for cooking and generating electricity.

(c) It does not cause pollution like CNG hence it can also be compressed.

(d) It does not release offensive odour

(e) It is ideal for small-scale production

(f) It improves health conditions and sanitation by checking the spread of faecal pathogens.

(g) It can be purified to yield a good grade of methane,

Disadvantages:

(a) It can be an ecological problem as methane absorbs infrared radiation and thus is a greenhouse gas. Its overproduction may cause global warming.

(b) Its production is very low.

(c) It has a low heat value

(d) It is expensive for large-scale production

(e) It contains some gases as impurities.

(f) It is quite difficult to store, transport and distribute methane

BIOFERTILIZERS:

These are microorganisms that bring about nutrient enrichment of soil by enhancing the availability of nutrients to crops. These are also known as microbial inoculants. These microorganisms act as biofertilisers in three ways:

1. **Nitrogen fixing biofertilisers:**
   - Free living asymbiotic bacteria: *Azotobacter, Bacillus, Clostridium*
   - Free living loose symbiotic bacteria: *Azospirillum*
   - Free living asymbiotic cyanobacteria: *Anabaena, Nostoc, Aulosira* (in rice field), *Cylindrospermum* (sugarcane and maize field), *Plectonema, Tolypothrix.*
   - Symbiotic nodule-forming bacteria in leguminous plants: *Rhizobium*
   - Symbiotic nodule-forming bacteria in non-leguminous plants: *Frankia*
   - Symbiotic cyanobacteria: *Anabaena azollae*
2. **Phosphate solubilising biofertilizers:** A fungus, *Glomus* develops a symbiotic association with roots of higher plants forming endomycorrhiza which converts insoluble phosphorus into soluble phosphorus for uptake of crop plants.
3. **Phosphate mobilising biofertilizers:** These facilitate unavailable and insoluble inorganic phosphorus into soluble and usable forms of phosphorus (phosphates) to crop plants. These may be
   - Bacteria: *Bacillus megatherium* var. *phosphaticum, Bacillus circulans, Pseudomonas striata*
   - Fungi: *Penicillium, Aspergillus awamori*

   Advantages:

   (a) It is cheaper and can be used by small and marginal farmers.

   (b) It increases the yield of crops by 25%.

   (c) It improves soil texture, water-holding capacity and organic matter.

   (d) It does not cause water pollution.

   (e) It does not allow pathogens to flourish.

   (f) It produces vitamins and growth-promoting chemicals

GREEN MANURING

It is a farming practice where a leguminous plant that has derived enough benefits from its association with species of *Rhizobium*, is ploughed into soil and then a nonlegume is grown and allowed to take the benefits of already fixed nitrogen. Some of the important leguminous plants used as green manure are as follows:

(a) Sunhemp (*Crotolaria juncea*)

(b) Dhaincha (*Sesbania aculeata*)

(c) Cluster bean (*Cyamopsis tetragonoloba*)

(d) Cow pea (*Vigna sinensis*)

(e) Sweet clover/Senji (*Melilotus pamniflora)*

(f) Egyptian clover/ Berseem *(Trifolium alexandrium*)

(g) Lentil (*Lens esculenta*)

(h) Horse gram (*Macrotyloma uniflorum*)

Importance:

(a) It adds organic matter and nitrogen to the soil.

(b) It controls the growth of weed.

(c) It helps in the reclamation of alkaline soil.

BIOCONTROL

It is the natural method of controlling pests (weeds, insects and pathogens) with the help of living organisms (viruses, bacteria, fungi, plants and animals). These organisms are called biopesticides.

Types of biopesticides:

1. **Bioherbicides**: These are living organisms or their products to either control or kill weeds or unwanted plants without harming useful plants. These are as such
   (a) Transgenic plants such as tobacco, and tomato show tolerance to specific herbicides.
   (b) Smoother crops such as sweet clover, soya bean, rye, alfalfa, and sunflower do not allow weeds to grow in the vicinity through chemicals. If crop rotation is done with smoother crops, then it will reduce the chance of growth of weeds.
   (c) Predatory herbivore such as Cochineal insect (*Cactoblastis cactorum*) checks the growth of cacti (*Opuntia*).
   (d) Predatory herbivore such as beetle (*Zygogramma bicolorata*) checks the growth of carrot grass or congress grass (*Parthenium hysterophorus.*
   (e) A fungus, *Cercospora rodmanii* controls the growth of water hyacinth (*Eicchornia*).
   (f) Mycoherbicide such as devine (a liquid suspension of chlamydospores of *Phytophthora palmivora* is sprayed to control Milkweed vine (*Morrenia odorata*).
   (g) Mycoherbicide such as collego (a dry powder formulation of a fungus, *Colletotrichum gloespirodes* is sprayed to control a leguminous weed, Northern joint Vetch (*Aeschyynomena*).
2. **Bioinsecticide**: These are living organisms that kill harmful insects. These are
   (a) Transgenic plants such as tomato show resistance to hornworm larvae.
   (b) Predators such as Dragonfly kill mosquitoes.
   (c) Predators such as Lady Bird (Ladybug beetle) and Praying mantis kill aphids, and scale insects of crop plants, respectively.
   (d) Fishes such as Gambusia and Guppy kill mosquito larvae.
   (e) A free-living fungus, *Trichoderma* kills several plant pathogens.
   (f) Baculoviruses (under the genus, Nucleopolyhedral virus or NPV) control many lepidopteran insects such as bollworms, gypsy moths, butterflies, beetle, ants, wasps of cotton, rice and vegetables. These are species-specific and have narrow-spectrum applications. These do not cause harm to plants, fish, birds, mammals or even non-target useful insects. The viruses after application enter into the mouth and digestive tract of insect pests and kill them.
   (g) **Natural insecticides** are formed which are as follows:
      - Rotenone: It is prepared from the root of *Derris elliptica*
      - Azadirachtin: It is prepared from the leaf of *Azadirachta indica.*
      - Nicotine: It is prepared from the leaf of *Nicotiana tabacci.*
      - Pyrethrum and Cinerin: These are prepared from flowers of *Chrysanthemum.*
   (h) A **commercial insecticide** such as sporeine is developed from *Bacillus thuringiensis.* It was the first such insecticide developed in France. It is species-specific and active against a wide variety of insects such as lepidopteran and dipteran larvae, cockroaches, mosquitoes, aphids, termites etc. This bacterium produces several exotoxins (α,β,δ) and endotoxin (δ) which have insecticidal properties.

SHORT ANSWER TYPE QUESTIONS

1. Why microbes are suitable for industrial processes?
2. Distinguish between distilled beverages and fermented beverages.
3. Mention the sources of five important enzymes.
4. What are the applications of streptokinase, cyclosporine and statin?
5. What are the features of a good antibiotic?

6. Distinguish between narrow-spectrum antibiotics and broad-spectrum antibiotics.
7. Discuss the mode of action of antibiotics.
8. What are the reasons for resistance against antibiotics?
9. Mention the sources of five important antibiotics.
10. Distinguish between primary sludge and secondary sludge.
11. Distinguish between phosphate solubilising biofertilizers and phosphate mobilising biofertilizers.
12. Mention five important nitrogen-fixing microbes.
13. Discuss the importance of biofertilisers.
14. Write notes on green manuring.
15. Write notes on natural insecticides.

LONG ANSWER TYPE QUESTIONS

1. Describe the role of microbes in producing household products.
2. Describe the role of microbes in the brewery.
3. Describe the role of microbes in producing organic acids.
4. Describe the role of microbes in sewage treatment
5. Describe the role of microbes in producing biogas.
6. Define biocontrol. Discuss the role of different organisms in the control of pests.

MULTIPLE CHOICE QUESTIONS

1. **During the formation of bread, it becomes porous due to the release of CO_2 by the action of**
   (a) Bacteria (b) Yeast
   (c) Protozoans (d) Archaebacteria
2. **In the manufacture of wine, as the alcohol concentration increases, yeast cells die. The cells can tolerate alcohol up to**
   (a) 5% (b) 10%
   (c) 15% (d) 30%
3. **When a huge amount of sewage is dumped into a river, the BOD will**
   (a) Decrease (b) Increase
   (c) Slightly decrease (d) Remain unchanged
4. **The group of bacteria used in biogas production is**
   (a) Methanotrophs
   (b) Methanogens
   (c) Thermoacidophiles
   (d) Organotrophs
5. **A common biocontrol agent for the control of plant disease is**
   (a) *Fusarium*
   (b) *Bacillus thuringiensis*
   (c) *Glomus*
   (d) *Trichoderma*
6. **Which of the following organisms helps in the absorption of phosphorus from the soil by plants?**
   (a) *Glomus* (b) *Frankia*
   (c) *Anabaena* (d) *Rhizobium*
7. **The nitrogen-fixing bacteria associated with maize roots belong to the genus**
   (a) *Azotobacter* (b) *Aulosira*
   (c) *Azospirillum* (d) *Azorhiza*
8. **Vesicular arbuscular mycorrhizae (VAM) are important in**
   (a) Nitrate nutrition of plants
   (b) Phosphate nutrition of plants
   (c) Carbonate nutrition of plants
   (d) Sulphate nutrition of plants

9. Green manuring increases crop yield by
(a) 5-10% (b) 20-25%
(c) 30-50% (d) 80-90%

10. Sporeine is based on
(a) Spores of algae
(b) Spores of fungi
(c) Bacteria
(d) Viruses

11. Which of the following N_2-fixing bacteria forms a loose association with the roots of grasses?
(a) *Azospirillum* (b) *Azotobacter*
(c) *Rhizobium* (d) *Bacillus*

12. Which of the following plants are used as green manure in crop fields and sandy soils?
(a) *Saccharum munja* and *Lantana camara*
(b) *Dichanthium annulatum* and *Azolla nilotica*
(c) *Crotolaria juncea* and *Alhagi camelorum*
(d) *Calotropis procera* and *Phyllanthus niruri*

13. Chloramphenicol antibiotic is isolated from
(a) *Streptomyces fradiae*
(b) *Streptomyces venezuelae*
(c) *Streptomyces rimosus*
(d) *Streptomyces aureus*

14. Genes for antibiotic resistance are located in
(a) Nucleus (b) Plasmid
(c) Chromosome (d) Cell wall

15. Antibiotics destroy
(a) Bacteria and viruses
(b) Bacteria and fungi
(c) Fungi and viruses
(d) Bacteria, viruses and fungi

16. Fermentation of sugar to yield alcohol is carried out by
(a) High temperature
(b) Microorganisms
(c) Zymase
(d) Cellulose

17. Distillation of wine is necessary because
(a) It decreases toxicity
(b) It has no action on the nature of wine
(c) It increases turbidity
(d) It prevents further fermentation

18. Cyclosporin produced from *Trichoderma* is useful because
(a) It lowers blood cholesterol
(b) It lowers blood glucose level
(c) It removes toxic materials in the body
(d) It prevents rejection reaction in transplantation

19. Mycorhiza improves the soil structure by
(a) Secreting carbonic acid
(b) Releasing glucosides
(c) Absorbing heavy metals
(d) Adding glycoprotein

20. Primary treatment of sewage involves
(a) Filtration and flocculation
(b) Flocculation and sedimentation
(c) Sedimentation and Homogenisation
(d) Filtration and sedimentation

21. Which of the following alcoholic drinks is produced without distillation?
(a) Brandy (b) Rum
(c) Whisky (d) Wine

22. In sewage treatment, the bacterial flocs are allowed to sediment in a settling tank. This sediment is called
(a) Primary sludge
(b) Secondary sludge
(c) Activated sludge
(d) Anaerobic sludge

23. BOD of wastewater is estimated by measuring the amount of
(a) Oxygen consumption
(b) Total organic matter
(c) Oxygen evolution
(d) Total dissolved solids

24. Methanogens are not found in
(a) Rumen of cattle
(b) Gobar gas plant
(c) Activated sludge
(d) Bottom of waterlogged paddy field

25. Biogas production technology was developed by
(a) IARI and CSIR
(b) IPRI and IARI
(c) CSIR and KVIC
(d) IARI and KVIC

26. Cheese is prepared by separating the curd from liquid part by
(a) *Bacillus megatherium*
(b) *Penicillium camemberti*
(c) *Lactobacillus acidophilus*
(d) *Propionibacterium shermanii*

27. In the preparation of bread, the sequence of enzymes involved is

(a) Amylase, maltase and zymase
(b) Maltase, amylase and zymase
(c) Zymase, maltase and amylase
(d) Amylase, zymase and maltase

28. The scientific name of wine yeast is

(a) *Saccharomyces cerevisiae*
(b) *Saccharomyces ellipsoideus*
(c) *Saccharomyces sake*
(d) *Torulopsis utilis*

29. Statin is produced from

(a) *Trichoderma polysporus*
(b) *Trichoderma viridiae*
(c) *Monoascus purpureus*
(d) *Saccharomyces cerevisiae*

30. Which of the following acids was first to be produced through fermentation?

(a) Acetic acid (b) Lactic acid
(c) Citric acid (d) Oxalic acid

31. The function of streptokinase is to

(a) Lower blood cholesterol
(b) Suppress the immune system
(c) Lower blood glucose level
(d) Dissolve blood clots

32. Which of the following vitamins was first to be produced through fermentation?

(a) Vitamin A (b) Vitamin B
(c) Vitamin C (d) Vitamin D

33. Which of the following antibiotics is broad-spectrum?

(a) Penicillin (b) Bacitracin
(c) Polymyxin (d) Chloramphenicol

34. The main composition of biogas is

(a) CH_4, CO_2 and O_3
(b) CH_4, O_3 and SO_2
(c) CH_4, NO and O_3
(d) CH_4, CO_2 and N_2

35. The production of biogas involves

(a) Solubilisation
(b) Acidogenesis
(c) Methanogenesis
(d) All of the above

36. An example of phosphate mobilising biofertilizer is

(a) *Bacillus megatherium*
(b) *Lactobacillus vulgaris*
(c) *Propionibacterium shermanii*
(d) *Clostridium butyricum*

37. Dextran is a plasma expander produced from

(a) *Bacillus megatherium*
(b) *Clostridium glutamicum*
(c) *Trichoderma polyporus*
(d) *Leuconostoc mesenteroides*

38. Which of the following fungi forms endomycorhiza?

(a) *Agaricus* (b) *Glomus*
(c) *Polyporus* (d) *Puccinia*

39. Which of the following is incorrect regarding baculovirus?

(a) It is species-specific
(b) It has narrow-spectrum applications
(c) It controls lepidopteran insects
(d) All of the above

40. Which of the following parts of plant produces a natural insecticide, Pyrethrum?

(a) Leaf (b) Flower
(c) Fruit (d) Seed

ANSWERS

1.a	11.a	21.d	31.d
2.c	12.c	22.c	32.c
3.b	13.b	23.a	33.d
4.b	14.b	24.d	34.d
5.d	15.b	25.d	35.d
6.a	16.c	26.d	36.a
7.c	17.c	27.a	37.d
8.b	18.d	28.b	38.b
9.b	19.d	29.c	39.d
10.c	20.d	30.b	40.b

CHAPTER-17

ORGANISMS AND POPULATION

Ecology is the study of the interrelationship between organisms and their environment. The term 'Ecology' (Gr. *Oikos*=home, *logos*-study) was coined by Reiter (1868). Ernst Haeckel (1886) first correctly defined ecology as 'the science which deals with the reciprocal relationship of organisms and the external world". Warming is regarded as the father of Ecology. He wrote a book" Oecology of Plants". Famous ecologists are Clarke, Warming, Clements, Odum, and Schimper. Ramdeo Misra is the father of Indian Ecology. Other famous Indian ecologists are G.S.Puri, S.C.Pandeya etc.

Terms:

1. **Species:** It is an organism (individual) of the same kind. The organism is considered a basic unit of study in ecology.
2. **Autecology:** It is the study of a single individual organism (individual species) with respect to its environment.
3. **Race:** It is a group of organisms within a species.
4. **Population:** It is a group of organisms belonging to the same species living in a particular geographical area at a particular time. These organisms interbreed freely. These do not establish a relationship between eating and being eaten. The local population of a species is called deme. The local population of a species of different areas is called the sister population. The local population adapted genetically to its particular environment is called Ecotype. The intermediate form between the two ecotypes is called ecoline. The entire group is called metapopulation. e.g., bacteria in a culture plate, fish in a pond, birds in a grove, elephants in a forest. e.g., Bacteria in a culture plate, Fish in a pond, Birds in a grove, Elephant in a forest etc.
5. **Synecology:** It is a study of a group of organisms (population) with respect to their environment.
6. **Community:** It is a group of organisms of different species (such as bacteria, fungi, plants, animals) in a particular geographical area interacting with one another through competition, predation, mutualism etc. e.g., Pond community
7. **Ecosystem:** It is an interacting system of a biotic community with the physical world (environment). e.g., Terrestrial ecosystem (Grassland and forest) and aquatic ecosystem (Pond and ocean).
8. **Biome:** It is a group of ecosystems stretching over a large physical area characterised by distinctive life forms or in other words, it is a community of plants and animals existing ina large area showing common features of the environment. e.g., Tropical rainforest, Temperate deciduous forest, Temperate grassland, Desert, Tundra, Taiga etc.
9. **Biosphere:** It is an entire inhabitable portion of the earth in which the ecosystem operates. It is the largest biological system. It receives solar energy for the maintenance of life on earth and releases heat energy to the environment by cellular respiration and decomposition of its dead organic remains. It includes
   (a) **Lithosphere:** Solid component (soils, rocks, minerals of earth crust)
   (b) **Hydrosphere:** Liquid component (water of ocean, sea, river, lake, pond)
   (c) **Atmosphere:** Gaseous cover around lithosphere and hydrosphere.
10. **Habitat**: It is a place where an organism lives. It is controlled by temperature, water, light, soil and other biotic components such as parasites, predators, competitors and pathogens.
11. **Niche**: It was coined by Grinnel (1917). It is a specific part of a habitat occupied by an individual limited by its range of tolerance, range of movement, microclimate, type and availability of food, shelter and type of predator. Every niche has its specific environment. It is the specific position and total role of the individual (species) in a habitat. It shows the profession of species. Each species has a distinct niche. Thus, no two species

occupy exactly the same niche for long period. For example, tadpoles and frog occupy different ecological niches because tadpoles are aquatic and herbivorous whereas frog is terrestrial and carnivorous.

12. **Biota**: It is a total collection of organisms of a geographical region.
13. **Environment:** It is a set of things and conditions around an organism.
14. **Weather**: It is a condition of the atmosphere at any place at a particular time.
15. **Climate:** It is long-term weather conditions and factors specific to an area.

POPULATION ATTRIBUTES

1. **Birth rate**: It is the number of individuals added in initial size of population.

$$B=\frac{\Delta N}{N}=\frac{\text{Number of individuals added}}{\text{Initial size of population}}$$

2. **Death rate**: It is the number of individuals removed in the initial size of the population. It often increases with age.

$$D=\frac{\Delta N}{N}=\frac{\text{Number of individuals removed}}{\text{Initial size of population}}$$

3. **Sex ratio**: It is the number of females in a population per 1000 males.

$$\text{Sex ratio}=\frac{\text{Number of females}}{\text{1000 males}}$$

4. **Age distribution**: It is the distribution of individuals of different age groups. There are three ecological groups in a population such as pre-reproductive, reproductive and post-reproductive individuals. These determine the three types of growth.
   (a) Positive growth: A population has a greater number of pre-reproductive individuals.
   (b) Zero growth: A population has an equal number of pre-reproductive and post-reproductive individuals.
   (c) Negative growth: A population has a greater number of post-reproductive individuals.
5. **Age pyramid**: It is a graphic representation of different age groups found in a population. It shows the following three different statuses of growth.
   (a) **Triangular:** It has more pre-reproductive (younger) individuals, moderate reproductive individuals and less post-reproductive individuals. It is an indicator of a growing population (positive growth) because the birth rate is high.
   (b) **Bell-shaped:** It has the same number of pre-reproductive and reproductive individuals, but the number of post-reproductive individuals is less. It is an indicator of a stable population (neither growing nor declining) because the birth rate is equal to the death rate hence there is zero population growth.
   (c) **Urn-shaped:** It has a lesser number of pre-reproductive individuals than both reproductive and post-reproductive individuals. It is an indicator of a declining population (negative growth) because the birth rate is low.
6. **Population density (size)**: It is the number of individuals of a species in a unit area (space) at a given time.

$$D=\frac{\text{Number of individuals}}{\text{Space}}$$

Space is expressed in two dimensions (m^2) for terrestrial organisms and in three dimensions (m^3) for aquatic organisms and the organisms suspended in space.

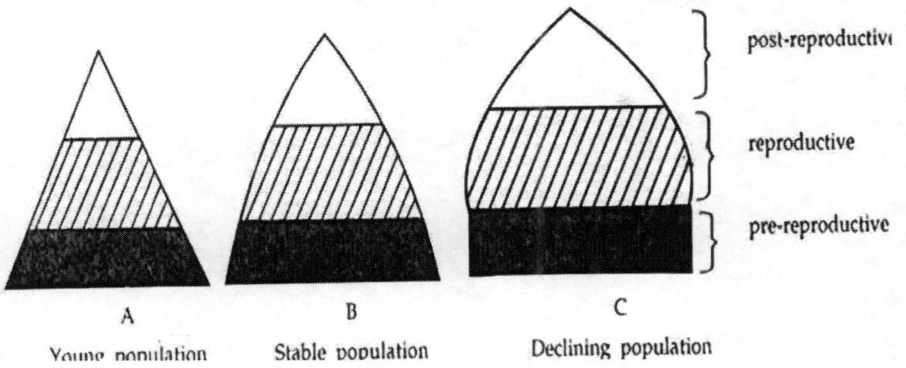

Population density depends on

(i) Natality: It is the number of births (or hatching/germination) per unit population per unit time. It adds a new number to the population hence increasing the density (size) of the population.

(ii) Mortality: It is the number of deaths per unit population. It removes the individuals from the population and hence decreases the density (size) of the population.

(iii) Immigration: It is the permanent inward movement of the same individuals into a local population. It increases the size of the local population.

(iv) Emigration: It is the permanent outward movement of the same individuals from a local population. It decreases the size of the local population. Under normal conditions, birth and death are important factors for influencing population density. Under special conditions, immigration and emigration are important factors for influencing population density.

Density= (Natality + Immigration) - (Mortality + Emigration)

Population density is determined by the following factors:

(a) Counting all the individuals in an area

(b) Measuring biomass

(c) Determining relative density

(d) Counting indirectly such as pug marks and faecal pellets in the case of tigers in national parks and reserves.

The growth of the population is controlled by the interaction between three factors which are as follows:

(a) **Biotic potential** (Reproductive potential): It is the natural capacity of a population to increase at its maximum rate when all the environmental conditions are favourable. It is very high for all animals. If it is unchecked, then the numbers of any species will quickly overrun the world.

(b) **Environmental resistance**: It is the sum of inhibitory factors such as unavailability of food, and shelter, drought, flood, fires, temperature fluctuations, accidents, pathogens, parasites and predators to check the biotic potential.

(c) **Carrying capacity**: It is the highest limit of population size or it is capacity beyond which growth is not possible. It is the number of individuals that a particular favourable environment can support over a long period (indefinitely). It is symbolised as K. It is property of the environment. It varies over space and time with the abundance of limiting resources. If a population reaches the carrying capacity of its environment, then there is zero growth rate of that population.

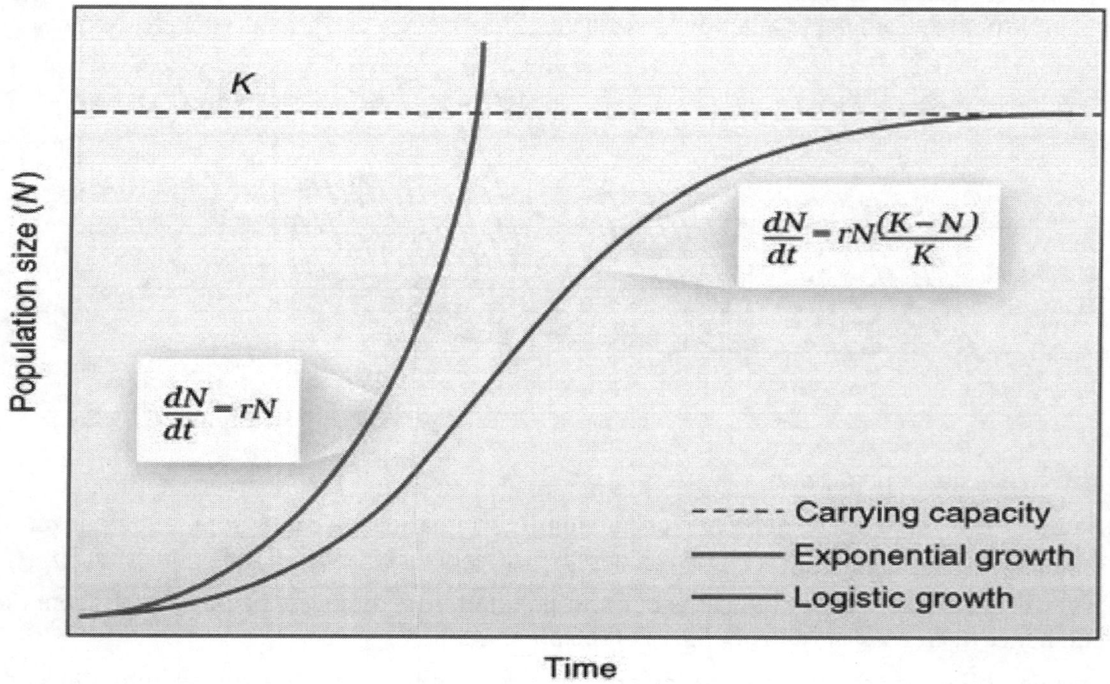

Population growth models: There are two models which are as such

(a) Exponential growth: It occurs when resource (food and space) is unlimited. It undergoes **lag phase** in which growth occurs slowly because individuals adapt themselves to the new environment and **log phase** in which growth occurs very rapidly and continues till enough food is available. With the excessive increase in population, the food is exhausted (deficiency of food) and environmental resistance increases hence there is mass mortality by which the population density (size) suddenly declines.

It shows J-shaped curve.

Biotic potential is realised because birth rate (natality) is higher than death rate (mortality).

It is shown by algal blooms, *Paramecium*, Insects, Lemming of Tundra and annual plants. The population of insects shows explosive increase in numbers during rainy season and disappear at the end of season. It is expressed as

$$dN/dT = rN$$
$$= (B-D)N$$

where dN /dt =change in population size per unit time,

r= intrinsic rate of natural increase

N= Population size, B= Birth rate, D=death rate

The integral form of exponential growth is represented as

$$N_t = N_o e^{rt}$$

where Nt=population size after time t;

No= population size in time t and

e= base of natural logarithm.

(b) Logistic growth: It occurs when resources (food and space) is limited hence it is more realistic than exponential growth. It is also called Verhulst-Pearl logistic growth. It undergoes the following phases:

Lag phase: Growth occurs slowly because individuals adapt themselves to the new environment.

Log phase: Growth occurs rapidly.

Inflexion point: Growth slows down as environmental resistance increases and finally

Stationary phase or plateau phase: Growth stops and the population stabilises around carrying capacity because the rate of mortality and natality become equal to each other.

It shows a sigmoid or S-shaped curve.

The equilibrium number of individuals or growth of organisms in a population is determined by carrying capacity (K) only.

It is shown by bacteria, and yeast cells in a culture medium and most organisms.

It is represented as

$$dN/dt = rN \left(\frac{1-N}{K}\right)$$

where dN/dt= Rate of change in population size

r= intrinsic rate of natural increase (natality-mortality)

N= size of population, K= carrying capacity

K-N/ K= environmental resistance

Exponential growth	Logistic growth
It is J-shaped.	It is S-shaped
It occurs when resources are abundant or unlimited.	It occurs when resources are limited or finite.
Population grows beyond carrying capacity.	The population cannot grow beyond carrying capacity.
The declining phase is not seen.	The declining phase is seen before reaching the equilibrium.
The exponential phase or log phase is of longer duration.	The exponential or log phase is of shorter duration.
Population crash occurs after growing to maximum population.	Population crash is generally not seen.

POPULATION INTERACTION

It is essential for the survival of different species and the community. It is categorised into neutral, positive and negative interactions depending upon the nature of the effect. There are three types of interaction:

Neutral interaction: It is neither beneficial nor harmful interaction.

1. **Neutralism**: It is the interaction between two individuals of two different species in which both organisms are neither benefited nor harmed. e.g., rats, shrews and rabbits live in the same area, but none will be affected.

Positive interaction: It is the interaction in which at least one partner is benefitted, and neither is harmed. It includes:

1. **Symbiosis** (obligate mutualism): It is the interaction between two individuals of two different species in which both organisms benefit from each other, and this association is obligatory (essential for the existence of both organisms). The term 'symbiosis' was coined by De Bary. It shows coevolution (both partners develop certain features to be successful).

   Examples:

   (a) Algae and fungi form lichen.

   (b) Algae and coelenterates form coral reefs.

   (c) Fungi and roots of higher plants form mycorrhiza.

   (d) *Rhizobium* and leguminous plants form root nodule.

   (e) Cyanobacteria and coralloid root of *Cycas* perform nitrogen fixation.

   (f) *Salvia* and bees perform pollination.

(g) Fig and gall wasp cause pollination.

(h) *Yucca* and pronuba moth cause pollination.

(i) Mango and ants cause pollination.

(j) *Chlorella vulgaris* (green alga) and *Hydra.* The alga provides O_2 and food to *Hydra* which in return provides shelter and nitrogenous components to alga.

(k) *Trichonymph* (Protozoa) and Termite. In this *Trichonymph* secretes cellulase enzyme which helps in the digestion of cellulose from wood by termite which in return provides shelter, a suitable environment and food to *Trichonymph.*

(l) *E. coli* and human beings. *E. coli* feeds on undigested food inside a human's intestine and in return, it releases vitamins and checks the growth of putrefying bacteria for humans.

2. **Protocooperation** (Facultative or Partial mutualism): It is the interaction of two individuals (organisms) of different species in which both organisms are mutually beneficial, but this association is non-obligatory (they can live independently).

Examples:

(a) Egret (bird) eats lice and ticks off the body of cattle and thus cattle gets relief.

(b) Egyptian plover (bird) eats leeches and decaying meat stuck in the teeth of crocodile when it opens its mouth and thus crocodile gets relief.

(c) Ox pecker (bird) eats lice and ticks off the body of rhinoceros and buffalo and thus these mammals gets relief.

(d) Hermit crab (*Eupagurus*) and sea anemone (*Adamsia*). Hermit crab takes sea anemones on its back to other feeding zones and in return, sea anemone protects crab from enemies.

3. **Commensalism**: It is the interaction of two individuals of different species in which one organism is benefited and the other is neither benefited nor harmed.

Examples:

(a) Orchid (epiphyte) grows on mango plant for space

(b) Algae grows animals (epizoic algae)

(c) Spider builds web on a tree.

(d) Lianas (vascular plants rooted in soil) get support from another plant for their erectness.

(e) Clownfish is protected from enemies by carrying sea anemones.

(f) Barnacle (an arthropod) is attached to a whale for space or shelter.

(g) Remora or *Echenesis* (sucker fish) are attached by dorsal fin to shark or whale for dispersal, protection from predators and obtaining nutrition from leftovers of shark or whale.

(h) Vulture eats the leftovers of a lion.

(i) Jackal eats the leftovers of a lion.

(j) Bacteria grow on the surface of root (Rhizoplanic bacteria).

(k) Bacteria grow on the surface of leaf (Phylloplanic bacteria).

Negative interaction: It is an interaction in which at least one partner is harmed.

1. **Predation:** It is the interaction between two organisms (individuals) of different species in which one organism (predator) captures, kills and eats the other (prey) to obtain food. The predators are mostly animals and have small population whereas preys are also animals that have large populations due to their high reproductive (biotic) potential. Prey will exponentially grow in the absence of predators.

Significance:

(a) Predators act as conduits between trophic levels for transfer of energy

(b) Predators maintain species diversity by reducing the competition among competing prey species (It was observed in the American pacific coast where all the starfish (*Pisater*) were removed from the particular area then several species of invertebrates became extinct within a year due to intraspecific competition).

(c) Predators keep the population of prey in check to maintain equilibrium.

(d) Predators help in the biological control of pests.

(e) Predators help in the growth of vegetation by checking the population of herbivores.

Examples:

(a) **Cannibalism**: It is a type of predation in which a larger individual of a species eats the smaller one of the same species. The large fish eats small fish.

(b) Predacious fungi such as *Dactylella, Dactylaria, Arthrobotrys* and *Zoophagus* kill annelids. These fungi have rings in mycelium which act as a noose and after that haustoria are released.

(c) Carnivores eat herbivores.

(d) Insectivorous plants such as *Nepenthes, Dionea, Utricularia* etc kill insects.

(e) Cochineal insect (*Cactoblastis canctorum*) kills *Opuntia*.

(f) Gambusia fish kills mosquito larvae.

(g) Ladybird beetle kills the aphids.

(h) Mynah kills red locust.

(i) Herbivores eat plants.

2. **Parasitism**: It is the interaction between two organisms (individuals) of different species in which one organism (parasite) benefits by obtaining nutrition from other organisms (host) which is harmed.

Parasitic adaptations:

(a) Loss of digestive system as in *Taenia*

(b) Occurrence of anaerobic respiration as in *Ascaris, Taenia* and *Fasciola.*

(c) Presence of adhesive organs(suckers) as in leech and *Taenia*

(d) Loss of sense organs

(e) Well-developed reproductive system

(f) Excessive multiplication

Types of parasites:

(a) **Ectoparasite**: These live on an external surface of the host. e.g., Lice, Tick, Flea, Aphid

(b) **Endoparasite**: These live inside the body of a host. It may be intracellular e.g., Plasmodium; body fluid e.g., *Trypanosoma*; Intestinal worms such as *Ascaris, Taenia.*

(c) **Temporary**: These live only for a brief period. e.g., Female mosquito, bedbug, Leech

(d) **Permanent**: These live for lifetime. e.g., *Ascaris, Entamoeba, Taenia*, Lice.

(e) **Partial or semi-parasite**: These are green and prepare their food but they depend on the host only for water and minerals.

Root parasites: These obtain nutrition from roots. e.g., *Santalum album* (largest parasite) found on *Albizzia, Dalbergia*, Morinda etc. *Striga* found on sorghum, sugarcane etc.

Stem parasite: These obtain nutrition from stems e.g., *Loranthus* (found on *Acacia, Albizzia, Dalbergia*, mango), *Viscum*.

(f) **Total or Holoparasite or obligate parasite**: These are non-green and fully dependent on the host for food and water. They obtain nutrition by haustoria.

Root parasites: These obtain nutrition from roots. e.g., *Rafflesia* (found on *Ficus*), *Orobanche* (found on crucifers, tomato, potato), *Epifagus*

Stem parasites: These obtain nutrition from the stem. e.g., *Cuscuta* (found on Citrus, *Zizyphus*); *Arceuthobium* (smallest parasite) found on *Pinus, Juniperus* etc.

(g) Pathogenic: These cause disease. e.g., Fungi (*Puccinia, Ustilag*) and Bacteria (*Mycobacterium, Vibrio, Corynebacterium*).

(h) Non-pathogenic: They do not cause disease but only they obtain a small amount of nutrition from the host. e.g., *Entamoeba coli*, Aphid eating sap from plants.

(i) Hyperparasite or Epiparasite. They live on another parasite e.g., Bacteriophage lives on a bacterium (parasite), *Pasteurella pestis* which also lives on another parasite rat flea (*Xenopsella cheopsis*).

(j) Brood parasitism: In fact, it is a case of co-evolution. In this cuckoo (a parasitic bird) lays eggs in the nest of the crow (a host) and lets the host incubate them. The eggs of the cuckoo develop to resemble the host's egg in size and colour to reduce the chance of the cuckoo locating the foreign eggs.

Significance:

(a) Parasite shows host specificity.

(b) Parasite has high reproductive or biotic potential.

(c) Parasite is smaller than host.

(d) Parasite reduces survival, growth and reproduction of the host.

(e) Parasite makes host weak, so it is prone to be attacked by predators.

(f) It is a case of co-evolution (if the host develops a certain mechanism to resist or avoid the parasite then the parasite will also develop certain features or mechanisms to counteract the host to be successful with the same host species.

3. **Amensalism**: It is the interaction of two individuals of different species in which one organism is harmed and the other is neither harmed nor benefited.

Examples:

(a) *Penicillium notatum* inhibits the growth of a bacterium, *Staphylococcus aureus* by secreting allelochemical (Penicillin).

(b) *Chlorella vulgaris* produces chlorellin which is toxic to other species.

(c) *Trichoderma* inhibits the growth of *Aspergillus.*

(d) *Convolvulus arvensis* checks the growth and germination of wheat.

(e) Roots of *Parthenium argentatum* check the growth of other plants by secreting trans-cinnamic acid.

(f) Roots of *Juglans regia* (Walnut) check the growth of apple, alfalfa tomato etc by secreting juglone.

(g) Roots of smoother crops (barley, sunflower and *Sorgum*) inhibit the growth of weeds by secreting allelochemicals.

4. **Competition:** It is the interaction (rivalry) between two or more individuals to obtain the same resource (food and shelter etc) in which one organism will die. It is of two types:

(a) Intraspecific: It is the competition between individuals of same species. It controls the size of the population. It is more acute and intense because all the individuals have similar requirements of food, light, water, space, shelter and mate.

(b) Interspecific: It is competition between individuals of different species for the same resources. It is essential for organic evolution.

Different terms related to competition:

1. **Competitive exclusion**: It was studied by Gause (1984). In this two closely related compatible species competing for the same limited resource cannot coexist in the same place hence one of them gets eliminated after some time due to strong competition. Thus, one species uses the resource more efficiently and reproduces more rapidly than others. It was shown by *Paramecium*. (*Paramecium aurelia* increases its number gradually and there is a decrease in the number of *Paramecium caudata*).

2. **Competition release**: Two closely related compatible species competing for the same limited resource cannot coexist in the same place hence one inferior species is restricted to a small geographical area due to the presence of competitively superior species. It was shown by Barnacle.
3. **Competition coexistence (Resource partitioning):** Two closely related compatible species competing for the same source coexist and avoid competition by choosing different feeding herbivores. It was shown by Warbler birds.

SHORT ANSWER TYPE QUESTIONS

1. Explain ecological niche.
2. Define biosphere.
3. Mention the major climatic regions.
4. Explain biotic potential and carrying capacity.
5. Distinguish between exponential growth and logistic growth.
6. Distinguish between partial parasite and total parasite.
7. Write notes on predation.
8. Write notes on amensalism.
9. Distinguish between intraspecific competition and interspecific competition.
10. Explain competitive exclusion.

LONG ANSWER TYPE QUESTIONS

1. Discuss the population growth models.
2. Discuss the different attributes of the population.
3. Explain positive population interaction with response to symbiosis, proto-cooperation and commensalism.
4. Describe the types of parasites and show their adaptations and significance.
5. Explain amensalism with examples.

MULTIPLE CHOICE QUESTIONS

1. Which of the following is incorrect?

(a) Exponential growth occurs when the resource is unlimited
(b) It shows J-shaped growth
(c) Population crash occurs
(d) Biotic potential is not realised

2. Population density of terrestrial organisms is measured in terms of individuals per

(a) Meter (b) Meter^2
(c) Meter ^{3} (d) Meter ^{4}

3. The carrying capacity of a population is determined by

(a) Natality rate
(b) Mortality rate
(c) Predation
(d) Limiting resources

4. Population termed r-strategists have

(a) J-shaped growth curve
(b) S-shaped growth curve
(c) Pioneer species
(d) All of the above

5. Population growth curve in most animals except human beings, is

(a) J-shaped
(b) S-shaped
(c) J-shaped with tail
(d) S-shaped with tail

6. Race is defined as

(a) Group of organisms of the same species
(b) Group of organisms within a species
(c) Group of organisms of different species
(d) Group of complex biotic community

7. If the birth rate is 50 and death rate is 40 and the population size is 80 then dN/dT is

(a) 200 (b) 400
(c) 800 (d) 1600

8. A person studied the population of cats in which the average natality was 100, average mortality 90, the average immigration was 10 and average emigration 15 then the net increase in population is

(a) 5 (b) 10
(c) 20 (d) 40

9. If the birth rate is 100 and the death rate is 10 in a population of 1000 individuals. The percentage natural growth rate is

(a) 5% (b) 9%
(c) 10% (d) 20%

10. In a pond there are 20 lotus plants. Due to reproduction 10 new plants were added last year taking the correct population of 30. The birth rate is

(a) 0.5 (b) 1
(c) 1.5 (d) 2

11. A person caught, marked and released 80 fish in a pond. Later 100 fish were caught randomly in which 40 were found to be marked. The number of fish in the pond is

(a) 120 (b) 15
(c) 180 (d) 200

12. The growth rate of a population according to the logistic model is zero when

(a) N/K is equal to zero
(b) N/K is exactly zero.
(c) N is close to carrying capacity
(d) Death rate is higher than birth rate

13. The ratio between mortality and natality is called

(a) Population ratio
(b) Vital index
(c) Density coefficient
(d) Carrying ratio

14. Which one of the following correctly represents an organism and its ecological niche?

(a) *Hydrilla* and pond
(b) Desert locust and desert forest
(c) Aphid and leaf
(d) Vulture and dense forest

15. Bell shaped pyramid is an indicator of

(a) Stable population
(b) Growing population
(c) Declining population
(d) None of the above

16. In logistic growth, the growth of organisms in a population is determined by

(a) Carrying capacity (K) only
(b) Size of population (N) only
(c) Biotic potential
(d) Both carrying capacity and Biotic potential

17. A cuckoo laying eggs in the nest of other species of birds shows

(a) Hyperparasitism
(b) Ectoparasitism
(c) Brood parasitism
(d) Commensalism

18. Which of the following interactions acts as a conduit for energy transfer across trophic level?

(a) Parasitism (b) Amensalism
(c) Protocooperation (d) Predation

19. Trichonymphs and termites exhibit

(a) Parasitism
(b) Symbiosis
(c) Protocooperation
(d) Commensalism

20. Remora and Shark exhibit

(a) Protocooperation
(b) Commensalism
(c) Amensalism
(d) Symbiosis

21. Which of the following is incorrect about parasitism?

(a) It is a case of co-evolution
(b) Plasmodium is intracellular endoparasite
(c) Parasite shows host specificity
(d) Parasitic adaptations do not occur in *Ascaris*, *Taenia* and *Hirudo*.

22. Which of the following is incorrect about competition?

(a) Intraspecific competition is more acute than interspecific competition.
(b) Competition exclusion in paramecium was explained by Gause

(c) Competition release was shown by Barnacle.

(d) Resource partitioning was not shown by Warbler birds.

23. The effective population size for completely monogamous species having 40 males and 10 females would be

(a) 10 (b) 20

(c) 30 (d) 40

24. Which of the following ecological factors is density-independent?

(a) Food (b) Predation

(c) Disease (d) Extreme temperature

25. Which of the following shows is most different from others?

(a) Algae embedded in coral tissue

(b) *Salmonella* in the human gastric tract

(c) Cellulolytic bacteria in termite gut

(d) Pollen collecting bees visiting flowers

26. The term 'biotype' means

(a) All individuals having the same phenotype

(b) All individuals having the same genotype

(c) All individuals having a different phenotype

(d) All individuals having a different genotype

27. The term 'ecotype' means

(a) Genetically different but phenotypically similar individual

(b) Genetically similar but ecologically different individuals

(c) Genetically adapted ecological race

(d) Genetically and phenotypically dissimilar individuals.

28. The characteristic population growth curve of bacteria is

(a) Straight (b) S-shaped

(c) Random (d) J-shaped

29. Which of the following is not a limiting factor to the growth of the population?

(a) Availability of food

(b) Increasing population of predators

(c) Introduction of disease

(d) Long-term abundance of space

30. Logistic growth is represented as

(a) dN/dt= rN

(b) dN/dt= rN(1-N/K)

(c) dN/dt= rN(1-K/N)

(d) dN/dt= r(1-N/K)

31. Competitive exclusion is best observed in

(a) Warbler birds (b) *Bonellia*

(c) *Paramecium* (d) *Euglena*

32. Which of the following acts as a conduit between trophic levels for the transfer of energy?

(a) Parasite (b) Symbiont

(c) Commensal (d) Predator

33. Hermit crabs and sea anemones show best examples of

(a) Symbiosis

(b) Protocooperation

(c) Commensalism

(d) Amensalism

ANSWERS

1. d	11. d	21. d	31. c
2. b	12. b	22. d	32. d
3. d	13. b	23. b	33. b
4. a	14. c	24. d	34.
5. a	15. a	25. b	35.
6. b	16. a	26. a	46.
7. c	17. c	27. c	37.
8. a	18. a	28. d	38.
9. b	19. b	29. d	39.
10. a	20. b	30. c	40.

CHAPTER-18

ECOSYSTEM

The ecosystem is an interactive system between living and nonliving components of the environment where both are interdependent. The term 'ecosystem' was coined by Tansley (1935).

It is an open system because there is continuous and variable entry and loss of energy and materials. It is the structural and functional unit of the environment.

Terms of Ecosystem:

1. Biocoenosis proposed by Carl Mobius
2. Microcosm proposed by Forbes
3. Biogeocoenosis proposed by Sukhachev
4. Holocoen proposed by Friederich

Types of ecosystem:

1. **Natural ecosystems:** These are naturally developed and operate without human intervention. These are of the following types
    (a) Terrestrial ecosystem: Grassland, forest, desert, garden
    (b) Aquatic ecosystem:
        (i) Freshwater (Limnetic): lake, pond (lentic or stagnant); river, stream (lotic or running) and
        (ii) Marine: Ocean, sea, estuary
2. **Artificial ecosystems**: These are developed by human beings hence called man-made. e.g., agriculture (largest), lake, dam, aquarium, garden, spacecraft etc.

Types of ecosystems based on size:

1. Megaecosystem: It is large. e.g., Ocean
2. Mesoecosystem: It is of medium size. e.g., Desert, forest, grassland
3. Microecosystem: It is of small size. e.g., Lake, pond
4. Nanoecosystem: It is of very small size. e.g., Aquarium, garden, log of wood etc.

COMPONENTS OF THE ECOSYSTEM

The ecosystem has two functional components which are as follows:

1. **Abiotic components**: These include kinds, quantity, quality and distribution of physical and chemical factors. These determine the maintenance of biotic components. These are temperature, light, water, soil, wind, humidity, topography, inorganic substances such as N, P, S, Ca etc. and organic substances such as carbohydrates, protein, lipids etc.
2. **Biotic components**: These include kinds, numbers and distribution of living organisms. These obtain inorganic nutrients and energy from abiotic components.
    (a) **Producers (Autotrophs)**: These are organisms that are capable of synthesising food in the presence of sunlight and chlorophyll by using inorganic compounds. These are sources of food for all other organisms which are incapable of synthesising their food. These are also called **converters or transducers** because they convert light energy to chemical energy. They take part in building their bodies by synthesising organic compounds as well as releasing energy for various physiological activities. They maintain CO_2/O_2 balance in nature. They pick up CO_2 from the atmosphere and release O_2 as a result of photosynthesis.

In aquatic ecosystems, major producers are phytoplanktons or rooted or large floating plants in freshwater ecosystems and algae in marine ecosystems but in terrestrial ecosystems, rooted plants (trees) are major producers.

Other producers are photoautotrophic and chemoautotrophic bacteria, cyanobacteria, and photosynthetic protists such as diatoms, dinoflagellates, and euglenoids.

(b) **Consumers** (Heterotrophs): These are organisms (animals) that are incapable of synthesising their food and depend on other organisms for food hence these are known as phagotrophs.

(i) **Primary consumers**: (Consumers of first order): These are organisms that directly feed on plants only. These are also called herbivores. These are also known as key industry animals because they convert plants into animals. e.g., grasshopper, deer, cattle (terrestrial ecosystem); molluscs, tadpoles and mosquito larvae (aquatic ecosystem).

(ii) **Secondary consumers** (Consumers of second order): These are organisms that feed upon herbivores. These are also called primary carnivores. e.g., Frog, spider, dog, cat, tiger and lion (in the terrestrial ecosystem); small fish, *Hydra* and frog (in the aquatic ecosystem).

(iii) **Tertiary consumers** (Consumers of third order): These are organisms that feed upon secondary consumers. These are also called secondary carnivores. e.g., snakes, hawks (of terrestrial ecosystem); sharks, and whales (of an aquatic ecosystem).

(iv) **Quaternary consumers** (Consumers of fourth order): These cannot be preyed upon by other animals. These are also called top carnivores. e.g., lions, vultures, falcons, peacocks.

(c) **Decomposers** (Saprotrophs): These are organisms that feed upon organic wastes of dead bodies of organisms. These are also called **osmotrophs** because they secrete extracellular enzymes to digest the organic matter hence the digested form of organic matter is partly absorbed by microbes for their assimilation. The remaining adds new materials and minerals back into the substratum. These are called **microconsumers** because of their small size. These are also called **reducers** because they remove or degrade the dead bodies of organisms. These may be called **mineralisers** because they make the nutrients available to the producers thus, they maintain a flow of nutrients in an ecosystem through the biogeochemical cycle.

(d) **Detrivores**: They break down organic matter as they feed upon dead organic remains (detritus). e.g., Earthworm, Termite.

(e) **Scavengers**: They feed on dead bodies of organisms. e.g., Crow, vulture, and Carrion beetle.

Different components of an ecosystem:

1. **Aquatic Ecosystem (Pond)**:

(a) Abiotic components: Water, temperature, light, pH of water, inorganic and organic substances

(b) Biotic components: These include Producers (plants), Consumers may be Primary (zooplanktons such as insect larvae, rotifers, molluscs), Secondary (Insects, medium fish, frog, snake), and Tertiary (large fish such as Shark) and Decomposers (Bacteria, Fungi).

2. **Terrestrial Ecosystem (Forest):**

(a) Abiotic components: include temperature, light, inorganic and organic substances.

(b) Biotic components: These include Producers plants (herbs, shrubs, trees); Consumers which may be Primary (Ants, Bugs, Birds, Herbivores), Secondary (Snake, lizard, large bird) and Tertiary (Tiger, lion) and Decomposers (Bacteria, Fungi).

Features of Ecosystem:

(a) **Species composition**: It is the variety of species present in an ecosystem. It depends upon geography, topography and location. It varies in different ecosystems. It is minimum in desert and arctic regions. It is maximum in tropical rainforests (in terrestrial regions) and coral reefs (in aquatic regions).

(b) **Stratification:** It is the presence of vertical layers that depend on the density of vegetation. It is a minimum (3 layers such as herbs, shrubs and trees) in desert and arctic regions and a maximum (7 layers such as tall trees, canopy trees, understorey trees, tree layer, shrub layer, herb layer, grass layer) in tropical rain forest.

(c) **Trophic structure:** It is the presence of a specific food chain and food web in the ecosystem. It is usually a grazing food chain in grassland.

(d) **Standing crop**: It is the amount of living matter or biomass (number of living organisms present in a unit area) in an ecosystem present at any time. It is usually measured in dry weight.

(e) **Standing state:** It is the amount of inorganic nutrients (materials present in soil and water) in an ecosystem per unit area at any time. It varies in different seasons and ecosystems.

FUNCTIONAL COMPONENTS OF ECOSYSTEM:

1. **Productivity**: It is the rate of production of organic matter (biomass) at any trophic level per unit area in a given period. It is measured as weight ($g/m^2/yr$) or energy ($Kcal/m^2/yr$).

   (a) **Primary productivity**: It is the amount of organic matter produced per unit area per unit time.

   - Gross primary productivity (GPP): It is the total amount of organic matter synthesised by producers as a result of photosynthesis. It depends on the photosynthetic efficiency of plants, the availability of solar energy and the availability of inorganic nutrients.
   - Net primary productivity (NPP): It is the net amount of organic matter stored by producers after loss in respiration (R) in plants. It is available biomass for the consumption of heterotrophs.

   $$NPP = GPP - R$$

   (b) **Secondary productivity**: It is the rate of accumulation of organic matter by consumer level per unit area per unit time. It depends on loss while transferring energy from the previous trophic level and consumption due to respiration and predation.

   (i) Gross Secondary productivity (GSP): It is the amount of food energy absorbed.

   (ii) Net secondary productivity (NSP): It is the amount of food energy stored.

   $$NSP = GSP - R$$

   (c) **Net community productivity**: It is the rate of storage of organic matter not used by heterotrophs (consumers) or it is the rate of net synthesis of organic matter by a community per unit area per unit time.

   Features:

   (a) The bulk of the earth's living mantle consists of plants and only a small fraction constitutes animals.

   (b) Primary productivity depends on plant species.

   (c) It varies in different ecosystems.

   (d) It is minimum in desert, arid land and deep sea.

   (e) It is moderate in temperate (coniferous) forests, crops, and grassland.

   (f) It is maximum in tropical rainforests, oceans, coral reefs, estuaries, and sugarcane fields. Coral reef grows in areas having sufficient light, enough warm water and abundant nutrients.

   Measurement of productivity: It is measured in the following ways:

   (a) Ecological efficiency: It is the percentage of energy transferred from one trophic level to another. It differs in different species, populations or communities.

   $$\text{Ecological efficiency} = \frac{\text{Energy converted into biomass at one trophic level}}{\text{Energy present in biomass at previous trophic level}} \times 100\%$$

   (b) Ecological Efficiency=Exploitation efficiency x Assimilation efficiency x Net production efficiency

   (c) Exploitation efficiency: It is prey biomass that is eaten by predator.

   (d) Assimilation efficiency: It is proportion of consumed energy that is assimilated.

   (e) Net production efficiency: It is proportion which assimilated energy is incorporated into growth, storage and reproduction.

(f) Net production efficiency $= \frac{NPP}{GPP} \times 100$

(g) Photosynthetic efficiency$= \frac{\text{Gross primary productivity}}{\text{Incident total solar radiation}} \times 100$

It is 0.8% in forest, 1.15% in grassland, 5% in cropland and 10% in sugarcane.

Factors affecting primary productivity:

(a) Temperature: Primary productivity is less in temperate plants where the temperature is minimum and more in tropical plants where the temperature is maximum.

(b) Light: It is less in the polar region because light is not available and more in the tropical region because light is abundant.

(c) Water: It is less in deserts because the water content is low but more in the tropical rain forests because the water content is high.

(d) Nutrients: It is less in deserts and seas because nutrient content is low but more in estuaries and coral reefs because nutrient content is high.

(e) Photosynthetic efficiency: It is less in forests and more in croplands. It is less in C_3 plants and more in C_4 plants. It is a maximum (10%) in sugarcane.

2. **Decomposition:** It is the breakdown of complex organic compounds by bacteria and fungi. The organic remains (dead plant parts, animal remains, and excreta) are called detritus. The upper layer of soil is the main site of decomposition in terrestrial ecosystems. It occurs as below-ground detritus (dead roots, underground dead animals) and above-ground detritus (dried plant parts, leaf litter, remains, and excreta of animals).

Process:

(a) **Fragmentation of detritus**: A large amount of detritus was fed upon by detrivores such as earthworms, termites and carrion flies and left small fragments. They cause fragmentation or pulverisation of detritus in the digestive tract of herbivores and a part comes out. The part digested by detrivores is immobilised. Due to fragmentation, leftover detritus comes to have a large surface area.

(b) **Catabolism**: The complex and insoluble organic substances are converted into simple and soluble organic and inorganic substances by extracellular enzymes released by decomposers such as bacteria and fungi. A part of broken-down food is taken up by decomposers and immobilised.

(c) **Leaching**: Soluble substances (sugars and inorganic nutrients) drain or enter in deeper layer of soil by percolating water.

Detrivores	Decomposers
They eat detritus (dead organic remains).	They obtain nutrition from detritus.
They do not decompose organic matter but ingest it	They decompose organic matter.
They do not release enzymes	They release enzymes.
They cause pulverisation of detritus	They cause humification and mineralisation.
e.g., earthworm, Termites, Carrion flies	e.g., Bacteria, Fungi

(d) **Humification**: It is the formation of humus by partial decomposition of detritus. It is partially decomposed, amorphous, dark-coloured, slightly acidic and colloidal. Humus is rich in cellulose and lignin. It is quite resistant to microbial action. It is a reservoir of NH^{+4}, Ca^{+2}, Mg^{+2} and K^{+} nutrients. These nutrients are released slowly as the humus is slowly decomposed. It maintains the soil moisture and aeration.

(e) **Mineralisation**: It is the release of inorganic substances such as minerals (NH_4^{+}, Ca^{+2}, Mg^{+2}, K^{+}) and non-minerals (CO_2 and H_2O) from organic matter. This process is slow due to immobilisation and trapping in the humus.

Detritus

↓Fragmentation, Catabolism and Leaching

Products

↓Humification

Humus

↓Mineralisation

NH_4^+, Ca^{+2}, Mg^{+2} (minerals) and CO_2, H_2O (non-minerals)

Factors affecting decomposition:

(a) Temperature: The optimum temperature for decomposition is 25°C. The rate of decomposition of detritus decreases if the temperature is less than 10°C and increases if the temperature is more than 25°C.

(b) Moisture: Optimum moisture is required. The rate of decomposition of detritus decreases if there is low and high moisture.

(c) pH of soil: The rate of decomposition of detritus decreases in acidic soil and increases in neutral or slightly alkaline soil.

(d) Oxygen content (aeration): The rate of decomposition of detritus decreases in anaerobic conditions and increases in aerobic conditions.

(e) Composition of detritus: The rate of decomposition of detritus is very slow if detritus is rich in cellulose, chitin, lignin etc and very rapid if detritus is rich in proteins, fatty acids and water-soluble sugars.

3. **Energy flow:** It is a continuous flow of energy starting from the sun to biotic components and then to the physical environment. It is governed by the first and second laws of thermodynamics.

Features:

(a) It is the basis of life. Living organisms use energy in the form of radiant energy (sunlight) and fixed energy (potential chemical energy bound in various organic substances).

(b) Its source is sunlight. Less than 50 % of solar energy incident over the earth is a photosynthetically active region (PAR) and the rest more than 50 % is lost (20% reflected by cloud +10% absorbed by cloud + 10% absorbed by gases/vapours + 10 % scattered by dust). Out of 50 % PAR, 95-99 % is lost from the plants by reflection, radiation or evaporation and the remaining 1-5% of incoming radiation is utilised by chlorophyll and used in photosynthesis (gross primary productivity). Out of 1-5 %, 0.8-4 % is utilised in net primary productivity and the rest 0.2-1% is lost in respiration.

(c) It is unidirectional. (Sunlight⟶Producers⟶Consumers). It does not pass back to autotrophs.

(d) It starts with the fall of sunlight on the green surface of plants, a part of it is transformed into chemical energy which is stored in various organic products in the plants. When herbivores consume plants as food and convert chemical energy accumulated in plant products into kinetic energy, degradation of energy will occur through its conversion into heat. When carnivores consume herbivores then further degradation will occur likewise.

(e) It decreases with a rise in trophic level (Energy of Producers > energy of primary consumer > energy of secondary consumer > top consumer).

(f) It determines the diversity and density of organisms.

(g) It determines the developmental and functional status of the ecosystem.

(h) It is either passed on to the higher trophic level or becomes available to detrivores or decomposers after the death of organisms.

Ten per cent law

It was proposed by Raymond Lindeman (1942). During the transfer of energy from one trophic level to the next, about 10 % of energy is stored in the flesh for bodybuilding and the remaining 90 % of energy is lost in ingestion, respiration, incomplete digestion and maintenance of body heat. The respiratory loss gradually increases in successive trophic levels. It is 20 % at the producer level, 30 % at the consumer level and 60 % at the top carnivore level.

Producer biomass (1000Kcal)
10% transferred ↓ 90 % lost
Primary consumer biomass (100 Kcal)
10% transferred ↓ 90 % lost
Secondary consumer biomass (10 kcal)
10% transferred ↓ 90 % lost
Top consumer biomass (1 Kcal)

Single channel energy flow model	Lindemann
Y-shaped energy flow model	E.P.Odum

FOOD CHAIN

It is the sequence of organisms of an ecosystem through which food and its energy continuously passes with each member. It can be seen in any ecosystem. It was first proposed by Charles Elton (1927).

Features:

(a) It is generally a straight and simple food chain that rarely occurs in nature. (Producers →Herbivores →Carnivores →Decomposers).

(b) It shows gradual decreases in biomass, number and energy of individual with rise in trophic level.

(c) It is sustained by producers and decomposers.

(d) The shorter the food chain, the greater the energy available.

Types of food chain:

1. **Grazing food chain** (Predatory food chain): It is most common. It directly depends on sunlight. It gets energy from sunlight. It starts from green plants (small organisms), goes to grazing herbivores and passes to carnivores (large animals). It supports the detritus food chain by providing organic matter. It is of two types:

   (a) **Terrestrial ecosystem:** It shows the following examples:
   - Grass→ Grasshopper→ Bird→ Hawk
   - Grass → Grasshopper → Lizard → Hawk
   - Grass → Grasshopper → Frog → snake → Peacock → Hawk
   - Grass → Rabbit → Fox → Wolf → Lion
   - Grass → Deer → lion
   - Grass → Insects → Insectivorous insect → Bird → Hawk

   (b) **Aquatic ecosystem:** It shows the following examples:
   - Phytoplankton → Zooplankton → Small fish → Large fish → Crocodile/Shark
   - Phytoplankton → Zooplankton → Frog → Snake
   - Phytoplankton → Zooplankton → Crustacean → Predatory insect → Small fish → King fisher

2. **Detritus food chain** (Saprophytic food chain): It is less common. It does not depend on sunlight hence producers are absent. It starts from detritus (organic remains of plants and animals). It gets energy from detritus. It supports the grazing food chain by providing inorganic matter.
   - Detritus → Earthworm (detrivore) → Sparrow → Falcon (large bird)
   - Detritus → Earthworm (detrivore) → Frog → Snake → Hawk

3. **Parasitic food chain:** It is also known as accessory or subsidiary food chain. It starts with the host (producer or large organism) and ends in parasite (small organism) without outright killing. It is often connected with the grazing food chain

- Grass → Cattle → Bacteria
- Tree → Fruit eating bird → Tick/lice → Bacteria/Fungi

FOOD WEB

It is a network of several food chains which become interconnected at different trophic levels forming a number of feeding connections amongst different organisms. One particular animal may feed upon organisms of different trophic levels. e.g., snakes may feed upon mice (herbivores) and frogs (carnivores). Likewise, Jackal is both a carnivore and a scavenger. Sparrow is a primary consumer when it eats seeds, fruits etc. as well as a secondary consumer when it eats insects and worms. Human consumes vegetables (producers), animals (primary consumers or herbivores) and Salmon (secondary consumers or carnivores).

Features:

(a) It is more real than a food chain.
(b) It consists of several food chains.
(c) It is not as straight as the food chain.
(d) It provides several alternative foods to consumers that become polyphagous.
(e) It maintains balance and stability in the ecosystem.
(f) It increases the adaptability and competitiveness of the organ.

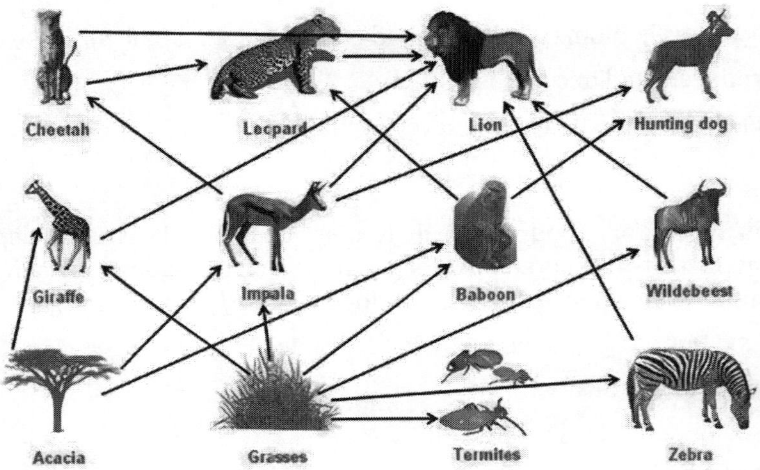

Fig: Terrestrial food web

Trophic level: It is a level in the food chain which is characterised by the method of obtaining its food. The parasites have no fixed trophic level since they feed on producers, herbivores as well as carnivores of various levels. Sparrow comes under more than one trophic level. Its number is 3 to 6. The number of trophic levels is less (3 in tropical forests, 4 in grassland) in terrestrial ecosystems. The number of trophic levels is more (5 in the coastal ocean ecosystem and 6 in the open ocean ecosystem. It is shown as such:

T_1--------(Producers)-------Plants
↓
T_2--------(Primary consumers)------Herbivores
↓
T_3--------(Secondary consumers)---Primary carnivores
↓
T_4---------(Tertiary consumers)-----Secondary carnivores
↓
T_5--------(Quaternary consumers)---Top carnivores
↓
T_6--------(Decomposers)

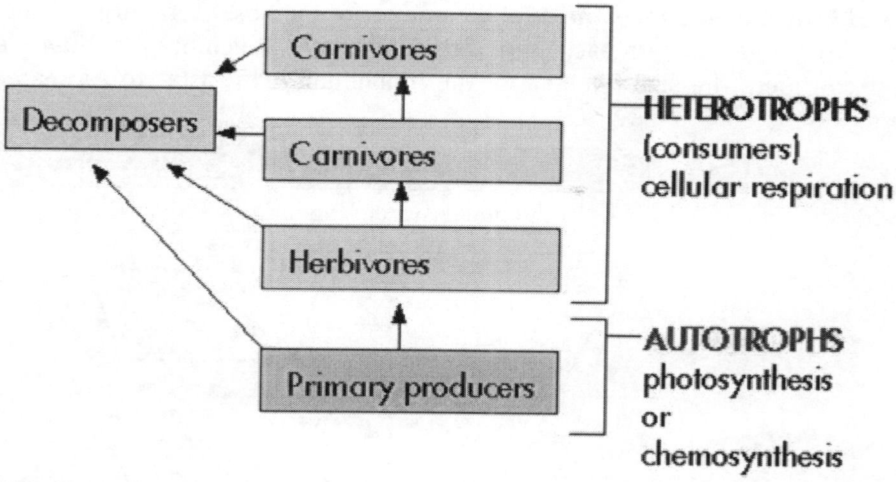

4. **Ecological pyramid:** It was proposed by Charles Elton (1927). It is a diagrammatic representation of the nutritional relationship of organisms of different trophic levels and the flow of energy from one trophic level to another in the ecosystem. It gives an idea about the structure of the ecosystem consisting of the interaction of food chains. The ecological pyramid is not suitable for detrivores and decomposers. It does not represent the food web. It does not include omnivores. It is of three types:

   (a) **Pyramid of Number**: It actually shows the number of individuals involved in each step of the food chain. It does not represent the true picture of the food chain because it does not show the relative effects of the food chain and the size of organisms.

Upright	Grassland ecosystem (terrestrial), Pond ecosystem (aquatic)
Inverted	Parasitic food chain
Spindle shaped	Forest or Tree ecosystem (terrestrial)

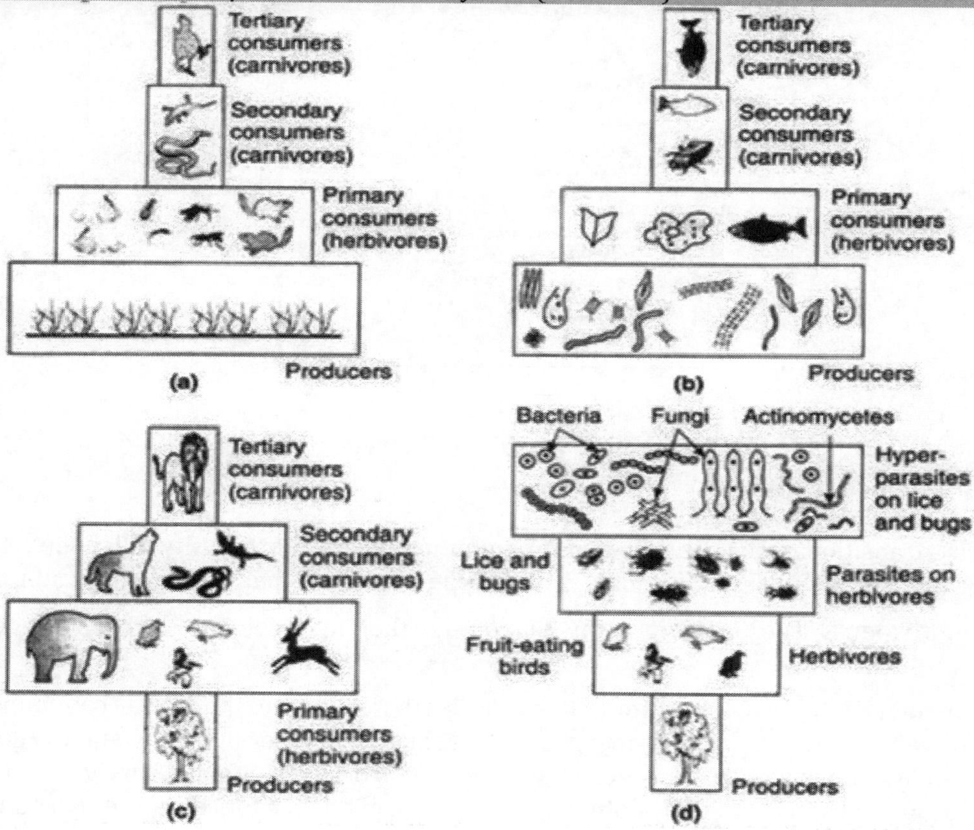

Fig: Pyramid of numbers: (a) Grassland ecosystem, (b) Pond ecosystem, (c) Forest ecosystem and (d) Parasitic food chain

(b) **Pyramid of biomass:** It shows the relative amount of biomass (total organic matter or total dry body weight of all the organisms) in each step of the food chain. It is more real than the pyramid of numbers because it considers the size (weight) of the individual. It is easier to estimate the dry weight in an ecosystem.

Upright	Grassland and forest ecosystem (terrestrial)
Inverted	Pond ecosystem (aquatic)

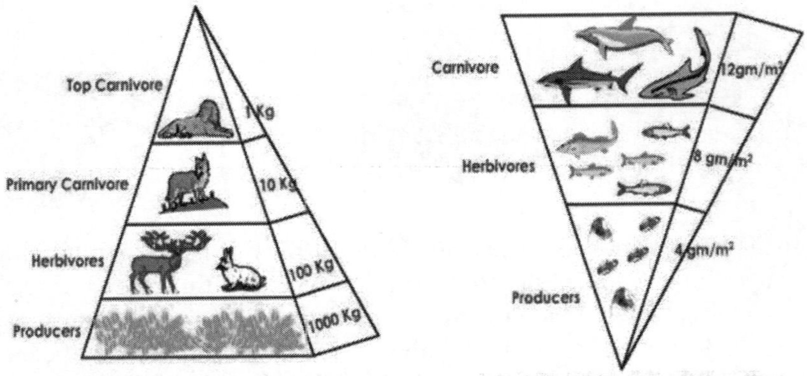

Upright Pyramid of biomass in a Terrestrial Ecosystem

Inverted Pyramid in an Aquatic Ecosystem

(c) **Pyramid of energy**: It shows the amount of energy in each step of the food chain. It gives the best picture of the overall nature of the ecosystem. It is always upright because the energy of producers is always more than primary consumer and so on hence energy goes on decreasing. It is clear that only a part of energy (10%) is transferred to the next trophic level from the previous trophic level and a major part (90%) is lost as heat.

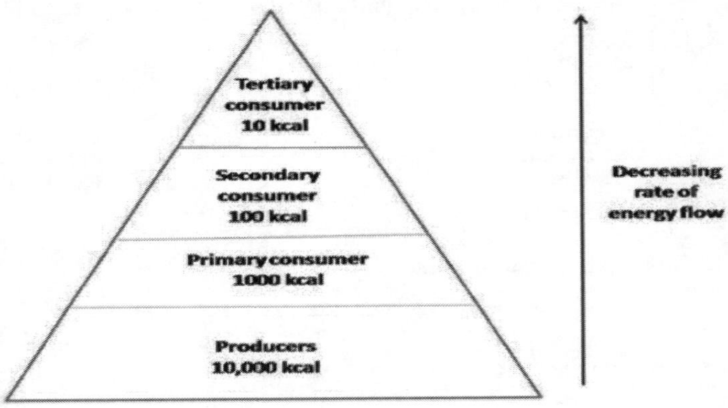

Fig. Pyramid of Energy

Terms related to the ecosystem:

(i) **Ecotone:** It is a transitional area between two communities. It is rich in diversity called the edge effect because it has a large number of species at the edge. Estuary is an ecotone of marine and freshwater habitats.

(ii) **Ecological equivalents:** These are organisms that occupy the same ecological niche in different geographical conditions.

(iii) **Keystone species (Ecological dominants):** It is a species that influences more on its community or ecosystem. It controls the relative abundance of other species. It is small in number but sufficient and it dominates other species in biomass. It has a huge impact on the community's organisation and survival. Its loss can lead to a sharp population drop and extinction of other species that depend on it for certain services and causes serious disruption in the functioning of a community. There are several examples which are as follows:

(a) **Fig** (*Ficus*): It bears fruits during seasons when no other fruits are available. If fig is removed, then many animals will die of starvation during periods of food scarcity.

(b) **Starfish** (*Pisaster*): It is a predator in the intertidal zone of the Pacific coast of North America. If it is removed, then there will be the overproduction of mussel (*Mytilus*) in the entire zone.

(c) **Kangaroo rat**: It is herbivorous and lives in the desert. It eats large seeds. If it is removed, then there will be an increase in a large number of herbs and shrubs thus whole desert will become green desert grassland.

(d) **African elephant**: It destroys shrubs and small trees by its feeding habit and converts entire woodland habitat to open grassland.

SHORT ANSWER TYPE QUESTIONS

1. Mention the several types of ecosystems based on size.
2. Distinguish between decomposers and detrivores.
3. Explain the terms standing crop and standing state.
4. Explain the ecological efficiency.
5. Write notes on the food web.
6. Explain the ten per cent law.
7. Write notes on the trophic level.
8. Explain keystone species.
9. Explain ecological equivalents.
10. What are the benefits of the ecosystem?

LONG ANSWER TYPE QUESTIONS

1. Describe the biotic components of the ecosystem.
2. Describe the functional component of an ecosystem with reference to decomposition.
3. Describe the functional component of an ecosystem with reference to productivity.
4. Discuss the energy flow in an ecosystem and comment upon the ten per cent law.
5. Describe the features and types of food chains in the ecosystem.
6. Discuss the three types of ecological pyramids.

MULTIPLE CHOICE QUESTIONS

1. Cropland ecosystem is characterised by

(a) Less diversity and less resilience
(b) Less diversity and high resilience
(c) High diversity and less resilience
(d) High diversity and high resilience

2. Which occupies more than one trophic level in the pond ecosystem?

(a) Phytoplankton (b) Fish
(c) Zooplankton (d) Frog

3. Maximum productivity is seen in

(a) Grassland ecosystem
(b) Forest ecosystem
(c) Desert ecosystem
(d) Aquatic ecosystem

4. Pitcher plant (*Nepenthes*) is an example of

(a) Primary producer
(b) Primary consumer
(c) Secondary consumer
(d) Both a and b

5. How much solar energy that falls on the leaves of a plant is converted into chemical energy by photosynthesis?

(a) Less than 1%
(b) 2-10%
(c) 10-20%
(d) More than 20%

6. **Which of the following must be preserved in an ecosystem if the system is to be maintained?**
(a) Producer
(b) Producer and Consumer
(c) Decomposer
(d) Producer and decompose

7. **The first plants to reappear in a badly burnt forest area will be**
(a) Grasses (b) Mosses
(c) Ferns (d) Liverworts

8. **Energy flow and energy transformation in the living system follow**
(a) Law of minimum
(b) Law of limiting factors
(c) Law of tolerance
(d) Law of thermodynamics

9. **Which one is most essential for recycling of materials?**
(a) Producers (b) Consumers
(c) Decomposers (d) All of the above

10. **Which of the following animals come under the category of detrivores?**
(a) Earthworm
(b) Termites
(c) Carrion flies
(d) All of the above

11. **Single channel energy flow model was proposed by**
(a) Clements (b) O.P Odum
(c) Lindemann (d) Charles Elton

12. **Energy for the detritus food chain comes from**
(a) Air (b) Water
(c) Sunlight (d) Organism

13. **The most stable ecosystem is**
(a) Forest (b) Ocean
(c) Desert (d) Mountain

14. **If phytoplanktons are destroyed by sea, then**
(a) Algae will grow heavily
(b) Primary consumers will grow luxuriantly
(c) The food chain will be affected
(d) There will be no effect

15. **How much of the NPP of a terrestrial ecosystem is eaten and digested by herbivores?**
(a) 1% (b) 10%
(c) 5% (d) 20%

16. **Which of the following ecological pyramids may be upright and inverted?**
(a) Pyramid of number only
(b) Pyramid of energy only
(c) Pyramid of number and energy
(d) Pyramid of number and biomass

17. **The cropland ecosystem is the largest manmade ecosystem characterised by**
(a) Less diversity and high productivity
(b) High diversity and high productivity
(c) Less diversity and less productivity
(d) High diversity and low productivity

18. **Extinction of a species in a food chain is compensated in**
(a) Food web
(b) Food chain
(c) Ecological pyramid
(d) All of the above

19. **The detritus food chain begins with**
(a) Viruses (b) Bacteria
(c) Algae (d) Protozoa

20. **The highest annual net productivity occurs in**
(a) Alpine forest
(b) Coniferous forest
(c) Subtropical deciduous forest
(d) Tropical rainforest

21. **The Great Barrier Reef of Australia is categorised as**
(a) Population (b) Community
(c) Ecosystem (d) Biome

22. **The biomass available for consumption by the herbivores and the decomposers is called**
(a) Gross primary productivity
(b) Net Primary productivity
(c) Secondary productivity
(d) Community productivity

23. **Which of the following is characteristic of a biological community?**
(a) Stratification (b) Mortality
(c) Natality (d) Sex ratio

24. **The pyramid of numbers in parasitic food chain is**
(a) Upright (b) Inverted
(c) Spindle-shaped (d) Dumb bell-shaped

25. The largest population in a food chain is found in

(a) Producers
(b) Primary consumers
(c) Top consumers
(d) Decomposers

26. What fraction of the assimilated energy is used in respiration by herbivores?

(a) 10% (b) 20%
(c) 30% (d) 50%

27. In a food chain such as grass--dear—lion, the energy cost of respiration as a proportion of total assimilated energy at each level would be

(a) 10-20-30% (b) 20-30-60%
(c) 10-20-40% (d) 20-40-80%

28. The process of decomposition does not involve

(a) Catabolism (b) Leaching
(c) Humification (d) Stratification

29. Which of the following is incorrect regarding keystone species?

(a) It is small in number
(b) It influences more on its community
(c) Its loss can cause extinction of dependent species
(d) It has no impact on the community's organisation and survival

30. The organisms which occupy the same ecological niche in different geographical conditions are known as

(a) Ecological dominants
(b) Ecotone
(c) Keystone species
(d) Ecological equivalents

31. In the recycling of mineral elements within an ecosystem, the responsible direct-acting organisms are called

(a) Producers
(b) Decomposers
(c) Primary consumers
(d) Secondary consumers

32. The pyramid of the number of parasitic food chains in forest ecosystems is

(a) Always inverted
(b) Always upright
(c) Mixture of upright and inverted
(d) Sometimes upright and sometimes inverted

33. The most stable ecosystem is

(a) Pond (b) Ocean
(c) Desert (d) Forest

34. The rate of storage of organic matter not used by heterotrophs is called

(a) Gross primary productivity
(b) Net primary productivity
(c) Net productivity
(d) Secondary productivity

35. Rapid nutritional cycling takes place in

(a) Coral (b) Desert
(c) Grassland (d) Ocean

36. The lowest productivity per sq.metre is seen in

(a) An open sea (b) A coral reef
(c) A small marsh (d) A grassland

37. Which of the following has the largest amount of free energy?

(a) Producer (b) Primary consumer
(c) Top consumer (d) Decomposer

38. The highest value in a grassland ecosystem is found in

(a) GPP (b) NPP
(c) SP (d) TP

39. All of the organisms in a particular area make up a food chain are called

(a) Population (b) Community
(c) Ecosystem (d) Niche

40. The net productivity of a closed ecosystem is

(a) Zero (b) Less than zero
(c) More than zero (d) +1/-1

41. Which of the following is not an example of terrestrial ecosystem?

(a) Garden pot (b) Human skin
(c) Volcano site (d) Rotten log

42. In a pond ecosystem, net productivity by zooplanktons is P and biomass consumed by small fish is C, then the ratio of C/P is termed as

(a) Consumption efficiency
(b) NPP
(c) Assimilation efficiency
(d) Conversion efficiency

43. Which of the following has the largest number of trophic levels?

(a) Grassland (b) Tropical forest
(c) Coastal ocean (d) Open ocean

44. How many times utilization of energy take place in the ecosystem?

(a) One (b) Two
(c) Four (d) Again, and again

45. The maximum number of trophic levels in any ecosystem is

(a) 4 (b) 5
(c) 6 (d) 7

ANSWERS

1.b	11.c	21.c	31.b	41.a
2.b	12.b	22.b	32.a	42.a
3.d	13.b	23.a	33.b	43.d
4.d	14.c	24.c	34.a	44.a
5.b	15.a	25.d	35.a	45. c
6.d	16.d	26.b	36.a	
7.b	17.a	27.b	37.a	
8.d	18.a	28.d	38.a	
9.c	19.b	29.d	39.d	
10.d	20.d	30.d	40.a	

CHAPTER-19

BIODIVERSITY AND CONSERVATION

Biodiversity is the number or variety of plants, animals and other organisms. The term biodiversity was coined by Walter Rosen but popularised by Edward Wilson.

According to the report of the International Union of Nature and Natural Resources (IUCN), 2004, the total number of plant and animal species described so far is more than 1.5 million. All species of plants taken together are known as flora and about 0.3 million of plants are known to date. All species of animals taken together are known as fauna and about 1.2 million animals are known to date. Among animals' 70 percent are insects. Animals constitute more than 70% of all species and the share of plants is about 22% on Earth.

India is the 7th largest country in the world with a geographical area of 32,87,591 sq. km, 15200 km long land frontiers and about 7,520 km of coastline. It accounts for only 2.4% of the total land area of the earth's surface. It contributes 8.1% of global diversity. India ranks the 6th among 12 mega-diversity regions of the world. In India, 45,000 plants (7% of world flora) and 81,000 animals (8% of world fauna) are found. Among invertebrates, arthropods are maximum and among vertebrates, fishes are maximum. In the case of plants, angiosperms are maximum (dicots are more than monocots).

Approximate Number of different species (animals) on earth

Animal groups	Number of species in world	Number of species in India
Mammals	5,416	390
Birds	9,956	1,232
Reptiles	8,240	456
Amphibians	6,199	209
Fish	30,000	2,546
Molluscs	81,000	5,000
Arthropods	990,000	68,389
Other invertebrates	1,30,000	8,400

Approximate Number of different species (plants) on earth

Plant Groups	Numbers in world	Numbers in India	Percentage in India
Angiosperms	2,60,000	17,500	7%
Gymnosperms	980	64	10%
Pteridophytes	13,000	1100	12%
Bryophytes	15,000	2850	17%
Algae	12,000	6500	18%
Lichens	10,000	2000	16%
Fungi	70,000	14500	20%

Levels of biodiversity: Based on hierarchy

1. **Genetic diversity**: It is a variety of genes within an individual (species) and among different species of population. It occurs in alleles, numbers and type of genes or in the structure of chromosomes by recombination during gamete formation, mutation and random mating (sexual fertilisation).

Features:

(a) It is a trait of species.

(b) It helps in speciation (evolution of species).

(c) It helps in producing varieties of plants and animals. Rice has more than 50,000 genetically different strains. Likewise, mangoes have thousands of varieties.

(d) It provides resistance to disease in species.

(e) It helps in adaptation to environmental conditions.

(f) It maintains diversity at the species and community levels.

2. **Species diversity**: It is the variety (number) of different species within an area.

   Features:

   (a) It is a trait of community.

   (b) It influences the biotic interaction and stability of the community.

   (c) It is more in western ghat than eastern ghat.

   (d) It is measured by the Shannon-Weiner index.

   (e) Its importance was described by the **Rivet-Popper hypothesis**.

   Rivet-Popper hypothesis: It was proposed by Paul Ehlrich. He suggested in an aeroplane (ecosystem) all parts are joined together using thousands of rivets (species). If every passenger starts popping a rivet (causing a species to become extinct), it may not affect flight safety (functioning of an ecosystem) initially but as more and more rivets are removed, the plane becomes dangerously weak. It shows significant loss of a certain species affects the function and diversity of an ecological community in an ecosystem.

3. **Ecosystem diversity**: It is the variety of ecosystems showing diversity in the number of niches, trophic levels, food webs, and nutrient cycles. It is high in India due to the presence of a number of ecosystems like rainforests, deserts, mangroves, coral reefs, estuaries, and wetlands.

Types of Ecological diversity

Whittaker (1972) divided it into three types:

1. α-diversity: It is the diversity of species in each community or habitat (within habitat diversity). It shows limited variation. There is lots of competition, adjustment and interrelationships among members of the same community.
2. β-diversity: It is the diversity of species between two communities at two habitats (between habitat diversity) due to the replacement of species with the change in the community. It arises due to the presence of microhabitats and niches, differences in environmental conditions and changes in habitat.
3. γ-**diversity**: It is the diversity of habitat (regions) over a large geographical area. It is also known as ecosystem diversity (regional diversity) due to the occurrence of large and different types of ecosystems such as rainforests, deserts, deciduous forests, coral reefs, mangroves, wetlands, estuaries, and alpine. It describes the number of niches, trophic levels and various other ecological processes.

Patterns of biodiversity

1. **Latitudinal gradient:** The diversity of plants and animals has been uneven at a global level.

   (a) Species diversity is maximum at the equator (tropical region) and minimum at the pole (arctic region). It decreases from the equator to the pole.

   (b) It is maximum in tropical rain forests (Amazon rain forest (lung of the planet) in South America) due to lack of glaciation (a type of disturbance), presence of uniform and stable climatic conditions (less seasonal variation), stability in environment and greater surface area.

   (c) It is maximum at plain and minimum at high altitude. It decreases as we move from the plain region to the top region due to a drop in temperature and greater seasonal variation.

(d) Insects are more numerous among animals.

(e) It shows the surface area relationship.

2. **Species-Area relationship**: Alexander Van Humboldt observed the relationship between species richness and area (without considering taxa or region) is considered a straight line (linear) on a log scale which is represented as

$$\text{Log } S = \log C + Z \log A$$

where, S= species richness; C= Y-intercept;

Z= slope of the line (regression coefficient); A= Area.

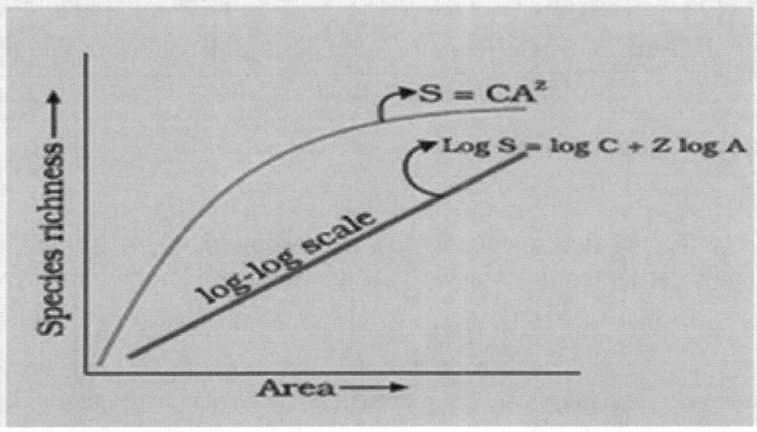

The relationship between species richness and area is rectangular hyperbola for a wide variety of organisms (taxa) where the slope of "Z" is much steeper.

$$S = CA^2$$

The value of Z lies in the range of 0.1-0.2 irrespective of the taxonomic group or the region. The value of "Z" is 0.6-1.2 for large geographical areas such as continents. It shows that species richness decreases with a decrease in area.

Importance of biodiversity:

(a) It stabilises the ecosystem by maintaining the gaseous composition of the atmosphere.

(b) It produces food in terms of plants and animals.

(c) It produces drugs and medicines.

(d) It protects land from soil erosion and loss of nutrients.

(e) It maintains soil fertility.

(f) It controls climate by forest.

(g) It checks flood by absorbing excess water.

(h) It produces a wide variety of useful products such as fibres, drugs and medicines.

(i) It promotes ecotourism.

Mega biodiversity in the world: There are 12 mega biodiversity regions which constitute 65% of all biodiversity. These are Australia, Brazil, China, Columbia Ecuador, India, Indonesia, Madagascar, Malaysia, Mexico, Peru and Zaire. The rank of India is sixth.

Biodiversity (Biogeographical region) in India: India is one of the 17th megadiversity richest regions in the world. It has 8.1% of species of the world and occupies 2.4% of the land of the world. India has 10 biogeographical regions which are as follows:

1. Western ghat: It is the richest in diversity. It is located parallel to the western coast of the Indian peninsula in Maharashtra, Karnataka, Kerala, and Tamil Nadu. It is mostly evergreen such as Silent Valley of Kerala (containing tropical rain forest) and Agasthyamalai hill.

2. Northeast: It is located in the region of Brahmaputra Valley and Northeast Hill. It is the second richest in diversity. It is called the cradle of civilisation
3. Deccan plateau: It is the central highland, Chhotanagpur plateau. It has the largest coverage area (42%).
4. Semiarid: It is seen in Punjab and Gujarat region. It includes the Aravalli hill range. It has the second-largest coverage area (16.6%).
5. Himalaya: It is found in the north, west, east and central Himalayan regions. It covers 7.2 % of the area.
6. Trans Himalaya: It is seen in the Ladakh region. It covers 5.7% of the area.
7. Gangetic plain: It is the northern region. It covers 11% of area.
8. Coast: It is seen on the western and eastern coast and the Lakshadweep region.
9. Island: It is seen in the Andaman and Nicobar region. It has the smallest coverage area (0.3%).
10. Desert: It is seen in Thar (Rajasthan) and Kutch (Gujarat) region. It covers 6.9 % of the area.

CAUSES FOR THE LOSS OF BIODIVERSITY

There are several causes which are as follows:

1. Destruction of habitat: It occurs due to cutting down trees, ploughing grassland, overgrazing, soil erosion, burning forest and human settlement by which plants are eliminated and animals move out into alien land.
2. Overexploitation: It occurs due to excessive hunting for food and shelter.
3. Invasion of alien or exotic species: The introduction of non-native (alien) species leads to the loss of native species because alien species are generally bigger, stronger, more aggressive and faster growing than native species. These alien species are as follows:
   (a) Water hyacinth (*Eichhornia crassipes*) causes the death of several aquatic plants and animals in water bodies. Initially, it was introduced in India to reduce water pollution. This plant is also known as the terror of Bengal.
   (b) Carrot grass or Congress grass (*Parthenium hysterophorus*) kills several herbs and shrubs.
   (c) *Eupatorium odoratum* reduces the population of *Tectona grandis.*
   (d) *Lantana* (a shrub) replaces several species in the forest of India.
   (e) *Periplaneta americana* replaces *Blatta orientalis.*
   (f) Nile perch (predatory fish) kills cichlid fish in Lake Victoria of East Central Africa.
   (g) African catfish (*Clarias gariepinus*) threatens native catfish (*Clarias bacterachus*).
4. Co-extinction: It occurs when the extinction of one species leads to the extinction of the other which was mutually associated with the former species.

   Note: The above four causes are called '**evil quartet**'.
5. Disturbances: It may be natural such as forest fire, tree fall, pest infestation, locust attack etc.
6. Mass extinction: It occurs due to volcanic eruption, drought, flooding, excessive heating and cooling.
7. Water Pollution: It occurs due to the use of chemical fertilisers, acid rain etc.
8. Eruption of disease: It occurs due to the breakout of viral, bacterial and fungal diseases from time to time.
9. Population attributes: It occurs due to large body size (e.g., Rhinoceros); small population and low reproductive potential (e.g., Blue whale, Giant panda), high status of trophic level in the food chain (e.g., Bengal tiger, Bald eagle), fixed habitat, fixed migratory route and inability to switch over alternative foods.

Impact of loss of biodiversity:

(a) Plant productivity will be reduced.
(b) Resistance to drought will be reduced.
(c) Food will be less available.
(d) Genetic and evolutionary variation will be reduced.

IUCN (INTERNATIONAL UNION OF CONSERVATION OF NATURE & NATURAL RESOURCES)

It is now called the World Conservation Union (WCU). Its headquarters is in Morges (Switzerland). It was founded in 1948 to maintain a complete record of every species that ever lived. It works in the field of nature conservation and sustainable use of natural resources. It prepares **a Red Data Book** or red list which provides information regarding taxa facing the risk of extinction. It was initiated in 1963. The Red Data Book is a public document that is created for recording endangered and rare species of plants, animals, fungi as well as some local subspecies which are present in a particular region.

Objectives of Red Data book:

(a) It provides information regarding the degree of threat.

(b) It prepares a conservation strategy.

(c) It provides a global index of the decline of diversity.

(d) It identifies the species at high risk of extinction.

(e) It provides complete information for research and analysis of different species.

(f) It is used to evaluate the population of a particular species.

Categories of species: The red data book has mentioned 8 categories of species based on rate of decline, population size, and area of geographical distribution. These include:

1. **Extinct**: Species do not exist, and it has been completely eliminated. e.g., Dodo (Mauritius), Passenger pigeon, Pink-headed duck, Mountain quail, Lesser 1- horned rhinoceros, Quagga=Zebra type (Africa), Thylacine =Tasmanian wolf (Australia), Steller's Sea Cow (Russia).
2. **Extinct in the wild**: Species survive only in cultivation (plants) or in captivity (animals). It is not found in natural habitat. e.g., *Ginkgo biloba* (plant), Alagoas curassow (bird).
3. **Critically endangered**: Species are at high risk of extinction and can become extinct at any time in future. e.g., Pigmy hog, Jerdon courser, Great Indian bustard, Gharial, Gangetic dolphin, Malabar civet, Kashmir stag.
4. **Endangered**: Species can become extinct if causal factors continue to operate. e.g., Red Panda, Snow leopard, Blackbuck, Bengal tiger, One-horned rhinoceros, wild Asiatic ass, Asiatic lion, Indian rock python, Blue whale, Lion-tailed macaque.

Endangered ornamental plant	*Rhododendron*
Endangered medicinal plants	*Atropa acuminata, Rauwolffia serpentina, Podophyllum hexandrium*
Endangered scientific plants	*Nepenthes, Balanophora, Dischidia, Sapria himalayensis*
Endangered economical plants	*Pinus gerardiana, Santalum album, Dioscorea*

5. **Vulnerable**: Species likely to move into the endangered category in the near future if causal factors continue to operate. e.g., Blackbuck

   (**Threatened category** includes three categories such as critically endangered, endangered and vulnerable which contains 272 species (maximum in mammals) in India).
6. **Lower risk**: Species are least concerned and do not come to any threatened categories.
7. **Data deficient**: There is insufficient information to assess the risk of extinction.
8. **Not evaluated**: Species have not been assessed for risk of extinction.

Two more categories were added:

9. **Rare**: These are species with small and localised populations. These are not at present endangered and vulnerable but are at risk. e.g., Clouded Leopard.
10. **Indeterminate**: Species is in danger but the reason is unknown. e.g., Mexican prairie dog, 3-banded Armadillo of Brazil

Green Databook: It is a catalogue of rare plants growing in protected areas and botanical gardens.

CONSERVATION APPROACHES

(a) **Narrow utilitarian**: Conservation is done because we get benefits from food (cereals, pulses, vegetables and fruits), fibres, firewood, and industrial and medicinal products.

(b) **Broadly utilitarian**: Conservation is done because we get the supply of oxygen from plants as a result of photosynthesis and the supply of fruits and seeds as a result of pollination by pollinators (insects, birds, bats etc).

(c) **Ethical**: Conservation of microbes, plants and animals is done because we share the planet and are morally bound together.

CONSERVATION STRATEGY

It is for the protection, restoration or preservation of wildlife (plants and animals) and natural resources such as forests and water. Among animals, 32% of amphibians followed by 23% of mammals are facing extinction. The world wildlife week is celebrated on the first week of October every year. These strategies are as follows:

(a) All threatened species should be protected on a priority basis.

(b) The natural habitat should be protected and restored.

(c) The reserves and protected areas should be developed

(d) Alien species should not be introduced

(e) Pollution should be controlled.

(f) National wildlife protection law (1972) should be enacted.

(g) Public awareness should be increased.

Types of conservation strategy

***In situ* conservation:** It is onsite conservation and protection of plants and animals in their native ecosystem (natural habitat). It includes:

1. **Hot spot**: It is a small area that contains the richest and most threatened species of plants and animals. It shows a high level of species diversity in which endemic species (present in a particular area not found anywhere else) are present. Maximum endemic species have been noticed in amphibians. Presently there are 34 hot spots in the world. In India, there are 3 hot spots such as the Himalayas, the Western Ghat, and the Indo-Burma region which cover less than 2% of the total area and contain 44% of species.
2. **National Park**: It is a protected area meant for the protection of both flora and fauna. In this, cultivation of land, private ownership and grazing of animals are not allowed. It occupies 1.23 % of a geographical area in India. Presently there are 109 national parks in India. The first national park established in India was Jim Corbett National Park (Uttaranchal). UNESCO has declared several National Parks (Kaziranga of Assam, Keoladeo of Rajasthan, Manas of Assam, Nanda Devi of Uttaranchal and Sundarbans of West Bengal) as world heritage sites.
3. **Wildlife Sanctuary**: It is a protected area meant for the protection of animals (fauna) only. In this, cultivation of land, private ownership and grazing of animals are allowed. It occupies 3.6% of the geographical area in India. There are around 450 sanctuaries in India.
4. **Biosphere Reserve**: It is a multi-purpose protected area to preserve biodiversity. It was launched in 1971 under the Man and Biosphere Programme (MAB) of UNESCO. The objectives of Biosphere reserves are as follows:

   (a) To conserve representative samples of the ecosystem of the world

   (b) To promote and facilitate ecological and environmental research

   (c) To promote appropriate sustainable management of the living resources

The first biosphere reserve was established in 1979 but around 600 biosphere reserves were formed in more than 105 countries. The Indian Government has established 18 biosphere reserves so far in India. Each biosphere reserve consists of the following 3 zones:

(a) **Core/Natural zone**: It is an innermost undisturbed area where there is no human activity.

(b) **Buffer zone:** It is a central area that is surrounded by a core zone where limited human activities such as education, research and training are allowed.

(c) **Transition/Manipulation zone**: It is an outermost zone where human settlement, cropping, recreation and forestry are permitted.

Biosphere reserves	Year	Regions
Nilgiri (First in India)	1986	Kerala, Karnataka,Tamil Nadu
Nanda Devi	1988	Uttaranchal
Nokrek	1988	Meghalaya
Manas	1989	Assam
Sunderbans	1989	West Bengal
Gulf of Mannar	1989	Tamil Nadu
Great Nicobar	1989	Andaman &Nicobar
Simlipal	1994	Orissa
Dibru-Saikhowa	1997	Assam
DihangDibang	1998	Arunachal Pradesh
Panchmarhi	1999	Madhya Pradesh
Khangchendzong	2000	Sikkim
Agasthyamalai hill	2001	Kerala
Achanakmar-Amarkantak	2005	Madhya Pradesh, Chattisgarh
Great Rann of Kutch	2008	Gujarat
Cold desert	2009	Himachal Pradesh
Seshachalam hill	2010	Andhra Pradesh
Panna	2011	Madhya Pradesh

Sacred groves and lakes: These are the most undisturbed forests without any human impact around places of worship. These are protected by tribal communities due to religious sanctity. e.g., Jaintia & khasi hills of Meghalaya; Aravalli hills of Rajasthan; Deodar groves in Kumaon; Pushkar lake in Ajmer (Rajasthan); Areas of Bastar in Madhya Pradesh.

5. **Cold desert**: Ladakh (Jammu and Kashmir)
6. **Hot desert**: Thar (Rajasthan)
7. **Wet land**: Assam and N-E states
8. **Forest**: Tropical moist evergreen forest in Assam
9. **Saline swampy area:**
10. **Coral reef:** Palk Bay, Gulf of Mannar, Gulf of Kutch, Andaman and Nicobar, Lakshadweep
11. **Mangroves**: Sundarbans in West Bengal

Ex situ conservation: It is off site conservation of rare or endangered plants and animals outside their natural habitat in artificial environment. It includes:

1. **Botanical garden**: It is an area of land where plants of different species and areas are grown.
2. **Zoological Park / Zoo / Zoological Garden**: It is a protected area of land where wild animals are kept under human care.

The largest zoo: Arignar Anna Zoo (Chennai)

The second largest zoo: Nandan Kanan Zoo (Bhubaneswar)

3. **Wildlife Safari Park**: It is a large area where animals are kept in semi-captivity under semi-natural conditions.
4. **Aquaria:** It is a tank, pool or bowl filled with water for keeping live fish and aquatic animals.
5. **Arboreta**: It is a garden with shrubs and trees.
6. **Orchard:** It is a home garden.
7. **Seed bank**: Seeds are stored in cold storage for many years. It has recalcitrant and orthodox seeds. A seed bank is the collection of seeds stored in a viable state. Here seeds are kept under controlled temperature (-20°C) and low humidity for storage and this is an efficient way to store the germplasm (genetic variety) of plants at low temperature. Seeds preserved under controlled conditions remain viable for long durations of time without substantial loss of vitality and many genetic changes.
8. **Tissue culture method**: It is a technique for growing cells/tissues/organs under aseptic conditions. Cryopreservation of disease-free meristem is very helpful. A long-term culture of excised roots and shoots is maintained. Meristem culture is very popular in plant propagation as it's a virus and disease-free method of multiplication.
9. **Cryopreservation**: It is the preservation of tissues, embryos, spores, seeds, and gametes at -196 °C under liquid nitrogen. Cryopreservation of disease-free meristem is very helpful. A long-term culture of excised roots and shoots is maintained. Meristem culture is very popular in plant propagation as it's a virus and disease-free method of multiplication.
10. **DNA bank** (gene bank): It is used to preserve the genetic variety of wild domesticated species. It is also used to store living sperm, eggs, and embryos of animal species. It is located in the National Bureau of Plant Genetic Resources (NBPGR), New Delhi.

Convention on Biological Diversity (CBD)

The United Nations Convention on Biological Diversity is intended to ensure international action to control the destruction of biological species, habitats and ecosystems. It was opened for signature at the UN Conference on Environment and Development- the **Earth Summit** in Rio de Janeiro, Brazil on June 5, 1992. It was signed by 152 countries. The convention subsequently came into force on December 29, 1993. India became a party to this convention in May, 1994. The CBD set three goals:

(a) Conservation of biodiversity

(b) Sustainable use of biological resources,

(c) Equitable sharing of benefits arising from commercial and other utilization of genetic resources.

World Summit: A world summit on sustainable development was held in Johannesburg (South Africa) in 2002.

Legislations by the Government of India:

National park Act	1955
Wildlife (protection) Act	1972 (amended in 1991)
Project Tiger	1972 but launched on 1st april,1973
Biodiversity Act	2002

Days to remember

21st March	World Forestry Day
22nd March	World Water Day
22nd April	World Earth Day
5th June	World Environment Day
1st July	Van Mahotsav Day
11th July	World Population Day

3rd October	World Animal Day
1st November	World Ecology Day
2nd December	National Pollution Control Day
3rd December	World Conservation day
29thDecember	World Biological Diversity Day

Symbolic animals:

India	Tiger
Australia	Kangaroo
New Zealand	Kiwi
USA	Bald eagle
China	Giant Panda
WWF	Red Panda

SHORT ANSWER TYPE QUESTIONS

1. Write notes on the Rivet-popper hypothesis.
2. Draw a graph between species richness and area.
3. Mention the biogeographical regions of India.
4. What are the objectives of the red data book?
5. Write notes on the threatened category of species.
6. Mention some of the endangered plants.
7. Write notes on biosphere reserve.
8. Mention any five biosphere reserves of India.
9. Write notes on the Earth summit of Brazil.
10. What are the legislations by the Government of India to conserve biodiversity?

LONG ANSWER TYPE QUESTIONS

1. Describe the levels, types and patterns of biodiversity.
2. Discuss the causes of the loss of biodiversity.
3. Describe the *in situ* and *ex-situ* conservation strategies.

MULTIPLE CHOICE QUESTIONS

1. The greatest biodiversity is found in

(a) India (b) South America
(c) England (d) Africa

2. Maximum number of species diversity occurs in

(a) Algae (b) Fungi
(c) Mosses (d) Lichens

3. The relationship between species richness and area for a wide variety of taxa shows

(a) Straight line
(b) Sigmoid
(c) J-shaped
(d) Rectangular hyperbola

4. **The relationship between the species richness and the area for a wide variety of taxa appears as**
(a) Straight line (b) Sigmoid
(c) Parabola(d) Rectangular hyperbola

5. **In Rivet-popper hypothesis, rivet and popper are concerned with, respectively**
(a) Ecosystem and species
(b) Species and community
(c) Community and species
(d) Species and ecosystem

6. **Rivet-Popper hypothesis explains**
(a) Role of species in an ecosystem
(b) Role of biological diversity in an ecosystem
(c) Role of disturbances in the ecosystem
(d) Role of pollutants in the environment

7. **Which of the following groups of vertebrates comprises the highest number of endangered species?**
(a) Fish (b) Reptiles
(c) Birds (d) Mammals

8. **A threatened category does not include**
(a) Critically endangered
(b) Endangered
(c) Vulnerable
(d) Rare

9. **If at high altitudes, birds become rare, the plants likely to disappear are**
(a) Orchids (b) Pines
(c) Oaks (d) *Rhododendron*

10. **Western ghat comes under the category of hot spot because of**
(a) High elevation (b) Tropical climate
(c) Good location (d) High endemism

11. **The exotic species introduced in India are**
(a) *Lantana* and *Rhododendron*
(b) *Lantana* and *Eichhornia*
(c) *Eichhornia* and *Vallisneria*
(d) *Apistia and Eichhornia*

12. **Which of the following is not observed in hot spots?**
(a) Species richness
(b) Lesser interspecific competition
(c) Endemism
(d) Accelerated species loss

13. **Maximum genetic diversity in India has been found in**
(a) Mango (b) Wheat
(c) Ground nut (d) Rice

14. **The contribution of India to global diversity is**
(a) 5% (b) 7.1%
(c) 8.1% (d) 10%

15. **Which of the following groups of plants comprises the least number of species?**
(a) Algae (b) Bryophyta
(c) Pteridophyta (d) Gymnosperms

16. **The Silent Valley of Kerala has**
(a) Temperate forest
(b) Alpine forest
(c) Tropical deciduous forest
(d) Tropical rainforest

17. **The Great Indian bustard comes under the category of**
(a) Extinct
(b) Extinct in wild
(c) Critically endangered
(d) Endangered

18. **World Biodiversity Day is celebrated on**
(a) 5th June (b) 29th December
(c) 11th July (d) 21st March

19. **Which of the following does not come under the strategy of *in situ* conservation?**
(a) Hot spot (b) Sacred groves
(c) Biosphere reserve (d) Botanical garden

20. **The Earth Summit took place in Rio de Janeiro, Brazil on**
(a) 5th June, 1990 (b) 5th June, 1991
(c) 5th June, 1992 (d) 5th June, 1993

21. **Which of the following has permanent status and is governed by central legislation?**
(a) National Park
(b) Forest
(c) Sport garden
(d) Wild sanctuary

22. **The most effective way to conserve the plant diversity of an area is**
(a) Biosphere reserve
(b) Botanical garden
(c) Seed bank
(d) Tissue culture method

23. Rivet popper hypothesis was proposed by

(a) Ernst Mayr (b) Ernst Haeckel
(c) O.P. Odum(d) Paul Ehlrich

24. Dodo is an extinct animal that belonged to

(a) Australia (b) Maldives
(c) Mauritius (d) Malaysia

25. Mangroves are found in

(a) Silent Valley of Kerala
(b) Sundarbans of West Bengal
(c) Thar of Rajasthan
(d) Ladakh of Jammu and Kashmir

26. Which of the following is known as between habitat diversity?

(a) Alpha diversity (b) Beta diversity
(c) Gamma diversity (d) Delta diversity

27. Amazon rain forest is found in

(a) North America (b) South America
(c) Africa (d) Europe

28. Human activity is not allowed in

(a) Core zone (b) Buffer zone
(c) Transition zone (d) Manipulation zone

29. Project Tiger was launched on

(a) Ist April,1970 (b) Ist April,1971
(c) Ist April,1972 (d) Ist April,1973

30. Which of the following is known as the cradle of civilisation?

(a) Western ghat (b) Desert
(c) Island (d) North-east

ANSWERS

1.b	6.a	11.b	16.d	21.a	26.b
2.b	7.a	12.b	17.c	22.c	27.b
3.d	8.d	13.d	18.b	23.d	28.a
4.d	9.d	14.c	19.d	24.c	29.d
5.a	10.d	15.d	20.c	25.b	30.d

MODEL PAPER-1

1. **Potato and sweet potato are similar at the level of**
(a) Order (b) Family
(c) Genus (d) Tribe

2. **Plaque is related to**
(a) T_2 phage (b) T_4 phage
(c) Prophage (d) Lysogenic phage

3. **In bacteriophage, the number of tail pins and tail fibres is**
(a) 4 and 4 (b) 4 and 6
(c) 6 and 4 (d) 6 and 6

4. **Which of the following groups of bacteria help in the recycling of nutrients?**
(a) Photoautotrophs
(b) Chemoautotrophs
(c) Photoheterotrophs
(d) Chemoheterotrophs

5. **Which of the following is odd with respect to homogametic organisms?**
(a) Fucus (b) *Cladophora*
(c) *Rhizopus* (d) *Synchytrium*

6. **The number of neck canal cells in archegonia of *Funaria, Dryopteris* and *Pinus* are, respectively.**
(a) 1, 6-10 and zero (b) 6-10, 1 and zero
(c) Zero, 1 and 6-10 (d) Zero, 6-10 and 1

7. **Male and female gametophytes do not have free-living and independent existence in**
(a) *Funaria* (b) *Pteris*
(c) *Cycas* (d) All of the above

8. **Which of the following plants shows opposite decussate phyllotaxy?**
(a) *Ficus* (b) *Calotropis*
(c) *Quisqualis* (d) *Syzygium*

9. **Which of the following structures found in the dicot seed will be genetically identical to its maternal plant?**
(a) Radicle (b) Cotyledon
(c) Endosperm (d) Testa

10. **The remnant of nucellus in the seed is known as**
(a) Caruncle (b) Aril
(c) Chalazosperm (d) Perisperm

11. **Which of the following compounds is used in the chemical analysis of living tissues?**
(a) Acetic acid (b) Trichloroacetic acid
(c) Glucose (d) Formaldehyde

12. **Suberised Casparian strips are found in**
(a) Dicot root (b) Dicot stem
(c) Monocot stem (d) All of the above

13. **Which of the following statements is wrong regarding smaller cells?**
(a) These are more efficient
(b) These are more metabolically active
(c) These have a high N: C ratio
(d) These have a low surface: volume ratio

14. **The periplasmic space in a prokaryotic cell is analogous to**
(a) Lysosome (b) Ribosome
(c) Mesosome (d) Nucleoid

15. **The rRNAs of the larger subunit of 80 S ribosome are**
(a) 28S, 5.8S and 5S (b) 23S, 5.8S and 5S
(c) 28S and 5S (d) 23S and 5.8S

16. **The lipid tails in the plasma membrane are**
(a) Polar and hydrophilic
(b) Polar and hydrophobic
(c) Nonpolar and hydrophilic
(d) Nonpolar and hydrophobic

17. **DNA is present in**
(a) Nucleus only (b) Mitochondria only
(c) Chloroplast only (d) All of the above

18. **Cells having secretory function have abundant**
(a) Lysosome (b) Dictyosome
(c) ER (d) Mitochondria

19. Maturation-promoting factor (PMF) formation triggers cells to cross

(a) G0 to G_1 (b) G_1 to S
(c) S to G_2 (d) G_2 to M

20. The anaphase 1 of meiosis is characterized by

(a) Division of centromere
(b) Separation of chromatids
(c) Separation of homologous pair of chromosomes
(d) All of the above

21. When sunlight is on the chloroplast, the pH is the lowest in

(a) Stroma (b) Thylakoid lumen
(c) Cytosol (d) Perichondrial space

22. The first stable product formed in the Calvin cycle of C_3 plants first undergoes process of

(a) Reduction (b) Dehydration
(c) Phosphorylation (d) Oxidation

23. Which of the following intermediates of respiration enters both dehydrogenation and decarboxylation?

(a) Pyruvic acid (b) Citric acid
(c) Isocitric acid (d) Oxaloacetic acid

24. Which of the following phytohormones is growth promoter as well as growth inhibitor?

(a) Auxin (b) ABA
(c) Ethylene (d) All of the above

25. How many oxygen atoms are required for complete oxidation of one molecule of phosphoglyceraldehyde?

(a) 2 (b) 3
(c) 4 (d) 6

26. A small protein attached to the outer surface of the inner membrane of mitochondria and acts as a mobile electron carrier for the transport of electron is

(a) Cytochrome c (b) Cytochrome b
(c) Cytochrome a (d) Cytochrome a3

27. Which of the following is surrounded by a callose wall?

(a) Microspore mother cell
(b) Pollen grain
(c) Male gamete
(d) Egg

28. Embryo doesn't show distinction of plumule, radicle and cotyledon in

(a) Orobanche (b) *Utricularia*
(c) Both a and b (d) *Cuscuta*

29. Elimination of bad traits from a plant is possible when it is reproducing

(a) Asexually by spores
(b) Asexually by propagule
(c) Vegetatively by grafting
(d) Sexually by meiosis and syngamy

30. A homozygous tall pistillate plant is crossed with a dwarf staminate plant then the genotype of endosperm will be

(a) TTt (b) TTT
(c) ttt (d) ttT

31. What would be the number of tall and pink plants if tall and pink plants (TtRr) are selfed if the first character shows complete dominance and the second character shows incomplete dominance?

(a) 3/16 (b) 6/16
(c) 9/16 (d) 1/16

32. How many gametes are formed if B and C are linked in genotype AaBbCcDDEe?

(a) 4 (b) 6
(c) 8 (d) 16

33. Blood group inheritance in human beings shows

(a) Dominance (b) Codominance
(c) Multiple alleles (d) All of the above

34. How many RNA polymerase enzymes are involved in eukaryotic transcription?

(a) One (b) Two
(c) Three (d) Four

35. The chain termination in the translation of protein synthesis involves

(a) Release factor
(b) Elongation factor
(c) Rho factor
(d) Sigma factor

36. Assuming that there are 6 types of nitrogenous bases available, and 40 types of amino acids are available for protein synthesis, then how many nitrogenous bases form each codon in the genetic code?

(a) 2 (b) 3
(c) 4 (d) 5

37. Which of the following fungi shows variation in universal genetic code?

(a) Yeast (b) *Rhizopus*
(c) *Agaricus* (d) *Neurospora*

38. **The human genome project was completed in**
(a) 2001 (b) 2002
(c) 2003 (d) 2005

39. **Which of the following varieties of maize is lysine-rich?**
(a) Proteins (b) Shakti
(c) Rattan (d) All of the above

40. **Somaclonal variation can be advantageous because**
(a) These produce monosomics
(b) These produce high genetic uniformity
(c) These produce chromosomal abnormalities
(d) These enrich genetic diversity

41. **Which of the following is left undegraded during the anaerobic digestion of organic waste for producing biogas?**
(a) Cellulose (b) Lipid
(c) Lignin (d) All of the above

42. **If recombinant DNA is introduced into animal cells then this process is known as**
(a) Transformation (b) Transduction
(c) Transfection (d) Conjugation

43. **Which of the following is the most efficient method of delivery of direct genes in the organism?**
(a) Electroporation (b) Gene gun
(c) Microinjection (d) Particle gun

44. **Which of the following conditions occurs in the midgut of an insect by which inactive toxin is converted to active toxin?**
(a) Acidic pH (b) Neutral pH
(c) Alkaline pH (d) At any pH

45. **The intermediate form between two ecosystems is called**
(a) Ecophene (b) Ecotone
(c) Edge effect (d) Ecocline

46. **The pyramid of energy is**
(a) Always upright (b) Always inverted
(c) Spindle-shaped (d) Sometimes upright

47. **If 50 kcal of energy is trapped at the producer level, then how much energy will be available to the peacock as food in the following chain: Plant > Mice > Snake > Peacock**
(a) 5 kcal (b) 0.5 kcal
(c) 0.05 kcal (d) 12.5 kcal

48. **Secondary productivity is available to**
(a) Producer level (b) Herbivore level
(c) Carnivore level (d) At all levels

49. **India became a party to the Convention on Biodiversity in**
(a) 1990 (b) 1992
(c) 1994 (d) 1996

50. **Which of the following is extinct now?**
(a) Dodo (b) Gangetic dolphin
(c) Palm Civet (d) All of the above

ANSWERS

1. (a)	2. (a)	3. (d)	4. (b)	5. (a)	6. (b)	7. (c)	8. (b)	9. (d)	10. (d)
11. (b)	12. (a)	13. (d)	14. (a)	15. (a)	16. (d)	17. (d)	18. (c)	19. (d)	20. (c)
21. (b)	22. (c)	23. (a)	24. (c)	25. (d)	26. (a)	27. (a)	28. (c)	29. (d)	30. (a)
31. (b)	32. (d)	33. (d)	34. (c)	35. (a)	36. (d)	37. (a)	38. (c)	39. (d)	40. (d)
41. (c)	42. (c)	43. (a)	44. (c)	45. (d)	46. (a)	47. (c)	48. (c)	49. (c)	50. (a)

MODEL PAPER-2

1. **Which of the following is an example of monotypic genus?**
   (a) *Solanum* (b) *Panthera*
   (c) *Homo* (d) All of the above

2. **Biological organisation starts with**
   (a) Submicroscopic molecular level
   (b) Atomic level
   (c) Cellular level
   (d) Organism level

3. **Murein and chitin are similar in**
   (a) NAG (b) NAM
   (c) Cellulose (d) Protein

4. **Which of the following protists lacks flagella?**
   (a) *Euglena* (b) *Gymnodinium*
   (c) *Gonyaulax* (d) *Pinnularia*

5. **Which of the following bryophytes represents algal ancestry in presence of pyrenoid?**
   (a) *Riccia* (b) *Marchantia*
   (c) *Anthoceros* (d) *Sphagnum*

6. **Megaphyllous pteridophytes belong to the class**
   (a) Psilotopsida (b) Lycopsida
   (c) Sphenopsida (d) Filicopsida

7. **There are three generations in the seed of gymnosperms but the future sporophyte generation is represented by all except**
   (a) Radicle (b) Cotyledon
   (c) Suspensor (d) Tegmen

8. **Multiple root cap is found in**
   (a) Pineapple (b) Screw pine
   (c) Blue pine (d) All of the above

9. **Which of the following plants shows cylindrical phylloclade?**
   (a) Cactus (b) Cocoloba
   (c) *Euphorbia* (d) All of the above

10. **The vegetative shoot apex is**
    (a) Dome shaped (b) Flat
    (c) Triangular (d) Dumb bell shaped

11. **Collenchymatous hypodermis is found in**
    (a) Monocot stem (b) Monocot root
    (c) Dicot stem (d) Dicot root

12. **The cork cambium in dicot root develops from**
    (a) Cortex (b) Pericycle
    (c) Endodermis (d) Medullary rays

13. **The nucleoid in bacterial cell is**
    (a) Single, circular and double stranded
    (b) Single, circular and single stranded
    (c) Double, linear and double stranded
    (d) Double, linear and double stranded

14. **The binucleate condition is seen in**
    (a) *Euglena* (b) *Ascaris*
    (c) *Paramecium* (d) *Mucor*

15. **Cilia and flagella contain**
    (a) 9 doublets and no singlet
    (b) 9 doublets and 1 singlet
    (c) 9 doublets and 2 singlets
    (d) 2 doublets and 9 singlets

16. **Condensation of chromatin in chromosome occurs at**
    (a) G1 phase (b) S phase
    (c) G2 phase (d) M phase

17. **Mustard gas acts as mitotic poison which causes**
    (a) Agglutination of chromosomes
    (b) Formation of chromatin
    (c) Dissolution of spindle fibres
    (d) All of the above

18. **The stroma lamellae lacks**
    (a) PS I and ATPase
    (b) PS II and NADP reductase
    (c) PS I and NADP reductase
    (d) PS II and ATPase

19. **Which of the following does not affect the rate of photosynthesis in plants growing in shade or in dense forest?**
    (a) Intensity of light (b) Duration of light
    (c) CO_2 (d) Temperature

20. The oxidative decarboxylation in complete respiration occurs at

(a) One place (b) Two places
(c) Three places (d) Four places

21. How many decarboxylation reactions in Krebs cycle are associated with the release of NADH2?

(a) One (b) Two
(c) Three (d) Four

22. The shape of graph plotted for absolute growth rate for various time of grand period of growth is

(a) Sigmoid (b) Straight line
(c) Hyperbola (d) Bell shaped

23. When a plant is not reproducing then most of the cytokinins are produced in its

(a) Root (b) Leaves
(c) Lateral bud (d) Shoot apex

24. How many ATPs are produced from one molecule of 1,3-diPGAL through electron transport chain?

(a) 12 (b) 15
(c) 17 (d) 22

25. Which of the following phytohormones is substitute for vernalisation?

(a) Auxin (b) Gibberellin
(c) Cytokinin (d) Ethylene

26. The protein in the pollen wall that causes allergy is contributed by

(a) Tapetum (b) Endothecium
(c) Exine of pollen (d) Intine of pollen

27. Funicle gets fused with the nucellus along one side and forms a ridge called

(a) Hilum (b) Epiblast
(c) Funicle (d) Raphe

28. A homozygous dwarf pistillate plant is crossed with tall staminate plant then the genotype of endosperm will be

(a) TTt (b) Ttt
(c) TTT (d) ttt

29. What is the sum total of phenotypes and genotypes in F2 generation if a character is controlled by two pairs of pooygene?

(a) 8 (b) 9
(c) 12 (d) 14

30. A carrier female for colour blindness is married to a haemophilic male. They produce eight children. How many of them will suffer from both haemophilic and colour blindness?

(a) 0% (b) 25%
(c) 50% (d) 100%

31. A woman whose father is colour blind but mother is normal is married to a haemophilic man with hypertrichosis. What percentage of their children will show genotypically any two of the traits out of three mentioned above at a given time?

(a) 0% (b) 25%
(c) 50% (d) 100%

32. It is known that there is large difference in DNA content between two organisms with similar developmental complexity. This is due to large differences in the number of

(a) Repetitive DNA and transposable elements
(b) Introns and transposable elements
(c) Exons and transposable elements
(d) Introns, exons and transposable elements

33. *E.coli* takes 40 min. to duplicate its genome using a bidirectional mode of replication. If *E.coli* were to use unidirectional mode of replication to synthesise a full copy of DNA complimentary to just one of the strands of the genome, it would take

(a) 40 minutes (b) 80 minutes
(c) 120 minutes (d) 160 minutes

34. The core of nucleosome contains

(a) Histone dimer (b) Histone tetramer
(c) Histone hexamer (d) Histone octamer

35. RNA polymerase II synthesises

(a) mRNA (b) tRNA
(c) 5 S rRNA (d) 5.8 S rRNA

36. Which of the following vectors is used in sequence annotations of Human genome project?

(a) Plasmid (b) Cosmid
(c) Phagemid
(d) Yeast artificial chromosome

37. Alexander Fleming discovered penicillin from a mould, *Penicillium notatum* while working on

(a) *Staphylococcus* (b) *Streptococcus*
(c) *Clostridium* (d) *Bacillus*

38. The anaerobic sludge digester in secondary treatment of sewage contains gases such as
(a) Methane and carbon dioxide
(b) Methane, hydrogen sulphide and carbon dioxide
(c) Methane and ammonia
(d) Methane, ammonia and carbon dioxide

39. Which of the following restriction enzymes produces blunt ends?
(a) *Eco* R I (b) *Hin* d III
(c) *Alu* I (d) *Bam* H I

40. Cosmid is an artificial vector which includes *cos* site having
(a) 6 bases of lambda phage
(b) 12 bases of lambda phage
(c) 20 bases of lambda phage
(d) 100 bases of lambda phage

41. Who first constructed recombinant DNA?
(a) Paul Berg (b) Korenberg
(c) Cohen and Boyer
(d) Arber, Nathans and Smith

42. Tissue plasminogen activator is a recombinant protein which is used in the treatment of
(a) Cancer (b) Haemophilia
(c) Cystic fibrosis (d) Heart disease

43. Which of the following is not the parasitic adaptations in the parasites?
(a) Anaerobic respiration
(b) Loss of digestive system
(c) Less developed reproductive system
(d) Excessive multiplication

44. Which of the following is the most limiting nutrient in marine ecosystem?
(a) Nitrogen (b) Carbon
(c) Phosphorus (d) Sulphur

45. Which of the following belongs to the category of primary, secondary and tertiary carnivore?
(a) Frog (b) Lizard
(c) Insectivorous bird (d) Peacock

46. Pyramid of biomass is inverted in
(a) Forest ecosystem
(b) Pond ecosystem
(c) Grassland ecosystem
(d) Parasitic food chain

47. The term ecological pyramid was coined by
(a) Charles Elton (b) Odum
(c) Reiter (d) Tansley

48. Maximum biodiversity is seen in
(a) Fish (b) Amphibians
(c) Reptiles (d) Mammals

49. The Rivet-Pauper hypothesis was given by
(a) Walter Rosen (b) Paul Ehlrich
(c) Edward Wilson (d) Weiner

50. The Chipkoo movement was started from
(a) Karnataka (b) Odisha
(c) Uttaranchal (d) Maharashtra

ANSWERS

1. (c)	2. (a)	3. (a)	4. (d)	5. (c)	6. (d)	7. (d)	8. (b)	9. (c)	10. (a)
11. (c)	12. (b)	13. (a)	14. (c)	15. (c)	16. (c)	17. (a)	18. (b)	19. (b)	20. (b)
21. (a)	22. (d)	23. (a)	24. (c)	25. (b)	26. (a)	27. (d)	28. (b)	29. (d)	30. (a)
31. (c)	32. (a)	33. (b)	34. (d)	35. (a)	36. (d)	37. (a)	38. (b)	39. (c)	40. (b)
41. (c)	42. (d)	43. (c)	44. (a)	45. (d)	46. (b)	47. (a)	48. (a)	49. (b)	50. (c)

MODEL PAPER-3

1. **The six kingdom classification was proposed by**
(a) Whittaker (b) Carl Woese
(c) Copeland (d) Haeckel

2. **In the five kingdom classification *Chlamydomonas* and *Chlorella* have been included in**
(a) Monera (b) Protista
(c) Fungi (d) Plantae

3. **Mycoplasma is not affected by penicillin because it lacks**
(a) DNA (b) RNA
(c) Ribosome (d) Cell wall

4. **Water is not required to transfer male gametes in**
(a) *Spirogyra* (b) *Ulothrix*
(c) *Chlamydomonas* (d) All of the above

5. **How many daughter plants will form from four detached gemmae in the body of** *Marchantia?*
(a) 2 (b) 4
(c) 8 (d) 16

6. **What is the ploidy of columella, calyptra, sporocyte, pseudoelater and egg in** *Funaria*
(a) n,2n,2n,2n,n (b) 2n,n,2n,2n,n
(c) 2n,n,2n,n,n (d) 2n,n,n,2n,n

7. **Wood of *Cycas* is**
(a) Monoxylic and pycnoxylic
(b) Monoxylic and manoxylic
(c) Polyxylic and pycnoxylic
(d) Polyxylic and manoxylic

8. **Polyadelphous condition of stamen is seen in**
(a) Cereals (b) Citrus plants
(c) Oil seeds (d) Nuts

9. **Among pea, potato,tulip, lily,petunia,lupin, and tobacco, how many plants show adhesion of stamens?**
(a) 3 (b) 4
(c) 5 (d) 6

10. **In cereals, the single cotyledon of embryo is represented by**
(a) Coleoptile (b) Coleorhiza
(c) Prophyll (d) Scutellum

11. **Medullary rays are tissues made up of**
(a) Phloem parenchyma
(b) Phloem fibres
(c) Xylem parenchyma
(d) Xylem fibres

12. **Bulliform cells are found in**
(a) Abaxial side of grass leaf
(b) Adaxial side of grass leaf
(c) Abaxial side of *Ficus* leaf
(d) Adaxial side of *Ficus* leaf

13. **Exodermis is formed in some dicots and most of monocot roots just below the epidermis which becomes suberised. It develops from**
(a) Cortex (b) Endodermis
(c) Pericycle (d) Vascular parts

14. **Plasma membrane in mammalian RBCs contains**
(a) 40% lipid and 52% protein
(b) 52% lipid and 40% protein
(c) 80% lipid and 20% protein
(d) 20% lipid and 80% protein

15. **How many meiotic divisions are required to produce 102 pollen grains?**
(a) 25 (b) 25.5
(c) 26 (d) 100

16. **The number of chromatids in 20 bivalents**
(a) 20 (b) 40
(c) 80 (d) 120

17. **Dinomitosis occurs in**
(a) Prokaryotic cells
(b) Eukaryotic cells
(c) Mesokaryotic cells
(d) All of the above

18. Meiosis in higher plants occurs in
(a) Apical meristem (b) Microspore
(c) Spore mother cell (d) Megaspore

19. Chemiosmosis was proposed by
(a) Calvin (b) Mitchell
(c) Bendal (d) Blackman

20. Photorespiration is also known as
(a) C_2 metabolism (b) C_3 metabolism
(c) C_4 metabolism (d) C_5 metabolism

21. The number of mobile electron carriers present in thylakoid is
(a) One (b) Two
(c) Three (d) Four

22. What will happen when an illuminated suspension of *Chlorella* is actively carrying out photosynthesis and suddenly light is switched off?
(a) The concentration of PGA will increase but that of RuBP will decrease
(b) The concentration of PGA will decrease but that of RuBP will increase
(c) The concentrations of both PGA and RuBP will decrease
(d) The concentrations of both PGA and RuBP will increase

23. Which of the following is called pacemaker enzyme?
(a) Hexokinase (b) Phosphofructokinase
(c) Aldolase (d) Enolase

24. Which of the following enzymes of Krebs cycle performs its catalytic activity similar to pyruvate dehydrogenase enzyme?
(a) Alpha ketoglutrate dehydrogenase
(b) Succinic dehydrogenase
(c) Malate dehydrogenase
(d) Thiokinase

25. Gibberellin causes flowering in non-inductive period in
(a) Rice (b) *Xanthium*
(c) Soyabean (d) Barley

26. How many molecules of metabolic water are formed when one molecule of alpha ketoglutaric acid enters Krebs cycle directly?
(a) 3 (b) 4
(c) 5 (d) 6

27. Which of the following enzymes of glycolysis can be regulated by feedback inhibition?
(a) Hexokinase (b) Phosphofructokinase
(c) Aldolase (d) Enolase

28. The first reaction in chemosynthesis is
(a) Reduction (b) Oxidation
(c) Dehydration (d) Decarboxylation

29. A perennial plant which bears flowers throughout the year is
(a) Banana (b) Mango
(c) China rose (d) Apple

30. How many total divisions are required to produce 4 trinucleate male gametophyte from sporophytic cell?
(a) 5 (b) 6
(c) 8 (d) 9

31. Which of the following layers of anther wall contains alpha cellulosic fibrous bands?
(a) Epidermis (b) Endothecium
(c) Middle layers (d) Tapetum

32. Attractants and rewards are essential for pollination by
(a) Water (b) Air
(c) Insects (d) All of the above

33. Genes for producing starch branching enzyme in pea is located on
(a) Chromosome number 1
(b) Chromosome number 4
(c) Chromosome number 5
(d) Chromosome number 7

34. Mr. Z is haemophilic and carrier for albinism. What percentage of sperm have defective alleles of both traits?
(a) 25 % (b) 50 %
(c) 75 % (d) 100 %

35. If maximum weight of orange is 60 gr and minimum weight is 30 gr. What will be the average weight of orange novariety with genotype Aabbcc?
(a) 30 gr (b) 40 gr
(c) 45 gr (d) 50 gr

36. The monohybrid ratio in haploid organisms is
(a) 1:1 (b) 3:1
(c) 1:2:1 (d) 1:1:1:1

37. **Which of the following carbon atoms is always occupied by an organic base?**
(a) First (b) Second
(c) Third (d) Fifth

38. **Unusual bases are not found in**
(a) mRNA (b) rRNA
(c) tRNA (d) Both a and b

39. **DNA absorbs maximum UV light at the wavelength of**
(a) 200 nm (b) 260 nm
(c) 300 nm (d) 340 nm

40. **Which of the blood cells is collected as sample in DNA fingerprinting?**
(a) RBCs (b) Platelets
(c) Both a and b (d) WBCs

41. **How many phosphodiester bonds are present in plasmid having 1000 base pairs?**
(a) 500 (b) 1998
(c) 250 (d) 1000

42. **During sewage treatment, flocs are masses of bacteria and fungi which are**
(a) Anaerobic and heterotrophic
(b) Anaerobic and autotrophic
(c) Aerobic and heterotrophic
(d) Aerobic and autotrophic

43. **The conversion of milk into curd improves its nutritional value by increasing the amount of**
(a) Vitamin-A (b) Vitamin-B1
(c) Vitamin-B12 (d) Vitamin-D

44. **Gentamicin is isolated from a bacterium which is**
(a) *Streptomyces griseus*
(b) *Bacillus subtilis*
(c) *Nocardia mediterranei*
(d) *Micromonospora purpurea*

45. **The introduction of rDNA in host cell is known as**
(a) Transduction (b) Transformation
(c) Conjugation (d) Transfection

46. **Which of the following restriction enzymes forms sticky ends?**
(a) *Hin* d II (b) *Hae* III
(c) *Alu* I (d) *Eco* RI

47. **GEAC stands for**
(a) Genetic Engineering Assessment Committee
(b) Genetic Engineering Approval Committee
(c) Genetic Engineering Action Committee
(d) Genetic Engineering Adhoc Committee

48. **Which of the age pyramids shows the less number of post reproductive individuals?**
(a) Triangular (b) Bell-shaped
(c) Urn shaped (d) Rectangular

49. **Pyramid of number is spindle shaped in**
(a) Grassland ecosystem
(b) Pond ecosystem
(c) Parasitic food chain
(d) Tree ecosystem

50. **Which of the following is second richest in biodiversity?**
(a) Western ghat (b) North-East
(c) Himalayas (d) Gangetic plain

ANSWERS

1. (b)	2. (b)	3. (d)	4. (a)	5. (c)	6. (b)	7. (d)	8. (b)	9. (c)	10. (d)
11. (c)	12. (b)	13. (a)	14. (a)	15. (c)	16. (c)	17. (c)	18. (c)	19. (b)	20. (a)
21. (b)	22. (a)	23. (b)	24. (c)	25. (d)	26. (a)	27. (a)	28. (b)	29. (c)	30. (d)
31. (b)	32. (c)	33. (d)	34. (a)	35. (b)	36. (a)	37. (a)	38. (d)	39. (b)	40. (d)
41. (b)	42. (c)	43. (c)	44. (d)	45. (b)	46. (d)	47. (b)	48. (b)	49. (d)	50. (b)

MODEL PAPER-4

1. **Protista was introduced in**
(a) 3-kingdom system
(b) 4-kingdom system
(b) 5-kingdom system
(d) 6-kingdom system

2. **A bacterial cell divides every two minutes and it takes 20 minutes to fill one fourth of cup. How much time will it take to fill the cup?**
(a) 20 minutes (b) 22 minutes
(c) 24 minutes (d) 40 minutes

3. ***Nitrobacter* takes energy for chemosynthesis from**
(a) Ammonia (b) Nitrate
(c) Nitrite (d) Protein

4. **The genetic material in viroid is**
(a) ss RNA (b) ds RNA
(c) ss DNA (d) ds DNA

5. **The pyrenoid is absent in**
(a) Green algae (b) Brown algae
(c) Red algae (d) Both b and c

6. **Heterospory in pteridophytes is not seen in**
(a) *Lycopodium* (b) *Selaginella*
(c) *Azolla* (d) *Marsilea*

7. **Racemose inflorescence is not characterised by**
(a) Indefinite growth of axis
(b) Centripetal opening of flowers
(c) Younger flowers are at base
(d) It does not end in terminal flowers

8. **Which of the following terms is associated with adhesion of stamens?**
(a) Syngenesius (d) Polyandrous
(c) Synandrous (d) Epipetalous

9. **Schizo-lysigenous cavity is found in**
(a) Monocot root (b) Dicot root
(c) Monocot stem (d) Dicot stem

10. **Asteraceae family is characterized by**
(a) Spike of spikelet inflorescence
(b) Vexillary aestivation
(c) Head inflorescence
(d) Parietal placentation

11. **Which of the following cells is routinely used to understand the structure of plasma membrane?**
(a) RBC (b) Stem cells
(c) Neuron (d) Myelin cell

12. **Nuclear spindle is made up of**
(a) One fibre (b) Two fibres
(c) Three fibres (d) Thirteen fibres

13. **The smallest functional unit of chloroplast is**
(a) Thylakoid (b) Stroma
(c) Quantasome (d) Granum

14. **How many SAT chromosomes exist in somatic cell of human beings?**
(a) 5 (b) 8
(c) 10 (d) 12

15. **Cell cycle was discovered by Howard and Pelc in**
(a) Broad bean (b) Kidney bean
(c) Soya bean (d) Hyacinth bean

16. **Single equatorial plate is formed at**
(a) Metaphase I (b) Metaphase II
(c) Anaphase I (d) Anaphase II

17. **If there are 20 bivalents then how many chromatids are found?**
(a) 10 (b) 20
(c) 40 (d) 80

18. **Terminalization of chiasma occurs at**
(a) Pachytene (b) Diplotene
(c) Diakinesis (d) Metaphase I

19. **Tonoplast of vacuole is an example of**
(a) Impermeable (b) Permeable
(c) Semipermeable (d) Selectively permeable

20. **Histone proteins are synthesised at**
(a) G_1 phase (b) S phase
(c) G_2 phase (d) M phase

21. **What is the first process of dark reaction?**
(a) Oxygenation
(b) Carboxylation
(c) Regeneration of RuBP
(d) Glycolytic reversal

22. **The example of a C_4 plant is**
(a) Wheat (b) Maize
(c) Sugarcane (d) Both b and c

23. **The cofactor /prosthetic group associated with oxidation-reduction reactions is**
(a) NADH (b) Zn+2
(c) Pyridoxal phosphate (d) Coenzyme A

24. **Which of the following is required for the functioning of the Calvin cycle?**
(a) Light (b) Darkness
(c) ATP alone (d) ATP and NADPH

25. **Photorespiration does not depend on**
(a) High temperature (b) Bright light
(c) Low O_2 (d) Low CO_2

26. **Which of the following reactions does not occur in glycolysis?**
(a) Phosphorylation (b) Isomerisation
(c) Dehydrogenation (d) Decarboxylation

27. **Which of the following intermediates of Krebs cycle is involved in synthesis of chlorophyll?**
(a) Citric acid (b) Succinyl coenzyme A
(c) Succinic acid (d) Malic acid

28. **How many protons are required to be transported to produce ATP from one molecule of NADH2?**
(a) 1 pair (b) 2 pairs
(c) 3 pairs (d) 4 pairs

29. **In plants, indole acetic acid causes cell elongation due to**
(a) Increase in pH of apoplast
(b) Increase in pH of cytoplasm
(c) Decrease in pH of apoplast
(d) Decrease in pH of cytoplasm

30. **Which of the following combinations is involved in morphogenesis?**
(a) Auxin: Gibberellin (b) Gibberellin: Cytokinin
(c) Auxin: Cytokinin (d) Cytokinin: Ethylene

31. **Grafting is not possible in**
(a) Apple (b) Mango
(c) Lemon (d) Banana

32. **Pollination is not required in**
(a) Gymnosperms
(b) Angiosperms
(c) Pteridophytes
(d) Phanerogams

33. **Which is most common dehiscence of anther?**
(a) Transverse (b) Porous
(c) Valvular (d) Longitudinal

34. **How many male gametes are formed from 2 germ cells?**
(a) Two (b) Four
(c) Six (d) Eight

35. **Obturator during fertilisation is formed from**
(a) Micropyle (b) Nucellus
(c) Funicle (d) Style of ovary

36. **During development of monocot embryo, the first division of terminal cell is**
(a) Longitudinal (b) Irregular
(c) Oblique (d) Transverse

37. **The term 'gene' was coined by**
(a) Mendel (b) Bateson
(c) Correns (d) Johannsen

38. **Coat colour in mice is an example of**
(a) Incomplete dominance
(b) Codominance
(c) Pleiotropy
(d) Epistasis

39. **ZO type of sex determination occurs in**
(a) Grass hopper (b) Bug
(c) Fish (d) Butterfly

40. **Antler in male deer is an example of**
(a) X-linked trait (b) Y-linked trait
(c) Sex limited trait (d) Sex influenced trait

41. **Which of the following enzymes removes supercoiling in replication of DNA ahead of replication fork?**
(a) Helicase (b) Primase
(c) DNA polymerase (d) Topoisomerase

42. **Which of the following also acts as catalyst in a bacterial cell?**
(a) 16 S rRNA (b) 23 S rRNA
(c) 5 S rRNA (d) All of the above

43. Which of the following amino acids is coded by maximum number of codons?

(a) Alanine (b) Tryptophan
(c) Leucine (d) Valine

44. In Lac operon, a mutation is created so that lactose cannot bind. Now if lactose is provided then what could happen?

(a) β galactosidase will be overexpressed
(b) Lac I repressor will remain inactivated
(c) Expresson of lac operon will remain inactivated
(d) Lac I repressor unaffected that could remain bound to oeron preventing expression

45. The speed of migration of a molecule in agarose gel electrophoresis depends on

(a) Charge alone (b) Charge and mass
(c) Mass alone (d) None of these

46. Pusa lerma is variety of

(a) Wheat (b) Rice
(c) Maize (d) Barley

47. Blood cholesterol is lowered by use of

(a) Cyclosporin (b) Statin
(c) Streptokinase (d) Glucose isomerise

48. Exponential growth does not involve

(a) Lag phase (b) Log phase
(c) Stationary phase (d) All of the above

49. Single channel energy flow model was proposed by

(a) Lindemann (b) Odum
(c) Charles Elton (d) R. Mishra

50. Pyramid of number in forest ecosystem is

(a) Upright (b) Inverted
(c) Spindle shaped (d) T-shaped

ANSWERS

1. (a)	2. (c)	3. (c)	4. (a)	5. (c)	6. (a)	7. (c)	8. (d)	9. (c)	10. (c)
11. (a)	12. (c)	13. (c)	14. (c)	15.(a)	16. (b)	17. (d)	18. (c)	19. (d)	20. (b)
21. (b)	22. (d)	23. (a)	24. (d)	25. (c)	26. (d)	27. (b)	28. (c)	29. (c)	30. (c)
31. (d)	32. (c)	33. (d)	34. (d)	35.(c)	36. (d)	37. (d)	38. (b)	39. (d)	40. (c)
41. (d)	42. (b)	43. (c)	44. (d)	45. (b)	46. (a)	47. (b)	48. (c)	49. (a)	50. (c)

MODEL PAPER-5

1. **Which of the following sets show different taxons at different categories?**
   (a) Plants, Dicots, Mammals
   (b) Polemoniales, Primata, Mammals
   (c) Angiospermae, Chordata, Primata
   (d) *Solanum*, Hominoidea, Dicotedonae

2. **Which of the following dinoflagellates is bioluminescent?**
   (a) *Gymnodinium* (b) *Gonyaulax*
   (c) *Noctiluca* (d) All of the above

6. **Gametangial contact leads to the formation of**
   (a) Zygospore (b) Chlamydospore
   (c) Zoospore (d) Oospore

4. **Which of the following is odd with respect to mosses?**
   (a) Absence of elaters
   (b) Presence of columella
   (c) Absence of gemmae
   (d) Presence of scales

5. **The coralloid root of *Cycas* shows symbiotic association with**
   (a) *Azolla* (b) *Aulosira*
   (c) *Nostoc* (d) *Anabaena*

6. **Which of the algae shows isogamy, anisogamy and oogamy?**
   (a) *Volvox* (b) *Chlamydomonas*
   (c) *Focus* (d) *Polysiphonia*

7. **Cruciform corolla is characteristic feature of**
   (a) *Ficus* (b) *Solanum*
   (c) *Salvia* (d) *Brassica*

8. **Multicarpellary and syncarpous ovary are found in**
   (a) *Ficus* (b) *Magnolia*
   (c) *Ranunculus* (d) *Papaver*

9. **Undifferentiated mesophyll cells occur in**
   (a) Leaf of *Ficus* (b) Leaf of *Solanum*
   (c) Leaf of *Salvia* (d) Leaf of *Smilax*

10. **Medullary rays are well developed in**
    (a) Monocot root (b) Dicot root
    (c) Monocot stem (d) Dicot stem

11. **Acrosome of sperm is formed by**
    (a) Endoplasmic reticulum
    (b) Mitochondria
    (c) Golgi apparatus
    (d) Lysosome

12. **The part of chromosome which does not take stain is**
    (a) Centromere (b) Chromomere
    (c) Chromatin (d) Chromatid

13. **The ratio of rRNA and protein in prokaryotic ribosome is**
    (a) 1:1 (b) 1:2
    (c) 2:1 (d) 3:2

14. **If there are 20 centromeres in a cell of anaphase then how many chromosomes are there in each daughter cell after cytokinesis?**
    (a) 10 (b) 20
    (c) 40 (d) 80

15. **Cells in G_0 phase**
    (a) Cannot re-enter the cell cycle
    (b) Can remain in that state for life time
    (c) Have entered from G1 or G2 check points
    (d) Do not divide in spite of favourable condition

16. **Lampbrush chromosome lasts for several months in**
    (a) Leptotene (b) Zygotene
    (c) Pachytene (d) Diplotene

17. **If one arm of a chromosome is missin then it is called**
    (a) Metacentric (b) Submetacentric
    (c) Acrocentric (d) Telocentric

18. **Cancer is uncontrolled**
    (a) Mitosis (b) Amitosis
    (c) Meiosis (d) Endomitosis

19. The Red drop phenomenon and Enhancement Effect were proposed by

(a) Robert Hill
(b) Emerson and Lewis
(c) Calvin
(d) Ingenhousz

20. For the fixation of 1 molecule of CO_2, how much ATP is required?

(a) 2 (b) 3
(c) 4 (d) 5

21. Assimilatory power does not include

(a) ATP (b) $NADH_2$
(c) $NADPH_2$ (d) All of the above

22. Which of the following pigment systems is absent in bacteria?

(a) PS I (b) PS II
(c) Both (d) None

23. How many additional ATPs are required to fix one molecule of carbon dioxide in C4 plants?

(a) 1 (b) 2
(c) 3 (d) 4

24. How many ATPs are produced from $FADH_2$ in one cycle of the Krebs cycle?

(a) 1 (b) 2
(c) 3 (d) 4

25. ATP synthase can bind to all of the following ions except

(a) NADH (b) ATP
(c) ADP (d) 3H+

26. Cyanides act on

(a) Complex II (b) Complex III
(c) Complex IV (d) Complex V

27. Synthesis of ethylene is induced by due to high concentration of

(a) Auxin (b) Gibberellin
(c) Cytokinin (d) ABA

28. Which of the following plant hormones is weak base?

(a) Auxin (b) Gibberellin
(c) Cytokinin (d) ABA

29. Photoperiodism was discovered in

(a) Tobacco (b) Potato
(c) Tomato (d) Maize

30. Which of the following plant is monocarpic?

(a) Radish (b) Carrot
(c) *Agave* (d) All of the above

31. Pollen viability is longest in

(a) *Typha* (b) *Cuscuta*
(c) *Mirabilis* (d) *Calotropis*

32. Pseudocopulation was observed in

(a) *Ophrys* (b) *Salvia*
(c) *Ficus* (d) *Ocimum*

33. The ploidy level of endosperm in gymnosperms is

(a) Haploid (b) Diploid
(c) Triploid (d) Tetraploid

34. If a normal male plant is crossed with tetraploid female plant then how many nuclei are involved in triple fusion?

(a) Three (b) Four
(c) Five (d) Six

35. The greatest distance among three genes is between a and c. It means that

(a) Gene a is close to b
(b) Genes are in the order: a-b-c
(c) Gene a is not recombining with c
(d) Gene a is between b and c

36. The genotypes of a husband and wife are IAIA × IAIO. How many genotypes and phenotypes are possible among the blood types of their children?

(a) 2 genotypes and 2 phenotypes
(b) 2 genotypes and 1 phenotype
(c) 1 genotype and 2 phenotypes
(d) 3 genotypes and 2 phenotypes

37. What proportion of offspring of a male carrying an X-linked dominant trait will be affected?

(a) All sons and no daughters
(b) All daughters and no sons
(c) 50% sons and 50% daughters
(d) All sons and daughters

38. In non-universal genetic code, UGA codes for

(a) Phenylalanine (b) Methionine
(c) Tryptophan (d) Glycine

39. Human genome project involved the construction of genomic library in

(a) Plasmid (b) Cosmid
(c) Bacteriophage (d) BAC

40. RNA polymerase was discovered by

(a) Kornberg (b) Severo Ochoa
(c) Okazaki (d) Khorana

41. Messelson and Stahl proved semiconservative replication of DNA in

(a) *E. coli* (b) *Salmonella*
(c) *Agrobacterium* (d) *Streptococcus*

42. The organellar DNA is

(a) Linear (b) Circular
(c) Zig-zag (d) Random

43. Which of the following restriction endonuclease enzymes does not require ATP?

(a) Type I (b) Type II
(c) Type III (d) All of the above

44. The most preferred choice for the development of hybrid plants from a male sterile line would be

(a) Ovary culture (b) Pollen culture
(c) Anther culture (d) Meristem culture

45. Pomato is a plant produced by

(a) Organ culture (b) Transgenic method
(c) Plant breeding (d) Protoplast fusion

46. If you want to use a plant tissue culture as a chemical factory for producing vitamins then you will choose

(a) Callus culture (b) Suspension culture
(c) Organ culture (d) Anther culture

47. Which of the following microbes can be effectively used as a biocontrol agent?

(a) *Psudomonas putida*
(b) *Bacillus megatherium*
(c) *Bacillus thuringiensis*
(d) *Agrobacterium tumefaciens*

48. Which of the following groups is most threatened by global extinction?

(a) Amphibians (b) Fish
(b) Birds (d) Mammals

49. Which of the following is not an example of terrestrial ecosystem?

(a) Volcano site (b) Rotten log
(c) Garden plot (d) Human skin

50. Aquatic ecosystems are least likely to be limited by

(a) Nitrogen (b) Phosphorus
(c) Carbon (d) Calcium

ANSWERS

1. (d)	2. (c)	3. (d)	4. (d)	5. (c)	6. (b)	7. (d)	8. (d)	9. (d)	10. (b)
11. (c)	12. (a)	13. (c)	14. (b)	15. (b)	16. (d)	17. (d)	18. (a)	19. (b)	20. (b)
21. (b)	22. (b)	23. (b)	24. (b)	25. (a)	26. (c)	27. (a)	28. (c)	29. (a)	30. (d)
31. (a)	32. (a)	33. (a)	34. (c)	35. (d)	36. (b)	37. (a)	38. (c)	39. (d)	40. (b)
41. (a)	42. (b)	43. (b)	44. (a)	45. (d)	46. (b)	47. (c)	48. (a)	49. (d)	50. (a)

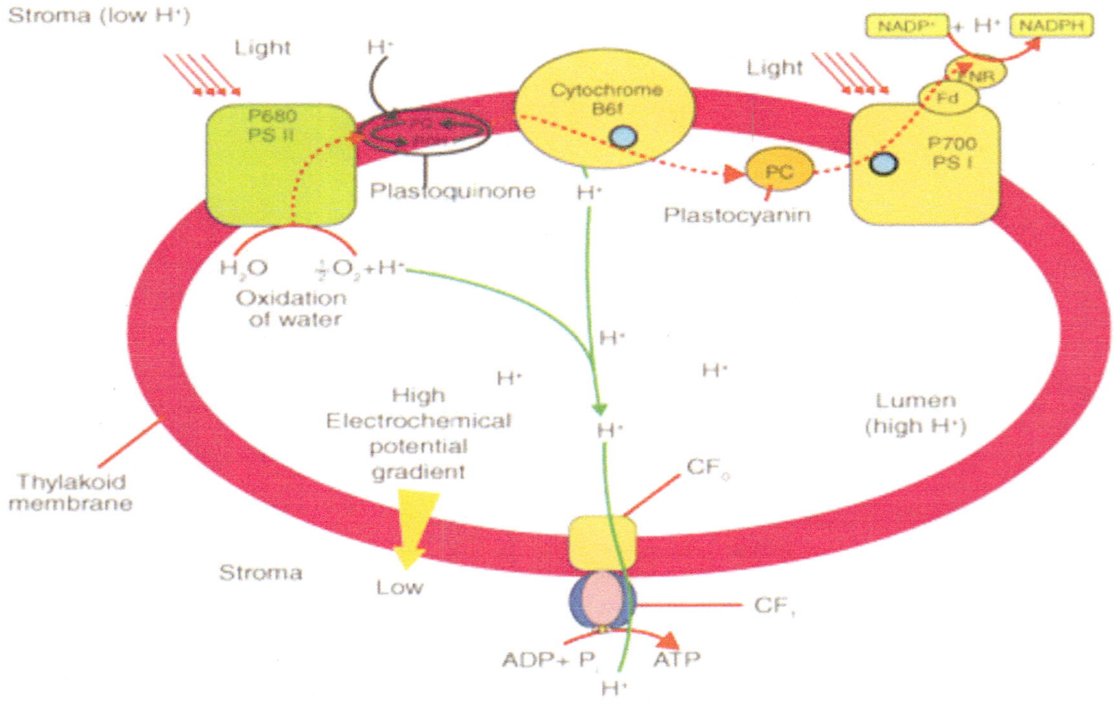

Mechanism of chemiosmosis (Page 182)

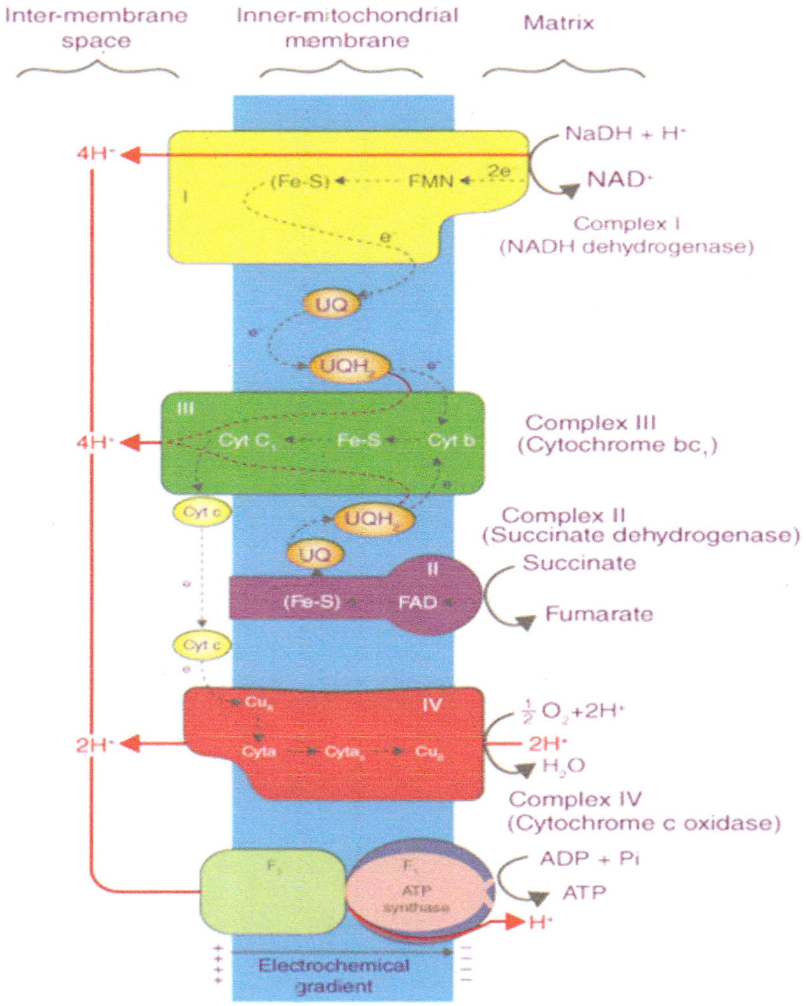

Mechanism of chemiosmosis (Page 208)

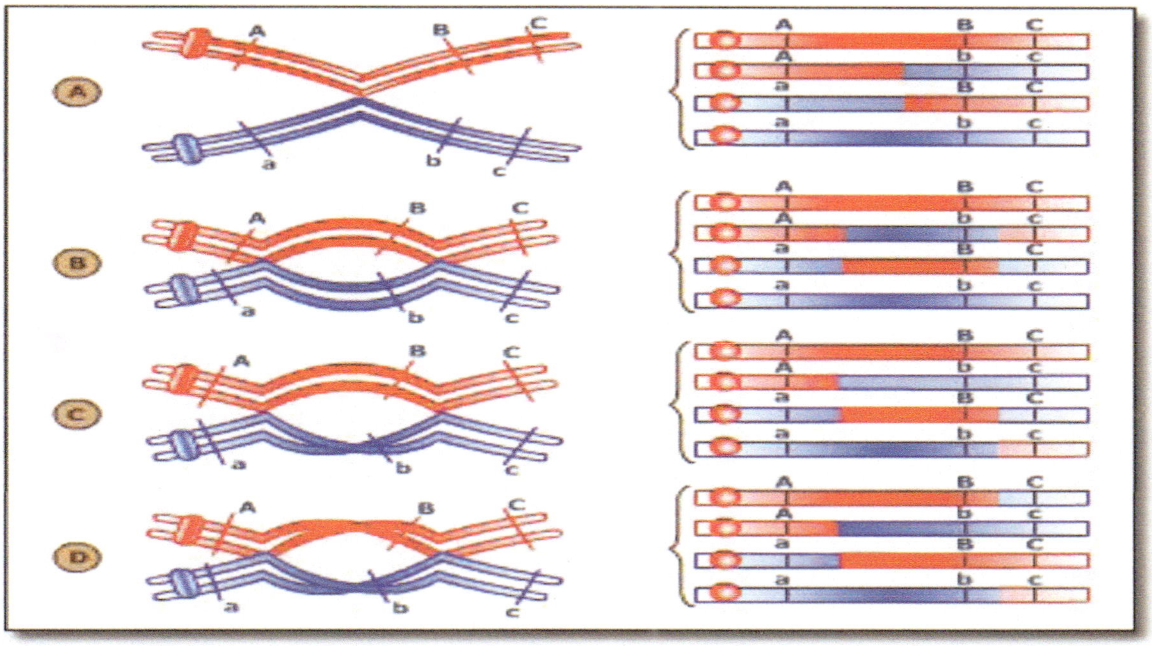

(A) Single crossing over; (B) Double crossing over (reciprocal); (C) and (D) Double crossing over (complimentary) (Page 288)

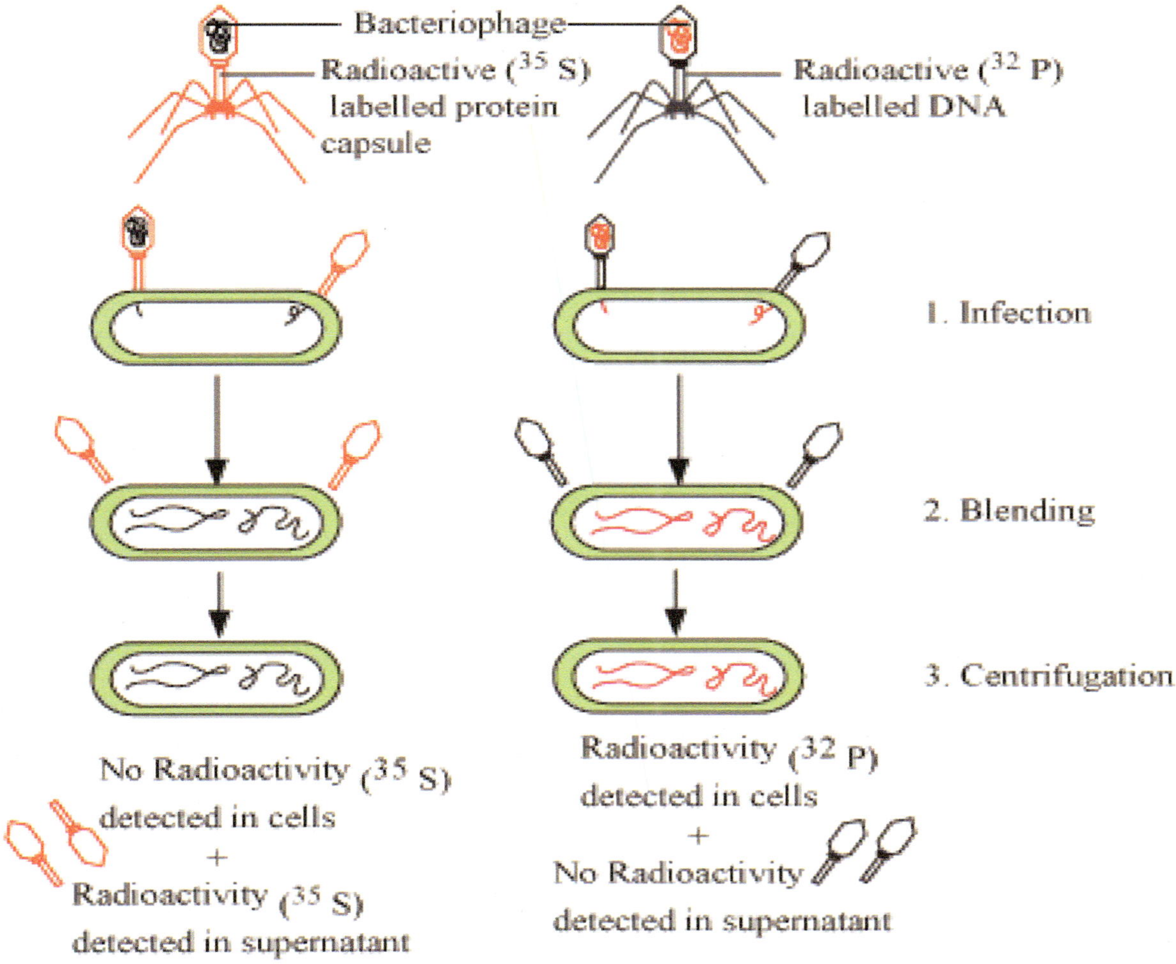

Blender experiment (Page 340)

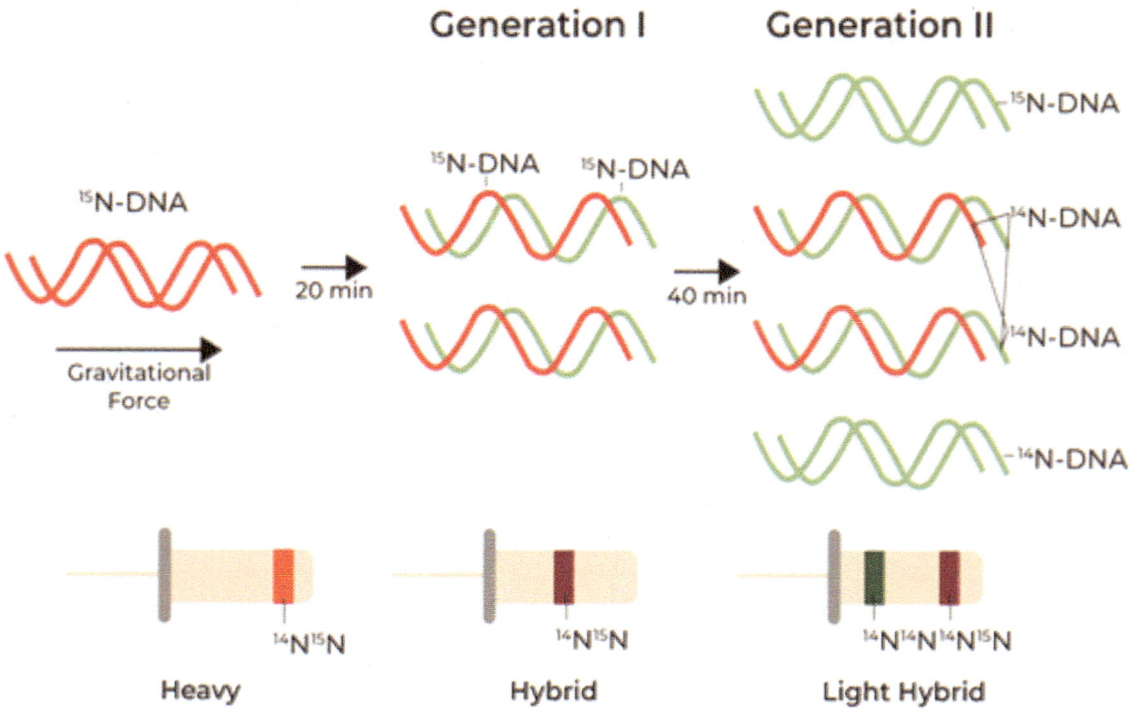

Experiment to prove semiconservative replication (Page 350)

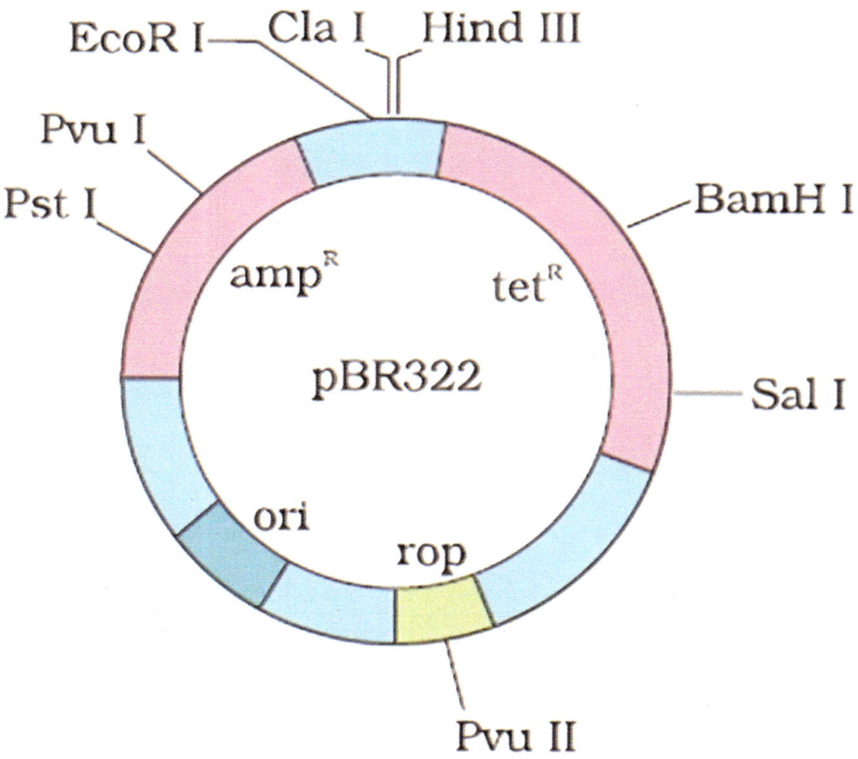

Artificial plasmids (Page 398)